# Companion to Environmental Studies

*Companion to Environmental Studies* presents a comprehensive and interdisciplinary overview of the key issues, debates, concepts, approaches and questions that together define environmental studies today. The intellectually wide-ranging volume covers approaches in environmental science all the way through to humanistic and post-natural perspectives on the biophysical world.

Though many academic disciplines have incorporated studying the environment as part of their curriculum, only in recent years has it become central to the social sciences and humanities rather than mainly the geosciences. 'The environment' is now a keyword in everything from fisheries science to international relations to philosophical ethics to cultural studies. The *Companion* brings these subject areas, and their distinctive perspectives and contributions, together in one accessible volume. Over 150 short chapters written by leading international experts provide concise, authoritative and easy-to-use summaries of all the major and emerging topics dominating the field, while the seven part introductions situate and provide context for section entries. A gateway to deeper understanding is provided via further reading and links to online resources.

*Companion to Environmental Studies* offers an essential one-stop reference to university students, academics, policy makers and others keenly interested in 'the environmental question', the answer to which will define the coming century.

**Noel Castree** is Professor of Geography at the University of Manchester, UK.

**Mike Hulme** is a Professor of Human Geography at the University of Cambridge, UK.

**James D. Proctor** is Professor of Environmental Studies at Lewis & Clark College, USA.

# Companion to Environmental Studies

**Edited by Noel Castree, Mike Hulme and James D. Proctor**

Routledge
Taylor & Francis Group
LONDON AND NEW YORK

First published 2018
by Routledge
2 Park Square, Milton Park, Abingdon, Oxon OX14 4RN

and by Routledge
711 Third Avenue, New York, NY 10017

*Routledge is an imprint of the Taylor & Francis Group, an informa business*



*British Library Cataloguing-in-Publication Data*
A catalogue record for this book is available from the British Library

*Library of Congress Cataloging-in-Publication Data*
A catalog record for this book has been requested

ISBN: 978-1-138-19219-5 (hbk)
ISBN: 978-1-138-19220-1 (pbk)
ISBN: 978-1-315-64005-1 (ebk)

Typeset in Stone Serif and Rockwell
by Apex CoVantage, LLC

# Contents

# Contributors

## Editors

**Noel Castree** is Professor of Geography at the University of Manchester, England. He has wide interests in human–environment research. He has published across the disciplines in the journals ranging from *Nature Climate Change* to *Environmental Humanities* to *New Political Economy*. His most recent book is a multidisciplinary exploration of how what we call 'nature' has been understood and its meanings socially deployed (*Making Sense of Nature* [2014], Routledge). He's co-author of the *Oxford Dictionary of Human Geography* (2013) and one of the senior editors of the 15 volume *The International Encyclopedia of Geography* (Wiley-Blackwell, 2017). He's also managing editor of *Progress in Human Geography* (Sage Publications) and a coeditor of the *Key Ideas in Geography* book series published by Routledge.

**Mike Hulme** is Professor of Human Geography at the University of Cambridge, UK. He's the author of *Can Science Fix Climate Change? A Case Against Climate Engineering* (Polity Press, 2014), *Exploring Climate Change Through Science and In Society* (Routledge, 2013), *Making Climate Change Work For Us* (co-edited, CUP, 2010) and *Why We Disagree About Climate Change* (CUP, 2009), chosen by *The Economist* magazine as one of its science and technology books of the year. From 2000 to 2007 he was the Founding Director of the Tyndall Centre for Climate Change Research and since 2007 has been the founding Editor-in-Chief of the review journal *Wiley Interdisciplinary Reviews (WIREs) Climate Change*. His most recent book is *Weathered* (Sage, 2016).

**James D. Proctor** is Professor of Environmental Studies at Lewis & Clark College. His recent work focuses on environmental theory, interdisciplinarity, and new learning technologies. He has a varied academic background including geography, religious studies, and environmental science/engineering. Proctor edited *Envisioning Nature, Science, and Religion* (2009) and *Science, Religion, and the Human Experience* (2005), and co-edited *Geography and Ethics: Journeys in a Moral Terrain* (1999). He helped form the Association for Environmental Studies & Sciences (AESS), the first national-scale environmental higher education professional association in the U.S.

# Contributors

**Joni Adamson** is Professor of Environmental Humanities, Department of English, Senior Sustainability Scholar, Wrigley Institute, Arizona State University, USA.

**Alison Anderson** is Professor of Sociology, Plymouth University, UK, and Adjunct Professor, School of Social Sciences, Monash University, Australia.

**Marco Armiero** is Director of the Environmental Humanities Laboratory at the KTH Royal Institute of Technology, Sweden, where he is also an Associate Professor of Environmental History.

**Susan Baker** is Professor of Environmental Policy, Cardiff School of Social Sciences and Co-Director, Sustainable Places Research Institute, Cardiff University, Wales.

**Greg Bankoff** is Professor of Environmental History at the University of Hull, UK.

**Edward B. Barbier** is the John S. Bugas Professor of Economics at the University of Wyoming, United States.

**Jon Barnett** is a Professor in the School of Geography at the University of Melbourne, Australia.

**Stewart Barr** is Professor of the School of Geography, University of Exeter, England.

**Chris Barrow** is Honorary Research Associate, Department of Geography, Swansea University, Swansea, UK.

**Megan Barry** is University Assistant at the Vienna University of Business and Economics, Austria.

**Simon Batterbury** is Professor of Political Ecology, Lancaster University and a Principal Fellow, University of Melbourne.

**David Benson** is Senior Lecturer, Department of Politics and the Environment and Sustainability Institute (ESI), University of Exeter, UK.

**Fikret Berkes** is Distinguished Professor Emeritus, Natural Resources Institute, University of Manitoba, Canada.

**Giovanni Bettini** is Lecturer at Lancaster University, United Kingdom.

**Frank Biermann** is Research Professor of Global Sustainability Governance, Copernicus Institute of Sustainable Development, Utrecht University, the Netherlands.

**Ingolfur Blühdorn** is Professor of Social Sustainability at the Vienna University of Business and Economics, Austria.

**Åsa Boholm** is Professor of Social Anthropology at the University of Gothenburg, Sweden.

**Frances Bowen** is professor of Innovation Studies and Head of School, School of Business and Management, Queen Mary University of London, UK.

**Sarah K. Brown** of Department of Earth Sciences, University of Bristol, UK.

**Katherine Browne** is a PhD researcher in the School of Natural Resources and Environment, the University of Michigan, USA.

**Christopher Buck** is Associate Professor of Government at St. Lawrence University in the United States.

**Sarah Burch** is Canada Research Chair in Sustainability Governance and Innovation and Associate Professor, Department of Geography and Environmental Management, University of Waterloo, Canada.

**Michael Carolan** is Professor of Sociology and the Associate Dean for Research for the College of Liberal Arts at Colorado State University, United States.

**Gemma Carr** is post-doctoral researcher at Vienna University of Technology, Austria.

**Eric D. Carter** is Edens Associate Professor of Geography and Global Health, Macalester College, Minnesota, USA.

**Wim Carton** is postdoctoral researcher at the Lund University Center for Sustainability Studies (LUCSUS), Sweden.

**Steve Carver** is Director, Wildland Research Institute, School of Geography, University of Leeds, UK.

**Ilan Chabay** is Head of Strategic Research Initiatives and Fellowships, Institute for Advanced Sustainability Studies (IASS), Potsdam, Germany.

**Jonathan Chapman** is Professor of Sustainable Design at the University of Brighton, UK.

**William W. L. Cheung** is Associate Professor in the Institute for the Oceans and Fisheries in the University of British Columbia, and Director (Science) of the Nippon Foundation-UBC Nereus Program studying climate change effects on marine ecosystems and fisheries.

**Jennifer Clapp** is Professor and Canada Research Chair in the School of Environment, Resources and Sustainability at the University of Waterloo, Canada.

**Melinda Cooper** is Associate Professor in Social and Political Sciences at the University of Sydney.

**Eileen Crist** is Associate Professor, Department of Science and Technology in Society, Virginia Tech, USA.

**Giorel Curran** is Senior Lecturer, School of Government and International Relations, Griffith University, Australia.

**Susan L. Cutter** is Carolina Distinguished Professor of Geography and Director of the Hazards and Vulnerability Research Institute at the University of South Carolina, USA.

**Anne Marie Dalton** is Professor Emerita of Religious Studies at Saint Mary's University, Halifax, Canada.

**Diana K. Davis**, DVM, PhD, Professor of History and Geography, University of California, Davis, USA.

**David Delaney** is Senior Lecturer in Law, Jurisprudence, and Social Thought, Amherst College, USA.

**Jessica Dempsey** is Assistant Professor, University of British Columbia, Department of Geography.

**Patrick Devine-Wright** is Professor of Geography at the University of Exeter, UK.

**Sidney I. Dobrin** is Professor and Chair, Director Trace Innovation Initiative, Department of English, University of Florida, USA.

**Thom van Dooren** is Associate Professor of Environmental Humanities at the University of New South Wales, Australia.

**Riley E. Dunlap** is Regents Professor of Sociology at Oklahoma State University, USA.

**Michael Ekers** is Assistant Professor, Department of Geography & Planning, University of Toronto, Canada.

**Erle C. Ellis** is Professor of Geography & Environmental Systems, University of Maryland, Baltimore County, USA.

**Kevin Ells** is Associate Professor of Mass Communication at Texas A&M University – Texarkana, USA.

**Jody Emel** is Professor of Geography, Clark University, USA.

**Richard Evanoff** is Professor in the School of International Politics, Economics, and Communication at Aoyama Gakuin University, Tokyo, Japan.

**Romain Felli** is a Lecturer in International Politics at the University of Lausanne, Switzerland.

**Robert Fletcher** is Associate Professor of Sociology of Development and Change, Wageningen University, the Netherlands.

**Tim Forsyth** is Professor of Environment and Development at the London School of Economics and Political Science.

**Robert A. Francis** is Senior Lecturer in Ecology in the Department of Geography at King's College, London.

**Tomas Frederiksen** is a Lecturer in International Development at the Global Development Institute at the University of Manchester.

**Silvio Funtowicz** is Professor at the Centre for the Study of the Sciences and the Humanities (SVT) of the University of Bergen (UiB), Norway.

**Greta Gaard** is Professor of English and Coordinator of the Sustainable Justice Minor at the University of Wisconsin, River Falls, USA.

**Greg Garrard** is Associate Professor, Sustainability, University of British Columbia.

**Frank W. Geels** is Professor of System Innovation and Sustainability at the University of Manchester, England.

**Mike Goodman** is Professor of Geography at the University of Reading, UK and Visiting Fellow in the Centre for Research in Spatial, Environmental and Cultural Politics (SECP), University of Brighton, England.

**David Griggs** is Professor Sustainable Development, Monash University (Australia) & Warwick University (UK).

**Matthias Gross** is Professor of Environmental Sociology at Helmholtz Centre for Environmental Research in Leipzig and the University of Jena, Germany.

**Kevin Grove** is an Assistant Professor of Human Geography at Florida International University, USA.

**Clive Hamilton** is Professor of Public Ethics, Charles Sturt University, Canberra, Australia.

**Philippe Hanna** is a Lecturer of International Studies at Leiden University, the Netherlands.

**Anders Hansen** is Associate Professor in the School of Media, Communication and Sociology at the University of Leicester, UK.

**Harriet Hawkins** is a Professor of GeoHumanities at Royal Holloway, University of London.

**Bronwyn Hayward** is Associate Professor of Political Science at the University of Canterbury, New Zealand.

**Ned Hettinger** is Professor Emeritus of Philosophy at the College of Charleston, USA.

**Steve Hinchliffe** is Professor of Geography at the University of Exeter, England.

**John G. Hintz** is Professor of Environmental, Geographical, and Geological Sciences at Bloomsburg University, USA.

**R. Anthony Hodge** is Professor, Department of Mining Engineering, Queen's University, Kingston, Canada.

**Joseph Holden** is Professor of Physical Geography at the University of Leeds, England.

**Alan Holland** is Emeritus Professor of Applied Philosophy at Lancaster University, UK.

**Alf Hornborg** is a Professor of Human Ecology, Lund University, Sweden.

**Peter Hough** is Assistant Professor of International Politics at Middlesex University, England.

**R. Bruce Hull** is Professor of Sustainability at Virginia Tech, USA.

**Raluca I. Iorgulescu** is Senior Researcher with the Institute for Economic Forecasting-NIER, Romanian Academy. Her areas of interest are socioeconomic metabolism, energy and sustainable agriculture.

**Michael Jefferson** is Professor, ESCP Europe Business School, London, and Honorary Senior Fellow, University of Buckingham.

**Susanna F. Jenkins** is Assistant Professor at the Earth Observatory of Singapore, Nanyang Technological University, Singapore.

**Willis Jenkins** is Professor of Religion, Ethics & Environment at the University of Virginia.

**Andrew Jordan** is a Professor in the Tyndall Centre for Climate Change Research, School of Environmental Sciences, University of East Anglia, Norwich, UK.

**Anja Kanngieser** is a Vice Chancellors Post-Doctoral Fellow in Geography at the University of Wollongong, Australia.

**Richard Kerridge** is a nature writer and ecocritic. He leads the MA in Creative Writing at Bath Spa University, and was founding Chair of ASLE-UKI. His nature writing memoir *Cold Blood* was published in 2014.

**Brian King** is Associate Professor of Geography at the Pennsylvania State University, USA.

**Brooke Lahneman** is an Instructor of Management at the University of Oregon, USA.

**Alice Larkin** (previously Bows) is Head of the School of Mechanical, Aerospace & Civil Engineering and Professor of Climate Science and Energy Policy in the Tyndall Centre, at the University of Manchester, England.

**Brendon M.H. Larson** is a Professor in the School Environment, Resources and Sustainability at the University of Waterloo, Canada.

**Maria Carmen Lemos** is Professor of the School of Natural Resources and Environment, The University of Michigan, USA.

**Les Levidow** is Senior Research Fellow at the Open University, UK.

**Martina K. Linnenluecke** is an Associate Professor at UQ Business School, The University of Queensland, Australia.

**Karen Litfin** is Associate Professor of Political Science at the University of Washington, USA.

**Jamie Lorimer** is Assistant Professor, School of Geography and the Environment, University of Oxford.

**Heather Lovell** is an Associate Professor of Sociology at the University of Tasmania, Australia.

**Michael Lynch** is President, Strategic Energy & Economic Research.

**Sherilyn MacGregor** is Associate Professor in Environmental Politics at the University of Manchester, UK.

**Deborah MacGregor** is Canada Research Chair in Indigenous Environmental Justice, Osgoode Hall Law School and Faculty of Environmental Studies, York University, Canada.

**Amanda Machin** is a post-doctoral researcher at Zeppelin University, Germany.

**Mark A. Maslin** is Professor of Physical Geography at University College London, UK.

**Charles Mather** is Professor of Geography at Memorial University, St John's, Canada.

**James Meadowcroft** is Professor and Canada Research Chair, School of Public Policy and Department of Political Science, Carleton University.

**Maurizio Meloni** is Senior Research Fellow at the Department of Sociological Studies, University of Sheffield, UK.

**John M. Meyer** is Professor and Chair, Department of Politics, Humboldt State University, USA.

**Manjana Milkoreit** is Assistant Professor of Public Policy, Purdue University, USA.

**Thaddeus R. Miller** is Assistant Professor, School for the Future of Innovation in Society, The Polytechnic School, Arizona State University, USA.

**Ben A. Minteer** is Professor of Environmental Ethics and Conservation in the School of Life Sciences at Arizona State University, USA.

**Róisín Moriarty** is Climate Change Mitigation Advisor to the Climate Change Advisory Council, Ireland and is affiliated to the Environmental Protection Agency, Ireland.

**Carol Morris** is Associate Professor of Rural Environmental Geography at the University of Nottingham, UK.

**Stephen Mosley** is Senior Lecturer in History at Leeds Beckett University, UK.

**Mark Mulligan** is Associate Professor of Geography, King's College, University of London.

**Harry W. Nelson** is Assistant Professor at the University of British Columbia, Canada.

**Sara Nelson** is Doctoral Candidate, Department of Geography, Environment and Society, University of Minnesota, USA.

**William Nikolakis** is a Lecturer at the University of British Columbia, Canada.

**Richard B. Norgaard** is Professor Emeritus of Ecological Economics in the Energy and Resources Group at the University of California, Berkeley, USA.

**Karen O'Brien** is a Professor in the Department of Sociology and Human Geography at the University of Oslo, Norway.

**Shannon O'Lear** is Ph.D. Professor of Geography, The University of Kansas, Lawrence, KS.

**John O'Neill** is Professor of Political Economy, University of Manchester, England.

**Tim O'Riordan** is Emeritus Professor of Environmental Sciences at the University of East Anglia, Norwich, UK.

**Yoshitaka Ota** is Research Assistant Professor, School of Marine and Environmental Studies, University of Washington, USA.

**Helen Pallett** is a Lecturer in Human Geography at the University of East Anglia, UK.

**James Palmer** is Departmental Lecturer (and MSc Course Director in Nature, Society and Environmental Governance) at the School of Geography and the Environment, University of Oxford.

**Adrian Peace** is an independent researcher living in America.

**John M. Polimeni** is Associate Professor of Economics, Albany College of Pharmacy and Health Sciences, Albany, USA.

**Jacques Pollini** is PhD Research Associate in the Department of Anthropology, McGill University, Canada.

**Jerome Ravetz** is an Associate Fellow of the Institute for Science, Innovation and Society at the University of Oxford. He is the co-author, with Silvio Funtowicz, of *Post-Normal Science* and also of the NUSAP notational system.

**Rupert Read** is a Reader in Philosophy at the University of East Anglia. His books include *Kuhn* (Polity, 2002), *Philosophy for Life* (Continuum, 2007), *There is No Such Thing as a Social Science* (Ashgate, 2008) and *Wittgenstein Among the Sciences* (Ashgate, 2012).

**William Rees**, PhD, FRSC is Professor Emeritus at the University of British Columbia, School of Community and Regional Planning.

**Katherine Richardson** is Professor in Biological Oceanography and Leader of the Sustainability Science Centre at the Danish Natural History Museum, University of Copenhagen, Denmark.

**Lauren Rickards** is Senior Lecturer, School of Global, Urban and Social Studies, RMIT University, Melbourne.

**Laura Rival** is an Associate Professor at the University of Oxford, where she teaches and researches the anthropology of nature, society and development.

**Magdalena Rogger** is post-doctoral researcher at Vienna University of Technology, Austria.

**Chris Russill** is Associate Professor in the School of Journalism and Communication, Carleton University, Ottawa, Canada.

**Stephanie Rutherford** is an Associate Professor in the School of the Environment at Trent University, Canada.

**Colin Sage** is Senior Lecturer in Geography at University College Cork, Republic of Ireland.

**Chris Sandbrook** is Senior Lecturer in Conservation Leadership at the UN Environment World Conservation Monitoring Centre and Affiliate Lecturer in Geography at the University of Cambridge, UK.

**Ivan R. Scales** is the Sir Harvey McGrath Lecturer in Geography at St Catharine's College, University of Cambridge.

**David Schlosberg** is Professor of Environmental Politics in the Department of Government and International Relations, and the Co-Director of the Sydney Environment Institute, at the University of Sydney, Australia.

**Tim Schwanen** is Associate Professor in Transport Studies, School of Geography and the Environment, University of Oxford, UK.

**Ian Scoones** is a Professorial Fellow at the Institute of Development Studies at the University of Sussex, England, and Director of the ESRC STEPS Centre.

**Morgan Scoville-Simonds** is Postdoc Fellow, Department of Sociology and Human Geography, University of Oslo, Norway.

**Nicole Seymour** is Assistant Professor of English at California State University, Fullerton, USA.

**Sergio Sismondo** is Professor of Philosophy at Queen's University, Canada.

**Vaclav Smil** is Distinguished Professor Emeritus at the University of Manitoba, Canada.

**Tom Smith** is the Frank Finn Professor in Finance at UQ Business School, The University of Queensland, Australia.

**Zoë Sofoulis** is Adjunct Fellow, Institute for Culture and Society, Western Sydney University, Australia.

**Sverker Sörlin** is Professor of Environmental History in the Division of History of Science, Technology and Environment, and a co-founder of its KTH Environmental Humanities Laboratory, at KTH Royal Institute of Technology, Stockholm, Sweden.

**Benjamin K. Sovacool** is Professor of Energy Policy at the University of Sussex in the United Kingdom, Director of the Sussex Energy Group, and Director of the Center on Innovation and Energy Demand. He is also Director of the Center for Energy Technologies, and Professor of Business and Social Sciences at Aarhus University in Denmark.

**David I. Stern** is Professor of Economics at the Crawford School of Public Policy, the Australian National University.

**Jack Stilgoe** is Senior Lecturer in Science and Technology Studies at University College London.

**Andy Stirling** is Professor of Science and Technology Policy in SPRU at Sussex University, England, and co-director there of the STEPS Centre.

**Daniel Chiu Suarez** is a Doctoral Candidate, University of California at Berkeley, Department of Environmental Science, Policy & Management.

**Jorge Daniel Taillant** is Executive Director of the Center for Human Rights and Environment (CHRE), a non-profit organization based in the USA.

**Ilanah Taves** is a master's student in International Development at Clark University, USA.

**Marcus Taylor** is Associate Professor in the Department of Global Development Studies, Queen's University, Canada.

**Peter J. Taylor** is Professor of Science in a Changing World at the University of Massachusetts, Boston.

**Michael Thompson** is a Senior Research Scholar at the International Institute for Applied Systems Analysis (IIASA), Laxenburg, Austria, and an Associate Fellow at the Institute for Science Innovation and Society (InSis), Oxford University, UK.

**Frank Vanclay** is Professor of Cultural Geography at the University of Groningen, The Netherlands.

**Phillip Vannini** is Professor of Communication and Culture at Royal Roads University, Canada.

**Alberto Viglione** is Assistant Professor at Vienna University of Technology, Austria.

**Peter Wade** is Professor of Social Anthropology at the University of Manchester, UK.

**Jeremy Walker** is Lecturer in Environment, Culture and Society at the University of Technology, Sydney.

**Samantha Walton** is Senior Lecturer in English Literature, Bath Spa University, UK.

**Stephen G. Warren** is Professor Emeritus of Atmospheric Science and Earth Science at the University of Washington in Seattle, USA.

**Michael J. Watts** is Class of 63 Professor, University of California, Berkeley, USA.

**Thomas Webler** is an Assistant Professor of Environmental Studies at Keene State College in Keene, New Hampshire, USA.

**Volker M. Welter** is Professor of Architectural History at the University of California at Santa Barbara, California, United States.

**Stephen M. Wheeler** is Professor, Department of Human Ecology, Landscape Architecture Program, U.C. Davis, USA.

**Iain White** is Professor of Environmental Planning at the University of Waikato, New Zealand.

**Mark Whitehead** is Professor of Human Geography at Aberystwyth University, Wales.

**Joshua Whittaker** is Research Fellow, Centre for Environmental Risk Management of Bushfires, University of Wollongong.

**Christopher Wright** is Professor of Organisational Studies at the University of Sydney Business School, Australia.

**Richard York** is Director and Professor of Environmental Studies and Professor of Sociology at the University of Oregon, USA.

# Preface

The systematic study of people and their natural and artificial environments is more popular than ever. Such study is evident across almost all academic disciplines and in many interdisciplinary enterprises, not least of which is 'environmental science'. This book aims to offer the most comprehensive introduction to the spectrum of ways that the non-human world impinges of the world of people, and vice versa. The lack of such an introduction is precisely what spurred its conception and completion. The book's contents amount less to a coherent whole and more to a kaleidoscope: this volume presents readers with different angles of vision on human–environment relations that reach across the natural sciences, the social sciences and the humanities.

This kaleidoscopic companion reflects the diversity of environmental studies. The diversity is less a sign of intellectual disarray than of the many legitimate perspectives that exist on a large, complex and dynamic world of plants, animals, ice, water and much more besides. Yet, if one thing holds these multiple perspectives together, it's a sense that human impacts on nature are of such scale, scope and significance that humans now need, through research and dialogue, to consider new and better ways of inhabiting the Earth. The challenge is not to reduce these perspectives to a single, overarching one in a grand act of synthesis. Instead, it is to learn from the diversity as we ask what sort of Earth we want to bequeath to future generations. Scientific approaches can tell us much about material causes, effects, changes and about technological problems and solutions. But vital questions arise about the human relationship with the non-human world that science alone cannot answer – questions that are moral, aesthetic, and spiritual, and that require answers based on often divergent values relating to concepts like rights, duties, entitlements, virtue, equity, justice, and so on. The social sciences and humanities are places to look for some of these answers, but so too are all sorts of places outside universities – such as major religions and minority cultures in countries like Brazil, Nepal or Canada.

We want to thank our many contributors for writing their chapters (more or less!) on time and for being responsive to editorial suggestions for change. Sadly, though, a number of commissioned chapters did not get written. If a second edition of this book is published in future we hope, as editors, to make amends. We'd like to acknowledge the superb assistance offered by the following people. Craig Thomas, at Manchester University, provided initial support in contacting contributors and fielding their questions.

Taneesha Amos-Hampson, at the University of Wollongong, then provided essential help when the book was almost ready to be handed over to the publisher. She undertook the prodigious task of formatting over 150 chapters, checking permissions for use of visual aids, collecting abstracts for chapters, soliciting ORCID identifiers, and doing much more besides – all this as a Dean's Scholar tasked to assist Noel Castree. The index was first drafted by Felicia Liu and Hannah Collis, Master's students at King's College, London.

Noel would like to dedicate this book to N-M B, THC, FMB and his mum, Christine. A very special mention goes to RO: *JMZ*. Mike would like to dedicate it to the many colleagues around the world who have helped him convert from an environmental scientist to an environmental humanist; he thanks in particular his Fellowship at the Rachel Carson Centre in Münich. Jim honours the last decade of courses and conversations with students at Lewis & Clark College that have both inspired and challenged him to help them build better frameworks for environmental studies; he also acknowledges brief sabbatical leave provided by Lewis & Clark College to work on this and related publications.

*Noel Castree, Mike Hulme and Jim Proctor*

# Introduction

## Environmental studies past, present and future

Noel Castree, Mike Hulme and James D. Proctor

### The environment and environmental studies

'The environment' refers to that which surrounds us in our day to day lives. Though natural environments have presented both threats and opportunities to humans for millennia, much of our 'habitat' is today artificial – it consists of buildings, transportation networks, power generation systems, urban parks and much more besides. As people utilise these constructed environments they often create a raft of unintended and unwanted environmental problems, such as warming of the global atmosphere because of greenhouse gas emissions from cars and cement factories. At the same time, we are intentionally transforming the natural environment to an unprecedented degree: deforestation, conversion of land to farming, urban expansion and other activities are 'denaturalising' the Earth's surface. This means that, for many of us alive today, our 'environment' is truly global: each time we fly in a plane or buy an imported product we are having an impact way beyond where we live. Indeed, some geologists believe that human effects on the non-human world are now such that this moment should be marked by declaring a new epoch of Earth history, the 'Anthropocene'. Such a naming has many interpretations. Some see it as an opportunity for contemporary humans to reflect on whether the world we are creating is one we want to bequeath our successors. Perhaps we can rethink our way of interacting with land, water and air; perhaps the Anthropocene is an overdue wake-up call for us to better conserve and restore the wonderful diversity of animate and inanimate phenomena created through natural evolution. Others see the Anthropocene as an incitement to take better control of the hydrosphere, atmosphere, cryosphere, lithosphere and biosphere. They envisage a technology-led future in which the division between the natural and the artificial disappears as humans exert more conscious and sophisticated influence on the planet. Still others question the undifferentiated status the Anthropocene gives to all humans; not all humans have equal power to shape the world-in-the-making and many remain disadvantaged, discriminated against and destitute. Social justice should be at the centre of the Anthropocene world.

'Environmental studies' describes the plethora of disciplines, sub-disciplines and inter-disciplines that together study the non-human world, both in its own right and the way humans affect (and are affected by) that world. These disciplines have existed for decades. For instance, in the 19th century, geographers like Mary Somerville and George Perkins Marsh sought to understand the world's physical configuration and how people were both adapting to it and exploiting it. However, since the early 1970s, the academic study of natural and constructed environments has grown prodigiously. In large part, this reflects both the increased power of many societies to alter the biophysical world, as well as escalating moral and political concern about how wise it is to exercise this power without restraint. Today, systematic studies of the non-human world, and how societies interact with it, can be found in disciplines as diverse as philosophy, theology and anthropology reflecting the humanities, sociology, cultural studies and economics from social sciences, and oceanography, archaeology and fisheries science from the natural sciences. And this diversity of engagements continues to grow.

In some countries, 'environmental studies' is distinguished from 'environmental science' and only encompasses disciplines that study human perceptions, uses of and responses to the non-human world. Environmental science came of age during the 1970s as an attempt to study the connections between elements of the earth surface too often previously studied in isolation. The 'earthrise' image, photographed by the Apollo 8 mission in 1968, offered a visual metaphor of this new interconnected science. In this volume, by contrast, we approach environmental studies more broadly, with an eye toward the full spectrum of disciplinary contributions to the field. Why so? Increasingly, it seems to us, it is limiting to study environments without reference to societies or to study societies without proper consideration of their biophysical underpinnings and affects. If we are, indeed, to declare the Anthropocene, then research (and policy informed by research) will need to reckon with the seamless character of our material and cultural lives. Today, it seems, everything from our diets to our clothing choices demands that we comprehend both their material dimensions and the value judgements, social relationships, historical legacies, economic practices and political regulations that they express and reproduce. Today we should talk not of societies *and* their environments but, increasingly, of 'socio-environments' wherein varied degrees of *co-constitution* apply to people and the biophysical worlds they inhabit.

## Key concepts, approaches, and topics

Given the considerable cross-disciplinary breadth of environmental studies, one may imagine that the most logical way to navigate the field would be discipline-by-discipline: economics, ecology, philosophy, and so forth. However, we have chosen not to organize the volume in this manner. Why? Disciplinary divisions of environmental studies are just one way to cut the metaphorical cake and do not, in fact, encompass all the pieces. In recent years, a multitude of sub-disciplines and cross-disciplinary approaches have emerged that address one or more aspects of environment and society. They mean that environmental studies is, to use another metaphor, an expanding ecosystem (of knowledge) where new niches are being created, yet not all the 'species' are in a direct or balanced relationship. This is not necessarily a 'problem': it does not merely reflect the gap between present day understanding and the 'total picture' some might one day like to paint of complex human–environment interactions. More positively, the 'heterodox' character of environmental studies arguably reflects the range of legitimate ways to comprehend its subject matter. For instance, while the sciences of environment can describe, explain and predict things like volcanic eruptions and El Niño events with relative accuracy, there are debates about how to value things like the Amazon rainforest, cows and freshwater. These debates cannot be resolved into one 'true' perspective. Environmental studies spans the 'fact-seeking' disciplines as well as those that try to foster intelligent discussion of how humans might best relate to nature according to diverse ethical, aesthetical and spiritual principles.

This book is the most comprehensive introduction to environmental studies published to date. It offers a 'wide angle' perspective on the field, ranging across environmental science, environmental social science and the environmental humanities. A fruitful way to navigate the field of environmental studies involves moving between its primary *concepts*, *approaches*, and *topics*. Concepts are the key ideas or organizing frameworks that circulate in the context of environment and inform our understandings; approaches are the ways we study environment, including disciplinary and cross-disciplinary tools and methods. For both concepts and approaches, we differentiate here between *classic* instances, corresponding to early developments of environmental studies from the mid-20th century on (though some are even older), and *contemporary* instances, which have arisen in the last few decades in response both to intellectual developments and the ever-changing world. It is important to keep in mind that the contemporary field of environmental studies includes *both* classic and contemporary influences – thus, rather than wholly replacing classic notions, contemporary notions often further diversify the field. Sections 1.1 through 4.2 include these four categories of classic/contemporary concepts and approaches.

This diversity of concepts and approaches is typically applied to a wide range of environmental topics, introduced in Sections 5.1 through 7.19. Here we consider three facets of these environmental topics. First, *Environmental challenges and changes* set the stage, helping us understand a dynamic Earth System replete with anthropogenic (human-induced) transformations. Next, *Human responses to environmental change* consider how individuals and institutions have addressed these changes in artistic, cultural, economic, political, scientific, and other arenas. Finally, *Key debates* considers topics related to environmental change and the human response for which there have been important, ongoing disagreements or challenges. Given the sheer breadth of the field, environmental studies has been and remains a meeting-ground of topics where sometimes widely differing concepts and approaches come into contact with each other; the ensuing dynamism of the field is an enduring strength, yet effectively guarantees that certain debates will persist.

## Navigating this book

You may be reading this volume to learn about a specific idea associated with environmental studies, perhaps in one discipline you are specializing in for your university degree. If so, navigating the volume via its table of contents, cross-references and index may be the best way to proceed. Yet bear in mind that there are well over 100 chapters, each introducing multiple ideas related to environmental studies, and so it may be difficult to assemble a complete, coherent picture of contemporary studies in this way.

You may wish to consider a more systematic approach, building on the notions of environmental studies *concepts*, *approaches*, and *topics* (and related older and contemporary periods) introduced above to help navigate this volume successfully, so that you can start to organize the highly diverse information you will encounter in this volume into a more coherent picture. You may wish to consider, for instance, how a particular concept is put into practice via various approaches, and/or in the context of various environmental topics; or, how classic and contemporary notions in environmental studies compare.

We expect you will find both synergies and differences in the chapter contents: in spite of noble attempts to unify the field (around, say, the idea of sustainability), environmental studies is incorrigibly plural. This, we feel, is one of its main strengths, but also a point of potential confusion for the would-be traveller simply hoping to chart the field successfully. We hope this volume helps you explore environmental studies in a fuller, clearer way, and wish you well on your journey in this exciting and deeply relevant field.

# Part 1

## Classic concepts

# Part 1 Introduction

A broad yet related set of ideas informed the field of environmental studies from its inception dating back to the 1960s and 1970s. Though many of these ideas have deeper roots, we shall call them 'classic concepts' here to distinguish them from ideas whose origin is even more recent. Many of these classic concepts have been refashioned in response to contemporary trends, and thus persist today. For instance, the prominent notion of 'sustainability' derives not only from the classic concept of sustainable development dating from the mid-1980s, but arguably channels the classic notion of 'limits to growth' from the early 1970s, itself built on ideas of the 'balance of nature' dating back centuries in Western thought. Then there is the more recent mark of 'neoliberalism', which between the 1980s and today transformed sustainable development as an international-scale notion into sustainability – typically expressed at the devolutionary scale of a building, business, or college campus to be managed efficiently in order to minimise waste. No matter what sorts of intellectual developments have unfolded in environmental studies since the 1960s and 1970s, then, classic concepts often remain in circulation, albeit in slightly different form.

Environmental studies is a cross-disciplinary academic field that emerged during a post-1960 period of global political turbulence. It was (and remains) motivated as much by concern over real-world issues as by intellectual curiosity. Concerns over pollution, ecological degradation, and the impacts of demographic and economic growth led to a search for guiding concepts not only to describe and explain, but ultimately to solve, the most prominent environmental problems of the time. Thus, in this section, you will find chapters on topics such as conservation and desertification, on environmental catastrophe and environmental footprint, and on scarcity and stewardship. Concepts such as these must have appeared eminently practical in the days of classic environmental thought, even though today they may sound somewhat dated – a product of their time.

Yet perhaps because of our nominal familiarity with many of these concepts (e.g., who hasn't heard the claim that 'overpopulation' will lead to major crisis?), they can as readily sound unarguable, as if classic concepts could continue to serve as solid ground for environmental studies today. But, in fact, many have been strongly contested in recent years – overpopulation for its neo-Malthusian logic, 'environmental catastrophe' for its overconfident apocalypticism, and so on. These contemporary challenges help us identify certain common traits among at least some classic concepts. First, the environment is largely understood as an entity under siege by humans. Second, the corrective actions humans should undertake are often understood as doing less – consuming less, polluting less, letting the natural world restore its balance rather than intervening in nature. Third, the near-impossibility of realistically imagining such corrective actions taking place on a widespread scale leads to gloomy future prognoses, perhaps nowhere more graphically conveyed than in the foreboding scenario of overshoot and collapse predicted under the (prototypically classic) banner of 'limits to growth'.

In many ways, then, classic concepts can somehow be both out of date and ubiquitous, in particular among popular environmental movements, if not current environmental scholarship. The foundational role they played in giving form to environmental studies at its point of formal inception nearly a half century ago, is itself sufficient justification for including them in this volume, as well as the fact noted above that, sometimes in altered form, they persist and continue to shape the field of environmental studies. For better or worse, many classic environmental concepts are perennial ideas we will continue to extend, transform, and/or challenge in future.

# 1.1 Adaptation

Marcus Taylor

Over the past decade and a half the idea of adaptation has moved from the periphery of academic thinking to become a key concept in the literature on societal responses to climatic change. This resurgence is notable in the steadily increasing volume of academic work dedicated to the concept and its recurrent use as a normative goal within contemporary policymaking. Strikingly, adaptation now acts as an encompassing framework that orientates the activities of international agencies, governments, corporations, non-governmental organisations and social movements alike (Ireland, 2012). The imperative to adapt has therein become firmly embedded as part of a collective institutional discourse on climate change in particular and environmental change more generally. Such cohesion, however, obscures that adaptation remains a fundamentally contested concept. There remain sharply diverging viewpoints on how to conceptualise adaptation and mobilise it as political practice.

## Evolutionary biology to cultural ecology

The concept of adaptation emerged within evolutionary biology in the nineteenth century where it was used to describe the process by which a species becomes better suited to its external environment from generation to generation. Individual members of a species that possess genetic traits most suited to their ecological niche will prosper and, through reproduction, pass these traits to the population at large. This process of natural selection will – over time – increase the overall fitness of the species. A change in the external environment, however, will require a species to adapt in new ways or potentially face extinction. Adaptation in evolutionary biology, therefore, refers to the process by which a species evolves gradually to better fit its environment. Notably, this adaptation occurs without purposeful design. It emerges through the random variations in genes occurring across a population in which more functional traits will slowly emerge, potentially taking many generations for a trait to become generalised.

Notwithstanding its foundation within the natural sciences, during the twentieth century a few social science frameworks drew explicitly upon the concept of adaptation. A branch of anthropology known as cultural ecology, for example, used the concept to explain the evolution of human cultural practices within traditional societies dependent upon the direct usage of natural resources. For cultural ecologists, human adaptation described not genetic shifts across a species but rather the emergence of belief systems and associated practices that enabled a social group to better fit the ecological niches that they inhabited (Rappaport, 1979). Their key argument was that cultural traditions helped regulate human–environment relationships in a manner that promoted social stability and enabled societies to better cope with persistent environmental change and uncertainty (Sponsel, 1986). This perspective, however, was critiqued sharply by political ecologists who accused it of advocating a form of environmental determinism that obscured the globalised drivers of cultural change including commercialisation, market penetration and (neo)colonialism (Watts, 1983).

## Climate change and the rebirth of adaptation

With the decline of cultural ecology in the 1970s the concept of adaptation lay dormant within the social sciences for several decades. It was a growing recognition in the 2000s that climate change posed new and challenging questions about society–nature relations that led to a swift resurrection of the concept (Head, 2010). As the Intergovernmental Panel on Climate Change (IPCC) asserted in 2007, some form of societal adjustment to climate shocks and stresses would be necessary to address impacts resulting from warming caused by past emissions (IPCC, 2007). Adaptation became the lynchpin concept under which to conceptualise and plan these social transformations. As the IPCC described it, adaptation is the adjustment in natural or human systems in response to actual or expected climatic stimuli or their effects that moderates harm or exploits beneficial opportunities (IPCC, 2007). The goal of adaptation was therein presented as a social imperative. The United Nations Development Programme (UNDP), for example, claimed that a failure to adapt is likely to consign the poorest 40 percent of the world's population to a future of sharply diminished opportunity (UNDP, 2007).

While the usage of adaptation unifies the literature, different frameworks build from diverging ideas about the normative goals of adaptation, its appropriate sites and scales, the rights and responsibilities of affected and contributor groups, and the necessary mechanisms of adjustment. Consequently, distinct approaches to adaptation legitimate contrasting policy responses. Three broad ways in which adaptation is conceptualised within the literature are particularly notable (Pelling, 2011).

## Adaptation as risk management

The mainstream of adaptation thinking represents adaptation as a process of risk management in which potential shocks and stresses are identified and adaptation measures are taken to reduce vulnerability. This typically involves three elements: (1) identifying and reducing exposure to shocks and stresses; (2) limiting the impact of any unavoidable shocks and stresses; and (3) building capacity to respond and recover in the medium term. Reducing exposure in a coastal region, for example, could take the form of building stronger flood defences while limiting impacts might involve improving existing drainage infrastructure. Improving capacity to respond to shocks typically involves improved emergency planning, educating populations to climate risks, and greater resources for disaster management.

This reactive approach to adaptation often adopts a strongly technocratic and managerial perspective wherein potential climate threats are seen as a series of identifiable risks that require a variety of institutional and technological solutions to preserve the status quo in the face of climatic change. As Bassett and Fogelman demonstrated in their extensive survey of the adaptation literature, over 70 percent of academic publications on the subject presented adaptation as a technical process of planned social engineering to guard against proximate climatic threats (Bassett and Fogelman, 2013). Rarely, however, do such approaches consider how present social inequalities shape how different social groups are unequally exposed to climate threats while also making them unevenly positioned to take advantage of any benefits provided by adaptation processes (Taylor, 2013). A key concern is that adaptation projects of this nature may come at the expense of marginalised groups, who may be ignored or even displaced in the name of fighting climate change. Such an outcome represents a case of 'maladaptation' wherein, for marginalised groups, the 'cure' of adaptation may well prove to be worse than the curse of climate change exposure (Marino and Ribot, 2012).

## Adaptation as building resilience

If the risk management approach sees adaptation as a process of planned adjustment led by centralised government actions, the resilience approach emphasises the need to foster autonomous and spontaneous adaptations at a more decentralised level. As developed by authors such as Carl Folke, the idea of resilience projects that coupled socio-ecological systems tend to move between multiple equilibrium states, and therefore display dynamic qualities in which a system is constantly changing through a process of perpetual re-organisation (Folke, 2006). Certain types of socio-ecological systems, however, are more able to cope with external shocks and stresses without losing core form and functions. Resilience thinking seeks to identify those key characteristics so as to facilitate appropriate governance strategies that can promote persistence within socio-ecological systems.

For some, resilience represents a profoundly neoliberal vision of society that is composed of self-adjusting individuals and communities able to 'bounce back' from shocks and stresses. It therefore can become a surrogate for imposing individual responsibility to collectively determined challenges, and therein blaming the victims for being insufficiently resilient (MacKinnon and Driscoll Derickson, 2013). While such politically reactionary renditions certainly exist, others have sought a recover a more radical essence within resilience (Leach et al., 2011; Brown, 2015). In this tradition, the incredible complexity of environmental change means that appropriate governance for ecosystem management must be decentralised, participatory and adaptable. It must be open to learning from diverse knowledge and practices about how humans interact with their environments (Biggs et al., 2015). These norms pose a significant challenge to adaptation as risk management because they profess a deep suspicion of top-down, managerial and technocratic approaches to organising society–nature relations. That said, whether such perspectives can adequately address questions of power and inequality remains an open question (Watts, 2014).

## Adaptation as transformation

The idea of adaptation as transformation is grounded within political ecology traditions that hold exposure to climatic shocks to be strongly determined by existing socio-economic inequality, disempowerment and marginality (Blaikie et al., 1994). As a result, adaptation should not focus solely on addressing the surface symptoms of vulnerability at the point of exposure, but rather target the underlying causes at a society-wide level (Eriksen et al., 2015). This turns adaptation from a process of risk management or resilience building into a process of empowering those most vulnerable. As a result, it can be achieved only through a fundamental process of socio-economic change. As Mark Pelling describes it, adaptation as transformation means prioritising actions that "have the reach to shift existing social systems (and their component structures, institutions and actor positions) onto alternative development pathways, even before the limits of existing adaptation choices are met" (Pelling et al., 2015). The compulsion to address adaptation as transformation appears to such authors to be particularly compelling given that climatic change is likely to sharpen the divisions between those with capacity to adapt and those without, therein accentuating the potential for the security of some to be attained by juxtaposing greater risks onto marginalised groups. From this perspective, adaptation must inherently involve new forms of redistributing resources in ways that address extant vulnerabilities, inequalities and power relations sharpened by present climatic change (Taylor, 2015).

## Learning resources

Within the academic literature, Mark Pelling's *Adaptation to Climate Change* (London: Routledge, 2011) provides the accessible introduction to the field.

One of the more comprehensive websites on adaptation is provided by the United Nations Environment Programme: http://www.unep.org/climatechange/adaptation/

The Intergovernmental Panel on Climate Change is the UN body that coordinates research on climate change impacts and potential adaptation options. It is extremely useful to familiarise yourself with its work: http://www.ipcc.ch/index.htm

The International Institute for Environment and Development provides a wealth of resources and briefing notes on adaptation related topics in developing country contexts: http://www.iied.org

## Bibliography

Bassett, T. and Fogelman, C. (2013) 'Déjà vu or Something New? The Adaptation Concept in the Climate Change Literature'. *Geoforum* 48(1): 42–53.

Biggs, R., Schlüter, M. and Schoon, M. (2015) *Principles for Building Resilience: Sustaining Ecosystem Services in Social-Ecological Systems*. Cambridge: Cambridge University Press.

Blaikie, P.M., Cannon, T., Davis, I. and Wisner, B. (1994) *At Risk: Natural Hazards, People's Vulnerability, and Disasters*. London and New York: Routledge.

Brown, K. (2015) *Resilience, Development and Social Change*. London: Routledge.

Eriksen, S., Nightingale, A. and Eakin, H. (2015) 'Reframing Adaptation: The Political Nature of Climate Change Adaptation'. *Global Environmental Change* 35(3): 523–33.

Folke, C. (2006) 'Resilience: The Emergence of a Perspective for Social-Ecological Systems Analyses'. *Global Environmental Change* 16(4): 253–67.

Head, L. (2010) 'Cultural Ecology: Adaptation – Retrofitting a Concept?'. *Progress in Human Geography* 34(2): 234–42.

IPCC (2007) *Climate Change 2007: The Scientific Basis. Wg I Contribution to IPCC 4th Assessment Report*. Cambridge: Cambridge University Press.

Ireland, Philip (2012) 'Climate Change Adaptation: Business-as-usual Aid and Development or an Emerging Discourse for Change?' *International Journal of Development Issues* 11(2): 92–110.

Leach, M., Scoones, I. and Stirling, A. (2011) *Dynamic Sustainabilities: Technology, Environment, Social Justice*. London: Earthscan.

MacKinnon, D. and Driscoll Derickson, K. (2013) 'From Resilience to Resourcefulness: A Critique of Resilience Policy and Activism'. *Progress in Human Geography* 37(2): 253–70.

Marino, E. and Ribot, J. (2012) 'Adding Insult to Injury: Climate Change and the Inequities of Climate Intervention'. *Global Environmental Change* 22(3): 323–8.

Pelling, M. (2011) *Adaptation to Climate Change*. London: Routledge.

Pelling, M., O'Brien, K. and Matyas, D. (2015) 'Adaptation and Transformation'. *Climate Change* 133(1): 113–27.

Rappaport, R.A. (1979) *Ecology, Meaning, and Religion*. Richmond, Calif.: North Atlantic Books.

Sponsel, L. (1986) 'Amazon Ecology and Adaptation'. *Annual Review of Anthropology* 15(1): 67–97.

Taylor, M. (2013) 'Climate Change, Relational Vulnerability and Human Security: Rethinking Sustainable Adaptation in Agrarian Environments'. *Climate and Development* 5(4): 318–27.

Taylor, M. (2015) *The Political Ecology of Climate Change Adaptation: Livelihoods, Agrarian Change and the Conflicts of Development*. London: Routledge.

UNDP (2007) *Human Development Report 2007/2008: Fighting Climate Change — Human Solidarity in a Divided World*. Geneva: United Nations.

Watts, M. (1983) *Silent Violence*. Athens: University of Georgia Press.

Watts, M. (2014) 'Resilience as a Way of Life: Biopolitical Security, Catastrophism, and the Food–Climate Change Question' in Chen, N. and Sharp, L. (eds) *Bioinsecurity and Vulnerability*. (pp 145–75). Santa Fe: SAR Press.

# Biorogionalism

1.2

## Richard Evanoff

While it can be argued that bioregional principles have been the norm for most of human history, contemporary bioregionalism emerged in the mid-1970s as a response to a growing recognition of the extent to which modern industrialization and consumerism contribute to ecological degradation, social alienation, and reduced levels of self-fulfillment. Bioregionalism can be briefly defined as a social movement which strives to recover a sense of place and a sense of community by revitalizing ecologically sustainable and culturally diverse societies in the context of their local geographical areas, or 'bioregions'.

## What is a bioregion?

The term *bioregion* (literally 'life-place') was originally coined by Allen Van Newkirk in 1975 and the concept was further elaborated by the field biologist, Raymond Dasmann, in his influential textbook, *Environmental Conservation* (1984). Peter Berg (2009, 2015) and Gary Snyder (1969, 1995) also contributed significantly to the development of bioregionalism as a distinct approach to environmentalism.

While acknowledging the role that the natural sciences could play in determining the boundaries of a bioregion, Berg proposed that the term be regarded primarily as a cultural rather than a scientific concept. In 'Reinhabiting California', an early manifesto of the bioregional movement, Berg and Dasmann wrote that a bioregion 'refers both to geographical terrain and a terrain of consciousness—to a place and the ideas that have developed about how to live in that place' (1977, p. 399). Bioregionalism attempts to overcome the 'nature vs. culture' dichotomy by seeing the two as forming a symbiotic relationship. Humans and their natural environments co-evolve dialectically, with each transforming the other, and the objective of bioregionalism is to achieve a co-adaptive fit between local cultures and local ecosystems.

Kirkpatrick Sale has suggested that bioregions can be distinguished by 'particular attributes of flora, fauna, water, climate, soils, and landforms, and by the human settlements and cultures those attributes have given rise to' (2000, p. 55). While granting that the borders of bioregions are overlapping and fluid rather than discrete, Sale identifies three different types, nested one within another: *ecoregions*, large territories sharing similar native vegetation and soil types; *georegions*, mid-size areas identified by mountain ranges, valleys, and river basins; and *morphoregions*, smaller units marked by changing life forms and human land use patterns. An ongoing interest among bioregionalists has been classifying and mapping regions with respect to their natural features rather than arbitrary political boundaries. For an example, see the national bioregion map for Australia (http://www.environment.gov.au/land/nrs/science/ibra/australias-bioregions-maps).

## Aims of bioregionalism

Jim Dodge identified three elements which he took to be fundamental to bioregional thought: 'a decentralized, self-determined mode of social organization; a culture predicated upon biological integrities and acting in respectful accord; and a society which honors and abets the spiritual development of its members' (1981, p. 10). Berg similarly suggests that bioregionalism endeavors to 'restore natural systems, satisfy basic human needs, and develop support for individuals' (2009, p. 162). A corresponding bioregional ethic seeks to integrate local environments and local communities in ways that are ecologically sustainable, socially just, and humanly satisfying (Evanoff, 2011).

A key bioregional concept is *reinhabitation*, which involves 'evolving social behavior that will enrich the life of [a] place, restore its life-supporting systems, and establish an ecologically and socially sustainable pattern of existence within it' (Berg & Dasmann, 1977, p. 399). Bioregionalism goes beyond both an ecocentric concern for preserving pristine wilderness areas and an anthropocentric focus on conserving natural systems for human use. Berg proposes that there are 'different zones of human interface with natural systems: urban, suburban, rural, and wilderness', each with 'a different appropriate reinhabitory approach' (2015, p. 139). Green cities, ecological restoration, and wilderness preservation are regarded as complementary rather than antagonist goals by bioregionalists.

## The bioregional paradigm

In contrast to what Sale (2000, p. 50) calls the 'industrio-scientific paradigm', which favors unlimited economic growth, centralized forms of decision making, and cultural homogenization at the national/global levels, the 'bioregional paradigm' espouses the devolution of economic, political, and social power to local communities. To achieve economic security, particularly in an age of peak oil and ecological limits to growth, bioregional communities aim at economic self-sufficiency in terms of food, clothing, shelter, energy, and other primary goods rather than depending on global markets to supply their basic needs (Cato, 2013). While complete self-reliance is neither necessary nor desirable, bioregionalism's advocacy of local production for local consumption lends support to grassroots movements within civil society advancing democratic alternatives to corporate globalization (Carr, 2004).

David Haenke's *Ecological Politics and Bioregionalism* (1984) provided the impetus for organizing the first North American Bioregional Congress (now the Continental Bioregional Congress) in 1984. In contrast with systems of governance that concentrate political power at the state, national, or international levels, bioregionalism champions more participatory forms of democracy that place ultimate decision-making power in the hands of local communities. Such communities can nonetheless be confederated at the appropriate levels 'through ecosystems, bioregions, and biomes' to resolve problems that cross local boundaries (Bookchin, 1982, p. 344). Following the principle of subsidiarity, issues are dealt with at the lowest appropriate level of social organization. Power flows not from the global to the local, but from the local to the global.

Distinguishing itself from mainstream environmentalism, bioregionalism adopts a proactive, prefigurative politics, which has less interest in protesting against the state, reforming laws and institutions, or taking control of the government than in acting directly to create practical alternatives to ecological devastation and social disintegration. Promoting biological and cultural diversity at the local level enables bioregional communities to both preserve their natural environments and prevent them from being exploited by others.

## Cultural diversity and global solidarity

Whereas globalization fosters what Berg (2009, pp. 129–37) refers to as a 'global monoculture', bioregionalism encourages learning and extending the lore of local cultures, including their customs, myths, and rituals. Bioregionalism's emphasis on localism has been criticized on the ground that it can lead to economic autarky, political isolationism, and cultural parochialism, with a corresponding inability to effectively address global environmental problems (Dudley, 1995). Bioregionalism fully embraces the Green slogan, 'think globally, act locally', however, and unequivocally rejects xenophobia, racism, and nativism.

Bioregionalists aspire to 'live regionally and yet learn from and contribute to planetary society' (Snyder, 1995, p. 247). It is only when individuals are rooted in a particular cultural tradition that they have something worthwhile to share with people from other cultures. Thomashow (1999) advocates a 'cosmopolitan bioregionalism', which recognizes that persons are simultaneously located not only in local landscapes (place) but also in global systems (space). Meredith similarly sees individuals as having overlapping relations at various scales, from micro-regions to macro-regions, with communities being 'interwoven between local and global affiliations' (2005, p. 93).

Local empowerment, rather than subservience to global forces beyond their control, is precisely what enables bioregional communities to engage in genuine acts of international (cross-bioregional) solidarity. As Berg writes, 'There are opportunities for life–place political alliances at all the levels from a local watershed to a continent (and eventually other continents' assemblies)' (2009, pp. 168–9). Bioregionalism has in fact become a global movement, with an influence that extends far beyond those who self-identify as bioregionalists.

## Learning resources

Pezzoli, K., 'Bioregional Theory,' Academic Computing and Media Services, 2013. Part 1: https://www.youtube.com/watch?v=9h6-ATaLGz4. Part 2: https://www.youtube.com/watch?v=0b2M8HF9ago. These two videos provide a comprehensive introduction to bioregional thought and practice.

Planet Drum Foundation: http://planetdrum.org. This website serves as a clearing house for information about bioregional activities, with links to projects, publications, and bioregional organizations worldwide.

Sale, K., 'Mother of All: An Introduction to Bioregionalism,' Third Annual E.F. Schumacher Lectures, 1983. http://www.centerforneweconomics.org/publications/lectures/sale/kirkpatric/mother-of-all. Kirkpatrick Sale introduces the key concepts of bioregionalism in this seminal lecture.

## Bibliography

Berg, P. (2009) *Envisioning Sustainability*. San Francisco: Subculture.

Berg, P. (2015) *The Biosphere and the Bioregion: Essential Writings of Peter Berg*. C. Glotfelty and E. Quesnel, eds. London: Routledge.

Berg, P. & Dasmann, R.F. (1977) 'Reinhabiting California'. *The Ecologist* 7(10): 399–401.

Bookchin, M. (1982) *The Ecology of Freedom: The Emergence and Dissolution of Hierarchy*. Palo Alto: Cheshire Books.

Carr, M. (2004) *Bioregionalism and Civil Society: Democratic Challenges to Corporate Globalism*. Vancouver: University of British Columbia Press.

Cato, M.S. (2013) *The Bioregional Economy: Land, Liberty, and the Pursuit of Happiness*. Oxford: Earthscan.

Dasmann, R.F. (1984) *Environmental Conservation*. 5th edition. New York: Wiley.

Dodge, J. (1981) 'Living by life: Some bioregional theory and practice'. *Coevolution Quarterly* 32: 6–12.

Dudley, J.P. (1995) 'Bioregional parochialism and global activism'. *Conservation Biology* 9(5): 1332–4.

Evanoff, R. (2011) *Bioregionalism and Global Ethics: A Transactional Approach to Achieving Ecological Sustainability, Social Justice, and Human Well-being*. New York: Routledge.

Haenke, D. (1984) *Ecological Politics and Bioregionalism*. Drury: New Life Farm.

Meredith, D. (2005) 'The bioregion as a communitarian micro-region (and its limitations)'. *Ethics, Place and Environment* 8(1): 83–94.

Sale, K. (2000) *Dwellers in the Land: The Bioregional Vision*. 2nd edition. Athens: University of Georgia Press.

Snyder, G. (1969) *Earth House Hold*. New York: New Directions.

Snyder, G. (1995) *A Place in Space: Ethics, Aesthetics, and Watersheds*. Washington, DC: Counterpoint.

Thomashow, M. (1999) 'Toward a cosmopolitan bioregionalism'. In *Bioregionalism*, pp. 121–32. M.V. McGinnis, ed. London: Routledge.

# Conservation

1.3

## Chris Sandbrook

### What is conservation?

The concept of conservation stems from concerns about how humans interact with, and ultimately harm, non-human nature, through processes such as overharvesting of natural resources, habitat conversion, and the effects of anthropogenic climate change (Primack, 2014). From this starting point, a wide range of different forms of conservation have emerged to tackle these perceived problems. This diversity makes conservation difficult to define, and in fact some definitions directly contradict one another. Seeking to identify shared themes running through all forms of conservation, Sandbrook (2015) defined it as 'actions that are intended to establish, improve or maintain good relations with nature' (p. 565). This high level definition helps to delineate the conceptual space within which conservation operates, but does little to unpack its diversity of ideas and practices. To do so, it is necessary to consider conservation's multiple competing visions for why, what, where and how to conserve.

### Why conserve?

There are various ways in which biodiversity (or more broadly non-human nature) can be considered to have value, and hence be worth conserving. These can be divided into 'use' values for people (such as for food, fuel and shelter) and 'non-use' values (such as cultural significance or benefits to future generations). Some argue that biodiversity has intrinsic value, irrespective of its use or even appreciation by humans. Indeed, some see a focus on intrinsic value as a definitive characteristic of conservation, distinguishing it from more anthropocentrically motivated environmentalism (e.g. Soulé, 2014). For others, conservation *should* be anthropocentric. This latter framing has gained popularity in recent years through the metaphor of 'ecosystem services' provided by nature to people, which are delivered by stocks of 'natural capital' (MA, 2005).

The latest manifestation of internal discussions within the conservation movement on the question of why to conserve is the so-called 'new conservation debate'. This has raged over the last few years, although many of its strands have a much longer history (Meine, 2014). Debates about why biodiversity should be conserved are not just academic. They affect profoundly subsequent choices about the objectives of conservation and how it should be practised.

## What to conserve and where?

Conservation effort is unevenly distributed both ecologically and spatially. In ecological terms, most attention over the history of conservation has been paid to the protection of species, and to large and charismatic animals in particular. These include the lions, pandas, elephants and gorillas that often feature in the names and logos of conservation organisations. More recently there has been a tendency to broaden the focus of conservation to include wider ecosystems, habitats and even land/sea-scapes, although species-based conservation remains widely practiced.

In spatial terms, many conservationists prioritise areas that contain high levels of biodiversity and/or that have low levels of human influence. Indeed, many are motivated to protect what they consider to be pristine 'wilderness' areas free (or almost free) from human influence. As more recent ideas around ecosystem services have emerged, there has been increasing conservation interest in areas that are heavily modified by people, including farmland and urban areas. This reflects an anthropocentric rationale for conservation, because biodiversity in such areas plays an especially important role in the delivery of value to people.

## How is conservation done?

Conservation has been practiced informally, by indigenous people, faith groups and others, for thousands of years through their cultural practices. What is now recognised as the modern conservation movement emerged in the mid-19th century in response to declines in the population of big game species on the plains of colonial Africa and harmful resource extraction in the frontier lands of the western United States (Adams, 2004). Since that time conservation has grown into a multimillion dollar industry, comprising a complex assemblage of actors including state agencies, civil society organisations, the private sector, international organisations and multilateral environmental agreements. These actors carry out many actions in the name of conservation, of which protected areas and market-based approaches deserve particular attention.

### *Protected areas*

Conservation throughout its modern history has had one really big idea – to create legally designated spaces for non-human nature within which human activities (or even presence) are forbidden. Beginning with the famous designation of Yellowstone National Park in the United States in 1872, protected areas now extend to a remarkable 14.7 per cent of the Earth's land, and 10 per cent of its territorial waters. The conservation community has set itself an ambitious target to further increase this coverage to 17 per cent of land and 10 per cent of all marine areas by 2020. Some conservationists go even further, and are calling for the designation of 50 per cent of Earth's surface into protected areas, under the slogans 'Nature Needs Half' and 'Half Earth' (e.g. Wilson, 2016).

Protected areas in many parts of the world were initially thought of as a tool for the complete separation of people and nature. This vision of human–non-human relations justified the forced removal of people already resident in designated areas, and the strict enforcement of boundaries through tools such as fencing and ranger patrols. This 'fortress conservation' or 'fences and fines' approach came in for much criticism in the latter part of the 20th century (Brockington, 2002). It was argued that it was harmful to the often poor and marginal communities living alongside parks, and ultimately counterproductive for conservation, which relies on cooperation and support from local stakeholders. An alternative 'community conservation' narrative emerged, emphasising the participation of local people and providing them with varying levels of decision-making power. It was often hoped that this approach would deliver 'win-win' outcomes for human development and conservation objectives (Adams et al., 2004).

In practice win-wins have proven hard to find, and in some areas there has been a 'back to the barriers' resurgence in the strict protection approach. In recent years this has been particularly associated with new forms of militarised conservation, deployed to protect charismatic species such as elephants and rhinos from the illegal wildlife trade. Links have been made between this trade and the financing of terrorist groups such as Al Shabaab, creating connections between conservation and the war on terror under an emerging paradigm of securitisation (Duffy, 2014).

## *Market-based conservation*

Conservation has always had a close relationship with capitalism and elite power. Early parks in North America were partly financed and justified by the tourists brought to them on newly built railroads, and those in colonial Africa were frequented by wealthy and powerful big game hunters such as Teddy Roosevelt (Adams, 2004). However, recent decades have seen a shift in gear in this relationship as wildlife has been increasingly expected to 'pay its own way' based on its (economic) value to people. This transition has taken place in response to the view that under the dominant neoliberal paradigm in the contemporary global political economy, only assets with market value will be taken into consideration by decision makers.

Various tools have emerged under this 'green economy' logic, including payments for ecosystem services, biodiversity offsetting, trading in carbon credits and the certification of biodiversity-friendly products (Pirard, 2012). At the same time, there has been a parallel trend to devolve conservation activity from state to non-state actors, including the private sector. For example, private protected areas are playing an ever-greater role in the delivery of conservation (Holmes, 2015). Noting the similarity of these trends to the broader process of neoliberalisation in society, some scholars have described the emergence, and rapid expansion, of 'neoliberal conservation' (Büscher et al., 2012).

Those arguing in favour of neoliberal forms of conservation say that they have brought (or perhaps, bought) conservation a seat at the table in the corporate boardrooms and ministries of finance where the most important decisions for the natural world are taken, and that market-based approaches deliver efficient allocation of scarce resources. Those criticising this approach say that neoliberal capitalism is hard-wired to pursue infinite economic growth, and therefore can never be fully reconciled with sustainability on a finite planet. They also point to weaknesses of market-based tools in practice, which can lead to perverse outcomes such as the erosion of non-material arguments for conservation. The relationship between conservation and capitalism remains an arena of intense theoretical and practical debate.

## Conclusion

'Conservation' today is an umbrella term for a wide, and often contradictory, set of ideas and practices. It might therefore make more sense to think of multiple conservations rather than a single movement. There is intense debate among conservationists regarding why, what, where and how to conserve, and this has always been the case. Some argue that such deliberations render conservation less effective in achieving societal change, whereas others hold that differences of opinion should be identified and debated rather than swept under the carpet (Matulis & Moyer, 2016). It seems unlikely that such disagreements will ever be resolved, because as Adams (2004) said, 'Conservation debates are not really arguments about nature, but rather about ourselves and the way we choose to live' (p. xiii).

## Learning resources

There are many online resources that can be used to find out more about key issues and contemporary debates in conservation:

International Union for Conservation of Nature (IUCN): a very large network of government and civil society organisations that provides much of the science and policy direction for conservation. Key resources include the IUCN Red List of Threatened Species: http://www.iucnredlist.org/ and the IUCN classification system for protected areas https://www.iucn.org/theme/protected-areas/about/protected-areas-categories.

Nature Needs Half: the online website for the (controversial) campaign to designate half of the planet's surface into protected areas for conservation: http://natureneedshalf.org.

Thinking Like a Human: a scholarly blog written by conservation social scientists on the relationship between nature and people: https://www.thinkinglikeahuman.com.

Conservation Bytes: an entertaining conservation biology blog with up to the minute reviews of new conservation research, as well as in depth analysis of some 'classics' https://conservationbytes.com.

United for Wildlife Online Course: this free course provides a basic introduction to a range of biological and social issues and conservation issues. It offers an excellent entry point into conservation, as well as insights into how a particular group thinks about conservation: https://learn.unitedforwildlife.org/course/index.php?categoryid=4.

## Bibliography

Adams, W.M. (2004) *Against Extinction: The Story of Conservation*. London: Earthscan.

Adams, W.M., Aveling, R., Brockington, D., Dickson, B., Elliot, J., Hutton, J., Roe, D., Vira, B. & Wolmer, W. (2004) 'Biodiversity conservation and the eradication of poverty', *Science*, 306, 1146–1149.

Brockington, D. (2002) *Fortress Conservation: The Preservation of the Mkomazi Game Reserve, Tanzania*. Oxford: James Currey.

Büscher, B., Sullivan, S., Neves, K., Igoe, J. & Brockington, D. (2012) 'Towards a synthesized critique of Neoliberal Biodiversity Conservation', *Capitalism Nature Socialism*, 23, 4–30.

Duffy, R. (2014) 'Waging a war to save biodiversity: The rise of militarized conservation', *International Affairs*, 90, 819–834.

Holmes, G. (2015) 'Markets, nature, neoliberalism and conservation through private protected areas in southern Chile', *Environment and Planning A*, 47, 850–866.

Matulis, B.S. & Moyer, J.R. (2016) 'Beyond inclusive conservation: The value of pluralism, the need for agonism, and the case for social instrumentalism', *Conservation Letters*, 10(3), 279–287.

Meine, C. (2014) 'What's so new about the new conservation?' In *Keeping the Wild: Against the Domestication of Earth* (Wuerthner, G., Crist, E. & Butler, T. eds) pp. 45–54. London: Island Press.

Millennium Ecosystem Assessment (MA) (2005) *Ecosystems and Human Well-being: Synthesis*. Washington, DC: Island Press.

Pirard, R. (2012) 'Market-based instruments for biodiversity and ecosystem services: A lexicon', *Environmental Science & Policy*, 19–20, 59–68.

Primack, R.B. (2014) *Essentials of Conservation Biology*, 6th edition. Sunderland, MA: Sinauer Associates.

Sandbrook, C. (2015) 'What is conservation?', *Oryx*, 49, 565–566.

Soulé, M. (2014) The 'new conservation'. In *Keeping the Wild: Against the Domestication of Earth* (Wuerthner, G., Crist, E. & Butler, T. eds) London: Island Press.

Wilson, E.O. (2016) *Half Earth: Our Planet's Fight for Life*. New York: Liverlight.

# 1.4 Desertification

## Diana K. Davis

Two thousand and sixteen saw the publication of two remarkable studies that make significant contributions to our knowledge about desertification. The first of these demonstrated that two centuries of vigorous afforestation in southern Europe has exacerbated climate warming rather than ameliorating it (Naudts et al., 2016). The second article detailed the sensitivity of different ecosystems to climate variability and showed that the drylands, including the Sahel, the Middle East, much of Australia, and the southwest US are the least sensitive of the terrestrial ecosystems – that is, they are the most ecologically resilient (Seddon et al., 2016). Both of these articles challenge us to reconsider ideas about the drylands and our common but outdated understandings of desertification.

They also follow a quarter century of scientific research on arid lands ecology that has repeatedly demonstrated that deserts are natural phenomena, not deforested, desertified wastelands, and that estimates of desertification have been significantly exaggerated. For centuries, deserts have been primarily conceived as previously vegetated land that has been deforested, burned or overgrazed by humans and their livestock. Although usually dated to the 1970s and the severe Sahelian drought and famine, concern about desertification actually became prominent during the colonial period and may be traced back several centuries (Davis, 2016). Many of the ecological misunderstandings of the drylands and some of the biases against their inhabitants have been retained since the colonial period and have negatively impacted development policies in the world's drylands. Understanding that legacy is crucial to developing new approaches to dryland development that will be more sustainable and equitable.

Before the 'age of exploration', Anglo-Europeans, when they considered deserts and arid lands, thought of them mostly as dry, hot, sandy places that presented difficulties for travel, often surprising riches, and rather strange peoples. The notion of the desert and arid lands as degraded landscapes, ruined by people and their livestock (early iterations of what would become known as desertification), though, appeared during the period of Western exploration and colonialism dating in large part to the seventeenth century. It was in the last half of that century that a significant connection began to be made between deforestation and decreased rainfall (desiccation theory) and accusations of environmental devastation by nomads began to be made by thinkers like Baron de Montesquieu. Montesquieu and others worried that deforestation in Europe would result in desiccation and civilizational collapse which they assumed had occurred in places like the region we know today as the Middle East. Reforestation at home in Europe and abroad was widely implemented in the face of these anxieties.

By the mid-nineteenth century it had become widely accepted in the Anglo-European world that deserts and drylands were not 'natural' but rather were the product of deforestation, burning and overgrazing. Indigenous peoples, especially nomads, were most often blamed for the dry landscapes assumed to be desiccated and ruined. Much of this kind of thinking about deserts and drylands was developed in the colonial territories of the Western countries, places like French Algeria and British India – but also in the western USA. Quite frequently, the assumption that the indigenous peoples had mismanaged the land, creating desert-like conditions, functioned as a justification for colonial appropriation of land and forests. It also justified imposing very specific land management tools like fire suppression, sedentarization of nomads and elimination or restriction of grazing, and reforestation in the name of 'bringing back the rains' and restoring a 'civilized and productive' climate.

This long intellectual legacy is evident in first use of the word 'desertification', for example, which was by the French colonial forester, Louis Lavauden, who wrote that "desertification is uniquely the act of humans. . . . [T]he nomad has created the pseudo-desert zone" nearly a century ago (Lavauden, 1927, pp. 267–337). He, like nearly all Anglo-European officials in the early twentieth century, recommended a standard package of techniques for the drylands, including sedentarizing nomads, curbing grazing, suppressing fire, reforestation and increasing irrigation. Over the course of the twentieth century, these colonial (mis)understandings of drylands and the techniques for their management were transferred and spread around the world by international organizations like the United Nations with their influential UNESCO Arid Zone Program.

Such knowledge was thus widespread by mid-century and when the devastating drought hit the African Sahel in the early 1970s, concern about desertification increased dramatically. This drought and the associated large famine stimulated international action to combat presumed desertification and the United Nations Conference on Desertification (UNCOD) was convened in 1977. Incorporating nearly all of the earlier colonial understandings about the drylands and their management, the UNCOD established global patterns of response to drought and management of the drylands that have been continued to the present via later international initiatives, especially the 1994 UN Convention to Combat Desertification (UNCCD) (Toulmin & Brock, 2016). Millions of people have been negatively affected by these anti-desertification policies, losing their land, livestock and other resources while these attempts to 'roll back the desert' all too often damaged the environment.

Much of the reason for these misguided policies and muddled understandings of the drylands that focus on 'desertification' lies in the fact that a large proportion of arid and semi-arid environments function as non-equilibrial systems, governed primarily by abiotic factors, due to the high inter-annual variability of rainfall (see Figure 1.4.1). Rainfall and drought in these zones are driven primarily by changes in oceanic temperatures,

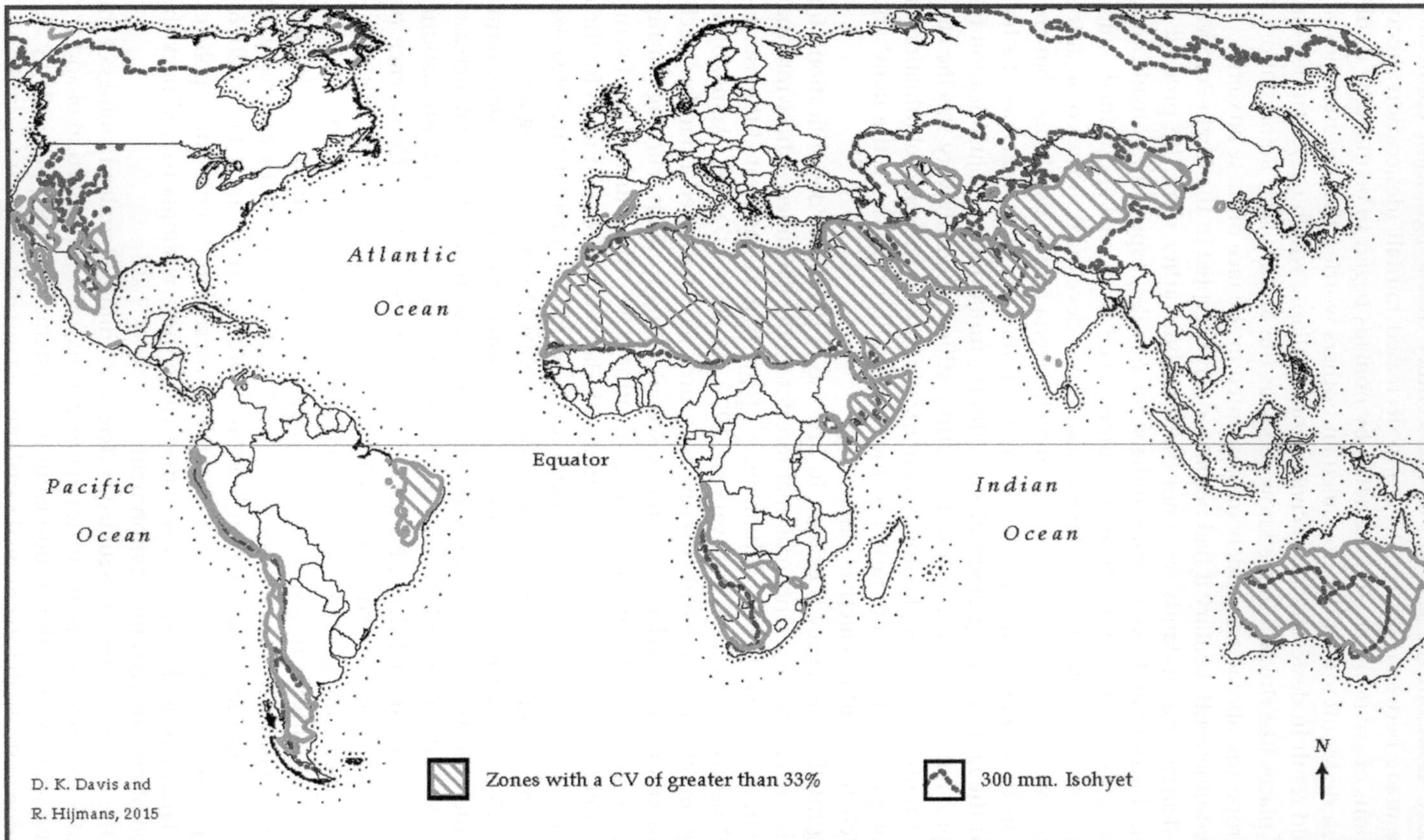


**Figure 1.4.1** Drylands variability map, showing 300 mm isohyet and the 33% CV (coefficient of variation of interannual rainfall).

Created by Diana K. Davis and Robert Hijmans. *Source*: D.K. Davis (2016) *The Arid Lands: History, Power, Knowledge*, The MIT Press, p. 16. Reproduced with the permission of The MIT Press. Higher resolution map available at: http://www.geovet.org/DrylandsMaps.html.

not by anthropogenic activities affecting vegetation such as overgrazing or deforestation (Giannini et al., 2008). Accordingly, most deserts are not continually expanding but they grow and shrink according to rainfall patterns more than any other single factor and have for millennia. Concern with desertification has been focused primarily on the semi-arid zone, that 20–25 per cent of the earth's surface that typically receives 200–500 mm of rainfall annually, where dramatic changes in vegetation are particularly obvious.

In these non-equilibrial regions, which comprise the majority of the arid and semi-arid lands, the vegetation is not governed by equilibrial dynamics that drive plant succession in more humid zones – dynamics controlled predominantly by biotic factors that are often understood to result in vegetation progressing from annuals to shrubs to trees. Moreover, in these dryland ecosystems, most of the vegetation is well-adapted to survive drought, heat and aridity, as well as fire and grazing. Many of the plants survive by existing for most of their lives as seeds in the soil, invisible to the naked eye, awaiting the next rainfall when they will germinate, grow quickly, set seed and senesce within just a few months. In such environments, it is nearly impossible to overgraze under the highly mobile herding systems of most societies indigenous to the drylands. Indeed, extensive grazing is frequently one of the most ecologically sustainable uses of the drylands. Given this complex ecology, "it is virtually impossible to separate the impact of drought from that of desertification" (Nicholson, 2011, p. 441), as is obvious in the recent 're-greening' of much of the African Sahel with higher rainfall.

Compounding confusion, more than 100 different definitions of desertification exist and there is no scientific consensus about how to measure or monitor desertification. Despite this, details of how degradation can occur in the drylands have been well documented and are known to include erosion, nutrient depletion, acidification, crusting, and salinization. Land degradation in arid and semi-arid lands that many categorize as desertification has taken place. Most of it, however, has resulted from over-irrigation, inappropriate afforestation, and the plowing of marginal soils, not 'traditional' forms of grazing or agriculture. These anthropogenic actions have resulted in serious land degradation in a relatively small number of discrete locations and have nearly always been driven by particular political, economic or social problems.

As prominent arid lands scientists have increasingly concluded, "we have a non-equilibrium world that is saddled with an overriding equilibrium mindset and [anti-desertification] policies that reflect it" (Hermann & Hutchinson, 2005, p. 38). This highlights the biggest problem for dryland development: the lack of scientific understanding of dryland ecologies among the policy community which facilitates the continuation of inappropriate and unsuccessful 'anti-desertification' policies. These policies frequently result in ecological harm, including lowered water tables from inappropriate afforestation, salinization from over-irrigation and poor drainage, and desiccation of soils and disruption of native vegetation from plowing soils marginal for farming. Better understanding the deep colonial roots of the underlying environmental imaginary of deserts and desertification which drives so much inappropriate dryland development provides multiple benefits. It will result in improved social outcomes, fewer environmental failures like that of afforestation driving desiccated soils in the drylands and more warming in Europe, and a more widespread appreciation of the important implications of the ecological resilience of the drylands.

## Learning resources

Books helpful for learning more about drylands and desertification include:

Behnke, R. & Mortimore, M. (Eds) (2016) *The End of Desertification: Disputing Environmental Change in the Drylands*. Dordrecht: Springer.

Middleton, N. (2009) *Deserts: A Very Short Introduction.* Oxford: Oxford University Press.

Nicholson, S. E. (2011) *Dryland Climatology*. Cambridge: Cambridge University Press.

A useful journal for many different issues related to drylands and desertification is the *Journal of Arid Environments.*

## Bibliography

Davis, D.K. (2016) *The Arid Lands: History, Power, Knowledge*, Cambridge, MA: The MIT Press.

Giannini, A., Biasutti, M. & Verstraete, M. (2008) 'A Climate Model-Based Review of Drought in the Sahel: Desertification, the Re-Greening and Climate Change', *Global Planetary Change*, 64(3–4): 119–128.

Hermann, S.M. & Hutchinson, C.F. (2005) 'The Changing Contexts of the Desertification Debate', *Journal of Arid Environments*, 63(3): 538–555.

Lavauden, L. (1927) 'Les Forêts du Sahara', *Revue des Eaux et Forêts*, 65(6): 265–277 and 329–341.

Naudts, K., Chen, Y., McGrath, M., Ryder, J. & Valade, A. (2016) 'Europe's Forest Management did not Mitigate Climate Warming', *Science*, 351(6273): 597–600.

Nicholson, S.E. (2011) *Dryland Climatology*. Cambridge: Cambridge University Press.

Seddon, A.W., Marcias-Fauria, M., Long, P.R., Benz, D. & Willis, K.J. (2016) 'Sensitivity of Global Terrestrial Ecosystems to Climate Variability', *Nature*, 531(7593): 229–232.

Toulmin, C. & Brock, K. (2016) 'Desertification in the Sahel: Local Practice Meets Global Narrative'. In *The End of Desertification: Disputing Environmental Change in the Drylands*, edited by R. Behnke and M. Mortimore. Dordrecht: Springer.

# Environment

1.5

*Sverker Sörlin*

Environment is a word with a long history but with a short and influential career as a modern concept to denote human–nature relationships, more often problematic than benign. The word was known in Medieval English and referred to an immediate vicinity, for example of a farm or village, rather than the more boundless 'nature'. Indeed, there was a verb, 'to environ' (Sörlin & Warde, 2009) which meant 'to form a ring around, to surround', originally French but in English usage from the 14th century. Environment would thus be more related to human action than nature, and it would expand with human presence and action. In a sense, if we did not name it as such, 'the environment' would not exist. The invention and use of the word says much about how some societies have come to understand their relationship with the biophysical world.

## Making impressions

With time the word became the carrier of an old idea, in the West known since Antiquity, that surrounding conditions impacted on humans and societies. Commonly the word used here was climate (*klima* in Greek), including temperature, humidity, character of landscape, but in the 19th century 'environment' was increasingly used to suggest such influences; early European examples included Thomas Carlyle (in the 1820s) and Herbert Spencer (from the 1850s onwards). Spencer suggested that environmental factors made imprints on thoughts, emotions and society but interestingly without attributing any specific properties to the environment itself; its effects were received but environment itself was not an object of investigation (Warde, 2016).

Although in current usage, environment tends to have a wider geography, in an extreme version including 'everything' around us (humans) and hence also being a 'global environment', 19th- and early 20th-century usage retained the old sense of an immediate surrounding. This became particularly evident in the way the word entered literature and art, where landscape and social atmosphere and setting were increasingly seen as important influences both on literary, fictional characters and on the authors and artists themselves. One could see this as a corresponding, more holistic environmental counterpoint to more distinctly sociological explanations of behavior, manners, morals and status of anything from criminals to schoolchildren, which was another rapidly growing practice at about the same time. Significantly, therefore, the early growth of 'environment' as a word in English takes off from about 1890, which coincides with the entry of 'sociology', 'ecology' and 'psychology' into common language, suggesting that concepts that tried to capture, and in due course quantify, external (or wider) causes behind human behavior and societal differences were *en vogue*.

## Determinist strands

This thinking became especially popular in strands of the discipline of geography. First articulated in Germany by Carl Ritter and Friedrich Ratzel, the idea that environmental conditions explained the 'level' of culture and civilization in a country or society spread widely. These ideas could have Darwinist justifications (although Darwin himself very rarely used environment as a scientific concept) and were readily compatible with hierarchical (including racist) understandings of human diversity, not least as Europeans were trying to make sense of their expanding overseas empires (Arnold, 1996). As they were systematized, they inspired attempts to calculate civilization as a function of certain environmental factors, for example by Ellen Churchill Semple in *Influences of Geographic Environment* (1911), or, perhaps most infamously, Ellsworth Huntington in his book *Civilization and Climate* (1915).

The notion of environment as an integrative concept for multiple factors and circumstances impacting on humans, and hence with a local or regional extent, also resulted in the word being used as a prefix, as in 'environmental medicine' or 'environmental health'. This referred to, for example, toxic elements in air, water, and soil, but also what has been called 'human engineering' to build a protective shield against a threatening environment. It could also mean an artificially managed micro-climate surrounding the individual human being, along similar lines as the capsules created for space flight with regulated air conditioning, temperature and tailored food rations; a micro climate surrounding the individual person. The soldier should thus perform and inhabit what could be seen as a micro version of the artificial landscapes that became part of technocratic landscaping and terraforming during the same period.

A key concept in this work was 'environment'. At the same time, starting in earnest in the 1930s, indoor environments became an issue, about energy efficiency, but also thermal comfort. The military and medical understanding of 'environment' as harsh or hostile conditions to which humans could more or less well adapt continued through the second half of the 20th century, however dwarfed by another branch of environmental health based in industrial medicine and occupational health.

## A crisis concept

The major turning point of 'environment' came after World War II. Always a relational concept, the environment 'of' something – any human being, any species, but also for example a particular city, a certain growing child – it now started to be talked about as 'the environment', a wider 'out there'. The word now referred not only to the impacts that environment had but, on the contrary, as the concept used for human impacts *on* their surroundings, often understood as 'nature' or some part thereof. Several significant books, reports and conferences over a period of a couple of decades after World War II established this pervasive change. Central to the change was a tone of concern, if not alarm. William Vogt in *The Road to Survival* (1948) and Fairfield Osborn, in *Our Plundered Planet* (1948), both ecologists, listed a range of problems that humans and their societies had wrought, all of them aggravating drastically: soil erosion, pollution, overuse of natural resources including fish and wildlife. Much of these resulted from rapid population growth, which then was defined, by Vogt, an avid Malthusian, as the 'mother' of all the other problems.

Human/societal treatment of *the* environment was as pervasive as it was destructive. This new understanding was not immediately heeded in the late 1940s, but it was underpinned by growing evidence across a broad range of scientific fields, for example the biological and the geological sciences, including geography, hydrology, oceanography, and others that environments were changing in most parts of the world. A seminal conference held in Princeton in 1955 made that change into its title, *Man's Role in Changing the Face of the Earth* (in book form 1956). At around the same time 'environmental sciences' became a concept, used by British primatologist and science advisor Solly Zuckerman in 1959, and spreading in the 1960s as a prefix word in titles of books, conferences, and soon also university departments.

## Predictive expertise

The success of 'environment' can be seen as an effect of its capacity to function as a unifying concept that could glue together almost any phenomenon or problem related to the human/nature, or human/outside world interface (Warde et al., 2018). Its growth as a concept was further reinforced by the fact that social and political interest was invested in the problems societies caused in the environment. To detect these issues, and subsequently to deal with them, required expertise and knowledge which quickly created institutionally well anchored interest groups, often siding with the broader environmentalism that started spreading from the 1960s as a civic movement and a political idea in many countries. These factors combined made environment a crisis concept, reflecting a perceived, hitherto neglected malfunctioning of modern societies.

Since it was the changes in the environment that were negative, the concept also soon became predictive, and so linked to interventions and solutions. In this there had been important forerunners in the decades before World War II that had worked out methods to understand and measure rate and direction of change in natural processes (Warde & Sörlin, 2015). Here belong for example: Alfred Lotka's equations in population ecology, early notions of anthropogenic climate change (Arrhenius, 1896;

Callendar, 1938), and Vladimir Vernadsky's work on humans as an earth-shaping 'geological' force affecting both 'the biosphere' (a concept he launched in a book 1926) and wider bio-geo-chemical processes. To this could be added a range of empirical observations of massive human impacts, and of course the establishment, also in the interwar years, of the demographic transition (Bashford, 2014). This in turn stimulated a population mathematics which projected future growth as the cause of further change in the environment.

Conceptually and historically the school of environmental determinism had suffered severe setbacks. A forceful empirical critique was waged by historians such as Lucien Febvre (1922) and others in the subsequently formed French *Annales* school of history, then from the disastrous uses of racism and various forms of biological determinism by especially the Nazi regime. Also, the war years had seen the enormous power of concerted human efforts, epitomized by the use of the atomic bomb. The terraforming capacity of humanity seemed almost endless, and while most interpretations of this fact were benign and fueled hopes of progress, the environment became the keyword for some of its undeniable and quantifiable downsides.

## The Age of Environment?

When it had crystallized as *the* integrative concept of choice for a troubled humanity/nature relationship its career was remarkable and it stands out as one of the most widely acknowledged and important scientific and political concepts of the 20th century, if not of modernity. It was used by Rachel Carson in *Silent Spring* (1962), either with a qualifier, 'man's total environment', or just as 'the environment', a usage that was now in some circulation. In 1965 it was actively used in a comprehensive report to President Lyndon B. Johnson on *Restoring the Quality of Our Environment*, covering anything from pest control and 'metropolitan problems' to atmospheric $CO_2$. As the title indicated, the environment was already considered in such bad shape it needed to be restored. This was an issue elevated to the highest possible level of importance: "The continued strength and welfare of our nation . . . depend on the quantity [sic] and quality of the environment in which our people live" (White House, 1965, p. 2).

By the 1970s, government departments for the environment started to form, and countless organizations, magazines, journals, think tanks and institutes were founded, starting in Europe and North America but soon literally everywhere. In June 1972 the United Nations Conference on the Human Environment was held in Stockholm, with that qualifiying 'human' prefix still there as a faint, but soon superfluous, emblem of the concept's human point of reference. But it was also there as a signal to those many nations where elementary health and wealth of humans was still considered much more important than issues that could easily be mixed up with conservation of nature, still perceived as a pet project of Western elites.

For some time, (the) environment has met some competition in the ever more crowded conceptual space around man/nature relations. Biodiversity and sustainability (both with a rise since 1980s), ecological modernization (1990s), and Anthropocene (2000s) are among the concepts that have been encroaching on the conceptual hegemony of 'environment', offering a mix of new narratives and chronologies, and identifying other causalities, complexities (and simplifications), and also different directions and destinies. Some of these offer more presence to humans, cultures, economies, and societies than did 'environment'. On the other hand 'environment' has proven remarkably elastic and resilient as a concept, providing new combinations that cater to diversities of all kinds, the humanities, social sciences, and to the human condition. Whether 'the environment' will in due course stand as a monumental term of a certain epoch that identified it as one of the chief challenges of the modern world – an 'Age of Environment' – or whether it will remain at center stage, will be the task of future generations to find out.

## Learning resources

A useful introduction to thinking reflexively about 'the environment' and how it has developed is the collection of seminal texts across a wide range of fields entitled *The Future of Nature: Documents of Global Change*, edited by Libby Robin, Sverker Sörlin & Paul Warde (New Haven: Yale University Press, 2013). It is organized and edited especially for classroom use. The same team have followed up with a full bodied, yet compact conceptual history, Warde, Robin & Sörlin, *The Environment – A History* (Baltimore MD: Johns Hopkins University Press, 2018).

Otherwise, the environment as a concept is complex enough to make most scholars cautious to approach it. Perhaps strangely, there are many book length studies of the concept of 'nature' (also complex indeed), or 'wilderness', but next to none on 'environment'. The exceptions are mostly works by historical geographers. Clarence Glacken's *Traces on the Rhodian Shore: Nature and Culture in Western Thought from Ancient Times to the End of the Eighteenth Century* (Berkeley, CA: University of California Press, 1967, and later editions) is, after more than half a century, still a very useful overview. The same goes for David N. Livingstone, *The Geographical Tradition: Episodes in the History of a Contested Enterprise* (Oxford: Basil Blackwell, 1992).

## Bibliography


Arnold, D. (1996) *The Problem of Nature: Environment, Culture and European Expansion.* Oxford: Blackwell.

Arrhenius, S. (1896) "On the Influence of Carbonic Acid on the Temperature on the Ground", *Philosophical Magazine and Journal of Science*, Series 5, Vol. 41, April 1896.

Bashford, A. (2014) *Global Population: History, Geopolitics, and Life on Earth.* New York: Columbia University Press.

Callendar, G.S. (1938) "The Artificial Production of Carbon Dioxide and its Influence on Temperature", *Quarterly Journal of the Royal Meteorological Society*, 64: 275.

Carson, R. (1962) *Silent Spring.* New York: Houghton Mifflin.

Febvre, L. (1922) *La Terre et l'Évolution Humaine.* Paris: Albin Michel, Engl. transl., *A Geographical Introduction to History*, The History of Civilization Series (1925/1932), 2nd ed. London: Kegan Paul, Trench, Trubner & Co.

Huntington, E. (1915) *Civilization and Climate.* New Haven, CT: Yale University Press.

Osborn, F. (1948) *Our Plundered Planet.* Boston & Toronto: Little, Brown and Company.

Robin, L., Sörlin, S. & Warde, P. (2013) "Introduction", in Robin, L., Sörlin, S. & Warde, P. (Eds) *The Future of Nature. Documents of Global Change* (pp. 1–14). New Haven, CT: Yale University Press.

Semple, E.C. (1911) *Influences of Geographic Environment: On the Basis of Ratzel's System of Anthropo-Geography*. New York: Henry Holt; London: Constable.

Sörlin, S. & Warde, P. (2009) "Making the Environment Historical – an Introduction", in Sörlin, S. & Warde, P. (Eds), *Nature's End: History and the Environment* (pp. 1–19). London: Palgrave Macmillan.

Vernadsky, V. (1926) *La Biosphère* (Russian orig. 1926), French transl. (1929). Paris: Félix Alcan.

Vogt, W. (1948) *The Road to Survival*. New York: William Sloane Associates.

Warde, P. (2016) "The Environment", in Coates, P., Moon, D. & Warde, P. (Eds), *Local Places, Global Processes* (pp. 32–46). Oxford: Windgather Press.

Warde, P. & Sörlin, S. (2015) "Expertise for the Future: The Emergence of Environmental Prediction c. 1920–1970", in Andersson, J., & Rindzeviciute, E. (Eds), *The Struggle for the Long Term in Transnational Science and Politics During the Cold War* (pp. 38–62). Abingdon: Routledge.

Warde, P., Robin, L. & Sörlin, S. (2018) *The Environment – A History*. Baltimore, MD: Johns Hopkins University Press.

White House (1965) *Restoring the Quality of Our Environment: The report of the Environmental Pollution Panel, President's Science Advisory Committee*. Washington, DC: The White House, November 1965.

# Ecosystems

1.6

Erle C. Ellis

An ecosystem is a community of organisms interacting with each other and with their abiotic environment such that energy is exchanged and system-level processes, such as the one-way flow of energy and the cycling of elements, emerge.

The ecosystem is a core concept in the fields of biology and ecology, serving as the level of biological organization in which organisms interact simultaneously with each other and with their immediate environment. As such, ecosystems are a level above that of the ecological community (organisms of different species interacting with each other) but are at a level below, or equal to, biomes and the biosphere. Essentially, biomes are regional ecosystems, and the biosphere is the largest of all possible ecosystems.

Ecosystems include living organisms, the dead organic matter produced by them, the abiotic environment within which the organisms live and exchange elements (soils, water, atmosphere), and the interactions among these components. Ecosystems embody the concept that living organisms continually interact with each other and with the environment to produce complex systems with emergent properties, such that "the whole is greater than the sum of its parts" and "everything is connected".

The spatial boundaries, component organisms and the matter and energy content and flux within ecosystems may be defined and measured. However, unlike organisms or energy, ecosystems are inherently conceptual, in that different observers may legitimately define their boundaries and components differently. For example, a single patch of trees together with the soil, organisms and atmosphere interacting with them may define a forest ecosystem, yet the entirety of all organisms, their environment, and their interactions across an entire forested region in the Amazon might also be defined as a single forest ecosystem. Moreover, the interacting system of organisms that live within the guts of most animals may also be seen as an ecosystem (the "microbiome"), despite their

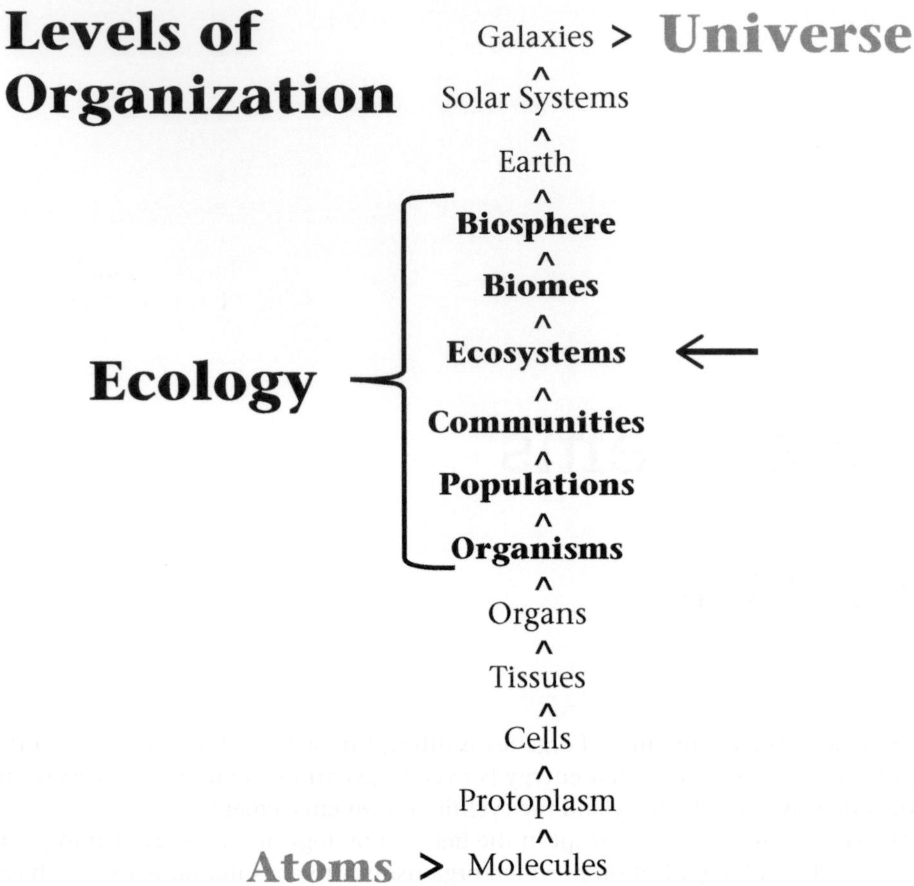


**Figure 1.6.1** Levels of organization in ecology, highlighting ecosystems.

Credit: Erle Ellis.

residence within a single organism, seemingly violating the level of organization-defining ecosystems. Note also that interactions among ecosystem components are as much a part of the definition of ecosystems as their constituent organisms, matter and energy. Despite the apparent contradictions that result from the flexibility of the ecosystem concept, it is just this flexibility that has made it such a useful and enduring concept.

## History of the ecosystem concept

The term "ecosystem" was first coined by Roy Clapham in 1930, but it was ecologist Arthur Tansley who fully defined the ecosystem concept. In a classic article in 1935, Tansley defined ecosystems as "The whole system, . . . including not only the organism-complex, but also the whole complex of physical factors forming what we call the environment". The ecosystem concept marked a critical advance in the science of ecology, as Tansley specifically used the term to replace the "superorganism" concept, which implied that communities of organisms formed something akin to a higher-level, more complex organism—a mistaken conception that formed a theoretical barrier to scientific research in ecology. Though Tansley and other ecologists also used the ecosystem concept in conjunction with the now defunct concept of the ecological "climax" (a "final", or "equilibrium" type of community or ecosystem arising under specific environmental conditions—as a "balance of nature"), the concept of ecosystem dynamics—the "fluxes of nature" has now replaced this. Eugene Odum, a major figure in advancing the science of ecology, made the ecosystem concept the central role in his seminal textbook on ecology, defining ecosystems as: "Any unit that includes all of the organisms (i.e. the "community") in a given area interacting with the physical environment so that a flow of energy leads to clearly defined trophic structure, biotic diversity (biodiversity), and material cycles (i.e. exchange of materials between living and non-living parts) within the system is an ecosystem."

## Ecosystem structure and function

### *Ecosystem components (structure)*

Ecosystems may be observed in many possible ways, so there is no one set of components that make up ecosystems. However, all ecosystems must include both biotic and abiotic components, their interactions, and some source of energy. The simplest (and least representative) of ecosystems might therefore contain just a single living plant (biotic component) within a small terrarium exposed to light to which a water solution containing essential nutrients for plant growth has been added (abiotic environment). The other extreme would be the biosphere, which comprises the totality of Earth's organisms and their interactions with each other, abiotic, "spheres" of the Earth system (the Atmosphere, Hydrosphere, Lithosphere, etc.). And of course, most ecosystems fall somewhere in between these extremes of complexity.

At a basic functional level, ecosystems generally contain primary producers capable of harvesting energy from the Sun by photosynthesis and of using this energy to convert carbon dioxide and other inorganic chemicals into the organic building blocks of life. Consumers feed on this captured energy, and decomposers not only feed on this energy, but also break organic matter back into its inorganic constituents, which can be used again by producers. These interactions among producers and the organisms that consume and decompose them are called trophic interactions, and are composed of trophic levels in an energy pyramid, with most energy and mass in the primary producers at the base, and higher levels of feeding on top of this, starting with primary consumers feeding on primary producers, secondary consumers feeding on these, and so on. Trophic interactions are also described in more detailed form as a food chain, which organizes specific organisms by their trophic distance from primary producers, and by food webs, which detail the feeding interactions among all organisms in an ecosystem. Together, these processes of energy transfer and matter cycling are essential in shaping ecosystem structure and function and in defining the types of interactions between organisms and their environment. It must also be noted that most ecosystems contain a wide diversity of species, and that this diversity should be considered part of ecosystem structure.

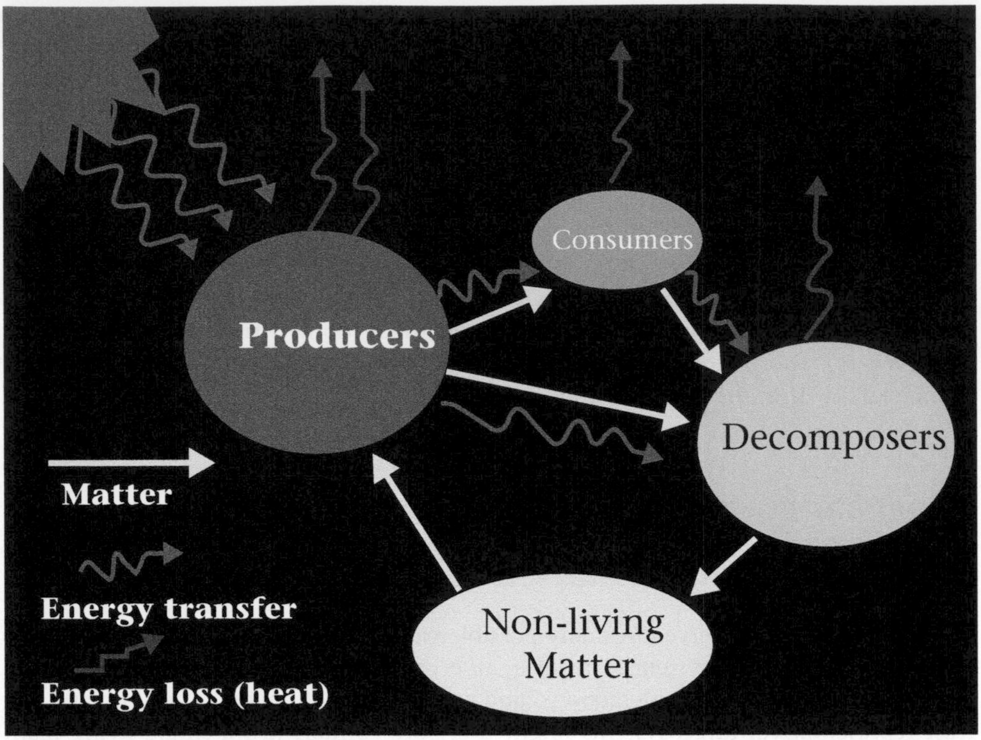


**Figure 1.6.2** Illustration of the flow of matter and energy in ecosystems.

Credit: Erle Ellis.

## *Ecosystem processes (function)*

By definition, ecosystems use energy and cycle matter, and these processes also define the basic ecosystem functions. Energetic processes in ecosystems are usually described in terms of trophic levels, which define the role of organisms based on their level of feeding relative to the original energy captured by primary producers. In keeping with thermodynamic laws, energy does not cycle, so ecosystems require a continuous flow of high-quality energy to maintain their structure and function. For this reason, all ecosystems are "open systems" requiring a net flow of energy to persist over time—without the Sun, the biosphere would soon run out of energy!

Energy input to ecosystems drives the flow of matter between organisms and the environment in a process known as biogeochemical cycling. The biosphere provides a good example of this, as it interacts with and exchanges matter with the lithosphere, hydrosphere and atmosphere, driving the global biogeochemical cycles of carbon, nitrogen, phosphorus, sulfur and other elements. Ecosystem processes are dynamic, undergoing strong seasonal and even daily cycles in response to changes in solar irradiation, causing fluctuations in primary productivity and varying the influx of energy from photosynthesis and the fixation of carbon dioxide into organic materials, driving remarkable annual variability in the carbon cycle—the largest of the global biogeochemical cycles. Fixed organic carbon in plants then becomes food for consumers and decomposers, who degrade the carbon to forms with lower energy, and ultimately releasing the carbon fixed by photosynthesis back into carbon dioxide in the atmosphere, producing the global carbon cycle. The biogeochemical cycling of nitrogen also uses energy, as bacteria fix nitrogen gas from the atmosphere into reactive forms useful for living organisms using energy obtained from organic materials and ultimately from plants and the Sun. Ecosystems also cycle phosphorus, sulfur and other elements. As biogeochemical cycles are defined by the exchange of matter between organisms and their environment, they are classic examples of ecosystem-level processes.

## Ecosystem research

Scientists who study entire ecosystems are generally called systems ecologists. However, most ecologists use the ecosystem concept and make measurements on ecosystem properties even if their work focuses on a single species or population.

### *Observing ecosystems*

Researchers can make direct observations on ecosystems in the field and indirect observations using remote sensing. Direct measurements include sampling and measurement of soils and vegetation, characterization of community structure and biodiversity, and the use of instruments for observing gas exchange and the fluxes of nutrients and water. As ecosystems can be very challenging to recreate under laboratory conditions, observational studies on existing ecosystems are a core methodology of ecosystem science.

### *Ecosystem experiments*

Though it has historically been difficult, ecosystems are now often studied using the classic experimental methods of science. For example, small- and meso-scale ecosystems containing a significant set of interacting organisms and their environment may be created in the laboratory, or in enclosures in the field. There are also methods for excluding organisms or altering environmental conditions in the field, such as the addition of nutrients and artificially enhancing carbon dioxide concentrations, temperature or moisture.

### *Modeling*

To better understand how ecosystems function and change, modeling is often used to simulate ecosystem dynamics, including the biogeochemical cycles of carbon and other elements, the role of specific species or functional groups in controlling ecosystem function, and even dynamic changes in ecosystem structure and function across landscapes and the entire biosphere.

## The future

Ecosystem science is evolving rapidly in both methodology and focus. Human alteration of ecosystems is now so pervasive globally that ecologists are working to integrate humans into ecosystem science at many levels—including the study of urban ecology, agroecology, social-ecological systems and global ecology and international efforts to assess the state of ecosystems globally—such as the "Millennium Ecosystem Assessment" and the Intergovernmental Panel on Biodiversity and Ecosystem Services. New techniques for ecosystem modeling are being developed all the time, as are new methods for observing ecosystems from space by remote sensing and aerial platforms, and even by networks of sensors embedded in soils and plants across ecosystems and on towers that can make observations on ecosystem exchanges with the atmosphere on a continuous basis. Examples of cutting edge ecosystem research are the Carnegie Airborne Observatory—an aerial remote sensing system capable of precisely mapping ecosystem carbon and species diversity, and the development of the National Ecological Observatory Network (NEON), a continental-scale research platform for discovering and understanding the impacts of climate change, land-use change, and invasive species on ecosystems.

## Learning resources

The Ecological Society of America http://www.esa.org/esa/
The National Ecological Observatory Network (NEON) http://www.neonscience.org/
The Millennium Ecosystem Assessment http://www.millenniumassessment.org/en/index.html
Intergovernmental Panel on Biodiversity and Ecosystem Services (IPBES) http://www.ipbes.net/

## Bibliography

Chapin III, F.S., Kofinas, G.P. & Folke, C. (2009) *Principles of Ecosystem Stewardship: Resilience-Based Natural Resource Management in a Changing World*. New York: Springer.

Chapin III, F.S., Matson, P.A. & Vitousek, P.M. (2012) *Principles of Terrestrial Ecosystem Ecology*. New York: Springer.

Golley, F.B. (1993) *A History of the Ecosystem Concept in Ecology: More Than the Sum of the Parts*. New Haven: Yale University Press.

Odum, E.P. (1971) *Fundamentals of Ecology*. 3rd edition. Philadelphia: Saunders, W.B.

Tansley, A.G. (1935) 'The use and abuse of vegetational concepts and terms', *Ecology*, 16: 284–307.

# Environmental catastrophe

1.7

Giovanni Bettini

The destructive powers and vagaries of 'nature' have constantly reminded humans of the contingency of their lifeworld. Thunders and plagues descending from the sky. Waves surging from the seas. Tremors, smoke and magma surfacing from the earth's deepest strata. Such cataclysms are recurring traumas in human history, inscribed in myth, religion, and politics (Clark, 2011; Ghosh, 2016; Hulme, 2009). But today 'environmental catastrophe' means also something else, as it alludes more to post-apocalyptic Mad Max-like scenarios than to biblical plagues. It casts a dystopic vista over the possibility, detected with the tools of modern science, of a human-induced planetary collapse. This is the case for all the catastrophes environmentalists have warned about and mobilized against (such as mass species extinction, nuclear contamination, the ozone hole, desertification, and more recently climate change). The fact that humans themselves are causing the very problem opens up an ethical dimension to the catastrophe, and poses divisive questions: Who exactly is to blame – can we really talk about humanity as a whole? And who should do (have done) what to avoid the materialization of the catastrophe, or at least to mitigate its fallout?

This chapter looks closer at how the idea of catastrophe has been inscribed in modern and contemporary discourses on the environment. It also introduces contrasting views on whether the invocation of environmental catastrophe can foster awareness and mobilize action, or obfuscates the underlying issues.

## A modernist nightmare: science, humanity and its planetary impacts

Contemporary discourses on environmental catastrophe are quintessentially 'modern'. They narrate, in a dystopic register, an alarming discovery made last century: since the European Industrial Revolution, human activities have had irreversible impacts on a planetary scale, pushing vital functions of the Earth system to the brink of collapse, with most serious implications for human societies and their survival. The signs of the looming disaster are detected by science – not superstition, religion, art. The dystopic scenarios that have permeated Western environmental consciousness since the 1960s are hyperbolic interpretations of discoveries made in labs, codified in computer models, illustrated with satellite pictures (Forsyth, 2003; Jamison, 2001; Latour, 2004; Lövbrand et al., 2009; Robbins, 2012). Key cultural milestones of environmentalism (such as Rachel Carson's *Silent Spring*, Paul Erlich's *Population Bomb*, and the Club of Rome's *Limits to Growth*) were all weaved within the frame of science. Environmental catastrophe – as a punishment for human hubris – has thus a Promethean character. It is slightly paradoxical that one of the tenets of the Enlightenment's reason (the idea of 'nature' and the planet being fundamentally indifferent to human action) is called into question by one of the expressions *par excellence* of that same reason (science), because of something (technology and industrialization) made possible by that very reason.

The concept of the Anthropocene is exemplary of this modernist knot, and helps illustrate another key ingredient of the environmental catastrophe: its planetary scale. The (problematic) definition of the Anthropocene as 'the époque of mankind' heralds the idea that humanity is becoming a geological force pushing the whole planet into a different state of operation. It thereby projects the question of human agency (and responsibility) onto a planetary level, empirically and ethically. This is key to environmental catastrophism, whose nightmares are not localized ecological degradation and disasters, but collapse on a planetary scale. In the Anthropocene – as in environmental catastrophe – the whole of humanity and the planet are at stake.

Another trope of catastrophism is the idea that humans are dangerously close to trespassing a limit beyond which abrupt and unpredictable shifts, unlikely benign for life on Earth, will materialize. The concepts of tipping points and planetary boundaries formalise such concerns. The existence of thresholds beyond which radical shifts will kick in has become more salient also in relation to 'runaway' climate change, which appears less unlikely every time scenarios and models are revised. An increasing number of observers (including scientists) warn over the possibility that, if a certain temperature increase limit is passed, we might be heading towards a 'climate catastrophe' that threatens the survival of human civilizations as we know them (Klein, 2014).

## A leap over the precipice

But does pointing to the magnitude and potentially catastrophic impacts of environmental change or the Anthropocene qualify as 'catastrophism'? Of course not. Awareness of the extreme dangers posed by environmental changes is widespread – being concerned does not mean being a catastrophist. And, in fact, concerns over global environmental change have profoundly influenced also mainstream political sensibilities and agendas (see the ecological modernization paradigm), and spurred the emersion of 'new' visions that are far from radical (such as those on Green Economy and Green Growth).

What qualifies catastrophism is rather the act of looking over the brink. Imagining to experience the downfall and to explore the ruins. Environmental catastrophe becomes a quasi-transcendental moment of revelation, an (imagined) encounter with radical dystopic change. A moment in which the worst 'finally' takes place and abruptly disrupts the present societal fabric – ending the world as we know it.

## Mobilizing, cathartic, or disempowering?

The place of the environmental catastrophe in (green) political agendas has been heavily scrutinized and challenged in green political theory and cognate fields. The point of contention is not the disruptive potential of global environmental changes. Rather, severe disagreements exist on whether waving the spectre of the catastrophe fosters awareness and enables action, or results in paralysing and/or serves as justification for conservative responses (Lilley et al., 2012).

One widespread assumption is that catastrophe is only a figure of the limit, a semi-transcendental, exceptional moment of rupture. As many have shown, however, catastrophe is also a stable – albeit paradoxical – component of mainstream modes of government. Aradau and van Munster (2011) offer a compelling genealogy of 'politics of catastrophe' in the realm of anticipatory governance and risk management, showing the extent to which catastrophe – with its peculiar epistemology, aesthetics and practices – is a structuring component of contemporary policy fields such as those on 'global terror' or climate change. Other authors have highlighted how the complex affective economies set in motion by catastrophic imaginaries are involved in the reproduction of existing configurations of power (Swyngedouw, 2010; Žižek, 2010). Put simply, catastrophe can be renormalized and regulate affective exchanges involved in mundane policy making. Nothing guarantees that the affective intensity that the spectre of catastrophe arouses is channelled into action or change.

To the contrary, a number of authors argue that framing socio-ecological crises as environmental catastrophe risks naturalizing and de-historicising the politics involved. The idea of catastrophe in green discourses has often resonated with determinism and the Malthusian idea that 'nature' places limits on human societies, thereby determining their fate. The invocation, often in the language of science, of planetary limits and natural laws risks overshadowing their politics and history (Harvey, 1974). And when disaster is made to look as unavoidable as the 'laws of nature', the responsibilities and inequalities involved are removed from the scene. We see this in that the catastrophe is often narrated as if 'we' were all equally responsible for it and faced the same fate. 'We are all in the same boat.' Using that metaphor, critics would argue that not everyone contributed equally to conjuring up the deluge, and that the boat does not have space for everyone. Put in more formalised terms, waving the spectre of the catastrophe risks conjuring up an un-reflexive universal, an undifferentiated human condition in the face of a common enemy – which contrasts to the histories of inequality and injustice that will lead (or have led) different groups to very different positions on the brink of the precipice (Chaturvedi & Doyle, 2015; Swyngedouw, 2010).

## Learning resources

'The Dark Mountain Project' is an influential albeit controversial network of writers, artists and thinkers who believe the world 'is entering an age of ecological collapse, material contraction and social and political unravelling'. http://dark-mountain.net/about/the-dark-mountain-project/

In the documentary 'When The Levees Broke: A Requiem in Four Parts', director Spike Lee explores how Hurricane Katrina became an uneven and racialized catastrophe when it hit New Orleans.

Oreskes, N. and Conway E.R. (2014). *The collapse of the western civilization: a view from the future*. New York: Columbia University Press. This is a science-based fiction book in which the authors, two acclaimed scholars, offer a 'history of the future' after catastrophic climate change has occurred.

## Bibliography

Aradau, C. & van Munster, R.V. (2011). *Politics of catastrophe: genealogies of the unknown*. London: Routledge.

Chaturvedi, S. & Doyle, T. (2015). *Climate terror: a critical geopolitics of climate change*. New York, NY: Palgrave Macmillan.

Clark, N. (2011). *Inhuman nature: sociable life on a dynamic planet*. Delhi: SAGE.

Forsyth, T. (2003). *Critical political ecology: the politics of environmental science*. London: Routledge.

Ghosh, A. (2016). *The great derangement: climate change and the unthinkable*. Chicago: University of Chicago Press.

Harvey, D. (1974). Population, resources, and ideology of science. *Economic Geography, 50*(3), 256–277.

Hulme, M. (2009). *Why we disagree about climate change: understanding controversy, inaction and opportunity*. Cambridge: Cambridge University Press.

Jamison, A. (2001). *The making of green knowledge: environmental politics and cultural transformation*. Cambridge: Cambridge University Press.

Klein, N. (2014). *This changes everything: capitalism vs. the climate*. London: Allen Lane.

Latour, B. (2004). *Politics of nature: how to bring the sciences into democracy*. Cambridge, Mass.: Harvard University Press.

Lilley, S., McNally, D., Yuen, E. & Davis, J. (2012). *Catastrophism: the apocalyptic politics of collapse and rebirth*. Oakland, Calif.: PM Press.

Lövbrand, E., Stripple, J. & Wiman, B. (2009). Earth System governmentality. *Global Environmental Change, 19*(1), 7–13.

Robbins, P. (2012). *Political ecology: A critical introduction*. New York: Wiley.

Swyngedouw, E. (2010). Apocalypse forever? Post-political populism and the spectre of climate change. *Theory, Culture & Society, 27*(2–3), 213–232.

Žižek, S. (2010). *Living in the end times*. London: Verso.

# Ecological footprint

1.8

## William Rees

Everyone has one – we couldn't live without them – yet most people remain unaware of their existence or why they matter. I refer here to peoples' *ecological footprints* which represent the biophysical connection each individual has with the planet. Think of your personal eco-footprint as the metaphorical placenta that binds you to Mother Earth.

## Concept and definition

Modern humans rarely consider themselves as ecological entities but, despite this cognitive lapse, we are just as much a part of nature as are wild species. In fact, *Homo sapiens* remains as dependent on the productive and assimilative capacity of ecosystems as any other consumer organism. The first question of human ecology should therefore be: just how much of the Earth's surface is dedicated to supporting just me in the lifestyle to which I have become accustomed?

We can answer such questions using ecological footprint analysis (EFA). EFA recognizes that, no matter how primitive or sophisticated an economy, every act of material production/consumption uses some product or process provided by terrestrial or aquatic ecosystems. It follows that if we can quantify the flows of energy and materials required to support a specified population, and we know the productivity/assimilative capacities of exploited ecosystems, we can convert the relevant flows into their corresponding ecosystem areas. Thus, we can define the ecological footprint of any study population (an individual or whole nation) as *the area of productive land and water ecosystems that the population requires on a continuous basis to produce the (bio)resources it consumes, and to assimilate its (mostly carbon) wastes, wherever on Earth the relevant ecosystems may be located* (Rees, 2013).

National eco-footprint estimates are based on each country's final consumption of energy, bulk commodities and manufactured goods, where consumption is defined as: domestic production plus imports minus exports. Several factors influence footprint size: national population, average material standard of living, the productivity of supportive ecosystems, and the efficiency of resource harvesting, processing, and use.

The Global Footprint Network estimates the national EFs of some 200 countries and territories using data provided by national statistical agencies and internationally recognized sources such as the United Nations' Food and Agriculture Organization, the International Energy Agency, and the UN Statistics Division. Since different ecosystems and management practices result in differing productivities, analysts generally convert EF estimates to their equivalents in global average hectares (gha) to facilitate international comparisons. (Details of current methods and results are available from WWF [2014] and on-line from the Global Footprint Network by following the links at http://www.footprintnetwork.org/en/index.php/GFN/).

## Turning carrying capacity on its head

If we think about eco-footprinting for even a moment, several implications emerge almost spontaneously. For example, if each of us requires an exclusive area of land and water just to live, then there must be an upper limit to the number of people our home region (or our country or the entire Earth) can support indefinitely; clearly, too, this upper limit will depend on lifestyles – greater resource consumption and waste production implies larger individual eco-footprints (and fewer people); finally, it follows that every person on this finite Earth is competing with everyone else (to say nothing of thousands of other species) for Earth's limited bio-capacity.

From this perspective, the EF concept is obviously closely related to that of 'carrying capacity' (CC). CC is normally defined as the average population of a particular species (e.g., elk, deer) that a given habitat can support more or less indefinitely but economists argue that the concept isn't applicable to humans. Why should the population of a given region or country be limited by its domestic resource endowments if it can import needed resources from elsewhere, become increasingly efficient in the use of resources, or invent human-made substitutes for natural resources? Trade dissolves constraints imposed by local resource shortages and technology promises to decouple the economy from nature entirely. Why shouldn't just about any region be free to grow indefinitely?

Indeed, the world community is in thrall to precisely this belief. Continuous economic growth is a primary goal of all modern governments, private corporations and international development agencies who promote it as the solution to contemporary problems ranging from poverty to pollution.

This is regrettable. By inverting the standard CC ratio, EFA shows the economists' rationale to be dangerous illusion: rather than asking how many people a given area can support, eco-footprinting asks how large an ecosystem area is effectively appropriated to support any given population. EFA recognizes that: a) whatever its technological base, the economy requires some physical inputs or life support services from ecosystems and; b) whether people consume locally-produced or traded goods, their material connection to 'the land' remains intact – trade merely shuffles bio-capacity around, enabling people to live in ecosystems located half a planet away.

Bottom line? EFA negates economists' principal objections to limits on growth and in the process shows the human enterprise to be in perilous overshoot. Even at current (inadequate) average material standards, humanity significantly exceeds the long-term carrying capacity of Earth. By enabling the unfettered expansion of humanity's collective metaphorical placenta, trade and technology have effectively turned *Homo sapiens* into a parasite on its planetary host.

## Eco-footprints, bio-capacity and overshoot

In 2010 the total area of productive land and water ecosystems (bio-capacity) on Earth was about 12 billion hectares (ha); divided among 6.9 billion people this amounts to 1.7 ha each. That same year, humanity's aggregate eco-footprint was actually 18 billion ha, or 2.6 gha per capita. In other words, human demand for bio-capacity was at least 50 per cent greater than available supply. We are using the planet as if it were half again as large as it is (data from WWF, 2014).

Such 'overshoot' is possible only temporarily because of the enormous accumulation of bio-resources in the ecosphere. For 40 years the humans have been harvesting so-called 'natural capital' faster than nature can regenerate and dumping wastes beyond its ability to assimilate. (In many nations the carbon footprint is half the total EF). In effect, the world community has been 'financing' its maintenance and growth partially by depleting tropical forests, fish stocks, biodiversity, soils, ground water, fossil fuels, carbon sink capacity and other resources essential to its own survival.

No one paying attention to global trends should be surprised. Propelled by the growth ethic and facilitated by technology, the human population increased fourfold to six billion and resource use by a factor of at least eight during the 20th century alone (Krausmann et al., 2009). This situation is unsustainable but in the absence of popular understanding and strong political action shows no sign of abating. Resource extraction ballooned by almost 50 per cent between 1980 and 2005 alone (SERI 2009); economic decoupling is simply not taking place (Giljum et al., 2014; Wiedmann et al., 2013).

## On ecological deficits and inequity

There are complicating factors. Personal eco-footprints are not equal, but vary widely – and unfairly – with income. People in high-income nations have average EFs ranging from 4 to 10 gha/capita, i.e., up to five times their equitable share of global biocapacity (1.7 gha); people in the poorest nations (e.g., Bangladesh, Zambia) live on under 1 gha, less than half their equal 'earth-share'.

Not surprisingly, many wealthy and most densely populated countries have EFs up to many times larger than their domestic bio-capacities (e.g., Japan and the Netherlands). Such countries are running ecological deficits with other countries and the global commons, i.e., their consumption of bio-resources exceeds 'income' from domestic bio-capacity (see Figure 1.8.1).

In 2010, per capita EFs exceeded per capita bio-capacity in 91 of 151 countries for which adequate data are available and this number is growing (WWF, 2014). (60 per cent of the world's bio-capacity is located within just ten countries, including the US and Canada).

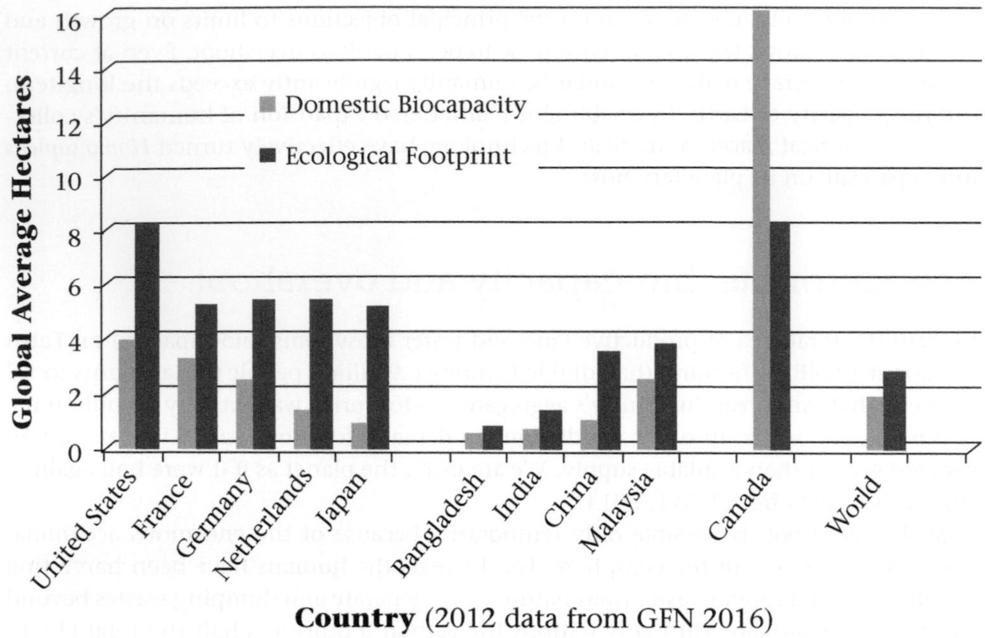


**Figure 1.8.1** Eco-footprints and bio-capacities of selected rich and poor countries compared to corresponding world averages. A few land-rich or low population countries like Canada have apparent bio-capacity surpluses but these are effectively incorporated into the eco-footprints of the majority of countries that run ecological deficits.

Kissinger and Rees (2009).

These data illustrate some of the hidden risks of globalization in an era of rapid ecological change – trade negates the 'negative feedback' that would normally come from exceeding carrying capacity thus facilitating uninterrupted population and consumption growth. This in turn: 1) creates unsustainable dependencies in both trading partners (what happens if climate change or geopolitical strife disrupts the exchange?) while; 2) accelerating the drawdown of remaining pockets of essential bio-resources all over the planet and placing all humanity at risk simultaneously (Kissinger and Rees 2009).

## Phantom planets and implications for sustainability

Overshoot is largely the result of overconsumption by the world's wealthy. EFA shows that we'd need nearly two additional Earth-like planets to support the entire human population at average European material standards; *three* more phantom Earths if everyone lived like North Americans. This already precarious situation is actually deteriorating because everyone's Earth-share shrinks with population growth and ecosystem degradation. (Remember, this finite planet is expected to accommodate the material demands of two *additional* billion people by 2050.)

Chronic poverty for billions of people is socio-politically unsustainable; accelerating overshoot is ecologically unsustainable. In the absence of additional planets, even the arithmetically challenged should recognize that today's growth-based approach to poverty alleviation is ultimately disastrous for everyone.

So, what is the alternative? Sustainability and social justice imply that we learn to live more equitably within the biophysical means of nature. EFA and similar analyses show that this requires reducing aggregate fossil energy and much material throughput (production and consumption) by 50 per cent globally by mid- to late century and 80 per cent in high-income countries (to free up 'ecological space' needed for justifiable increased consumption among the poor).

The good news is that such reductions are theoretically achievable with available technologies and modest lifestyle changes that might actually increase well-being (von Weizsäcker et al., 2009; Wilkinson and Pickett, 2010); the bad news is that society is doing little to make it happen.

The reasons are anything but theoretical. Human nature is the greatest barrier to sustainability. People are naturally myopic, preferring the here and now over future possibilities and distant events; we react defensively to any change that might threaten our creature comforts or social status; corruption and greed – all but sanctioned by contemporary morality – overshadow the public interest (governing political and economic elites are already vigorously resisting reform). To top things off, the world community is arguably too fractious and too mutually suspicious to respond effectively to the sustainability challenge. The issue then becomes whether *Homo sapiens* will be able to rise above reptilian instincts to exercise uniquely human attributes associated with high intelligence. Avoiding collapse demands unprecedented international cooperation in devising long-term plans for collective survival, plans consistent with both the biophysical evidence and the requirements of social justice (Rees, 2014). Is anyone taking bets on the likelihood of success?

## Learning resources

Ecological footprint analysis (EFA) was introduced in 1992 and elaborated in 1996; see:

Rees, W.E. (1992) 'Ecological footprints and appropriated carrying capacity: What urban economics leaves out'. *Environment and Urbanization* (4) 121–130.

Wackernagel, M. and Rees, W.E. (1996) *Our Ecological Footprint*. Gabriola, BC: New Society Publishers.

For current EFA practice, including data on the eco-footprints of nations, see the various links at: http://www.footprintnetwork.org/en/index.php/GFN/

For clear and insightful analyses with models illustrating how modern countries could 'downsize' or learn to live well without growth see:

Jackson, T. (2009) *Prosperity Without Growth: Economics for a Finite Planet*. London: Earthscan.

Victor, P. (2008) *Managing Without Growth: Slower by Design, not Disaster*. Cheltenham: Edward Elgar.

## Bibliography

GFN (2016) *Ecological Wealth of Nations* (interactive map). Available at http://www.footprintnetwork.org/ecological_footprint_nations. (Accessed 31 August 2013).

Giljum, S., Dittrich, M., Lieber, M. & Lutter, S. (2014) 'Global patterns of material flows and their socio-economic and environmental implications: A MFA study on all countries world-wide from 1980 to 2009', *Resources*, 3: 319–339.

Kissinger, M. & Rees, W.E. (2009) 'Footprints on the prairies: Degradation and sustainability of Canadian agriculture in a globalizing world', *Ecological Economics*, 68: 2309–2315.

Krausmann, F., Gingrich, S., Eisenmenger, N., Erb, K-H., Haberl, H. & Fischer-Kowalski, M. (2009) 'Growth in global materials use, GDP and population during the 20th century', *Ecological Economics*, 68(10): 2696–2705.

Rees, W.E. (2013) Ecological Footprint, Concept of. In Simon Levin (ed.), *Encyclopedia of Biodiversity* (2nd edn.). Salt Lake City, UT: Academic Press/Elsevier.

Rees, W.E. (2014) *Avoiding Collapse – An Agenda for Sustainable Degrowth and Relocalizing The Economy*. Vancouver: Canadian Centre for Policy Alternatives. Available at https://www.policyalternatives.ca/sites/default/files/uploads/publications/BC per cent20Office/2014/06/ccpa-bc_AvoidingCollapse_Rees.pdf. (Accessed 1 August 2016).

SERI et al. (2009) *Overconsumption? Our use of the World's Natural Resources*. Sustainable Europe Research Institute (SERI), Friends of the Earth Europe and GLOBAL 2000 (Friends of the Earth Austria). Available at https://www.foe.co.uk/sites/default/files/downloads/overconsumption.pdf. (Accessed 31 August 2016).

von Weizsäcker, E., Hargroves, K., Smith, M., Desha, C. & Stasinopoulos, P. (2009) *Factor 5: Transforming the Global Economy through 80 Per Cent Increase in Resource Productivity*. London: Earthscan.

Wiedmann, T.O., Schandl, H., Lenzen, M., Moran, D., Suh, S., West, J. & Kanemoto, K. (2013) 'The material footprint of nations', *Proceedings of the National Academy of Sciences of the United States of America*, 112: 6271–6276.

Wilkinson, R. & Pickett, K. (2010) *The Spirit Level: Why Equality is Better for Everyone*, London: Penguin Books.

WWF (2014) *Living Planet Report 2014*. Gland, Switzerland: World Wide Fund for Nature. Available at http://www.worldwildlife.org/pages/living-planet-report-2014. (Accessed 1 August 2016).

# The environmental Kuznets curve

1.9

David I. Stern

The environmental Kuznets curve (EKC) is a hypothesized relationship between various indicators of environmental degradation and countries' gross domestic product (GDP) per capita. In the early stages of a country's economic growth, environmental impacts and pollution increase, but beyond some level of GDP per capita (which will vary for different environmental impacts) economic growth leads to environmental improvement. This implies that environmental impacts or emissions per capita are an inverted U-shaped function of GDP per capita, whose parameters can be statistically estimated. Figure 1.9.1 shows a very early example of an EKC. A large number of studies have estimated such curves for a wide variety of environmental impacts ranging from threatened species to nitrogen fertilizers, though atmospheric pollutants such as sulfur dioxide and carbon dioxide have been most commonly investigated. The name Kuznets refers to the similar relationship between income inequality and economic development proposed by Nobel Laureate Simon Kuznets and known as the Kuznets curve.

The EKC has been the dominant approach among economists to modeling ambient pollution concentrations and aggregate emissions since Gene Grossman and Alan Krueger (1991) introduced it in an analysis of the potential environmental effects of the North American Free Trade Agreement. The EKC also featured prominently in the 1992 *World Development Report* published by the World Bank and has since become very popular in policy and academic circles and is even found in introductory economics textbooks.

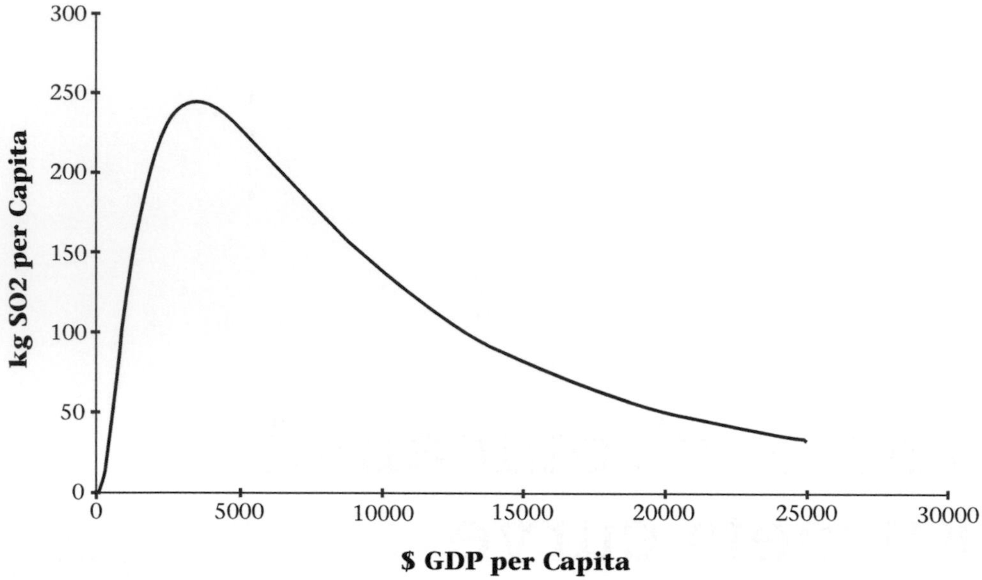


**Figure 1.9.1** An environmental Kuznets curve estimated by Panayotou (1993).

*Source*: see Stern et al. (1996) for details.

## Critique

Despite this, the EKC was criticized almost from the start on empirical and policy grounds, and debate continues. It is undoubtedly true that some dimensions of environmental quality have improved in developed countries as they have become richer. City air and rivers in these countries have become cleaner since the mid-20th century and in some countries forests have expanded. Emissions of some pollutants such as sulfur dioxide have clearly declined in most developed countries in recent decades. But there is less evidence that other pollutants such as carbon dioxide ultimately decline as a result of economic growth. There is also evidence that emerging countries take action to reduce severe pollution. For example, Japan cut sulfur dioxide emissions in the early 1970s following a rapid increase in pollution when its income was still below that of the developed countries and China has also acted to reduce sulfur emissions in recent years.

As further studies were conducted and better data accumulated, many of the econometric studies that supported the EKC were found to be statistically fragile. Figure 1.9.2 presents much higher quality data with a much more comprehensive coverage of countries than that used in Figure 1.9.1. In both 1971 and 2005 sulfur emissions tended to be higher in richer countries and the curve seems to have shifted down and to the right. A cluster of mostly European countries had succeeded in sharply cutting emissions by 2005 but other wealthy countries reduced their emissions by much less.

Initially, many understood the EKC to imply that environmental problems might be due to a lack of sufficient economic development rather than the reverse, as was conventionally thought, and some argued that the best way for developing countries to

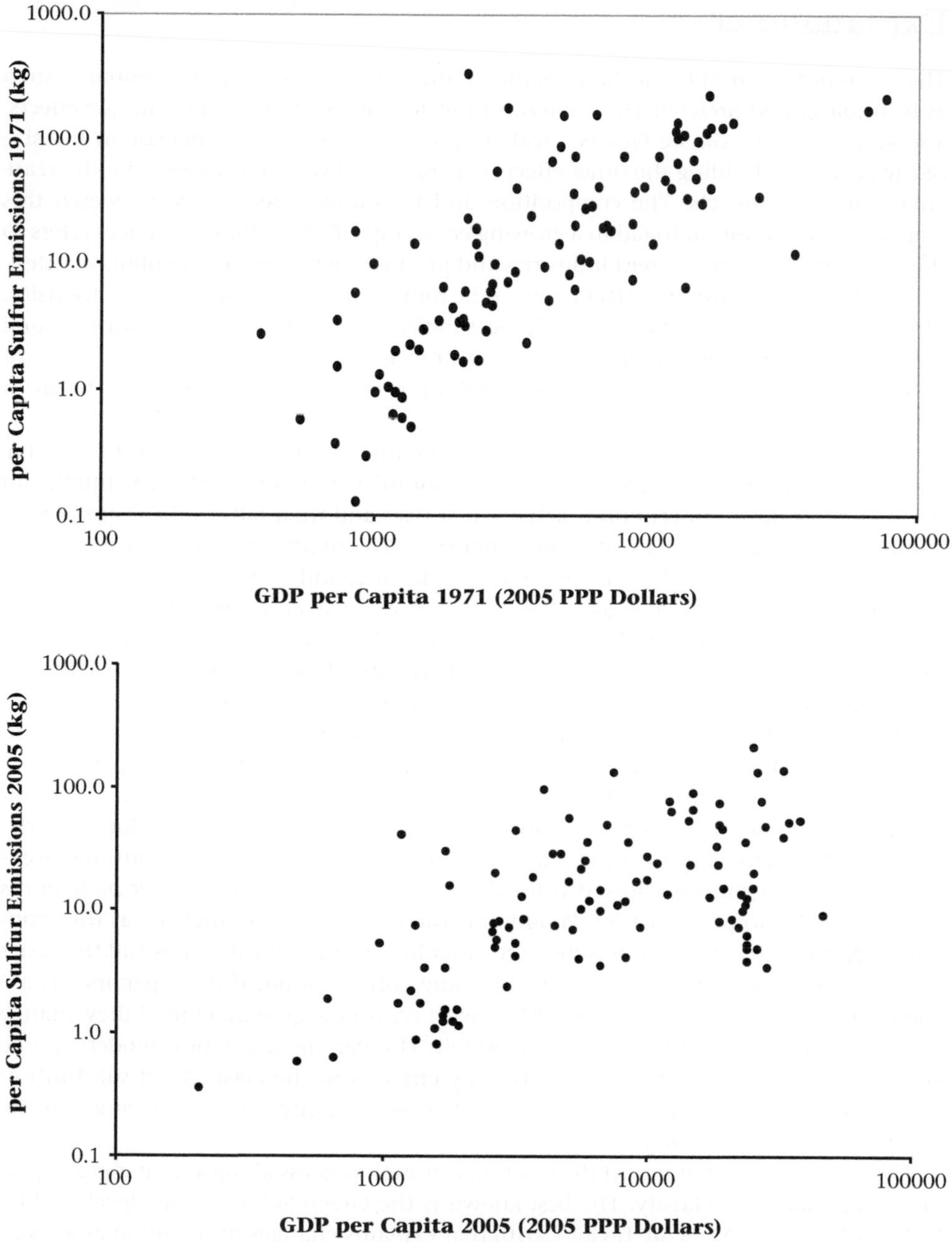


**Figure 1.9.2** Sulfur emissions and GDP per Capita 1971 and 2005.

*Source*: Sulfur emissions data from Smith et al. (2011) and GDP data from the Penn World Table.

improve their environment was to get rich. This alarmed others, as while this might address some issues like deforestation or local air pollution, it would likely exacerbate other environmental problems such as climate change.

## Explanations

The existence of an EKC can be explained either in terms of deep determinants such as technology and preferences or in terms of scale, composition, and technique effects, also known as "proximate factors". Scale refers to the effect of an increase in the size of the economy, holding the other effects constant, and would be expected to increase environmental impacts. The composition and technique effects must outweigh this scale effect for pollution to fall in a growing economy.[1] The composition effect refers to the economy's mix of different industries and products, which differ in pollution intensities. Finally the technique effect refers to the remaining change in pollution intensity. This will include contributions from changes in the input mix – e.g. substituting natural gas for coal; changes in productivity that result in less use, everything else constant, of polluting inputs per unit of output; and pollution control technologies that result in less pollutant being emitted per unit of input.

Over the course of economic development the mix of energy sources and economic outputs tends to evolve in predictable ways. Economies start out mostly agricultural and the share of industry in economic activity first rises and then falls as the share of agriculture declines and the share of services increases. We might expect the impacts associated with agriculture, such as deforestation, to decline, and naively expect the impacts associated with industry such as pollution would first rise and then fall. However, the absolute size of industry rarely does decline and it is improvement in productivity in industry, a shift to cleaner energy sources, such as natural gas and hydro-electricity, and pollution control that eventually reduce some industrial emissions.

Static theoretical economic models of deep determinants, that do not try to also model the economic growth process, can be summarized in terms of two parameters: the elasticity of substitution between dirty and clean inputs or between pollution control and pollution, which summarizes how difficult it is to cut pollution; and the elasticity of marginal utility, which summarizes how hard it is to increase consumer well-being with more consumption. It is usually assumed that these consumer preferences are translated into policy action. Pollution is then more likely to increase as the economy expands, the harder it is to substitute other inputs for polluting ones and the easier it is to increase consumer well-being with more consumption. If these parameters are constant then either pollution rises or falls with economic growth. Only if they change over time will pollution first rise and then fall. The various theoretical models can be classified as ones where the EKC is driven by changes in the elasticity of substitution as the economy grows or models where the EKC is primarily driven by changes in the elasticity of marginal utility.

Dynamic models that model the economic growth process alongside changes in pollution, are harder to classify. The best known is the Green Solow Model developed by William Brock and M. Scott Taylor (2010) that explains changes in pollution as a result of the competing effects of economic growth and a constant rate of improvement in pollution control. Fast-growing middle-income countries, such as China, then having rising pollution, and slower growing developed economies, falling pollution. An alternative model developed by Ordás Criado et al. (2011) also suggests that pollution rises faster in faster growing economies but that there is also convergence so that countries with higher levels of pollution are more likely to reduce pollution faster than countries with low levels of pollution.

## Recent empirical research

Recent empirical research builds on these dynamic models painting a subtler picture than did early EKC studies.[2] We can distinguish between the impact of economic growth on the environment and the effect of the level of GDP per capita, irrespective of whether an economy is growing or not, on reducing environmental impacts. Economic growth usually increases environmental impacts but the size of this effect varies across impacts and the impact of growth often declines as countries get richer. However, richer countries are often likely to make more rapid progress in reducing environmental impacts. Finally, there is often convergence among countries, so that those that have relatively high levels of impacts reduce them faster or increase them slower. These combined effects explain more of the variation in pollution emissions or concentrations than either the classic EKC model or models that assume that either only convergence or growth effects alone are important. Therefore, while being rich means a country might do more to clean up its environment, getting rich is likely to be environmentally damaging and the simplistic policy prescriptions that some early proponents of the EKC put forward should be disregarded.

## Notes

1 We use "pollution" in the following discussion for concreteness, but a similar analysis can be applied to other environmental impacts too.
2 See Stern (2015) for details.

## Learning resources

Gapminder is a website where you can create and watch animations of the development of many environmental and development indicators across the world over time:

https://www.gapminder.org/
These include the sulfur emissions data shown in Figure 1.9.2 and carbon dioxide emissions. For a more in depth discussion of the issues covered in this chapter, see: https://ideas.repec.org/p/een/ccepwp/1514.html

## Bibliography

Brock, W.A. and Taylor, M.S. (2010) The green Solow model. *Journal of Economic Growth*, 15, 127–153.
Grossman, G.M. and Krueger, A.B. (1991) Environmental impacts of a North American Free Trade Agreement. *NBER Working Papers*, 3914.
Ordás Criado, C., Valente, S. and Stengos, T. (2011) Growth and pollution convergence: Theory and evidence. *Journal of Environmental Economics and Management*, 62, 199–214.
Panayotou, T. (1993) Empirical tests and policy analysis of environmental degradation at different stages of economic development. *Working Paper, Technology and Employment Programme, International Labour Office, Geneva*, WP238.

Smith, S.J., van Ardenne, J., Klimont, Z., Andres, R.J., Volke, A. and Delgado Arias, S. (2011) Anthropogenic sulfur dioxide emissions: 1850–2005. *Atmospheric Chemistry and Physics*, 11, 1101–1116.

Stern, D.I. (2015) The environmental Kuznets curve after 25 years. *CCEP Working Papers*, 1514.

Stern, D.I., Common, M.S. and Barbier, E.B. (1996) Economic growth and environmental degradation: The environmental Kuznets curve and sustainable development. *World Development*, 24, 1151–1160.

# Gaia

1.10

## Karen Litfin

### From planetary science to cultural symbol

Gaia theory began as an interdisciplinary hypothesis that Earth functions as a self-regulating biogeochemical system that sustains living organisms. In other words, the biosphere interacts with the atmosphere, lithosphere and hydrosphere in ways that create conditions conducive to the furtherance of life (Lovelock, 1972, 1979). Working with NASA in the 1960s to study the possibility of life on other planets, James Lovelock, an atmospheric chemist, observed that the dynamic chemical balance of Earth's atmosphere was a product of the biosphere. His unorthodox view became even more so with its christening by Lovelock's friend, the novelist William Golding. Perceiving the resonance of Lovelock's ideas with the ancient notion of *anima mundi* (or living earth), he suggested Gaia, the primordial Greek deity who created herself out of chaos and gave birth to much of the Greek pantheon. This mythological name only heightened the scientific controversy surrounding the hypothesis for decades.

In contrast to conventional evolutionary theory, Gaia theory postulates that life, rather than merely adapting to a fixed environment, co-creates its environment. Gaia theory represents a paradigmatic shift from disciplinary reductionism to scientific holism, most evident in its contribution to the integrative field of Earth system science. Gaia theory is influencing the full panoply of academic disciplines (Schneider et al., 2004; Crist & Rinker, 2010): ecology (Harding, 2006), economics (Ekins, 1992), philosophy (Latour, 2013; Ruse, 2013), theology (Primavesi, 2000), political science (Litfin, 2012), and astrobiology (Andrulis, 2012).

Gaia's cultural allure was evident at the outset. James Lovelock (2000) was astonished to receive twice as many letters in response to his first book from people interested in its religious aspects as from those with a more scientific bent. Had Lovelock named his theory "homeostatic Earth systems theory", he surely would not have received such a response. This cultural resonance, however, stood in stark contrast to the response of the scientific community. As philosopher of science Michael Ruse notes, some biologists were particularly offended, dismissing Lovelock's ideas as "a metaphor, not a mechanism" (Ruse, 2013).

One biologist, however, took a special interest in the Gaia hypothesis, and became Lovelock's principal collaborator. Lynn Margulis' trailblazing research in microbiology led her to believe that atmospheric trace gases (primarily methane, carbon dioxide, ammonia and hydrogen sulfide) are produced not just by life as a whole but specifically by bacteria traceable to the origins of life itself. Margulis' work both complemented and augmented Lovelock's claim for the coevolution of life and Earth (see Lovelock & Margulis, 1974; Margulis, 1999). Facing intense criticism, misunderstandings, and unanswered questions, Lovelock and Margulis spent decades refining the Gaia hypothesis. In the meantime, many of their insights were incorporated into the emerging integrative field of Earth system science.

## Strong and weak Gaia

At least four general readings of Gaia theory are possible. First, by substantially impacting (or even regulating) Earth's geochemistry, living organisms contribute to the habitability of the biosphere. Stated in this way, Gaia theory qualifies as a falsifiable scientific hypothesis. Second, Earth itself may be understood as a complex, bounded, self-organizing, adaptive living system with emergent properties. This perspective is entirely consistent with living systems theory, with Gaia constituting the largest known living system. Third, Gaia functions not only as a living system but as a kind of animate super-organism. Fourth, and most radically, Gaia is conceived by some as an evolutionary planetary intelligence that operates to the benefit of the whole, or perhaps even according to some larger purpose. Of the four readings, the first two are broadly accepted among scientists. Yet even scientists who disavow the stronger versions of Gaia use language that personifies "her"—for instance, the titles of Lovelock's book, *The Revenge of Gaia* (2007) and Margulis' (1996) essay 'Gaia is a Tough Bitch'. While Gaian sentience and teleology implicit in the third and fourth versions are beyond the purview of natural science, these perspectives resonate with elements of the broader culture.

## Humanity and Gaia

In displacing humans from the center or apex of creation, Gaia theory offers both scientific and metaphysical support for an alternative to modernity's anthropocentrism. For some social theorists, this is its primary contribution to green politics (Dobson, 1990). Yet because "on a planetary scale, life is near immortal" (Lovelock, 1986: 28), this biocentrism does not translate easily into a strategy for social and political change. If Gaia will inevitably re-establish a new dynamic equilibrium, then perhaps we can continue with business as usual. This logic, however, is deeply flawed: whereas Gaian time scales are on the order of aeons, human systems rarely consider anything longer than a generation. Gaia's resilience says nothing about that of civilization, which developed in the context of an optimal interglacial period and peak biodiversity.[1] Yet an understanding of Gaia's great biogeochemical subsystems (the global carbon, nitrogen, oxygen, sulphur, and water cycles) can point to broad thresholds for sustaining human systems within a planetary equilibrium. In calling for a 1.5°C threshold for global warming, for instance, the 2015 Paris Agreement gestures in this direction—but, as they say, the devil is in the details.

Gaia's big-picture perspective might also tempt us to ignore questions of justice and equity and get on with "saving the planet." Yet, while Gaia theory undercuts anthropocentrism, the pragmatic requirements of sustainability have the paradoxical effect of highlighting questions of racism, sexism and North–South inequity. As Dobson suggests, a return to a weak anthropocentrism is needed to transform Gaian thinking from either pure science or mystified philosophy into a practical worldview with a strategy for social change (Dobson, 1990). If, for instance, over-consumption cannot be globalized without destabilizing Gaia, then importing Gaian insights into the socio-political arena requires attending human needs and aspirations—especially those in developing countries who represent the wave of the global future.

While Gaia theory's political implications are necessarily unclear at this juncture, the utter embeddedness of human systems within Gaia suggests a need for harmonizing them with Earth's biogeochemical cycles. The global economy, which drives matter and energy on a linear trajectory from resource extraction to production to consumption, appears profoundly at odds with living systems, for which one species' "waste" is always another species' food. Gaian governance would regulate production and consumption within the homeostatic equilibrium of the Earth system.

Likewise, Gaian governance highlights the perennial procedural question—*who decides and how?*—on a planetary scale. Natural and social scientists are collaborating on all of these questions in the emerging field of Earth-system governance (Biermann et al., 2010). Because the Earth system is the wider context in which social, political, and economic systems operate, we are increasingly compelled to develop institutions and ways of living compatible with a favourable Gaian equilibrium.

## Gaia and the political imagination

An Internet search for "Gaia" yields over a million websites, with most being about environmentalism and various forms of spirituality. The popular embrace of Gaian imagery in rationalistic and technologically advanced societies seems surprising. Yet, because the image of a living Earth has deep mythopoetic roots, perhaps this return to a cosmology of human embeddedness rather than human exceptionalism makes sense. Some see in Gaia the rebirth of paganism and others an ally in the politics of ecofeminism. For others, environmentalism and feminism have given Gaia theory "the emotional and moral force it may need to become politically relevant" (Roszak, 1992).

For Vaclav Havel, former president of the Czech Republic, Gaia theory serves as inspiration for an alternative discourse of human rights, one rooted in "the revitalized authority of the universe." The authority of a world democratic order can be built on Gaia theory, he suggests, because it fosters "the awareness of being anchored in the Earth and in the universe. . . . Only someone who submits to the authority of the universal order of creation, who values the right to be a participant in it, can genuinely value himself and his neighbours, and thus honour their rights as well" (Havel, 1997). Gaia need not inform only questions of environmental governance; it can also inspire a wider context for envisioning human rights.

In a time when anxiety and despair threaten our capacity for positive action, Gaia reminds us that we are an integral part—and an astonishing result—of an evolutionary process that has been unfolding on our home planet for over 4 billion years. We are a species with the same bacterial ancestry as all other species, yet we may also be the means by which Gaia is growing into self-awareness. The ramifications of this insight will no doubt suffuse human culture and institutions for generations to come.

## Note

1 For this reason, astrobiologists make the crucial distinction between habitability, a property of planets, and sustainability, a property of human systems (Ogle, 2013).

## Learning resources

Martin Ogle has compiled an excellent compendium of readings, videos, quotes, and presentations at www.gaiatheory.org.

James Lovelock explains Gaia Theory in this excerpt from David Suzuki's *The Sacred Balance*: https://www.youtube.com/watch?time_continue=95&v=44yiTg7cOVI.

Ian Enting's "Gaia Theory: Is it Science Yet?" offers a critical analysis of the strong and weak variants of Gaia theory outlined at the beginning of this chapter. https://theconversation.com/gaia-theory-is-it-science-yet-4901.

Along with a host of eminent scientists, James Lovelock has compiled a highly accessible illustrated anthology, The Earth and I. https://www.taschen.com/pages/en/catalogue/graphic_design/all/02888/facts.james_lovelock_et_al_the_earth_and_i.htm.

## Bibliography

Andrulis, E. (2012) "Theory of the Origin, Evolution, and Nature of Life," *Life*, 2: 1–105.

Biermann, F., Betsill, M., Gupta, J., Kanie, N., Lebel, L., Liverman, D., Schroeder, H. & Siebenhuener, B. (2010) "Earth System Governance: Navigating the Anthropocene," *International Environmental Agreements: Politics, Law and Economics*, 10(4): 277–298.

Crist, E. & Rinker, H.B. (2010) *Gaia in Turmoil: Climate Change, Biodepletion, and Earth Ethics in an Age of Crisis*. Cambridge, MA: MIT Press.

Dobson, A. (1990) *Green Political Thought: An Introduction*. London: Unwin Hyman.

Ekins, P. (1992) *The GAIA Atlas of Green Economics*. Lanham, MD: Rowman and Littlefeld.

Harding, S. (2006) *Animate Earth: Science, Intuition and Gaia*. Totnes, UK: Green Books.

Havel, V. (1997) *The Art of the Impossible: Politics as Morality in Practice*, transl. by P. Wilson. New York: Knopf.

Latour, B. (2013) "Facing Gaia: Six Lectures on the Theology of Nature". Online. Available at: http://macaulay.cuny.edu/eportfolios/wakefield15/files/2015/01/LATOUR-GIFFORD-SIX-LECTURES_1.pdf (accessed 16 October 2016).

Litfin, K. (2009) "Principles of Gaian Governance: A Rough Sketch," in Crist and Rinker (eds) *Gaia in Turmoil: Climate Change, Biodepletion, and Earth Ethics in an Age of Crisis*. Cambridge, MA: MIT Press.

Litfin, K. (2012) "Thinking Like a Planet: Integrating the World Food System into the Earth System," in Peter Dauvergne (ed.) *International Handbook of Environmental Politics*, 2nd edition. Cheltenham: Edward Elgar.

Lovelock, J. (1972) "Gaia As Seen Through the Atmosphere," *Atmospheric Environment*. Vol. 6: 579–580.

Lovelock, J. (1979) *Gaia: A New Look at Life on Earth*. Oxford: Oxford University Press.

Lovelock, J. (1986) "Gaia: The World as Living Organism," *New Scientist*, 112(1539): 25–28.

Lovelock, J. (2000) "Foreword" in Primavesi, A. *Sacred Gaia: Holistic Theology and Earth System Science*. London: Routledge.

Lovelock, J. (2007) *The Revenge of Gaia: Earth's Climate Crisis and the Fate of Humanity*. New York: Basic Books.

Lovelock, J. & Margulis, L. (1974) "Atmospheric Homeostasis By and For the Biosphere: The Gaia Hypothesis," *Tellus*. Vol. 26: 2–10.

Madron, R, & Jopling, J. (2003) *Gaian Democracies: Redefining Globalisation and People-Power*. Totnes, UK: Green Books.

Margulis, L. (1996) "Gaia is a Tough Bitch," in John Brockman (ed.), *The Third Culture: Beyond the Scientific Revolution*. New York: Simon & Schuster, pp. 129–150.

Margulis, L. (1999) *Symbiotic Planet: A New Look at Evolution*. New York: Basic Books.

Ogle, M. (2013) "Sustainability and Habitability on a Living Planet," available here: http://entrepreneurialearth.com/sustainability-and-habitability-on-a-living-planet/ (accessed October 16, 2016).

Primavesi, A. (2000) *Sacred Gaia: Holistic Theology and Earth System Science*. London, Routledge.

Roszak, T. (1992) *The Voice of the Earth*. New York: Simon & Schuster.

Ruse, M. (2013) *The Gaia Hypothesis: Science on a Pagan Planet*. Chicago: The University of Chicago Press.

Schneider, S.H., Miller, J., Crist, E., Boston, P., Torres, P., Lovelock, J. & Margulis, L. (eds) (2004) *Scientists Debate Gaia: The Next Century*. Cambridge, MA: MIT Press.

# 1.11 The Jevons Paradox

## John M. Polimeni and Raluca I. Iorgulescu

In his book published in 1865, *The Coal Question,* the British economist William Stanley Jevons promoted the idea that energy efficiency leads to an equal or greater amount of energy consumption rather than the expected reduction. In the seventh chapter of his book, Jevons postulated that less fuel consumption per unit causes greater total consumption (Polimeni et al., 2008; Alcott, 2005). At the time, the British empire was powered by coal and the improvements to steam engines increasing the scale of production. Jevons wrote:

> Every such improvement of the engine, when affected, does but accelerate anew the consumption of coal. Every branch of manufacture receives a fresh impulse – hand labor is still further replaced by mechanical labor.
>
> (Jevons, 1865)

This efficiency-consumption dynamic is known as the *Jevons Paradox*. Over one hundred years later, the paradox was revisited by modern economists. Nicholas Georgescu-Roegen (1975) observed that technological enhancements are usually energy-using and labor-saving through the use of more powerful energy converters. As a result, the efficiency gains lead to a decrease in prices, which instigates greater demand since the consumer can purchase more with the same budget constraint.

## The Khazzoom-Brookes postulate

Years later, Leonard Brookes (1978, 1979) and Daniel Khazzoom (1980) examined the paradox. Their research has come to be known as the Khazzoom-Brookes postulate. The postulate states that energy efficiency improvements will increase energy consumption more than what the increase would have been without the efficiency advances (Saunders, 1992; Herring and Sorrell, 2009). Furthermore, Khazzoom (1980) found that with a Cobb-Douglas production, which represents the relationship between output and the combination of inputs used for production (typically capital and labor), energy efficiency improvements for any factor of production will lead to an increase in energy consumption; and advancements in energy efficiency increase energy consumption as long as the elasticity of substitution, a measure of how easily inputs can be substituted for one another, is greater than one (Saunders, 1992).

## The rebound effect

This research on the Jevons Paradox was expanded to enable the estimation of the size of the energy consumption increase. This quantification of the energy efficiency improvement resulting in an increase in energy consumption dynamic has come to be known as *the rebound effect*. The rebound has been measured as insignificant (Lovins, 1988, pp. 156–157) to various levels of confidence (Schipper and Grubb, 2000, pp. 367–368).

The research literature has identified three types of rebound effects (Herring, 2006; Bhattacharyya, 2011):

1. Direct rebound effect: increased energy consumption results in lower energy prices because of energy efficiency improvements.
2. Indirect rebound effect: occurs when energy efficiency improvements lead to cost savings due to a reduction in energy consumption, allowing the consumer to spend more of their disposable income on other goods and services, which require energy to make. Furthermore, they will also be able to spend more money on energy.
3. General rebound effect: ensues as a result of adjustments in supply and demand by all producers and consumers in all sectors of the economy.

Based upon these definitions and findings, the research on the Jevons Paradox has mostly focused on the rebound effect. In particular, considerable effort has been expended trying to measure the magnitude of the rebound effect, largely on a micro level for individual economic sectors. If the size is relatively small, for example 10 per cent, then plans for increasing energy efficiency could be viable. However, if the magnitude of the rebound effect is 80–90 per cent, then the feasibility of the energy efficiency program is questionable (Bhattacharyya, 2011, p. 158).

This increase in demand is not confined only to more efficient products, but can also involve other end-uses because they compete for the same budget allocation (Khazzoom, 1980; Polimeni, et al., 2008). As a result, there is a direct micro rebound and an indirect macro rebound. A micro rebound is specific to one product, sector, or end-use and is the subject of the literature on the rebound effect because the elasticities of demand and substitution can be calculated for specific energy uses to estimate direct rebound effects. However, energy efficient improvements tend to impact the economy as a whole, the macro economy. When a macro rebound occurs, there is an income effect which creates an increase in real income enabling consumers to upgrade in quality, as well as an increase in demand (Wirl, 1997, pp. 20, 26–27, 31, 41, 197; Saunders, 2000a). However, these studies (for a list of articles see the learning resources section at the end of the chapter) did not attempt to ascertain which factor or factors cause the rebound.

## Jevons Paradox and the environmental connection

The Jevons Paradox has significant implications for the environment. Since the environment is a complex system, less resource consumption, as a result of energy efficiency improvements, is good because there is less pollution. However, less pollution only occurs if the complex system was not able to adapt. Complex systems generally adapt quickly, and when technological improvements are initiated, one or both of the following can occur: (1) an expansion of current levels of activity within the original setting, and (2) an increase in the option space with additional activities (Giampietro and Mayumi, 2005; Polimeni et al., 2008; Giampietro and Mayumi, 2008). Consequently, as efficiency improves, the resource becomes less expensive to use, and the current technology will be used more or a new technology will be developed with more options and features. For an illustration of the Jevons Paradox we encourage the reader to read the article at http://thebreakthrough.org/archive/jevons-paradox-illustrated.

Therefore, it is important to explore and understand the relationship between energy consumption and energy efficiency for policy makers developing energy policy. A starting point is to use the I=PAT model by Ehrlich and Holdren (1971) which examines the environmental impact (I) of population (P), affluence (A), and technology (T) (Alcott, 2005). Under this formula, energy consumption can be inserted on the left hand side of the equation as a proxy for environmental impact. The right hand side of the equation can then represent the three main macro factors influencing energy consumption:

1. Population measures (P): total population, population density, urban population, rural population, population growth rate;
2. Affluence measures (A): gross domestic product and its individual components (except for investment), gross national income, GDP per capita;
3. Efficiency improvements (T): energy intensity, a proxy for technological improvements.

Table 1 maps the variables in the I=PAT equation.

Using this mapping, a macro level examination of the Jevons Paradox can be undertaken to determine whether there is some evidence that the paradox exists, and if there is proof that suggests its presence then what factors contribute to its existence and which factors are most important in its existence. While seemingly these are basic questions, they are essential. Separate arguments could be made that energy efficiency improvements are not the cause of increased energy consumption but rather a growing population or an expanding economy. Therefore, using the I=PAT mapping can be transferred into a macro level regression model to understand how these factors and others influence energy consumption. This model specification, shown in equation 1, provides the information on possible existence and the magnitude of the variables inform us as to their importance in the energy consumption relationship.

$$\textit{Energy Consumption} = \beta_1 + \beta_2 \textit{Population} + \beta_3 \textit{Affluence} + \beta_4 \textit{Technology} \qquad [1]$$

where the variables in the map can be used in the appropriate place in the regression.

The Jevons Paradox is important given the current status of the world. Climate change, pollution, higher energy costs, and a diminishing supply of natural resources

**Table 1.11.1** The variables in the I=PAT equation

| *Environmental impact (I)* | *Population (P)* | *Affluence (A)* | *Technology (T)* |
|---|---|---|---|
| Total primary energy consumption | Total population | Gross domestic product (GDP) | Energy intensity |
| | Urban population | Exports | |
| | Rural population | Imports | |
| | Population density | Household consumption | |
| | Population growth rate | Government consumption | |
| | | GDP per capita | |
| | | Gross national income | |

*Source*: Adapted from Polimeni et al. (2008, p. 148).

to use for energy production impact everyone. Policy makers, scientists, economists, and others state that new technological improvements will be developed as a result of price signals. However, empirical research on the Jevons Paradox (see the Learning Aid section at the end of the chapter) and the rebound effect indicate that the policy makers and the rest are probably wrong. As a result, efficiency improvements are likely not the answer, rather the solution will require behavioral consumption changes.

## Learning resources

We encourage the readers to follow their own curiosity on the rebound effect and Jevons Paradox and read the articles listed below.

### *Rebound effect research articles*

Saunders (1992), Greening and Greene (1998), Greening et al. (2000), Schipper (2000), Saunders (2000a, 2000b), Allan et al. (2006), Hanley et al. (2006, 2009), Barker et al. (2007a, 2007b), Sorrell (2007), Sorrell and Dimitropoulos (2007), Wei (2007, 2010), Evans and Hunt (2009), Herring and Sorrell (2009), Turner (2009), Anson and Turner (2009), Saunders (2009), and Jenkins et al. (2010).

### *Quantitative Jevons Paradox articles*

Iorgulescu Polimeni and Polimeni (2007), Polimeni (2007a, 2007b, 2008), Polimeni and Iorgulescu Polimeni (2006, 2007), Clement (2011), Wolfe (2012) and Freire-González and Puig-Ventosa (2015).

## Videos on Jevons Paradox

https://www.youtube.com/watch?v=dbLUXwES-yI and https://www.youtube.com/watch?v=tylBCUoki8k

## Bibliography

Alcott, B. (2005) 'Jevons' Paradox', *Ecological Economics*, 54: 9–21.

Allan, G., Hanley, N., McGregor, P., Swales, K. & Turner, K. (2006) *The Macroeconomic Rebound Effect and the UK Economy*. Final Report to DEFRA.

Anson, S. & Turner, K. (2009) 'Rebound and Disinvestment Effects in Refined Oil Consumption and Supply Resulting from an Increase in Energy Efficiency in the Scottish Commercial Transport Sector', *Energy Policy*, 37(9): 3608–3620.

Barker, T., Ekins, T. & Foxon, T. (2007a) 'Macroeconomic Effects of Efficiency Policies for Energy-intensive Industries: The Case of the UK Climate Change Agreements, 2000–2010', *Energy Economics*, 29(5): 760–778.

Barker, T., Ekins, T. & Foxon, T. (2007b) 'The Macroeconomic Rebound Effect and the UK Economy', *Energy Policy*, 35: 4935–4946.

Bhattacharyya, S.C. (2011) *Energy Economics: Concepts, Issues, Markets and Governance*. London: Springer.

Brookes, L. (1978) 'Energy Policy, the Energy Price Fallacy and the Role of Nuclear Energy in the UK', *Energy Policy*, 6(1): 94–106.

Brookes, L. (1979) 'A Low Energy Strategy for the UK', *Atom*, 269: 73–78.

Clement, M.T. (2011) 'The Jevons Paradox and Anthropogenic Global Warming: A Panel Analysis of State-level Carbon Emissions in the United States, 1963–1997', *Society & Natural Resources*, 24(9): 951–961.

Ehrlich, P.R. & Holdren, J.P. (1971) 'Impact of Population Growth', *Science*, 171: 1212–1217.

Evans, J. & Hunt, L. (2009) *International Handbook on the Economics of Energy*. Cheltenham: Edward Elgar.

Freire-González, J. & Puig-Ventosa, I. (2015) 'Energy Efficiency Policies and the Jevons Paradox', *International Journal of Energy Economics and Policy*, 5(1): 69.

Georgescu-Roegen, N. (1975) 'Energy and Economic Myths', *Southern Economic Journal*, XLI: 347–381.

Giampietro, M. & Mayumi, K. (2005) *Jevons Paradox and Complex Adaptive Systems: Exploring the Epistemological Conundrum When Modeling the Evolution of Hierarchical Systems*, unpublished manuscript, http://www.webmeets.com/files/papers/ERE/WC3/1155/JevonsPaper.pdf

Giampietro, M. & Mayumi, K. (2008) 'The Jevons Paradox: The Evolution of Complex Adaptive Systems and the Challenge for Scientific Analysis', In *The Jevons Paradox and the Myth of Resource Efficiency Improvements*, pp.79–140. London: Earthscan.

Greening, L. & Greene, D. (1998) *Energy Use, Technical Efficiency, and the Rebound Effect: A Review of the Literature*. Oak Ridge, TN: Center for Transportation Analysis, Oak Ridge National Laboratory.

Greening, L., Greene, D. & Difiglio, C. (2000) 'Energy Efficiency and Consumption: The Rebound Effect, a Survey', *Energy Policy*, 28(6–7): 389–401.

Hanley, N., McGregor, P., Swales, J. & Turner, K. (2006) 'The Impact of a Stimulus to Energy Efficiency on the Economy and the Environment: A Regional Computable General Equilibrium Analysis', *Renewable Energy*, 31: 161–171.

Hanley, N., McGregor, P., Swales, J. & Turner, K. (2009) 'Do Increases in Energy Efficiency Improve Environmental Quality and Sustainability?', *Ecological Economics*, 68: 692–709.

Herring, H. (2006) 'Energy Efficiency: A Critical Review', *Energy*, 31(1): 10–20.

Herring, H. & Sorrell, S. (2009) *Energy Efficiency and Sustainable Consumption: Dealing with the Rebound Effect*. Basingstoke: Palgrave Macmillan.

Iorgulescu Polimeni, R. & Polimeni, J.M. (2007) 'Multi-scale Integrated Analysis of Societal Metabolism and Jevons' Paradox for Romania, Bulgaria, Hungary, and Poland', *Romanian Journal of Economic Forecasting*, 4(4): 61–76.

Jenkins, J., Nordhaus, T. & Schellenberger, M. (2010) *Energy Demand Backfire and Rebound as Emergent Phenomena: A Review of the Literature*. Oakland, CA: Breakthrough Institute.

Jevons, W.S. (1865) *The Coal Question: An Inquiry Concerning the Progress of the Nation and the Probable Exhaustion of our Coal-Mines*. 3rd Edition revised by A.W. Flux (1965). New York: Augustus M. Kelley.

Khazzoom, J.D. (1980) 'Economic Implications of Mandated Efficiency in Standards for Household Appliances', *Energy Journal*, 1: 21–40.

Lovins, A.B. (1988) 'Energy Saving from More Efficient Appliances: Another View', *Energy Journal*, 9: 155–162.

Polimeni, J.M. (2007a) 'Jevons' Paradox: A Case Study of China', *International Journal of Interdisciplinary Social Sciences*, 2(2): 383–394.

Polimeni, J.M. (2007b) 'Jevons' Paradox and the Economic Implications for Europe', *International Business and Economics Research Journal*, 6(10): 109–119.

Polimeni, J.M. (2008) 'Jevons' Paradox: The Cases of India and China', *Journal of Interdisciplinary Social Sciences*, 3(8): 145–157.

Polimeni, J.M. & Iorgulescu Polimeni R. (2006) 'Jevons' Paradox and the Myth of Technological Liberation', *Ecological Complexity*, 3(4): 344–353.

Polimeni, J.M. & Iorgulescu Polimeni R. (2007) 'Energy Consumption in Transitional Economies (Part I): Jevons' Paradox for Romania, Bulgaria, Hungary, and Poland', *Romanian Journal of Economic Forecasting*, 3: 63–80.

Polimeni, J.M., Mayumi, K., Giampietro, M. & Alcott, B. (2008) *The Jevons Paradox and the Myth of Resource Efficiency Improvements*. London: Earthscan.

Saunders, H. (1992) 'The Khazzoom-Brookes Postulate and Neoclassical Growth', *The Energy Journal*, 13(4): 131–148.

Saunders, H. (2000a) 'A View from the Macro Side: Rebound, Backfire and Khazzoom-Brookes', *Energy Policy*, 28(6–7): 439–449.

Saunders, H. (2000b) 'Does Predicted Rebound Depend on Distinguishing Between Energy and Energy Services?', *Energy Policy*, 28(6–7): 497–500.

Saunders, H. (2009) 'Theoretical Foundations of the Rebound Effect'. In: Herring, H., Sorrell, S. (Eds) *Energy Efficiency and Sustainable Consumption: Dealing with the Rebound Effect*. Basingstoke: Palgrave Macmillan.

Schipper, L. (Ed.) (2000) 'On the Rebound: The Interaction of Energy Efficiency, Energy Use and Economic Activity', *Energy Policy*, 28(6–7): 351–353.

Schipper, L. & Grubb, M. (2000) 'On the Rebound? Feedback between Energy Intensities and Energy Uses in IEA Countries', *Energy Policy*, 28(6–7) 367–388.

Sorrell, S. (2007) *The Rebound Effect: An Assessment of the Evidence for Economy-wide Energy Savings from Improved Energy Efficiency*. London: UK Energy Research Centre.

Sorrell, S. & Dimitropoulos, J. (2007) 'Evidence from Energy, Productivity and Economic Growth Studies'. In: Sorrell, S. (Ed.) *The Rebound Effect: An Assessment of the Evidence for Economy-wide Energy Savings from Improved Energy Efficiency*. London: UK Energy Research Centre.

Turner, K. (2009) 'Negative Rebound and Disinvestment Effects in Response to an Improvement in Energy Efficiency in the UK Economy', *Energy Economics*, 31: 648–666.

Wei, T. (2007) 'Impact of Energy Efficiency Gains on Output and Energy Use with Cobb-Douglas Production Function', *Energy Policy*, 35: 2023–2030.

Wei, T. (2010) 'A General Equilibrium View of Global Rebound Effects', *Energy Economics*, 32: 661–672.

Wirl, F. (1997) *The Economics of Conservation Programs*. Boston, MA: Kluwer Academic.

Wolfe, M. (2012) 'Beyond "Green Buildings:" Exploring the Effects of Jevons' Paradox on the Sustainability of Archival Practices', *Archival Science*, 12(1): 35–50.

# 1.12 Nature

## R. Bruce Hull

What is nature? How should humans relate to it? These two questions have perplexed philosophers for centuries. This chapter focuses mostly on answering the first question, but will speculate about the more political and consequential second question.

People purchase medicines made of natural ingredients. Foodies eat natural foods. Recreationists retreat to nature parks. Conservationists protect nature's health. But, what is it that makes something natural?

Nature is a conveniently pliable term. People use it to describe and justify many things, which is why the term is both powerful and problematic: it does not have a single "true meaning" (Hull, 2006). Five different definitions are explored below: economic, aesthetic, moral, scientific, and the Anthropocene.

## Nature as an economic ideal

Capitalism shapes our understanding of most things. Nature is no exception. Throughout history, humans have extracted, harnessed, or otherwise converted nature to "resources" that fuel the economy and build civilization. Consider two examples: factory and service.

### *Nature as factory*

Nature can be understood as a means of producing consumer goods such as wood, water, and food. Sometimes nature's processes appear inefficient and in need of human management. For example, a wild forest grows species that don't have economic value, thus wasting the productive potential of soil and sun. Worse, old trees grow slowly and rot before they can be converted into lumber, paper, or chemicals. Forest managers plant only merchantable trees, space them to maximize growth, and harvest them before their growth slows.

## *Nature as ecosystem service*

Nature provides free life support services. Oxygen is inhaled and $CO_2$ exhaled. Evaporation, rain, and gravity purify and distribute water. Microbes build and fertilize soil. Ozone blocks ultraviolet radiation that mutates genes. And insects pollinate crops. The list of services is long and their combined economic value surpasses the market economy of all countries combined (Millennium Ecosystem Assessment, 2005). Efforts are underway to privatize and monetize ecosystem services so that they can be traded and subjected to the efficiencies of markets.

## Nature as an aesthetic ideal

In 1757, Edmund Burke published *A Philosophical Enquiry into the Origin of Our Ideas of the Sublime and Beautiful,* an essay defining the natural aesthetic. Sublime scenery impresses the viewer with God's grandeur. It promotes feeling humble and insignificant relative to the glory of Creation. Such landscapes have rugged cliffs, sharp and angular lines, towering mountains, and dark, foreboding gorges. Beautiful landscapes, in

**Figure 1.12.1** Economic nature: a forest plantation.

*Source:* © CSIRO Land and Water.

contrast, with their gentle, curving, soothing features, promote feelings of tranquility and contemplation. Artists and landscape architects popularized sublime and beautiful views, so they are familiar to us because of coffee-table books, environmental club calendars, and parkway drives.

Urban people feel nostalgia for an idealized rural past and yearn for nature and solitude as contrasts to urban bustle. Parks and open spaces were established to meet this growing demand, often by forcibly evicting the people native to those areas whose lifestyles and economy did not jive with the aesthetic ideal of an untrammeled, pristine,

**Figure 1.12.2** Aesthetic nature: a sublime view. 'Wanderer above the sea of fog' by Caspar David Friedrich (1818).

*Source*: Public domain.

static, and safe nature. Millions of hectares of nature are now managed for the billions of recreation visits that occur every year. Whole government agencies and professional societies service these aesthetic experiences. Closer to home, people spend billions of dollars and countless hours on lawns and gardens trying to recreate and reconnect to this aesthetic.

## Nature as a moral ideal

Humanity has a long history of looking to nature for inspiration. Jesus, Mohammad, and Buddha retreated to nature to find lessons. Modern philosophers such as John Locke, Thomas Jefferson, and Henry David Thoreau repeatedly invoked nature as justification and rationale for their arguments. America's wild frontier supposedly forged the new nation's character. Rather than looking backward, toward customs and tools that worked in the past, Americans lunged headfirst into the frontier and into the future, confident that a little ingenuity and a lot of brute force could solve most any problem (Marx, 1964).

Some believe nature teaches children essential life lessons. As an example, in the early 20th century, when Europe and the US transitioned from a rural to urban society, people worried that "city children" were losing the "earnest, self-reliant and high minded" qualities forged by nature. Exposing children to nature became a priority. Textbooks about nature were introduced to schools and after-school nature clubs connected children to nature. We see the same concerns expressed today, over 100 years later, with the movement titled *No Child Left Indoors* (Louv, 2008). Perhaps, in response to the uncertainties of globalization, people are again looking to nature for moral guidance.

## Nature as scientific ideal

"The balance of nature" is a phrase often invoked during environmental debates. Some speakers use it to argue against human interventions that they worry will disrupt Earth's fragile ecology. Other speakers use it to defend the opposite conclusion: that nature's robust and redundant systems can withstand aggressive resource exploitation. Both arguments are problematic. Contemporary science confirms that nature is not balanced—no stable, ideal, original, authentic condition exists that can be said to be objectively best or good or healthy. Instead, science shows variability, openness, and dynamism driven by positive and negative feedback loops accelerating and moderating nearly constant change. Ecological systems are hierarchical clusters of fragmented and dynamic patches constantly fluctuating due to internal evolution and external perturbation (Worster, 1994).

The "balance" argument is an example of how people misuse appeals to science to justify their arguments for environmental conditions they actually prefer for aesthetic, economic, moral or other reasons (Cronon, 1996). Popular scientific terms such as biodiversity, integrity, and ecological health are similarly problematic. Ecological science cannot be used to set policy goals—to claim ought from is (Sagoff, 1988). We certainly can learn from studies of undisturbed natural areas how environmental systems function, but we can't then claim that those functions *ought* to occur elsewhere. Constructing goals (oughts) for environmental policy requires the hard work of collaboration among stakeholders, each with their own values and understandings of desired future conditions.

## Conclusion

So, then, what is nature? The centuries-old debate is not resolved here except to suggest that "nature" can be defined many ways, each definition depending upon the different values, cultures, and purposes of the people using the term.

What about the second, more political question asked in the introduction: how should humans relate to nature? The outlines of a possible answer to this question emerge when examining the Anthropocene and the supposed "end of nature" (Davies, 2016). The Holocene is the just-ending ~10,000-year geological epoch characterized by stable biophysical conditions that nurtured agriculture, urbanization, literacy, science, democracy, organized religion, global capitalism, and most other institutions that support human civilization. The Anthropocene—a world made by humanity, by both accident and design—is replacing it. Humans now manage more than half of Earth's land, harvest most of the ocean fisheries, use most of the fresh water, and otherwise impact most planetary processes. These trends are likely to accelerate over the coming decades as we add several billion people to the global middle class, rapidly urbanize, and double energy and food consumption. The accelerating changes giving rise to the new epoch are well documented and widely accepted (Steffen et al., 2015), but the implications are hotly debated.

Some worry that the Anthropocene could be used to rationalize risky technical interventions like geo-engineering of climate or abandoning efforts to conserve and respect things natural. But others, and here the chorus is growing, argue the Anthropocene demands a new way of thinking. Perhaps a new worldview is needed that moves beyond defining nature independently of humans, a view that accepts that humans and nature form a tightly coupled, interconnected, and interdependent system. Perhaps our economic model should change from exploitation of capital to partnership for resilience, our conception of beauty change from wonder at wildness to appreciation of tending, and our sciences evolve from seeking control to adapting to change (Purdy, 2015).

## Learning resources

More information about the five lenses that reveal different aspects of nature (i.e., economic, aesthetic, moral, scientific, and Anthropocene), as well as additional lenses such as health, equity, religion, evolution, biorights, and anthropology, can be found in Hull (2006).

"No child left indoors" programs and "Nature Deficit Disorder" prevention efforts near you can be found with quick internet searches. These efforts seem similar to movements 100 years ago that birthed nature camps, scouting clubs, and environmental education programs to get children back in touch with nature. Explore the connection between today's efforts and those of yesteryear, and speculate on the causes.

Walk the aisles of a grocery store looking for product labels that claim "100% natural," "pure," "sustainable," and "organic." Compare and contrast the claims using Good Guide and reviews of "natural" food labels by Consumer Reports.

www.GoodGuide.com

http://www.consumerreports.org/content/dam/cro/magazine-articles/2016/March/Consumer_Reports_Natural_Food_Labels_Survey_2015.pdf

Explore the Anthropocene, beginning with a review of rapidly accelerating trends in most key social and environmental indicators:

Welcome to the Anthropocene: http://www.anthropocene.info/great-acceleration.php

## Bibliography

Cronon, W. (1996) 'The trouble with wilderness: or, getting back to the wrong nature', *Environmental History* 1:7–55.

Davies, J. (2016) *The Birth of the Anthropocene*. Oakland: University of California Press.

Hull, R. (2006) *Infinite Nature*. Chicago: University of Chicago Press.

Louv, R. (2008) *Last Child in The Woods: Saving Our Children From Nature-Deficit Disorder*. New York: Algonquin Books.

Marx, L. (1964) *The Machine in the Garden: Technology and the Pastoral Ideal in America*. London: Oxford University Press.

Millennium Ecosystem Assessment. (2005) *Ecosystems and Human Well-being: Synthesis*. Washington, DC: Island Press.

Purdy, J. (2015) *After Nature: A Politics of the Anthropocene*. Cambridge, Mass: Harvard University Press.

Sagoff, M. (1988) 'Ethics, ecology, and the environment: Integrating science and law', *Tennessee Law Review* 56:77–229.

Steffen, W., Broadgate, W., Deutsch, L., Gaffney, O., Ludwig, C. (2015) 'The trajectory of the Anthropocene: The great acceleration', *The Anthropocene Review* 2:81–98.

Worster, D. (1994) *Nature's Economy: A History of Ecological Ideas*. Cambridge, England: Cambridge University Press.

# 1.13 One world

## Volker M. Welter

The term 'one world' indicates an awareness that humanity lives on one planet, which requires environmental care on a global scale. The expression rose to prominence with the United States' outer-space exploration from the 1960s and 1970s because it was closely tied to the emergence of whole Earth photographs (Cosgrove, 1994). On 24 December 1968 astronauts of Apollo 8 captured 'Earth Rise' and on 7 December 1972 their colleagues from Apollo 17 'Blue Marble'; two widely published color photographs depicting planet Earth as it appeared from outer space. Suddenly, humans could see that they indeed lived in one world, regardless of all other criteria that divided them along, for example, geographical, national, ethnic, or cultural lines. For centuries, humanity had attempted to visualize Earth with the help of maps and globes. Now images of the Earth were available that humans had photographed after physically leaving the planet and looking at it from outer space. Pictures of the Earth within outer space acquired various symbolic meanings, focusing most notably on the physical limits of the planet. An alternative view recognized the same images as a call to shift the care for the environment from a multitude of local levels to a singular global scale, adequate to the interconnectedness of all human beings and their environment. Historically, the roots of the symbolic rise of planet Earth from being merely the physical home of humanity to the conceptual focal point of the notion of one world stretch back much further than the twentieth century.

Knowledge that humanity existed on some celestial body is ancient. In Greek mythology, the story of the Titan Atlas, who was condemned to hold up eternally the sky above the western edge of the earth, suggests an awareness that humanity itself lived on a distinct celestial body.

Equally old is the wish of humans to comprehend their environment by visualizing its extent and overall form. The Farnese Atlas in the National Archaeological Museum in Naples, Italy, an antique Roman copy of an older, lost Greek sculpture, portrays the Titan as carrying the universe in the form of a celestial sphere. The Greek philosopher Anaximander (c. 610–c. 546 BC) created one, if not the first map of the world. Common to such visualizations of humanity's environment is the adoption of an imagined viewpoint that is located outside of the depicted geographical area or entity. The prevalence of such abstract viewpoints only changed with the invention of air balloons in the late eighteenth century and, approximately one hundred years later, of airplanes. Now it was possible (at least for the pilots and their passengers) to see with one's own eyes sections of the earth from high above rather than having to imagine what these may look like. This advance eventually culminated in those aforementioned outer-space photographs of the whole earth.

Throughout the nineteenth and twentieth centuries, many biologists, geographers and environmentalists hoped that visualizing the single world of humanity would further their cause of comprehending and caring for the environment. In 1897, the French geographer Élisée Reclus (1830–1905) envisioned a giant globe for the Universal Exposition in Paris in 1900. The globe was never constructed, but Reclus argued that being able to look at any location on the earth's surface from a ramp that was to wrap around the sphere would initiate feelings of empathy for the environment.

In the early decades of the twentieth century, the Scottish biologist Patrick Geddes (1854–1932) stated that viewing models, maps, engravings, and photographs of the earth, or parts of it, would stir viewers into greater efforts to plan responsibly the use of the natural environment.

In 1966, the American environmentalist Stewart Brand (born 1938), the creator of the countercultural *Whole Earth Catalog*, demanded that NASA should release images of the whole earth. Brand assumed that such photographs must have existed, but were withheld deliberately. The contemporary fears of imminent ecological catastrophes drove Brand's interest in the images. He argued that outer-space photographs showing the earth as a sphere or as a flat disk in case there was no shadow on its surface—two finite geometric shapes that closed on themselves—would make tangible the physical limits both of planet Earth and of the resources it harbored.

Finally, images of the whole Earth raised expectations for a renewed emotional relationship between humanity and its planet. In 1966, the Californian counter-cultural initiative, New Games, developed games using an earth ball in order to channel human aggression into non-competitive tournaments. After initially experimenting with a painted, 6 ft rubber ball, lightweight plastic Earth balls became standard that had to be inflated. As *The New Games Book* explains, this pre-game activity was understood as "donating one's breath" to the earth (Fluegelman 1976, 9). From the participants' "chance to play with the planet, whether . . . pushing, passing or throwing it; kicking or hugging it; on top, beneath or against it" (134) an emotionally closer involvement with the environment was to emerge. Yet the ease with which this form of play turned the earth into a cuddly rubber toy points towards a different strand in environmental thinking. This tradition had developed in the nineteenth century and likewise relied on a notion of one world, albeit one that understood the later as a call for actively remodeling the planet on a global scale.

In 1881, the British Arts & Crafts designer, and socialist, William Morris (1834–1896) defined in 'The Prospect of Architecture in Civilization' as the major task of the coming modern age "the moulding and altering to human needs of the very face of the earth itself, except in the outermost desert" (119). This call for designing almost the entire physical world found an echo in the 1955 conference *Man's Role in Changing the Face of the Earth*. The meeting convened in Princeton, New Jersey, in order to take stock of humanity's attempts to shape the entire globe according to its needs. The world games Richard Buckminster Fuller (1895–1983) developed during the latter half of the last century were rooted in a comparable notion of a large-scale, technocratic makeover of the earth's entire environment. Relying on visual representations of the earth in the form of acrylic-glass globes and three-dimensional maps complete with colored lights indicating the misaligned distribution of scarce resources and plentiful humans, Fuller envisioned the technocratic planner both as the administrator of earth's limited resources and the captain of spaceship earth.

The British economist Barbara M. Ward (1914–1981) popularized the latter term when she highlighted in her book *Spaceship Earth* (1966) the unequal distribution of wealth and resources across the globe. Together with the American microbiologist René Dubos, Ward wrote *Only One Earth* in 1972 for the United Nations Conference on the Human Environment. This was one of the first books arguing that environmental problems ought to be addressed on the global scale of one world rather than on national or even smaller scales. The question, however, which scale is appropriate to address environmental issues, was debated already from the nineteenth century onwards.

In 1877, the British Darwinist Thomas H. Huxley (1825–1895) raised doubts in his book *Physiography* whether explaining to a child the earth as a celestial body would instill a sense of understanding and, accordingly, responsibility for the natural environment. Huxley proposed to explain the world instead by concentrating on the small scale of the immediate, local surroundings of the child. Subsequently, Geddes emphasized that knowledge about the earth at large was only a stepping-stone from which to zoom in on the much smaller region around one's home as the best framework for engaging with the environment.

German philosopher Edmund Husserl (1859–1938) stated in 1934 that while he may not know what lies beyond the horizon, other humans live beyond that border. They will know their corner of the larger world, which ends, in turn, at their respective horizons. Beyond the latter, more human beings live, and more even further out. From the knowledge each human has of his or her place within the world, a collectively created image of the one world emerges as the habitat of all of humanity. In short, Husserl suggests that in order to understand the world all we need to know is the many worlds of all its inhabitants.

The debate continues today. Bioregionalism, for example, views the world in the tradition of Geddes as composed of smaller regions to which successful environmental intervention should adhere. By comparison, the presumed scale of coming ecological disasters such as the possible consequences of global warming makes some scientists call for climate engineering or geoengineering on a global scale.

## Learning resources

Gateway to Astronaut Photography of Earth eol.jsc.nasa.gov

An official NASA website that allows one to explore and view historic and contemporary earth photographs taken by astronauts and satellites.

*"For the Sake of the Prospect": Experiencing the World from Above in the Late 18th Century* (publicdomainreview.org/2016/07/20/for-the-sake-of-the-prospect-experiencing-the-

world-from-above-in-the-late-18th-century/) and *"Unlimiting the Bounds": The Panorama and the Balloon View* (publicdomainreview.org/2016/08/03/unlimiting-the-bounds-the-panorama-and-the-balloon-view). These are two essays, including a great number of historic images, by Lily Ford on how during the 18th and 19th centuries humanity gradually learned viewing the world from above, and the consequences this perspective had on humankind's understanding of the earth.

## Bibliography


Brand, S. (1977) 'Why haven't We Seen the Whole Earth?' In Lynda Rosen Obst *The Sixties: The Decade Remembered Now, By the People Who Lived It Then* (pp. 168–70). New York: Random House/Rolling Stone Press.

Cosgrove, D. (1994) 'Contested Global Visions: One World, Whole Earth, and the Apollo Space Photographs', *Annals of the Association of American Geographers*, 48: 270–94.

Fluegelman, A. (1976) *The New Games Book*. Garden City, NY: Headlands Press.

Husserl, E. (1934) 'Foundational Investigations of the Phenomenological Origin of the Spatiality of Nature: The Originary Ark, the Earth Does not Move'. In Maurice Merleau-Ponty (2002) *Husserl at the Limits of Phenomenology* (pp. 117–131). Edited by Leonard Lawlor. Evanston, IL: Northwestern University Press.

Huxley, T.H. (1877) *Physiography: An Introduction to the Study of Nature*. London: Macmillan & Co.

Morris, W. (1881) 'The Prospect of Architecture in Civilization'. In William Morris (1966) *The Collected Works of William Morris with Introductions by his Daughter May Morris*. Vol. xxii (pp. 119–52). New York: Russell & Russell.

Poole, R. (2008) *Earthrise: How Man First Saw the Earth*. Oxford: Oxford University Press.

Thomas, W.L. (ed.) (1956) *Man's Role in Changing the Face of the Earth*. Chicago: University of Chicago Press.

Ward, B. (1966) *Spaceship Earth*. New York: Columbia University Press.

Ward, B. & Dubos, R.J. (1972) *Only One Earth: The Care and Maintenance of a Small Planet*. New York: W.W. Norton.

Welter, V.M. (2010/11) 'From Disc to Sphere', *Cabinet: A Quarterly of Art and Culture*, 40: 19–25.

# 1.14 Overpopulation

Eric D. Carter

The global population continues to grow, from about 7.3 billion people today, to an expected 8.5 billion in 2030, and over 11 billion by 2100. While the rate of growth has slowed, overpopulation is still widely seen as a driving cause of current environmental problems and the seed of impending catastrophe. Blaming overpopulation for all kinds of social and environmental troubles has a long and layered history, going back to at least the eighteenth century, and it is hard to understand modern environmental thought and environmental science without full consideration of the debates over the overpopulation issue.

## History of the population question

Modern thought on the population question begins with Thomas Malthus, one of the notable classical British economists, whose *Essay on the Principle of Population* was first published in 1798. In it Malthus famously asserted that the population of a given country grows at a geometrical rate while food supply expands only arithmetically; thus, eventually, the means of subsistence would reach a natural limit and the result would be, inevitably, shortages, hunger, famine, and epidemics. Even in normal times, population growth depressed wages and deepened the misery of the working classes, leading Malthus to argue that the Poor Laws of England and other forms of economic relief for the poor only stimulated rampant population growth and delayed inevitable crisis. Over the years, Malthus's ideas were used to justify conservative class interests, the eugenics movement, and callous imperial responses to hunger and famine (Bashford, 2014; Davis, 2001). Yet, Malthusianism was also entangled with struggles for women's reproductive rights, led by feminists such as Emma Goldman, who proclaimed that the working class could achieve its own emancipation through 'conscious procreation', and Margaret Sanger, the founder of Planned Parenthood (Connelly, 2008; Masjuan & Martinez-Alier, 2004).

In the middle of the twentieth century, the so-called neo-Malthusians adapted Malthus's demographic determinism to warn of the exhaustion of land and natural resources more generally, not just the food supply. Best-selling books such as William Vogt's *The Road to Survival* and Fairfield Osborn's *Our Plundered Planet* (both from 1948) stirred fears that rampant population growth and industrialization were bringing the Earth to the limits of its 'carrying capacity' (Robertson, 2012). The overpopulation thesis became one pillar of the modern environmental movement with the work of Paul Ehrlich, a Stanford University population biologist, whose book *The Population Bomb* (1968) predicted global environmental crisis. Philosophically, the neo-Malthusians stood out for their conviction that natural laws—the principles of demography and ecology, specifically—governed the fate of a global human society. In imposing a natural-science framework on the dynamics of human political economy, Ehrlich actually brought Malthusian thinking 'full circle', since Malthus had inspired Darwin's evolutionary theory, which in turn served as the foundation for modern population biology (Worster, 1994).

After Ehrlich, population growth became a focal point in systematic studies of the causes of environmental degradation (mainly in economics and sociology). Ehrlich helped to develop the IPAT formula, where environmental impact (I) is calculated as the product of population growth (P), income or affluence (A), and technology (T). This formula is intrinsically related to the concept of 'ecological footprint', which introduced an accounting method to assess the impacts of consumption beyond national borders. In all, this kind of impacts accounting research has shown that rising consumption in countries with relatively high income and low population growth (mainly, the Global North), has far-reaching environmental impacts in developing countries (the Global South)—since, for example, "the footprint of the typical American is nearly 25 times greater than that of the typical Bangladeshi" (York et al., 2003, p. 295). However, more recent "changes in patterns of consumption have blurred this binary distinction between rich and poor societies" (Rosa et al., 2015, p. 37).

## Critical voices

One strand of critique of the overpopulation thesis, from the 'economic optimists', emphasizes the power of markets and technological innovations to respond to scarcity. This notion is exemplified by the economist Julian Simon, who famously wagered with Ehrlich in the 1970s against the prospect of a global ecological calamity. From the political left, other critics take a 'distributionist' angle, seeing famine and environmental degradation as produced by unequal political-economic structures (Newbold, 2014). Karl Marx and Friedrich Engels initiated this critique of Malthus, which was further elaborated by Josué de Castro of Brazil, in his book *The Geography of Hunger* (1952) and the geographer David Harvey (1974). The economist Amartya Sen—hardly a Marxist—also argued that population pressure on food supply is seldom the cause of hunger and famine, which can instead be traced to widespread deprivation, market failures, and unresponsive governments; that is, political-economic, not natural, causes (Sen, 1994).

The effect of population growth on land use change is also controversial. In the 1960s, Danish economist Ester Boserup effectively turned Malthusian logic on its head, arguing that agricultural production has ecologically flexible limits and land managers respond to localized demographic pressure with intensification of land use, as demonstrated by countless examples including the well-known case of the Machakos district in Kenya (Mortimore & Tiffen, 1994). Conversely, local population decline (for example, from labor outmigration) may actually lead to environmental degradation, due to the decay of landesque capital, such as agricultural terraces, or a turn towards less labor-intensive yet destructive practices, such as deforestation for cattle grazing. In all, research shows that land use change has complex drivers, and proximate population pressure on resources is seldom the key driver of land degradation (Blaikie & Brookfield, 1987; Lambin et al., 2001).

There has always been resistance to the policy ramifications of neo-Malthusianism. Political ecologists have placed overpopulation discourse in the company of mutually reinforcing environmentally deterministic theories (e.g., Garrett Hardin's Tragedy of the Commons thesis and the desertification narrative) that dominated environment-and-development work well into the 1990s, mainly to the disadvantage of smallholders and traditional resource users in developing countries (Moseley & Laris, 2008). Fears of population crisis propelled the Green Revolution in agriculture, which arguably worsened rural poverty in developing countries as technologically intensive practices displaced traditional farmers (Cullather, 2010).

But perhaps the clearest political consequence of neo-Malthusianism was a reorientation of international development policy toward fertility control and 'family planning'. To some critics of this broad effort, international development planners used faulty premises and a narrative of pending environmental and humanitarian crisis to push a racist agenda with a long heritage, dating back to nineteenth-century European imperialism, scientific racism, and the eugenics movement (Connelly, 2008). Looming crisis led major development organizations (such as UNFPA, the United Nations Fund for Population Activities, the Ford and Rockefeller Foundations, and USAID) to support or condone coercive efforts, such as the One-Child Policy in China and compulsory sterilization in India in the 1970s. Early in this decade of stringent population control efforts, David Harvey (1974) pointedly explained the political implications of the overpopulation argument: "Somebody, somewhere, is redundant, and there is not enough to go round. Am *I* redundant? Of course not. Are *you* redundant? Of course not. So who is redundant? Of course, it must be *them.*" Thus, he continued, "The overpopulation argument is easily used as part of an elaborate apologetic through which class, ethnic, or (neo-) colonial repression may be justified". Along similar lines, Sen (1994) perceived a reactionary political agenda behind the then-popular overpopulation argument: "fears [in the Global North] of being engulfed" by migrants and refugees from the poorer countries of the Global South.

## Population and environmentalism today

Today, the politics of overpopulation are vexed. Malthusian thought was a vital underpinning of an international environmental movement in the 1970s, and we can see its traces in 'lifeboat ethics,' Spaceship Earth, the 'limits to growth' stance, concerns over desertification, and other tropes of environmental discourse from that era. However, by now population control is, at best, a back-burner issue for most major environmental organizations (Cafaro, 2015). Conservatives reject neo-Malthusianism for its anti-growth ethic and coercive family planning efforts while progressives cringe at its racist and neo-colonial overtones (Hoff & Robertson, 2016). Meanwhile, fertility rates have generally declined, in the US and around the world, also deflating the urgency of population control. This 'baby bust' or population stagnation has implications for all kinds of issues, from immigration to sustaining welfare state benefits to resurgent ethnic nationalism (Robbins & Smith, 2016). Today, there is a broad consensus, consistent with demographic transition theory, that mortality-reducing improvements in public health, along with poverty reduction and female empowerment, generally lead toward lower fertility, diminishing the need for coercion in reproductive matters.

In environmental studies, the overpopulation question may seem dead, a thing of the past. And yet, the question remains, can population *and* consumption continue to grow together at current rates without severe impact on natural resource stocks, ecosystem services, and global climate? As Crist and Cafaro (2012, p. 4) frame it, technology and a mostly unchallenged dedication to a capitalist growth paradigm have enabled humanity to push well past the limits Malthus or Ehrlich foresaw, in the process "turning the whole world in Resource World"—yet, is this really an achievement to be celebrated, or rather, "an ignoble and unjust goal"? Perhaps it is time for a rapprochement between Malthusian and anti-Malthusian camps, to deal squarely with the social, environmental, and ethical challenges of this Anthropocene era.

## Learning resources

*Imagine the Population of Tomorrow*. Hosted by INED, France's national demographic institute, this interactive simulation game lets you create your own future population growth scenarios by manipulating fertility rates and other variables, for the world and for specific countries. http://www.ined.fr/en/everything_about_population/population-games/tomorrow-population/

*New York Times Retro Report: The Population Bomb?* This 13-minute documentary video vividly details the rise and fall of neo-Malthusian concerns in the 1960s and 1970s, in the United States and globally. https://www.youtube.com/watch?v=W8XOF3SOu8I

*United Nations Population Division*. A reliable source for information on population, mortality, fertility, and migration trends internationally. http://www.un.org/en/development/desa/population/index.shtml

## Bibliography

Bashford, A. (2014) *Global population: history, geopolitics, and life on earth*. New York: Columbia University Press.

Blaikie, P.M. & Brookfield, H.C. (1987) *Land degradation and society*. London; New York: Methuen.

Cafaro, P. (2015) *How many is too many?: The progressive argument for reducing immigration into the United States*. Chicago: University of Chicago Press.

de Castro, J. (1952) *The Geography of Hunger*. Boston: Little, Brown.

Connelly, M.J. (2008) *Fatal misconception: The struggle to control world population*. Cambridge, MA: Harvard University Press.

Crist, E. & Cafaro, P. (2012) Human population growth as if the rest of life mattered. In Cafaro, P. & Crist, E. (Eds), *Life on the brink: environmentalists confront overpopulation* (pp. 3–15). Athens: University of Georgia Press.

Cullather, N. (2010) *The hungry world: America's Cold War battle against poverty in Asia*. Cambridge, MA: Harvard University Press.

Davis, M. (2001) *Late Victorian holocausts: El Niño famines and the making of the Third World*. New York: Verso.

Ehrlich, P.R. (1968) *The Population Bomb*. New York: Ballantine Books.

Harvey, D. (1974) 'Population, resources, and the ideology of science', *Economic Geography*, 50(3): 256–277.

Hoff, D.S. & Robertson, T. (2016) 'Malthus today'. In R.J. Mayhew (Ed.), *New perspectives on Malthus* (pp. 267–293). Cambridge: Cambridge University Press.

Lambin, E.F., Turner, B.L., Geist, H.J., Agbola, S.B., Angelsen, A., Bruce, J.W. & Folke, C. (2001) 'The causes of land-use and land-cover change: moving beyond the myths', *Global Environmental Change*, 11(4): 261–269.

Masjuan, E. & Martinez-Alier, J. (2004) *'Conscious procreation': Neo-Malthusianism in Southern Europe and Latin America in around 1900*. Paper presented at the International Society for Ecological Economics, Montréal 11–15 July 2004.

Mortimore, M. & Tiffen, M. (1994) 'Population growth and a sustainable environment: the Machakos story', *Environment: Science and Policy for Sustainable Development*, 36(8): 10–32.

Moseley, W.G. & Laris, P. (2008) 'West African environmental narratives and development-volunteer praxis', *Geographical Review*, 98(1): 59–81.

Newbold, K.B. (2014) *Population geography: tools and issues* (2nd edn.). Lanham, Maryland: Rowman & Littlefield.

Robbins, P. & Smith, S.H. (2016) 'Baby bust towards political demography', *Progress in Human Geography*. https://doi.org/10.1177/0309132516633321.

Robertson, T. (2012) *The Malthusian moment: Global population growth and the birth of American environmentalism*. New Brunswick: Rutgers University Press.

Rosa, E.A., Rudel, T.K., York, R., Jorgenson, A.K. & Dietz, T. (2015) 'The human (anthropogenic) driving forces of global climate change'. In R.E. Dunlap & R.J. Brulle (Eds), *Climate change and society: Sociological perspectives*. New York: Oxford University Press.

Sen, A. (1994) 'Population: delusion and reality'. *New York Review of Books, 41*(15).

Worster, D. (1994) *Nature's economy: a history of ecological ideas*. Cambridge: Cambridge University Press.

York, R., Rosa, E.A. & Dietz, T. (2003) 'Footprints on the earth: The environmental consequences of modernity', *American Sociological Review*, 68: 279–300.

# Precaution

1.15

## Tim O'Riordan and Rupert Read

Precaution emerged as a central tenet of European environmental policy in the mid-1970s (Boehmer-Christiansen, 1994; Jordan, 2000; Gee, 2013). Its regulatory companions were carefully juxtaposed: stimulation of the best available technology for prevention and efficiency of resource use, safeguarding of nurturing environmental space, proportionality of action in relation to gains and losses, sharing the burden of regulatory responsibilities, ensuring the interests of forthcoming lives of people and biota, making the polluter pay, and placing the burden of proof of additionality to sustainability on those proposing to introduce new products, technology or developments. Its genius lies in making it clear that the good cause need not necessarily be certainty, fully known knowledge, or even evidence-based analysis. It may be simply *exposure* to a grave potential harm.

In essence, precaution imposes a duty of planetary care on the human will. It seeks to deepen, to widen and to lengthen all manner of so-called impact assessments. Above all, it provides a basis for public discussion and deliberation over what kind of society and moral accountability we collectively should choose to adopt.

In this formulation, precaution fits in with many, if not all, of the themes encompassed by the companion chapters. It certainly addresses the notions of safe planetary boundaries, of participatory science, of the 'Anthropocene', of bioethics, of political power and corporate manipulation, and of cultural sensitivities. Because of its universality of agenda, precaution is cherished, tolerated, ignored and despised in equal measure. It reached the heady status of a Rio Principle agreed by political leaders who attended the first Earth Summit in 1992:

> In order to protect the environment, the precautionary approach shall be widely applied by States according to their capabilities. Where there are threats of serious . . .

> irreversible damage, lack of full scientific certainty shall not be used as a reason for postponing cost-effective measures to prevent environmental degradation.
>
> (Principle 15)

A key role of precaution is to slow the implementation of innovations that may have cause to be seriously destructive. This Rio formulation brings out clearly the other side of this coin: that rapid intervention is called for to *prevent* such destruction, even before research may have demonstrated that such action is necessary. Precaution calls upon us to take care, and first of all to do no harm; it simultaneously calls upon us to act swiftly in order to head off great harm. Since 1992 its star has waned. Here we examine its strengths, its weaknesses, and its pathways to restorative salvation.

## The power of precaution in its initial formulation

The precautionary principle is an heir to the German *Vorsorgeprinzip* (Boehmer-Christiansen, 1994: 33–43). Though often interpreted as 'foresight', *Vorsorge* applies to responsible and equitable appraisal of future *action*. 'Fore-care' is a useful literal translation. It was developed in the mid-1970s in the aftermath of the UN Conference on the Human Environment held in Stockholm in 1972, and in the furore following the publication of *The Limits to Growth* (Meadows et al., 1972); (see also O'Riordan, 2012). Lying behind the evolution of this interpretation were both a sense of provable responsibility for public, private and civic actions, and a sharing of accountability between the state, the change-agent and the electorate. In this sense, precaution connects with more recent trends in the deliberative approaches to uncertainty and forecasting, as well as to the question of methods of calculating toxicity and biodiversity damage.

Today, there is a greater willingness (arguably resulting from corporate capture of public discourse) to accept regulatory voluntarism as an expression of self-styled corporate social responsibility. This is despite continuing skepticism amongst wary public interest groups. The early rise of precaution relied more on formal rules of engagement and consensus amongst business, unions and civil society brokered by the state. This arrangement coincided with the emergence of powerful state-run health and safety organizations, powerful pollution prevention agencies, and a heightened regulatory legal regime. By the mid-1980s, in its regulatory heyday in many countries, the precautionary principle guided many of the procedures for assessing health and safety generally, chemical toxicity, for protecting endangered species, and for providing regulatory space in the face of scientific ignorance.

In the UK the emergence of the prestigious and multidisciplinary Royal Commission on Environmental Pollution in 1970 (Owens, 2015) coincided with a deepening ethical and even theological interpretation of precaution. In its path-breaking reports on radioactive waste, setting environmental standards, restricting marine pollution, avoiding possible toxic effects of pesticides and incineration, removing lead from gasoline, and exploring the scope of genetically modified organisms, the Commission imaginatively explored the widening of the precautionary principle.

In all of this flurry of mid-1980s legislation, the precautionary principle was invoked to ensure that the integrity of ecosystems-functioning was protected, that primary biodiversity was enhanced, and that human health and well-being were given attention, especially where minority rights and future generations were concerned. In its early formulation(s), precaution championed the interests of the two unseen and unheard components of the planet: future generations, especially of the disadvantaged, and ecological assemblages and interconnections, which combine to retain life and habitability.

## Mid-life crisis

The original ideals for precaution, especially regarding shifting the burden of proof onto the promoter, together with strong and social justice driven regulatory arrangements, were steadily eroded. This was in part a direct result of deregulatory pressure. In the case of toxic pesticides, for example, the evidence is instructive. A group of pesticides known as neonicotinoids, closely associated with damage to insect pollinators, especially bees, should have been banned on the basis of precaution. Instead, after extensive lobbying by the farming and agro-chemical industries, the European Commission announced a 3-year temporary ban to explore the outcomes of the trial removal. It is impossible for ecological science to prove one way or another that such a 'fallow period' would show the effects on bee populations which are under attack from many sources. But it was, seemingly, equally (politically) impossible for the Commission to ban these formulations simply on the basis of precaution.

On the front of genetically modified organisms, the role of precaution has coincided with widespread public disquiet throughout Europe (but not in North or South America) over the introduction of non-natural genes into living organisms, especially food crops. Here the application of precaution coincided and helped to buttress that massive opposition by consumers. So it is possible for precaution still to have a role. But such a role probably has to be subsumed into a wider and more potent political anger. Acting on its own, precaution is nowadays a weakened scientific and political weapon.

The Precautionary Principle has been targeted in recent years by those who seem determined to prevent anything preventing the free rein of 'the free market' (American Enterprise Institute, 2016). Moreover, much of what is called 'evidence' by those who want to downgrade the precautionary principle is not statistically significant, in relation to the potential for catastrophic events (Taleb, 2007; see also Taleb et al., 2014). Such events are by definition rare, and usually barely-evidenced. Where substantial evidence in the true sense of the word is lacking, the precautionary principle ought to fill the breach. In practice, it will only do so by a deliberate act of political or legal will.

## Precaution, future generations and the ethics of sustainability

Thus advocates of precaution are fighting back using such routes as mass citizen awareness and/or legal intervention. In today's world of active social media, where political opinion can be explosive if well-targeted and directed; and where there are increasing legislative demands for formal consultation, there is perhaps a new lease of life for precaution. The emergence of potentially-devastating crises in *other* areas of human concern beside 'the environment' has also, arguably, underscored the contemporary salience of precautionary reasoning. Examples here include *finance* (Taleb, 2007); and *public health*, where potential super-epidemics such as Ebola (Bar-Yam and Hardcastle,

2014), or the potential ineffectiveness of antibiotics, echo the same kinds of non-calculable risk of serious, potentially irreversible harm.

The emergence of 'deliberative science', as creatively explored in an official UK science report (Wolport, 2014), has come of age. No longer is it possible to introduce new technology, or products, onto an unsuspecting public. Increasingly it is also necessary for creators of innovative products or technology formally and publicly to take into account the interests of both the disadvantaged and the next generations (social and intergenerational justice tests).

The precautionary principle is particularly relevant here. Where there is a clear and sharp *asymmetry* where one party, which will make the decision, stands potentially to benefit, while another party, which has no say in the decision, stands catastrophically to lose out – then the application of the precautionary principle arguably ought to be decisive.

As many of the companion chapters will reveal, environmental concepts are metaphors for social ills and moral triumphs. All of the themes in this volume are constantly being reinterpreted in the context of changing cultures and new environmental bedfellows. The precautionary principle epitomizes this feature. It will rise and fall in social salience and in our consciences according to the swells of all of the other notions which adorn this volume. Thus it speaks for the age and for the uniqueness of the human condition.

## Learning resources

The following websites contain much of relevance and interest:

Bar-Yam and Hardcastle (2014):
http://necsi.edu/research/social/pandemics/ebola3.pdf

Denis Campbell (2014):
http://www.theguardian.com/society/2016/mar/15/antibiotics-becoming-ineffective-at-treating-some-child-infections

WHO factsheet:
http://www.who.int/mediacentre/factsheets/fs194/en/

American Enterprise Institute (2016):
https://www.aei.org/publication/the-problems-with-precaution-a-principle-without-principle/

Taleb, Read, Douady, Norman and Bar-Yam (2014):
http://www.fooledbyrandomness.com/pp2

## Bibliography

Boehmer-Christiansen, S. (1994) The precautionary principle in Germany – enabling government. In O'Riordan, T. and Cameron, J. (eds), *Interpreting the Precautionary Principle* (pp. 31–60). London: Cameron and May.

Gee, D. (ed.) (2013) *Late Lessons from Early Warnings: Science, Precaution, Innovation.* Copenhagen: European Environment Agency.

Jordan, A. (2000) The precautionary principle in the European Union. In O'Riordan, T., Cameron, J. and Jordan, A. (eds), *Reinterpreting the Precautionary Principle* (pp. 143–162). London: Cameron and May.

Meadows, D.H., Meadows, D.L., Randers, J. and Behrens III, W.W. (1972) *The Limits to Growth: A Report for the Club of Rome's Project on the Predicament of Mankind.* New York: Universe Books.

O'Riordan, T. (2012) *The Limits to Growth* revisited. In Vaz, S.G. (ed.) *Environment: Why Read the Classics?* (pp. 93–113). Sheffield: Greenleaf Publishing.

O'Riordan T. and Cameron, J. (eds) (1994) *Interpreting the Precautionary Principle.* London: Cameron and May.

O'Riordan, T., Cameron, J. and Jordan, A. (2000) *Reinterpreting the Precautionary Principle.* London: Cameron and May.

Owens, S. (2015) *Knowledge, Policy and Expertise: The UK Royal Commission on Environmental Pollution 1970–2011.* Oxford: Oxford University Press.

Taleb, N.N. (2007) *The Black Swan: The Impact of the Highly Improbable.* New York: Random House.

Wolport, M. (ed.) (2014) *Innovation: Managing Risk not Avoiding it.* Annual Report of the Government Chief Science Advisor. London: Department of Business, Innovation and Skills.

# 1.16 Risk

Susan L. Cutter

Risk is a very old concept and has always been part of the human experience. At its most basic level, risk is the chance that something bad will happen. It is the threat of danger or the potential exposure to harm. The transition from an industrial world to one with increasing reliance on technology created the "Risk Society" (Beck, 1992). The reflexive nature of risk—human agency both producing and managing risk—coupled with globalization and increasing complexity and interconnectivity of societal systems are key elements in our contemporary understanding of risk. Simply put, most risks now facing the world are of our own making—acts of people, not acts of some deity nor only of nature.

## The changing nature of risk

Modernity has improved the human condition, but it has also led to the production of risks never before seen or experienced such as climate change, pandemics, or cyber threats. New or unknown risks are dreaded more than old or known ones; there is less worry about risks we can control versus those we cannot; and there is more fear of those risks whose effects are hardest to observe such as airborne toxins (Slovic, 2000). Significant advancements in our understanding of how individuals perceive risks are due to the evolution of the interdisciplinary field of risk perception—the study of why people make different estimates of danger—drawing upon theories, concepts, and methods from psychology, decision sciences, communications, anthropology, sociology, and geography.

Differences in risk perception (based on cognitive influences, personality, and emotions), experience, culture, social and political organizational processes, demographics, proximity, and mass media help to explain why some risks are judged as acceptable and others are not. The social construction of risk is a key component in our present understanding of risk and helps explain why specific risks become unacceptable and others highly politicized. 'Social construction' here means that risks are no longer viewed in objective isolation, but are interpreted through the context in which they occur. For example, experts may have one view on the risk and its outcome such as the use of genetically modified food, but the public may have an alternative perspective that does not align with the expert judgement on the risks and benefits. In effect, they are less accepting of the risks and desire greater regulation or protection from private (insurance) and public institutions.

## Managing risks

The management of risks is a process whereby society (and individuals) makes choices about which risks to take responsibility for and regulate or mitigate and which ones to essentially ignore. Risk management includes a four-step assessment process that includes hazard identification, the statistical relationship between the level of exposure and adverse outcomes (dose-response), the probability of anticipated exposures (exposure assessment), and the estimated incidence of the known adverse effect in a given population (risk characterization). Once the assessment process is completed, risks are prioritized and options for their reduction identified. Depending on the field (medicine, public health, engineering, business, public safety and security or disasters) risk assessment outputs are either qualitative, quantitative, or some combination of the two.

There are many techniques for managing risks but they generally fall into one of the following categories: legislation, governmental regulation, insurance, common law adjudication, and self-regulation (Covello and Mumpower, 1985; Cutter, 1993). Let's consider each in turn.

*Legislation* is one of the most important techniques for managing risks from local to national scales using legally binding and enforceable laws. Internationally, multinational treaties and directives accomplish the same goal but are not as legally binding or enforceable. The US Clean Air Act (regulating air emissions by setting ambient air quality standards to public health and welfare) is one example of a risk management law at the national level. The 2015 Paris Agreement (UN Framework Convention on Climate Change) is an example of an international treaty designed to reduce the risks and impacts of climate change through greenhouse gas emissions reductions.

*Government regulation* is another approach to risk management. Government regulation derives its mandate from laws or legal statutes and establishes regulatory structures to administer and enforce the laws. There is considerable variability in how the legal mandates are interpreted by the regulating agencies and rule makers, and this has often led to the contested nature of risk and increased debates about how safe is safe enough. The so-called risk managers govern most aspects of modern life from our transportation to the safety of our buildings, to workplace regulation, to the food we eat, and the consumer products we purchase. For some, the intrusion of governmental regulation into daily life is viewed negatively not as a protective or precautionary action, again illustrating the highly politicized nature of the social construction of risk.

*Insurance* is a time-honored approach to managing risk. The basic premise for insurance is to share the loss burden over a wide segment of the public. The collection of premiums from individual policies (homeowners insurance, crop insurance, flood insurance) is then redistributed to affected parties in times of loss. Insurers may leave certain market areas (e.g. canceling or not renewing policies in high-hazard areas such as the coast or near rivers), or increase deductibles while reducing what is covered to remain profitable. Re-insurers (companies that insure property and casualty insurance companies) are now leading climate change risk reduction efforts because of the lack of governmental action and the exponential increase in insured losses from extreme weather events which threaten their future financial solvency and bottom line.

The judicial system also plays a part in the management of risks through *adjudication*. Litigation based on common law or 'toxic torts' allows people to claim injuries from the activities of someone else or an industrial facility. In this civil action, the injured party (plaintiff) claims that the defendant (industrial plant) is liable for damages. Litigation has become a preferred mechanism for managing risks, especially in North America. The most recent case of toxic torts (the largest product liability settlement ever) is the $7.8 billion settlement between British Petroleum and claimants from the 2010 Deepwater Horizon oil spill in the Gulf of Mexico.

The last general technique for managing risks is *self-regulation*. Private sector self-regulation only works when the risks (and technologies) are well known and pose significant liabilities which force a responsible industry approach to risk reduction. While some view the self-regulation as contributing even more to the risk society, the establishment of standard-setting organizations and professional certification and licensing helps mediate those concerns. Organizations such as the International Organization for Standardization (ISO) produce consensus-based voluntary standards that ensure the safety, reliability and quality of goods and services in a number of important sectors—health and safety, food, water, transport, energy, and development. ISO standards cover such areas as principles and guidelines for risk management for companies and organizations (ISO 31000) or standards for environmental management systems for companies worldwide (ISO 14000) life-cycle analyses and environmental impact labeling.

## What the future portends?

The distribution of risk within society is uneven as are their impacts. In some instances the benefits of production accrue to one social group, community, or nation, while the risks are borne by another group, community, or nation. The transfer of risk from rich to poor neighborhoods or from developed to developing nations creates the uneven riskscape of the modern world. In the risk society, disaster and industrial failures now are routine with worst cases increasingly normal (Clarke, 2011). Mounting public concern about risks and the ability of science and technology to control them has led to a loss of confidence and trust in many risk management institutions.

To complicate matters, every day brings new challenges about emergent risks, risks that are less understood, have ambiguous consequences, and lack data about exposure and impact. Emergent risks are generally systemic in nature (failure in the entire system, not just one component of it) and hard to anticipate. Similarly, black swan events (Taleb, 2007) are equally unexpected, random, and rare events with extreme impacts that in hindsight could have been anticipated. Future risk managers will have to learn to foresee emergent risks (e.g. large-scale involuntary migrations, freshwater water crises, new pandemics), worst cases, and black swan events. At the same time they will be challenged in understanding the connectivity between them and the likely cascade of impacts as the risk society becomes more interdependent and reflexive than ever before.

## Learning resources

A sampling of resources about risk and risk reduction:

United Nations Office for Disaster Risk Reduction (UNISDR): This website provides information on international risk reduction policy and practice: http://www.unisdr.org

Envirofacts: A US Environmental Protection Agency website that allows explorations of toxic substances in communities: https://www3.epa.gov/enviro/

## Bibliography

Beck, U. (1992) *Risk Society: Towards a New Modernity*. Los Angeles, CA: SAGE Publications.

Clarke, L. (2011) *Worst Cases: Terror and Catastrophe in the Popular Imagination*. Chicago: University of Chicago Press.

Covello, V.T. & Mumpower, J. (1985) 'Risk Analysis and Risk Management: An Historical Perspective', *Risk Analysis*, 5(2): 103–120.

Cutter, S.L. (1993) *Living with Risk: The Geography of Technological Hazards*. London: Edward Arnold.

Slovic, P. (2000) *The Perception of Risk*. Sterling, VA: Earthscan.

Taleb, N.N. (2007) *The Black Swan: The Impact of the Highly Improbable*. New York: Random House.

Tierney, K. (2014) *The Social Roots of Risk: Producing Disasters, Promoting Resilience*. Stanford, CA: Stanford University Press.

# 1.17 Resilience

## Jeremy Walker and Melinda Cooper

'Resilience', a quality of 'complex adaptive systems', has moved far beyond its original applications in environmental science to become a ubiquitous term in contemporary practices of crisis management. Developed within ecology in the 1970s, complex systems theory served to conceptualise the stable or resilient dynamics of ecosystems subject to extractive industry. The concept of resilience has in recent years rapidly infiltrated vast areas of the social sciences, becoming a key term in finance, central banking, corporate strategy, psychology, development, urban planning, public health, education and national security. Malleable and capacious enough to encompass human and non-human 'systems' within a single analytic, the concept of resilience is well established in the lexicon of global governance.

## Origins

The ecologist Crawford Holling did important work to renovate classical systems ecology in terms of the new 'complexity sciences'. His work on resilience marked a shift away from the mechanistic assertions of equilibrium typical of post-war cybernetics. The key image of science that propelled the formalisation of economics (in the 1870s) and ecology (in the 1950s), was of smooth and continuous returns to equilibrium after a shock, one derived from different vintages of classical mechanics and thermodynamics. Holling's classic (1973) paper 'Resilience and Stability of Ecological Systems' exemplifies the destabilisation of the notion of 'equilibrium' as the core of the ecosystem concept and the normal terminus of ecosystem trajectories. It initiated a retreat amongst ecologists from the idea that there exists a 'balance of nature' to which ecosystems will return if left to self-repair. Speaking as an experienced resource manager and conservation ecologist, Holling began his paper noting that:

> traditions of analysis in theoretical and empirical ecology have been largely inherited from developments in classical physics and its applied variants [. . .] But this orientation may simply reflect an analytic approach developed in one [field] because it was useful and then transferred to another where it may not be.
>
> (1973: 1)

He went on to distinguish between an existing notion he calls 'engineering resilience' and his alternative, a properly 'ecological' resilience. Engineering resilience, associated with mathematical models, is an abstract variable, simply the time (*t*) it takes an ecosystem to return to a stable maximum (or equilibrium position) after disturbance. The return is simply assumed, and the equilibrium state conflated with long-term persistence. Holling articulated instead a complex version of resilience which can account for the ability of an ecosystem to remain cohesive even amidst extreme perturbations. 'Ecological' resilience designates the complex biotic interactions that underpin "the persistence of relationships within a system", thus resilience is "a measure of the ability of these systems to absorb changes of state variables, driving variables, and parameters, and still persist" (1973: 17).

Holling criticised the management theory of 'maximum sustained yield' (MSY), long dominant in industrial forestry and fisheries, with its claims to quantify the 'surplus' portion of a population that can be harvested year in year out, without undermining the regenerative capacity of the ecosystem. Holling's argument here was that the long-term expectation of stability may be inherently destabilising. Acting on assumptions of a permanent resource yield may undermine the complex interdependencies comprising the resilience of the system as a whole, rendering its organisation ever more fragile and vulnerable to collapse.

> The very approach . . . that assures a stable maximum sustained yield of a renewable resource might so change these deterministic conditions that the resilience is lost or reduced so that a chance and rare event that previously could be absorbed can trigger a sudden dramatic change and loss of structural integrity of the system.
>
> (1973: 21)

Holling's perspective reflects emerging critical voices which, in the early 1970s, insisted that intensive maximisation of agricultural and industrial production would at some point meet inherent limits, resulting in mass extinctions and intolerable over-pollution. For him, the equilibrium analysis was dangerously abstract: glossing over the unknowably complex interdependencies of ecosystems pressed into the conditions of maximised yield, it accelerated the process of fragilisation, potentially leading to irreversible losses of biodiversity, thus "resilience is concerned with probabilities of extinction . . ." (1973: 20). By contrast, Holling's perspective opens up a management approach capable of sustaining productivity even under conditions of extreme instability. Its ability to adapt to crisis events derives from the fact that it has abandoned long-term expectations:

> A management approach based on resilience . . . would emphasize the need to keep options open, [. . .] and the need to emphasize heterogeneity. Flowing from this

> would be not the presumption of sufficient knowledge, but the recognition of our ignorance: not the assumption that future events are expected, but that they will be unexpected.
>
> (1973: 21)

This passage is significant because it so clearly anticipated the guiding ideas of contemporary complex systems theory and its practical applications in crisis response. 'Resilience' now denotes an approach to risk management which foregrounds the limits to predictive knowledge and preventative action, insisting on the prevalence of the unexpected, seeking to 'absorb and accommodate future events in whatever unexpected form they may take'.

## Evolution

In the late 1990s, Holling formed the Resilience Alliance, a consortium which would build consensus with mainstream economists and ambitiously expand the insights of the resilience perspective well beyond ecology. These initiatives were brought together within the Stockholm Resilience Centre, a high-profile think tank which promotes and applies 'resilience thinking' in international environment and development policy. Holling and his colleagues were now concerned to advance resilience as an integral property linking societies *and* ecosystems, reconceptualised as unified coevolutionary systems. This research into 'social-ecological resilience' aspires to become a general systems theory integrating society, the economy, and the biosphere in a totality dubbed the 'Panarchy':

> the structure in which systems, including those of nature (e.g., forests) and of humans (e.g., capitalism), as well as combined human-natural systems (e.g., institutions that govern natural resource use such as the Forest Service), are interlinked in continual adaptive cycles of growth, accumulation, restructuring, and renewal.
>
> (Gunderson & Holling, 2002, cover)

There are significant differences between this account of socio-ecological resilience and Holling's earlier work. Holling was no longer arguing that ecological communities are at risk of irreversible extinction under the stress conditions of maximum sustained yield. Now, resilience is proposed as a perspective capable of analysing *all* socio-ecological systems in terms of an 'adaptive cycle' of recurring events, characterised by phases of rapid growth (*r*) toward a temporary stable maximum (K), then collapse (Ω), and spontaneous reorganisation for a new growth phase (α).

Having emerged as a critical perspective on modernist theories of economic growth in the post-WW II era, resilience theory today presents itself as an alternative theory of growth in far-from-equilibrium conditions, in which the capacity of systems to spontaneously re-organise through catastrophic events is denoted as 'capital'.

Arguably, the proliferation of 'resilence' across so many spheres of governance can be traced to its formal, political and ontological resonance with the influential philosophy of the Austrian economist and arch-neoliberal protagonist, Friedrich Hayek. From the 1970s, Hayek's radical critique of socialist, Keynesian and neoclassical economics would increasingly take the form of an account of market society as an evolving 'spontaneous order',

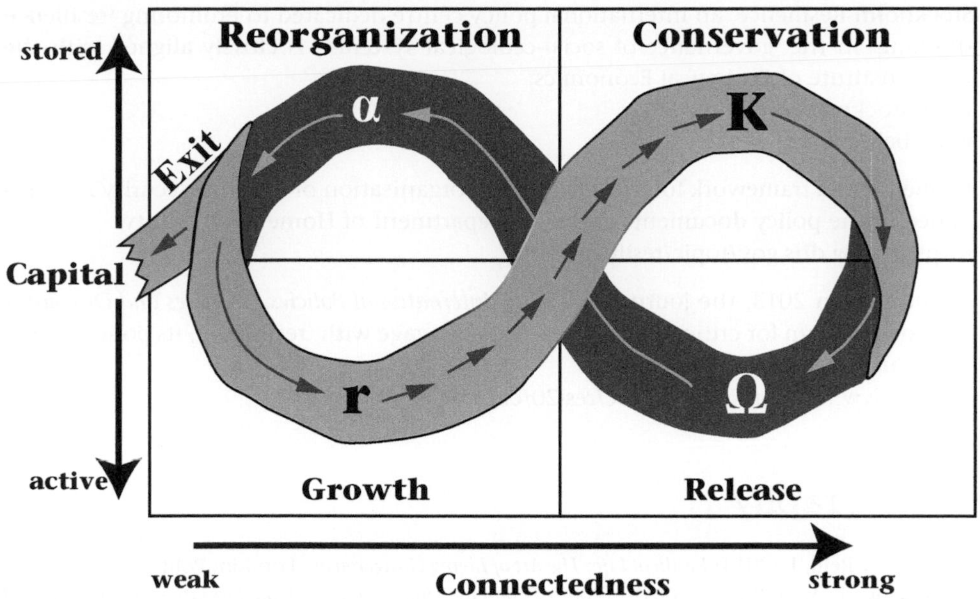


**Figure 1.17.1** A stylised representation of the four ecosystem functions (r, K, Ω, and α) organised into an adaptive cycle.

*Source*: adapted from Gunderson and Holling (2002) and reprinted with permission of Island Press, copyright 2002.

far too complex for any individual or government to understand or predict, much less to regulate for social objectives, including long-term ecological stability (Hayek, 1974).

From the 1990s, the International Monetary Fund, the World Bank and other global institutions began incorporating strategies of 'resilience' into their logistics of crisis management, financial (de)regulation and development (eg. World Bank 2006). With the post-9/11 revolution in 'homeland security', resilience has become a byword among agencies charged with coordinating security responses to climate change, critical infrastructure protection, natural disasters, pandemics and terrorism (Evans and Reid, 2014), reorienting these once distinct policy arenas toward a horizon of critical future events which (we are told) we cannot predict or prevent, but merely adapt to by 'building resilience'. In the process, resilience has largely ceased to operate as a critique emphasising the fragility of complex communal life, and now arguably functions to naturalise power and normalise catastrophe. This can be seen when prominent ecologists propose that global financial markets should be understood as complex adaptive ecosystems (May et al., 2008), or argue against conservationists that "Nature is so resilient that it can recover rapidly from even the most powerful human disturbances" (Kareiva et al., 2012).

## Learning resources

*Ecology and Society*, previously named *Conservation and Society*, is the house journal of the Resilience Alliance.
www.ecologyandsociety.org/

Stockholm Resilience, an international policy centre dedicated to promoting 'resilience thinking' in the 'governance of social-ecological systems', is closely aligned with the Beijer Institute of Ecological Economics.
www.stockholmresilience.org
www.beijer.kva.se

Resilience as a framework for the analysis and organisation of national security is exemplified in the policy documents of the US Department of Homeland Security.
https://www.dhs.gov/topic/resilience

Established in 2013, the journal *Resilience: International Policies, Practices and Discourses* provides a forum for critical social scientists to engage with 'resilience', its policy frameworks and their consequences.
http://www.tandfonline.com/toc/resi20/current

## Bibliography

Evans, B. & Reid, J. (2014) *Resilient Life: The Art of Living Dangerously*, London: Polity.

Gunderson, L. & Holling, C.S (eds) (2002) *Panarchy: Understanding Transformations in Human and Natural Systems*, Washington, DC: Island Press.

Hayek, F. (1974) *The Pretence of Knowledge*. Lecture upon the award of the Swedish Bank Prize in Economics in Memory of Alfred Nobel.

Holling, C.S. (1973) 'Resilience and Stability of Ecological Systems', *Annual Review of Ecology and Systematics*, 4: 1–23.

Kareiva, P., Lalasz, R. & Marvier, M. (2012) 'Conservation in the Anthropocene: Beyond Fragility and Solitude', *Breakthrough Journal*. Available at https://thebreakthrough.org/index.php/journal/past-issues/issue-2/conservation-in-the-anthropocene.

May, R., Levin, S. & Sugihara, G. (2008) 'Complex Systems: Ecology for Bankers', *Nature*, 451: 893–895.

World Bank (2006) *Social Resilience and State Fragility in Haiti: A Country Social Analysis*, Caribbean Country Management Unit, Report No. 36069–HT, April 27.

# The resource curse

1.18

Michael J. Watts

Can too much of a good thing produce decidedly bad outcomes? Friar Lawrence in Shakespeare's *Romeo and Juliet*, invoking the plenitude of the earth, warned precisely of such a prospect: "Nor aught so good but, strained from that fair use revolts from true birth, stumbling on abuse" (Act 2, Scene 3). In a quite different register, Honore de Balzac in *Father Goriot* suggested that personal abundance and riches always contained a sinister underbelly: "The secret of great wealth with no obvious source is some forgotten crime, forgotten because it was done neatly." Both of these sentiments stand at the heart of the 'resource curse', a body of scholarly, policy and practical work resting on the claim – in effect a paradox – that there are "adverse effects of a country's natural resource wealth on its economic, social, or political well-being" (Ross, 2015: 239). The "paradox of plenty" (Karl, 1997) now encompasses a vast literature – dominated by economists, political scientists and development practitioners (for a review see Ross, 2015; Rosser, 2006; Frankel, 2012) – which purports to show how natural resource commodity wealth, particularly in the Global South (so-called resource-dependent economies), leads to substandard economic performance, state pathologies (governance failures such as corruption and lack of accountability), non-democratic politics (patrimonialism and/or despotic institutions), and civil conflict and war. How might an abundance of hydrocarbons, diamonds or coltan turn out to be a curse? Harvard economist Jeffrey Frankel (2012: 17), for example, identifies seven core aspects of commodity wealth: "they are: long-term trends in world commodity prices, volatility, permanent crowding out of manufacturing, weak institutions, unsustainability, war, and cyclical Dutch disease." All of this leads to an equal and opposite reaction, namely a set of prescriptions to 'escape the resource curse' (see Humphreys et al., 2007): "The most promising ideas include indexation of contracts, hedging of export proceeds, denomination of debt in terms of the export commodity, Chile-style fiscal rules, a monetary target that emphasizes product prices, transparent commodity funds, and lump-sum distribution" (ibid.). The pathologies of the resource curse seem to assume their most archetypical and pernicious forms when the resource in question is oil, and the political economy the 'petro-state' (see Karl, 1997; Ross, 2012; Le Billon, 2012; Collier, 2010). To say that these sorts of curse claims have proven to be contentious is a considerable understatement: Stanford economists Gavin Wright and Jesse Czelusta (2004), for example, simply refer to the curse argument as "a myth". Such disputation – or mythology – might not have surprised the great Polish journalist Ryzard Kapuściński (1982: 35) who in describing the political economy of oil-rich Iran during the reign of Shah Mohammed Reza Pahlavi wrote: "Oil creates the illusion of a completely changed life, life without work, life for free . . . oil is a fairy tale and like every fairy tale a bit of a lie."

The idea that nature or nature resources confer advantages for human well-being is of course of great antiquity. Within modern (post-1945) mainstream development theory there is a long lineage in which natural resource endowment is seen to confer major advantages for economic growth. Modernization (Rostow, 1961) and neoliberal theorists (Balassa, 1980) alike argued that 'take off' or industrial development could be facilitated by resources providing investment capital and domestic markets. At the same time there was also a structuralist account – most associated with Raoul Prebisch (1950) and his Economic Commission for Latin America (ECLA) colleagues – that trade and international commodity markets could undercut the 'resources as blessing' argument. While a minority view (Rosser, 2006), during the 1980s it gained momentum curiously from some of the multi-lateral development institutions such as the IMF and the World Bank (Sala-i-Martin and Subramanian, 2013). The origins of the term resource curse is usually traced to economic geographer Richard Auty (see Auty, 2001 and Auty and Gelb, 2001). While at Princeton (1982–83) he had met Sir Arthur Lewis who recommended him to World Bank researcher Allan Gelb who included him in a 1988 volume entitled *Oil Windfalls: Blessing or Curse?*, and was subsequently elaborated in his 1993 book *Sustainable Development in Mineral Economies: The Resource Curse Thesis*. There is an echo here with respect to earlier work on extraction and the deleterious impact of mineral booms (during the 1970s E.L. Dean Kohrs developed the notion of the Gillette Syndrome to refer to the higher rewards and "great pains" (alcoholism, accidents and absenteeism) associated with boom conditions – in his case in Gillette, Wyoming. During the 1990's the curse theme was elaborated along three axes. First, Jeffrey Sachs (Sachs and Warner, 1995) and others using large data sets suggested that natural resource abundance was negatively correlated with economic growth. Second, Paul Collier (2008) in his hugely influential book *The Bottom Billion* argued that the 'natural resource trap' (the ration of natural resources to GDP) accounted for the calamitous conditions of the poorest of the poor not only through poor growth but through the likely – albeit curvilinear – relation between resources and the outbreak, onset and duration of civil war. And third Michael Ross (2012) and others argued that resource abundance is associated with low levels of democracy (regime type) and governance failures (corruption, lack of transparency, patrimonialism). Over the same period, and in part driving the expansion of policy and scholarly debate, a number of activist organizations (most prominently Global Witness and Revenue Watch) began to advocate for public action around not only the plundering of resources in the Global South (timber in Cambodia, diamonds in Sierra Leone, Oil in Angola) but also to regulate and transform corporate practice. The disclosure of large off-book payments by BP in Angola in 2001 proved to be a turning point. Since that time resource curse analysis and prescription has produced a small industry of regulatory institutions and practices many operating under the sign of multi-stakeholder initiatives (MSI) and civic engagement (Brockmeyer and Fox, 2015).

An overwhelming proportion of the resource curse literature deploys a large-n approach – the work of geographers (Watts, 2012; Le Billon, 2012) and anthropologists (Coronil, 1997; Welker, 2014) is an exception since they tend to make use of ethnographic and case study approaches. One consequence of the use of statistical measures of association is that the calibration of the model can have important consequences and results in often wildly different outcomes. For example 'resource dependence' can be measured in very different ways: the resource as a percentage of GDP, as a percentage of exports, the proportion of large (oil) fields, the percentage of government revenues, or on a per capita basis. Some of these measures – for example the size of government rents – can be very difficult to obtain (or are largely fictional). These differing measures tend to produce a diversity of trends and forms of association. Take for example the purported relations between resource abundance and economic performance. Richard Auty claims that "since the 1960s, the resource-poor countries have outperformed the resource-rich countries compared by a considerable margin" (2001: 840). Sachs and Warner (1995) presented evidence of an inverse statistical relationship between natural resource-based exports (agriculture, minerals, and fuels) and growth rates during the period 1970–90 and argued subsequently that

> what the studies based on the post-war experience have argued is that the curse of natural resources is a demonstrable empirical fact, even after controlling for trends in commodity prices. . . . Almost without exception, the resource-abundant countries have stagnated in economic growth since the early 1970s, inspiring the term 'curse of natural resources.' Empirical studies have shown that this curse is a reasonably solid fact.
>
> (2001: 828, 837)

As Wright and Czelusta (2004) note, the resource-curse literature pays little attention to the economic character of mineral resources or to the concept of resource abundance and tend to interpret booms, busts, windfalls in an uncritical way. They conversely point to the generative relations between resources and the 'knowledge economy'. Studies that use other measures of mineral abundance (such as reserves per capita or the level of natural resource exports per worker) do not find that these variables are negatively associated with growth rates (Maloney, 2002). Similar problems are associated with the putative effect of resources on institutions: rents contribute to corruption and booms to the lack of state capacity, revenue volatility (the boom and bust cycle) compromises long-term planning and institutional 'quality' and may have inflationary effects which damage non-oil sectors (the 'Dutch Disease'). Yet if one examines for example subnational (state-level) variation in complex federal states with resource abundance there is enormous variability in 'institutional performance' – all of which demands a fine-grained and granular approach to political economy, institutional history and political dynamics (Porter and Watts, 2016).

Central to how resources matter is the question of the conditions of possibility: under what conditions is oil or gold consequential? (Or to put it differently what are the precise causal relations linking resource exploitation and political conflict or authoritarianism.) The work on the relations between oil wealth and democracy, for example, turns out to be profoundly ambiguous: scholars might emphasize the degree to which oil rents purchase middle class consent, or promote authoritarianism, or help stabilize democratic regimes, or is conditional upon pre-oil levels of inequality (see Dunning, 2008; Ross, 2012). Many political scientists turn to the 'rentier effect' and how the centralization and capture of oil revenues by state apparatuses and the political mechanisms by which rents are distributed mitigates against effective taxation systems (rupturing the logic of taxation-produces-political representation). Collier (2008) in his account of civil war – which turns on a simplistic dichotomy of combatant 'greed or grievance' – argues that as the value of resource wealth increases the risk of conflict rises then falls. Others turn to the location of the resources (proximate or distant from centralized state authority, onshore or offshore) or the properties of the resource (portability, lootability point or diffuse) (for a review see Le Billon, 2012). In all of this the causal mechanisms tend to be blunt, singular and linear: rents are predated or looted; there is a logic of the survival of the fattest ('patrimonialism'), combatants are criminals. What is striking in so much of this work is the degree to which certain causal powers are almost entirely missing. In Collier's account large transnational oil companies only appear to the extent that they are looted! Until the recent work on corporate mining practices (Welker, 2014), the complex powers exercised by capital have been shockingly absent from the resource curse literature.

What is required in resource curse studies is an understanding that first a resource has material and biophysical properties and qualities (the oiliness of oil matters); second that the resource is a commodity and enters into commodity circuits through determinate relations of production; third that as a commodity a resource has use, exchange and fetishistic properties (oil is a fairy tale as Kapuściński put it); and not least oil is exploited in particular places at particular times and is inserted into a ready-existing political economy (say in 1960 in the Niger delta region of Nigeria within a new post-colonial multi-ethnic federal system). Collectively this means that many of the sorts of 'variables' identified by the large-n studies operate in complex configurations and congeries – the outcomes are multiply determined – which will require precisely the sorts of granular and historicized analyses offered by geographers (see Le Billon, 2012; Watts, 2012).

## Bibliography

Auty, R.M. (1993) *Sustaining Development in the Mineral Economies: The Resource Curse Thesis*. London: Routledge.

Auty, R.M. (2001) 'The Political Economy of Resource-Driven Growth', *European Economic Review*, 45: 839–946.

Auty, R. & Gelb, A. (2001) 'The Political Economy of Resource-Abundant States', in R. Auty (ed.), *Resource Abundance and Economic Development* (pp. 126–144). Oxford: Oxford University Press.

Balassa, B. (1980) *The Process of Industrial Development and Alternative Development Strategies*. Princeton: Princeton University.

Brockmeyer, M. & Fox, J. (2015) *Assessing the Evidence*. London: Transparency and Accountability Initiative.

Collier, P. (2008) *The Bottom Billion*. New York: Oxford University Press.

Collier, P. (2010) *The Plundered Planet*. New York: Oxford University Press.

Coronil, F. (1997) *The Magical State: Nature, Money, and Modernity in Venezuela*. Chicago: University of Chicago Press.

Dunning, C. (2008) *Crude Democracy: Natural Resource Wealth and Political Regimes*. New York: Cambridge University Press.

Frankel, J.A. (2012) *The Natural Resource Curse: A Survey of Diagnoses and Some Prescriptions*. HKS Faculty Research Working Paper Series RWP12–014, John F. Kennedy School of Government. Cambridge, MA: Harvard University.

Gelb, A. (1988) *Oil Windfalls: Blessing or Curse?* New York: Oxford University Press.

Humphreys, M., Sachs, J. & Stiglitz, J.E. (2007) *Escaping the Resource Curse*. New York: Columbia University Press.

Kapuściński, R. (1982) *Shah of Shahs*. New York: Harcourt.

Karl, T.L. (1997) *The Paradox of Plenty: Oil Booms and Petro-States*. Berkeley, LA and London: California University Press.

Le Billon, P. (2012) *Wars of Plunder: Conflicts, Profits, and the Politics of Resources*. New York: Columbia University Press.

Maloney, W.F. (2002) 'Missed Opportunities: Innovation and Resource-Based Growth in Latin America', *Economia*, 3: 111–150.

Porter, D. & Watts, M. (2016) 'Righting the Resource Curse: Institutional Politics and State Capabilities in Edo State, Nigeria', *Journal of Development Studies*, 53(2): 249–263.

Prebisch, R. (1950) *The Economic Development of Latin America and its Principal Problems*. Lake Success, NY: United Nations.

Ross, M.L. (2012) *The Oil Curse: How Petroleum Wealth Shapes the Development of Nations*. Princeton, NJ: Princeton University Press.

Ross, M.L. (2015) 'What Have We Learned About the Resource Curse?', *Annual Review of Political Science*, 18: 239–259.

Rosser, A. (2006) *The Political Economy of the Resource Curse*. Working Paper 268. Brighton: Institute of Development Studies, Sussex University.

Rostow, W. (1961) *The Stages of Economic Growth: A Non-communist Manifesto*, Cambridge: Cambridge University Press.

Sachs, J. (2001) 'The Curse of Natural Resources', *European Economic Review*, 45: 827–838.

Sachs, J. & Warner A. (1995) *Natural Resource Abundance and Economic Growth*. Dev. Disc. Pap. 517a, Cambridge, MA: Harvard Institute for International Development.

Sala-i-Martin, X. & Subramanian, A. (2013) 'Addressing the Natural Resource Curse: an Illustration from Nigeria', *Journal of African Economics*, 22(4): 570–615.

Watts, M. (2012) 'Blood Oil'. In S. Reyna, A. Behrends and G. Schlee (eds), *Crude Domination: An Anthropology of Oil* (pp.49–71). Oxford: Berghahn.

Welker, M. (2014) *Enacting the Corporation: An American Mining Firm in Post-authoritarian Indonesia*. Berkeley, LA and London: University of California Press.

Wright, G. & Czelusta, J. (2004) 'Why Economies Slow: The Myth of the Resource Curse', *Challenge*, 47(2): 6–38.

# 1.19 Scarcity and environmental limits

## Sara Nelson

This chapter provides an overview of debates about environmental limits from the 18th century to the present. It will demonstrate that, historically, these debates tend to be linked to concerns over the availability of key resources. These debates have been characterized by two general perspectives: those which see limits as *absolute*, or immutable to social change; and those which see them as *relative* to a given socio-economic system. The terms of this debate have shifted, however, with climate change and the proposed advent of the Anthropocene. Whereas previous notions of limits were tied to resource scarcity, new concepts like 'planetary boundaries' stress critical thresholds of environmental change. And whereas progressive thinkers have tended to eschew notions of limits that constrain political futures, many recognize that anthropogenic global change poses new challenges to political thought and action. The problem of environmental limits is therefore charged with renewed import in the contemporary moment.

## Classical political economy

For classical political economists, natural resources were non-substitutable inputs to the production process. This idea underpinned David Ricardo's notion of diminishing returns to land, and John S. Mill's notion that the economy would eventually reach a 'steady state' equilibrium without growth (Gómez-Baggethun et al., 2010: 1211). The quintessential thinker of environmental limits in this tradition was Thomas Malthus (1798/1998). He argued in his *Essay on the Principle of Population* that resource scarcity was inevitable due to the inherent tendency of population growth to outrun food supplies. Because population grows geometrically (or exponentially) while food production grows arithmetically, population would always exceed food production, resulting in misery and starvation (see Figure 1.19.1).

Malthus was explicit about the political implications: if misery and starvation acted as natural 'checks' on population, the poor served as a kind of buffer between these 'checks' and the rest of society. An increase in social equity would therefore only extend this misery to more people. For this reason, he argued against the expansion of social support for the underclasses.

Malthus's analysis was refuted by Karl Marx. He argued that the supposedly 'natural' law of population was really a law inherent to capitalism. Marx (1973: 607) argued that environmental limits were not transhistorical, but always *relative* to a given mode of production: "The overpopulation among hunting peoples, which shows itself in the warfare between the tribes, proves not that the Earth could not support their small numbers, but rather that the condition of their reproduction required a great amount of territory for few people." The underemployed 'surplus population' that Malthus described was a necessary result, Marx argued, of the contradictions of capitalist competition. Political ecologists and resource geographers have built on this approach, insisting that 'resources' are not naturally given but historically produced.

Malthus's arguments were taken up, decades later, by William Stanley Jevons in his pioneering work *The Coal Question*. Accepting that "exterior nature presents a certain absolute and inexorable limit," Jevons (1866: 173) argued that rather than grain (with

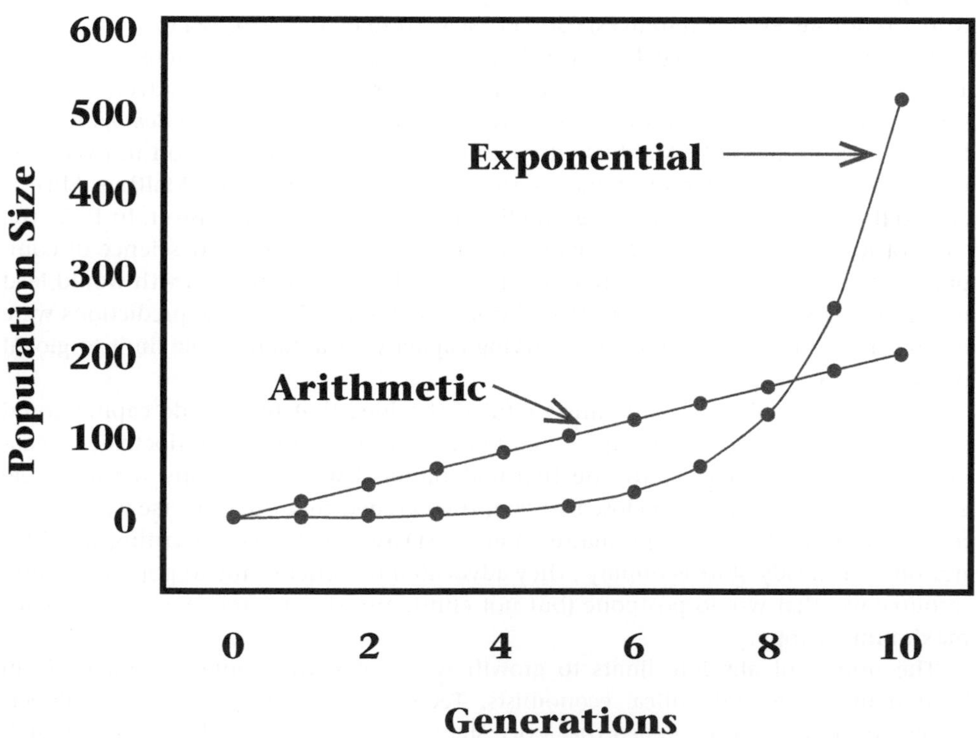


**Figure 1.19.1** Model of Malthusian growth: exponential (population) vs. arithmetic (food production).

which Malthus had been concerned), coal now represented "the staple produce of the country." As population increased, so would the use of coal, but not in a linear manner – rather, with technological innovation, coal use promised to become *more* intensive per capita. This idea became distilled in the 'Jevons paradox,' influential to modern environmental economics, which posits that increasing energy efficiency tends ironically to increase overall energy consumption. Likening coal to the colonial conquests that had expanded agricultural production beyond Malthus's predictions, Jevons argued that the exhaustion of coal reserves was now the chief limit to population growth and to British hegemony. His predictions of energy scarcity, however, were dramatically wrong. Positing the limits to coal reserves as an absolute limit to Britain's economic growth, Jevons did not anticipate the rise of petroleum power that would transform the global economy (and Britain's position within it) in ways that he could not foresee.

## From neoclassicism to neo-Malthusianism

If questions of scarcity and environmental limits were prescient for classical political economy, they were all but forgotten with the rise of neoclassical economics in the 1870s. With a shift in emphasis from land to capital as the key production input alongside labor, neoclassical economists sidelined concern with natural resources throughout the 1910s and 1930s, and began instead to theorize their 'substitutability' with capital through technological innovation (an exception to this was Harold Hotelling's (1931) work on non-renewable resources) (Gómez-Baggethun et al., 2010: 1211).

This trend was challenged by two linked developments in the 1960s and 1970s: 1) the return of Malthusian fears over population growth and 2) the advent of environmental and ecological economics. In his 1968 Book *The Population Bomb*, biologist Paul Ehrlich warned, like Malthus, of a natural tendency for population to overcome resources, which would result in mass starvation in the 1970s. Unlike Malthus, Ehrlich argued that birth control – coercively applied if necessary – was the answer. In 1972, the Club of Rome's best-selling *The Limits to Growth* report used the new science of computer modeling to demonstrate that, if left unchecked, population growth would lead to a global crisis in the next 100 years (Meadows et al., 1972). These predictions were based on a static understanding of 'carrying capacity' as a quantifiable limit to global population (Sayre, 2008).

Meanwhile, ecological economists refuted the idea that man-made capital could substitute for scarce 'natural capital,' and criticized the mainstream doctrine of economic growth. Influenced by the thermodynamic law that entropy (or unusable energy) never decreases in a closed system, they argued that the universe was limited in its supply of "low-entropy matter-energy" (Daly, 1992: 21). Returning to Mill's notion of a 'steady-state economy', they advocated for strict limits to population and resource use that would postpone (but not eliminate) the inevitable encounter with maximum entropy.

The notion of absolute limits to growth was met with counter-arguments from both mainstream and critical economists. Technological optimists such as Robert Solow and Julian Simon argued that price signals in markets would catalyze a switch to less resource-intensive products ahead of resource exhaustion, and that most natural resources could be substituted with man-made inputs. In contrast, many ecological Marxists accepted some of the insights of ecological economics, while rejecting both Malthusian alarmism and market optimism. They argued that environmental limits were the result of capitalism's inherent environmental contradictions and its drive toward growth, justifying a shift toward an ecologically-attuned socialism (Benton, 1996).

As in previous moments, this debate was shaped by its political context. Some critics pointed out the geopolitical anxieties underpinning fears of Third World population growth, in a moment when formerly-colonized states were asserting sovereign control over resources (Golub and Townsend, 1977). The 1973 'oil shock,' widely interpreted as a manifestation of neo-Malthusian fears, was more accurately a result of these geopolitical changes. As oil companies worked to regain control of prices, they seized on new ways of measuring oil reserves (notably M. King Hubbert's 'peak' theory), encouraging popular associations of 'peak oil' with widespread energy crisis (Mitchell, 2012). In this context, Marxist geographer David Harvey (1974) argued, discussions of population and resources were never politically neutral. And as feminist demographers like Betsy Hartmann (1987) showed, catastrophist fears about population growth informed coercive population policies that were arguably counterproductive to limiting birth rates in the long term.

## Contemporary debates

While popular fears of peak oil persist (Schneider-Mayerson, 2015), technological and market changes have rendered unconventional deposits such as oil shale economically viable, making it increasingly evident that the problem is one of too much oil, not too little: it has been estimated that more than half of global fossil fuel reserves are 'unburnable' if we are to avoid more than 2°C of global warming (Jakob and Hilaire, 2015). In other words, it is now greenhouse gas emissions – not fossil fuels – that are treated as scarce resources, and debates over climate policy tend to concern the equitable distribution of emissions among nations (Chakrabarty, 2015).

A key recent innovation is the concept of "planetary boundaries" (Rockström et al., 2009). Unlike earlier notions of limits, planetary boundaries do not focus on resource exhaustion or carrying capacity, but on "tipping points" of irreversible environmental change (see Figure 2.18.1). There is also renewed attention to environmental limits in Marxian literature, notably from Jason Moore (2015: 12), who analyzes the history of capitalism as a series of encounters with limits to particular "historical natures". In the *longue durée,* Moore (ibid.: 112) argues, capitalism's expansionary tendencies are limited by both the "finite character of the biosphere" and a tendency to appropriate nature faster than the latter can reproduce itself.

Apart from problems of resources, historian Dipesh Chakrabarty (2015: 109) has argued that anthropogenic environmental change presents "a possible limit to our very human-centered thinking about justice, and thus to our political thought as well." In other words, problems such as ocean acidification, climate change, and biodiversity loss challenge the human exceptionalism implicit in progressive thought (Stengers, 2017). From this posthumanist perspective, environmental limits in the Anthropocene do not only refer to resource scarcity but also to the limits of the human – both in terms of its usefulness as a universal category, and its possible extinction.

## Learning resources

For an account of the debate over peak oil and environmental futures since the 1950s, see:

Priest, T. (2014) 'Hubbert's Peak: The Great Debate over the End of Oil', *Historical Studies in the Natural Sciences*, 44(1): 37–79. Available at: https://typriest.files.wordpress.com/2012/07/hubberts-peak-hsns.pdf

For a response to neo-Malthusian theories of population, see:

Sen, A. (1994) 'Population: Delusion and Reality', *The New York Review of Books*, 22 September. Available at: http://www.nybooks.com/articles/1994/09/22/population-delusion-and-reality/

This lecture provides an introduction to the theory of planetary boundaries:

Rockström, Johan. (2010) 'Let the Environment Guide our Development', TED talk, July. Oxford. Available at: https://www.youtube.com/watch?v=RgqtrlixYR4

For an example of the contemporary debate around planetary boundaries, see this summary:

Lalasz, B. (2013) 'Debate: What Good are Planetary Boundaries?' *The Nature Conservancy* blog. Available at: http://blog.nature.org/science/2013/03/25/debate-what-good-are-planetary-boundaries/

## Bibliography

Benton, T. (ed.) (1996) *The Greening of Marxism*, New York: The Guilford Press.

Chakrabarty, D. (2015) 'Whose Anthropocene? A Response', in Robert Emmet and Thomas Lekan (eds) *Whose Anthropocene? Revisiting Dipesh Chakrabarty's 'Four Theses'*. RCC Perspectives – Transformations in Environment and Society: 103–114. Available at: http://www.environmentandsociety.org/sites/default/files/2015_new_final.pdf

Daly, H.E. (1992) *Steady-State Economics*, 2nd edn. London: Earthscan.

Ehrlich, P. (1968) *The Population Bomb*, Binghamton, NY: Ballantine Books.

Golub, R. & Townsend, J. (1977) 'Malthus, Multinationals and the Club of Rome', *Social Studies of Science*, 7(2): 201–222.

Gómez-Baggethun, E., De Groot, R., Lomas, P.L. & Montes, C. (2010) 'The History of Ecosystem Services in Economic Theory and Practice: From Early Notions to Markets and Payment Schemes', *Ecological Economics*, 69(6): 1209–1218.

Hartmann, B. (1987) *Reproductive Rights and Wrongs: The Global Politics of Population Control and Contraceptive Choice*, New York: Harper & Row.

Harvey, D. (1974) 'Population, Resources, and the Ideology of Science', *Economic Geography*, 50(3): 256–277.

Hotelling, H. (1931) 'The Economics of Exhaustible Resources', *Journal of Political Economy*, 39(2): 137–175.

Jakob, M. & Hilaire, J. (2015) 'Climate Science: Unburnable Fossil Fuel Reserves', *Nature*, 517(7533): 150.

Jevons, W.S. (1866) *The Coal Question*, 2nd ed. London: Macmillan and Co.

Malthus, T. (1798/1998) *An Essay on the Principle of Population*. Electronic Scholarly Publishing Project. Available at: http://www.esp.org.

Marx, K. (1973) *Grundrisse*, New York: Pelican Books.

Meadows, D.H, Meadows, D.L., Randers, J. & Behrens III, W.W. (1972) *The Limits to Growth: A Report for the Club of Rome's Project on the Predicament of Mankind*, New York: Universe Books.
Mitchell, T. (2012) *Carbon Democracy*, New York: Verso.
Moore, J.W. (2015) *Capitalism in the Web of Life: Ecology and the Accumulation of Capital*, New York: Verso.
Rockström, J., Steffen, W., Noone, K., Persson, Å., Chapin, F.S., Lambin, E.F., Lenton, T.M., Scheffer, M., Folke, C., Schellnhuber, H.J., et al. (2009) 'A Safe Operating Space for Humanity', *Nature*, 461: 472–475.
Sayre, N. (2008) 'The Genesis, History and Limits of Carrying Capacity', *Annals of the Association of American Geographers*, 98: 120–134.
Schneider-Mayerson, M. (2015) *Peak Oil: Apocalyptic Environmentalism and Libertarian Political Culture*, Chicago: University of Chicago Press.
Stengers, I. (2017) 'The Intrusion of Gaia', *South Atlantic Quarterly*, 116(2): 381–400.

# 1.20 Stewardship

## Willis Jenkins

Stewardship means an entrusted responsibility for something belonging to another. While the etymology of the English word "steward" is somewhat murky, tracing to an Old English term early in the Middle Ages, the concept generally refers to someone tasked with keeping a household on behalf of its owner or managing a trust on behalf of its beneficiary. The key notion of this office, which has many ancient precedents and cultural analogs, is accountability to another for something given into one's care on trust.

In the late 20th century this minor managerial concept acquired new cultural life as a term for environmental responsibility. Stewardship serves as the organizing concept in many environmental statements from governments, corporations, institutions, and civic organizations. There are, however, reasons to be sceptical of its popularity: stewardship may reinforce assumptions about the human relationship to nature that perpetuate ecological problems.

## Entrusted responsibility

What exactly environmental stewardship entails depends on how its several parties and their relations are imagined. First, who is the steward? It sometimes refers to individual persons, who should exercise stewardship in the context of their own lives. This notion of the steward appears in approaches to environmental issues that focus on individual freedom and accountability for the choices one makes about lifestyle and property, including any land or creatures in one's care. An individualist conception of the steward is especially attractive to those who are sceptical of collective environmental management or who dissent from the way nature is imagined in collective policies. However, the steward is not always an individual; it might also be a community, city, nation, corporation, or even the human species as a whole.

Second, what is the object of stewardship? Sometimes its "environment" refers to a set of resources that should be managed prudently. Much as a fiduciary manages an investment trust for the sake of future beneficiaries, stewards might think of nature as a trust made up of resource stocks whose overall value they should preserve or increase. However, on another conception, the trust might be imagined as a living community made up of organisms, species, and relations that are intrinsically valuable. Exercising stewardship with that view may entail protecting each particular member of the trust. Or, on yet another conception, if nature is imagined as a dynamic evolutionary system, then caring for it might entail protecting qualities like creativity or resilience.

Third, to whom is the steward accountable? Grounding the claim of responsibility carried by stewardship is its notion that the object entrusted to a steward properly belongs to someone else, to whom the steward is accountable. One major way of answering that question holds that, because the object is creation, stewardship is accountable to the Creator. However, it is important to recognize that stewardship does not require theism; the steward might be accountable to a larger community like a nation, to an intergenerational membership like a people, or to Earth itself. One of the most common ways to imagine stewardship outside of religious discourse is in terms of responsibility to future generations. Perhaps in recognition that we have inherited gifts of care from previous generations, we may have an obligation to care for the ecological conditions we pass on to future generations (Page, 2007). In any case the question for the steward is: what are the criteria of good stewardship set by this authority? How will future generations evaluate our environmental care or what does the Creator expect?

Answers to those questions yield a wide range of environmental imaginaries. Stewardship may name a relation of humanity to Earth as indifferent as that of a servant managing a house for an absentee landlord, yet it can also depict humans cherishing the bioregional membership to which they belong. It may reproduce views of anthropocentric mastery over nature, yet it can also constrain human freedom in respect of nature as a gift or in care for a particular watershed.

## Religious varieties

Stewardship ideas are prominent in modern Jewish, Christian, and Islamic environmental thought, probably because the concept is so compatible with theistic worldviews in which humans are accountable to a Creator for their use of creation. Whereas some members of Abrahamic traditions are uneasy about environmental thought that begins from the sacredness of nature or seems to elide humanity's special status, stewardship allows an approach to environmental responsibility that begins from divine commands for human behavior. One reason stewardship ideas are prominent in Abrahamic communities, in other words, is because it can help maintain a traditional structure of moral reasoning.

Nonetheless, stewardship was not a central moral trope in Judaism or Christianity until the 20th century, and has only an indirect basis in their scriptures. "Steward" is often the translation of the *oikonomos* who appears in the New Testament in several of Jesus' parables as an enslaved household manager dealing with a difficult master – not obvious ground for an environmental ethic. More important are the two creation accounts at the beginning of the Hebrew scriptures. In Genesis 1, God tells humans to "fill the earth and subdue it, and have dominion over every living thing." In Genesis 2, God puts Adam in a garden "to cultivate and guard it." The dominionist stance of the first is emphasized in theologies that present stewardship as a partnership with God, in which humans are assigned to rule over creation as God's proxy, sometimes contributing to it as co-creators with God, and sometimes even helping to redeem it from evil. (The latter notion may have figured in the rise of science; Harrison, 2007). The agrarian humility of the second creation story features in theologies that present stewardship in terms of care, wherein humans learn proper use of creation through service to it (Wirzba, 2003). A major question for stewardship in those traditions, therefore, is which text takes interpretive priority.

In Islam, stewardship as entrusted responsibility is more explicitly grounded in scripture. In the Qur'an, Allah says "I am placing on the earth one that shall rule as my deputy (*khalifa*)" (2.30). Deputy rulership seems to align with a dominionist view of stewardship, yet the Qur'an also expresses scepticism about human ability to fulfill that office by relating a story (33.72) in which the trusteeship for creation was offered to the heavens, the earth, and the mountains, which all refused for fear that they could not do it justice. Humans then accept the offer, but the Qur'an suggests they do so foolishly and unfairly. The powerful deputies thus require a great deal of legal and moral instruction in order to care properly (Haq, 2003).

Shared across the Abrahamic traditions is a sense that humans bear a special responsibility for creation, the pattern of which is found by submitting to God's guidance (which may be found in scripture, or creation, or both). A different version of stewardship sometimes appears in some modern indigenous environmental thought. Indigenous leaders sometimes describe the biocultural practices of their people as a form of responsible membership in earth and an intergenerational trust which they have an obligation to pass on. That takes the form of stewardship but, in contrast with the typical Abrahamic conception, here the pattern of responsibility is often described as learned through interdependence and reciprocity with all one's relations. (See *Renewing Relatives* in the learning resources section.)

## Planetary stewardship

The notion that humans have exceptional responsibilities for earth is also compatible with secular views of the human role in nature. For example, if the Anthropocene refers to a new geological epoch characterized by pervasive human influence throughout planetary systems, then, some scientists argue, humans must learn to become "active stewards of the Earth System as a whole" (Folke et al., 2011: 719). In this context, appealing to stewardship shifts conventional ecological science in two ways. First, it makes cultural transformation an explicit goal of environmental policy, which should seek not only to ameliorate ecological problems but also to develop cultural connections with ecological systems likely to foster responsibility for earth. Second, it shifts focus from managing problems in reference to benchmarks of nature toward "the active shaping of trajectories of change in coupled social-ecological systems . . . to enhance ecosystem resilience and promote human well-being" (Chapin et al., 2011: 45). The stewardship frame thus shifts attention from protecting what nature would be without human interference toward managing what nature is becoming under conditions of constant human influence.

The idea of planetary stewardship is controversial on both counts. Some reject the idea that ecological policy can or should drive ethical transformation in how humans understand their responsibilities. Others criticize the framework for implicitly perpetuating anthropocentrism by assigning humans to manage planetary ecological systems. Nonetheless, the idea seems apt for the challenges of the Anthropocene, if only because a moral metaphor in which humans are entrusted with a special role responsibility for earth fits a reality in which human power has forced itself into earth systems. How humans imagine their role on Earth and enact their responsibilities will be a major factor in the future of life's evolution on Earth, and indeed the future of humanity's understanding of itself.

## Learning resources

*Renewing Relatives: Nmé Stewardship in a Shared Watershed,* available at: http://hfe-observatories.org/project/renewing-relatives-nme-stewardship-in-a-shared-watershed

The following handbook offers an introduction to many intersections of religion and environment: Jenkins, Tucker, and Grim (eds) (2016) *The Routledge Handbook of Religion and Ecology*. Abingdon: Routledge.

## Bibliography

Chapin, F., Pickett, S., Power, M.E., Jackson, R.B., Carter, D.M. and Duke, C. (2011) 'Earth Stewardship: A Strategy For Social-Ecological Transformation To Reverse Planetary Degradation', *Journal of Environmental Studies & Sciences*, 1: 44–53.

Folk, C., Jansson, Å, Rockström, J., Olsson, P., Carpenter, S.R., Chapin, F.S., Crépin, A-S., Daily, G., Danell, K., Ebbesson, J., et al. (2011) 'Reconnecting to the Biosphere', *Ambio*, 40: 719–738.

Haq, N. (2003) 'Islam and Ecology: Toward Retrieval and Reconstruction', in Foltz, Denny and Baharuddin (eds) *Islam and Ecology: A Bestowed Trust* (pp. 121–154). Cambridge: Harvard University Press.

Harrison, P. (2007) *The Fall of Man and the Foundations of Science*. New York: Cambridge University Press.

Page, E (2007) 'Fairness on *The Day After Tomorrow*: Justice, Reciprocity, and Climate Change', *Political Studies*, 55(1), 225–242.

Steffen, W., Persson, Å., Deutsch, L., Zalasiewicz, J., Williams, M., Richardson, K., Crumley, C., Crutzen, P., Folke, C., Gordon, L., et al. (2011) 'The Anthropocene: From Global Change to Planetary Stewardship', *Ambio*, 40: 739–761.

Wirzba, N. (2003) *The Paradise of God: Renewing Religion in an Ecological Age*. New York: Oxford University Press.

# 1.21 Sustainable development

Mark Whitehead

Over the last 30 years sustainable development has emerged as a core principle of local, national and international governance. Secretary-General of the United Nations Ban Ki-moon recently claimed that achieving 'sustainable development is the central challenge of our time', while leading economist Jeffrey Sachs has described our current era as the *age of sustainable development* (Sachs, 2015). Despite its political prominence, the notion of sustainable development is often a subject of misinterpretation and misapplication. It was famously described by Gro Harlem Brundtland as 'development that meets needs of the present without compromising the ability of future generations to meet their own needs' (WCED, 1987: 43). Sustainable development asserts that decisions about human development made in the here-and-now should anticipate and address the long-term consequences of related actions. Advocates of sustainable development argue that the key to delivering effective forms of development in the present and future is integrated thinking. Accordingly, it is claimed that development can only be sustainable if it finds ways of balancing social, economic, and environmental needs. The social (or more precisely human welfare), economic (usually defined in relation to the production and distribution of goods and services), and the environment (normally used to refer to ecological systems, and resources), are often referred to as the *three pillars of sustainable development*. As this chapter demonstrates, sustainable development policies tend to approach the harmonization of social, economic, and environmental needs in two main ways. First, certain forms of sustainable development seek ways of regulating economic growth while optimizing the provision of social welfare and environmental protection. Second, other sustainable development strategies are based upon the search for creative means to fuse relatively unlimited forms of economic growth with the redistribution of wealth and ecological conservation. This production of so-called *win-win-win* forms of sustainability—in which economic growth is hitched to social development and environmental protection—has become the dominant model of sustainable development throughout the world.

## Scientific background

According to Sachs (2015) sustainable development is both an *analytical* and *normative* category. As an *analytical* category sustainable development provides a framework in and through which we can study how the world operates. As a *normative* category sustainable development offers a blueprint for how the world should be. More will be said about the normative aspects of sustainable development in the following section.

As an analytical category sustainable development has origins in the forest management sciences that emerged in seventeenth-century Europe. The science of forest management recognized that in order to be sustainable the extraction of timber should not exceed the levels at which surrounding ecological systems could replenish the supply. Through the careful study of woodlands, the sciences of forest management were able to determine the sustainable yields at which timber resources could be used. The analysis of sustainable yields, or thresholds, is now central to the goals of the sciences that are associated with sustainable development. A sustainable yield establishes the level at which any resources—including timber from forests; water from aquifers; or fish from the oceans—can be extracted while allowing that resource to naturally replenish. A sustainable threshold refers to the level at which an activity can continue without causing damage to underlying ecological systems.

While based upon the analytical insights of sustainable yields and thresholds, modern forms of sustainable development science explore sustainability at much larger scales of complexity than those associated with forest management. As a 'science of complex systems' sustainable development considers sustainable yields and thresholds in the context of diverse forms of interaction between economic, social, and environmental systems (Sachs, 2015). There are two particular aspects of the sustainable science of complex systems that are important to note. First, it recognizes that sustainable yields and thresholds are not necessarily stable states and may fluctuate according underlying changes in both the operation of environmental systems and the technologies that are used to extract resources from the natural world. Second, it draws analytical attention to the complex relationships and feedback systems that exist between levels of economic development and the (un)sustainable use of the environment. Economic growth is often seen as presenting the greatest threat to the sustainability of environmental systems. Scientific studies of sustainable development have, however, revealed the damaging consequences of poverty and under-development on the sustainable use of the environment. Studies of agricultural communities around the world have thus shown that poverty places a great constraint on the ability of people to plan for a sustainable future, particularly when their day-to-day survival is under threat.

## Geopolitical origins

In addition to being grounded in science, sustainable development is also a product of geopolitical struggle and compromise. Many trace the origins of sustainable development as a policy goal to 1972 and the United Nations Conference on the Human

**Table 1.21.1** United Nations Sustainable Goals

| *Sustainable Development Goal* | *Description of Goal* |
|---|---|
| 1. No poverty | The eradication of all forms of poverty everywhere. |
| 2. Zero hunger | Improved food security, promotion of sustainable agriculture, and improved nutrition. |
| 3. Good health and well-being | Promotion of healthy lifestyles and improved physical and mental health. |
| 4. Quality education | Support for equal access to education and promotion of lifelong learning opportunities. |
| 5. Gender equality | Achieving equality of treatment and opportunity for women and girls. |
| 6. Clean water and sanitation | Achieving access to clean water and sanitation for all, and establishment of sustainable water management practices. |
| 7. Affordable and clean energy | Ensuring access to clean and affordable energy for all. |
| 8. Decent work and economic growth | Pursuit of sustainable and inclusive forms of economic growth. |
| 9. Industry, innovation, and infrastructure | Support development of resilient infrastructure and pursue sustainable and inclusive forms of industrialization. |
| 10. Reduced inequalities | Reduce inequalities of opportunity and income both within and between countries. |
| 11. Sustainable cities and communities | Make human settlements safe, sustainable and resilient. |
| 12. Responsible production and consumption | Establishment of sustainable patterns of consumption and production. |
| 13. Climate action | Take immediate action to address climate change mitigation and adaptation. |
| 14. Life below water | Support sustainable use of marine resources. |
| 15. Life on land | Support the sustainable use of terrestrial resources and avoid unsustainable patterns of deforestation, desertification, and biodiversity loss. |
| 16. Peace, justice and strong institutions | Promote the building of peaceful and just societies, with access to justice for all and the presence of accountable institutions. |
| 17. Partnerships for the Goals | Strengthen the global partnerships through which sustainable development is being pursued. |

*Source*: Adapted from The Sustainable Development Knowledge Platform: https://sustainabledevelopment.un.org/?menu=1300

Environment, which was convened in Stockholm. The conference was convened in order to instigate and coordinate international action on a range of emerging environmental issues. While the Stockholm conference was, in many ways, a landmark event in the history of sustainable development, the concept was not used at the event. Sustainable development actually emerged as a direct consequence of the geopolitical divisions that emerged in Stockholm. On one side of these divisions were More Economically Developed Countries (particularly in Western Europe) who were keen to forge international agreements that would help protect the local and global environment. On the other side were many Less Economically Developed Countries who were concerned that the ratification of international agreements on environmental protection would undermine much needed development in their territories (Whitehead, 2006). It was precisely in the context of this geopolitical stalemate that sustainable development emerged as a potential international compromise.

Although the emergence of sustainable development as a concept in the 1970s was in part based upon the scientific principles of sustainable yield and threshold, it was also strongly influenced by emerging economic ideas and principles (Bernstein, 2002). These economic ideas suggested that not only was effectively oriented economic growth not harmful to the environment, but that it was a crucial prerequisite of environmental protection and enhancement. In 1983 the United Nations established the World Commission on Environment and Development (WCED) in order to develop a new framework upon which multilateral international action on environmental protection and socio-economic development could be delivered. Drawing on the insights of sustainability science and economics, the WCED introduced sustainable development as a form of 'breakthrough idea' that could pave the way for new patterns of global development in the future (Bernstein, 2002). In 1987 the WCED produced the so-called Brundtland Report, which set out how sustainable development could be deployed in a range of different sectors (WCED, 1987). In 1992, the United Nations convened the UN Conference on Environment and Development in Rio de Janeiro in order to internationally ratify the principles of sustainable development. The most recent United Nations summit on sustainable development was again held in Rio in 2012 (it become known as the Rio+20 conference). The Rio+20 Summit produced a series of 169 sustainable development targets and 17 goals that are to be met by 2030.

## The sustainable development debate

There is much debate about the value and effectiveness of sustainable development thinking and policies. These debates often revolve around the extent to which sustainable development is providing a sustainable balance between economic growth, social need and environmental protection. For some, mainstream sustainable development policy reflects a form of *weak sustainability*, which wrongly assumes that the technological developments associated with free markets and economic growth will be able to ensure that we can take more from nature without compromising the sustainability of environmental systems. For others, sustainability reflects the only viable way of achieving effective forms of collective action on pressing environmental issues while also addressing global inequality.

It appears that only time will demonstrate for sure if we really can effectively combine the goals of environmental protection, a reduction in social inequality, and economic growth. What is undeniable, however, is that challenges identified within the sustainable development debate will continue to be a defining feature of our collective futures.

## Learning resources

For more information on emerging initiatives in the sustainable development policy area visit the Sustainable Development Solutions Network:

http://unsdsn.org

For up to date information on United Nations actions on sustainable development go to the Sustainable Development Knowledge Platform: https://sustainabledevelopment.un.org

Details of the UN's current Sustainable Development Goals can be accessed here:

https://sustainabledevelopment.un.org/?menu=1300

## Bibliography

Adams, W.M. (2001) *Green Development: Environment and Sustainability in the Third World*, London: Routledge.

Bernstein, S. (2002) *The Compromise of Liberal Environmentalism*, New York: Columbia University Press.

Sachs, J.D. (2015) *The Age of Sustainable Development*, New York: Columbia University Press.

WCED (1987) World Commission on Environment and Development.

Whitehead, M. (2006) *Spaces of Sustainability: Geographical Perspectives on the Sustainable Society*, Abingdon: Routledge.

# The tragedy of the commons

1.22

Kevin Ells

The "tragedy of the commons" is the name the biologist Garrett Hardin gave to a thought experiment in a now famous 1968 *Science* article. It predicted global resource degradation and societal ruin from the net consequences of individuals acting in their short-term interests but at a long-term cost to the environment. This chapter will outline the history of the concept and, after presenting some compelling critiques of it, demonstrate its continuing utility in helping us understand environmental problems.

## History

Hardin's principal source was an 1832 Oxford lecture W. F. Lloyd published in rebuttal to Thomas Malthus's influential analysis (in six editions from 1798 to 1830) of unsustainable countervailing growth trends in human population versus agricultural resources. Malthus predicted famine and accelerated death rates in England once geometric growth in human population outpaced linear arithmetic growth in the food supply.

Lloyd argued rational self-interest would undermine Malthus's appeal to "moral restraint" in procreation with reference to the degraded state of England's common grazing areas. Herdsmen each add cattle to a common area because the long-term loss in food supply is shared by all herdsmen, with only a small net loss for each. By analogy, Malthus's solutions to limit human population growth, voluntary delayed marriage and sexual abstinence, would fail, and the common labor market was doomed to overstocking to the point of saturation.

Hardin reiterated Lloyd's hypothetical example in a thought experiment intended to demonstrate that global overpopulation would lead to the collapse of Earth's carrying capacity.

> Picture a pasture open to all. . . . Explicitly or implicitly, . . . [each herdsman] asks, "What is the utility to me of adding one more animal to my herd?" . . . Since [he] receives all the proceeds from the sale of the additional animal, the positive utility is nearly +1. . . . Since, however, the effects of overgrazing are shared by all . . ., the negative utility for any particular decision-making herdsman is only a fraction of 1. . . . [T]he only sensible course . . . is to add another animal to his herd. . . . Therein is the tragedy. Each man is locked into a system that compels him to increase his herd without limit—in a world that is limited. Ruin is the destination toward which all men rush, each pursuing his own best interest in a society that believes in the freedom of the commons. Freedom in a commons brings ruin to all.

Hardin substituted Malthus's voluntary moral restraint with mandatory restraint by the state, "mutual coercion, mutually agreed upon." Malthus was of course wrong that the population of England would exhaust its food supply (as the global population has not worldwide), but reiterations of his thesis appeared in later social Darwinist writing, and persisted even after the Third Reich tamped down public support for eugenics. For example, American ornithologist William Vogt and sociologist (and eugenicist) Elmer Pendell warned in their 1948 books against incompatible trends in human fertility and environmental sustainability, publishing detailed updates in 1960.

Hardin's article appeared during a high watermark in public fear of overpopulation. During the late 1960s and early 1970s, biologist Paul Ehrlich's *The Population Bomb* (also 1968) and *The Limits to Growth* (Meadows et al., 1972) sold millions of copies, and *Soylent Green*, a science fiction film depicting a diabolical solution to preventing famine in an overpopulated future society, was a box-office hit in 1973.

## Critique

Criticism of Hardin's thesis appeared as early as readers' letters in a 1968 issue of *Science*. They offered alternative solutions (contraception, agronomics), or found Hardin's appeal to the social good contradicted his assertion that people are motivated only by economic self-interest. David Harvey demonstrated that curbing population growth presumes our cultural perspectives on nature, social organization of scarcity, and attitude toward material goods are fixed and immutable (1974, pp. 272– 273). Thousands of people have since proven complex problems have multiple points of engagement by agreeing to buy environmentally neutral or beneficial products, organizing to manage commons areas, and so on.

Certainly, if profits are distributed privately, but the costs of resource loss or toxic waste are disbursed among millions of taxpayers, corporate or state entities may find themselves caught up in a system wherein they must pollute. So the "tragedy of the commons" fable clarifies the need for legislative intervention to manage or preserve a wide range of areas or resources analogous to a public common.

However, in a post-agrarian society, we cannot logically conflate this narrative with the decision of parents to bear children. The herdsman adds cattle to the commons as capital. Parents' income from children, if any, does not recapitulate, let alone exceed, the costs incurred in raising them. Even considering the intangible rewards of child-rearing, doubling household population does not double the love within it, nor do three children in addition to the first quadruple security in old age. Ironically, Hardin's fable remains directly relevant to virtually every environmental issue *except* overpopulation.

Hardin did concede thirty years after publishing "The tragedy of the commons" that he should have specified he was writing about "an unmanaged commons" (1998, p. 682), a crucial distinction, as it leaves open the political choice of management scheme (from enforced decree through free market incentives), as well as widely varying responses to local conditions.

Elinor Ostrom and her colleagues refer instead to "the drama of the commons" (2002). Ostrom's extensive research program has documented numerous case studies of local resource commons use ending *not* in tragedy (though some do) but in perpetual sustainability for the benefit of the communities that manage them.

## Utility

Social commentator Friedrich Engels optimistically argued in 1844 that science would discover a solution to Malthus's conundrum (the 1950s Green Revolution in high-yield agricultural productivity is the most prominent confirmation of this prediction). But Engels deemed Malthus's work of great use in posing the problem so clearly. In a sense, the present chapter takes the same stance toward Hardin's rhetorical argument.

Multiple case studies in the half century since Hardin raised his alarm have disproven his claim that the fate of a resource commons is inexorably tragic. Some have suggested retiring the concept for good ("Malthusianism" and "The tragedy of the commons" appear in the anthology *This Idea Must Die*, 175 short essays answering the question "What scientific idea is ready for retirement?").

Yet the structure of Hardin's argument retains its utility for environmental rhetoric and advocacy. It is a vicarious narrative in which readers can imagine themselves in the place of the fictional herdsman, subject to his economic temptation to overuse a community resource. Even those who hope they would find workable alternatives can easily infer that numerous neighbors might not.

Hardin's thought experiment proves an implied two-part deductive argument. First,

1. In reading about someone doing X, you can imagine yourself doing X
2. If you can imagine yourself doing X, you can imagine anyone doing X
3. You can therefore imagine everyone doing X.

This X is an individual act, relatively harmless in itself to the system in which one is embedded, which would destroy the whole if everyone else did it, too. Here is the second part:

1. Everyone doing X will bring about Y
2. You can imagine everyone doing X (proven in part one)
3. Y will happen.

Hardin's vicarious narrative is not a story narrated in the second person, but a hypothetical example he invites the reader to interpret vicariously ("Picture a pasture open to all"). Rather than depicting a horrifying future that *might* occur "if this goes on," Hardin makes a future that logically *must* occur a new aspect of the reader's sense of the present. His rhetorical accomplishment is that an audience for his story believes its global implications must occur *precisely because they attended to the story and understood it.*

## Conclusion

"The tragedy of the commons" remains a compelling rhetorical trope, since it applies with full logical force to multiple environmental issues such as overfishing, deforestation, watershed pollution, and soil erosion. Writers and advocates may still use the structure of Hardin's argument to demonstrate the danger of relaxing regulations that manage any natural resource commons, or persuade people to see themselves as influential agents in systems in which they are embedded. This ready applicability could account for the persistent popularity of Hardin's article even for readers opposed politically or ethically to Hardin's prescriptions for curbing human population, or for whom global overpopulation has cooled as a front-burner environmental concern.

When Erlich's dire scenarios did not materialize, his bestseller passed into obscurity. Hardin remains relevant because the structure of his thought experiment still cogently expresses the potential intractability of many environmental problems. One may plausibly assert that Hardin takes a narrow view of human nature, or call his herdsman the product of a particular, perhaps peculiar, culture. But those engaged in environmental journalism or education may set aside Hardin the policy analyst while imitating Hardin the rhetorician, as have readers familiar only with excerpts of his article. One need not throw out the persuasive baby with the political bath water.

## Learning resources

Here is a compelling and detailed interactive map of twenty-first-century population projections (net births, life expectancy, income, and more) by region:
http://www.prb.org/Publications/Datasheets/2015/2015-world-population-data-sheet/world-map.aspx#map/world/population/2015

A generally balanced 10-minute survey of population concerns and misconceptions, using the UK as a case study applicable in varying degrees to overcrowded or sparsely populated areas:
https://www.youtube.com/watch?v=VUTP93qWV7I

A study in contrasts – two thoroughly cited and richly-linked articles on opposite ends of the current spectrum of concern about human population growth:
http://www.pewresearch.org/fact-tank/2015/06/08/scientists-more-worried-than-public-about-worlds-growing-population/
http://www.slate.com/articles/technology/future_tense/2013/01/world_population_may_actually_start_declining_not_exploding.single.html

## Bibliography

Brockman, J. (2015) *This idea must die: Scientific theories that are blocking progress*. New York: Harper Perennial.

Dietz, T., Dolšak, N., Ostrom, E. & Stern, P.C. (2002) The drama of the commons. In Ostrom, E., Dietz, T., Dolšak, N., Stern, P.C., Stonich, S. and Weber, E.U. (Eds) *The drama of the commons* (pp. 3–36). Washington, D.C.: National Academy Press.

Ehrlich, P. (1968) *The population bomb*. New York: Ballantine Books.

Engels, F. (2010) Outlines of a critique of political economy. In *Marx and Engels: Collected works, volume 3*. London: Lawrence and Wishart.

Hardin, G. (1968) The tragedy of the commons. *Science*, 162: 1243–1248.

Hardin, G. (1998) Extensions of 'The tragedy of the commons'. *Science*, 280, 682.

Harvey, D. (1974) Population, resources, and the ideology of science. *Economic Geography*, 50(3), 256–277.

Lloyd, W.F. (1980) W.F. Lloyd on the checks to population. *Population and Development Review*, 6: 473–496.

Meadows, D.H., Meadows, D.L., Randers, J. & Behrens III, W.W. (1972) *The limits to growth: a report for the Club of Rome's project on the predicament of mankind*. New York: New American Library.

Ostrom, E., Dietz, T., Dolšak, N., Stern, P.C., Stonich, S. and Weber, E.U. (Eds) (2002) *The drama of the commons*. Washington, D.C.: National Academy Press.

# 1.23 Uncertainty

## Andy Stirling

Threats to climate, biodiversity, soils, air and water join long-standing issues of poverty and vulnerability in demanding urgent action. Radical transformations are required in global institutions and infrastructures for provision of energy, food, water, mobility and livelihoods. So, political, economic and social – as well as environmental – stakes are high. Yet, it is rarely the case that all details of the issues in question can be definitively pinned down (Gee et al., 2013). There usually remains significant scope for questions over: appropriate knowledges; causal processes; possible implications; and relevant actions. These are the dilemmas of what is often ambiguously called 'uncertainty' – but more accurately described as 'incertitude' (Harremoës et al., 2001). As we will see, specialist usages of 'uncertainty' can elide crucially different features of context – allowing irresponsible "pretence of knowledge" (Hayek, 1978). An overarching term like 'incertitude' helps avoid this.

Whatever it is called, what makes incertitude more tricky are the political realities in which it is set. Despite the value of high quality evidence and analysis, even the best available policy-relevant science is prone to delivering divergent pictures of salient problems and solutions – for instance in energy (Sundqvist et al., 2004), chemicals (Saltelli et al., 2008), biotechnology (Stirling & Mayer, 2001) and industrial safety (Amendola, 2001). Indeed, the levels of incertitude associated with a relevant peer-reviewed literature are (when attended to) often sufficient to support many contrasting possible courses of action (Funtowicz & Ravetz, 1990). The most important feature of 'sound science' or 'evidence based policy', then, is that these disciplines are rarely adequate, definitely to justify any single particular intervention. In other words, dilemmas of incertitude typically mean that no particular policy can be uniquely validated by the available evidence. The idea of a single 'evidence based policy' is an oxymoron.

Sadly, these challenges are seriously neglected – sometimes even denied – in mainstream environmental policy. This can be seen even in a field where 'science based' comparative appraisal methods are arguably at their most sophisticated and mature: in the energy sector. This is an area where climate change as well as many other environmental issues present some of their most formidable policy challenges. So, an impressive range of techniques address questions over which energy strategy looks on balance most favourable. Focusing on results expressed as monetary 'environmental externalities', Figure 1.23.1 summarises a problem that is also common to environmental risk assessment, multicriteria appraisal or life cycle analysis (Stirling, 2010). Shown on the right, is the number of peer-reviewed studies for each energy option (all with results cited in official regulatory interventions). Particular findings of one indicative study are shown in grey. Results obtained in the entire literature are shown as black bars, the thickest parts of which give the range for the central half of all studies.

Figure 1.23.1 shows that the incertitude expressed in an individual study of comparative environmental implications of different policies, tends to be very small compared to the corresponding range in the relevant peer-reviewed literature as a whole. So policy debates informed by just a subset of studies, typically only get a partial impression of the issues at stake. It is therefore routinely possible to justify on grounds of selected studies, any one among many different policy choices. And what is true in the field of energy policy, also applies in other areas of environmental governance – like agriculture, transport, or water. In interactions between different sectors, dilemmas of incertitude are correspondingly amplified.

Policy challenges are even further compounded, when it is taken into account how pressures for closure in the 'real world' of policy making conflict with actual levels of

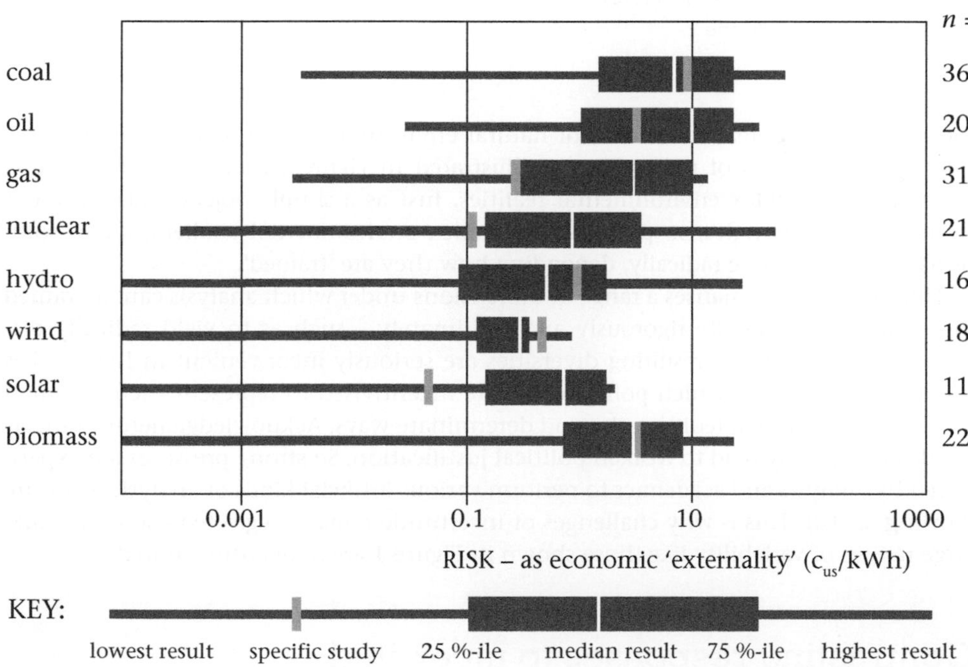


**Figure 1.23.1** Neglected variability in environmental appraisal – the example of energy assessments.

*Source*: Adapted from Sundqvist et al. (2004).

**Equally relevant to quantitative and qualitative approaches**

| | | |
|---|---|---|
| setting agendas | defining problems | posing questions |
| prioritising issues | deciding context | choosing methods |
| addressing power | definition of options | selecting alternatives |
| handling dissensus | designing process | drawing boundaries |

**More relevant to expert and calculative approaches**

| | | |
|---|---|---|
| discounting time | formulating criteria | characterising metrics |
| setting baselines | deriving probabilities | including disciplines |
| expressing uncertainties | recruiting expertise | commissioning research |
| constituting proof | exploring sensitivities | interpreting results |

**More relevant to participatory and discursive approaches**

| | | |
|---|---|---|
| identifying stakeholders | phrasing questions | bounding remits |
| recruiting participants | providing information | focusing attention |
| engaging personalities | conducting discourse | facilitating interactions |
| documenting findings | persuading critics | adopting norms |

**Figure 1.23.2** Examples of different kinds of framing assumptions implicated in environmental incertitude.

incertitude in the 'real real world' of natural environments and societies themselves. The open-endedness of policy choices illustrated in Figure 1.23.1 is an unavoidable reflection of complex environmental realities. Just as a simple object looks different from contrasting angles, so pictures of complex environmental challenges and solutions differ even more radically, depending how they are 'framed'.

Figure 1.23.2 summarises a range of dimensions under which analysis can be framed differently – but equally rigorously and legitimately – such as to yield radically distinctive answers. These resulting diversities are seriously inconvenient in high stakes political processes, in which policy actors are incentivised to represent their favoured arguments in exaggeratedly precise and determinate ways. Acknowledgements of incertitude are typically held to weaken political justification. So strong pressures are experienced by analysts and academics to perform various artificial kinds of analytical closure (Stirling, 2010). This is why challenges of incertitude remain neglected – and typically large ranges of variability like those shown in Figure 1 are frequently ignored.

## Contrasting responses to incertitude

What can be done about these formidable challenges of incertitude in environmental governance? Although there are no panaceas, there exist many practical responses. Like incertitude itself, many of these also remain neglected. In limited space, only an indicative sample can be covered here. For a start, there is the language of environmental appraisal. Terms like 'risk', 'uncertainty', 'ambiguity' and 'ignorance' are bandied around in many diverse ways in this field. It is not realistic or useful to try to assert only one kind of usage. But what is essential in interests of effective decisions, is to resist instrumental pressures to use such words in ways that systematically exaggerate the tractability of incertitude.

Why this matters is summarised in Figure 1.23.3. Long central to environmental appraisal (EPA 1998; UK Department of Environment, 1995; Suter, 2006), the word 'risk' has for more than a century referred to a condition under which there is confidence that problematic knowledge can satisfactorily be addressed by assigning probabilities to reflect perceived relative *likelihoods* for each of a defined range of possible *outcomes* (NRC, 1983; PCCRARM, 1997; Byrd & Cothern, 2000). Accordingly, the dimensions of Figure 1.23.3 represent confidence in the quality of knowledge experienced under each of these twin constituting parameters of risk: probability and outcomes (Stirling, 2010). In opening up these dimensions, a range of complementary methods become visible beyond risk assessment.

For a century or so, economists have explicitly contrasted a strict definition of 'uncertainty' with the state of 'risk' – uncertainty in these terms being a condition under which there *cannot* be confidence in any single representation of probabilities (Knight, 1921; Keynes, 1922). As such, uncertainty is a situation in which (in the words of the

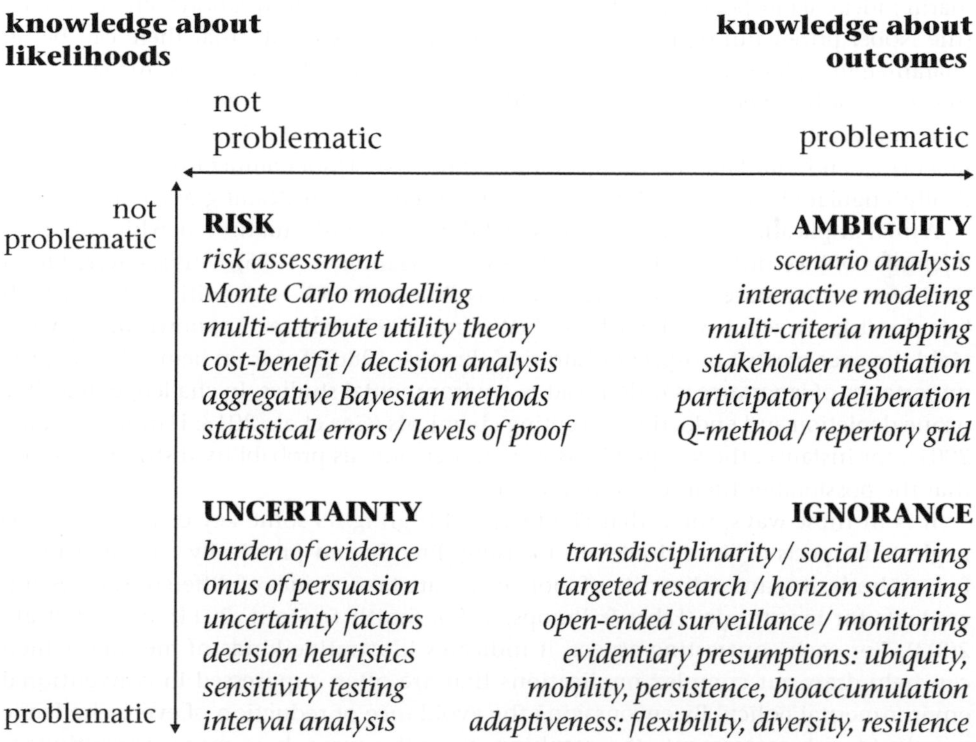


**Figure 1.23.3** Different aspects of incertitude, as distinguished in relation to two dimensions of knowledge.

Adapted from Stirling (2010).

celebrated probability theorist de Finetti (1974), probabilities simply "do not exist". Yet this same term 'uncertainty' is now routinely used in environmental studies, with exactly the *opposite* meaning – where political pressures discussed above nonetheless force assertion of singular representations of probabilities. Suppressing the open-endedness of colloquial ideas of uncertainty as well as clashing with the strict definition (Funtowicz & Ravetz, 1990), it is this that constitutes the "pretence of knowledge" referred to above (Hayek, 1978). So, it is to avoid such misleading confusions, that the term 'incertitude' can be used in a more general fashion, clearly covering all senses of the colloquial general implications of 'uncertainty'. But the point here is not about words, but whether academic or policy imaginations even acknowledge at all, a condition under which there is little confidence in any single picture of probabilities (Wynne, 1992).

However, Figure 1.23.3 also shows that this is not the only reason for thinking about 'incertitude' in this general way. Even if suppression of 'uncertainty' (in a strict sense) is avoided, other deep challenges remain. For instance, colloquial use of the word 'ambiguity' refers to a parallel dilemma, in which it is not the likelihood of different outcomes that is at issue, but even more fundamental problems around defining, measuring, partitioning or bounding the possibilities themselves (Stirling, 2010). This can occur even for historical events that have already occurred (which are in this sense 'certain'), but where questions still arise over 'what happened?' In other words, ambiguity is about 'contradictory certainties' (Thompson & Warburton, 1985).

Ambiguity involves many conflicts, in: interests or values; notions of 'benefit' or 'harm'; ideas about policy options. Here, it is well established in rational choice theory (the Nobel prize-winning paradigm underlying risk assessment), that there can be no guarantee in a plural society, of any single ordering of social preferences (Arrow, 1963). In other words – even aside from difficulties with probabilities – the idea of calculating a uniquely optimal environmental response in 'the real world' – is not only difficult in practice . . . it is fundamentally impossible in principle. Under ambiguity, then, simplistically singular ideas of 'sound science' are also a seriously misleading oxymoron.

A final implication of Figure 1.23.3, is that there also exist situations in which all these dilemmas of incertitude apply together (Dovers & Handmer, 1995). This is referred to in the literature as a state of 'ignorance' (Faber & Proops, 1990) – a condition under which "we don't know what we don't know" (Wynne, 1992), where "unknown unknowns" yield the ever-present prospect of 'surprise' (Brooks, 1986). Far from being abstract, the importance of ignorance is widespread in environmental studies. In challenges like BSE, ozone depletion and endocrine disrupting chemicals (Gee et al., 2013; Harremoës et al., 2001), for instance, the key problems were not erroneous probability distributions, but that the possibilities themselves were surprises.

It is in these ways, then, that Figure 1.23.3 highlights some key challenging-but-neglected aspects of 'incertitude'. In focusing directly and practically on implications for methods, it spans a diversity of more elaborate taxonomies of different sources and contexts for incertitude (Faber & Proops, 1990; Smith & Stern, 2011; Walker et al., 2003). For each contrasting aspect, it indicates illustrative kinds of method, which can help draw out complex implications that are often suppressed in conventional environmental policy. By encouraging the avoidance of reduction of every challenge merely to risk assessment, the resulting general approach is more precautionary (Harremoës et al., 2001). But the point here is not to suggest that each specific instance of environmental problem can be assigned to a particular category. Instead, the framework is heuristic – aiding thinking about a wider diversity of methods beyond reductive forms of risk assessment or optimisation.

Taken together, the methods shown in Figure 1.23.3 offer ways to be more realistic about the inconvenient fact that contrasting framings (Figure 1.23.2), often lead to enormous discrepancies in environmental appraisal (Figure 1.23.1). To ignore this in governance can leave the most vulnerable people exposed to the consequences – a syndrome the social theorist Beck called "organised irresponsibility" (Beck, 1992). By making greater use of these practical 'Cinderella methods' for addressing different aspects of incertitude, environmental appraisal can become more robust about specific conditions under which contrasting conclusions arise over the best policy actions. Methods can help reshape institutions. So environmental policy can hope at the same time to become more rigorous and more accountable. Instead of forcing technocratic closure of a kind that can reinforce entrenched interests, this can help open crucial space for what has so often proven essential in achieving the necessary environmental transformations: democratic struggle (Scoones et al., 2015).

## Learning resources

A classic early articulation of challenges of incertitude can be found in reference (Funtowicz & Ravetz, 1990). A wealth of case studies and practical lessons are deeply analysed in references (Gee et al., 2013) and (Harremoës et al., 2001).

A website that builds on rich accumulated experience in exactly the tradition of analysis of incertitude introduced in reference (Funtowicz & Ravetz, 1990), to point to a variety of practical tools for analysing different aspects of incertitude can be found here: http://www.nusap.net/

Another website that discusses broader methods likes those shown in Figure 1.23.3 for generally 'broadening out' and 'opening up' policy appraisal – with links to many other web resources – can be found at the STEPS Centre: http://steps-centre.org/methods/

A particular web-based interactive tool that practically illustrates what is meant by 'opening up' incertitude as discussed here, can be found at: http://www.sussex.ac.uk/mcm/index

## Bibliography

Amendola, A. (2001) Recent paradigms for risk informed decision making. *Safety Science*, 40, 17–30.

Arrow, K.J. (1963) *Social Choice and Individual Values*, New Haven: Yale University Press.

Beck, U. (1992) *Risk Society: Towards a New Modernity*, London: SAGE.

Brooks, H. (1986) The typology of surprises in technology, institutions and development, in Clark, W.C. and R.E. Munn, (eds) *Sustainable Development of the Biosphere* (pp. 235–348), Cambridge: Cambridge University Press.

Byrd, D.M. & Cothern, C.R. (2000) *Introduction to Risk Analysis: A Systematic Approach to Science Based Decision Making*, Oxford: Government Institutes Press.

de Finetti, B. (1974) *Theory of Probability – A Critical Introductory Treatment*, Chichester: Wiley.

Dovers, S. & Handmer, J.W. (1995) Ignorance, the precautionary principle, and sustainability. *Ambio*, 24(2), 92–97.

Environmental Protection Agency (EPA) (1998) *Guidelines for Ecological Risk Assessment*, Washington DC: EPA.

Faber, M. & Proops, J.L.R. (1990) *Evolution, Time, Production and the Environment*, Berlin: Springer.

Funtowicz, S. & Ravetz, J.R. (1990) *Uncertainty and Quality in Science for Policy*, Dordrecht: Kluwer Academic Publishers.

Gee, D., Grandjean, P., Foss Hansen, S., van den Hove, S., MacGarvin, M., Martin, J., Nielsen, G., Quist, D. & Stanners, D. (eds) (2013) *Late Lessons from Early Warnings: Science, Precaution, Innovation*, Copenhagen: European Environment Agency.

Harremoës, P., Gee, D., MacGarvin, M., Stirling, A., Keys, J., Wynne, B. & Vaz, S.G. (eds) (2001) *Late Lessons from Early Warnings: The Precautionary Principle 1896–2000*, Copenhagen: European Environment Agency.

Hayek, F.A. (1978) *New Studies in Philosophy, Politics, Economics and the History of Ideas*, London: Routledge.

Keynes, J.M. (1922) A treatise on probability. *The Philosophical Review*, 31(2): 180.

Knight, F.H. (1921) *Risk, Uncertainty and Profit*, Boston: Houghton Mifflin.

National Research Coucil (NRC) (1983) *Risk Assessment in the Federal Government: Managing the Process*, Washington, DC: The National Academies Press.

Presidential/Congressional Commission on Risk Assessment and Risk Management (PCCRARM) (1997) *Risk Assessment and Risk Management in Regulatory Decision-Making*, Washington, DC: US Government Printing Office.

Saltelli, A., Ratto, M., Andres, T., Campolongo, F., Cariboni, J., Gatelli, D., Saisana, M. & Tarantola, S. (2008) *Global Sensitivity Analysis: The Primer*, Chichester: John Wiley.

Scoones, I., Leach, M. & Newell, P. (eds) (2015) *The Politics of Green Transformations*, London: Earthscan Routledge.

Smith, L. & Stern, N. (2011) Uncertainty in science and its role in climate policy. *Philosophical Transactions. Series A, Mathematical, Physical, and Engineering Sciences*, 369(1956): 4818–4841.

Stirling, A. (2010) Keep it complex. *Nature*, 468: 1029–1031.

Stirling, A. & Mayer, S. (2001) A novel approach to the appraisal of technological risk: A multicriteria mapping study of a genetically modified crop. *Environment and Planning C-Government and Policy*, 19(4): 529–555.

Sundqvist, T., Stirling, A. & Soderholm, P. (2004) Electric power generation: valuation of environmental costs. In C.J. Cleveland (ed.) *Encyclopedia of Energy* (pp. 229–243). Stockholm: Elsevier Science, Luleå University of Technology.

Suter, G.W. (2006) *Ecological Risk Assessment*, Boca Raton: CRC Press.

Thompson, M. & Warburton, M. (1985) Decision making under contradictory certainties: How to save the Himalayas when you can't find what's wrong with them. *Journal of Applied Systems Analysis*, 12: 3–34.

UK Department of Environment (1995) *A Guide to Risk Assessment and Risk Management for Environmental Protection*, London: Stationery Office Books.

Walker, W.E., Harremoës, P., Rotmans, J., van der Sluijs, J.P., van Asselt, M.B.A., Janssen, P. & Krayer von Krauss, M.P. (2003) Defining uncertainty: A conceptual basis for uncertainty management. *Integrated Assessment*, 4(1): 5–17.

Wynne, B. (1992) Uncertainty and environmental learning: Reconceiving science and policy in the preventive paradigm. *Global Environmental Change*, 2(2): 111–127.

# Vulnerability

## Morgan Scoville-Simonds and Karen O'Brien

The concept of vulnerability is widely used in the environmental sciences to describe the potential to be harmed by a process or event. Vulnerability is also used in the social sciences, such as human geography and psychology. Although vulnerability thinking has a long history, the concept continues to demonstrate its relevance to contemporary environmental problems and the strategies and policies people design to address them. While global environmental changes will have widespread impacts across all societies, some will be affected disproportionately more than others, and in different ways. To understand the 'who, what and why' of vulnerability calls for attention to the interplay of social and biophysical factors, including their relationship to macro-scale and global processes.

Vulnerability has proven to be a valuable concept for understanding the uneven outcomes and equity dimensions related to a variety of environmental challenges, including climate change, disaster risk management, pollution and food security. However, interpretations of vulnerability vary across scientific traditions. This leads to different approaches to identifying vulnerable individuals, groups, systems, sectors, or regions, as well as to different strategies for reducing vulnerability. This chapter seeks to identify how academic thinking about vulnerability has evolved and explain why it is a useful concept. We will also point to some of the limits to how the term tends to be employed, and identify some emerging challenges in vulnerability-related research. Below we start with a brief overview of how the concept has evolved over time.

## Vulnerability research

Vulnerability is derived from the Latin word *vulnerabilis*, which relates to the potential to be wounded or harmed (Kelly and Adger, 2000). Before about the 1970s, scientific development of the term (and the related concept of resilience) was particularly strong in the field of psychology, where it related to psychological disorders, particularly maladaptive responses to stressful events. It was also (and remains) related to childhood development, particularly relationships and attachment issues. Since then, however, the term has been used increasingly in diverse fields related to environmental studies.

Ian Burton et al.'s (1978) book *The Environment as Hazard* represents one early contribution to the vulnerability literature that recognizes the 'social' nature of so-called 'natural' hazards. This work analyzed the various ways that people in different locations manage specific risks, pointing out that hazard outcomes (e.g. loss of life, homes, assets, or livelihoods) not only depend on the magnitude or frequency of the hazard itself (e.g. the strength of a hurricane, the height of floodwaters), but on how human populations are distributed (e.g. settlements on floodplains) and the strategies they use to prepare for and respond to such hazards. In the policy sphere, this marked a shift from one of post-disaster recovery to supporting awareness and preparedness.

While this was an important contribution to understandings of hazards and risk, critics called for a 'thicker' understanding of the social dimensions of vulnerability and a greater emphasis on the root causes (see Blaikie et al., 1994). Research in the field of famine and food security was particularly influential in the early history of the evolution of thinking on vulnerability. Amaryta Sen's (1981) early work, although not focusing explicitly on the concept of vulnerability, argued that famine must be understood not simply as food shortage but as 'entitlement' failure. This argument, and the concept of entitlements in particular, drew attention to the analytical and practical importance of focusing on the social institutions that determine people's access to natural resources.

Building on this type of research, Watts and Bohle (1993) drew attention to the role of the political economy (that is the social organization of goods and services) and entitlements in creating spaces of vulnerability. This work represents one of the first attempts to integrate social-ecological with political-economic analysis into a conceptual vulnerability framework, and it inspired further critical thinking on vulnerability. Kelly and Adger (2000) similarly sought to move the analysis from not only identifying the level of vulnerability of individuals or groups (e.g. through indicators of resource access and distribution), but also to examining the determinants of social vulnerability. Their work suggests in particular that understanding the 'architecture of entitlements' requires examining the role of institutions (understood to include formal political structures as well informal 'rules of the game' or norms) through which resource access is determined and contested. Geography in particular has been an important field for the discussion of vulnerability, as exemplified by the 'hazards of place' model that situated vulnerable places in both geographical and social spaces (Cutter, 1996).

Political ecology approaches have been particularly useful in understanding why and how some groups are more vulnerable than others. It adds an environmental dimension to the study of political economy. Piers Blaikie et al.'s (1994) book *At Risk: Natural Hazards, People's Vulnerability and Disasters*, presented a 'Pressure and Release' framework in which disasters (or the risk thereof) were understood as the interaction of a natural hazard (e.g. cyclones) with 'unsafe conditions' in the local physical, social, and economic context (e.g. low income, unprotected infrastructure, endemic disease). The novelty of the approach was to point out that to understand how 'unsafe conditions' are produced in particular places and times for particular groups, these conditions must be traced back through the dynamic processes that sustain them, and ultimately, to root causes. The root causes of the vulnerability, they argue, are to be found in the limits in access to power and resources, and in the structure of political and economic systems themselves.

Others have built on this approach and on the concept of 'causal chains' from political ecology, which similarly suggests that vulnerability outcomes resulting from climate impacts must be traced back to not only environmental factors, but to social and political processes (e.g. Ribot, 2014). More recently, Marcus Taylor (2013) has drawn on political ecology to demonstrate how dynamic social, economic and political relationships guarantee security for some groups while producing vulnerability for others. This relational quality of vulnerability in particular calls into question whether vulnerability can be reduced by simply augmenting the capacities or resources of vulnerable groups (i.e., changing 'unsafe conditions'). It suggests instead that power relationships must also be addressed.

More generally, there has been an emerging recognition that vulnerability is determined by environmental, social, and political factors operating at not only local but also global scales. For example, it has been demonstrated how vulnerability in particular contexts can be examined as the product of multiple processes of change, or 'multiple stressors', including the effects of agricultural trade liberalization and other factors linked to globalization (O'Brien et al., 2004; Eakin, 2005). In terms of environmental studies, this has led to calls for a much stronger integration of the social sciences into global environmental change research.

Alongside this evolution in thinking, a number of authors have pointed out that competing conceptions of vulnerability continue to exist in the scientific and policy-oriented literature related to vulnerability (e.g. Bassett and Fogelman, 2013). Different interpretations of the concept of vulnerability reflect not only different disciplinary approaches but different value-based assumptions and worldviews, which ultimately influence the range of response options that are emphasized (e.g. technical, social, political) and at what scale (e.g. local, national, global). Broadly speaking, analyses that treat vulnerability as the 'end-point' or 'outcome' of environmental impacts tend to lead to infrastructural and technological fixes, while those taking social vulnerability as a 'starting point' highlight the social, economic and political processes that create a contextual basis for vulnerability (Kelly and Adger, 2000; O'Brien et al., 2007). This suggests a need for changes or transformations that may be more overtly political in nature than the technical adjustments that address vulnerability as an outcome of environmental impacts. In short, how we think about vulnerability affects the range of technical, infrastructural, social, or political factors that are eventually addressed, and which remain unaddressed. Although diverse approaches are employed as reviewed above, it is generally recognized that vulnerability analysis is fundamentally different from projecting the outcomes of specific impacts. Specifically, whereas impact analysis begins with a specific environmental change and projects the potential impacts into the future (Figure 1.24.1), vulnerability analysis traces particular outcomes to multiple causal factors, both biophysical and social in origin.

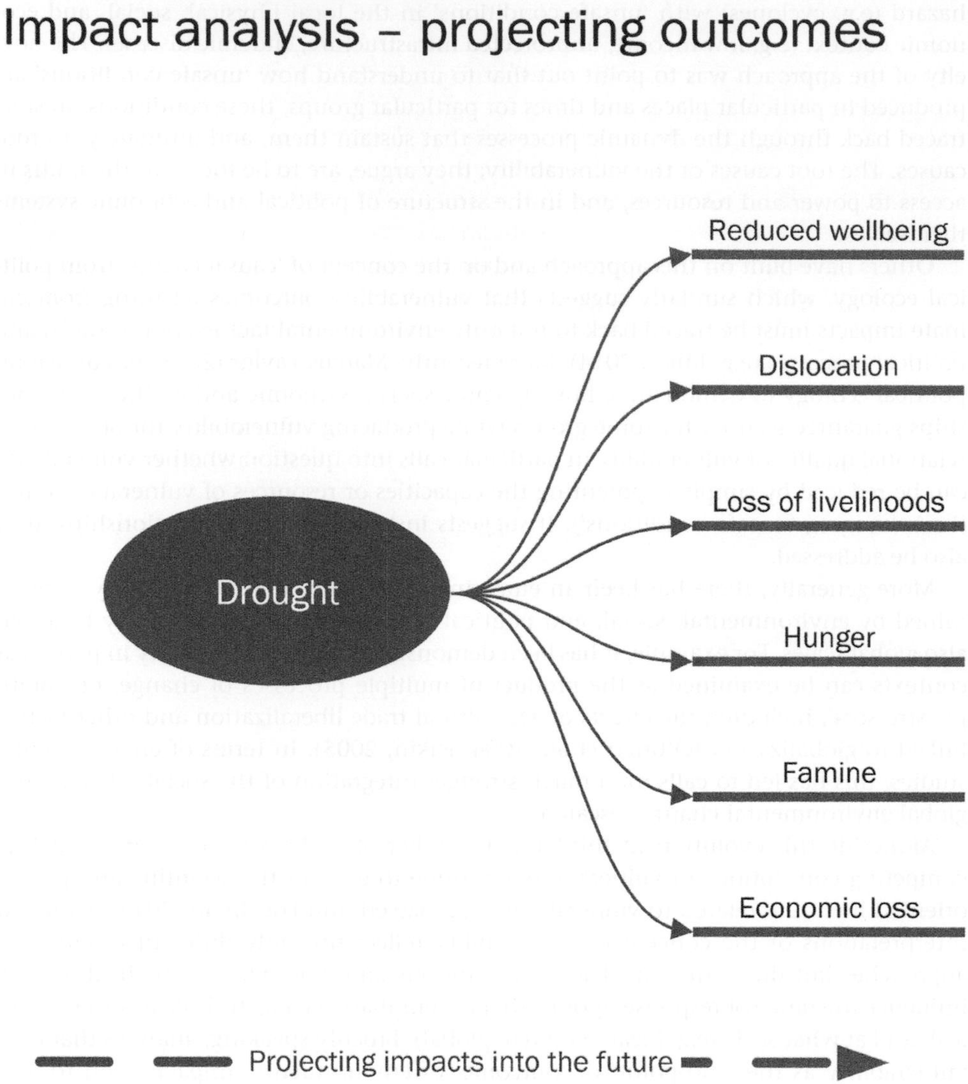


**Figure 1.24.1** Projecting the impacts of environmental change. Starting with a specific environmental impact, multiple specific outcomes (hypothetical examples given here) are projected into the future (moving to the right on the diagram). The environmental impact itself is considered the primary causal force in generating vulnerability.

*Source*: After Ribot (2014).

## From vulnerability to resilience?

There is no doubt that the concept of vulnerability has proven a useful term for research across diverse fields. This research has also had implications for environmental policy, evidenced in the growing use of vulnerability assessments in planning processes. It is increasingly recognized that vulnerability is related not only to biophysical processes that may pose the immediate threat, but also to local and global, social and political processes that put people in harm's way to begin with, and constrain or enable the material resources and range of choices available, with different currents emphasizing the importance of different factors.

More generally, the broad thrust of the evolution of thinking on vulnerability, whether this is understood through the language of political economy, structures, entitlements, or causal chains, has been increased attention to the underlying politics of vulnerability. This broadens vulnerability analysis to identifying not only *who* is vulnerable, but *how* and *why*. This involves examining the processes involved in the production of vulnerability in specific contexts and for particular individuals and groups, linking processes at the local to global scales. This can be considered one of the main contributions and important emphases in the evolution of thinking within vulnerability research.

In recent years, the concept of resilience has also been used to analyze differential capacities to respond to change, including shocks and stressors. Resilience analyses

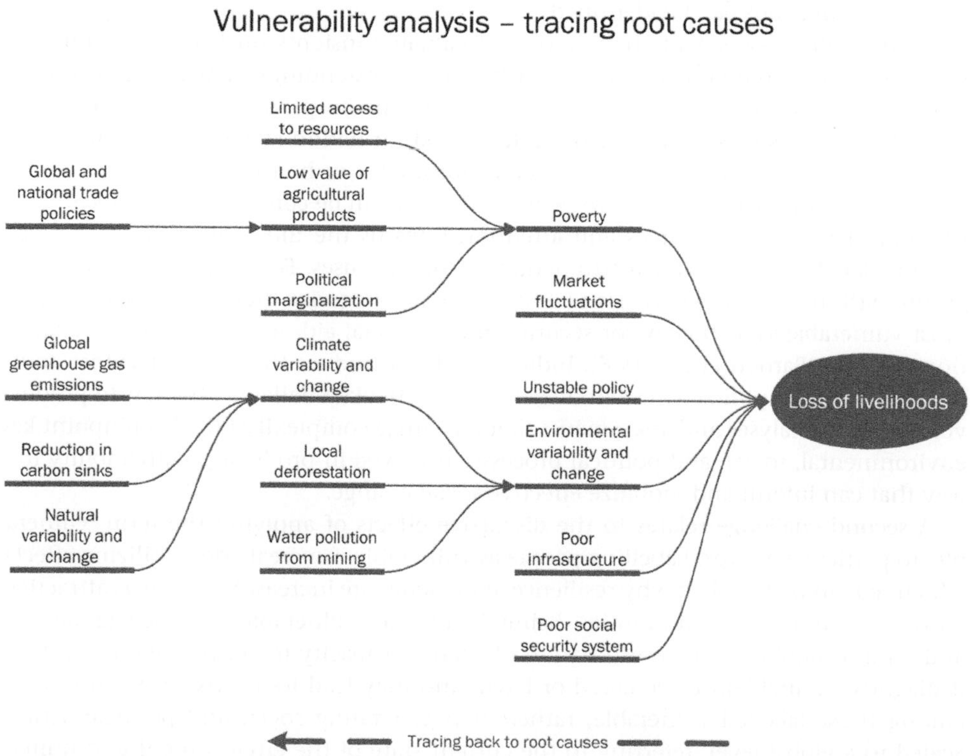


**Figure 1.24.2** Tracing the root causes of vulnerability. Example of how specific outcomes have multiple causes. These relate not only to local contextual conditions, but can also be traced back in time and outwards in space from the local to the global (moving left on the diagram). Rich contextual studies are required to identify which are the most salient conditions, processes, and root causes for analysis and possible intervention (hypothetical examples given here).

*Source*: After Ribot (2014).

tend to take a systems approach to social-ecological relations, emphasizing thresholds, feedbacks and environmental consequences, whereas vulnerability analyses tend to focus on social and political factors, such as values, power, agency, and social differentiation (Miller et al., 2010). More recently, frameworks that combine vulnerability and resilience narratives have been developed to identify resilience-enhancing and vulnerability-reducing responses that contribute to adaptation processes, while avoiding maladaptive outcomes (Maru et al., 2014). This research makes it clear that resilience approaches are not alternatives to vulnerability approaches, but complementary.

## Challenges and future directions

This introduction to vulnerability has sought to point out not only the concept's rich research tradition but also the unique conceptual contributions of vulnerability research to environmental studies. Though vulnerability research continues to be a vibrant and evolving field of scientific study and debate, a number of existing challenges and emerging future directions can be identified.

Vulnerability assessments have provided valuable insights on how environmental impacts are distributed, contributing to better understandings of winners and losers from global change processes. Some assessments aim to produce specific vulnerability indicators, whether they have been developed for nation-states, cities, households, social groups, ecosystems or species. Such indicators can be useful in policy-oriented contexts as they provide quantitative measures that can be compared and ranked, with the aim of targeting resources and attention towards the most vulnerable. The 'vulnerable' label may also have instrumental or strategic uses. For example, countries (or groups within countries) may have an incentive to categorize themselves as among the most vulnerable as a strategy for securing international aid or climate change adaptation funding (Barnett et al., 2008). Indicators also introduce the risk that vulnerability may be overly simplified or 'essentialized'. One particular challenge, then, is to produce vulnerability analyses and assessments that recognize complexity but also pinpoint key environmental, social, and political processes involved in producing vulnerability in a way that can inform and mobilize effective social change.

A second challenge relates to the discursive effects of applying the term 'vulnerable' to particular groups. Labelling people as vulnerable can create demobilizing effects, which may in part explain why resilience approaches are increasingly seen as attractive, particularly to policy communities. Being denoted as 'vulnerable' may be stigmatizing and tend to limit one's own estimation of agency (capacity to act positively). Further, it diagnoses a problem to be cured or fixed, and may lead to a sense of victimization among those labelled vulnerable, rather than generating social and political action. Scaled to a global level, referring to the current state of the environment and human condition as 'vulnerable' and characterized by notions of insecurity, uncertainty, risk, hazard, or crisis can potentially paralyze rather than motivate effective responses. Again, vulnerability research must meet the challenge of communicating effectively. While researchers have an important role to play in pointing out social justice issues in the production of vulnerability globally, they must also be attendant to the demobilizing effects such terms can have and the way they may be strategically employed.

The subjective dimensions of vulnerability should not be underestimated, and this is largely an area for future research. It is clear that people, places, species, sectors, or systems may be vulnerable to environmental changes, especially when stressors and shocks influence material conditions or outcomes. However, the question of how and why individuals and groups may *feel* vulnerable to these changes, whether or not material losses are experienced, has been underexplored. Ultimately, and particularly in the case of projected or predicted rather than already-experienced environmental changes, it is perceptions and feelings, not the changes themselves, that influence how vulnerability is experienced and that may or may not motivate action.

Although to be psychologically and emotionally vulnerable is part of the human condition, the changing nature of environmental risks related to human impacts on the environment contributes to existential threats, which introduces a super-subjective and collective dimension to vulnerability. The capacity or agency to deal with the diversity of factors contributing to vulnerability (such as war or pandemics) may be limited if problems are seen as separate rather than inter-linked.

In a related way, the role of culture and values is a further dimension for future research on vulnerability. Cultural and religious beliefs and practices may influence perceptions and experiences of vulnerability on individual and collective levels (Schipper, 2010). For example, fatalism and determinism may limit commitments to vulnerability reduction, whereas solidarity and a shared concern for the future may engage people in action. What is valued within and across communities may differ, and it is not surprising that communities experience and deal with changes, stressors and shocks in different ways (O'Brien and Wolf, 2010). Fundamentally, if vulnerability relates to the potential to experience harm or loss, then this must depend on what is *valued* (what it would hurt to lose) within different societies. This point in particular calls on vulnerability research to take into account not only contextual biophysical and socio-economic but also *cultural* factors. Further, given that 'to assess' means to judge or estimate *value* or *quality*, the next generation of vulnerability assessments may need to take particular groups' values as a starting point, likely requiring the use of in-depth qualitative methods.

Finally, we would like to suggest the potential positive implications of recognizing our shared vulnerability. To be vulnerable is to recognize that we live in relationship with others and with the environment. Without denying the serious risks to life and well-being that vulnerability involves, it should be recognized that vulnerability is fundamentally an essential part of our relationships – to each other and to the earth. It thus also reflects, and calls on us to reflect upon our relationships, our shared humanity and our shared home. Rather than obscuring the profound inequalities in the differential production of vulnerability globally, shared feelings of vulnerability may be strong sources and motivators for social change.

## Learning resources

Some of the most relevant scientific publications on vulnerability have been cited above and are listed in the bibliography. Learning resources that may be more relevant to a wider audience are provided here. A wide variety of perspectives and approaches to vulnerability exist, and this diversity is reflected in these different resources.

*weADAPT* is a platform for information and collaboration related to climate change adaptation and is supported by the Stockholm Environment Institute. It includes

a variety of resources on vulnerability oriented towards researchers, practitioners, and policy-makers.
https://www.weadapt.org/knowledge-base/vulnerability.

The Global Programme of Research on Climate Change Vulnerability, Impacts and Adaptation (PROVIA) is an initiative led by several United Nations organizations to consolidate research on climate change adaptation. Its latest guidance document, oriented to researchers, practitioners and policy-makers, covers a diversity of approaches and assessment and planning tools.
https://www.sei-international.org/mediamanager/documents/Publications/Climate/PROVIA-guidance-Nov2013-summary-low-res.pdf

A presentation by Lori Peek at Colorado State University introduces the Pressure and Release model discussed above and presents a case study of research and community action with communities affected by oil spills on the US Gulf Coast.
https://youtu.be/jlXWot27Omk.

UKCIP at the University of Oxford's Environmental Change Institute provides, among other resources, a simple 'wizard' for vulnerability assessments. It is oriented primarily towards businesses and other organizations.
http://www.ukcip.org.uk/wizard.

## Bibliography

Barnett, J., Lambert, S. & Fry, I. (2008) 'The Hazards of Indicators: Insights from the Environmental Vulnerability Index', *Annals of the Association of American Geographers* 98(1): 102–119.

Bassett, T.J. & Fogelman, C. (2013) 'Déjà vu or Something New? The Adaptation Concept in the Climate Change Literature', *Geoforum* 48: 42–53.

Blaikie, P., Cannon, T., Davis, I. & Wisner, B. (1994) *At Risk: Natural Hazards, People's Vulnerability and Disasters*. London and New York: Routledge.

Burton, I., Kates, R.W. & White, G.F. (1978) *The Environment as Hazard*. New York: Guilford Press.

Burton, I., Kates, R.W. & White, G.F. (1993) *The Environment as Hazard*, 2nd edition. New York: Guilford Press.

Cutter, S.L. (1996) 'Vulnerability to Environmental Hazards', *Progress in Human Geography* 20(4): 529–539.

Eakin, H. (2005) 'Institutional Change, Climate Risk, and Rural Vulnerability: Cases from Central Mexico', *World Development* 33(11): 1923–1938.

Kelly, P.M. & Adger, W.N. (2000) 'Theory and Practice in Assessing Vulnerability to Climate Change and Facilitating Adaptation', *Climatic Change* 47(4): 325–352.

Maru, Y.T., Stafford Smith, M., Sparrow, A., Pinho, P.F. & Dube, O.P. (2014) 'A Linked Vulnerability and Resilience Framework for Adaptation Pathways in Remote Disadvantaged Communities', *Global Environmental Change* 28: 337–350.

Miller, F., Osbahr, H., Boyd, E., Thomalla, F., Bharwani, S., Ziervogel, G., Walker, B., Birkmann, J., van der Leeuw, S., Rockström, J. et al. (2010) 'Resilience and Vulnerability: Complementary or Conflicting Concepts?', *Ecology and Society* 15(3): 11.

O'Brien, K.L. & Wolf, J. (2010) 'A Values-Based Approach to Vulnerability and Adaptation to Climate Change', *Wiley Interdisciplinary Reviews: Climate Change* 1(2): 232–242.

O'Brien, K.L., Eriksen, S., Nygaard, L.P. & Schjolden, A. (2007) 'Why Different Interpretations of Vulnerability Matter in Climate Change Discourses', *Climate Policy* 7(1): 73–88.

O'Brien, K.L., Leichenko, R., Kelkar, U., Venema, H., Aandahl, G., Tompkins, H., Javed, A., Bhadwal, S., Barg, S., Nygaard, L. et al. (2004) 'Mapping Vulnerability to Multiple Stressors: Climate Change and Globalization in India', *Global Environmental Change* 14(4): 303–313.

Ribot, J. (2014) 'Cause and Response: Vulnerability and Climate in the Anthropocene', *The Journal of Peasant Studies* 41(5): 667–705.

Schipper, E.L.F. (2010) 'Religion as an Integral Part of Determining and Reducing Climate Change and Disaster Risk: An Agenda for Research', in Voss, M. (ed.) *Der Klimawandel* (pp. 377–393). Berlin: VS Verlag für Sozialwissenschaften.

Sen, A. (1981) *Poverty and Famines: An Essay on Entitlement and Deprivation*. Oxford: Oxford University Press.

Taylor, M. (2013) 'Climate Change, Relational Vulnerability and Human Security: Rethinking Sustainable Adaptation in Agrarian Environments', *Climate and Development* 5(4): 318–327.

Watts, M.J. & Bohle, H.G. (1993) 'The Space of Vulnerability: The Causal Structure of Hunger and Famine', *Progress in Human Geography* 17(1): 43–67.

# 1.25 Wilderness

Phillip Vannini

## The concept

The word "wilderness" originates in twelfth-century English. The root of the word "wild" is "will" and "willed," as in will-force and willpower. *Deor* means deer, or in general an animal, whereas "*Ness*" – derived from Old Norse, Swedish, Danish, and Low German – refers to a promontory headland or cape. The word "wilderness" then simply denotes a place that is self-willed. Etymologies aside, the most common contemporary understanding of wilderness is that of a wild, pristine, and uncultivated natural region that is inhabited by animals but not by humans. Such understanding fuels the popular imagination and is widely manifested in media culture, tourism, law, science, and conservation. For the last two decades, however, this conceptualization of wilderness has undergone extensive criticism.

The notion of wilderness as a pristine land free of human interference is problematic and paradoxical. Wilderness in practical terms is an area officially designed and protected by public authorities aiming to conserve a place's wild character through a variety of legal measures and conservation practices. Several countries gazette and protect wilderness areas following the International Union for Conservation of Nature's (IUCN) influential definition. For the IUCN, wilderness areas are "free of modern infrastructure, development and industrial extractive activity", free of "other permanent structures" and "agriculture including intensive livestock grazing, commercial fishing." They are "preferably with highly restricted or no motorized access" and "characterized by a high degree of intactness" "of sufficient size to protect biodiversity" "and to maintain evolutionary processes." Furthermore wilderness areas "offer outstanding opportunities for solitude." An example is shown in Figure 1. So, in essence wilderness is an entity protected by law, informed by political priorities, and managed through science-based frameworks. It is hardly a self-willed subject (IUCN).

**Figure 1.25.1** An aerial view of Lowell Glacier, within Canada's Kluane National Park and Preserve, which together with Tatshenshini-Alsek Park, and Alaska's Wrangell-St. Elias and Glacier Bay Parks constitutes the world's largest contiguous UNESCO World Heritage Site.

*Source*: author.

## The debate

Central to the popular understanding of wilderness is the dualist notion of nature. Humans – following this ideology – are somehow outside of nature and their social and cultural practices end up spoiling otherwise pristine natural environments. This is not only a short-sighted understanding, but also a dangerous one. Critics have argued that the idea of wilderness as pristine nature is a powerful discursive construction which hides how nature is always subject to contesting and constantly-shifting social forces. The entire world's surface, even its most spectacular parks and preserves, has inevitably been touched in one way or another by people. In the words of environmental historian William Cronon: "far from being the one place on earth that stands apart from humanity, [wilderness] is quite profoundly a human creation – indeed the creation of very particular human cultures at very particular moments in human history [. . .]. Wilderness hides its unnaturalness behind a mask that is all the more beguiling because it seems so natural" (1996, p. 69).

The trouble with the view of wilderness as pristine nature thickens if we take a broader historical perspective. Max Oelschlaeger (1991) and Roderick Nash (2001) have shown that wilderness is a relatively recent idea: an outcome of worldwide population growth and residential and industrial development. When extensive tracts of forested land were widely available wilderness was largely seen in part as menace and in part as yet-unexploited resource for expansion, or simply as useless wasteland. But as more and more people began to lament the urban condition and the domination of humans over nature, wilderness grew in the popular imagination as a place of refuge and escape, a ground for moral and physical fortitude, and a site for the manifestation of the sublime.

Much writing has recently contributed to unsettling the idea of wilderness as a pristine, "natural" environment. Many scholars have found that the legal designation of an area as wilderness has often resulted in the displacement of human dwellers, especially aboriginal people. Such situation has taken particularly dramatic tones outside the Western world where the imperial visions of European colonizers first, and the conservation pressure of some international environmental NGOs later, have resulted in the formation of people-free parks and nature reserves that have deeply reshaped indigenous ways of life.

The sustained popularity of leisure practices in wilderness and protected areas – ranging from lifestyle sports to family camping, and from nature-based tourism to hunting and fishing – further contributes to the construction of wilderness as a space outside of permanent human control, yet a place awaiting adventurous exploration and conquest. Cultural studies of wilderness-based leisure practices have shown that de-constructing wilderness as a domain separate from culture and society results in the exposure of many anthropocentric, ethnocentric, nationalist, regionalist, and androcentric ideologies.

Nevertheless, the dominant idea of wilderness continues to exercise a powerful pull over the social imagination. By denying that the wild exists, by relegating "nature" to a product of human action, and by resigning to the fact that pristine environments no longer exist, we are essentially giving up on the value of conservation. Why bother to protect nature, after all, it if has already been spoiled? Many conservation advocates believe that in spite of their philosophical acuity critiques of wilderness may simply end up throwing the baby away with the bath water. Wilderness as a concept may indeed be in need of a serious update, but wild places themselves should not be jettisoned with an old idea.

## The value of wild areas

There are multiple reasons why preserving wild places might both provide valuable "services" to people and might also be the right moral course of action for the sake of life's inherent worth. According to Nelson (1998, pp. 156–193) wilderness is a repository of natural resources; a rich hunting and fishing ground; a vast source of species for natural medicinal use; a self-regulating service "industry" for humankind (e.g. by working as a carbon sink); a life-supporting ecosystem; a place to seek physical and mental health; an arena for athletic and recreational pursuits; an aesthetically-rewarding environment and source of inspiration; a place for spiritual encounters with the divine; a scientific field "laboratory" for the observation of natural processes; a base-datum, or comparative standard measure, for land health; a reservoir of bio-diversity; an outdoors "classroom"; a memorial site commemorating human evolution; a cultural landscape; a symbol of national or regional character and identity; a place for self-realization; a buffer-zone protecting humans from unknown viruses and bacteria; a bastion of individual freedom; the birth ground of many myths; a metaphysical necessity for our understanding of civilization; a site for the practice of minority rights; a liminal space for the formation of social bonds; the prime site for animal welfare and for the smooth functioning of the earth as an organism; an inheritance for future generations; a place full of enchantment, wonder, and unknowns; a place that has its own intrinsic value.

Given its value and despite its problematic historical baggage, Callicott (2008) argues that we would be better off by replacing the term "wilderness" with the more scientifically precise "biodiversity." The expression "biodiversity" – which has indeed grown in popularity worldwide – is more neutral and alerts us to the value of preserving habitats for species who find them suitable for their survival. However, abandoning the word "wilderness" would be the wrong solution, according to others. Plumwood (1998), for example, wishes to preserve the term wilderness because of its potential to remind us of important historical lessons. Wilderness has been traditionally defined as "virgin" land to be conquered, mastered, and controlled. These narratives reveal a great deal about androcentric and colonial attitudes toward nature and about the Western tendency to separate culture and nature and therefore to view nature as the absence and denial of culture. Instead of continuing to view wilderness as "empty," Plumwood argues, we simply need to re-conceptualize it as a site for the *presence* of the Other.

## Learning resources

The Wilderness Institute: The Wilderness Institute is housed within the University of Montana. Its website provides information on scholarships, programs of study, research projects, and reading resources. http://www.cfc.umt.edu/wi/

Wilderness.net: A website focused on educating users about the value of public lands preservation. Contains information on management, policy, training, education, research, and maps. http://www.wilderness.net/NWPS/ALWRI

Vannini, P. and A. Vannini (2016). *Wilderness*. London: Routledge. This book offers an in depth exploration of the wilderness idea and its use in environmental policy.

## Bibliography

Callicott, J.B. (2008) 'Contemporary criticisms of the received wilderness idea', pp. 355–377 in Nelson, M. & Baird Callicott, J. (eds), *The wilderness debate rages on.* Athens: University of Georgia Press.

Cronon, W. (1996) 'The trouble with wilderness; or, getting back to the wrong nature', pp. 69–90 in Cronon, W. (ed.), *Uncommon ground: rethinking the human place in nature.* New York: W.W. Norton & Co.

International Union for Conservation of Nature (IUCN) (n.d.). Protected areas: Category 1b: wilderness. http://www.iucn.org/theme/protected-areas/about/protected-area-categories/category-ib-wilderness-area. Accessed September 15 2016.

Nash, R.F. (2001) *Wilderness and the American mind.* New Haven, CT: Yale University Press.

Nelson, M. (1998) 'An amalgamation of wilderness preservation arguments', pp. 154–198 in Callicott, J. & Nelson, M. (eds), *The great new wilderness debate*. Athens, GA: University of Georgia Press.

Oelschlaeger, M. (1991) *The idea of wilderness: from prehistory to the age of ecology*. New Haven, CT: Yale University Press.

Plumwood, V. (1998) 'Wilderness skepticism and wilderness dualism', pp. 652–690 in Callicott, J. and Nelson, M. (eds), *The great new wilderness debate*. Athens, GA: University of Georgia Press.

# Part 2

## Contemporary concepts

# Part 2 Introduction

Environmental studies has, over the last several decades, responded to developments both within and beyond environmental scholarship, resulting in a set of contemporary concepts that range from logical extensions to outright repudiations of classic environmental ideas dating from the 1960s and 1970s.

One development within environmental scholarship has been the increased volume and specificity of data related to environmental problems. For instance, whereas the classic 1970s *The Limits to Growth* (Meadows et al., 1972) global-scale computer model was strikingly simple, contemporary global climate models offer far greater spatial and temporal resolution, benefitting from the tremendous increase in remotely sensed data and computational power. One would, then, expect environmental concepts to similarly evolve, as global generalizations give way to spatiotemporal specificities. For instance, the contemporary version of *Our Common Future*'s (1986) One World notion may be the highly data-driven, scientifically grounded, process-oriented notion of the Earth System (see Clive Hamilton's chapter on the latter, this volume).

Other important developments have occurred outside the field of environmental studies proper. In the broadest sense, the continued human transformation of the earth has, at least in some scholarly circles, tended to marginalize classic concepts such as wilderness, which may seem quaint set against contemporary concepts such as the Anthropocene that describe, perhaps even embrace, this heightened human role on earth. Unfolding realities have thus presented environmental studies with the challenge to build concepts that better fit the world as it is coming to be.

More specifically, the expanding range of environmental issues emerging in our world has led to the recognition that not all environmental problems are the same. In particular, the classic problems that motivated important environmental legislation in the 1960s and 1970s such as the U.S. Clean Air Act and Clean Water Act were viewed as bounded, often limited to point-source pollution and identifiable polluters, and thus in theory readily solvable by imposing tougher regulations on irresponsible parties. These well-defined environmental problems remain, but environmental scholars now recognize a wholly different class of 'wicked' environmental problems (the most prominent being climate change), whose unbounded nature makes it effectively impossible to agree on what these problems are, let alone how – indeed, whether – they can be solved. In short, environmental issues at the heart of the field may be more complex than they were thought to be during the heyday of classic environmental thought – again motivating new concepts to address new realities.

The contemporary concepts that arise from the last several decades of change within and outside the field of environmental studies are at times elaborations of classic concepts – understandable developments that could be expected in any field. One example would be the notion of 'planetary boundaries', a more sophisticated and data-driven idea that nonetheless resonates strongly with the classic notion of limits to growth. At other times, and perhaps most significantly, contemporary concepts serve as major departures from ideas of the past. One good example is 'hybridity', a concept that revels in the entanglement of human and nonhuman in a manner that would have minimally confused and possibly enraged the nature-protecting environmentalist of classic times. Another example, with a distancing from classic environmental thought clearly implied in the name itself, is 'post-environmentalism', built on the contention that classic environmental concepts and practices are effectively useless today.

This rich mix of contemporary concepts provides environmental studies with a dynamism and relevance that speak to our current and future socio-ecological realities, and ensures that environmental studies will remain fresh and provocative as the field continues to evolve.

# 2.1 The Anthropocene

Mark A. Maslin

## What is the Anthropocene?

The Anthropocene is currently an informal term used to mark when human actions begin to have a global impact on Earth's environment (Crutzen & Stoermer, 2000; Bonneuil & Fressoz, 2013; Castree, 2014; Lewis & Maslin, 2015a; Waters et al., 2016). There is general scientific agreement that human activity has been a geologically recent, yet profound, influence on the Earth System. The magnitude, variety and longevity of human-induced changes, to the lithosphere, hydrosphere, cryosphere, biosphere and atmosphere suggests that we should refer to the present, not as within the Holocene Epoch (as it is currently formally referred to), but instead as within the Anthropocene Epoch. Considerable debate exists about how to define the Anthropocene, including when it began, ranging from many thousands of years ago to a few decades ago (Lewis & Maslin, 2015a). It is desirable to have an agreed scientific definition of the Anthropocene Epoch, because clear and precise definitions aid understanding. Surprisingly, some scientists argue that the Anthropocene should be deliberately left undefined (Ruddiman et al., 2015). Applied broadly, this route may lead to confusion and should, many specialists argue, be avoided. However, differing disciplinary definitions, particularly outside the physical sciences, should be recognised (Chakrabarty, 2009, 2015; Maslin & Lewis, 2015; Latour, 2015; Barry & Maslin, 2016).

## Formal definition

Geologists are discussing a formal definition of the Anthropocene Epoch to ensure the integrity and internal consistency of the Geologic Time Scale (GTS). Geologists have uncovered the major events of Earth's 4.6 billion-year history, dividing this history into a hierarchical series of ever-finer units, with stages nested within epochs, nested within periods, nested within eras, nested within eons (Gradstein et al., 2012). Divisions represent changes in the functioning of Earth as a system and the concomitant changes in the resident life-forms. Larger differences result in classifications at higher unit-levels. The question, in terms of geology and the GTS, is whether human activity has altered the Earth as a system, with permanent or extremely long-lived impacts, to such an extent that defining a unit of geological time is logically obvious. A strict adherence to the previously defined norms of defining epochs and other higher units is thus necessary to ensure that the categorisation of Earth's history is consistent across time (Smith et al., 2014; Walker et al., 2015). Furthermore, given the clearly contentious nature of the idea of the Anthropocene, strict adherence to the requirements generated *before* the modern usage of the Anthropocene term at the turn of this century, provides a relatively clear and objective geological test of the impact of the human activity on the Earth and its future trajectory (Lewis & Maslin, 2015b). The alternative is to argue that the time closest to the present day within the GTS is somehow in need of exceptions to agreed norms. This would obviously present challenges to the process of defining the Anthropocene as any definition would be open to easy criticism that the process is biased and 'ideologically driven' (Maslin & Lewis, 2015).

Geological time is divided into a hierarchical series of ever-finer units, the fines being stages, with stages nested within epochs (Smith et al., 2014). The present, according to *The Geologic Time Scale* (2012), is in the Holocene Epoch (Greek for 'entirely recent'; started 11,650 BP, 'before present' where present is defined as 1950), within the Quaternary period (started 2.588 million years ago), within the Cenozoic era ('recent life'; started 66 million years ago) of the Phanerozoic eon ('revealed life'; started 541 million years ago). Divisions represent differences in the functioning of Earth as a system and the concomitant changes in the resident life-forms. Larger differences result in classifications at higher unit-levels.

Formally, geological time units are defined by their lower boundary, that is, their beginning. Boundaries are demarcated using a GSSP (Global Stratotype Section & Point), or by an agreed date, termed a GSSA (Global Standard Stratigraphic Age). For a GSSP, a 'stratotype section' refers to a portion of material that develops over time (rock, sediment, glacier ice), and 'point' refers to the location of the marker within the stratotype. These 'golden spikes' are a single physical manifestation of a change recorded in a stratigraphic section, often reflecting a global-change phenomenon. These are then complemented by other stratigraphic records showing a global change to the Earth system (Smith et al., 2014). Thus, for a long-term change to the Earth from one state to another, a single boundary time is chosen at a specific point within that long-term change. Suggested GSSPs include (Figure 2.1.1): 11,700 BP, replacing the Holocene boundary with the Anthropocene (Walker et al., 2009), 5,010 BP the rise in atmospheric methane ascribed to early agriculture (Ruddiman et al., 2015), 1610 ACE with the marked dip in atmospheric $CO_2$ (Lewis & Maslin, 2015a) and the peak in radiocarbon in 1964 due to radionuclide fallout from above ground nuclear weapons testing (Rakowski et al., 2013). The precise definition of each GSSP differs in how the geological requirements are combined, depending upon the time period, sediment types that are available and the types of change occurring at that point within Earth's history.

It is also possible, following a survey of the stratigraphic evidence, that a GSSA date may be agreed by committee to mark a time unit boundary. GSSAs are common in the Precambrian (>630 million years ago) because well-defined geological markers and clear events are less obvious further back in time. Suggested GSSAs include 1778, 1800, 1945 and post-1950 (Zalasiewicz et al., 2015). However, Walker et al. (2015) eloquently argue

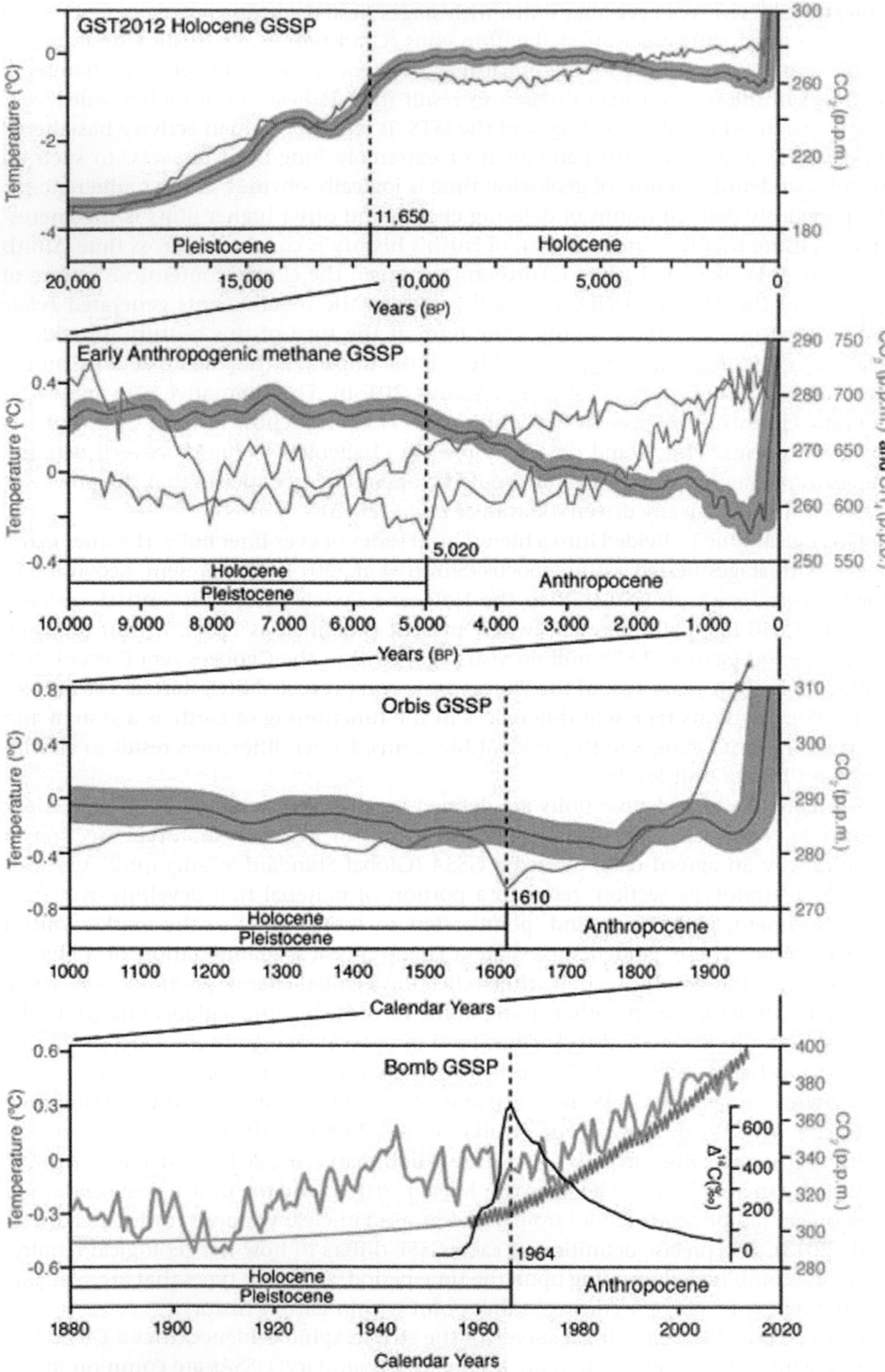


**Figure 2.2.1** Potential GSSPs to define the Anthropocene.

From Lewis and Maslin (2015a).

that geological convention clearly shows that GSSAs are not appropriate for defining the Anthropocene Epoch.

The formal definition of any geological stage is a long and bureaucratic process that has been followed for all geological boundary definitions (Smith et al., 2014). The still young Anthropocene Working Group of the Subcommission of Quaternary Stratigraphy would need to make a formal recommendation. This would be followed by a supermajority vote of the International Commission on Stratigraphy (ICS), and finally ratification by the International Union of Geological Sciences (see Finney, 2014 for full details).

## Potential golden spikes

Lewis and Maslin (2015a) assessed each suggested start date for the Anthropocene against the formal criteria used to define previous epochs. They concluded that most previously proposed start dates, including the earliest detectable human impacts through farming and historic events such as the start of the Industrial Revolution, should be rejected. While they are important events they have not produced *globally* synchronous markers and few of these human impacts will discernibly change Earth for the next few million years, the time span of a typical epoch. In contrast, Ellis et al. (2016) argued that the Anthropocene was a long-term human process and all the stages of human development should be recognised (Figure 2.1.2).

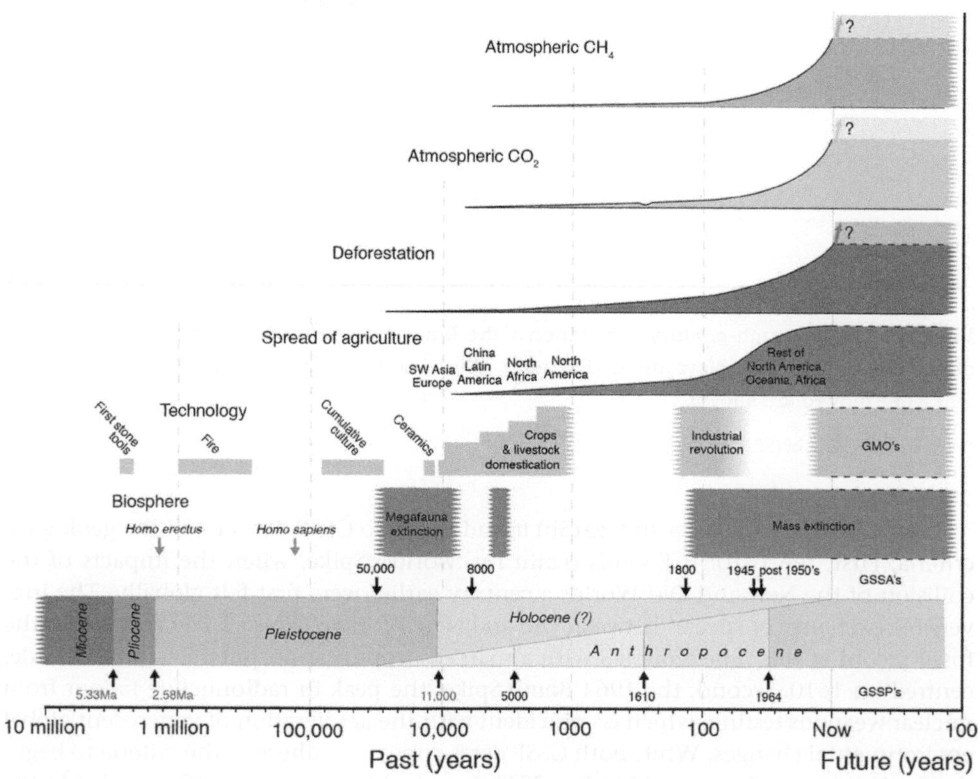


**Figure 2.1.2** Summary of the long-term impacts that humanity has had on the global environment and potential markers for the Anthropocene.

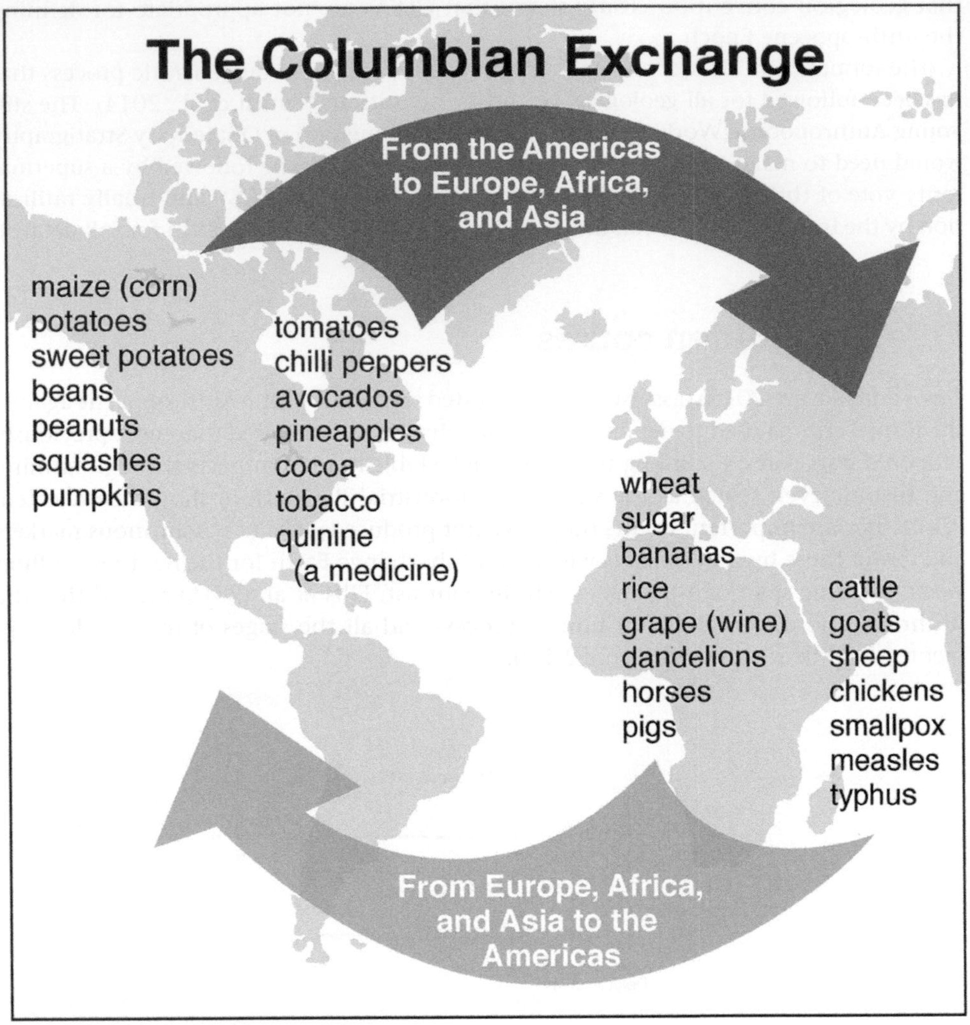


**Figure 2.1.3** Sixteenth-century annexation of the Americas started the irreversible exchange of species between the New and Old Worlds called the Columbian Exchange – key species that were moved are shown.

From Lewis & Maslin (2015b).

Lewis and Maslin (2015a and 2015b) found that two GSSP dates fit all the geological criteria. First, the 1610 ACE Orbis (Latin for 'world') Spike, when the impacts of the collision of the New and Old Worlds a century earlier were first felt globally. The irreversible exchange of species between Old and New Worlds (Figure 2.1.3) is noted in the fossil record at this time, coupled with a marked drop in atmospheric carbon dioxide, centred on 1610. Second, the 1964 Bomb Spike, the peak in radionuclide fallout from nuclear weapons testing, which is coincident with the acceleration of very recent global environmental changes. While both GSSP dates appear to adhere to the criteria to begin the Anthropocene, Lewis and Maslin (2015a) suggest that overall 1610 is probably the strongest contender while Waters et al. (2016) argue for the 1964 date.

## Beyond the geosciences

A formal definition of the Anthropocene Epoch is not *the* only definition of the Anthropocene. In fact, it is the role of geologists officially to define geological time, and have the institutions that enable them to do so. But they have no ability, nor legitimacy, to define the beginning of historic periods or political systems. It is therefore incumbent on other subjects such as history, politics, anthropology, geography etc. to have their own definitions of the Anthropocene. Moreover, if the term Anthropocene is not fit for purpose within the disciplines, then other terms such as Capitalocene or Anglocene (Bonneuil & Fressoz, 2013) should be used and defined. For example Haraway (2015) has proposed the 'chthulucene', named from the Greek 'chthonic' meaning literally 'subterranean' to represent all those human and non-human entities intertwined on Earth. This term is deliberately spelt different to "Cthulhu" the nightmare monster imagined by H.P. Lovecraft. Instead it steals some of its symbolism as Cthulhu awaits at the edge of the Solar System for its opportunity to devour humanity. But maybe Cthulhuocene would be more apt if it is we that are the monster devouring the planet. With all these multiple definitions the AWG should review all contributions to this debate from all disciplines (Ellis et al., 2016). But the recommendation they make in their final report to ICS has to be based on the fundamental principles of stratigraphy and be able to be defended scientifically against accusations of political bias or agenda. If it does not follow these principles, it is very clear that the ICS will not ratify the Anthropocene Epoch. If a narrow geological defined Anthropocene Epoch is agreed then I suggest this would generate more debate and, indeed, it may push critical social scientists to re-evaluate their definitions of past political and social constructions.

## Why does the Anthropocene matter?

Formally ratifying the Anthropocene as an epoch requires the agreement of a series of geological committees. If it is adopted, it will be a defining moment in human history, fundamentally changing our relationship with our environment, changing the story that humans tell about the societal development. Embracing the Anthropocene reverses 500 years of scientific discoveries that have moved humans to a position of ever increasing insignificance. From the sixteenth-century Copernican revolution to the modern understanding that our Sun is one of $10^{24}$ stars in the Universe; to the nineteenth-century Darwinian revolution establishing humans as merely a twig on the tree of life with no special origin. A twenty-first-century adoption of the Anthropocene reverses this insignificance: humans are not passive observers of Earth. We argue that *Homo sapiens* are central as the future of the only place where life is known to exist is being determined by the actions of humans.

A more widespread recognition that human actions are driving far-reaching changes to the life-supporting infrastructure of Earth will have implications for our philosophical, social, economic and political views of our environment. By defining the Anthropocene we are providing a deep insight to the current relationship between humanity and the global environment. Because the power that humans wield is unlike any other force of nature, it is reflexive and therefore can be used, withdrawn or modified. The first stage of solving our damaging relationship with our environment is recognising it.

## Learning resources

To find out more about the key debates on the Anthropocene, the following are very accessible and succinct reviews:

https://en.wikipedia.org/wiki/Anthropocene

Lewis, S.L. and Maslin, M.A. (2015a) Defining the Anthropocene. *Nature*, 519: 171–180.

Steffen, W., Leinfelder, R., Zalasiewicz, J., Waters, C.N., Williams, M., Summerhayes, C., Barnosky, A.D., Cearreta, A., Crutzen, P., Edgeworth, M., et al. (2016) Stratigraphic and Earth System approaches to defining the Anthropocene. *Earth's Future*, 4: 324–345.

Waters, C.N., Zalasiewicz, J., Summerhayes, C., Barnosky, A.D., Poirier, C., Gałuszka, A., Cearreta, A., Edgeworth, M., Ellis, E.C., Ellis, M. et al. (2016) The Anthropocene is functionally and stratigraphically distinct from the Holocene, *Science*, 351(6269), DOI: 10.1126/science.aad2622.

## Bibliography

Barry, A. and Maslin M.A. (2016) The politics of the Anthropocene: A dialogue, *Geo: Geography and Environment*, DOI: 10.1002/geo2.22

Bonneuil, C. and Fressoz, J-B. (2013) *L'événement anthropocène: La Terre, l'histoire et nous*, Paris: Seuil.

Castree, N. (2014) Geography and the Anthropocene 2: Current Contributions, *Geography Compass* 8(7): 450–463.

Chakrabarty, D. (2009) The climate of history: Four theses. *Critical Inquiry* 35(Winter): 197–222.

Chakrabarty, D. (2015) The Anthropocene and the convergence of histories. In Hamilton, C., Bonneuil, C. and Gemenne, F. (eds) *The Anthropocene and the Global Environmental Crisis: Rethinking Modernity in a New Epoch*. Abingdon: Routledge.

Crutzen, P.J. and Stoermer, E.F. (2000) The Anthropocene. *IGBP Global Change Newsletter*, 41: 17–18.

Ellis, E., Maslin, M., Boivin, N. and Bauer, A. (2016) Involve social scientists in defining the Anthropocene, *Nature* 540(7632): 192–193.

Finney, S.C. (2014) The 'Anthropocene' as a ratified unit in the ICS International Chronostratigraphic Chart: Fundamental issues that must be addressed by the Task Group. *Geographical Society of London Special Publications*, 395, 23–28.

Gradstein, F.M., Ogg, J.G., Schmitz, M.D. and Ogg, G. (2012) *The Geologic Time Scale 2012*, The Netherlands: Elsevier BV.

Haraway, D. (2015) Anthropocene, Capitalocene, Plantationocene, Chthulucene: Making kin. *Environmental Humanities*, 6, 159–165.

Latour, B. (2015) Telling friends from foes in the time of the Anthropocene. In Hamilton, C., Bonneuil, C. and Gemenne, F. (eds) *The Anthropocene and the Global Environmental Crisis: Rethinking Modernity in a New Epoch*. Abingdon: Routledge.

Lewis, S.L. and Maslin, M.A. (2015a) Defining the Anthropocene. *Nature*, 519: 171–180.

Lewis, S.L. and Maslin, M.A. (2015b) A transparent framework for defining the Anthropocene Epoch. *Anthropocene Review*, 2(2): 128–146.

Maslin, M.A. and Lewis, S.L. (2015) Earth system, geological, philosophical and political paradigm shifts. *The Anthropocene Review*, 2(2): 108–116.

Rakowski, A.Z., Nadeau, M-J., Nakamura, T., Pazdur, A., Pawelczyk, S. and Piotrowska, N. (2013) Radiocarbon method in environmental monitoring of $CO^2$ emission. *Nuclear Instruments & Methods in Physics Research Section B-Beam Interactions with Materials and Atoms*, 294: 503–507.

Ruddiman, W.F., Ellis, E.C., Kaplan, J.O. and Fuller, D.Q. (2015) Defining the epoch we live in. *Science*, 348: 38–39.

Smith, A.G., Barry, T., Bown, P., Cope, J., Gale, A., Gibbard, P., Gregory, J., Hounslow, M., Kemp, D., Knox, R. et al. (2014) GSSPs, global stratigraphy and correlation. *Geological Society, London, Special Publications*, 404(1): 37.

Steffen, W., Broadgate, W., Deutsch, L., Gaffney, O. and Ludwig, C. (2015) The trajectory of the Anthropocene: The Great Acceleration. *The Anthropocene Review*, 2(1): 81–98.

Walker, M., Gibbard, P. and Lowe, J. (2015) Comment on "When did the Anthropocene begin? A mid-twentieth century boundary is stratigraphically optimal" by Jan Zalasiewicz et al. *Quaternary International*, 383: 196–203.

Walker, M., Johnsen, S., Rasmussen, S.O., Popp, T., Steffensen, J-P., Gibbard, P., Hoek, W., Lowe, J., Andrews, J., Björck, et al. (2009) Formal definition and dating of the GSSP (Global Stratotype Section and Point) for the base of the Holocene using the Greenland NGRIP ice core, and selected auxiliary records. *Journal of Quaternary Science*, 24: 3–17.

Waters, C.N., Zalasiewicz, J., Summerhayes, C., Barnosky, A.D., Poirier, C., Gałuszka, A., Cearreta, A., Edgeworth, M., Ellis, E.C., Ellis, M. et al. (2016) The Anthropocene is functionally and stratigraphically distinct from the Holocene, *Science*, 351(6269), DOI: 10.1126/science.aad2622.

Zalasiewicz, J., Waters, C.N., Williams, M., Barnosky, A.D., Cearreta, A., Crutzen, P.J., Ellis, E.C., Ellis, M., Fairchild, R.J., Grinevald, J. et al. (2015) When did the Anthropocene begin? A mid-twentieth century boundary level is stratigraphically optimal. *Quaternary International*, 383: 196–203.

# 2.2 Biopolitics

## Kevin Grove

Few concepts have impacted critical thought in the past decade quite like *biopolitics* and its corollary, biopower. In brief, the concept of biopolitics enables us to study how life and power intertwine. It focuses attention on the way power in modern liberal societies operates through seemingly banal and apolitical efforts to "make life live", or to promote the security, development and well-being of individuals and collectives. In other words, it recognizes that strategies and techniques of power refer back to a vision of the life that needs to be secured and developed – whether this is "life" thought of as an individual citizen, a biological population, or a complex social and ecological system. This chapter will clarify biopolitics and flag up some of the concept's implications for environmental studies. Perhaps most importantly, biopolitics challenges us to think differently about the way we think about and study power and politics, even as they relate to the non-human environment. Rather than a distinct concept with clear meaning that can be used to deductively describe particular situations, biopolitics is more appropriately thought of as a problem space for thinking how life and power intertwine in contextually specific ways. Biopolitics *complexifies rather than clarifies* a messy empirical field. In the case of environmental studies, this involves folding in thorny cultural, political and ethical dimensions to problems such as climate change adaptation, resilience, conservation, hazards studies or disaster management that are typically approached as clearly bounded ecological or (behavioral) economic topics.

## A new model of power

Scholarly interest in biopolitics is indebted to French historian Michel Foucault. Foucault initially deployed the term to rethink the relation between power, the subject and the state. In the influential final chapter of his *History of Sexuality, Volume 1* (Foucault,

1990) as well as the final lecture in his 1975–1976 Collège de France lectures (Foucault, 2003), he characterizes biopower as a distinct form of power that operates through "making live" and "letting die". Biopower contrasts with sovereign power, which operates through "making die" and "letting live". These phrases succinctly capture the differences between the two forms of power: if sovereign power is a negative power of appropriation that operates through techniques such as taxation, imprisonment or execution (appropriating money, mobility and life, respectively), then biopower is a positive power of security that operates through techniques such as normalizing training and education, statistics and actuarial calculations (securing a population's continued circulation, development and growth in the face of uncertainty).

This seemingly straightforward definition has profound implications for how we can understand power and analyze power relations. Indeed, Foucault set his critical sights on what he saw to be a narrow and confining modernist political imaginary that limits our ability to recognize how power operates within contemporary liberal societies. A modernist political imaginary reifies power: it becomes an entity that can be held by a sovereign subject (individual or state) and wielded to impose the individual's will on others. Power here can only be negative: the negation of another individual's will and agency. From this perspective, the question of power becomes a matter of adjudicating the contest between individual wills: what institutional forms (representative democracy, parliament, etc.) can best limit the arbitrary exercise of power? This is, of course, the basis of modernist understandings of sovereignty and the state: Hobbes' famous social contract embodies the recognition that individuals voluntarily cede their sovereignty to a king and subject themselves to a system of rights and responsibilities in exchange for freedom. In contrast to this "contractual" model, biopower offers a model of power that is not centered on the individual subject of rights, responsibilities and agency. Instead, the concept directs attention to what Maruzio Lazzarato (2006: 12) calls "a multitude of forces acting and reacting in relation to each other." This biopolitical prism distorts key categories of a modernist political imaginary. First, power is no longer a thing. Instead, it is a *relation* that exists as such only as it is enacted through interaction. Second, power is no longer negative. Instead, it is now productive: power *produces* objects of power – abnormal human and non-human others requiring regulation, and subjects with certain kinds of desires, interests and capacities. From this perspective, the sovereign individual is an *effect* of power relations that need to be analyzed and explained, rather than the essential foundation of political analysis. Third, government does not refer to an institution that holds and exercises sovereign power. Instead, it is a process of regulating the conduct of others designated as objects of power. Government involves the "conduct of conduct" (Foucault, 2007), action on the action of others that produces power effects.

But there is more. Because power is a relation that produces subjects and objects of government, it never exists outside of interactions. A topographical analysis of power concerned with the location of power (who or what holds power) and the limits of power (the territorial extent of sovereignty or the design of institutions to limit the arbitrary exercise of sovereign power) tells us little about these interactions and their effects. Understanding power relations and the kinds of subjects they produce requires a *topological* analysis attuned to the ways different techniques of power (such as statistics, observation and evaluation, instructional training, physical punishment, or actuarial calculation) become combined and re-combined with each other in response to contextually-specific problems of government (Collier, 2009). A biopolitical analysis thus focuses on the *strategic* dimensions of power relations. Here, strategy refers to set objectives, or visions of proper forms of conduct and desired forms of order, and mechanisms of power deployed to realize these objectives. Strategies are always deployed in response to specific transgressions, against people and things that refuse to be governed as intended. This means that power is no longer the antithesis of freedom. Instead, it relies on freedom – not the liberal freedom of individual choice, but a more radical freedom to live beyond any externally-imposed determinant of meaning and value. Transgressing the limits of the present provokes new kinds of strategic governmental intervention designed to bring life under the sway of a calculative, controlling rationality.

Taken together, these conceptual innovations involve a radical shift in how we can think the relation between power and life. Biopower is not an abstract concept that can be deductively applied to explain an empirical field. Instead, it is a provocation to inductively examine the contextually specific ways that power attempts and fails to regulate life itself. Rather than an explanatory variable, it is more productively thought of as a problem space in which life is made amenable to governmental intervention.

## Biopolitics in action: producing resilient subjects

Critical research on resilience helps illustrate these somewhat abstract claims. In recent years, resilience has become an increasingly prominent organizing principle

**Figure 2.2.1** Biopolitics in action: community-based disaster resilience programming in Trinityville, Jamaica. The author carrying out transect walk interviews with community members; the data from interviews became part of a community disaster response plan.

**Figure 2.2.2** The launch of the plan, which was rolled out to great fanfare at an evening event held in a local school, and publicized through the Jamaican state's media agency (see also https://societyandspace.com/material/article-extras/seeing-resilience-like-a-state-kevin-grove/).

for governing social and ecological relations (Chandler, 2014). Proponents assert that building resilience will enable individuals, communities and even nations to survive and thrive in a world of complex interconnection and abrupt, non-linear emergence (e.g., Walker and Salt, 2012; see Figures 2.2.1 and 2.2.2).

However, research into the biopolitical effects of resilience interventions offers a very different story. This work has shown how resilience mobilizes a critique of centralized environmental management that is topologically indistinct from neoliberal economists' critique of centralized economic planning. In both critiques, centralization fails to capture the inherent complexity of socio-economic-ecological life. The solution in each case involves decentralizing decision-making authority to individuals, who are most in tune with the contextually-specific dynamics of complexity. There is thus an "intuitive ideological fit" between resilience and neoliberal economic thought: both provide justification for rolling back state regulations and the provision of social and ecological security (Walker and Cooper, 2011). Despite coming from diametrically opposed political leanings, and despite resilience proponents' good *intentions*, both produce the same biopolitical *effects*: "resilient" individuals, communities, and even nations that *embrace* risk, *live with* vulnerability and do not challenge the uneven political economic relations that produce this vulnerability in the first place (Evans and Reid, 2014). Thus, for Mark Duffield (2011), resilience is the "official policy response" to the "fabricated uncertainty" of neoliberal development. It does not empower individuals to adapt to change as much as it depoliticizes adaptation and turns it into a way of simply "surviving the after-effects of industrial modernization, the green revolution and the Washington Consensus" (Walker and Cooper, 2011: 155).

## Conclusion

The ability of biopolitics to re-orient thought towards the under-acknowledged biopolitical effects of seemingly apolitical, technical interventions makes it an absolutely essential concept for analyzing contemporary efforts to regulate socio-environmental conditions. It encourages scholars to recognize how power operates and flows within *and* beyond the state. Seemingly banal practices such as cataloguing environmental conditions or devising new techniques for managing environmental resources, building resilience, enhancing biodiversity conservation, and so forth, are potentially vectors along which power relations can flow (Braun, 2002). Thus, biopolitics presents environmental scholars with a new ethical challenge to reflect on how their knowledge might *resist*, *subvert*, or *destabilize* rather than shore up an uneven political economic status quo.

## Learning resources

Because the concept of biopolitics emerged out of critical theory, engaging with it can at first glance appear to be a daunting challenge. The most straightforward introduction to the concept remains Michel Foucault's Collège de France lectures. In contrast to the dense writing in Foucault's polished texts, his lectures are considerably more dialogical, instructive and accessible. The following lecture series remain the single most important resource for readers looking for an introduction to the concept:

Foucault, M. (2003) *'Society Must be Defended': Lectures at the Collège de France, 1975–1976*. New York: Picador.

Foucault, M. (2007) *Security, Territory, Population: Lectures at the Collège de France, 1977–1978*. New York: Picador.

Foucault, M. (2008) *The Birth of Biopolitics: Lectures at the Collège de France, 1978–1979*. New York: Picador.

The 2009 special issue of *Theory, Culture & Society* (volume 26, issue 6) offers a number of essays that discuss these lectures' impacts. Further details on the political and philosophical debates surrounding biopolitics can also be found in these excellent texts:

Campbell, T. (2011) *Improper Life: Technology and Biopolitics from Heidegger to Agamben*. Minneapolis: University of Minnesota Press.

Lemke, T. (2011) *Biopolitics: An Advanced Introduction*. New York: New York University Press.

## Bibliography

Braun, B. (2002) *The Intemperate Rainforest: Nature, Culture and Power on Canada's West Coast*. Minneapolis: University of Minnesota Press.

Chandler, D. (2014) *Resilience: The Governance of Complexity*. London: Routledge.

Collier, S. (2009) 'Topologies of power: Foucault's analysis of political government beyond "governmentality"', *Theory, Culture & Society*, 26(6): 78–108.

Duffield, M. (2011) 'Total war as environmental terror: Linking liberalism, resilience, and the bunker', *South Atlantic Quarterly*, 110(3): 757–769.

Evans, B. & Reid, J. (2014) *Resilient Life: The Art of Living Dangerously*. Cambridge: Polity.

Foucault, M. (1990) *The History of Sexuality, Volume 1*. New York: Vintage.

Foucault, M. (2003) *'Society Must be Defended': Lectures at the Collège de France, 1975–1976*. New York: Picador.

Foucault, M. (2007) *Security, Territory, Population: Lectures at the Collège de France, 1977–1978*. New York: Picador.

Foucault, M. (2008) *The Birth of Biopolitics: Lectures at the Collège de France, 1978–1979*. New York: Picador.

Lazzarato, M. (2006) 'From biopower to biopolitics', *Tailoring Biotechnologies*, 2(2): 11–20.

Walker, B. & Salt, D. (2012) *Resilience Practice*. Washington, D.C.: Island Press.

Walker, J. & Cooper, M. (2011) 'Genealogies of resilience: From systems ecology to the political economy of crisis adaptation', *Security Dialogue*, 42(2): 143–160.

# 2.3 Biosecurity

## Steve Hinchliffe

Biosecurity is a relatively recent term to describe the efforts that are (or, it is argued, should be) taken to reduce the dangerous effects of biological organisms that are on the move. It is used in response to the fears that surround the spread of epidemics and pandemics (like Ebola and Zika), livestock diseases (like foot and mouth disease), crop and tree diseases (like Ash dieback), food borne illnesses (like *E. coli*), threats to infrastructures (or vital systems) and the 'perfect storms' that are generated through inter-system failures that may include promiscuous and exuberant microbial life as one of a number of triggers.

The organisms in question tend to be microorganisms associated with infectious diseases but can also include larger organisms, 'invasive' plants and animals, and disease vectors, like mosquitos, that can carry parasites. Some of these life forms, and particularly viruses, bacteria and fungi, are constantly and rapidly evolving as a result of mutations, responses to their changing environments (including developing resistance to human control), opportunities for mixing with new hosts or other microbial life, or even as a result of modifications developed within laboratories. In this sense, biosecurity is an attempt to manage biological movements in a world that seems less and less biologically stable.

The feared instabilities are often associated with social changes – the human-induced rise in global traffic and intensifying connections (the four 'T's of travel, transport, trade and tourism), and the consequences of accelerated ecological disturbance (leading to greater mixing of people and nonhuman animals). Added to this are the vulnerabilities associated with impoverishment and wars (prompting further habitat change, human displacement and reduced resilience), climate change (affecting host ranges of insects and plants), and the large-scale expansion of animal livestock operations that increase animal movements and multiply the number of bodies that may be susceptible to circulating disease organisms.

## Four approaches to biosecurity

Traditionally, the response to all of this activity and movement has been to attempt to contain, exclude, or 'keep out' dangerous organisms, to keep biological matters in their place. In Table 2.3.1 this is the exclusion model, a biblical and medieval approach involving spatial division and banishment of the ill. It's a model that can still be seen in the control of many livestock diseases, like foot and mouth and avian influenza, where the world is often divided for trade purposes into those countries or regions with endemic problems (often within the Global South) and those where the disease is thought to be controlled or eliminated through so-called modern livestock systems. A geography emerges of so-called diseased and non-diseased spaces that are often coterminous with uneven power relations and access to political resources (Law and Mol, 2008)

*Exclusion* tends to be supplemented and coexist with several other ways of organising or securing the biological. There are, for example, efforts to police and discipline populations such that disease becomes more manageable. In Europe, from the middle ages onwards, efforts were increasingly made to *incorporate*, through registration and ordering, town and city populations as a means to limit the effects of plague. With formalisation of medical interventions around the eighteenth century, including the use of vaccinations, the model was supplemented again to one that not only tried to keep things out, and manage or regulate people, but also develop healthier populations through public health (aka 'normalisation'). Finally, these diagrams have been supplemented again in more recent years by a concern with inter-system vulnerabilities and the feared onset of terrorist, climate, disease and economic disasters that seem to demand their own form of security associated with *emergency measures*, contingency planning and more military-style interventions.

Biosecurity in this sense is clearly wrapped up in the religious, economic and geopolitical debates of the time. These can define who and what is to be secured and how it is to be achieved. The ways in which these various approaches work together, their relative prominence, will depend on the social setting and priorities of the organisations involved.

**Table 2.3.1** Four approaches to biosecurity

| | *Exclusion* | *Incorporation* | *Normalisation* | *Emergency* |
|---|---|---|---|---|
| *Action* | Divides | Organises | Intervenes | Contingency planning |
| *Technique* | Banishment | Quarantine | Vaccination | Scenarios |
| *Organising principles* | Religion/law | Political economy | Public health | Operations management |
| *Power* | Law/ sovereignty | Discipline | Security | Military |

*Source*: After Thacker (2009).

## A geography of biosecurity approaches

In the past two decades, a crude geography of biosecurity may well approximate to the following: a concern with emergencies, triggered by emerging diseases, terrorism and laboratory science dominates in the US, while Europeans have been particularly concerned with food safety (discipline), and Australasians have been most active in the area of exclusion of invasive species (Lakoff and Collier, 2008). Within Asia and Africa, concern has often focused on the threats to livelihoods posed by diseases and trade restrictions, and what some might see as the disproportionate responses of western-based interests in policing emerging diseases while neglecting less mobile but more significant health concerns for people living in impoverished conditions.

There are also important variations in emphasis across different sectors. Bodies like the World Health Organisation (WHO) have tended to advocate emergency preparedness for pandemics, while veterinary and animal health organisations have advanced exclusion of threats (often as a means to conform to international trading rules). International development and food organisations (like for example Food and Agricultural Organization (FAO)), meanwhile, have tended to juggle the concern to modernise or improve production systems with the need to safeguard livelihoods and sustenance. Given these historical and geographic differences, it is useful to map out what factors shape these slightly different versions of biosecurity and to critically examine their effects.

## What shapes biosecurity and its outcomes?

There are four key drivers that seem to shape the mix of biosecurity approaches (exclusion, inclusion, normalisation, emergency) that are implemented (Hinchliffe et al., 2016) (see Figure 2.3.1). Each seems to push biosecurity in different directions. Legal sanction, for example, is a powerful means by which biosecurity is pushed towards the exclusion model. Food businesses, laboratories and hospitals, for example, fear being held responsible for lapses in biosecurity that lead to food scares, release of pathogens or the emergence of medicine resistant microbes. Their model is to pursue hygiene across their organisation. The result can sometimes be paradoxical – the pursuit of control can inadvertently create the conditions for new or unexpected problems. Examples include the tendency of avian influenza and some food borne diseases to flourish in intensive, highly biosecure concentrated livestock operations (Wallace, 2009). Another would be the tendency of anti-microbial treatments to produce resistant organisms and contribute to the current threat of antimicrobial resistance (Landecker, 2015).

If legal sanction pulls biosecurity towards exclusion, then biological emergence tends to drive us to the opposite end of Table 2.3.1, to emergency planning. So the WHO's *Safer Planet* (World Health Organisation, 2007) for example, argued that stopping disease at borders was no longer feasible in a highly mobile and biologically promiscuous world. The emergence of new diseases as well as globalisation rendered control and exclusion nonsensical. Instead, nation states needed to be prepared for unpredictable though highly likely disease events. Preparation included priming health and emergency services for the next pandemic.

The degree of exclusion or prepared openness are in turn shaped by two current modes of economic and political organisation (security and liberalism). Biosecurity is

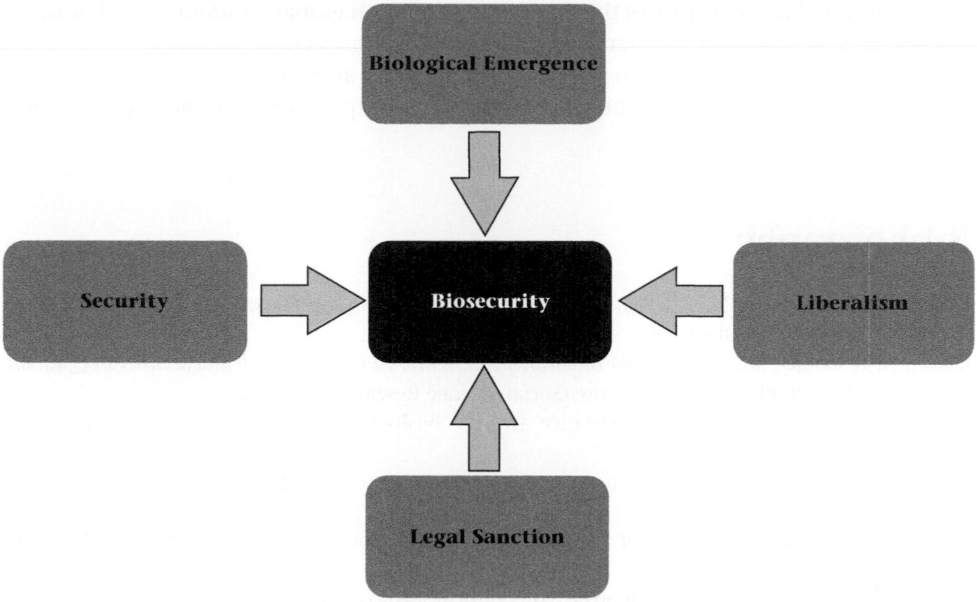


**Figure 2.3.1** The four dimensions of biosecurity.

clearly related to current understandings of what it means to be secure. The latter is not easily reducible to defence of existing borders and tends to speak to a much broader, expanding and relentless set of activities. Security in this sense can tend to involve the regulation of lives and of the networks that connect those lives to others (and into the future). So securing freedoms to trade, or to move, often involve attempts to minimise the risks that such movements can incur. This is not so much exclusion, or interdiction, but the detailed management of activities to desired ends (air travel is a good example) and a tendency to anticipate and pre-empt risk.

Finally, all of this takes place within what many characterise as liberal or neoliberal approaches to governance. The latter tend to be less concerned with state provision and sanction and more oriented to the expansion of circulation (or traded goods), the 'responsibilisation' of private sector actors and the marketisation of risk and its management. The result tends to be a heady mix of logics that attempt to reduce state responsibility at the same time as playing up bio-social uncertainties and insecurities in order to generate greater powers to develop more business opportunities for security. As was the case for the legal sanction approach, the effects can be paradoxical. Pursuing biosecurity by pre-empting the emergence of new diseases tends for example to create new problems of controlling scientific information. Indeed, it could be said that biosecurity exists to manage the risks created by its existence.

## Learning resources

A very good introduction to biosecurity is Dobson, A., Barker, K. & Taylor, S. L. (eds) *Biosecurity: The socio-politics of invasive species and infectious diseases*. London and New York: Routledge.

A fictional film that captures the fears associated with global pandemics and biosecurity is *Contagion* (Warner Brothers) http://www.warnerbros.com/contagion.

A game that introduces you to the public health and other security issues associated with emergent microbes is *Plague Inc.* http://www.ndemiccreations.com/en/22-plague-inc

## Bibliography

Hinchliffe, S., Bingham, N., Allen, J. & Carter, S. (2016) *Pathological Lives: Disease, Space and Biopolitics*, London: Wiley Blackwell.

Lakoff, A. & Collier, S.J. (eds) (2008) *Biosecurity Interventions: Global Health and Security in Question*, New York: Columbia University Press/Social Science Research Council.

Landecker, H. (2015) 'Antibiotic resistance and the biology of history', *Body and Society*, 22(4): 19–52.

Law, J. & Mol, A. (2008) 'Globalisation in practice: On the politics of boiling pigswill', *Geoforum*, 39: 133–143.

Thacker, E. (2009). 'The shadows of atheology: Epidemics, power and life after Foucault', *Theory, Culture and Society*, 26: 134–152.

Wallace, R.G. (2009) 'Breeding influenza: The political virology of offshore farming', *Antipode*, 41: 916–951.

World Health Organisation (2007) *The World Health Report 2007 – A safer future: Global public health security in the 21st century*. Geneva, Switzerland: World Health Organisation.

# Corporate environmental responsibility 2.4

## Christopher Wright

The environmental challenges now facing humanity are profound in their scope and complexity. Climate change, ocean acidification, biodiversity decline, and disruptions to the nitrogen and phosphorus cycles all illustrate how modern human civilization is now challenging basic planetary boundaries (Steffen et al., 2015). Business corporations are central to this process. As the engines of the modern global economy, large businesses underpin the production and consumption of an ever-distending cornucopia of products and services. They play a dominant role in an economic system driven by the need for infinite growth and fossil-fuel based energy (Wright and Nyberg, 2015). However, the environmental damage resulting from industrialization has also resulted in social criticism, political mobilization and government regulation. In response, a new business narrative has emerged which promotes an image of business corporations as environmentally responsible actors who can harness their innovative capacities in ways that deliver both returns to shareholders, as well as environmental benefits. Indeed, business corporations are increasingly seen as central to humanity's response to 'grand challenges' such as climate change. This chapter explores the concept of corporate environmental responsibility by firstly examining its origins, what this has involved in practice, and then reviewing normative and more critical accounts of its impact in an era of climate crisis.

## The origins of corporate environmental responsibility

Social concern about environmental decline in response to industrialization has a long history. Although this led to limited regulatory constraints on business and the emergence of early conservation movements in developed economies, in most countries the unfettered power of private firms to exploit nature went largely unchallenged for much of the twentieth century (Guha, 2000). During the 1960s and 1970s this dominant view was challenged by a wave of environmental critique highlighting a deeper questioning of industry's impact upon nature. Symbolic of the shift was zoologist Rachel Carson's influential book *Silent Spring* (1962), which revealed the devastating environmental impact of pesticides and humanity's vulnerability within nature. Environmental awareness duly gathered pace, signified through events such as the first Earth Day in 1970, the publication of the Club of Rome's *The Limits to Growth* (Meadows et al., 1972), and growing media focus on industrial accidents and environmental catastrophes. The emergence of 'Green' political parties in Europe, Australia and elsewhere further underscored the growing environmental critique of industrialization (Dunlap and Mertig, 1992; Jermier et al., 2006).

During the 1970s many countries initiated regulatory reforms which challenged industry's environmental impact. In the US this included legislation such as the National Environmental Policy Act (1970), the Clean Air Act (1970) and the Clean Water Act (1972). In response to these regulatory pressures, many large businesses adopted a more strategic approach to the environment that extended beyond legislative compliance. As Hoffman (2001) argues, this involved an emerging view that businesses could be responsible custodians of the natural environment and moreover, that environmental protection could also deliver business benefits (Porter and van der Linde, 1995). The institutionalization of 'corporate environmentalism' was evident in the development of new capabilities within businesses around concepts such as pollution prevention, industrial ecology, life-cycle analysis, carbon footprinting, and most recently, business sustainability (Hoffman and Bansal, 2012).

## Corporate environmental responsibility in practice

Jermier et al. (2006) argue that the rhetoric of corporate environmentalism prioritises the central role of business in achieving economic growth *and* ecological benefit (a so-called 'win-win' outcome) through voluntary actions which go beyond regulatory requirements. Indeed, related concepts such as 'creating shared value' (Porter and Kramer, 2011) promote a vision of global corporations as 'saviours'; reinventing capitalism and providing environmental and social benefit through innovation and economic growth. Here, there are links to the broader idea of ecological modernization, which argues that increasing economic development, technological innovation and environmental reform have minimised pollution and environmental harm over time (Hajer, 1995; Mol, 2002).

Within large corporations the uptake of corporate environmentalism has become increasingly evident over the last 20 years in response to growing social and political concern over climate change. This has included businesses seeking to minimize their

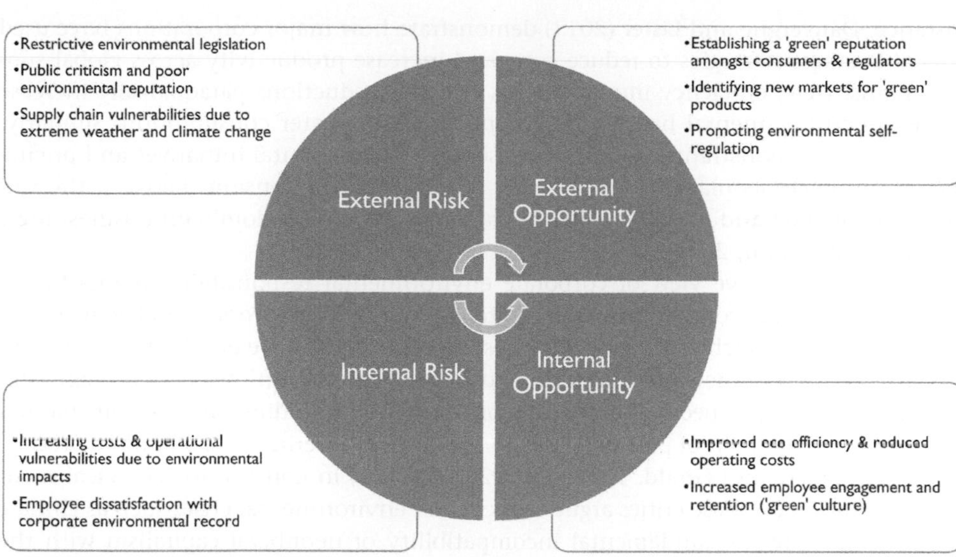


**Figure 2.4.1** Business risks and opportunities of corporate environmentalism.

*Source*: Based on Wright, C. and Nyberg, D. (2015) *Climate Change, Capitalism and Corporations: Processes of Creative Self-Destruction*. Cambridge: Cambridge University Press, pp.14–23.

exposure to regulatory and physical climate risks, as well as identifying new market opportunities (Hoffman, 2005; Lash and Wellington, 2007). The practice of corporate environmentalism is diverse including both internally and externally-focused initiatives. These include: reducing operational costs through improved eco-efficiency and waste reduction; promoting 'green' and environmentally responsible products and services; marketing and branding as 'green' companies to consumers and the public; and developing corporate cultures and engaging employees around environmental activities (Dauvergne and Lister, 2013; Wright and Nyberg, 2015). Corporate environmentalism also extends to political action in lobbying government over environmental regulations, and developing political coalitions that promote the vision of business as environmentally and economically progressive (Nyberg et al., 2013). As evident in companies such as GE, Walmart, Virgin, Nestlé and Unilever, corporate environmentalism thus presents an image of business leadership on the environment, and promotes the merits of corporate self-regulation, innovation and technology in solving key environmental challenges (Esty and Winston, 2006; Hart, 1997).

## Critiques of corporate environmentalism

However, critics argue that the practice of corporate environmentalism often fails to live up to its pro-environmental rhetoric (Banerjee, 2001). At one level, critics point to the potential for corporate environmental claims to amount to little more than 'greenwashing'; fictitious and misleading marketing which masks a darker reality of environmental degradation (Ramus and Montiel, 2005). More substantively, others point to the way in which corporate environmentalism extends beyond rhetoric and involves practices which incorporate the environment more completely within business activities. For instance, Dauvergne and Lister (2013) demonstrate how major corporations have used 'green' business strategies to reduce costs and increase productivity across global supply chains. Here, efficiency improvements and cost reductions paradoxically increase aggregate environmental harm through encouraging greater consumption. Similarly, the focus on demonstrating a 'business case' for environmental initiatives and pricing natural resources as market commodities, reveals how the tensions between the values of the market and nature are all but erased in favour of a dominant business logic (Nyberg and Wright, 2013).

This more negative view of corporate environmental responsibility points to the more fundamental conflict between capitalism and the environment. For instance, ecosocialist critics such as Foster (1999; Foster et al., 2010) have emphasized how capitalism as an economic system has resulted in a 'metabolic rift' between humans and nature. From this perspective capitalism continually commodifies and exploits natural resources in the pursuit of profit and economic growth; a veritable 'treadmill of production' (Schnaiberg and Gould, 1994; York, 2004). Thus, in contrast to ecological modernist perspectives, these critics argue that current environmental crises such as climate change, illustrate the fundamental incompatibility of neoliberal capitalism with the sustainability of nature and human society. Rather than dramatically reducing environmental impacts, corporate environmentalism provides cover for an economic system busily focused on 'extractivism' and the exploitation of new carbon frontiers such as tar sands, gas fracking and deep-water and Arctic oil drilling (Klein, 2014).

From this perspective, corporate environmental responsibility is viewed less as a normative model for pro-environmental action, and more as a form of ideological justification in the increasingly contested politics surrounding climate and environment (Levy, 1997). By presenting an image of the business corporation as a force for good, the rhetoric and practice of corporate environmental responsibility aims to forestall social criticism, more stringent regulation of business activities, and the possibilities for radical change. Pro-environmental actions are thus constrained within system-compliant boundaries. In an era of 'creative self-destruction', the human response to existential threat is thus limited within 'business as usual' trajectories (Wright and Nyberg, 2015).

## Learning resources

A useful introduction to the way in which patterns of production and consumption underpin environmental destruction is the video *The Story of Stuff*: http://storyofstuff.org/movies/story-of-stuff/ There is an associated website which provides further details on specific problems like the privatization of water and micro-bead pollution, as well as solutions to environmental problems here: http://storyofstuff.org/

Ecomagination: GE's website outlining its focus on creating profit by responding to environmental problems. A good example of corporate environmentalism in practice: https://www.ge.com/about-us/ecomagination

This Changes Everything: a website by Canadian journalist and activist Naomi Klein including links to her best-selling book on climate change and capitalism and the associated film: https://thischangeseverything.org/

You can listen to a podcast interview with Naomi Klein on the topic of the climate crisis and business here: https://theconversation.com/speaking-with-naomi-klein-on-capitalism-and-climate-change-47037

Climate, People & Organizations: a website that explores the role of corporations in shaping the social, organizational and political implications of climate change https://climatepeopleorg.wordpress.com/

350.org: a leading environmental organization focused on ending our reliance on fossil fuels through social mobilisation: https://350.org/

## Bibliography


Banerjee, S.B. (2001) 'Managerial Perceptions of Corporate Environmentalism: Interpretations from Industry and Strategic Implications for Organizations', *Journal of Management Studies*, 38(4): 489–513.

Carson, R. (1962) *Silent Spring*. Boston: Houghton Mifflin.

Dauvergne, P. & Lister, J. (2013) *Eco-Business: A Big-Brand Takeover of Sustainability*. Cambridge, MA: MIT Press.

Dunlap, R.E. & Mertig, A.G. (1992) *American Environmentalism: The US Environmental Movement, 1970–1990*. Washington, DC: Taylor & Francis.

Esty, D.C. & Winston, A.S. (2006) *Green to Gold: How Smart Companies Use Environmental Strategy to Innovate, Create Value, and Build Competitive Advantage*. Hoboken, NJ: John Wiley & Sons.

Foster, J.B. (1999) 'Marx's Theory of Metabolic Rift: Classical Foundations for Environmental Sociology', *American Journal of Sociology*, 105(2): 366–405.

Foster, J.B., Clark, B. & York, R. (2010) *The Ecological Rift: Capitalism's War on the Earth*. New York: Monthly Review Press.

Guha, R. (2000) *Environmentalism: A Global History*. New York: Longman.

Hajer, M. (1995) *The Politics of Environmental Discourse: Ecological Modernization and the Policy Process*. Oxford: Oxford University Press.

Hart, S.L. (1997) 'Beyond Greening: Strategies for a Sustainable World', *Harvard Business Review*, 75(1): 66–77.

Hoffman, A.J. (2001) *From Heresy to Dogma: An Institutional History of Corporate Environmentalism*. Stanford, CA: Stanford University Press.

Hoffman, A.J. (2005) 'Climate Change Strategy: The Business Logic Behind Voluntary Greenhouse Gas Reductions', *California Management Review*, 47(3): 21–46.

Hoffman, A.J. & Bansal, P. (2012) 'Retrospective, Perspective, and Prospective: Introduction to the Oxford Handbook on Business and the Natural Environment', in Bansal, P. & Hoffman, A.J. (eds) *The Oxford Handbook of Business and the Natural Environment* (pp. 3–25). Oxford: Oxford University Press.

Jermier, J.M., Forbes, L.C., Benn, S. & Orsato, R.J. (2006) 'The New Corporate Environmentalism and Green Politics', in S.R. Clegg, C. Hardy, T.B. Lawrence & W.R. Nord (eds) *The Sage Handbook of Organization Studies* (pp. 618–50). London: Sage.

Klein, N. (2014) *This Changes Everything: Capitalism Vs. The Climate*. New York: Simon & Schuster.

Lash, J. & Wellington, F. (2007) 'Competitive Advantage on a Warming Planet', *Harvard Business Review*, 85(3): 94–102.

Levy, D.L. (1997) 'Environmental Management as Political Sustainability', *Organization & Environment*, 10(2): 126–47.

Meadows, D.H., Meadows, D.L., Randers, J. & Behrens III, W.W. (1972) *The Limits to Growth: A Report for the Club of Rome's Project on the Predicament of Mankind*. New York: Potomac Associates.

Mol, A.P.J. (2002) 'Ecological Modernization and the Global Economy', *Global Environmental Politics* 2(2): 92–115.

Nyberg, D. & Wright, C. (2013) 'Corporate Corruption of the Environment: Sustainability as a Process of Compromise', *The British Journal of Sociology*, 64(3): 405–24.

Nyberg, D., Spicer, A. & Wright, C. (2013) 'Incorporating Citizens: Corporate Political Engagement with Climate Change in Australia', *Organization*, 20(3): 433–53.

Porter, M.E. & Kramer, M.R. (2011) 'Creating Shared Value', *Harvard Business Review*, 89(1/2): 62–77.

Porter, M.E. & van der Linde, C. (1995) 'Green and Competitive: Ending the Stalemate', *Harvard Business Review*, 73(5): 120–34.

Ramus, C.A. & Montiel, I. (2005) 'When Are Corporate Environmental Policies a Form of Greenwashing?', *Business & Society*, 44(4): 377–414.

Schnaiberg, A. & Gould, K.A. (1994) *Environment and Society: The Enduring Conflict*. New York: St. Martin's.

Steffen, W., Richardson, K., Rockström, J., Cornell, S.E., Fetzer, I., Bennett, E.M., Biggs, R., Carpenter, S.R., de Vries, W., de Wit, C.A., Folke, C., Gerten, D., Heinke, J., Mace, G.M., Persson, L.M., Ramanathan, V., Reyers, B. & Sörlin, S. (2015) 'Planetary Boundaries: Guiding Human Development on a Changing Planet', *Science* 347(6223): 1259855.

Wright, C. & Nyberg, D. (2015) *Climate Change, Capitalism and Corporations: Processes of Creative Self-Destruction*. Cambridge: Cambridge University Press.

York, R. (2004) 'The Treadmill of (Diversifying) Production', *Organization & Environment*, 17(3): 355–62.

# The Earth System

2.5

## Clive Hamilton

Only in the last three of four decades have scientists begun to think of the Earth as a 'system'. Systems thinking emerged in the 1950s and 1960s as a better way to understand certain engineering and social problems. It was a reaction against *reductionism*, which separates (or 'reduces') the thing being studied into its component parts. A systems approach focuses instead on the relationships between the components. In short, the whole is more (a lot more) than the sum of its parts. We can't do without anatomy, but anatomy cannot tell us how a human being works.

Reductionism is especially inadequate when trying to understand how natural systems, including the Earth as a whole, function. Natural systems are characterized by the continued recycling of materials, powered by external energy sources (ultimately the Sun), which means they are in constant transformation (Langmuir and Broecker, 2012: 20–22). In fact, the Earth can be thought of not as a thing or an object but as an entity composed entirely of processes.

Natural systems embody *feedback mechanisms*: when a change in one element of the system affects another it then 'feeds back' to change the first (Steffen et al., 2004: 258). That change then affects the second again, and so on. Feedback responses can be 'negative', so that they dampen down the interactions, or positive, so that they amplify them, leading to a runaway process unless something else intervenes. As an example of a positive feedback effect, Arctic sea-ice reflects solar radiation back into space because it is white, but as the Earth warms the sea-ice melts and the newly exposed dark-coloured ocean absorbs more heat from the Sun, warming the planet further and melting more ice.

Natural systems also have *emergent properties*, that is, properties that belong to the system but cannot be found in any individual element of it. They evade all cause-and-effect explanations. The unpredictability of emergent properties means the future evolution of natural systems always holds surprises.

Systems thinking applied to the environment goes back to the 1930s, but it was only in the 1990s that it began to be applied to the Earth as a whole, with much of the work centered around the International Geosphere-Biosphere Programme, formed in 1986 and based in Stockholm. The new science of the Earth System grew from a number of events, including the creation of global observation systems and databases that suggested planet-wide linkages, and the enhanced computing power capable of seeing the patterns (Steffen et al., 2004). More abstractly, the 'blue planet' images of the Earth taken in the 1970s by the Apollo space missions were highly suggestive, as was James Lovelock's Gaia hypotheses that characterized the Earth as a single living organism (Lenton, 2016: 5).

More importantly, in the 1980s scientists became increasingly concerned about the impact of human activity on the planet as a whole, beyond the damage being done to local or regional environments. Evidence of the dangers to life from the expanding hole in the stratosphere's ozone layer and of global warming due to burning fossil fuels led scientists to build global monitoring networks and to explore the interactions of the various components of planetary processes through elaborate models that attempted to capture the connections between the major processes governing the Earth.

In the case of global warming, it was becoming apparent that human-sources of greenhouse gases were not just warming the atmosphere, but also causing acidification of the oceans, melting of ice masses, changes in vegetation cover and extinction of species. In time, geologists would begin to realize that climate change is also affecting the ground beneath our feet through landslides, seismic shocks and even volcanic activity (McGuire, 2012). In short, the Earth System as a whole is disrupted.

## The secrets of ice cores

The most decisive evidence for the idea that the Earth behaves as a single, complex, dynamic entity emerged in 1999 with the Vostok ice core drilled in Antarctica. The 4-kilometre ice core revealed the history of the Earth System going back over 420,000 years. Close analysis showed the Earth operating as a single system, with three especially noteworthy features (Steffen et al., 2015: 3).

1. It was revealed that the Earth's average global temperature and the concentration of carbon dioxide in the atmosphere, both planet-wide phenomena, move in tandem (driven mainly by changes in the way the Earth moves around the Sun).
2. The arrival of glacial conditions and warm (inter-glacial) periods occur on a regular cyclical pattern, telling us that the entire Earth moves through major phases that affected *all* of its components—atmosphere, oceans, land cover, biota. It is also prone to very abrupt shifts in its functioning as positive feedback effects kick in.
3. Despite these abrupt and dramatic changes, the temperature of the Earth has remained within distinct upper and lower bounds, which means that the Earth displays 'a high degree of self-regulation'. In other words, it functions as a single entity.

These insights into the deep history of the planet confirmed the conception of the Earth as a complex, dynamic, self-regulating totality—as the Earth System. One of the more remarkable insights into how it functions is that life itself has been intimately involved in the evolution of the Earth System. Perhaps the two most spectacular instances have been the role of cyanobacteria in creating an atmosphere rich in oxygen

2.3 billion years ago, and the role of humans over the last 70 years in bringing on, with unprecedented speed, a warming episode.

All of this raises the question of where the Earth System starts and finishes. The Earth System extends outwards to the limit of the Earth's atmosphere, and perhaps includes the Moon, whose influence is felt on the tides. The lower boundary may be placed at the upper layer of the thin crust on its surface (known as the lithosphere) that life depends on. However, this definition adopts a relatively short-term perspective. The lithosphere is divided into plates that are in constant movement, and these continental shifts are related to movements in the mantle beneath. In deep time the crust of the Earth changes with the movement of tectonic plates and they in turn are influenced by the movements in the Earth's mantle and indeed by its hot, dense inner core (Langmuir and Broecker, 2012: Chapter 11). The source of some volcanic eruptions (which can transform the Earth's atmosphere, at times radically) lies in the boundary between the inner and outer mantles. So in deep time the Earth System may be said to begin at the centre of the planet and extend to the top of the atmosphere or out to the Moon.

## A phase-shift in the Earth System

Constant movement and feedback effects mean that natural systems behave in a 'non-linear' way, that is, a small change in one variable can give rise to very large changes in others. In some circumstances, the system can cross a *tipping point* to flip it from one state to another as a result of a quite small perturbation. This has happened often in the history of the Earth, plunging the planet into a deep freeze or driving it back up into a hot phase.

This ability to shift, often abruptly, from one state to a quite different one means that natural systems are generally not in a state of equilibrium in which the various forces balance each other out keeping the system stable with no tendency to change. They are better described as existing in a state of steady-state disequilibrium, 'a balance of forces and fluxes . . . where the natural system remains within narrow bounds, perched at a disequilibrium state' (Langmuir and Broecker, 2012: 16). That is, until something happens that tips them into a new and quite different state. Earth System scientists believe that anthropogenic warming may shift the Earth System into a new state this century.

The idea of the Anthropocene, first put forward in 2000, was not possible before systems thinking was applied to the Earth as a whole. The new geological epoch is defined by the fact that the 'human imprint on the global environment has now become so large and active that it rivals some of the great forces of Nature in its impact on the functioning of the Earth system' (Steffen et al., 2011: 843). Before the development of the concept of the Earth as a complex, dynamic system it was not possible to speak of the 'functioning' of the Earth system. Various attempts to locate the onset of the new epoch several thousand or even several hundred years ago (such as in 1610) fail the basic test of being able to show that anthropogenic forces had changed the way the Earth System functions in a way that is detectable (Hamilton, 2016).

## Learning resources

The website of the (now-closed) International Geosphere-Biosphere Programme is an excellent resource to explore. See http://www.igbp.net.

Another is a website maintained by Carleton University http://serc.carleton.edu/introgeo/earthsystem/index.html.

## Bibliography

Hamilton, C. and Grinevald, J. (2015) 'Was the Anthropocene anticipated', *The Anthropocene Review*, 2(1): 59–72.

Hamilton, C. (2016) 'The Anthropocene as rupture', *The Anthropocene Review*, 3(2): 93–106.

Langmuir, C. and Broecker, W. (2012) *How to Build a Habitable Planet: The Story of Earth from Big Bang to Humankind* (revised and expanded edition), Princeton: Princeton University Press.

Lenton, T. (2016) *Earth System Science: A Very Short Introduction*, Oxford: Oxford University Press.

McGuire, B. (2012) *Waking the Giant*, Oxford: Oxford University Press.

Steffen, W., Sanderson, A., Tyson, P.D., Jäger, J., Matson, P.A., Moore III, B., Oldfield, F., Richardson, K., Schellnhuber, H.J., Turner II, B.L. et al. (2004) *Global Change and the Earth System: A Planet Under Pressure*, Berlin: Springer.

Steffen, W., Grinevald, J., Crutzen, P. and McNeil, J. (2011) 'The Anthropocene: conceptual and historical perspectives'. *Philosophical Transactions of The Royal Society A*, 369: 842–67.

Steffen W. et al. (2015) The Trajectory of the Anthropocene: The Great Acceleration. *The Anthropocene Review* 2(1): 81–98.

# Ecosystem services

2.6

Daniel Chiu Suarez and
Jessica Dempsey

## What are ecosystem services?

Across virtually all indicators, biodiversity loss continues to worsen on a planetary scale (Secretariat of the Convention on Biological Diversity, 2014). From species extinction and habitat destruction to overharvesting and climate change, the situation is dire. This growing sense of socio-ecological crisis and widespread institutional failure has prompted many in the environmental movement to look with renewed urgency toward new ideas, new strategies, and ultimately a new way forward for biodiversity conservation. Over the past two decades, one such framework has been vigorously promoted as offering this new way forward: ecosystem services.

The Millennium Ecosystem Assessment defines ecosystem services simply as 'the benefits people obtain from ecosystems' (MA, 2005). The concept expresses in measurable, often monetary terms the economic values encompassed in biodiversity: mangos, clean drinking water, weekend hikes, flood protection, pharmaceuticals, two-by-fours, and so on ad infinitum. In a set of foundational reports authored by over 1,300 experts from around the world, the MA sorted these 'services' into what became a classic four-category classification: (i) *provisioning services* such as food, freshwater, and timber, (ii) *regulating services* such as climate, flood, and disease attenuation, (iii) *supporting services* such as nutrient cycling, primary productivity, and soil formation, and (iv) *cultural services* such as educational, aesthetic, or spiritual fulfilment (see Figure 2.6.1). While typologies for ecosystem services vary, they all express a particular way of making sense of human–non-human relationships—a framework for systematizing and quantifying the many benefits that people derive from nature. They articulate a strategy for making nature *valuable*, recognizably worthwhile to society, and investable by the state and finance.

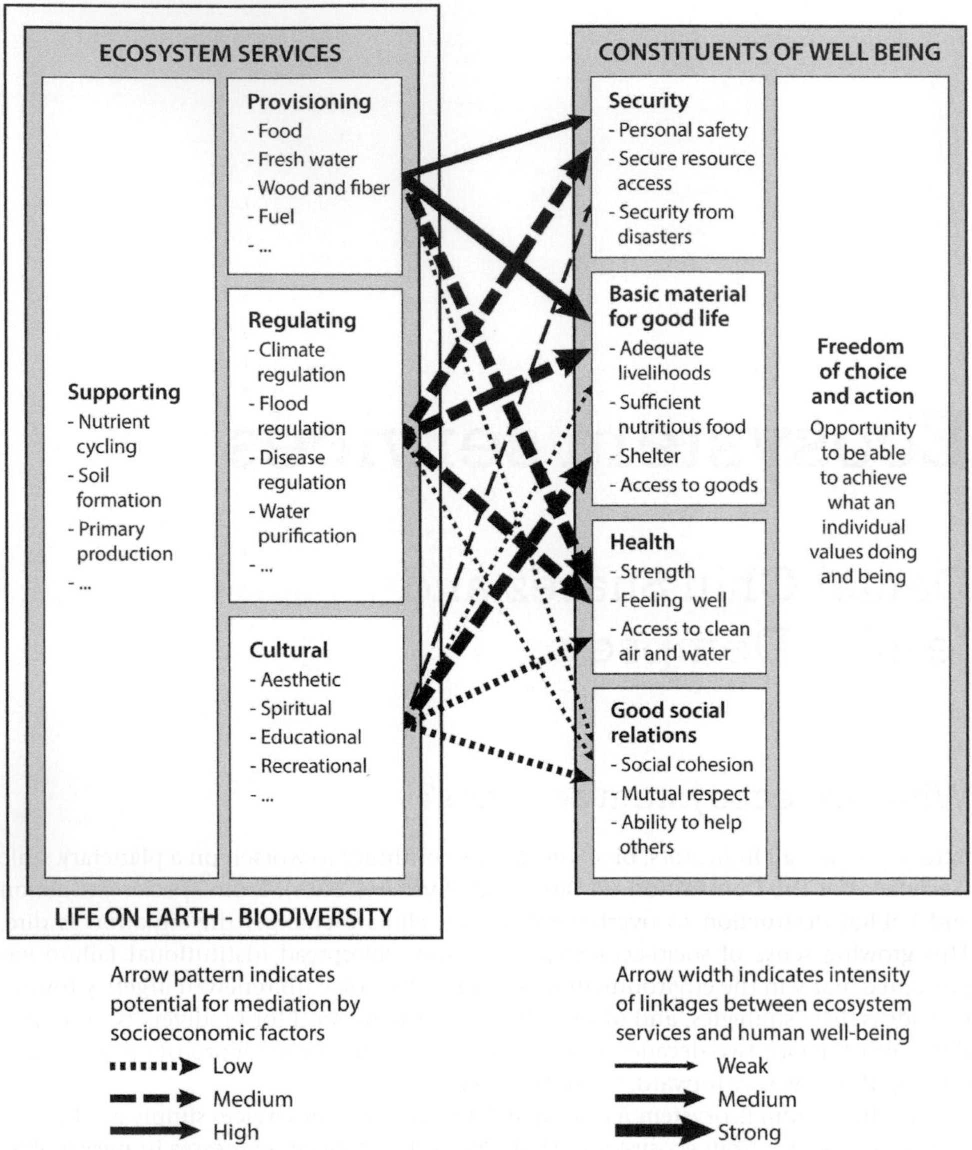


**Figure 2.6.1** Model of ecosystem service framework employed by the Millennium Ecosystem Assessment



The concept has, since its inception, been subject to vigorous debate and confronted its researchers and practitioners with an array of definitional, methodological and theoretical challenges, particularly related to valuation (for overview see Dempsey and Robertson, 2012). It is a fundamentally slippery concept. Moreover, it has become a site of contestation. As with other environmental discourses, such as 'wilderness' or 'biodiversity', the idea of ecosystem services prescribes norms for environmental governance: it re-shapes how we understand what nature means, what it is worth, and how we ought to manage it, with potentially far-reaching implications for how we structure present and future socio-ecological relationships.

## The ecosystem services gambit

The central question embedded in ecosystem services—i.e. *how much is nature worth?*—has increasingly come to preoccupy the organized conservation movement. While proponents of ecosystem services express a variety of aspirations for the concept, they typically highlight some combination of three main functions performed by ecosystem services valuation (TEEB, 2010; Suarez & Corson, 2013): (a) *recognizing* value, thereby legitimating and strengthening rationales for protecting nature, (b) *demonstrating* value, to provide estimates for supporting rational decision-making frameworks such as cost-benefit analysis, and (c) *capturing* value, by operationalizing new policy instruments, institutional arrangements, and various mechanisms for managing nature such as payment for ecosystem services (PES) programs.

For its supporters, the idea of ecosystem services promises new arguments (e.g. the 'business case for nature', in contrast to ethical appeals), new allies (e.g. powerful constituencies amenable to market discourse), new resources (e.g. public and private conservation finance streams enabled by ecosystem services arrangements), and a powerful overarching framework for aligning conservation with the multiple, competing priorities of sustainable development. By translating what nature does into economic language, proponents of ecosystem services endeavour to make previously 'invisible', taken-for-granted environmental costs and benefits related to major societal decisions finally 'visible' (TEEB, 2010). In the absence of such values, they suggest, nature is implicitly assigned one by default: zero. Thus, properly accounting for nature—in other words, bringing biodiversity into the fold of decision-making realms typically dominated by narrow economic considerations—requires that nature, too, be made economically legible and considered worthwhile in commensurable terms. According to this reasoning, such commensurability allows decision-makers to more accurately parse 'trade-offs' between different land-use choices: a standing mangrove forest, for example, provides greater quantified ecosystem service benefits through flood protection and as a fish nursery compared with the short-term revenues and environmental costs of building a shrimp farm over it.

In 1997, researchers famously applied this logic to the global scale by estimating the total economic value of the world's ecosystems—from wetlands and tropical forests to tundra and coral reefs—at US $33 trillion: nearly double global economic output that year (Costanza et al., 1997). The publication of this widely cited figure, followed by the major international undertaking of the Millennium Ecosystem Assessment (2001–2005), heralded exponential growth in ecosystem services research (Figure 2.6.2).

Understanding this graph (Figure 2.6.2), which charts the rise of ecosystem services, requires situating the concept in its wider historical, cultural, and political-economic context. Throughout the 1980s and 1990s, governments, particularly in the United States and United Kingdom, began a dramatic and well-documented 'neoliberal' shift toward markets as preferred mechanisms for addressing a wide range of governance questions including in the domains of environmental policy and management. During this period, the United States innovated market-based pollution trading to address acid rain, developed wetland banking to offset environmental impacts from development, and, most famously, advocated for the inclusion of carbon markets in the 1997 Kyoto

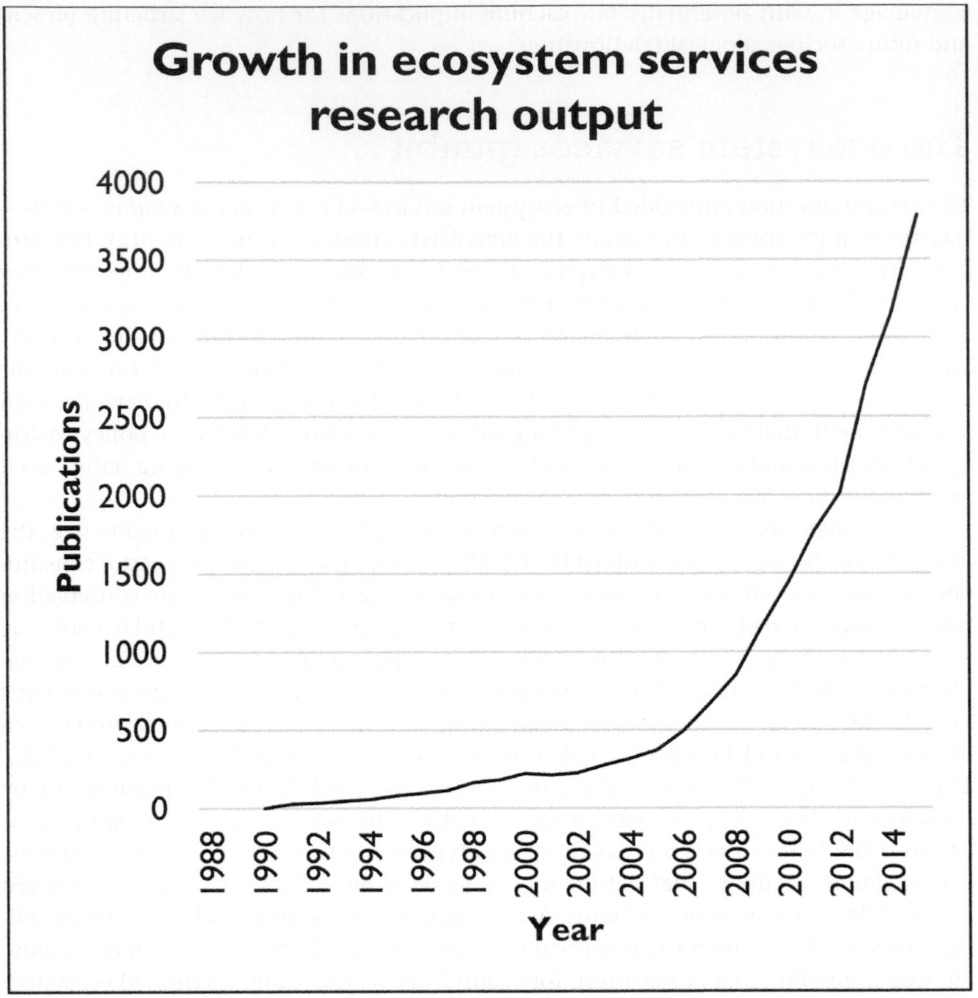


**Figure 2.6.2** Graph showing yearly record counts of publications catalogued in Web of Science using the topics 'ecosystem services' or 'ecological services'. Research output dates back to 1977, increases in the 1990s, and accelerates after 2005.

Protocol as a means of pursuing emissions reductions. The turn to ecosystem services emerges from and contributes to this broader trajectory.

Today, as conservationists Redford and Adams (2009: 785) write, the concept of ecosystem services now 'increasingly structures the way conservationists think, the ways they explain the importance of nature to often sceptical policy makers, and the ways they propose to promote its conservation.' The emergence of ecosystem services has sustained, and been sustained by, a diverse and rapidly growing community of research and practice disseminating its discourse. Ecologists and economists, policymakers, activists of varying stripes, governmental and intergovernmental bureaucrats, celebrities, business leaders, and many others have joined together to re-imagine 'nature' as a stock of 'natural capital' assets generating flows of valuable services. One can watch Hollywood actor Adrian Grenier (of *Entourage* fame) explain the 'hidden value in nature' in a short video (https://www.youtube.com/watch?v=qcbDdId-6vw) or play the 'eco earner' (http://eco-earner.com/) video game which aims to educate people about ecosystem services. The concept has crystallized in a range of institutionalized forms including dozens of PES programs; natural capital accounting standards (e.g. the UN's System for Environmental-Economic Accounting, SEEA); private sector coalitions (e.g. the business-led Natural Capital Protocol); new academic journals (e.g. *Ecosystem Services*); and a proliferation of new research initiatives and communities of practice (e.g. the Natural Capital Project, Ecosystem Services Partnership, and many others).

## The ecosystem services road to ruin

The economistic nature of ecosystem services, along with its conspicuously market-oriented connotations and applications, have provoked tremendous debate among conservationists and academics. Critics have introduced a diverse broadside of methodological, ethical, political, and strategic concerns about the concept. Some see ecosystem services as a dangerously narrow re-conceptualization of the aims of biodiversity conservation, which, as one critic argued, 'is to imply—intentionally or otherwise—that nature is only worth conserving when it can be made profitable' (McCauley, 2006: 28). The idea, these critics emphasize, threatens to brush aside intrinsic values, ethical duties, and the sense of aesthetic or spiritual connection to living things historically ingrained in conservation's ideals. Others argue that the concept fails to adequately capture the often difficult-to-quantify complexities of ecosystems (Norgaard, 2010).

In broader political terms, critics point to the colonizing conceit of the notion, adding it to a longer list of concepts that Westerners and elites have tried to apply to cultures and communities who have their own ways of understanding human–nonhuman relations and practicing land and water management (e.g. Sullivan, 2009). The political-economic implications of ecosystem services have also been subject to intense criticism from analysts and activists who disparage the concept as complicit with the commodification of nature and ongoing neoliberalization of environmental governance (e.g. Robertson, 2012; IPMSDL, 2013). From such a perspective, ecosystem services seem not only somewhat but *archetypally* neoliberal: preoccupied with extending economistic, market-mediated relations to any and all aspects of nature. In this way, critics emphasize how ecosystem services both reflects and serves to reinforce the continuing subordination of conservation to a hegemonic discursive and political economic order. Thus, conservationists are now compelled to speak in economic language, think in market categories, and create arrangements that either do not challenge or outright perform capitalist imperatives—arguably, the very processes implicated in driving environmental degradation and biodiversity loss.

## So, what *are* ecosystem services?

Clearly, ecosystem services have come to represent many *different* things, variously auspicious and ominous. In this respect, ecosystem services represents a kind of *chimera*: a shifting, hybrid reflection of the specific activity of the various actors envisioning it, interpreting it, developing it, using it, instituting it, responding to it, and contesting it. Its meaning remains subject to continuing re-negotiation, its form and function contingent 'on the hands through which the concept and policies pass' (Dempsey and Robertson, 2012: 760).

## Learning resources

The recent emergence of ecosystem services as a 'Big Idea' in conservation and development has been attended not only by considerable research output but also by a proliferation of publicly disseminated media, a sample of which is provided below. These clips, variously critical and supportive of the idea, provide revealing glimpses into the kinds of narratives, policy discourses, and strategic sensibilities that have characterized debates around ecosystem services.

Pavan Sukhdev TED talk: 'What's the price of nature?' https://www.ted.com/talks/pavan_sukhdev_what_s_the_price_of_nature?language=en

George Monbiot: 'Put a price on nature? We must stop this neoliberal road to ruin' https://www.theguardian.com/environment/georgemonbiot/2014/jul/24/price-nature-neoliberal-capital-road-ruin

Sir David Attenborough: 'Forests: Nature at Your Service' https://www.youtube.com/watch?v=BCH1Gre3Mg0

Natural Capital Project: 'Big Question: What Is Nature Worth?' https://www.youtube.com/watch?v=TartoYpK1yI

OPERAs Project: 'Ecosystem Services in brief' https://www.youtube.com/watch?v=Y2KdM9zoF8E

California Academy of Sciences: 'Ecosystem Services' https://www.youtube.com/watch?v=BCH1Gre3Mg0

## Bibliography

Costanza, R., d'Arge, R., de Groot, R., Farber, S., Grasso, M., Hannon, B., Limberg, K., Naeem, S., O'Neill, R.V., Paruelo, J. et al. (1997) 'The value of the world's ecosystem services and natural capital', *Nature*, 387(6630): 253–260.

Dempsey, J. & Robertson, M. (2012) 'Ecosystem services: Tensions, impurities, and points of engagement within neoliberalism', *Progress in Human Geography*, 36: 758–779.

The International Indigenous Peoples Movement for Self-Determination and Liberation (IPMSDL) (2013) Declaration of Indigenous Peoples on the World Trade Organization. Available at http://ipmsdl.wordpress.com/2013/12/08/declaration-of-indigenous-peoples-on-the-world-trade-organization-wto, last accessed November 2014.

McCauley, D.J. (2006) 'Selling out on nature', *Nature*, 443(7107): 27–28.

Millennium Ecosystem Assessment (MA) (2005) *Ecosystems and human well-being: Synthesis*. Washington, DC: Island Press.

Norgaard, R.B. (2010) 'Ecosystem services: from eye-opening metaphor to complexity blinder', *Ecological Economics*, 69(6): 1219–1227.

Redford, K.H. & Adams, W.M. (2009) 'Payment for ecosystem services and the challenge of saving nature', *Conservation Biology*, 23(4): 785–787.

Robertson, M. (2012) 'Measurement and alienation: Making a world of ecosystem services', *Transactions of the Institute of British Geographers*, 37(3): 386–401.

Secretariat of the Convention on Biological Diversity (2014) *Global Biodiversity Outlook 4*. Montreal: SCBD.

Suarez, D.C. & Corson, C. (2013) 'Seizing center stage: Ecosystem services, live, at the Convention on Biological Diversity!', *Human Geography*, 6(1): 64–79.

Sullivan, S. (2009) 'Green capitalism, and the cultural poverty of constructing nature as service-provider', *Radical Anthropology*, 3: 18–27.

The Economics of Ecosystems and Biodiversity (TEEB) (2010) *The economics of ecosystems and biodiversity: The ecological and economic foundations*. London: Earthscan.

# Environmental governance

Susan Baker

2.7

## What is governance?

Traditionally, governing was associated with governments, that is, the formal institutions of the state. However, towards the end of the twentieth century new ways of governing emerged, involving greater use of non-state actors. The term governance is now employed to capture governing that does not rest on the authority and sanctions of government alone (Gouldson and Bebbington, 2007). The state, rather, governs with and through non-state actors.

A mix of hierarchies, markets and networks are used to govern. Hierarchy involves top-down regulation by the state through legal rules backed by (often criminal) sanctions (Black, 2008). In the environmental arena this sees the state, among other things, pass legislation, ensure compliance with international agreements, set regulatory standards in relation to pollution prevention and control and oversee the application of environmental law. Market governance uses markets tools, such as prices, taxes, subsidies or permits. The European Union Emission Trading Scheme and Payments for Ecosystem Services (PES) provide examples. Finally, there is increased use of policy networks (Pierre and Peters, 2000), that is, organized society groups involved in either formal or informal institutionalized relations with the state. Networks can form around economic interests or among groups from civil society, such as non-governmental organizations (NGOs) like charities and voluntary groups. This results in various forms of public-private collaboration, including the use of private market actors for the delivery of social goods. The mix of these governance styles can be different across policy sectors and countries, and can change over time. Ideological views, such as neoliberal belief in the effectiveness and efficiency of markets, also shapes the mix used.

## Why change?

Traditional methods of governance have come under pressure from three key forces: internationalization, globalization and Europeanization. These interact in complex ways to reduce the problem-solving capacity of the state when acting alone (Black, 2008). This is occurring within the context of global environmental change, especially climate change and biodiversity loss, bringing an array of governance challenges that cannot be dealt with within a national territory or through top-down steering alone.

Internationalization refers to the increasing importance of international relations, treaties and alliances between or among nations. The United Nations (UN) is a key actor, bringing order to the governance of environmental change at the international level. Under its auspices regulatory regimes for the global management of climate change (the UN Framework Convention on Climate Change) and biodiversity (the Convention on Biological Diversity CBD) have developed. It has also elaborated a set of organizing principles for governance. Sustainable development, a form of development that mutually reinforces societal, economic and ecological well-being over time, is one key organizing principle. Promoting sustainable development requires complex systems of multi-level governance, open to participatory processes (Farrell et al., 2005). Reflexive governance, using participation, experiments and learning, offers one way to governance for sustainable development (Voss et al., 2006). Another approach is proposed by the Earth System Governance Project, linking governance for sustainable development to the issue of political legitimacy and social justice (Biermann, 2007). Promoting sustainable development also requires acknowledging the interrelationship between the natural and social worlds. The governance of this 'coupled' social-ecological system means ensuring that institutions 'fit' with ecological systems, for example, a river that crosses several country borders requires transboundary governance (Ostrom, 2009).

The UN also promotes the principle of 'good governance', which includes effectiveness and efficiency, rule of law, participation, accountability, transparency, respect for human rights, absence of corruption, tolerance and gender equality (Esty, 2006). The concept of 'metagovernance' refers to the ethical principles, or norms, that shape the governing process (Sørensen, 2006). Through this principle the UN has supported the participation of economic and social actors in governance arrangements. However, there are deep divisions, including over the role of the private sector and of markets. Many Latin American countries, for example, voice strong objections to the use by the CBD of market tools like PES, seeing it as a 'commodification of nature' (IIED, 2012).

Globalization is the latest stage of a process whereby technological, economic, ecological, cultural, and military trends, traditionally operating on a geographically limited scale and scope, are extended to the entire globe (Daly, 1999). Globalization sees new players, such as multinational corporations, undermine traditional political actors. For governments, their capacity to govern is increasingly dependent upon their ability to mobilize the support of these new actors (Finger, 1999).

The rise of *private governance systems* provides a useful illustration of the impact of globalization (Cashore, 2002). The United Nations Global Compact, the Forest Stewardship Council, Fair Trade, and Food Alliances have developed their own private regulation schemes, establishing governance structures and rules dealing with the production and sale of various products and services. These private governance schemes act like governments (Brammer et al., 2012). This raises concerns about whether private governance is democratic and whose interests its serves (Rudder, 2008).

The European Union plays a central role in environmental governance, within its member states, near neighbours, and at a global level. Europeanization has shifted policy-making competence upwards to the EU, while its multi-level governance system makes greater use of subnational, regional and local authorities, as well as networks (Painter, 2000). Influenced by the spread of neoliberalism in many European states in the latter half of the twentieth century, the EU also uses market-based instruments, like taxes, tradable permits, such as carbon trading, and voluntary agreements (Jordan et al., 2003). Neoliberalism holds that market-led approaches are needed to improve the efficiency and effectiveness of public service provision. Changes in the style of governance were also driven by public scepticism about state-centred approaches to environmental risk regulation and management (Gunningham, 2009). This often erupted into civil protests, for example, about nuclear power, road building and airport extensions, and pollution.

In summary, the current period sees environmental governance rely less on hierarchical steering by the state alone and more on processes that operate across multiple-levels and involve multiple actors. The state acts as both a co-ordinator and facilitator, while retaining its traditional regulatory and oversight roles. Foucault refers to this as a change in 'governmentality', the rationalities and tactics of government. He argues that this does not diminish the power of the state, but rather that the state now uses new 'technologies of governing' to respond to changing social, economic and cultural conditions (Foucault, 1979).

## The significance of governance

Contemporary methods of governing are seen by some as more effective, compared with earlier hierarchical and bureaucratic styles. The use of market instruments gives the state more ways in which it can steer. Market actors can also deliver on policy when the state has limited capacity or resources. Participation of stakeholders increases the range of solutions available to policy makers, forges co-operation and learning that produces greater stakeholder 'buy in', and results in policy responses that better take account of local needs. All this reduces the risk of policy failure. Over the longer term, this style of governance promotes new forms of participatory democracy (Dryzek, 2005) and socializes people into an ethic of citizenship.

However, some point to the undemocratic character of contemporary governance (Swyngedouw, 2005). The growing use of markets undermines the state's traditional welfare functions, such as supporting groups in society that are vulnerable to climate change. As a result, inequalities increase. Markets work on the short term and can over-use resources, to the detriment of future generations. Networks wield power and influence but the absence of clear channels of access makes it difficult for certain social groups to participate. If the aim of society is not just to manage the environment, but to promote sustainable developments then state steering is needed to address distributional issues, including intergenerational (between generations) and intra-generational (across generations) equity. The rise of network governance poses problems for traditional democratic values and there is need for the state to ensure the accountability and transparency of network arrangements and that they include broad societal interests (Gouldson, 2009).

We should not exaggerate the changes, overemphasizing the 'original' story of the state as unified, coherent and territorially sovereign (Painter, 2000). We have to be careful not to see a separation of the social world into distinct 'spheres' (the state and the economy, or governments and markets) because the state is always involved in the operation of the market (Painter, 2000). The three governance styles are 'ideal types' and in reality the boundaries between them are not clear cut: market governance can see the delegation of traditional governmental functions to private players and thus works closely with network governance; similarly, hierarchical governance can be strongly influenced by lobbying from networks; networks themselves can become closed and elitist and lose ties with civil society and participatory practices. Empirical research on environmental governance shows that markets and networks do not replace but *co-mingle* with traditional, hierarchical governance (Baker and Eckerberg, 2008). Nevertheless, there are grounds for concern as a growing array of practices are used to steer society's future and where the mechanisms to ensure that these remain open, inclusive and democratic remain underdeveloped.

## Conclusion

The term 'governance' captures important changes in how policy is made, but is also used in normative discussion on the style of steering that ought to be adopted to achieve desired outcomes. Differences abound: some argue for enhancing the role of markets; others for stronger state steering; some argue that new forms of governance are needed, such as more reflexive methods, while others focus on the fit of institutions across multi-level, even polycentric scales. Some argue that contemporary governance practices reconfigure the relationship between the state, civil society and markets. Others caution against overstating the changes, arguing that the state is not governing less, but is governing differently and in collaboration with new agents and actors. Changing patterns of governance raise important questions about the nature of democracy and the meaning of citizenship: who is in charge here and what are the consequences for our collective future? Contemporary environmental governance confronts an age old challenge – how to ensure voice for the people who are marginalized (and how to govern in nature's interest when nature cannot, as it were, 'speak'). At the same time it confronts a new challenge: how to steer society in ways that protect the natural environment upon which all our future depends.

## Learning resources

For an introduction to governance and a wide range of examples of different styles see:

Baker, S. (2010) 'The governance dimensions of sustainable development', in P. Glasbergen and I. Niestroy (Eds) *Sustainable Development and Environmental Policy in the European Union: A Governance Perspective*, Den Haag: Open University Press.

For an introduction to the specific challenges associated with the governance of sustainable development see:

Meadowcroft, J. (2007) 'Who is in charge here? Governance for sustainable development in a complex world', *Journal of Environmental Policy & Planning*, 9, 3–4: 299–314.

For discussion on the role of the state in environmental governance see:

Barry, J. & Eckersley, R. (Eds) (2005) *The State and the Global Ecological Crisis*, London: MIT Press.

# Bibliography


Baker, S. & Eckerberg, K. (Eds) (2008) *In Pursuit of Sustainable Development: New Governance Practices at the Sub-National Level in Europe*, New York: Routledge.

Biermann, F. (2007) '"Earth system governance" as a crosscutting theme of global change research', *Global Environmental Change*, 17(3–4): 326–337.

Black, J. (2008) 'Constructing and contesting legitimacy and accountability in polycentric regulatory regimes', *Regulation & Governance*, 2: 137–164.

Brammer, S., Jackson G. & Matten, D. (2012) 'Corporate social responsibility and institutional theory: new perspectives on private governance', *Socio-Economic Review*, 10: 3–28.

Cashore, B. (2002) 'Legitimacy and the privatization of environmental governance: How non-state market-driven (NSMD) governance systems gain rule-making authority', *Governance*, 15(4): 503–529.

Daly, H.E. (1999) *Globalization Versus Internationalization*, available online at: https://www.globalpolicy.org/component/content/article/162/27995.html.

Dryzek, J. (2005) *The Politics of the Earth*, Oxford: Oxford University Press.

Esty, D.C. (2006) 'Good governance at the supranational scale: globalizing administrative law', *Yale Law Journal*, 115: 1490.

Farrell, K., Kemp, R., Hinterberg, F., Rammel, C. & Ziegler, R. (2005) 'From *for* to governance for sustainable development in Europe: what is at stake for future research?', *International Journal of Sustainable Development*, 8(1–2): 127–151.

Finger, M. (1999) 'Globalisation and governance' *Newsletter of the IUCN Commission on Environmental, Economic and Social Policy*, Issue no. 6, available online at: https://www.iucn.org/downloads/pm6.pdf.

Foucault, M. (1979) 'On governmentality', *Ideology and Consciousness*, 6: 5–21.

Gouldson, A. (2009) 'Advances in environmental policy and governance', *Environmental Policy and Governance*, 19: 1–2.

Gouldson, A. & Bebbington, J. (2007) 'Corporations and the governance of environmental risk', *Environment and Planning C: Government and Policy*, 25: 4–20.

Gunningham, N. (2009) 'Environment law, regulation and governance: shifting architectures', *Journal of Environmental Law*, 21(2): 179–212.

International Institute for Environment and Development (IIED) (2012) 'Nature has values, and markets can be governed', available online at: http://www.iied.org/nature-has-values-markets-can-be-governed.

Jordan, A. Wurzel, R.K.W. & Zito, A. (Eds) (2003) *New Instruments of Environmental Governance?: National Experiences and Prospects*, London: Frank Cass.

Ostrom, E. (2009) 'A general framework for analyzing sustainability of social-ecological systems', *Science*, 325: 419–422.

Painter, J. (2000) 'The State and Governance', in E. Sheppard and T.J. Barnes (Eds) *The Companion to Economic Geography* (pp. 359–376), Oxford: Blackwell.

Pierre, J. & Peters, B.G. (2000) *Governance, Politics, and the State*, Basingstoke: Macmillan.

Rudder, C.E. (2008) 'Private governance as public policy: A paradigmatic shift', *The Journal of Politics*, 70(04): 899–913.

Sørensen, E. (2006) 'Metagovernance: The changing role of politicians in processes of democratic governance', *American Review of Public Administration*, 36(1): 98–114.

Swyngedouw, E. (2005) 'Governance innovation and the citizen: the Janus face of governance-beyond-the-state', *Urban Studies*, 42(11): 1991–2006.

Voss, J-P., Bauknecht, D. & Kemp, R. (Eds) (2006) *Reflexive Governance for Sustainable Development*, Cheltenham: Edward Elgar.

# 2.8 Green democracy

Amanda Machin

This chapter addresses the connection between environmentalism and democracy. Traditionally, green parties and movements have promoted a participatory, grass-roots democracy. Green political theorist Robert Goodin believes, however, that this should not be their priority. For him, commitment to environmental values is commitment to substantive *outcomes*. What matters is the achieving of green *ends* not the *means* of achieving them (Goodin, 1992). Others disagree. There is alarm at the assumption that the 'right' outcomes can be determined in advance of democratic discussion – as if environmental problems were technical issues with straightforward solutions rather than matters for political debate (Swyngedouw, 2011). Many thinkers and activists are motivated by the challenge of creating institutions that care for the environmental and democratic values. Their aim is not only to strengthen environmental policy outcomes but also to rejuvenate – perhaps even reinvent – democratic institutions and practices. In the face of ecological crisis, they believe, democracy is more important than ever. This claim underpins the concept of 'green democracy'.

## Green ends and democratic means

The argument for green democracy can be made from two different directions. On the one hand, democracy can be seen as the best political system for reaching the objective of environmental sustainability. Advocates of 'green democracy' point to the rich variety of perspectives that provoke informed and creative environmental policy making. For example, in deciding what to do about a local habitat, scientific expertise is complemented by the knowledge the local community has accumulated over generations; the latter is a crucial resource for identifying new methods and anticipating their potential detrimental side effects. But such knowledge would be left untapped in a technocracy:

democratic institutions are open to new ideas and perspectives, whereas technocratic rule by experts is not. Furthermore, environmental policies made democratically claim far more legitimacy and are more likely to gain the support of the public than policies made by a remote authority, who ultimately cannot be held accountable.

On the other hand, advocates of green democracy draw attention to the fact that democracy can only function within an environment that is clean and stable enough for human beings to be able to formulate, register and communicate their preferences (Mills, 1996). A healthy democracy requires a healthy environment, and a situation of environmental hazard would destabilise the measures and commitment to values that underpin democracy. In short, then, democracy and environmentalism are mutually reinforcing (Mason, 1999).

## Towards green democracy

Green democrats agree that democratic and environmental goals are connected. But they have different ideas about the form that democracy should take in order to incorporate environmentalist insights. There are three main pathways towards a 'green democracy'.

### *Extending rights*

Environmentalists are concerned that contemporary democratic systems overlook non-human nature and future generations. One suggestion, therefore, is that democratic rights are extended to these constituents. These rights are not *substantive,* but *procedural;* the only guarantee here is that non-human nature and future generations are included in the decision making process, not that they are necessarily the beneficiaries of that process. Terence Ball, for example, promotes a 'greatly expanded democracy' or what he calls 'biocracy' (2006: 139). In this 'democracy of the affected' the boundaries of the community are extended to include animals, ecosystems and future generations. It is not that these new members are given voting rights, but rather that their *interests are represented,* perhaps by designated positions for spokespeople in legislative bodies (2006: 144).

### *Crossing boundaries*

The allocation and extension of rights would normally be undertaken by the state. For sure, democracy is conventionally understood today as the concern of nation-states. But environmental issues such as acid rain, biodiversity loss and climate change do not heed state borders. Not only do the environmental policies of states affect those outside their borders, but actions to combat these issues demand cooperation across these borders. Some promote a *global* democracy in which supranational bodies and institutions such as the United Nations are expected to play a crucial role (Holden, 2002). Actors in global civil society such as international non-governmental organisations are also seen as nurturing green democracy. Questions arise here, however, regarding whether such organisations are internally democratic, and how 'the people' is established at this global level. The state has always been the bastion of democratic legitimacy. Should it really be jettisoned? Robyn Eckersley (2004) defends the role of the national-state as a crucial 'node' in ecological transnational governance.

Other green democrats demand not an expansion but rather a *reduction* of scale. The state has been criticised for its inability to be fully responsive to the environmental concerns and knowledge of local communities. Some green democrats therefore argue for more decentralised decision making. At the local level, they claim, expertise can be translated into relevant and effective policy strategies and include previously excluded or 'silenced' participants. The various levels at work in both the causes and the solutions of environmental problems seem to require 'multi-level' governance – but it is crucial to consider how these levels *interconnect* in order to ensure that decisions are green, inclusive and legitimate.

## *Deepening participation*

Green democrats have joined other voices in advocating greater participation in politics. Their aim is to deepen participation by engaging people beyond voting in elections. In particular, greens have been attracted to a discursive or 'deliberative' form of democratic participation. Deliberative green democrats argue that by introducing real opportunities for discussion and debate, participants may come to formulate more informed, socially and environmentally responsible opinions. As advocates such as Graham Smith (2003) and John Barry (1999) suggest, deliberation educates citizens and stake-holders, nurtures a sense of community and exposes environmentally unsustainable practices to public scrutiny. This strategy suggests that by engaging citizens with different types of knowledge and interests in an inclusive, authentic and consequential discussion, complex socio-environmental issues are tackled more effectively and legitimately. The aim of deliberative forums such as citizen juries, citizen councils, town meetings and consensus conferences is to improve not only the *quantity* but also the *quality* of political participation. John Dryzek (2002) suggests extending deliberation to include non-human nature; by 'building sensitivity' towards the signals coming from natural systems – for example, he believes that democracy can be made more ecologically rational. Indeed, although the focus of deliberation has tended to be upon *speaking*, also critical is the practice of 'good *listening*' (Dobson, 2012). The deliberative approach has been criticised, too, for its blindness to power imbalances, its focus upon reaching rational consensus and its lack of clarity regarding the relationship of deliberative forums to formal political institutions (Irwin, 2015). Other forms of participation, such as debate and protest are equally important in deepening democracy.

## Conclusion

Green democrats argue that democracy and environmentalism are tightly connected. Challenges posed by the looming ecological crisis offer an opportunity to reinvent political systems. Environmental challenges can alert us to new ways of extending and enhancing political mechanisms and practices that may make them more robust, inclusive and legitimate. By extending rights to new participants, by functioning at various levels, and by encouraging new forms of participation, democracy can become greener *and* more democratic.

## Learning resources

www.opendemocracy.com is an independent global media platform that aims at cultivating democratic debate. Its environment section publishes interesting short articles. www.opendemocracy.net/transformation/environment

*Participedia* is an online database of experiments in participatory political processes. It provides a tool for students, researchers and practitioners to search and compare organisations and methods for participatory democracy. www.participedia.net

## Bibliography

Ball, T. (2006) "Democracy" In Dobson and Eckersely (eds) *Political Theory and the Ecological Challenge.* Cambridge and New York: Cambridge University Press.

Barry, J. (1999) *Rethinking Green Politics: Nature, Virtue and Progress.* London, Thousand Oaks and New Delhi: Sage.

Dobson, A. (2012) "Listening: the new democratic deficit", *Political Studies*, 60(4): 843–859.

Dryzek, J. (2002) *Deliberative Democracy and Beyond: Liberals, Critics, Contestations.* Oxford: Oxford University Press.

Eckersley, R. (2004) *The Green State: Rethinking Democracy and Sovereignty.* London: MIT Press.

Goodin, R.E. (1992) *Green Political Theory.* Cambridge and Oxford: Polity Press.

Holden, B. (2002) *Democracy and Global Warming.* London and New York: Continuum.

Irwin, A. (2015) "On the local constitution of global futures: scientific and democratic engagement in a decentred world", *Nordic Journal of Science and Technology Studies*, 3(2): 24–33.

Mason, M. (1999) *Environmental Democracy.* Abingdon and New York: Earthscan.

Mills, M. (1996) "Green democracy: the search for an ethical solution." In B. Doherty and M. de Geus (eds) *Democracy and Green Political Thought: Sustainability, Rights and Citizenship.* London and New York: Routledge.

Saward, M. (1996) "Must Democrats be environmentalists?" In B. Doherty and M. de Geus (eds) *Democracy and Green Political Thought: Sustainability, Rights and Citizenship.* London and New York: Routledge.

Smith, G. (2003) *Deliberative Democracy and the Environment.* London and New York: Routledge.

Swyngedouw, E. (2011) "Whose environment? The end of nature, climate change and the process of post-politicization", *Ambiente & Sociedade*, 14(2): 69–87.

# 2.9 Environmental security

Jon Barnett

Environmental change is a political issue because it impacts on things that people and societies value. But not all things are valued equally: people value their bicycles, but usually not as much as they value their health; and societies value their sporting teams, but usually not as much as they value peace. When things that people and societies value very highly are at risk then this becomes a matter of 'security'. When environmental change puts at risk things that people and societies value highly then the issue is one of 'environmental security'.

The concept of environmental security ties together processes of 'securitisation' with those of environmental change. Securitisation is the process whereby ordinary political issues are elevated to matters of extraordinary importance by virtue of being described as matters of security (Waever, 1995). It is a process that draws on the power of 'national security' to justify the extraordinary policies traditionally associated with that issue, including large budgetary allocations to the military, conscription, constraints on political freedoms, and declarations of war. So, when environmental issues are described as 'security' issues a claim for similarly extraordinary measures to tackle the problem is implied.

Environmental security is an ambiguous concept that can be applied in many different ways. This is because the values that underpin securitisation are heterogeneous, for example not everyone in the United States cares so much about the demise of Hawaii's reefs or permafrost thawing in Alaska that they consider these to be security issues that justify an extraordinary response from their government. It is also ambiguous because the environmental processes that make valued entities vulnerable are not perfectly understood, for example there is uncertainty and debate about the causal processes that might lead climate change to cause violent conflict, which is a much cited example of environmental (in)security (Gemenne et al., 2014).

## Origins and key ideas

The idea that environmental change is in some way a security issue is as old as contemporary environmentalism, and emerges from its early concerns with scarcity and the limits to growth. Early classic works in environmental politics drew on arguments from the peace movement to juxtapose the dangers of environmental change with the limits of contemporary national security practices to address these more pervasive dangers. Gwyn Prins (1990) was later to capture this argument eloquently when he said 'you can't shoot a hole in the ozone layer'. Peace scholars extended this argument to show how national security institutions were indeed powerful agents of environmental degradation – thus casting national security practices as the drivers of harm, and the environment as the object to be secured (Westing, 1984).

The national security policy community was not slow to appropriate the risk to its legitimacy posed by these early critical uses of the concept of environmental security. The idea that environmental change is a risk to national security and to peace was initially proposed by Richard Ullman (1983) who argued that poverty and environmental scarcity in developing countries was likely to drive armed conflict and migrants and refugees, making developing country governments more militarily confrontational in their relations with developed countries, as well as degrading the quality of life of people in developed countries. This argument that environmental change is a risk to national security justifies national security institutions to secure the state from exogenous risks, and has since remained the dominant and popular understanding of environmental security.

Whereas these interpretations of environmental security provided by the environmental movement and security institutions are somewhat exclusive, there has since emerged a middle ground that derives instead from the international development community and its concern for human security (UNDP, 1994). For the development community human security is a state of well-being that is threatened by various economic and political processes, which can include national security institutions and violence, and environmental changes that undermine human health and well-being. Such an interpretation grounds environmental security in local people's vulnerability and adaptation. It draws on environmental geography and political ecology to explain the way environmental change threatens human security in vastly different ways, highlighting that environmental security is very much a matter of injustice.

## Conflict or cooperation?

The idea that environmental change can cause or exacerbate the risk of armed conflict is central to the concept of environmental security (Homer-Dixon, 1999). It is an argument that has been propagated by both the national security policy community as well as environmentalists. At the heart of many such arguments is the simple assumption that people will fight over increasingly scarce resources. In such accounts scarcity is taken as a natural rather than a socially constructed condition, and as being resolved principally through exclusion and violence rather than markets, cooperation, and ingenuity.

Scarcity of water resources is often said to be a cause of armed conflict. This discussion initially focused on conflict over the waters of some of the world's 261 transboundary river systems. The issue was highlighted by Boutrous Boutrous-Ghali when he was the Egyptian Foreign Minister in the late 1980s, who said "the next war in our region will be over the waters of the Nile, not politics" (cited in Gleick, 1991: 20). As it happened, the next war in the Middle East was the first Gulf War, which was about oil and territory and not at all about water. Indeed, almost all the evidence shows that shared waters are a basis for international cooperation: there are over 400 treaties over shared water and no examples of significant armed conflict (Priscoli and Wolf, 2009).

More recently there has been some debate about the extent to which water scarcity drives conflict among pastoralists, particularly in the horn of Africa. Though water scarcity may be a factor in such forms of violence, it seems to be a weak influence relative to political and economic variables (Raleigh and Urdal, 2007).

Earlier debates about the influence of environmental change on conflict have more recently been repeated with respect to climate change. If there is a difference, it is that much of the more recent debates about climate change and conflict hinge on the appropriate use and analysis of statistical techniques to infer causality (Adger et al., 2014). However, the issue is perhaps less one of method and more one of underlying theory, because armed conflict is complex and multi-causal, and the significance of environmental variables is very hard – if not impossible – to discern. Therefore few scholars suggest that climate change is a primary or even a major cause of armed conflict, but there is some agreement that it increases the risk in indirect ways, leading to some consensus that it is a 'threat multiplier'.

Rather than focussing on the risks environmental change might pose to armed conflict, some researchers focus on the potential of shared environmental problems to be a basis for peacebuilding. This research takes its lead from the analysis that shared water resources is a basis for cooperation, and extends its analysis into cases such as peace parks, shared fisheries, and mainstreaming environmental issues into post-conflict processes (Conca and Dabelko, 2002). Its focus on the environment as a basis for conflict prevention and the transnational networks forming around the environmental peacebuilding initiative provide a shift in focus towards practices that can sustain peace despite environmental change (Unruh and Williams, 2013).

## Conclusion

The meaning of environmental security is therefore ambiguous (Barnett, 2001). More than any other concept, it points to the challenge of maintaining peace and prosperity through the Anthropocene. It performs a valuable function in bringing together environmental, development and security policy communities into a novel environmental security policy community. The ultimate value of the concept may well be judged by whether this new environmental policy community leads to militarised and divisive, or peaceful and fair responses to environmental change.

## Learning resources

The following contain much relevant information: The Institute for Environmental Security: http://www.envirosecurity.org

The Environmental Change and Security Program at the Woodrow Wilson International Center for Scholars: http://www.wilsoncenter.org/ecsp

The environmental peace-building initiative: http://www.environmentalpeacebuilding.org

## Bibliography

Adger, W.N., Pulhin, J., Barnett, J., Dabelko, G., Hovelsrud, G., Levy, M., Oswald Spring, Ú. & Vogel, C. (2014) 'Human security'. In Field, C., Barros, V., Mastrandrea, M., Mach, K., Abdrabo, M., Adger, W., Anokhin, Y., Anisimov, O., Arent, D., Barnett, J. et al. (eds) *Climate Change 2014: Impacts, Adaptation, and Vulnerability. Part A: Global and Sectoral Aspects. Contribution of Working Group II to the Fifth Assessment Report of the Intergovernmental Panel on Climate Change* (pp. 755–791). Cambridge: Cambridge University Press.

Barnett, J. (2001) *The Meaning of Environmental Security: Ecological Politics and Policy in the New Security Era*. London: Zed Books.

Conca, K. & Dabelko, G.D. (eds) (2002) *Environmental Peacemaking*. Woodrow Wilson Center Press: Washington DC.

Falk, R. (1971) *This Endangered Planet: Prospects and Proposals for Human Survival*. New York: Random House.

Gemenne, F., Barnett, J., Adger, N. & Dabelko, G. (2014) 'Climate and security: evidence, emerging risks, and a new agenda', *Climatic Change*, 123(1): 1–9.

Gleick, P. (1991) 'Environment and security: the clear connections', *The Bulletin of the Atomic Scientists*, 47(3): 17–21.

Homer-Dixon, T. (1999) *Environment, Scarcity, and Violence*, Princeton: Princeton University Press.

Prins, G. (1990) 'Politics and the environment', *International Affairs*, 66(4): 711–730.

Priscoli, J. & Wolf, A. (2009) *Managing and Transforming Water Conflicts*. Cambridge: Cambridge University Press.

Raleigh, C. & Urdal, H. (2007) 'Climate change, environmental degradation and armed conflict', *Political Geography*, 26(6): 674–694.

Ullman, R. (1983) 'Redefining security', *International Security*, 8(1): 129–153.

United Nations Development Program (UNDP) (1994) *Human Development Report 1994*. New York: Oxford University Press.

Unruh, J. & Williams, R. (eds) (2013) *Land and Post-Conflict Peacebuilding*. Abingdon: Routledge.

Waever, O. (1995) 'Securitisation and desecuritisation', in Lipschutz, R. (ed) *On Security* (pp. 46–86). New York: Columbia University Press.

Westing, A. (ed) (1984) *Environmental Warfare: A Technical, Legal, and Policy Appraisal*. London: Taylor and Francis.

# 2.10 Food systems

## Michael Carolan

The term food system denotes all those activities, and the beliefs, knowledge systems, and cultural routines they reflect, involved in the production, processing, transportation and consumption of food. This is sometimes called taking a "farm-to-fork" approach to food, or "seed to shelf", recognizing the role that the input sector (seed, fertilizers, biotechnology, etc.) plays in shaping how and what we eat at home or in restaurants. And we must not forget about food waste either, reminding us that food systems reach far beyond "forks" and "shelves". The following offers a brief journey from seed to landfill.

## Food system concentration

The world's top four seed firms control 56 percent of the global brand-name seed market. Concentration is even greater for genetically engineered seed. In the US, for example, the largest three seed firms control 85 percent of transgenic corn patents and 70 percent of non-corn transgenic plant patents. Of all the land cultivated in the US in genetically engineered seed, 85 percent of the corn and 92 percent of soybean acreage holds seeds that are the proprietary technology of Monsanto.

A popular measure of market concentration is the four firm concentration ratio—or simply CR4. The CR4 is defined as the sum of market shares of the top four firms for a given industry. A standard rule of thumb is that when the CR4 reaches 20 percent a market is considered concentrated, 40 percent is highly concentrated, and anything past 60 percent indicates a significantly distorted market. With this rule of thumb in mind, note in Table 2.10.1 the CR4 (unless otherwise noted) for certain agrifood markets in the US.

**Table 1** Concentration of US agricultural markets

| *Sector* | *CR4** |
|---|---|
| Beef Slaughter | 82% |
| Beef Production (feedlots) | Top four have one-time feeding capacity of 1.983 million head |
| Pork Slaughter | 63% |
| Pork Production | Top four have 1.62 million sows in production |
| Broiler Slaughter | 53% |
| Turkey Slaughter | 58% |
| Animal Feed | 44% |
| Flour Milling | 52% |
| Wet Corn Milling | 87% |
| Soybean Processing | 85% |
| Rice Milling | 85% |
| Cane Sugar Refining | 95% |
| Corn Seed | CR1 80% |
| Soybean Seed | CR1 93% |
| GE Cotton Seed | CR1 96% |

*unless otherwise stated

Developed from Harvey et al. (2013) and Wilde (2013).

The CR4 statistic is a measure of horizontal concentration; concentration at one "link" in the food commodity chain. Horizontal integration occurs when firms in the same industry and at the same stage of production merge and dominate a market. Yet the food system has also undergone tremendous vertical concentration—a phenomena describing when companies are united throughout the supply chain by a single owner. Smithfield Farms, for example, has captured (CR1) a 31 percent market share of the US pork packing industry, additionally raising 19.7 percent of all hogs in the US. This dual concentration—horizontal and vertical—gives firms unique advantages that cannot be had in more open markets.

Retail sector concentration varies considerably around the world. Independent grocers, for example, still account for 85 percent of all retail sales in Vietnam and for 77 percent in India (Carolan, 2016). At the other extreme is Australia, whose retail sector has a CR2 statistic of 80 percent. CR3 ratios for the retail sectors in Sweden, the Netherlands, France, Spain, Greece, and Italy are 95 percent, 83 percent, 64 percent, 44 percent, 32 percent, and 32 percent, respectively (ibid).

Large retail firms wield tremendous buyer power due to the sheer volume they deal in. Walmart's global sales for 2015 topped US$482.1 billion (up from approximately US$250 billion in 2003). The second largest food retailer is Tesco, with sales of a little over US$100 billion. This size gives retailers like Walmart enormous market leverage. General Mills and Kraft Foods, for instance, generate approximately 20 percent of their revenue through Walmart retail sales. When so much revenue is dependent upon one contract this gives the buyer tremendous negotiating power.

When looked at from farm-to-fork, the demographics of the food system take on an hourglass figure. The system is relatively wide at both "ends", representing farmers and consumers, with a severely truncated "middle", where the food processors and manufacturers and food and beverage retailers reside. Adding the even less competitive seed industry, a better metaphor might be an hourglass hanging by a thread. Figure 2.10.1 illustrates this market concentration for the US.

## Tastes: growing homogeneity in global food supplies

A rule held throughout the sciences is that diversity enhances the health, functioning, and resiliency of systems. Yet modern agriculture and global food commodity chains have pointed us in the other direction, toward a *shrinking* of biodiversity, cultural diversity, knowledge diversity and even taste diversity. A recent study reveals an emerging standard global food supply consisting of such energy-dense foods as soybeans, sunflower oil and palm oil, along with more historically familiar staples like rice and wheat (Khoury et al., 2014). Wheat was found to be a major staple in 97.4 percent of all countries and rice in 90.8 percent, whereas soybean has become significant in 74.3 percent of countries. Meanwhile, many crops with long held regional and cultural importance—cereals like sorghum, millet and rye as well as root crops such as sweet potato, cassava and yam—are disappearing from fields and diets. For example, a nutritious tuber crop known as *Oca*, once grown throughout the Andean highlands, has declined significantly in this region both in cultivation and consumption. Figure 2.10.2 illustrates these trends. Soybean consumption has increased 282 percent in developing countries between 1969 and 2009, while the consumption of millet and sorghum has more than halved during that same time period. This is an interesting nutritional disparity between the foods at the top and bottom of the figure. Those being eaten at

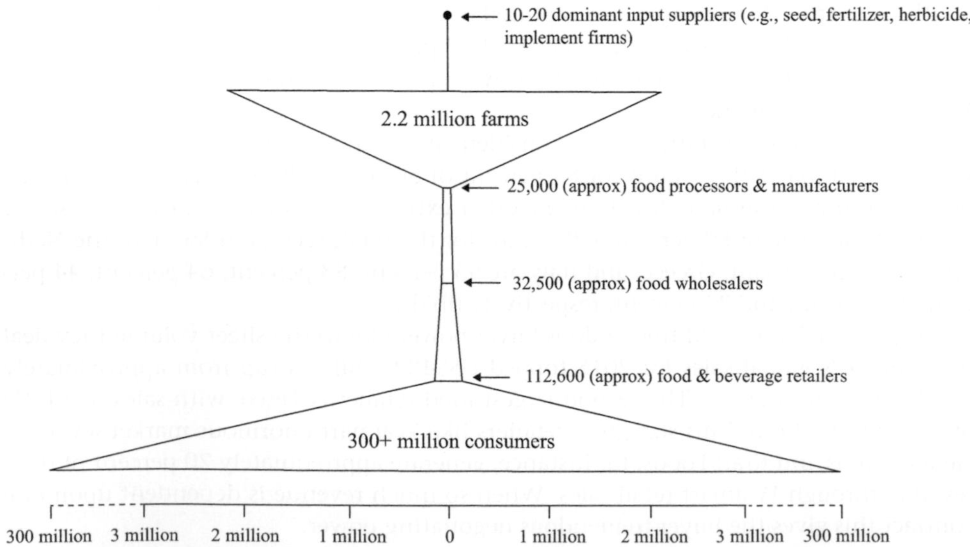


**Figure 2.10.1** US food system "hourglass" (hanging by a thread).

greater rates tend to be richer in macronutrients, while those disappearing are micronutrient rich. Not only that, many of the crops near the bottom of the figure are stress tolerant, which make them particularly valuable when we talk about sustainably feeding future populations in the context of climate change.

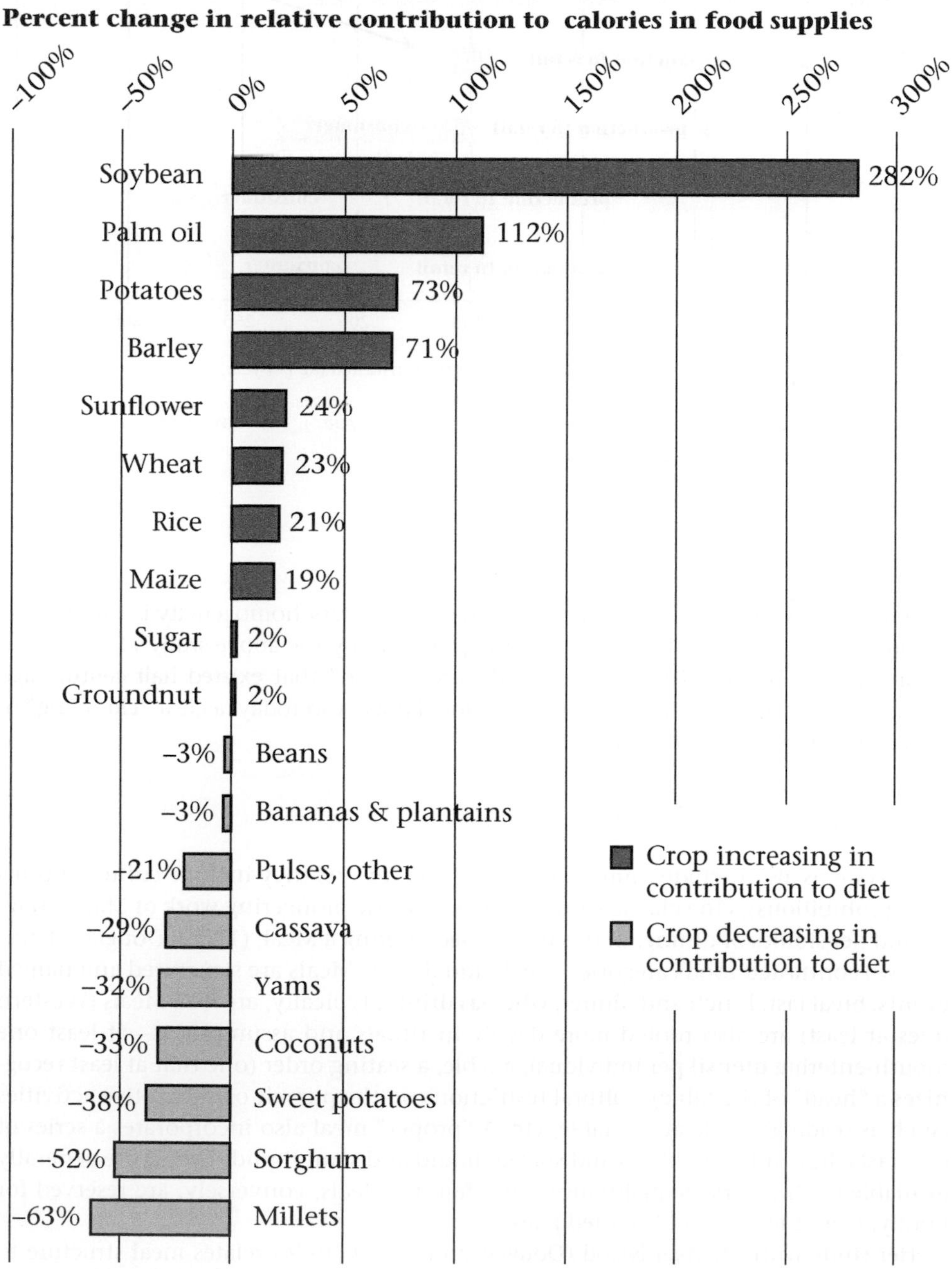


**Figure 2.10.2** Median change in the relative contribution to calories from crops of interest to the CGIAR in national diets in developing countries, 1969–2009.

*Source*: Adapted from Khoury et al. 2014. *Proc. Natl. Acad. Sci.* USA.

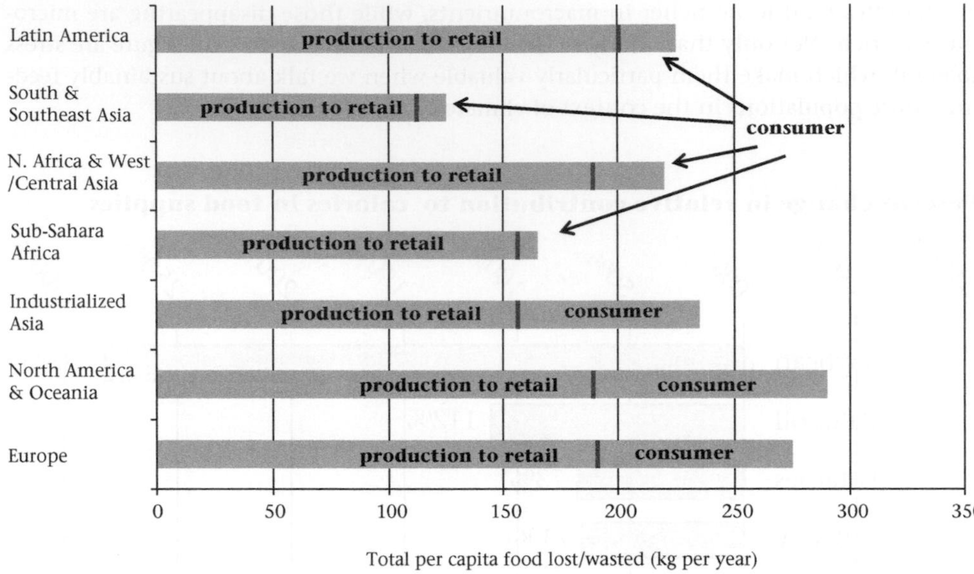


**Figure 2.10.3** Per capita food lost/wasted (kg per year), at consumption and pre-consumption stages, for select regions.

*Source*: Gustavsson et al. (2011).

Another way to visualize this move toward global dietary homogeneity is through the following image constructed (and graciously provided) by the above study's lead author, Colin Khoury: Figure 2.10.4. Notice the dietary "spread" that existed half century ago around the world, by the dispersal of light colored dots. And today: a clear "clumping" of darker dots near the middle of the figure.

## The cooked (Western) meal

Food systems also include human systems, which means they include cultural norms (and prohibitions). One classic work in this area is the pioneering work of Mary Douglas and her studies of dining. In the article 'Deciphering a Meal' (1972), Douglas identifies two contrasted food categories: meals and drinks. Meals are structured and named events, breakfast, lunch and dinner, whereas drinks, typically, are not. Meals (Western ones at least) are also rooted more deeply in rituals and assumptions—at least one mouth-entering utensil per individual, a table, a seating order (one that at least recognizes a "head" of the table), cultural restrictions on the pursuit of alternative activities (such as reading) while at the table, etc. A "proper" meal also incorporates a series of contrasts: hot and cold, bland and spiced, liquid and semi-liquid. Drinks are generally available to strangers, acquaintances, and family. Meals, conversely, are reserved for family, close friends, and honored guests.

Her study with Michael Nicod (Douglas and Nicod, 1974) relates meal structure to meal content. Examining the dining patterns of English working-class families, they note their diets centered on two staple carbohydrates—potatoes and cereals. This was in contrast to upper- and middle-class diets, which made greater use of a range of cereals, beans, and roots. Societal norms around the "proper" meal structure, for example, help us understand cultural barriers to vegetarian diets, and why vegetarian items mimicking meat (tofu *burger*, lentil *hotdogs*, etc.) are so popular.

## Food waste

British consumers, for example, discard approximately 7 million tons of food, which is about a third of what they purchase. At a retail cost of around £10.2 billion (US$19.5

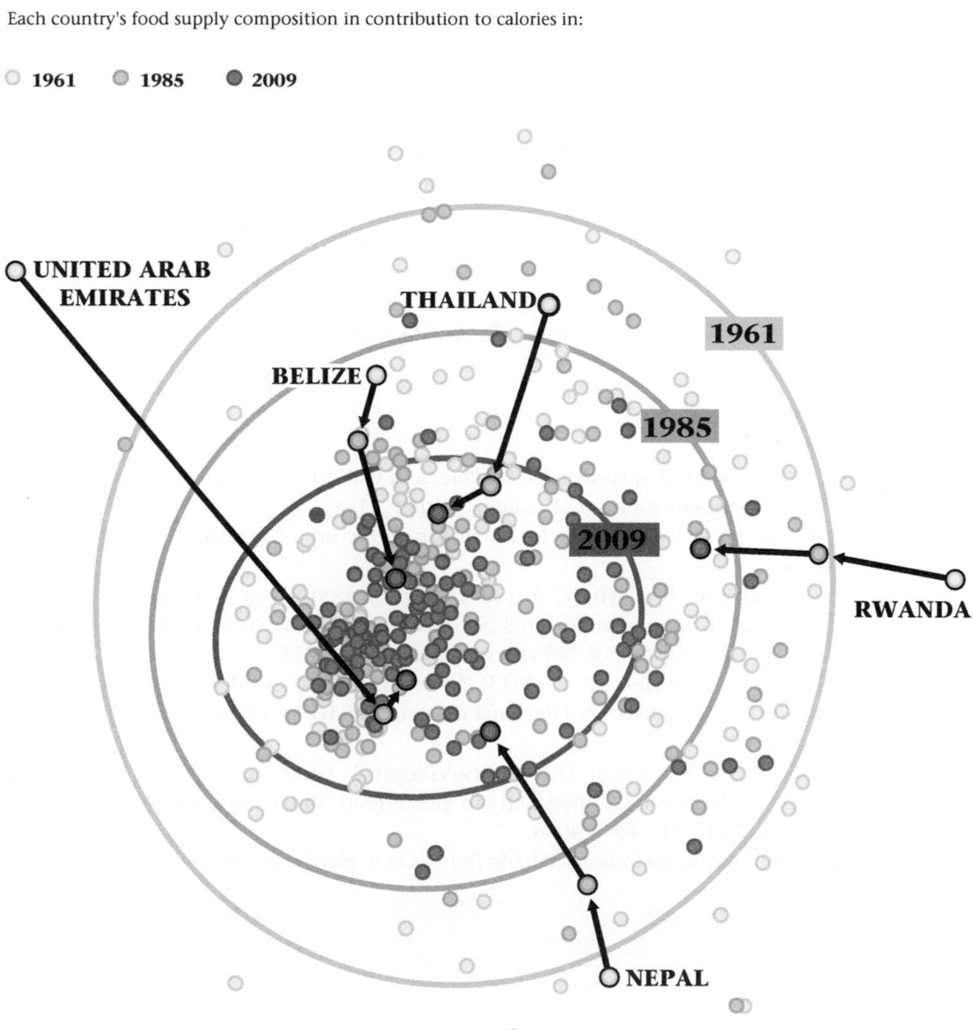


**Figure 2.10.4** A study of the world's countries finds that over the last 50 years, diets have become ever more similar.

*Source*: Khoury et al. (2014)

billion), this waste has a $CO_2$ equivalent of 18 million tonnes—an amount equal to the annual emissions of one-fifth of Britain's total car fleet. Canadians toss out about C$27 billion worth of food annually. Japan's household and food industries together waste about 17 million metric tonnes of edible food each year. In India, roughly 30 percent of the country's fruits and vegetables and 30 percent of its grain are lost due to poor storage facilities. In China, food waste has increased exponentially in recent years and now accounts for close to 70 percent of all household and commercial waste (Carolan, 2016). As a percentage of their total food, developing countries do not fare much better in terms of wasting food. What varies is where in the food system losses occur. As illustrated in Figure 2.10.3, whereas high-income regions incur considerable losses at the consumer end, most of the food lost in lower income regions occurs prior to its arrival into people's homes.

## Learning resources

Those wanting to learn more about the subject can turn to the following publications, which in turn can direct readers to additional resources:

Carolan, M. (2016) *The Sociology of Food and Agriculture*, 2nd edition. New York: Routledge.
Guptill, A., Copelton, D. & Lucal, B. (2013) *Food and Society: Principles and Paradoxes*. Malden, MA: Polity Press.
Lang, T. (2015) *Food Wars: The Global Battle for Mouths, Minds and Markets*. London: Routledge.

## Bibliography

Carolan, M. (2016) *The Sociology of Food and Agriculture*, 2nd edition. New York: Routledge.
Douglas, M. (1972) 'Deciphering a meal', *Daedalus*, Winter: 61–81.
Douglas, M. & Nicod, M. (1974) 'Taking the biscuit: the structure of British meals', *New Society*, 30(637): 744–747.
Gustavsson, J., Otterdijk, R. & Meybeck, A. (2011) 'Global food losses and food waste', Food and Agriculture Organization of the United Nations, Rome, available at: http://www.fao.org/docrep/014/mb060e/mb060e00.pdf, last accessed January 30, 2015.
Harvey, J., Hendrickson, M. & Howard, P. (2013) 'Networks, power and dependency in the agri-food industry', in J. Harvey (ed) *The Ethics and Economics of Agrifood Competition* (pp. 99–126). Dordrecht: Springer.
Khoury, C., Bjorkman, A., Dempewolf, H., Ramirez-Villegas, J., Guarino, L., Jarvis, A., Rieseberg, L. & Stuik, P. (2014) 'Increasing homogeneity in global food supplies and the implications for food security', *PNAS*, 111(11): 4001–4006.
Wilde, P. (2013) *Food Policy in the United States: An Introduction*. New York: Routledge.

# Green economy

2.11

## Les Levidow

'Green economy' became a high-profile idea when promoted by the United Nations Environment Programme (UNEP) in preparing the 2012 Rio+20 Earth Summit. Proponents have celebrated the widespread global take-up of this idea. Yet the green economy concept encompasses diverse agendas and priorities. Such differences warrant close attention through critical perspectives, as outlined here.

## Economic growth versus resource limits

For a couple of decades there have been efforts to go beyond 'sustainable development', a concept implying resource limits on economic growth. Policymakers have sought to identify and promote types of economic growth that would be environmentally beneficial. According to an early proponent of a green economy, Michael Jacobs, 'Green growth not only insists on that compatibility, but claims that protecting the environment can actually yield *better* growth' (Jacobs, 2013: 6).

UNEP began to promote Green Economy in 2008, initially as a Global Green New Deal (UNEP, 2009). This had resonances with various national proposals of a similar name, as promoted by Green and social-democratic parties. This agenda would expand public services, regulate private-sector activities, distribute resources more fairly and promote less resource-intensive patterns. This Green New Deal somewhat overlapped with proposals to overcome energy poverty and to provide a socially equitable distribution of benefits, e.g. through transition assistance (ILO, 2012). Likewise NGOs have promoted a community-based model of equitable access to natural resources and state investment in public goods (Green Economy Coalition, 2012: 33; WDM, 2012).

However, UNEP soon abandoned its early version, instead favouring market instruments and public-private partnerships (UNEP, 2011a: 1). Greater resource-efficiency in production methods would achieve sustainable growth. The new focus on a green economy reflects the 'growing recognition that achieving sustainability rests almost entirely on getting the economy right' (UNEP, 2011b: 17). This agenda aims to decouple economic growth from resource burdens. For example, green economy can achieve a faster rate of economic growth than business as usual, even 'while enhancing stocks of renewable resources, reducing environmental risks, and rebuilding capacity to generate future prosperity' (UNEP, 2011b: 24). Sustainable development is linked with eco-efficient technoscientific innovation, especially through North-South international cooperation (UN, 2012).

This agenda overlaps with 'green growth' proposals of the World Economic Forum (Green Growth Action Alliance, 2013). The OECD has promoted several variants – 'green economy', 'growing green' and 'green growth' – as if they were interchangeable ways to achieve sustainable development. A key means is 'identifying cleaner sources of growth' (OECD, 2010: 9), thus stretching resource limits. Likewise a European Commission report expects more efficient technology to reconcile conflicting aims:

> The current technological potential, if intelligently and appropriately supported by shifting the playing field towards favouring 'green' economic growth, could accelerate that path and result in the creation of a "European Way of Life", a new, sustainable and profitable ideal for middle class aspirations.
>
> (Perez, 2016: 4)

Sceptics have raised doubts about expectations to decouple economic growth from resource burdens through greater efficiency. For example, 'growth that does not recognise its roots will turn into a cancer' and 'will be corrupted by greed', according to the Director of the Green Economy Coalition (Greenfield, 2012). UNEP reports feature 'over-optimism and over-simplification of the challenges to greening growth' (Borel-Saladin & Turok, 2013: 216). A green economy cannot readily achieve GHG savings, or close the gap between rich and poor, 'without some curtailment of ambitions for economic growth' (Victor & Jackson, 2012: 11).

Since the 2007 financial crisis, many economic stimulus packages have been promoted as 'green' but have a doubtful basis for such a claim. Some schemes operate in a structurally conservative way; for example, Germany's car scrappage premium has benefited companies and employees in politically strong sectors, stimulating energy consumption (Brand, 2012a: 22). Rather than a win-win game, green economy brings new conflicts: 'The currently dominant interest is in expanding capitalist market structures', especially through economic growth (ibid.: 38).

Towards a fairer resource-allocation, the UNEP agenda proposes market instruments. As a rationale, monetising natural resources can more wisely manage them to benefit the poor (UNEP, 2011a: 1). Policy measures are necessary 'to enable green markets and ensure more efficient use of the environment and natural resources', especially by incentivising private-sector investment, according to the UNEP report on *Enabling Measures* (Wooders, 2011).

Many NGOs have suspected this agenda of commoditising nature rather than benefiting the poor (FoEE, 2012; WDM, 2012). In the name of making nature more visible, this agenda makes commons less visible and more vulnerable than before. Advocates of monetising nature ignore the social actors who help to maintain 'ecosystem services'; indeed, 'such terms all but obscure the social context' (Unmüßig et al., 2012: 28).

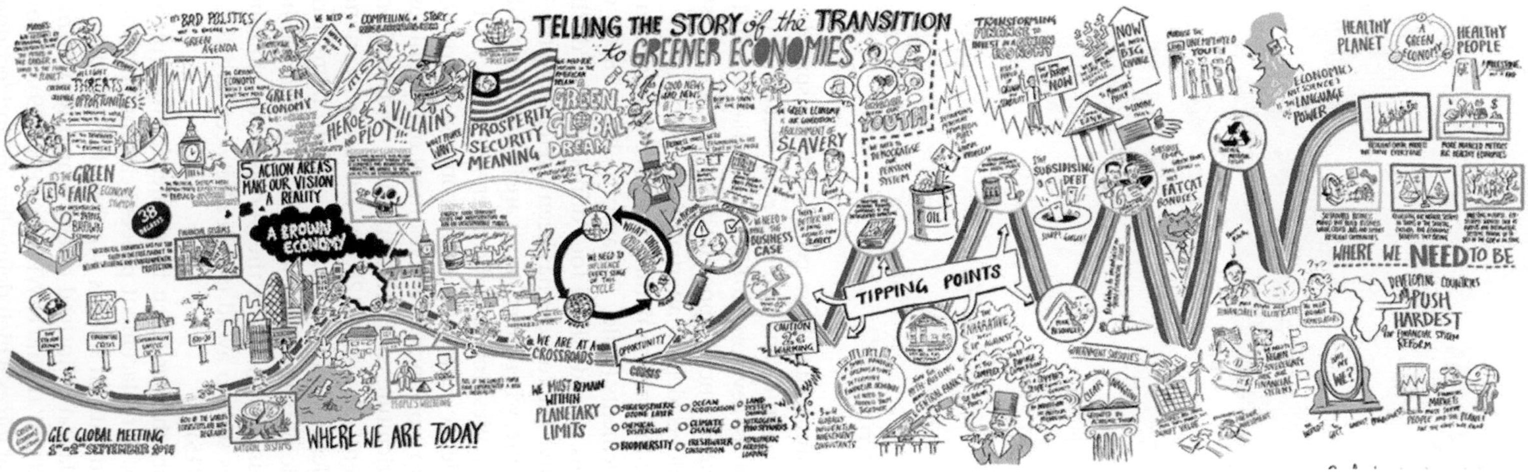


**Figure 2.11.1** Taking the Green Economy into the mainstream: telling the story of the transition.

Green Economy Coalition (2014) with permission.

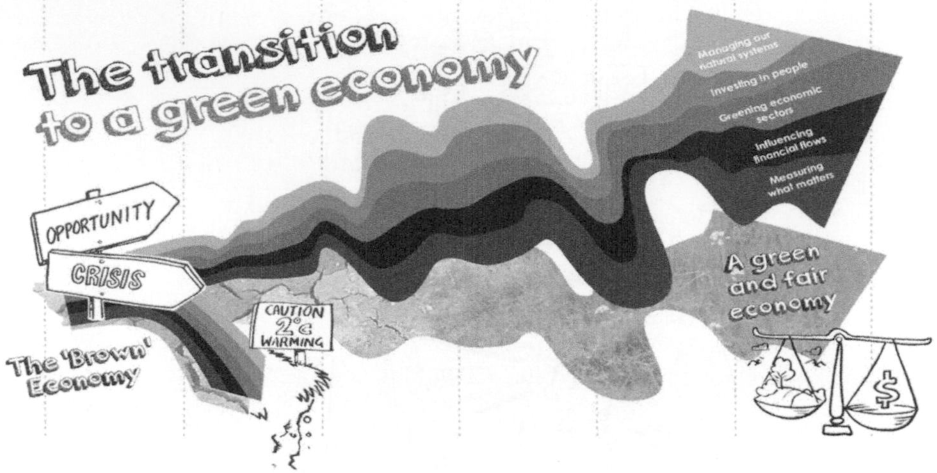


**Figure 2.11.2** Transition to a Green Economy.

Green Economy Coalition (2016) with permission.

## Which green economy?

Some proponents and analysts alike have celebrated the spread of 'green economy' agendas, thus downplaying significant differences in meanings of 'green', 'economy' and their linkages (Forsyth & Levidow, 2015). Any initiative can be analysed as four functional domains: global financial opportunities through investment; transitional policies whereby stakeholders debate alternative socioeconomic futures; regulatory standards and metrologies for valuing natural resources; and green cultural economy, including opposition movements (Bailey & Caprotti, 2014: 1800). Competing normative and distributional aims pervade those four domains, thus distinguishing any specific form of green economy. The concept has been stretched to the informal economy, e.g. in smallholder agriculture, where livelihoods are increasingly vulnerable to resource degradation; economic activities recycle and conserve resources (IIED, 2016).

According to some academic critics of a green economy, however, prevalent agendas prioritise means to open up more lucrative investment opportunities. Indeed, 'financial capital has discovered agriculture, soil, infrastructure, and environmental protection as a new field of investment' (Brand, 2012b: 1). The green economy agenda attempts 'to turn ecological obstacles to capital accumulation into mere barriers that can open the door to new economic opportunities', while depoliticising inherent conflicts and trade-offs (Kenis & Lievens, 2016). Given the pervasive tension amongst aims, some 'green economy' strategies seek to avoid conflict between growth versus post-growth perspectives (Ferguson, 2015).

There are doubts about whether green investment truly replaces brown economy or instead supplements it. A replacement process cannot be achieved or incentivised simply by financial instruments. At stake is whether this agenda generates a new green regime of capital accumulation or instead offers 'a mainly presentational greening of the old finance-dominated one' (Brand & Wissen: 2015: 519). The outcome will depend on the interplay of political forces which define 'green' and 'economy' in practice.

# Learning resources

Al Jazeera (2015) 'Pricing the planet: can a "green" economy save the planet?'. Online. Available at: <http://www.aljazeera.com/programmes/specialseries/2015/11/pricing-planet-151119111851963.html>.

Dale, G., Mathai, M.V. & de Oliveira, P. (2016) *Green Growth, Ideology, Political Economy and the Alternatives*. London: Pluto.

Scoones, I., Leach, M. and Newell, P. (eds) (2015) *The Politics of Green Transformations*. London: Routledge/Earthscan.

# Bibliography


Bailey, I. & Caprotti, F. (2014) 'The green economy: functional domains and theoretical directions of enquiry', *Environment and Planning A*, 46: 1797–1813.

Bina, O. (2013) 'The green economy and sustainable development: an uneasy balance?', *Environment and Planning C: Government and Policy*, 31(6): 1023–1047.

Borel-Saladin, J. & Turok, I. (2013) 'The green economy: incremental change or transformation?', *Environmental Policy and Governance*, 23(4): 209–220.

Brand, U. (2012a) *Beautiful Green World: On the Myths of a Green Economy*, Berlin: Rosa Luxembourg Foundation.

Brand, U. (2012b) 'Green economy and green capitalism: some theoretical considerations', *Journal für Entwicklungspolitik*, 28(3): 118–136.

Brand, U. & Wissen, M. (2015) 'Strategies of a green economy, contours of a green capitalism', in K. van der Pijl (ed.), *Handbook of the International Political Economy of Production* (pp. 508–523). Cheltenham: Edward Elgar.

Bullard, N. & Muller, T. (2012) 'Beyond the "Green Economy": system change, not climate change?', *Development*, 55(1): 54–62.

Ferguson, P. (2015) 'The green economy agenda: business as usual or transformational discourse?', *Environmental Politics*, 24(1): 17–37.

Forsyth, T. & Levidow, L. (2015) 'Towards an ontological politics of comparative environmental analysis', *Global Environmental Politics*, 15(3): 140–151.

Friends of the Earth Europe (FoEE) (2012) Pursuit of the green economy in the Rio+20 debate. Online. Available at: <http://www.foeeurope.org/green-economy-rio-conference>.

Green Economy Coalition (2012) *The Green Economy Pocketbook: The Case for Action* Online. Available at: <http://www.accaglobal.com/content/dam/acca/global/PFD-memberscpd/AFF/Green-economy-pocketbook.pdf>.

Green Growth Action Alliance (2013) *The Green Investment Report: The Ways and Means to Unlock Private Finance for Green Growth*. Cologny, Switzerland: World Economic Forum.

Greenfield, O. (2012) *December Update*. London: Green Economy Coalition.

Greens in EP (2012) *Rio + 20: Green Priorities for the Earth Summit*, Greens/European Free Alliance in the European Parliament. Online. Available at: <http://www.greens-efa.eu/the-future-we-really-want-7181.html>

International Institute for Environment and Development (IIED) (2016) *Informality and Inclusive Green Growth: Evidence from the 'Biggest Private Sector' Event*. London: IIED, UK Aid, Center for International Forestry Research (CIFOR).

International Labour Organisation (ILO) (2012) *Working Towards Sustainable Development: Opportunities for Decent Work and Social Inclusion in a Green Economy*. Geneva: International Labour Organisation.

Jacobs, M. (1991) *The Green Economy: Environment, Sustainable Development, and the Politics of the Future*. London: Pluto Press.

Jacobs, M. (2013) Green growth: economic theory and political discourse, in R. Falkner (ed.), *Handbook of Global Climate and Environmental Policy*. Oxford: Wiley Blackwell. A similar argument is available at <http://www.lse.ac.uk/GranthamInstitute/publications/WorkingPapers/Papers/90–99/WP92-green-growth-economic-theory-political-discourse.pdf>.

Kenis, A. & Lievens, M. (2016) 'Greening the economy or economizing the green project? When environmental concerns are turned into a means to save the market', *Review of Radical Political Economics*, 48(2): 217–234.

Levidow, L. (2014) *What Green Economy? Diverse Agendas, Their Tensions and Potential Futures*, IKD Working Paper no.71 Online. Available at: <www.open.ac.uk/ikd/publications/working-papers>.

Organisation for Economic Co-operation and Development (OECD) (2010) *Interim Report of the Green Growth Strategy: Implementing our Commitment for a Sustainable Future*. Meeting of the OECD Council at Ministerial Level. Paris: OECD.

Organisation for Economic Co-operation and Development (OECD) (2013) *Putting Green Growth at the Heart of Development*. Paris: OECD.

Perez, C. (ed.) (2016) *Changing Gear in R&I: Green Growth for Jobs and Prosperity*. Online. Available at: <http://www.kowi.de/Portaldata/2/Resources/horizon2020/coop/Expert_Group_Green_Growth.pdf>.

United Nations (UN) (2012) *The Future We Want*, Geneva: UN. Online. Available at: <http://www.uncsd2012.org/content/documents/727The%20Future%20We%20Want%2019%20June%201230pm.pdf>.

United Nations Environment Programme (UNEP) (2009) *Global Green New Deal: Policy Brief*. Geneva: UN Environment Programme.

United Nations Environment Programme (UNEP) (2011a) *Towards a Green Economy: Pathways to Sustainable Development and Poverty Eradication – A Synthesis for Policy Makers*. Online. Available at: <http://www.unep.org/greeneconomy/Portals/88/documents/ger/GER_summary_en.pdf>.

United Nations Environment Programme (UNEP) (2011b) *Towards a Green Economy: Pathways to Sustainable Development and Poverty Eradication*. Online. Available at: <http://www.unep.org/greeneconomy/Portals/88/documents/ger/ger_final_dec_2011/Green%20EconomyReport_Final_Dec2011.pdf#>.

Unmüßig, B., Sachs, W. & Fatheuer, T. (2012) *Critique of the Green Economy*. Brussels: Heinrich Böll Foundation.

Victor, P.A. & Jackson, T. (2012) 'Commentary on UNEP's green economy scenarios', *Ecological Economics*, 77(1): 11–15.

Wooders, P. (2011) *Enabling Conditions: Supporting the Transition to a Global Green Economy*. Online. Available at: <http://www.unep.org/greeneconomy/Portals/88/documents/ger/14.0_EnablingConditions.pdf>.

World Development Movement (WDM) (2012) *Rio + 20: Whose Green Economy?* London: World Development Movement/Global Justice Now.

# Green governmentality

2.12

Stephanie Rutherford

## The origin of the concept

Green governmentality emerges out of Michel Foucault's articulation of governmentality, an exploration of how power operates in modern society. For Foucault, power is neither monolithic nor static. The way it is practiced changes through both time and context, with different historical periods more closely associated with different kinds of power. For Foucault, governmentality is the characteristic system of power in modern (neo)liberal democracies. Named the "conduct of conduct" or the "art of government" (Foucault, 1991), the framework of governmentality seeks to understand how various state and non-state authorities (governments, universities, corporations, non-governmental organizations, and so on) together promote and encourage specific systems of knowledge (like health care, statistics, insurance, demographics, census taking) which can, in turn, then foster particular practices in the populations they govern. In essence, authorities encourage people to govern themselves.

The "how" is important here: studies in governmentality take aim at the various discourses, techniques, and practices that are employed by various authorities to direct human behavior. In staying with the how of government, governmentality offers a window into the "positive" mechanisms of power. We are used to thinking about power as repressive or about domination; the governing power of the legal system functioning through prohibition and punishment offers a clear example. But what governmentality alerts us to is that power is often productive, generative of particular ways of seeing the world, strategies of intervention, and practices of self-government. Moreover, power is diffused across the social body, instead of a thing to be held only by specific people or within institutions. Put another way, power is not a zero-sum game, with some hording power and others having none at all. Instead, every interaction, even at a micro-scale, is inflected with differential and uneven power relations. Power is always a product of relational engagement with the world.

The 1990s and 2000s saw a remarkable uptake of governmentality as a conceptual approach, particularly in Anglo-American scholarship. All manner of disciplines—from political science to environmental studies—sought to apply this way of thinking to a range of societal problems. The concept of green governmentality (also known under the monikers of eco-governmentality or environmentality) emerged within this period. It offers an extension of Michel Foucault's original articulation, utilizing this analytical toolkit to study environmental politics. Scholars who take this approach begin from the starting point that when we speak of the environment, we are talking about more than just the biophysical world; the environment is also a site of power where meanings, discourses, and subjectivities are constantly circulated, negotiated, and renegotiated.

## Three key aspects

To make this more comprehensible, this chapter will break down the discussion of green governmentality into three constitutive parts: environmental discourses, technologies of rule, and environmental subjectivities. These divisions are meant as neither mutually exclusive nor exhaustive. Rather, the chapter puts these aspects of governmentality forward as a framework through which one might grasp the larger concept. It fleshes these ideas out by drawing on the instructive example of climate change. Of course, climate change is not the only issue to which green governmentality has been applied. However, it is one with which readers will have some familiarity and also represents an area where some of the most current scholarship on green governmentality is taking place.

### *Environmental discourses*

For Foucault, the world we encounter is made in and though power. There is no underlying truth or objective reality outside of discourse that simply needs to be revealed by the right person or ideology. Instead, through discourse, he asserts, "power produces; it produces reality, it produces domains of objects and rituals of truth" (Foucault, 1995, p. 194). Thinking about how this relates to the natural world, the ways we talk about the environment generate its truth, structuring how we can think about it as an object of knowledge. As a result, the discourses around, for example, biodiversity, sustainable development, ecosystem services, or deforestation coalesce into what Foucault would call a knowledge/power nexus, where some notions become possible while others are rendered incoherent by the rules and practices of its respective discourse. Along these same lines, some people are authorized to speak the truth of each specific discourse, to articulate its contours, to ensure its circulation and re-inscription.

With reference to climate governmentality, a certain way of thinking about climate has emerged over time that was not pre-given. Indeed, thinking about climate as a global system—as we do today—was only one way we could have imagined and acted upon the issue of anthropogenic climate change. Building upon decades of thinking about environmental problems as planetary, the United Nations Framework Convention on Climate Change (UNFCCC) was born in 1992 as the vehicle that made this global concern manifest. As Oels suggests, "Viewing the planet as a 'global' ecosystem is the important step of framing the problem of climate change as one that requires global solutions, while obscuring the scope for regional and local action" (2005, p. 198). It rendered the planetary climate a domain of political action.

In so doing, particularly located experts and bureaucracies were authorized to speak its truth. Pre-eminent here, of course, is the Intergovernmental Panel on Climate Change (IPCC), an international scientific body made of recognized natural and social science experts who "provide policymakers with regular assessments of the scientific basis of climate change, its impacts and future risks, and options for adaptation and mitigation" (IPCC Factsheet). Oels contends that the IPCC, as both the creator and disseminator of definitive knowledge about the planetary climate, facilitated the understanding of climate as a global problem, one that was in need of "natural science expertise and a technological fix" (2005, p. 198).

## Technologies of rule

The discourse around climate—or any environmental issue—has consequences; it does not leave the world untouched. Power/knowledge frameworks not only problematize particular environmental issues, but also offer strategies to calculate and ameliorate environmental harm, a hallmark of governmentality. Technologies of rule, or strategies of intervention, produce the knowledge required to act on a particular issue while at the same time circumscribing its potential solutions. They feed into and support the discursive framing of the environmental problem.

In the case of climate change, we are witness to a whole variety of calculative strategies that render the issue visible as a planetary problem: computer modeling and forecasting, remote sensing, carbon inventories, and the counting of carbon sinks. All of these technologies operate through the lens of science to produce authoritative accounts of the state of the global climate and the possible scenarios to fix it. However, the translation into policy not only has to do with science, but also politics and economics. The Kyoto Protocol (1997–2002), the global effort to reduce greenhouse gas emissions through a cap-and-trade system, took note of the information produced by science, but also worked along the lines of targets and flexibility mechanisms, which meant that even in the face of successive IPCC reports, many countries in the Global North, like the United States and Canada, avoided implementing their recommendations. In a sense, the economics trumped science as a discourse of truth, and a market-based approach to climate change became normalized as the only way to deal with the problem, rather than, for example, a carbon tax. Initiating a global cap-and-trade system became the common sense solution to the problem of climate change.

## Environmental subjectivities

The final aspect of green governmentality is the governance of the self—the ways that individuals constitute themselves through discursive frameworks. Studies in governmentality tell us that one of its most salient features is that it functions by governing at a distance, encouraging individuals to make choices for themselves that also work to the best end of the state. As a result, people become entrepreneurs of the self, choosing among a range of subject positions and practices that demonstrate who they wish to be in the world.

Governing at a distance is clear in the range of options that individuals have to ameliorate their carbon consumption. In what Matt Paterson and Johannes Stripple call the "conduct of carbon conduct", practices like calculating your ecological footprint, purchasing carbon offsets for a flight, or engaging in self-sacrifice through carbon diets offer the possibility for people to answer the question "what will you do to save the climate" (2010, p. 355)? These practices shape our understandings of ourselves and offer the possibility that an individual can, in fact, change the course of global climate change. And they are also highly visible, such that a person can publicize their efforts and encourage (or shame) others into counting their carbon as well. As such, a new subjectivity is born, what Skogland (2014) calls *Homo Clima,* a person defined by climate insecurity but engaging in moral practice, "transparency about one's changeability, a thorough assessment and fine-tuning of everyday mundane practices and the global spread of all this as the correct way of living" (p. 166).

## Learning resources

Internet Encyclopedia of Philosophy – "Michel Foucault: Political Thought": This is an online learning tool that explores various branches in philosophy. Here you will get a brief overview of Michel Foucault's analytic of power.
http://www.iep.utm.edu/fouc-pol/

Timothy W. Luke – "Generating Green Governmentality": This online resource is an accessible explanation of the key facets of green governmentality, written by one of the key scholars who coined the term.
http://www.cddc.vt.edu/tim/tims/Tim514a.PDF

Shauna Barnhart – "Environmentality": This website presents an introduction to the concept of green governmentality, and applies it to the case of waste governance.
https://discardstudies.com/2016/07/27/environmentality

## Bibliography

Foucault, M. (1991) 'Governmentality' in G. Burchell, C. Gordon and P. Miller (eds) *The Foucault Effect: Studies in Governmentality,* (pp. 87–104). Chicago: University of Chicago Press.

Foucault, M. (1995)[1977] *Discipline and Punish: The Birth of the Prison.* New York: Vintage Books.

Intergovernmental Panel on Climate Change (IPCC) (n.d.) IPCC factsheet: What is the IPCC? Online. Available at: http://www.ipcc.ch/news_and_events/docs/factsheets/FS_what_ipcc.pdf

Oels, A. (2005) 'Rendering climate change governable: from biopower to advanced liberal government', *Journal of Environmental Policy & Planning,* 7(3): 185–207.

Paterson, M. & Stripple, J. (2010) 'My space: Governing individuals' carbon emissions', *Environment and Planning D: Society and Space,* 28: 341–362.

Skogland, A. (2014) 'Homo Clima: the overdeveloped resilience facilitator', *Resilience: International Policies, Practices and Discourses,* 2(3): 151–167.

# Hybridity

2.13

Jacques Pollini

According to the Meriam Webster dictionary online, a hybrid is: (1) an offspring of two animals or plants of different 'races', breeds, varieties, species, or genera; (2) a person whose background is a blend of two diverse cultures or traditions.

The term hybrid has been widely used in biology since the discovery of the Mendelian laws of biological inheritance, in 1865. For biologists, a hybrid is an organism that is made out of the combination of two parent categories. The parents, as well as the hybrid that is obtained, are distinct types that are given a name and whose specific characteristics are recognized. If the concept of hybrid was expanded to any mix of origins, then it would have no value as virtually all organisms could be called hybrids. Pure varieties that serve as parents in the process of hybridization would be hybrid in that loose sense as they are the outcome of an evolution that involved various origins and strains. Those organisms that result from the random crossing of non-pure varieties would be hybrid too as they are mixtures of various strains, even though these strains could not be identified. For biologists, thus, hybrids are categories or types defined with the same rigor than the parent categories that are used to create them. Hybrids are consistent with a so-called 'essentialist' view of the world. They are new essences created through the mixing of other essences that pre-existed them. They are new entities in a world of closed objects that can be changed only through careful engineering that follows explicit purposes.

In the social sciences, the term hybrid is widely used in discussions about people's identities, especially in post-colonial studies. In this context, it is synonym for 'creolization'. Historically, this was the preferred word when dealing mostly with the mixing of black and white people's identities, before the generalization of the concept to all human groups and mixes. The studies of creolization and hybridity contributed to the development of intellectual and social movements against racism and ethnocentrism, against the 'essentialization' and hierarchical ranking of identities from a Eurocentric

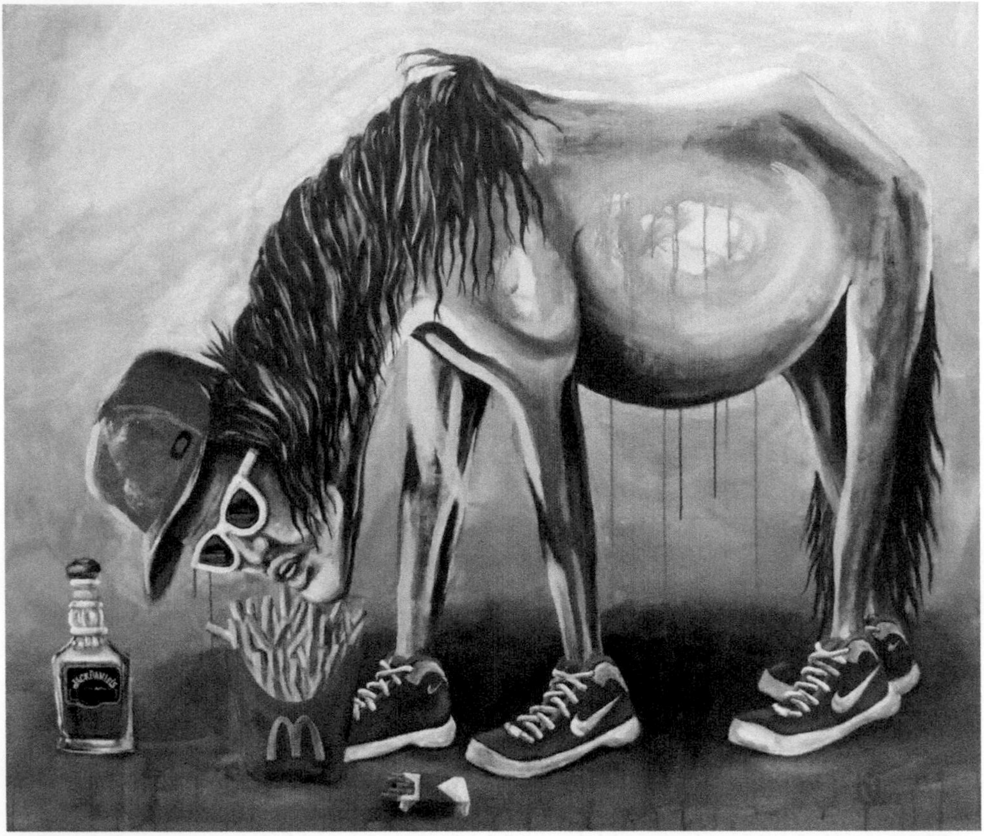

**Figure 2.13.1** 'A consumerist man animal hybrid eats MacDonalds': a painting by street artist Dinho Bento, reproduced with permission. Though MacDonald's food is often criticized for being 'processed', this painting provokes us to consider that all organic things are, these days, mixtures of various substances by virtue of human agriculture, science and industry.

*Source*: www.streetsonart.com

perspective. These studies formed a critique of the use of science by the geo-economic world 'center' to subordinate the 'periphery'. More recently, this critique is associated with post-modernism, post-structuralism, and constructivism, all movements that challenge the canon of Western sciences and the hegemonies of social and natural orders that it produces. Homi Bhabha's (1994) influential book *The location of culture* conceptualizes a third space, in between Western hegemonies and subaltern cultures, which enables the emergence of hybrid, new identities. Cultural critic Gayatri Spivak (1999) criticized this model, arguing that hybridization implies the creation of pure entities of which the hybrids are made. Her critique reveals that the use of the hybridity concept in the social sciences may have ontological implications similar to those that arise from its use in biology after all. She raises the question of what new hybrid entities are and who defines them. Imperialism, she argues, may still operate in the process that leads to the emergence of hybrids and may bias the characterization of their identities as 'good' or 'regressive'.

Today, hybridity is pervasive to the natural and social sciences literature dealing with society and the environment, especially since the publication of influential post-modern (Haraway, 1991) and constructivist (Latour, 1993) books on the subject. Things that constitute the world are said to be natural/cultural hybrids coproduced by nature and culture. Although used in new and broader domains of knowledge, the concept of hybrid still carries with it its rebellious content. Since the European Enlightenment, Western science became hegemonic in defining what nature is and how the environment should be managed. The colonized, subordinate periphery absorbs the canon of Western science and its recommendations regarding environmental management through the training of its elites in Western universities and the design and funding of its policies by international aid agencies. If the purpose of science is to know what nature is and how it works, then there should be normative definitions of how nature should be managed and what it should look like. Against this essentialization of nature by what is sometimes called 'positivist' science, the hybridity argument claims that there are as many natures as there are cultures that interact with it, that nature is not separate from culture, that it may even be misguiding to distinguish the two realms, and that all cultures have the same legitimacy to define what nature should look like and how it should be managed. Hybrid forms of knowledge would then be needed to account for these varieties of views. In the end, hybridity characterizes cultural identities, things co-produced by nature and culture and with which groups of various identities interact, but also knowledge co-produced by these hybrid groups about these hybrid things.

The hybridity argument is of great political importance to counteract the hegemony of Western science and of the representations it produces about a world that Western people do not occupy and use alone. This resistance is required to oppose Western subjugation of the others and control of the things they use. But it poses an ontological problem as it leads to questioning the relevance of any norms, or at least fails to define criteria to determine which norms are operational when the social and natural worlds need to be changed, and when the management of our common environment needs to be improved. Obviously, and post-modernists would not contest it, several options are always available to manage the environment even for a specific, given purpose, and some options fail while others succeed. Scientists and ordinary people alike know that their knowledge can be right or wrong, and that they can take right or wrong decisions and achieve (or not) expected outcomes, regarding which variety to sow in a field to obtain a better harvest, how to handle a locust invasion, or what strategy to adopt when climate patterns change. In a non-essentialist world, what are the categories or types to which one can refer to characterize features and processes that shape the environment we live in? If there are as many valid representations of things as there are identities capable of seeing these things, and if all these representations have the same legitimacy to claim they are correct, on what ground can the notions of right and wrong be rebuilt? If they cannot, how to make decisions that deliver expected outcomes, and how to explain that expected outcomes do not always realize? How to combine the legitimation of any hybrid form of knowledge with the necessity to distinguish between true and false statements about the world, without falling back into the trap of adopting the norms of those social groups that have the most power and influence in the debate?

It is worthwhile at this stage to go back to Bhabha's work. For him (Bhabha, 1994), hybridity emerges in moving spaces located on borderlines. It is the outcome of a "moment of processes," a concept that evokes continuum rather than discrete categories. Entities move across space and change during that process, an idea that is similar to Michel Serre's (1991) argument that learning occurs through the confusion generated by passing through unknown ground to reach a new location, like passing across a river and being challenged by rapids. The hybridity concept facilitates the mapping of entities around which we revolve, by raising the question: hybrid of what? But the "what" question leads to envisioning the world as being made of discrete entities, whereas complexity is better reflected by a multiplicity of continuum within which these entities are dissolved. Mapping is a mental process that is required to find directions and muddle through complex spaces. Dualism facilitates this mental process by generating contrasting categories such as black and white, natural or cultural, good and bad, but the poles created by this dualism may not exist because things may never be pure forms. Purity can only be an approximation of reality, and the same could be said of hybrids. Mapping and categories just create a mental space that has no concrete existence. The things the reality is made of are not the features of the map. They have no boundaries. They are not pure forms. They constitute dialogies (Morin, 1992) engaged in a dialogue, rather than dual categories defined by their contrasting features.

One of today's key cognitive challenges to account for the world we live in and face its greatest challenges, like addressing the environmental crisis, may be to move beyond seeing the world as made of hybrids of known categories to seeing it as a noisy mess from which some order emerges in unexpected places; as an open system filled with an undecipherable web of continua rather than a closed system populated with discrete, pure or hybrid (this would not after all make such a difference ontologically speaking) entities. To achieve this, we may need to mobilize all forms of knowledge wherever they are located, whatever canon they use to be produced, and create a constructive dialogue between these forms of knowledge to solve problems that matter for the people, involving these people in this dialogue and taking seriously their knowledge. We may need to find ways to settle controversies whenever they arise. How? A consensus exists already: experience. There is no culture and there are no people for whom experience is not taken seriously. For anybody, the unexpected outcomes of any action leads to a questioning of the validity of that action and of the knowledge that sustains it, and to experiment new ways to proceed. Participatory action research (Greenwood & Levin, 2007) already developed methods aimed at producing knowledge about hybrid, should we say 'open', or 'moving', objects. Participatory action research may not cancel power asymmetries among those involved in the production of knowledge, but by putting experience in real world setting at the core of scientific enquiry, it renders more difficult the production of false statements by the powerful and gives a concrete weapon to the powerless. By adopting such methods, hybrid forms of knowledge will be produced about hybrid natural/cultural objects, but the question of what these hybrids are made of, although worthy to be raised as part of a mapping or philosophical exercise, or for the purpose of reminding the dangers of power imbalances, may not matter so much or may even be misguiding ontologically; the things themselves and the validity of knowledge produced about these things, tested through experience and recognized by all participants, will matter the most.

## Learning resources

The following explore different facets of hybridity in interesting ways in different arenas of contemporary life:

A hybrid artistic/scientific project on hybridity: http://hybridmatters.net/pages/about
'Soft' hybridity – the future of American faces: http://ngm.nationalgeographic.com/2013/10/changing-faces/funderburg-text
'Hard' hybridity – trans-human futures: http://humanityplus.org/
Cool hybrid stuff: http://www.designboom.com/art/matthias-jung-surreal-architecture-collage-09–22–2015/

## Bibliography


Bhabha, H.K. (1994) *The location of culture*. London: Routledge.

Greenwood, D.J. & Levin, M. (2007) *Introduction to action research: social research for social change*. Thousand Oaks, California: Sage Publications.

Haraway, D.J. (1991) *Simians, cyborgs, and women: the reinvention of nature*. New York: Routledge.

Latour, B. (1993) *We have never been modern*. Cambridge, Massachusetts: Harvard University Press.

Morin, E. (1992) *The nature of nature*. New York: P. Lang.

Serres, M. (1991) *Le tiers-instruit*. Paris: François Bourin.

Spivak, G.C. (1999) *A critique of postcolonial reason: toward a history of the vanishing present*. Cambridge, Massachusetts: Harvard University Press.

# 2.14 Land grabbing

Ian Scoones

Land grabbing emerged as a global phenomenon in the period following the global financial crisis of 2007–08. Investors in search of financial returns looked to land across the world, but particularly in parts of Africa and south-east Asia. Large-scale investments in land for agriculture, biofuels or conservation uses are not new, and the recent dynamics must be seen in an historical perspective. There has been much debate about the extent and impacts of land deals, and ongoing discussion about whether there are alternatives that facilitate overseas investment, while protecting local rights and enhancing smallholder livelihoods.

## The global land rush

There have been a number of drivers of global land deals. These include (White et al., 2012): the anticipation of food insecurity in certain parts of the world, due to growing populations and urbanisation; response to fuel scarcity in a period of 'peak oil', and the push for biofuels; the commodification of resources, including carbon and biodiversity, in response to the environmental and climate crises; the development of infrastructure corridors and economic zones as sites for investment and extraction; the creation of new financial instruments to channel investment, enhancing speculative behaviour; and global development narratives that emphasise the role of private, foreign investment, set within a neoliberal philosophy.

Through the enclosures of past centuries to the expropriations of the colonial era to more recent expansions of agribusiness fuelled by the economic reforms of 'structural adjustment', land grabs have been a recurrent phenomenon. What, then, is new about the recent period? Some novel features stand out, including the role of financialised capital in a globalised context; the importance of market environmentalism driving 'green grabbing' (Fairhead et al., 2012), and the extent and rapidity of the recent unfolding of land deals across the globe (Edelman et al., 2013).

However, data on land deals are scarce, and contracts with investors often not transparent (Scoones et al., 2013; Cotula, 2015). It is therefore very difficult to know how many deals have taken place, where, with whom and with what consequences. Early estimates of land acquisitions varied between 40 million hectares to over 200 million hectares, with at least 60 percent in sub-Saharan Africa. Land Matrix estimates in mid-2016 suggest that 1,265 deals have been concluded across 44 million hectares, with a further 206 deals proposed across 18 million hectares (see the learning resources section, below).

Detailed field-level research has exposed much complexity, however. For example in a study of three African countries, Cotula et al. (2015) show how land deals make up a small proportion of agricultural land; deals involve national investors as well as foreign players; there is much diversity in state involvement across deals and countries; and levels of implementation vary considerably. Hall et al. (2015a) show similar diversity across eight African cases. While deals were struck, in many cases investment never followed. Shifts in commodity prices since the late 2000s also meant that incentives to invest, for example in biofuels, have plummeted. Despite claims that many areas are 'unutilised', research has shown that such locations are often of low potential productivity, far from markets and used by pastoralists or other mobile peoples. Many land deals occurred on state land, where farms had been abandoned, bringing new investment, jobs and markets. Land remains an important asset, and through a process of globalised financial investment, there is much speculative activity and multiple players involved (Fairbairn, 2015). Research shows that state elites and community leaders are often heavily involved, benefiting from deals, resulting in a variety of forms of resistance (Wolford et al., 2013; Hall et al., 2015b). Significant displacements continue to occur, often without effective consultation and 'free prior informed consent' (Vermeulen and Cotula, 2010).

## Governing land deals

It is a mixed picture, but it is the governance of land deals that really matters. Where conditions of consultation, transparency and appropriate compensation are adhered to, then land deals may sometimes generate positive benefits. But in many settings land governance is weak and corruption is rife. Deals may benefit a few, but not the majority due to adverse 'terms of incorporation' in a deal (McCarthy et al., 2012). This depends crucially on the 'powers of exclusion' (Hall et al., 2011) and patterns of 'land control' (Peluso and Lund, 2011) – the interacting social, political and economic processes that include or exclude people from land. Central here is the agrarian question of labour, and whether new capitalist enterprises cause dispossessions, and a 'surplus population' of labour (Li, 2011), or whether large-scale agriculture can be labour-absorbing, even if offering poor working conditions (Gibbon, 2011).

With development agencies supporting private sector investment in commercial agriculture, many also recognise the risks of land 'grabbing', as resistance and protest can undermine the extension of market relations. As a result, there has been a proliferation of guidelines, principles and frameworks for addressing such issues, and a parallel set of toolkits and resources for implementing them. The most significant is the 'Voluntary Guidelines for the Responsible Governance of Tenure of Land, Fisheries and Forests in the Context of National Food Security', agreed through a long consultation process. While voluntary, the guidelines are framed in terms of international human rights obligations and link to national constitutional provisions in some regions. As an agreed framework by national governments, businesses, NGOs and social movements, the guidelines provide an important platform (Hall et al., 2016).

Given the long history of land struggles – from the colonial era to the present – challenges of land governance in many countries are not surprising. This is why an historical political economy perspective on changing land relations is essential. Shifting configurations of power in a globalised, financialised economy results in capital seeking profit in new ways. The 'land grab' phenomenon must therefore been seen in relation to longer-term trends in agribusiness expansion (Amanor, 2012), and wider agrarian political economy in a globalised world (Moyo et al., 2012). Bargains between capital, states and local elites continue to frame outcomes, as has been the case since the enclosures of the past (Alden Wily, 2012).

In the African context, the limits of state capacity in the post-adjustment era mean that effective regulation and governance of land deals in the interests of citizens is constrained, such that investors can often run roughshod over unimplemented laws. In some countries – for instance Mozambique – highly progressive land laws exist, seen by some as a model. But in practice, local elites, in alliance with investors, are able to subvert these and encourage exclusionary appropriation of land, undermining local livelihoods (Cabral and Norfolk, 2016).

## Commercial land investments: multiple models

There is no single type of land deal. The standard image of a huge plantation or estate, growing an export mono-crop on a massive scale with a highly mechanised operation exists, but is not standard. There are many different types of commercial agriculture, and it is important to differentiate the type of 'business model' or 'labour regime', as well as the type of crop and scale of activity (Cotula and Leonard, 2010). The relationship between agricultural enterprises of different scales is crucial too, as this has an impact on the 'multiplier' and 'linkage' effects. An enclave arrangement contrasts with a dualistic system with a more integrated, variegated pattern. This has a huge influence on who benefits and who does not (Smalley, 2013).

For example, large estates may employ significant numbers of people in some operations, including women. If they have processing plants, for example, this may add to employment opportunities. Others may operate a nucleus estate and be reliant on a network of small-scale out-growers to supply commodities. The estate may provide support, ranging from inputs to extension advice for outgrowers. Contract farming too takes on many forms, with some arrangements allowing for dividend returns to smallholders who hand over land to the estate, while in others smallholders may produce themselves, but be bound to the contracting company with very restrictive terms. In other instances, more flexible arrangements may be possible (Little and Watts, 1994; Hall et al., 2017).

Investments in land may not come just from large-scale companies. An important, poorly understood phenomenon is the growth of medium-scale farms. Studies in Africa show their significance across many countries (Jayne et al., 2015). Such farms may be the result of accumulation and consolidation at a local level, or may represent investments from outside, as former urban-based workers invest retirement packages or relatives oversee investment of remittance incomes. The net effect is a changing agrarian structure, with a different pattern of farm sizes, and with this class relations. Such farms are more likely to be embedded in the local economy and employ labour locally, but their owners may not be as compliant with regulations governing land use or employment rights as larger companies.

## Conclusion

In order to assess the impacts of land grabs and commercial agriculture on income poverty, food security, labour, gender relations or the environment, for example, particular contexts need to be taken into account. There are diverse pathways of agricultural commercialisation, and different land deals have different impacts in different places.

Understanding the contextual dynamics of agrarian change and the extension of capitalism through commercial agriculture – and the class, gender and other social relations that underpin this – is vital if a more rounded assessment of global land grabbing is to be achieved.

Improving the governance of land deals is therefore essential, whether through reform of investment treaties to generate greater balance and accountability; increased transparency of land deal contracts; more effective consultation and consent mechanisms in planning; improved tenure security and land rights of affected communities; and greater legal empowerment and political voice for local communities influenced by land investments.

## Learning resources

- Land Matrix – An online database on land deals (http://landmatrix.org/en).
- Farmlandgrab – A multilingual news compilation site, with excellent search facilities (http://www.farmlandgrab.org).
- Land Portal – A site useful for keeping track of international aid agency-led initiatives (https://landportal.info).
- International Institute for Environment and Development (IIED) – A source of numerous downloadable papers, and materials on legal resources (http://www.iied.org/land-acquisitions-rights).
- Future Agricultures Consortium (FAC) – A research consortium focusing on Africa, including material on land and agricultural commercialisation (http://www.future-agricultures.org/research/land).

## Bibliography

Alden Wily, L. (2012) 'Looking back to see forward: the legal niceties of land theft in land rushes', *The Journal of Peasant Studies*, 39(3–4): 751–775.

Amanor, K.S. (2012) 'Global resource grabs, agribusiness concentration and the smallholder: two West African case studies', *The Journal of Peasant Studies*, 39(3–4): 731–749.

Cabral, L. & Norfolk, S. (2016) 'Inclusive land governance in Mozambique: good law, bad politics?', *IDS Working Paper*, 478. Brighton: Institute of Development Studies.

Cotula, L. (2015) *Land Rights and Investment Treaties: Exploring the Interface*. London: International Institute for Environment and Development.

Cotula, L. & Leonard, R. (2010) *Alternatives to Land Acquisitions: Agricultural Investment and Collaborative Business Models*. London: International Institute for Environment and Development.

Cotula, L., Oya, C., Codjo, E., Eid, A., Kakraba-Ampeh, M., Keeley, J., Kidewa, A.L., Makwarimba, M., Seide, W.M., Nasha, W.O., Asare, R.O. & Rizzo, M. (2015) 'Testing claims about large land deals in Africa: findings from a multi-country study', *Third World Quarterly*, 50(7): 903–925.

Edelman, M., Oya, C. & Borras Jr, S.M. (2013) 'Global land grabs: historical processes, theoretical and methodological implications and current trajectories', *Third World Quarterly*, 34(9): 1517–1531.

Fairbairn, M. (2015) 'Foreignization, financialization and land grab regulation', *Journal of Agrarian Change*, 15(4): 581–591.

Fairhead, J., Leach, M. & Scoones, I. (2012) 'Green grabbing: a new appropriation of nature?', *The Journal of Peasant Studies*, 39(2): 237–261.

Gibbon, P. (2011) 'Experiences of plantation and large-scale farming in 20th century Africa', *DIIS Working Paper*, 20. Copenhagen: Danish Institute of International Studies.

Hall, D., Hirsch, P. & Li, T. (2011) *Powers of Exclusion: Land Dilemmas in Southeast Asia*. Singapore: NUS Press.

Hall, R., Scoones, I. & Henley, G. (2016) 'Strengthening land governance: lessons from implementing the voluntary guidelines', *LEGEND State of the Debate Report*. London: Overseas Development Institute. Online. Available at: https://landportal.info/sites/landportal.info/files/Strengthening%20Land%20Governance.pdf

Hall, R., Scoones, I. & Tsikata, D. (eds) (2015a) *Africa's Land Rush: Rural Livelihoods and Agrarian Change*. Woodbridge: James Currey.

Hall, R., Tsikata, D. & Scoones. I. (2017) 'Plantations, outgrowers and commercial farming in Africa: agricultural commercialization and implications for agrarian change', *The Journal of Peasant Studies*, 44(3): 515–537.

Hall, R., Edelman, M., Borras Jr, S.M., Scoones, I., White, B. & Wolford, W. (2015b) 'Resistance, acquiescence or incorporation? An introduction to land grabbing and political reactions from below', *The Journal of Peasant Studies*, 42(3–4): 467–488.

Jayne, T.S., Chamberlin, J., Traub, L., Sitko, N., Muyanga, M., Yeboah, F.K., Nkonde, C., Anseeuw, C., Chapoto, A. & Kachule, R. (2015) 'Africa's changing farmland ownership: the rise of the emergent investor farmer', Plenary paper presented at the 29th Triennial International Conference of Agricultural Economists, 13 August 2015, Milan, Italy.

Li, T. (2011) 'Centring labour in the land grab debate', *The Journal of Peasant Studies*, 38(2): 281–298.

Little, P. & Watts, M. (eds) (1994) *Living Under Contract: Contract Farming and Agrarian Transformation in Sub-Saharan Africa*. Madison: University of Wisconsin Press.

McCarthy, J.F., Vel, J.A. & Afiff, S. (2012) 'Trajectories of land acquisition and enclosure: development schemes, virtual land grabs, and green acquisitions in Indonesia's Outer Islands', *The Journal of Peasant Studies*, 39(2): 521–549.

Moyo, S., Yeros, P. & Jha, P. (2012) 'Imperialism and primitive accumulation: notes on the new scramble for Africa', *Agrarian South: Journal of Political Economy*, 1(2), 181–203.

Peluso, N. & Lund, C. (2011) 'New frontiers of land control: introduction', *The Journal of Peasant Studies*, 38(4): 667–681.

Scoones, I., Hall, R., Borras Jr, S.M., White, B. & Wolford, W. (2013) 'The politics of evidence: methodologies for understanding the global land rush', *The Journal of Peasant Studies*, 40(3): 469–483.

Smalley, R. (2013) 'Plantations, contract farming and commercial farming areas in Africa: a comparative review'. *FAC Working Paper*, 55. Land and Agricultural Commercialization in Africa project. Brighton: Future Agricultures Consortium.

Vermeulen, S. & Cotula, L. (2010) 'Over the heads of local people: consultation, consent, and recompense in large-scale land deals for biofuels projects in Africa', *The Journal of Peasant Studies*, 37(4): 899–916.

White, B., Borras Jr, S.M., Hall, R., Scoones, I. & Wolford, W. (2012) 'The new enclosures: critical perspectives on corporate land deals', *The Journal of Peasant Studies*, 39(3–4): 619–647.

Wolford, W., Borras Jr, S.M., Hall, R., Scoones, I. & White, B. (2013). 'Governing global land deals: the role of the state in the rush for land', *Development and Change*, 44(2): 189–210.

2.15

# Metabolic rift

Richard York

The concept of metabolic rift originated with Karl Marx's (1867/1976) analysis of capitalist agricultural practices. Sociologist John Bellamy Foster (1999, 2000) explicates Marx's conceptualization, providing the term "metabolic rift" and arguing that it is central to understanding the modern environmental crisis. Foster, Clark, and York (2010) build on this work and use "ecological rift" as a broader term than metabolic rift (although often the terms are used synonymously) to go beyond Marx's focus on agriculture and encompass the full array of environmental processes that have been disrupted by global capitalism. This chapter recounts that conceptual journey from Marx to the present.

## Marx's conceptualization

Marx carefully studied agricultural practices and followed developments in soil science, especially the work of the German chemist Justus von Liebig, who made important contributions to the scientific understanding of the role of nitrogen and minerals as nutrients in crop production. Marx observed that before the capitalist-industrial era, traditional agriculture to a large extent entailed local production for local consumption. In this system, the nutrients extracted from the earth to grow crops are typically recycled back into the soil in the form of waste (human, animal, and agricultural) stemming from production and consumption proximate to the site of agricultural production. In many traditional societies, this nutrient cycle or metabolic cycle – a process of exchange of nutrients from soil to food and other agricultural products to waste and back to the earth – ensured the enduring fertility of the soil, sustaining the land and people.

In contrast to the traditional system of local production, capitalism led to a separation of the sites of production from the sites of consumption. This separation was driven by the rapid urbanization of the population in many European countries that came in the wake of industrialization, as displaced peasants moved to the cities to work in factories. Additionally, the enclosure movement combined many small tracts of land into a few larger ones that were generally controlled by wealthy owners. The affluent landowners, following capitalist principles, structured production to maximize short-term profits, rather than to sustain the long-term fertility of the soil or provide for human needs. In this new system, soil nutrients in the form of food and other agricultural products were moved from the country to the markets of the city. In the city the nutrients ended up as pollution, being dumped into rivers and other bodies of water and accumulating in garbage heaps and in the streets, instead of being returned to the country where they could fertilize farmland. This disruption in the traditional nutrient cycle constituted a metabolic rift – a break in the process of exchange – that undermined the sustainability of agricultural production, harming the environment and threatening the food supply for humans.

## Extensions of metabolic rift

Clark and York (2005) developed and expanded the applicability of metabolic rift analysis by showing how it could be used to help understand environmental crises other than the degradation of soil, by applying Foster's Marxian insights to an analysis of the global carbon cycle and the climate change crisis. They showed how the dynamics of capitalism disrupt the global carbon cycle by greatly increasing fossil-fuel use, which adds long-stored carbon to the atmosphere and oceans, while also impairing carbon sinks, such as by driving deforestation around the world. Following this work, multiple authors used the metabolic rift conceptualization to analyze a variety of other processes of anthropogenic environmental change, including the alteration of the nitrogen cycle (Mancus, 2007) and the destruction of fisheries (Clausen and Clark, 2005; Longo et al., 2015). Ariel Salleh (2010) has done important work building on metabolic rift theory, showing the importance of grassroots movements and people marginalized in the capitalist world-system (indigenous people, peasants, and unpaid caregivers) in creating metabolic value and countering the rift creation of capitalism.

One important feature of the metabolic rift concept is that it highlights a qualitative dimension of environmental problems. Quantitative dimensions of environmental degradation are widely recognized by ecological Marxists and other environmental scholars, such as the fundamental unsustainability of exponential economic and population growth (Schnaiberg, 1980; York et al., 2003). Metabolic rift analyses bring to light how even well before the scale of production becomes unsustainable due to the growth dynamics of capitalism, qualitative features of economic systems can undermine the conditions of production and life. The depletion of agricultural soil fertility due to the practices Marx identified is the most obvious example of how the qualitative aspects of processes can undermine sustainability even when production is at a small scale.

Metabolic rift analyses have shown that the generation of rifts in ecological processes has been common throughout the era of capitalist development. Clark and York (2008) characterize the capitalist-ecological regime as in part one of "rifts and shifts," where capitalist development successively disrupts one ecological process, undermining the availability of some natural resource or ecosystem function, thereby creating a metabolic rift, then shifts to using different resources and rupturing other ecological processes. In agriculture this can be seen where the exploitation of the soil under capitalism led to declining fertility of farmland, a rift, which was in part addressed by mining guano from oceanic islands, a shift, which disrupted the ecology of these islands, especially their seabird colonies and the fisheries surrounding them that relied on the nutrients from the guano (Clark and Foster, 2009). Subsequently, these ecological disruptions were shifted to those stemming from chemical fertilizers and pesticides. In this fashion, capitalism creates a chain of ecological crises, where each one is partially "solved" by creating a new crisis. This can be seen in the energy sector, where early industrial development led to massive deforestation to provide fuel wood for factories, then to fossil fuels to replace biomass as the primary fuel source, which, of course, led to global climate change and a variety of other pollution problems stemming from fossil-fuel use. The process of rifts and shifts highlights how ecological problems cannot properly be understood as simply about prevailing technologies, but rather about the capitalist system, which continually pushes for private profits at any cost.

Some scholars have criticized some of the theoretical aspects of the metabolic rift. Jason W. Moore's (2011, 2015) work to a large degree is a reaction to Foster's formulation of the metabolic rift. Moore claims that Foster's theorization, in making distinctions between the ecological and social, is grounded in Cartesian dualism. Schneider and McMichael (2010) recognize metabolic rift analysis as important, but criticize Foster's conceptualization as based on an outmoded understanding of agricultural ecological processes and labor practices. Following Moore, they argue that the metabolic rift tradition tends to separate the ecological from the social, and that the rift is in part due to historically changing modes of labor practice and thought that go back to the origins of capitalism. Foster (2016) provides a response to these critiques, noting the limitations of the monist view of Moore and his followers, where the insistence on making no distinction between biophysical and social processes undermines dialectical analysis and denies ecological science and the material reality of the world. As Foster (2016) has noted, Moore's analysis is focused on the language used in discourse rather than on the material world, making it at odds with Marx's approach.

## Continuing importance of metabolic rift theory

The metabolic rift has taken its place as an important, even essential, part of Marxian ecological analyses, especially following Foster, Clark, and York's (2010) further integration of the concept into political-economic theory on environmental crises. A number of Marxian scholars now draw on the concept in a variety of ways as part of efforts to extend and refine socio-ecological theory (e.g., Burkett, 2006; Dickens, 2004; Malm, 2016; York and Mancus, 2009). The central strength of metabolic rift analysis is the instance on recognizing how socio-ecological material processes are dialectically shaped by political-economic regimes, and in particular how the modern global environmental crisis needs to be understood as a product of the dynamics of capitalism.

## Learning resources

The web-based journal *Climate & Capitalism* is a good source of ecosocialist analyses of ecological crises. A recently posted interview of John Bellamy Foster about the metabolic rift theory and critiques of it provides a helpful introduction to the topic:

http://climateandcapitalism.com/2016/06/06/in-defense-of-ecological-marxism-john-bellamy-foster-responds-to-a-critic.

*Monthly Review*, edited by John Bellamy Foster, is one of the longest standing Marxian journals in the United States and frequently has articles on ecological Marxism. A recent article there explains metabolic rift in a clear and accessible manner:

http://monthlyreview.org/2013/12/01/marx-rift-universal-metabolism-nature.

## Bibliography


Burkett, P. (2006) *Marxism and Ecological Economics*, Boston: Brill.

Clark, B. & Foster, J.B. (2009) 'Ecological imperialism and the global metabolic rift: unequal exchange and the guano/nitrates trade', *International Journal of Comparative Sociology*, 50(3–4): 311–334.

Clark, B. & York, R. (2005) 'Carbon metabolism: global capitalism, climate change, and the biospheric rift', *Theory and Society*, 34(4): 391–428.

Clark, B. and York, R. (2008) 'Rifts and shifts: getting to the root of environmental crises', *Monthly Review*, 60(6): 13–24.

Clausen, R. & Clark, B. (2005) 'The metabolic rift and marine ecology: an analysis of the ocean crisis within capitalist production', *Organization & Environment*, 18(4): 422–444.

Dickens, P. (2004) *Society and Nature: Changing Our Environment, Changing Ourselves*, Cambridge: Polity.

Foster, J.B. (1999) 'Marx's theory of metabolic rift: classical foundation for environmental sociology', *American Journal of Sociology*, 105(2): 366–405.

Foster, J.B. (2000) *Marx's Ecology: Materialism and Nature*, New York: Monthly Review Press.

Foster, J.B. (2016) 'Marxism in the anthropocene: dialectical rifts on the left', *International Critical Thought*, 6(3): 393–421.

Foster, J.B., Clark, B. & York, R. (2010) *The Ecological Rift: Capitalism's War on the Earth*, New York: Monthly Review Press.

Longo, S.B., Clausen, R. & Clark, B. (2015) *The Tragedy of the Commodity: Oceans, Fisheries, and Aquaculture*, New Brunswick, NJ: Rutgers University Press.

Malm, A. (2016) *Fossil Capital*, London: Verso.

Mancus, P. (2007) 'Nitrogen fertilizer dependency and its contradictions: a theoretical explanation of socio-ecological metabolism', *Rural Sociology*, 272(2): 269–288.

Marx, K. (1867/1976) *Capital*, vol. 1. New York: Vintage.

Moore, J.W. (2011) 'Transcending the metabolic rift: a theory of crises in the capitalist world-system', *The Journal of Peasant Studies*, 38(1): 1–46.

Moore, J.W. (2015) *Capitalism in the Web of Life: Ecology and the Accumulation of Capital*, London: Verso.

Salleh, A. (2010) 'From metabolic rift to "metabolic value": reflections on environmental sociology and the alternative globalization movement', *Organization & Environment*, 23(2): 205–219.

Schnaiberg, A. (1980) *The Environment: From Surplus to Scarcity*. New York: Oxford University Press.

Schneider, M. & McMichael, P. (2010) 'Deepening, and repairing, the metabolic rift', *The Journal of Peasant Studies*, 37(3): 461–484.

York, R. & Mancus, P. (2009) 'Critical human ecology: historical materialism and natural laws', *Sociological Theory*, 27(2): 122–149.

York, R., Rosa, E.A. & Dietz, T. (2003) 'A rift in modernity? Assessing the anthropogenic sources of global climate change with the STIRPAT model', *International Journal of Sociology and Social Policy*, 23(10): 31–51.

# Offsetting

2.16

## Heather Lovell

Offsetting in relation to the environment involves undertaking additional environmentally beneficial activity in order to compensate for known environmental damage elsewhere. The aim is to achieve 'no net loss' of environmental assets, so offsetting is about counteracting or neutralising environmental harm – for example caused by a new road or housing development. Offsetting is therefore conditional on being able to calculate the amount of environmental harm that has been done, as well as the benefit (in order that remedial activity can be undertaken). For this reason much of the day-to-day practice of offsetting has centred on valuation and measurement. A fundamental belief underpinning environmental offsetting is that different environmental activities are commensurable, i.e. that they can be made to be the same or equivalent.

An early application of environmental offsetting was in relation to wetlands. It was first proposed in the USA during the 1970s – through the 1972 Clean Water Act – that new development could take place in an area with valuable wetlands as long as the wetlands could be recreated in another location, within the same catchment. In this way the environmental harm caused by a new development would be cancelled out or 'offset'. But it was when offsetting was introduced to mitigate greenhouse gas atmospheric pollution within the United Nations Framework Convention on Climate Change (UNFCCC) Kyoto Protocol, through what is known as the 'Clean Development Mechanism' (CDM), that it gained much wider recognition and uptake. Carbon offsetting has taken place at an international scale between nation states since 2006. Countries in the Global South are able to register projects which save or sequester carbon in order to counteract emissions from countries in the Global North ('Annex 1' countries, as defined in the Kyoto Protocol). The value of investment in greenhouse gas emission reduction projects in developing countries through the CDM is calculated to be $US90 billion (2006–2014), which equates to 13 per cent of the total renewable energy investment in these countries (World Bank & Ecofys, 2015: 35).

However, as is discussed below, offsetting has not had unanimous acceptance as a mechanism to protect the environment. Offsetting has in many cases been implemented as a market-based response to environmental problems, and as such sits within a wider set of neoliberal ideas, debates and practices, including privatising environmental assets, natural capital and greening capitalism. There have been vocal opponents to offsetting who disagree with any type of response to environmental problems that places a financial value on the environment, viewing it as a 'licence to pollute'.

In the remainder of this short chapter the following topics are covered: different types of offsetting, illustrated with examples; and the main theoretical and practical disadvantages and advantages of offsetting.

## Types of offsetting

The two main types of environmental offsetting are biodiversity and carbon offsetting. Here their practices, institutional structures, and scope and size of activity are briefly outlined.

### *Biodiversity offsetting*

Biodiversity offsetting has grown in popularity of late – with legislation requiring or encouraging use of biodiversity offsets now in operation in 45 countries, and under development in another 27 (Madsen et al., 2011). But biodiversity offsetting is not a new concept – it has been practiced in some form in countries such as the USA, Australia and the UK for decades. Biodiversity offsets are defined as:

> measurable conservation outcomes resulting from actions designed to compensate for significant residual adverse biodiversity impacts arising from project development after appropriate prevention and mitigation measures have been taken. The goal of biodiversity offsets is to achieve no net loss and preferably a net gain of biodiversity on the ground with respect to species composition, habitat structure, ecosystem function and people's use and cultural values associated with biodiversity.
> (BBOP, 2009: 4)

A number of key principles of biodiversity offsetting are captured in this comprehensive definition. First, the notion of a hierarchy of actions, with offsetting as a 'last resort', used only when measures to prevent and mitigate environmental harm have already been implemented. Second, the principle of 'no net loss' of biodiversity, i.e. that offsetting at a minimum brings biodiversity back up to its original pre-development level.

It is difficult to obtain accurate data on the size of biodiversity offsets worldwide, as many offset schemes are small, and not necessarily formally registered. The 2011 State of the Biodiversity Market Report estimated the annual global market size as US$2 to 4 billion, with at least 187,000 hectares of land under some sort of conservation management or permanent legal protection (Madsen et al., 2011: v).

Common ecosystem types where biodiversity offsetting is practiced include wetlands (e.g. Wetland Banking in the USA), marine (fish habitat compensation, e.g. in Canada), and forests (e.g. Brazilian Forest Code). Biodiversity offsets are also used to protect particular species or groups of species, such as native vegetation (e.g. Australian State of Victoria's BushBroker scheme; see also the USA Endangered Species program (POST, 2011)).

## Carbon offsetting

Carbon offsets are perhaps what many people think of when the term 'offsetting' is used in relation to the environment. A carbon offset allows emission reduction targets to be met in one location by purchasing emission reductions from a climate mitigation project based elsewhere. A distinctive feature of carbon offsetting is its international sphere of operation, enabled by the uniform global mixing of greenhouse gases in the atmosphere. In other words, because of a shared atmosphere (there is one global level of $CO_2$ in the atmosphere, currently c. 400 parts per million) carbon emissions can effectively be offset by other carbon saving or carbon sequestering projects anywhere else in the world. Thus it is that carbon offsets became a formal policy tool at the international level as part of the 1997 UNFCCC Kyoto Protocol 'flexibility mechanisms', finally agreed on in 2001 at the 7th Conference of the Parties (COP) in Marrakesh. The CDM brought into being a highly regulated and complex set of measurement, legal and financial mechanisms to enable the global production and trading of carbon offsets (Bumpus & Liverman, 2008). In recent years, because of prolonged uncertainty about the UNFCCC agreement, the CDM has lost value, both financially and politically.

There is also a second type of global carbon offset market known as the voluntary offset market. This is an informal or 'parallel' market governed by a mix of non-governmental and private sector organisations. The voluntary offset market allows companies and individuals who wish to offset their greenhouse gas emissions to directly compensate for them. The voluntary carbon offset market is much smaller than the UNFCCC compliance market, with 87 MtCO2e transacted by voluntary buyers in 2014 (Hamrick & Goldstein, 2015: 2) compared with 60 million MtCO2e (CERs) in the CDM (World Bank & Ecofys, 2015: 36). It was in the voluntary sector though that the original idea of carbon offsets was first put into practice, with a project in 1988 by a US company called Applied Energy Services that offset carbon emissions from a coal-fired power plant through a forest plantation project in Guatemala (Moura-Costa & Stuart, 1998).

## Disadvantages of offsetting

The main disadvantages of offsetting can be classified into two categories: moral and practical. First, moral concerns are about the broader ethics of valuing the environment. In these arguments there is a fundamental objection to a whole raft of market-based environmental management approaches, including offsetting. It is believed that putting any financial value on environmental assets – commodifying them – devalues them as it undermines their intrinsic value (Bond & Dada, 2004; O'Neill, 2007).

Practical concerns about offsetting are more specific and focus on *how* offsetting has been done (in contrast to the more fundamental 'why' questions posed by those with moral objections). An example is concern over the amount of resources involved in setting up a system of governance for offsets (within the CDM this has been labelled 'CDM bureaucracy' (see Lovell, 2010)). Objectors argue that the overall environmental benefit of offsetting is relatively modest, and that scarce resources to tackle environmental problems could be more usefully deployed elsewhere (Lohmann, 2005).

Another key practical disadvantage of offsets that is commonly stated is their narrowness, or limited scope. Again it is perhaps easiest to illustrate this with reference to the CDM, where a key governance challenge has been the relative weighting given to sustainable development issues in CDM projects in the Global South versus emission reductions. The CDM has been criticised for not delivering on its sustainable development objectives, and focusing instead too narrowly on carbon (Olsen, 2007).

## Advantages of offsetting

There are a number of advantages that are commonly discussed in relation to environmental offsetting. First, it is suggested that offsetting allows for the productive engagement of industry and business in finding solutions to environmental problems through adapting practices and approaches that they understand, e.g. financial valuation, accounting, and auditing. Such arguments are part of a wider debate about natural capital and business and the environment. These are pragmatic viewpoints which see benefit in diversity: offsetting is one of a number of tools and approaches used to help protect the environment, which can be embraced alongside other approaches. Second, as a counter to criticisms outlined above about the amount of resources spent on creating and managing offsets, it is suggested that these can be solved by good governance mechanisms. In other words, it is the detail of how contemporary environmental mitigation practices are put into practice that is important (MacKenzie, 2009).

## Learning resources

There are numerous websites that provide information on offsets:

Business and Biodiversity Offsets Programme: this website is about biodiversity offsets and includes a useful summary of what biodiversity offsets are, and a Publications Library: http://bbop.forest-trends.org.

*The Guardian*: this UK newspaper has published a comprehensive and easy-to-read article about carbon offsetting, which also has links to other relevant resources "Complete guide to carbon offsetting": https://www.theguardian.com/environment/2011/sep/16/carbon-offset-projects-carbon-emissions.

Two good academic journal papers that summarise key thinking in the field, across the topics of carbon and biodiversity offsets, include:

Bumpus, A. & Liverman, D. (2008) "Accumulation by decarbonisation and the governance of carbon offsets", *Economic Geography*, 84: 127–156.

Bull, J.W., Suttle, K.B., Gordon, A., Singh, N.J. & Milner-Gulland, E. (2013) "Biodiversity offsets in theory and practice", *Oryx*, 47: 369–380.

## Bibliography


Bond, P. & Dada, R. (2004) *Trouble in the Air: Global Warming and the Privatised Atmosphere*. Durban, South Africa: Center for Civil Society, University of Natal.

Bumpus, A. & Liverman, D. (2008) "Accumulation by decarbonisation and the governance of carbon offsets", *Economic Geography*, 84: 127–156.

Business and Biodiversity Offsets Programme (BBOP) (2009) "Business, biodiversity offsets and BBOP: an overview", Washington, DC: Business and Biodiversity Offsets Programme, Forest Trends.

Hamrick, K. & Goldstein, A. (2015) "Ahead of the curve: state of the voluntary carbon market", Washington, DC: Ecosystem Marketplace & Forest Trends.

Lohmann, L. (2005) "Marketing and making carbon dumps: commodification, calculation and counterfactuals in climate change mitigation" *Science as Culture*, 14: 203–235.

Lovell, H. C. (2010), "Governing the carbon offset market" *Wiley Interdisciplinary Reviews: Climate Change*, 1: 353–362.

MacKenzie, D. (2009) "Making things the same: gases, emission rights and the politics of carbon markets" *Accounting, Organizations and Society*, 34: 440–455.

Madsen, B., Carroll, N., Kandy, D. & Bennett, G. (2011) "Update: state of The Biodiversity Market Report", Washington, DC: Forest Trends.

Moura-Costa, P. & Stuart, M.D. (1998) "Forestry-based greenhouse gas mitigation: a story of market evolution", *The Commonwealth Forestry Review*, 77(3): 191–202.

Olsen, K. (2007) "The clean development mechanism's contribution to sustainable development: a review of the literature", *Climatic Change*, 84: 59–73.

O'Neill, J. (2007) *Markets, Deliberation and Environment*, London: Routledge.

Parliamentary Office of Science and Technology (POST) (2011) "POSTNote 369: Biodiversity offsetting", London: Parliamentary Office of Science and Technology.

World Bank & Ecofys (2015) "State and trends of carbon pricing", Washington, DC: World Bank Climate Group.

# 2.17 Peak oil

## Michael Lynch

The past decade saw a near obsession about the prospect of 'peak oil'. Dozens of books and innumerable articles (print and internet) have predicted an imminent peak and decline in global oil production with a variety of economic and environmental impacts. Aside from widespread coverage in the business press, the idea has entered popular culture, even being mentioned in an episode of *The Simpsons*.

The actual amount of research in support of the theory was fairly slim, with most writers either repeating the arguments of others or focusing on the impacts of peak oil. David Goodstein (2005), for example, argued that peak oil necessitated more nuclear power, Michael J. Lynch (no relation to the present author) wrote about peak oil and prison reform, while Jeff Rubin talked about the end of long-distance transport of foodstuffs. Some were more apocalyptic, suggesting peak oil meant the 'possible extinction of mankind.' Needless to say, promoters of renewable energy used the peak oil theory to bolster their arguments for increased government support for their products.

Partly because of the post-World War II economic boom, concerns about rapid population growth and the soaring demand for resources that resulted led to a revival of long-standing fears about 'over-population' going back to nineteenth-century English economist Thomas Malthus (who proposed that there are periodically too many people and too few natural resources available to sustain them). This neo-Malthusianism was exemplified by biologist Paul Ehrlich's 1968 *The Population Bomb* and the Club of Rome project that, in 1972, produced the best-selling book *The Limits to Growth* (Meadows et al., 1972). Both texts were immensely influential and their arguments were bolstered by the oil crises of the 1970s (when prices soared) as well as inflation in prices for food and other resources. By 1980, a broad consensus existed that oil prices, having tripled twice in the 1970s, would continue rising due to inadequate resource supply.

These predictions all proved incorrect. More importantly, they relied on bad theory. Ehrlich's apocalyptic predictions of looming global starvation were based almost entirely on the assumption that agricultural productivity could not keep up with population growth, even though in many parts of the world rural land was still exploited using low-tech methods. The Club of Rome's computer model, used in *The Limits to Growth,* was flawed primarily in its conservative assumptions about future resource availability and omission of technical innovation and price effects on demand. Both have seen their warnings proved incorrect, although many still cling to them.

The 1980s and 1990s saw a return to more normal commodity prices, and fears of resource scarcity receded, with only a few analysts still writing on the subject. However, Jean Laherrere and Colin Campbell, two retired petroleum geologists, revived the so-called Hubbert curve. Their 1998 *Scientific American* article 'The End of Cheap Oil' gained much notoriety. (Campbell's 1989 article which claimed the peak was occurring that year did not receive much attention.) When prices began rising after supply disruptions in Venezuela and Iraq, they seized on that as evidence of the validity of their research and attention to the subject soared.

Unfortunately, this view mirrored the way many analysts misinterpreted the 1970s transient oil supply problems as evidence of permanent, geological scarcity. An extraordinary amount of disruptions in oil producing nations, including the strike and mass layoffs at Petroleos de Venezuela in 2002/3, the second Gulf War in 2003, hurricane Katrina in 2005, attacks on Nigerian oil infrastructure beginning in 2006, the Arab Spring in 2010, and economic sanctions against Iran, took well over a billion barrels of oil off the market. Yet the consensus pivoted from interpreting high prices as a cycle to the belief that \$100 a barrel was a new price floor.

But peak oil theory reflected not just a simple misinterpretation, but a number of serious analytical flaws. 'Presentism' was driving peak oil advocates, and two particular cases demonstrate this. When a brief, sharp spike in oil prices occurred in 2008, passing \$140/barrel, prominent peak oil advocates saw it not as a temporary, speculative bubble but rather as the beginning of an extended period of even higher prices, insisting that \$200/barrel was imminent and inevitable.

In terms of supply, predictions by peak oil advocates like Campbell and Laherrere were notable for predicting looming production peaks in nearly every country, and having to repeatedly revise their estimates of production and resources upwards. Both were notably pessimistic about Russian oil, ignoring its ample resource base and immature technology, and insisting that production levels couldn't be maintained even as new highs were repeatedly set.

Aside from bias (cherry-picking data that supported a pessimistic view), the theory behind the peak oil argument was deficient, but few examined the actual research. The use of the Hubbert curve, a bell-shaped curve said to represent national or global production paths, was not only not based on any coherent theory, but could easily be seen to be violated in most cases. Treating it as a natural scientific fact was in practice contradictory to the evidence. Indeed, rather than natural science, nearly all of the work constituted curve-fitting and simple extrapolation, and again and again, real world developments violated the supposedly scientifically-determined production paths, such as the assertion that production, once it peaked, would never recover.

Tellingly, a number of arguments were made that defied belief, including that technological advances did not increase supply but simply accelerated production and that estimates of field sizes known as P50 or 50 per cent probability were on average accurate and did not increase over time. Both were at the core of the mathematical research, but both were easily shown to be incorrect.

As critics pointed out these flaws, some peak oil advocates shifted their position to focus on above-ground problems, arguing that resource scarcity was not a threat, but rather the inability to add enough capacity to meet demand would result in a peak in supply. A combination of political constraints, rapid decline in existing fields, and high costs were said to be challenges unlikely to be overcome.

But most of these problems are long-standing and have been overcome throughout the industry's history. Here, the inexperience of most peak oil advocates came into play, as their ignorance of the industry and failure to examine most of the available data led them to overstate the problems they were describing. When discussing decline in old fields, for example, there is rarely mention of previous decline levels; the industry has always had to replace capacity lost to depletion, and current levels are not significantly different from historical amounts. And political risk has been a challenge for the industry for over a century—Stalin got his political start as an oil worker organizer in Baku in the early twentieth century.

Some problems are real but cyclical. Costs have risen, tripling in the past decade (before prices dropped) but not because 'the easy oil is gone' so much as due to price inflation caused by high investment levels by big oil companies. (The same thing occurred in the early 1980s.) Also expensive resources were developed not because of a lack of alternatives, but the fact that high prices allowed them to be profitably developed. The cost of the marginal barrel rose, but with lower prices costs are now falling again because of cyclically lower input costs and abandonment of expensive projects.

Peak oil advocates have viewed the rise of shale oil and gas exploitation with the same pessimistic bias they have shown towards conventional oil. They argue that the resource was too expensive to produce and high well decline rates would make it difficult if not impossible to produce significant amounts. Just as in 1989 Colin Campbell dismissed deepwater resources as inconsequential, Arthur Berman in 2010 insisted that the Marcellus shale would never be a major producer. (It is now the largest single source of gas in the U.S.).

The entire episode supports recent theories about the formation of beliefs, namely that people tend to adopt beliefs first, then seek confirming evidence, ignoring anything contradictory. The mathematical models and the assumptions used in the initial peak oil publications were clearly invalid, yet advocates continued to embrace them for years. The fact that recent concerns about supply adequacy mirrored those of the 1970s was apparently unknown to a new generation of neo-Malthusians and the repeated failure of their models and predictions has caused few if any to re-evaluate their views, which is an unfortunate commentary on the analytical abilities of so many.

## Learning resources


Adelman, M.A. (2003) *The Economics of Petroleum Supply*, Cambridge: MIT Press.

Aguilera, R.F. & Radetzki, M. (2015) *The Price of Oil*, Cambridge: Cambridge University Press.

Dahl, C. (2015) *International Energy Markets: Understanding Pricing, Policies and Profits*, Tulsa: Pennwell Publishing.

Lynch, M.C. (2015) *The Peak Oil Scare and the Coming Oil Flood*, Santa Barbara, CA: Praeger.

## Bibliography

Campbell, C.J. (1989) 'Oil Price Leap in the 90s', *Noroil*, 17(12): 35–38.
Campbell, C.J. & Laherrere, J. (1998) 'The End of Cheap Oil', *Scientific American*, 278(3) 78–83.
Ehrlich, P. (1968) *The Population Bomb*, New York: Ballantine.
Goodstein, D. (2005) *Out of Gas: The End of the Age of Oil*, New York: Norton Books.
Meadows, D.H., Meadows, D.L., Randers, J. & Behrens III, W.W. (1972) *The Limits to Growth: A Report for the Club of Rome's Project on the Predicament of Mankind*, New York: New American Library.

# 2.18 Planetary boundaries

## Katherine Richardson

The conditions we recognise as being the environment of the Earth are created by the interactions of a number of different global processes or 'sub-systems', i.e., the hydrological cycle, climate, biological activity, etc. These sub-systems are interlinked and, together, they constitute a single, complex Earth System (ES). The planetary boundaries framework attempts to identify scientifically-based levels of human perturbation of them beyond which its functioning may be significantly altered. The framework was developed within the relatively new discipline of Earth System Science that emerged in the late twentieth century. This discipline aims not only to develop an understanding of how the Earth functions as a system but also to describe and predict how changes occurring within and between components in the Earth System will influence the state of the system as a whole. The Amsterdam Declaration on Global Change in 2001 recognised that the Earth System 'behaves as a single, self-regulating system comprised of physical, biological, chemical and human components' (Earth System Science Partnership, 2001), i.e., that the influence of human activities can be detected at the planetary level and can potentially alter the state of the system as a whole.

The planetary boundaries concept was conceived at a 2008 workshop in Sweden convened by Johan Rockström, who at the time was Executive Director of Stockholm Environment Institute & Stockholm Resilience Centre. The Swedish workshop brought together experts from the contributory fields to ES science and considered whether the then contemporary understanding of ES function was sufficient to identify processes or components within the ES where human perturbation might have the potential to ultimately change the state of the system as a whole and, furthermore, to identify levels of perturbation of these processes/components beyond which the risk of human activities catalyzing a state change becomes substantially increased.

It is well established from paleo records that the ES has existed in a number of different 'states', i.e., much colder or warmer than is the case today. Over the last 100,000 years alone, ice-cores have documented large fluctuations in temperature, and large-scale shifts in ecosystems and environmental processes. Workshop participants noted that, while humans have existed in their current biological form for tens or hundreds of thousands of years, all great civilisations and everything we associate with modern humanity, i.e., development of agricultural practices, written language, etc., have developed within the last ~12,000 years. This is the geological epoch known as the Holocene. It is a prolonged period in human history where the climate has been unusually stable and relatively warm, enabling societies to make reliable plans for agriculture, trade and so on. The workshop deliberations therefore took as their starting point the fact that these 'Holocene-like' conditions represent the only ES state where we know for certain in which modern humanity can thrive. They concluded that it would, therefore, be unwise of humanity to allow its activities to substantially increase the risk of triggering a state change outside of Earth's current relatively stable interglacial conditions.

Based on the workshop deliberations, the first peer reviewed, published planetary boundaries paper (Rockström et al., 2009) identified nine processes within the ES that are critical for maintaining system state and that are also substantially impacted by anthropogenic activities:

- climate change
- ocean acidification
- biodiversity loss
- global freshwater use (hydrological cycle)
- atmospheric aerosol loading
- stratospheric ozone depletion
- change in land use
- biogeochemical flow (release of reactive nitrogen (N) and phosphorus (P))
- chemical pollution

Control variables that would serve as a robust global measure for these processes were suggested and discussed. For seven of the identified processes (not chemical pollution or atmospheric aerosol loading), a potential quantitative limit for human perturbation was proposed, based on an analysis of the available scientific literature. These proposed limits are the 'planetary boundaries'.

Thus, the planetary boundaries framework argues that these boundaries define a 'safe operating space' for humanity. This original planetary boundaries paper argued that three of the boundaries are exceeded: biodiversity loss, climate change, and the biogeochemical flow boundary for nitrogen. Note, however, that exceeding the planetary boundaries is not the same as crossing a threshold into environmental catastrophe. Rather, each of the planetary boundaries can be seen as analogous to the use of blood pressure measurement in health care. Having high blood pressure does not necessarily mean the patient will have a heart attack, but if blood pressure is not reduced, the patient does have an increased risk of serious and even life-threatening cardiovascular problems. Similarly, if every effort is made to reduce the human pressure on the global environment so it remains within the planetary boundaries, this reduces the risk of an undesirable change in critical processes that could potentially change the state of the ES as a whole.

The introduction of the planetary boundaries framework attracted considerable attention in civil society, within the business and policy communities, and within the scientific community. In light of the more recent scientific discussions about the nine critical Earth system processes, a second peer reviewed planetary boundaries paper appeared in 2015 (Steffen et al., 2015). For several of the planetary boundaries, this paper proposed changes in control variables (Table 2.18.1). This later paper focuses more on the functioning of the Earth system. Thus, it argues that the state of the ES is, ultimately, the product of the interaction between, and co-evolution of, the climate and the biosphere, i.e., the totality of living organisms and their interactions. Climate change and 'biosphere integrity' (which had been referred to as 'biodiversity' in the earlier paper) are argued to represent 'core' boundaries. The change in terminology from biodiversity to biosphere integrity was made to emphasise that it is not the numbers of individual species *per se* that is important for ES function but, rather, the activities of the biosphere as a whole. All of the other identified planetary boundaries are then perceived to act on the ES through their influence(s) on these two core boundaries.

Another change in the second paper is that 'chemical pollution' was redefined as the introduction of 'novel entities'. This was an effort to signify that chemical substances are not the only anthropogenic introductions to the environment of potential concern at the planetary level.

The 2015 planetary boundaries paper concluded (Figure 2.18.1) that four of the nine boundaries are currently exceeded: climate change, loss of biosphere integrity, land-system change and biogeochemical flows (both N and P).

While the prospect provided by the planetary boundaries framework of potentially being able to identify 'safe limits' for human perturbation of the global environment has been found attractive by many, the concept has also generated some criticism (*The*

**Table 2.18.1** Control variables for the nine planetary boundaries identified in the second planetary boundaries paper. The control variables for biosphere integrity are considered to be provisional and to be used only until more appropriate metrics are developed. No control variable for the introduction of novel entities was proposed.

| *ES process* | *Control variable* |
|---|---|
| Climate change | Atmospheric $CO_2$ concentration<br>Increase in radiative forcing relative to pre-industrial level |
| Loss of biosphere integrity | *Genetic diversity*: Extinction rate<br>*Functional diversity*: Mean species abundance (MSA) |
| Land-system change | Area of forested land as % of original forest cover |
| Freshwater use | Maximum amount of consumptive blue water use ($km^3$/yr) |
| Biogeochemical flows (P and N cycles) | *P cycle*: P flow from freshwater systems into the ocean<br>*N cycle*: Industrial fixation of N |
| Atmospheric aerosol loading | Aerosol Optical Depth (AOD) suggested, but there is much regional variation and no global value is suggested |
| Ocean acidification | Aragonite saturation state ($\Omega_{arag}$) |
| Stratospheric ozone depletion | $O_3$ concentration in DU (Dobson Units) |
| Novel entity | No global boundary yet identified |

From Steffen et al. (2015).

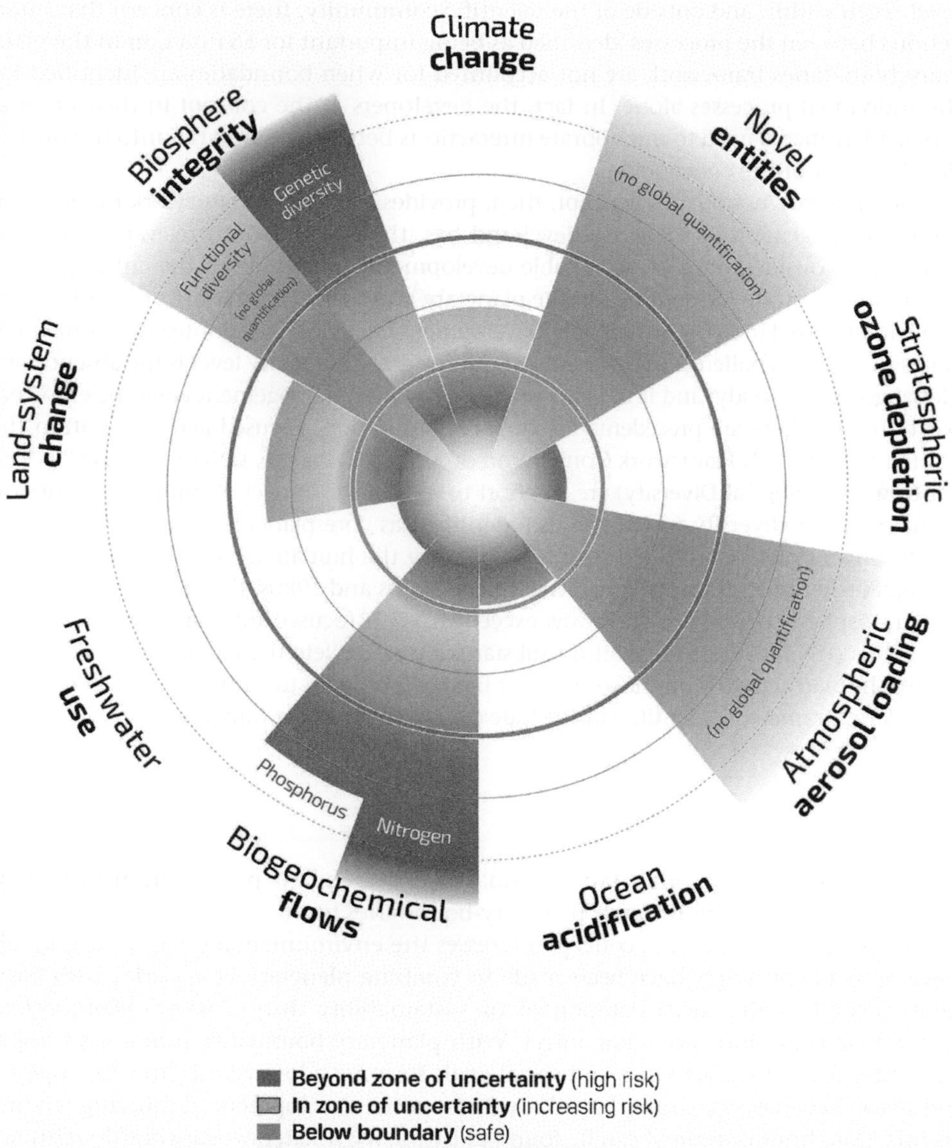


**Figure 2.18.1** Planetary boundaries: a safe operating space for humanity.
*Source*: Steffen et al. (2015). Design: Globaïa.

*Economist*, 2012). The criticisms most often voiced include questions regarding the validity of establishing such boundaries based on the assumption that Holocene-like conditions represent the only ES state where it is known for certain that modern humanity can thrive, and concern that it may not be appropriate to assemble the diverse processes identified as being important for determining ES state in the same framework as some (i.e., climate change) clearly operate at the global level while the most obvious impacts of others (i.e. release of reactive nitrogen and phosphorous) appear at the local/regional level. Both within and outside of the scientific community, there is concern that interactions between the processes identified as being important for ES function in the planetary boundaries framework are not accounted for when boundaries are identified for the individual processes alone. In fact, the developers of the concept in their original article identified a need to incorporate interactions between boundaries into the further development of the concept.

The planetary boundaries concept, then, provides a potential framework for environmental management at the global level and has, therefore, obvious relevance for environmental considerations of sustainable development. Indeed, all of the anthropogenic perturbations of the ES identified in the planetary boundaries framework are implicitly or explicitly referred to in the Sustainable Development Goals (SDGs) adopted by UN member states in 2015. A challenge in governing resources at the planetary level is the absence of a global governing body, and it remains unclear how such management could be enforced. Nevertheless, there are precedents for global coordination. Focused activities within the United Nations (UN Framework Convention on Climate Change, UNFCCC, and UN Convention on Biological Diversity) are an effort to reduce the impact of human activities on climate and biodiversity which are also identified as core planetary boundaries. Within these conventions, new mechanisms for managing the human–ES relationship are developing. Furthermore, it can be noted that in the 1970s and 1980s, the planetary boundary for stratospheric ozone depletion was exceeded. With focused international effort since the 1990s (UN Montreal Protocol on substances that deplete the ozone layer), the influence of human activities on the ozone layer has been brought to within what the planetary boundaries framework identifies as the 'safe operating space for humanity'.

## Learning resources

New developments in planetary boundaries research are posted at: http://www.stockholmresilience.org/research/planetary-boundaries.html

The planetary boundaries concept addresses the environmental pillar of sustainable development. Attempts have been made to combine planetary boundaries with basic goals related to the social component of sustainability (https://www.oxfam.org/en/research/safe-and-just-space-humanity). With planetary boundaries forming an upper boundary for human activities and social goals forming a lower limit, the 'safe operating space' becomes doughnut shaped. Updates on the development of thinking relating to this 'doughnut economy' can be found at http://www.kateraworth.com/doughnut.

The concept of planetary boundaries and management of environmental resources at the global level raises a number of governance issues. These have been reviewed by Galez, V., de Zeeuw, A., Shiroyama, H. & Tripley, D. (2016) 'Planetary boundaries – governing emerging risks and opportunities', *The Solutions Journal*, 7(3): 46–52. Online. Available at: https://www.thesolutionsjournal.com/article/planetary-boundaries-governing-emerging-risks-opportunities.

## Bibliography

Earth System Science Partnership (2001) *The Amsterdam Declaration on Global Change*. Online. Available from: https://web.archive.org/web/20070721122243/http:/www.essp.org/en/integrated-regional-studies/open-science-conferences/the-amsterdam-declaration.html.

*The Economist* (2012) 'Boundary conditions'. Online. Available from: http://www.economist.com/node/21556897.

Rockström, J., Steffen, W., Noone, K., Persson, Å., Chapin, S., Lambin, E.F., Lenton, T.M., Scheffer, M., Folke, C., Schellnhuber, J., Nykvist, B., de Wit, C.A., Hughes, T., van der Leeuw, S., Rodhe, H., Sörlin, S., Snyder, P.K., Costanza, R., Svedin, U., Falkenmark, M., Karlberg, L., Corell, R.W., Fabry, V.J., Hansen, J., Walker, B., Liverman, D., Richardson, K., Crutzen, P. & Foley, J. (2009) 'A safe operating space for humanity', *Nature*, 461: 472–475.

Steffen, W., Richardson, K., Rockström, J., Cornell, S.E., Fetzer, I., Bennett, E.M., Biggs, R., Carpenter, S.R., de Vries, W., de Wit, C.A., Folke, C., Gerten, D., Heinke, J., Mace, G.M., Persson, L.M., Ramanathan, V., Reyers, B. & Sörlin, S. (2015) Planetary boundaries: guiding human development on a changing planet. *Science*, 347(6223). doi: 10.1126/science.1259855.

# 2.19 Post-environmentalism

Christopher Buck

Post-environmentalism is a concept that describes new understandings of the biophysical world as well as humans' ethical and political relations to it. The prefix 'post' highlights how post-environmentalism resembles the postmodern project of questioning predominant narratives (Lyotard, 1984), in this case by describing (or calling for) a break with conventional accounts of the environment and conventional approaches to environmental ethics and politics.

## Reconceptualizing the nature of the environment

More often than not, the environment of environmentalism refers to the 'natural' world that provides the raw materials and resources that sustain human life on planet Earth. In the 1970s radical ecological thinkers such as Arne Naess (1995) and Murray Bookchin (1980) criticized this portrayal of the environment in favor of one that appreciates the complex interactions that occur between humans, other species, nutrient cycles, and the abiotic world. While these ecological accounts emphasize the relations that occur between the human species and other components of ecosystems, they continue to treat the dualism between humanity and nature as more or less fixed. Post-environmentalism, however, calls into question not only the stability of this dualism but also the stability of the identities of the entities that comprise ecosystems (Castree, 2003).

One way in which to challenge the dualism between humanity and nature is by suggesting that the former has physically transformed the latter to such an extent that the distinction between the two is no longer meaningful. In *The End of Nature*, for example,

American environmentalist Bill McKibben (1989) argued that pollution on a global scale in the form of greenhouse gas emissions responsible for climate change leaves no part of the planet unaffected by human activity. From this perspective, the idea of nature as a realm that exists independently of human societies loses its force. While embracing the term post-naturalism as opposed to post-environmentalism, philosopher Steven Vogel (2015) also highlights the ways in which humans have transformed the environment, but claims that this is not a recent development because nature (as McKibben uses the term) ends whenever humans encounter a new landscape, as they have done for millennia.

Another approach to post-environmentalism emerges from the field of Science and Technology Studies, most notably in the work of Donna Haraway and Bruno Latour. Whereas environmentalists have a reputation for blaming technology for accelerating ecological degradation, Haraway (1991) celebrates the revolutionary potential of technology insofar as it facilitates transgressing the boundaries between humans, other animals, and machines, which in turn disrupts the dualism between nature and human culture. Similarly, Latour (1993, 2004) emphasizes the role that hybrid entities play in shaping the ecological issues that affect contemporary public life. Rather than maintain a rigid distinction between active human subjects and an environment consisting of more or less passive objects, Latour draws attention to the ways in which networks comprised of both human and nonhuman entities function as actors in the world.

## Reconceptualizing environmental ethics

While challenging a strict dichotomy between humanity and nature might not appear to be particularly controversial, the perceived ethical implications of post-environmentalism have led some scholars to portray it as a 'social siege of nature' that provides justification for ecological degradation (Soulé, 1995, p. 137). This perspective seems to presuppose that the environment is worthy of ethical consideration solely on the basis of its otherness to humanity. In a similar vein, McKibben laments the end of nature because he sees the idea of nature that has an existence independent of human activity as a central source of ethical inspiration. Post-environnmentalists, however, do not claim that nonhuman entities lack ethical value but rather point towards different sources of this value other than the autonomy of nature.

If the environment is literally socially constructed as Vogel contends, insofar as social relations organize the labor that transforms the biophysical world to make it more amendable to human wants and needs, then people are causally responsible for the environment they inhabit. This causal responsibility can serve as a starting point for taking moral responsibility for the condition of the environment, especially when ecological degradation is an unintended consequence of economic production. Unlike approaches to environmental ethics that prioritize the preservation of wild nature, Vogel's post-natural environmental ethic has the advantage of encouraging people to take care of the artifacts in their lives by engaging in more sustainable forms of production and consumption.

Inspired by Latour's refusal to grant humanity a monopoly on the ability to act in the world, political theorist Jane Bennett (2010) draws attention to what she refers to as the vital materiality of things, which can serve as the basis for a post-environmental ethics. Much like Naess's deep ecology, which refuses to insist that human interests always take priority over those of other species, Bennett's 'vital materialism' undermines hierarchical ethical systems that place humanity at the pinnacle of creation by emphasizing the properties that humans share in common with other beings as opposed to what makes them unique. In addition, acknowledging how humans depend upon a multitude of nonhuman organisms that reside within our bodies, such as the gut bacteria that facilitate digestion, can foster a broader notion of self-interest that considers the well-being of other forms of life.

## Reconceptualizing environmental politics

Just as Bennett's vital materialism invites one to rethink what it means to be human, post-environmentalism encourages reflection on the identity of the environmental movement. Historian John Young coined the term in his attempt to weave together different strands of radical ecology into a 'post environmentalist movement' that 'recognize[s] the mix of soft energy, appropriate technology, sustainable economics, communal autonomy and cultural identity as the central ingredients of consistent ideology which can harness the diverse energies of a crisis-conscious generation' (Young, 1990, p. 165). Ironically, when the concept of post-environmentalism reemerged nearly 15 years later with the publication of Michael Shellenberger and Ted Nordhaus's *The Death of Environmentalism* (2004), it would become associated with two of the most vocal proponents for the increased use of nuclear power as a 'clean' energy source to combat climate change.

Shellenberger and Nordhaus call for the death of environmentalism because the movement is too attached to its identity as the protector of the environment, which has the consequence of framing major issues such as climate change too narrowly at the expense of building a broad coalition to address it. Public support for environmental causes tends to be widespread but not strongly felt, such that other concerns like economic security take precedence. For this reason, Shellenberger and Nordhaus argue that political responses to global warming are more likely to resonate with people if they are framed in ways that emphasize the job-creating potential of investments in clean energy rather than the benefits of reduced greenhouse gas emissions for charismatic megafauna like the polar bears inhabiting the icy regions of the Arctic Circle.

Another problem with environmentalism from a post-environmental perspective is that it tends to call for placing constraints on human activity to protect the environment rather than encouraging human ingenuity in the form of scientific and technological innovations to mitigate ecological degradation (Nordhaus and Shellenberger, 2007). Sociologist Frederick H. Buttel (2005), for example, argues that blanket condemnations of genetic engineering by environmentalists alienate potential allies in the scientific community who see a role for biotechnology in sustainable agriculture. While some environmentalists embrace the precautionary principle as a justification for prohibiting the implementation of new technologies until proven ecologically sound, Latour (2011) offers a post-environmental interpretation of the principle as encouraging policymakers to pay greater attention to the unintended ecological consequences of human actions.

Taking greater responsibility for the co-production of the environment also involves reconsidering the nature of democracy. Whereas some environmentalists look to nature for guidance when it comes to organizing political life, Vogel dismisses such appeals to nature as an attempt to avoid engaging in democratic deliberation, which he sees as the most rational means for political decision making. For Vogel, ecological degradation is not the result of a failure to heed nature's lessons but rather the result of leaving significant decisions regarding economic production up to the market instead of deciding them through democratic debate. Bennett offers a more provocative proposal for democratic reform by redefining the *demos* itself to include not only the human community but all forms of vital materiality and by entertaining the possibility of representing the 'voices' of nonhuman entities in political institutions (Bennett, 2010, p. 150).

## Conclusion

Post-environmentalism is difficult to define because only a handful of activists and scholars explicitly identify themselves as post-environnmentalists and they do not all agree over the meaning of the concept. That said, they all share an interest in challenging worldviews that reinforce a strict separation between human society on the one hand and the 'natural' environment on the other. At first glance it might seem as though post-environmentalism is an attempt to dismiss or reject the concerns of the environmental movement, but this initial impression is misleading. Rather, post-environnmentalists maintain a commitment to the goal of ecological sustainability but believe that this goal can best be achieved by appreciating how the natures of both humans and the environment are not as straightforward as they once seemed.

## Learning resources

For a classic critique of environmentalism's tendency to perpetuate the dualism between humanity and nature, see:

Cronon, W. (1996) 'The Trouble with Wilderness; Or, Getting Back to the Wrong Nature,' in Cronon, W. (ed.) *Uncommon Ground: Rethinking the Human Place in Nature* (pp. 69–90). New York: W.W. Norton & Co. Available at: http://www.williamcronon.net/writing/Cronon_Trouble_with_Wilderness_1995.pdf

For more information about the research agendas and political issues prioritized by Nordhaus and Shellenberger's approach to post-environmentalism, see:

*The Breakthrough Institute*: This is the website of the organization they co-founded: http://thebreakthrough.org

For a critique of some of Nordhaus and Shellenberger's policy proposals from a perspective that is nonetheless sympathetic to the broader aims of post-environmentalism, see:

Buck, C.D. (2013) 'Post-Environmentalism: An Internal Critique', *Environmental Politics*, 22.6: 883–900. Available at: https://www.researchgate.net/publication/263598998_Post-environmentalism_An_internal_critique

## Bibliography

Bennett, J. (2010) *Vibrant Matter: A Political Ecology of Things*, Durham, NC: Duke University Press.

Bookchin, M. (1980) *Toward an Ecological Society*, Montreal: Black Rose Books.

Buttel, F.H. (2005) 'The Environmental and Post-Environmental Politics of Genetically Modified Crops and Foods', *Environmental Politics*, 14.3: 309–323.

Castree, N. (2003) 'A Post-environmental Ethics?, *Ethics, Place, and the Environment*, 6.1 (March): 3–12.

Haraway, D.J. (1991) *Simians, Cyborgs, and Women: The Reinvention of Nature*, London: Routledge.

Latour, B. (1993) *We Have Never Been Modern*, Cambridge, MA: Harvard University Press.

Latour, B. (2004) *Politics of Nature: How to Bring the Sciences into Democracy*, Cambridge, MA: Harvard University Press.

Latour, B. (2011) 'Love Your Monsters: Why We Must Care for Our Technologies as We Do Our Children', in Shellenberger, M. and Nordhaus, T. (eds) *Love Your Monsters: Postenvironmentalism and the Anthropocene*, Oakland, CA: The Breakthrough Institute, pp. 16–23.

Lyotard, J-F. (1984) *The Postmodern Condition: A Report on Knowledge*, Minneapolis: University of Minnesota Press.

McKibben, B. (1989) *The End of Nature*. New York: Anchor Books.

Naess, A. (1995) 'The Shallow and the Deep, Long-Range Ecology Movement: A Summary', in Drengson, A. and Inoue, Y. (eds) *The Deep Ecology Movement: An Introductory Anthology*, Berkeley, CA: North Atlantic Books.

Nordhaus, T. and Shellenberger, M. (2007) *Break Through: From the Death of Environmentalism to the Politics of Possibility*. Boston: Houghton Mifflin.

Shellenberger, M. and Nordhaus, T. (2004) *The Death of Environmentalism: Global Warming Politics in a Post-Environmental World*, http://www.thebreakthrough.org/images/Death_of_Environmentalism.pdf (Accessed on September 14, 2016).

Soulé, M.E. (1995) 'The Social Siege of Nature', in Soulé, M.E. and Lease, G. (eds.), *Reinventing Nature? Responses to Postmodern Deconstruction*, Washington, DC: Island Press.

Vogel, S. (2015) *Thinking Like a Mall: Environmental Philosophy After the End of Nature*, Cambridge, MA: MIT Press.

Young, J. (1990) *Post Environmentalism*, London: Belhaven.

# The social construction of nature

## Michael Ekers

## Nature is 'constructed', really?

How can we speak of the 'construction' of nature when the word nature evokes such a strong sense of naturalness, as a realm free from human activities?

There are a number of risks associated with accepting the naturalness of nature. To begin, as soon as we consider a wilderness park to be natural we lose sight of how it was made. In most cases, the creation of parks has been premised on the historical removal of certain people from the land that becomes the protected area thereby contributing to its apparent 'naturalness' (Adams and Hutton, 2007). More broadly, given the assumption that environmentalism is about protecting the quality of nature (pristine versus degraded), there is a tendency to focus on ecological questions rather than the social factors that may either damage or enhance ecological processes (Cronon, 1996; Smith, 1984). Finally, the stakes involved in determining whether particular groups of people are 'natural' or not are immense and troubling. Europeans, in establishing colonial settlements in North America, dehumanized an entire continent of Indigenous people by suggesting that they were less civilized and more natural than settlers, which in part legitimized the colonial project (Umeek of Ahousaht, 2002; King, 2012).

Dating back 40 years now, scholars advanced the notion that nature is 'socially constructed' as a response to how notions of naturalness can legitimize injustices and obscure the root causes of environmental change. The phrase social construction refers to how different social groups *create* and *fabricate* the worlds they inhabit, experience and represent (Evans, 2008). Various social processes, including peoples' labour, the search for profits, scientific research and cultural creations like films and photographs, contribute to the active construction of such things as the economy, social norms and even nature, despite the assumption that it is the very thing that can't be constructed (see Wilson, 1992).

At its most basic level, the social construction of nature refers to the *making* of the world both *physically* and *imaginatively*, whether that is urban landscapes replete with infrastructure and parks or even children's books that constitute the ideas we have about what *counts* as nature (Demeritt, 2001a; Gandy, 2002). In the social sciences and humanities these approaches achieved influence and momentum through the 1980s, 1990s and early 2000s (Braun and Castree, 1998; Proctor, 1998; Castree and Braun, 2001). Their legacy lives on today in both research and university teaching. Below, I explore different approaches to considering nature as socially constructed while also touching on the implications, potential and possible limits to these approaches.

## Origins of the 'social constructionist' perspective

In 1984, the geographer Neil Smith (1984: 32) suggested 'when [the] immediate appearance of nature is placed in historical context, the development of the material landscape presents itself as a process of the production of nature.'[1] Smith argued that we can look at any landscape, agricultural field or park space, and work backwards analytically to understand the various ways in which they were constructed. Smith and others concentrated on the role of labour (both physical and imaginative) in making new natures (Smith, 1998; Mitchell, 1996; Ekers and Loftus, 2013). For instance, the large-scale agricultural landscapes of California are carefully flattened and cultivated through people's exertions to facilitate the intensive production of crops. However, it is not simply agrarian ideals that bring people to transform landscapes and produce agricultural commodities. Rather, it is the circulation of money that structures investments and the construction of nature in the hope of generating a profitable return (see Walker, 2004).

Whether it was Smith's original writing, or how others have carried forward this agenda, the emphasis has been on the role of capitalism (as the dominant economic system of our time) in constructing nature (Harvey, 1996; Swyngedouw, 2004). The endless search for profits means that monetary values come to be attached to, and transform, various entities including organisms, resources, infrastructure and commodities. For instance, bitumen (tar sand) is extracted from Northern Alberta, refined and transported by rail or pipeline and shipped to global markets. This oil may be refined into gasoline or consumed in industrial processes but in all cases ends up increasing atmospheric carbon dioxide concentrations thus remaking the world over. In this example, nature is socially constructed at particular and planetary scales in ways that are intentional but also in ways that were never fathomed when oil was first discovered (on a related point see Castree, 2002).

Alongside Smith, who focused on landscape change, were others such as Donna Haraway (1991, 1997) and Bruno Latour (1993, 2004) who concentrated on the role of science, scientists and material artifacts (test tubes and mosquitos, for instance) in constructing various networks of human and non-human life. Haraway (1991) deployed the trope of the 'cyborg' to describe how contemporary life (both human and non-human) was being transformed by techno-science in ways that defied any neat separation of nature and culture. Contemporary use of smart phones, avatars in online worlds and algorithms governing social media plainly demonstrate what was once controversial, specifically that humans indeed lead a cyborg existence.

One of the key motivations for scholars working in this tradition has been to disable the authority often associated with scientific claims about nature, which are often entangled (Haraway, 1989; Magubane, 2003). Through studying scientific researchers as they conduct experiments, Latour sought to relativize scientific facts. He argued that scientists don't simply observe and record 'truths' about the world but rather enroll all sorts of material artifacts (test tubes, microbes, soil, water, Bunsen burners etc.) in constructing facts, including facts regarding what is often described as nature 'out there' (Latour, 1999, 2004). The point is that scientific knowledge is not absolute and, therefore, claims to nature couched in scientific terms are often more subjective and socially constructed than is often assumed. Debates between progressive climate change campaigners and deniers regarding the scientific truth of whether or not our world is warming (which it is!) demonstrate how the establishment and contestation of scientific facts is central to the politics of managing nature (Demeritt, 2001b, 2006). In Latour's view, the planet is indeed warming, but scientists work hard to 'construct' their knowledge about climate change and we need to be cognizant of how this knowledge is not an infallible 'mirror' reflecting inviolate truths about the atmosphere.

Building on the debates noted above, a group of writers have focused on interrogating the relations of power inherent in representations of nature within the mediums of novels, films, and photography among others (Wilson, 1992; Sandilands, 2000; Braun, 2002). Representations including discourses, stories, images and symbols construct natures and render them legible in particular ways, which means there are no innocent or culturally unmediated views of nature. For instance, the Government of Canada is currently 'celebrating' the 150th anniversary of Confederation and has offered Discovery Passes which the Minister responsible for Parks Canada describes as 'the gift of free admission to all Parks Canada places to Canadians' (McKenna, as quoted by Parks Canada, 2016). In the context of the United States, the historian William Cronon (1996) suggests that the establishment of wilderness parks romanced earlier moments of colonial settlement by preserving areas that resembled unsettled frontiers. 'Discovery Passes' in their very name and purpose function similarly: Canada's 'gift of free admission' encourages 'Canadians' to celebrate the history and geography of the nation through visiting wilderness parks and historic sites.

How might Indigenous people feel about 'Discovery Passes' being granted to Canadians so that they can visit what was and is Indigenous land? It should be noted that First Nations communities have been vocal critics of the Canada 150 celebrations (Harp, 2017; Manuel, 2017). The Government of Canada's call to discover wilderness parks is couched in nationalist rhetoric that obscures the voices and perspectives of Indigenous people for whom the land in question is not a wilderness landscape but rather a site of culture, sovereignty and sustenance (Coulthard, 2014). This example is useful in highlighting the power dynamics of settlement and indigeneity inherent in how nature is constructed and regarding who gets to 'discover' what landscapes. At its best this literature marries an examination of representational practices with a so-called 'postcolonial' approach centered on how colonial relations infuse notions of 'the public', 'the nation' and 'wilderness' (Said, 1993; Braun, 2002).

## The limits and possibilities of the construction of nature

Is it fair to question the value of different perspectives on the social construction of nature in light of the enormous ecological challenges we face in the twenty-first century (the effects of anthropogenic climate change being a key example)? Is it not risky to traffic in theoretical niceties regarding the construction of nature in intellectual forums, such as this chapter, while urgent action is needed every time we put down a book? Does talk of 'construction' not at some level deny or trivialize the serious material realities of what is happening to what we have learnt to call 'nature'?

By way of conclusion there are two obvious ways in which the perspectives outlined above remain as relevant as ever. First, we live in a world in which politics is now waged on the terrain of 'alternative facts' and 'fake news'. Environmental scientists working within the United States government are worried that the Trump administration may compromise research data regarding climate change and impede their ability to complete their research. While we must remember that this data is constructed through the very work that climate scientists do, debates outlined in this chapter are a vital resource in understanding how struggles over climate change in part are about the construction and contestation of environmental knowledge, which is one way in which nature is socially constructed.

Second, it is easy to feel paralysed in light of the magnitude at which capitalism has transformed landscapes and ecological processes the world-over. My students, in response to a question I ask, overwhelmingly suggest that it is easier to imagine catastrophic ecological collapse rather than the end of capitalism. In light of this, Smith's work is of a vital importance as he points to the possibility of constructing nature differently. He writes:

> The gloom of 'the end of nature' . . . is relinquished for a perspective which . . . makes a future of nature plausible. The political question becomes this: how, by what social means and through what social institutions, is the production of nature to be organized? How are we to create democratic means for producing nature? What kind of nature do we want?
>
> Smith (1996: 50)

Over 20 years later, these are the questions that need to be taken up with more urgency and force than ever.

## Learning resource

The book *Next Nature*, edited by Koert van Mensvoort and Hendrik-Jan Grievink (2015, Actar: New York), offers a very accessible and visually arresting set of examples of the social construction of nature, its causes and effects. A related website is www.nextnature.net

## Note

1 Smith uses the term production to understand the making of nature in an effort to stress the physical making of nature and the role of labour therein. In the context of this piece I will use

the term 'construction' in relation to his work but it should be noted that there are different conceptual and analytical distinctions to be made between the production and construction of nature (see Castree and Braun,1998; Demeritt, 2001a).

## Bibliography

Adams, W.M. & Hutton, J. (2007) 'People, parks and poverty: Political ecology and biodiversity conservation', *Conservation & Society*, 52(2): 147–183.

Braun, B. (2002) *The intemperate rainforest: Nature, culture and power on Canada's west coast*, Minneapolis: University of Minnesota Press.

Braun, B. & Castree, N. (1998) *Remaking reality: Nature at the millennium*, London: Routledge.

Castree, N. (2002) 'False antithesis? Marxism, nature and actor networks', *Antipode*, 34(1): 111–146.

Castree, N. & Braun, B. (1998) 'The construction of nature and the nature of construction: Analyical and political tools for building survivable futures' in B. Braun & N. Castree (eds) *Remaking reality: Nature and the millennium* (pp. 3–42), London: Routledge.

Castree, N. & Braun, B. (2001) *Social nature: theory, practice, and politics*, Oxford: Blackwell Publishers.

Coulthard, G. (2014) *Red skin, white masks: Rejecting the colonial politics of recognition*, Minneapolis: University of Minnesota Press.

Cronon, W. (1996) 'The trouble with wilderness; Or, getting back to the wrong nature' in W. Cronon (ed.) *Uncommon ground: Rethinking the human place in nature* (pp. 69–90), New York: W.W. Norton & Co.

Demeritt, D. (2001a) 'Being constructive about nature' in N. Castree & B. Braun (eds) *Social nature: Theory, practice, and politics* (pp. 22–40), Oxford: Blackwell.

Demeritt, D. (2001b) 'The construction of global warming and the politics of science', *Annals of the Association of American Geographers*, 91(2): 307–337.

Demeritt, D. (2006) 'Science studies, climate change and the prospects for constructivist critique', *Economy and Society*, 35(3): 453–479.

Ekers, M. & Loftus, A. (2013) 'Revitalizing the production of nature thesis: A Gramscian turn?', *Progress in Human Geography*, 37(2): 234–252.

Evans, J. (2008) 'The social construction of nature.' in P. Daniels, M. Bradshaw, D. Shaw, & J. Sidaway (eds) *An introduction to human geography*, 3rd Edition (pp. 256–272), Essex: Pearson.

Gandy, M. (2002) *Concrete and clay: reworking nature in New York City*, Cambridge, MA: MIT Press.

Haraway, D. (1989) *Primate visions: gender, race, and nature in the world of modern science*, New York: Routledge.

Haraway, D. (1991) *Simians, cyborgs, and women: The reinvention of nature*, New York: Routledge.

Haraway, D. (1997) *Witness@Second_Millennium. FemaleMan©_Meets_OncoMouse™*, London: Routledge.

Harp, R. (2017) 'Crashing the Canada 150 birthday party', *Radio Indigena with Rick Harp* [podcast]. Online. Available at: http://www.mediaindigena.com/rick-harp [Accessed 8th March 2017]

Harvey, D. (1996) *Justice, nature, and the geography of difference*, Cambridge, MA: Blackwell Publishers.

King, T. (2012) *The Inconvenient Indian: A curious account of Native people in North America*, Toronto: Doubleday Canada.

Latour, B. (1993) *We have never been modern*, Cambridge, MA: Harvard University Press.

Latour, B. (1999) *Pandora's hope: Essays on the reality of science studies*, Cambridge, MA: Harvard University Press.

Latour, B. (2004) *Politics of nature: How to bring the sciences into democracy*, Cambridge, MA: Harvard University Press.

Magubane, Z. (2003) 'Simians, savages, skulls, and sex: Science and colonial militarism in nineteenth-century South Africa', in D. Moore, J. Kosek & A. Pandian (eds) *Race, Nature, and the Politics of Difference* (pp. 99–121), Durham, NC: Duke University Press.

Manuel, A. (2017) 'Until Canada gives Indigenous people their land back, there can never be reconciliation', *Rabble*. Online. Available from: <http://news.gc.ca/web/article-en.do?nid=1167879>. [Accessed 8th March 2017]

Mitchell, D. (1996) *The lie of the land: migrant workers and the California landscape*, Minneapolis: University of Minnesota Press.

Parks Canada (2016) 'Free Parks Canada discovery pass', Government of Canada Online. Available from: http://news.gc.ca/web/article-en.do?nid=1167879 [Accessed 8th March 2017].

Proctor, J. (1998) 'The social construction of nature: relativist accusations, pragmatist and critical realist responses', *Annals of the American Association of Geographers* 88(3): 352.

Said, E. (1993) *Culture and imperialism*, New York: Vintage Press.

Sandilands, C. (2000) 'Observing nature, observing Canada: national parks, discipline, and spectacle', in J. Kleist & P. Lang (eds) *Canada Observed* (pp. 99–108), New York: Peter Lang Publishing.

Smith, N. (1984) *Uneven development: nature, capital, and the production of space*, New York: Blackwell.

Smith, N. (1996) 'The production of nature' in G. Robertson (ed.) *FutureNatural: nature science and culture* (pp. 35–52), London: Routledge.

Smith, N. (1998) 'Nature at the millennium: Production and reenchantment' in B. Braun & N. Castree (eds) *Remaking Reality* (pp. 271–285), New York: Routledge.

Swyngedouw, E. (2004) *Social power and the urbanization of water: flows of power*, Oxford: Oxford University Press.

Umeek of Ahousaht (Richard Atleo) (2002) 'Discourses in and about Clayoquot Sound: A First Nations perspective' in W. Magnusson & K. Shaw (eds) *A political space: Reading the global through Clayoquot Sound* (pp. 199–208), Montreal: McGill-Queen's University Press.

Walker, R. (2004) *The conquest of bread: 150 years of agribusiness in California*, New York: New Press.

Wilson, A. (1992) *The culture of nature: North American landscape from Disney to the Exxon Valdez*, Cambridge, MA: Blackwell.

# Symbolic environmental politics

2.21

## Megan Barry and Ingolfur Blühdorn

The negative social and ecological impacts of the lifestyles of privileged countries or social groups around the globe are becoming ever more visible. As their patterns of self-realisation, notions of the good life and particular interpretations of their inalienable rights are dangerously disrupting natural ecosystems and militating against core values of human dignity, well-being and equality, a profound transformation of the socio-economic order seems more urgent than ever. Yet among political decision makers half-heartedness, pretending and delay seem endemic. Exactly this is what critics often refer to as *symbolic politics*. But symbolism and symbolic action actually entail more than false promises and the deception of disempowered citizens by the rich and powerful. In fact, symbols are an indispensable ingredient of all political communication; and in environmental politics, too, their strategic use is a practice that all actors are engaged in.

The picture of planet Earth as seen from outer space has become a symbol of the unity and vulnerability of the terrestrial ecosystem. Pictures of modern wind-turbines or photovoltaic panels have become a symbol of technological innovation helping to protect the natural environment. International climate summits are a symbol of global cooperation towards the realisation of shared eco-political goals. In private households, bottle recycling and the purchase of particular products are symbolic practices which stand for and anticipate much more comprehensive changes in the way we live and relate to our environment. And the provision of vegan appetisers at a private prosecco-reception may be the symbolic expression of a particular lifestyle or personal image.

Symbols are signs that refer to something, the signified, which may be so large and comprehensive that it transcends human cognitive and practical capacities. Symbols then help us to experience the unimaginable – or make present what remains absent. In that they reduce and organise complexity, symbols help to make sense of the world and facilitate communication. They transport narratives of meaning which are jointly produced by those trying to package something into a symbol and those reading something out of it. In highly differentiated and fast-changing societies, they can help forge agreement and assist decision making. In contexts of uncertainty or lacking resources, symbolic action may maintain political momentum until more information has become available, or more resources and support have been mobilised for the implementation of more satisfactory solutions. And in the media- and information-society, where an ever increasing number of actors are competing for ever more limited public attention, and where *communicated, mediated* reality seem set to gain priority over so-called *primary* or *authentic* reality, symbols and symbolic action are becoming ever more important, still.

## Deception, manipulation and power

Thus, symbols and symbolic action fulfil a range of indispensable functions and their use is not exclusive to only some political actors. Yet, the prevailing understanding of the term symbolic politics is – in environmental discourses as elsewhere – still what the American political scientist Murray Edelman described and criticised already in the 1960s: a strategic tool in the hands of ruthless power-elites used to deceive, manipulate and control the disempowered masses (Edelman, 1964). In fact, as after a long period of confidence in an educated citizenry and democratic self-governance, populist actors and rhetoric are profoundly reshaping public political discourse, this understanding of symbolic politics is today very prominent again. It portrays modern societies as being divided into small, self-interested, corrupt elites and the masses of disenfranchised and alienated citizens, deceived by their leaders and denied their right to political self-determination. For Edelman, this mass public was confined to the role of the spectator, with a limited understanding of, and no genuine influence on, the political process, but coopted into it by means of political rituals (such as democratic elections) and myths (such as the narrative of representation) carefully designed to secure public acquiescence and stabilise established power relations.

Edelman himself did not make specific reference to environmental issues, but his notion of symbolic politics became constitutive for eco-political mobilisation throughout the 1970s and 1980s. In particular, eco-emancipatory movements continued to see modern society as divided into a powerful, self-interested elite and the disempowered citizenry. Yet, this mass public now evolved into an increasingly educated and self-confident political actor, 'civil society', which challenged the established order. It conceived of itself as the subject and voice of the public environmental interest and nature's right to integrity. Political and economic elites, in contrast, were perceived as anti-environmental and interested only in their own privileges. To demands for the extension of democratic rights and structural changes to the established socio-economic order, policy makers appeared to respond only with measures which they knew were ineffective and which were designed to deliver no more than the minimum required to mollify public unrest. These policy measures were seen to articulate a false commitment and make promises that the elites had no intention to ever fulfill. This symbolic politics was perceived as dishonest and immoral, an alibi indicative of a political system that refuses to respond to legitimate public demands, places elite interests over

**Table 2.21.1** The ideal of authentic politics versus the critique of symbolic politics

| | *Authentic politics* | *Symbolic politics* |
|---|---|---|
| **Primary objective** | to resolve problems, to effect change | to stabilise established structures, to mollify public unrest |
| **Moral quality** | genuine commitment | false pretence, dishonest |
| **Policy effectiveness** | high | low |
| **Democratic quality** | representative, responsive, nurtures trust | elitist, manipulative, undermines trust |

the common good, and denies citizens their right to political self-determination. But the emancipatory movements were confident that once this symbolic politics had been exposed and the self-serving elites removed, the true public interest and environmental good could swiftly be implemented.

## Post-ecologism and simulative politics

Up to the present, this conceptualisation and critique of symbolic environmental politics has its uses and legitimacy. Yet, modern societies have evolved beyond the conditions which once underpinned this notion. Indeed, in the contemporary context, the simplistic narrative it evokes contributes as much to obstructing a differentiated understanding of modern societies' eco-politics as to shedding light on it. Today, the idea of the dualistic divide of society into an anti-environmental elite and an ecologically committed civil society is – although popular – untenable: as issues of climate change and sustainability are anchored in all policy agendas, elites cannot collectively be categorised as anti-environmental; and as the differentiation of modern societies has given rise to many different ideas of nature and diverse views of what ought to be sustained, for whom, and for what reasons, civil society is not the unified voice of an unambiguous environmental and public interest. Furthermore, even if a consensus about the ecologically necessary was achieved, there is no reason to assume that it could easily be implemented. Not only are major parts of the societal mainstream – despite all environmental awareness and commitment – equally committed to values and lifestyles which they know to be socially and ecologically destructive, but the development of modern societies is significantly shaped by forces which are beyond their governments' control, and which restrict their capacity for coordinated, sustained and effective (eco-) political action.

This condition where ambiguity is proliferating, political agency eroding and major parts of the well-educated citizenry are environmentally committed but also deeply attached to values and lifestyles which are known to be socially exclusive and ecologically destructive, has been conceptualised as the *post-ecologist* constellation (Blühdorn, 2000, 2014). It breeds disillusionment and frustration that overshadows the optimism of the earlier environmental movement. The lack of uncontroversial ecological imperatives is disorientating and politically disabling. For individual citizens and collective actors, the tension between the evidence that established values and behaviours can be sustained only at significant social and ecological costs, and the desire to hold on

**Table 2.21.2** The concepts of symbolic politics versus simulative politics

| | *Symbolic politics* | *Simulative politics* |
|---|---|---|
| **Perspective** | normative, activist | descriptive, analytical |
| **Aim of concept** | critique of elites, improvement of policy | diagnosis of and explanation for societal impasse |
| **Actor infocus** | government institutions, economic and political elites | many actors across society |
| **Suggested function** | deception and manipulation of public, to stabilise established power relations | management of contradictions, coping strategy for irresolvable value conflicts |
| **Availability of alternatives** | effective politics is perfectly feasible | caught up in conundrum |

to them anyway, leads into irresolvable contradictions. In this situation, all kinds of symbolic action become an attractive coping mechanism: they provide opportunities to articulate ecological values and experience social commitment, while postponing any major revision of established value preferences and lifestyles (Blühdorn, 2017). And the old narrative of symbolic politics gains new significance because it offers simple explanations for the bleak and schizophrenic condition.

Building on the work of the French sociologist Jean Baudrillard, this inherently paradoxical eco-politics in the post-ecologist constellation has been described as *simulative politics* (Blühdorn, 2007). Baudrillard diagnosed a condition where citizens have ever less first-hand experience of what they are talking about and refer ever more to a reality constructed via images and signs. In this scenario, the world of signs, he believed, develops its own life, and images and symbols no longer organise and *make sense of* primary reality but – as the referent of societal communication – *replace* it. Truth then turns into a *simulacrum*, and societal conduct into *simulation* (Baudrillard, 1981/1994). In this same sense, simulative environmental politics has been said to discursively perform a societal – and eco-political – condition unaffected by the paradoxes outlined above. In contrast to the notion of symbolic politics, this concept shifts the emphasis away from the critique of established power-relations to a socio-cultural diagnosis of advanced modern societies. Its focus is not on the manipulative practices of particular elites, but on procedures of self-deception that are pervasive throughout society. In contemporary societies' eco-politics, both symbolic and simulative politics are ever-present and closely intertwined. Together the two concepts facilitate a richer understanding of the reality of *sustained unsustainability* (Blühdorn, 2011, 2013).

## Learning resources

The resources listed below provide analyses of current sustainability and environmental politics through the lens of symbolic and simulative politics.

Baker, S. (2007) 'Sustainable Development as Symbolic Commitment: Declaratory Politics and the Seductive Appeal of Ecological Modernisation in the European Union', *Environmental Politics*, 16(2): 297–317.

Blatter, J. (2009) 'Performing Symbolic Politics and International Environmental Regulations: Tracing and Theorizing a Causal Mechanism beyond Regime Theory', *Global Environmental Politics*, 9(4): 81–110.
Blühdorn, I. (2015) 'A Much-needed Renewal of Environmentalism? Eco-politics in the Anthropocene', in Hamilton, C., Gemenne, F. & Bonneuil, C. (eds) *The Anthropocene and the Global Environmental Crisis: Rethinking Modernity in a New Epoch* (pp. 156–167), London: Routledge.
Death, C. (2011) 'Summit Theatre: Exemplary Governmentality and Environmental Diplomacy in Johannesburg and Copenhagen', *Environmental Politics*, 20(1): 1–19.
Sander, J. (1997) *The Symbolic Politics of the Political Systems of the United States and Great Britain: An Analysis of the Reagan Presidency and the Thatcher Premiership*, Ann Arbor: ProQuest.

## Bibliography

Baudrillard, J. (1981/1994) *Simulacra and Simulation*, Ann Arbor: University of Michigan Press.
Blühdorn, I. (2000) *Post-ecologist Politics: Social Theory and the Abdication of the Ecologist Paradigm*, London: Routledge.
Blühdorn, I. (2007) 'Sustaining the Unsustainable: Symbolic Politics and the Politics of Simulation', *Environmental Politics*, 16(2): 251–275.
Blühdorn, I. (2011) 'The Politics of Unsustainability: COP15, Post-Ecologism and the Ecological Paradox', *Organization & Environment*, 24 (1): 34–53.
Blühdorn, I. (2013) 'The Governance of Unsustainability: Ecology and Democracy beyond the Post-Democratic Turn', *Environmental Politics*, 22 (1): 16–36.
Blühdorn, I. (2014) 'Post-ecologist Governmentality: Post-democracy, Post-politics and the Politics of Unsustainability', in Swyngedouw, E. & Wilson, J. (eds) *The Post-Political and its Discontents: Spaces of Depoliticisation, Spectres of Radical Politics* (pp. 146–166), Edinburgh: Edinburgh University Press.
Blühdorn, I. (2017) 'Post-capitalism, Post-growth, Post-Consumerism? Eco-political Hopes Beyond Sustainability', *Global Discourse*, 7(2): 42–61.
Edelman, M. (1964) *The Symbolic Uses of Politics*, Champaign: University of Illinois Press.

# 2.22 Tipping points

## Chris Russill

In the last 15 years, the concept of 'tipping points' has circulated widely in academic, policy and media circles. It describes threatening forms of environmental change, particularly the dangers of abrupt climate change. Tipping points were adopted as a media savvy way for high profile scientists to communicate a different conception of planetary change to non-scientists. While many believe the Earth is stable, resilient, and slow to change, these scientists argue that the climate and other Earth systems have different possible states. Transition between these states is not always gradual or predictable. The stability and resilience of environments can be eroded by small and gradual changes that suggest few serious consequences until a threshold is crossed and rapid, chaotic, and usually irreversible change takes place. In this respect, a tipping point is "a critical threshold at which a tiny perturbation can qualitatively alter the state or development of a system" (Lenton et al., 2008: 1786). The transitions precipitated by tipping points are often experienced as surprising and unmanageable, if not catastrophic, as the recent past of an environment becomes an unreliable guide to its future.

### Origins

The tipping point approach gained traction in the 1980s with the surprising appearance of an ozone hole, a sudden and disconcerting shift in the chemical composition of the atmosphere caused by humans. Given global concern with the ozone crisis, scientists and policy makers became more curious about other sorts of abrupt and unpredictable change. Wallace Broecker, a geochemist specializing in oceanic dynamics, gained wide notoriety for depicting climate change in this way. While often represented as a steady increase in global temperature that results from the gradual build up of $CO_2$ in the atmosphere, Broecker argued that climate change occurred in sudden and unpredictable jumps that no civilization was equipped to manage. By substituting analogy and vivid metaphor for the usual emphasis on model simulations, which he found biased toward gradualist types of change, Broecker (1987) called for a fundamental reorganization of research into the 'jumps,' 'great leaps,' and 'flips' of climate and related Earth systems.

Broecker's emphasis on the dangers of abrupt environmental change circulated during a resurgence of interest in dynamical systems theories, particularly chaos and complexity theory. A fleet of popularizing books had emerged to discuss the significance of chaotic change generally, including Prigogine and Stengers' (1985) *Order out of Chaos*, James Gleick's (1987) *Chaos: Making a New Science*, Gregoire Nicolis and Ilya Prigogine's (1989) *Exploring Complexity: An Introduction*, Ian Stewart's (1989) *Does God Play Dice?*, Katherine Hayles' (1990) *Chaos Bound*, Al Gore's (1992) *Earth in the Balance*, Mitchell Waldrop's (1992) *Complexity*, and of course Michael Crichton's (1990) popular novel, *Jurassic Park*, which featured a chaos theorist predicting imminent collapse of a human-dominated ecosystem. Ian Malcolm, the fictional mathematician of the book, forecasts the failure of a human designed environment that precipitates an unexpected return to the dangers from our paleo-geological past (represented in this case by extinct dinosaurs rather than past climate change). The flips and switches characteristic of abrupt change were not yet called tipping points; still, there was clear interest in modeling their dynamics conceptually and mathematically to inform environmental policy making. The phrase would enter popular consciousness when Malcolm Gladwell (1996, 2000) synthesized how social scientists and marketing mavens understand dynamical systems change using his 'tipping point' framework.

## A new paradigm

By 2002, scientists were willing to embrace Broecker's framework for climate change as a "paradigm shift for the research community", one "well established by research over the last decade . . . but little known and scarcely appreciated in the wider community of natural and social scientists and policy-makers" (National Research Council, p. 16, p. 1). The lack of recognition did not last long. Gary Comer, the billionaire founder of Land's End, provided millions of dollars for research and fellowships on abrupt change, much of it distributed on the advice of Broecker and Sherwood Rowland (known for his work relating to the ozone hole). Broecker's iconic example of a 'trigger' for abrupt climate change also found its way into popular culture. In brief, Broecker conceptualized oceanic circulation as a global 'conveyor' having on/off modes. If the conveyor were switched off in the North Atlantic, the abrupt climate shifts characteristic of the pre-Holocene could recur. It is an alarming scenario that figured prominently in the Hollywood film, *The Day After Tomorrow* (2004), in Gore's (2006) *An Inconvenient Truth*, and in a report commissioned by the Pentagon on the national security implications of such events.

High profile scientists began warning overtly of the dangers of tipping points to popularize the abrupt change paradigm to policymakers and to engender wider societal concern. The sensationalist representations of Broecker's 'trigger' circulated widely and raised concerns about undue alarmism. Increasing emphasis was placed on the disappearance of Arctic sea-ice as a more likely example of tipping points that could be observed visibly in the near-term. Additionally, dynamical systems theorists incorporated the concept to propose a holistic 'Earth systems science' for managing planetary tipping points of all sorts. Tipping points were no longer a colloquial metaphor for a particular sort of change, but used to reframe the risk management frameworks governing how environmental science and policymaking were brought together at the global scale.

## Criticism and enduring problems

The decision to reorganize scientific and public understanding of environmental change in terms of tipping points is not without contention. The use of a buzzword popularized by a journalist was criticized as alarmist, overwrought, and confusing. Communication scholars warned that tipping points represented a new alarmism and undermined the credibility of scientists while fostering fatalism in citizens about their ability to ameliorate climate change. Tipping points, in this respect, reproduced a tendency to emphasize apocalyptic and catastrophic futures in public discourse. Many scientists suggested that tipping points were simply "old wine in new bottles" (*Nature*, 2006) and that the novelty of this conceptual approach to environmental change was overblown and of limited use in describing geophysical phenomena. Confusion has been a long-standing concern. The holistic scope and complexity of dynamical systems science is difficult and non-intuitive in its approach to change; the use of a popular buzzword for scientific purposes introduced a range of popular connotations that conflicted with the desires of its proponents to circumscribe its technical meanings.

A broader problem with the tipping point approach is the managerial politics of its main proponents. The urgency involved in detecting, observing, and preparing for tipping point dangers has tended to overshadow debates over the conceptions of political agency that its risk management systems imply. The narrow conception of social science and politics drawn upon in this approach tends to retain the glibness of Gladwell's approach and remains insensitive to other ways of envisioning how the futures of societal dynamics and planetary systems are linked. In this respect, tipping points are not simply a scientific idea for understanding environmental change, but a crucial site through which struggles over the environmental conditions and security of our existing socio-economic order should be engaged.

## Learning resources

'The Tipping Points': This is a television series and website introducing the notion of planetary tipping points: http://www.thetippingpoints.com/about-the-tv-series.

'Wake up, freak out – then get a grip': This is an animation produced by Leo Murray (2009) to communicate the urgency often associated with tipping points: https://vimeo.com/1709110?pg=embed&sec=1709110.

Scheffer, M. 2009. *Critical Transitions in Nature and Society*. Princeton: Princeton University Press. This is a good and an accessible introduction to dynamic systems theory authored by one of its main proponents.

# Bibliography


Broecker, W. (1987) 'Unpleasant surprises in the greenhouse?', *Nature*, 328: 123–126.

Crichton, M. (1990) *Jurassic Park*. New York: Alfred K. Knopf.

Gladwell, M. (1996) 'The tipping point', *New Yorker*, June, 3. Available https://www.newyorker.com/magazine/1996/06/03/the-tipping-point

Gladwell, M. (2000) *The Tipping Point: How Little Things Can Make a Big Difference*. Boston, MA: Back Bay.

Gleick, J. (1987) *Chaos: Making a New Science*. New York: Viking Press.

Gore, A. (1992) *Earth in the Balance: Ecology and the Human Spirit*. Boston, MA: Houghton Mifflin.

Hayles, K. (1991) Chaos and Order: Complex Dynamics in Literature and Science. Chicago, IL: University of Chicago Press.

Lenton, T.M., Held, H., Kriegler, E., Hall, J.W., Lucht, W., Rahmstorf, S. & Schellnhuber, H.J. (2008) 'Tipping elements in the Earth's climate system', *Proceedings of the National Academy of Sciences*, 105: 1786–1793.

National Research Council (2002) *Abrupt Climate Change: Inevitable Surprises*. Washington, DC: The National Academies Press.

*Nature* (2006) 'Editor's summary', *Nature*, 441: 785.

Nicolis, G. and Prigogine, I. (1989) *Exploring Complexity: An Introduction*. London: W. H. Freeman and Company.

Prigogine, I. and Stengers, I. (1985) *Order out of Chaos: Man's New Dialogue with Nature*. New York: Bantam Books.

Stewart, I. (1989) *Does God Play Dice? The New Mathematics of Chaos*. Hoboken, NJ: Wiley Blackwell.

Waldrop, M. (1992) *Complexity: The Emerging Science at the Edge of Order and Chaos*. Simon and Schuster.

# 2.23 Wicked environmental problems

Michael Thompson

There are some – those who talk of "planetary boundaries", for instance, and then use those boundaries to define a "safe operating space for humanity" (Rockström et al., 2009) – who see nature as manageable: *stable within limits* that can be determined by certified experts (Royal Societies, National Academies of Science and so on). Others, however, are adamant that there are *no limits*; "if something is unsustainable", they reassure one another, "it will stop". Yet others – like the environmental movement Earth First!, with its bleak vision of a future consisting of "just cockroaches and Norway rats" – insist that there are *no safe limits*, and that all of us must therefore learn to "tread lightly on the earth", before it is too late. Still others are sceptical of all these contradictory certainties, arguing that, since nature operates without rhyme or reason, and since people are everywhere fickle and untrustworthy, "nothing you can do will make much difference".

In other words, there are four contending "voices", each rejecting much of what the others are saying. But it does not follow from this that one must be right and the others wrong. Rather, each is distilling elements of wisdom and experience that are being missed by the others: the world, at times and in places, can be each of these four ways. This, a little more formally set out, is the *theory of plural rationality* (also called *cultural theory*) and it tells us that, whenever we encounter this sort of "contested terrain" – multiple and mutually incompatible certainties – we are faced with a *wicked problem*.

# The plural rationality framing: wicked problems, uncomfortable knowledge, clumsy solutions

With wicked problems (climate change is currently the prime example), and in marked contrast to *tame problems* (the hole in the ozone layer, for instance, to which climate change is often, and erroneously, compared), there are contending and mutually contradictory definitions of the problem-and-solution, and these do not converge as the policy process gets under way. If they are treated as tame problems then the assumption of a *single* definition imposes *elegance*: in the orthodox practice of policy analysis, for instance, one establishes a "single metric" – dollars, lives saved, quality-adjusted life years or whatever – which can then be used to evaluate options, select the best one, and then optimize around it. But elegance comes at a cost: the exclusion of those actors who subscribe to the other definitions. The valid and useful knowledge generated by these excluded actors, since it inevitably calls into question the knowledge that is generated by the "hegemonic" actor, is then seen as *uncomfortable* and is ignored or marginalized. If that is to be avoided then things will have to be arranged institutionally so that each of the voices is able to make itself heard and is then responsive to, rather than disdainful of, the others. Only then will we see the emergence of those more robust, consent-preserving, surprising-lessening and inherently democratic outcomes: *clumsy solutions*.

## *Beyond tame problems*

Wicked problems were first delineated more than 40 years ago by two University of California political scientists, Horst Rittel and Melvin Webber (Rittel and Webber, 1973). These problems have several inter-related characteristics and, as a result of these characteristics, people typically clash over how to *define* them and over how to *resolve* them:

- The range of possible causes is large and uncertain (as are the possible interactions of those causes);
- The range of possible solutions is equally large and uncertain;
- Every solution is a "one-shot operation" and will have serious consequences (there are, in other words, considerable "sunk costs", and this means that a decision – to build a super-sewer, say, or a high-speed rail link – cannot easily be backed out of if things do not go quite as expected);
- Many people, organizations and social domains are involved;
- Wicked problems are essentially unique and novel;
- They have no "stopping rule" (every attempt at resolution leads to new problems);
- There are no absolutely right solutions.

Marco Verweij (2011), in his book *Clumsy Solutions for a Wicked World*, checks climate change and the ozone hole against these seven characteristics, thereby confirming that the former is a wicked problem and the latter a tame one. That is the copper-bottomed test for this crucial distinction; usually the existence of plural and mutually incompatible definitions of problem and solution, together with their non-convergence as the policy process gets under way, suffices.

## Uncomfortable knowledge

The well-known academic response "New not true; true not new" nicely captures the way in which the members of a scientific establishment tend to deal with knowledge that threatens the paradigm around which they are stabilized. The philosopher of science, Imré Lakatos (1976), in his book *Proofs and Refutations*, showed that "monsters" – pieces of knowledge that cannot be accommodated within the prevailing paradigm – can occur even in mathematics, and he went on to tease out the various ways in which this sort of uncomfortable knowledge can be handled: by *monster-adjusting*, for instance (in which both the paradigm and the offending piece of knowledge are progressively modified until a fit, of sorts, is achieved) or by *monster-barring* (in which the offending knowledge, and its carriers, are rejected out of hand, as happened with the first attempts to publish Ohm's law: "these preposterous theories of Professor Ohm" was the response, and Ohm lost his university position). An environmental example is the response, back in 2010, by the then head of the IPCC (the Intergovernmental Panel on Climate Change), Rajendra Pachauri, to the glaciologists who helpfully pointed out that a recent IPCC report had made a serious mistake in its predictions about the rate at which Himalayan glaciers are retreating. He called them "voodoo scientists" (others have even found themselves stigmatized as "climate change deniers").

Quite independently of Lakatos, the anthropologist Mary Douglas (1970), in her book *Natural Symbols*, also homed-in on monsters: in her case these were animals – such as the pangolin, among the Lele in the former Belgian Congo – that simply could not be fitted into the indigenous typology. A few years later David Bloor, a philosopher, in a celebrated article "Polyhedra and the abominations of Leviticus" (Bloor, 1982), synthesized these approaches, thereby establishing a thorough-going theory of uncomfortable knowledge.

> The books [*Proofs and Refutations* and *Natural Symbols*] have a common theme: they deal with the ways men [sic] respond to things which do not fit into the boxes and boundaries of accepted ways of thinking; they are about anomalies to publicly-accepted schemes of classification. Whether it be a counter-example to a proof; an animal that does not fit into the local taxonomy; or a deviant who violates the current moral norms, the same range of reactions is generated.
>
> (Bloor, 1982: 191)

## Clumsy solutions

As a concept, this originated back in 1988, with Michael Shapiro, a lawyer at the University of Southern California. To be precise, he used the term "clumsy institution" (Shapiro, 1988) in order to stress the good sense inherent in the seemingly messy way in which new members of the US Supreme Court are chosen. It is a way of escaping from the commonsensical prescription that, when faced with contradictory definitions of problem-and-solution, we must choose one and reject the rest. The idea was subsequently picked up by a number of anthropologists and policy analysts (see, for instance, the case studies of elegant failures and clumsy successes in Verweij and Thompson, 2011) and the term clumsy institution is now used to characterize the sort of "policy sub-system" in which those who speak with the four voices all enjoy both accessibility and responsiveness. Clumsy institution is thus the polar opposite of what political theorist Robert Dahl (1971) in his *theory of pluralist democracy*, called "closed hegemony": the hyper-elegant situation in which just one voice drowns out the others. But, where Dahl had just his dualistic distinction – closed hegemony versus pluralist democracy – our typology of four voices gives us four distinct varieties of closed hegemony. It also enables us to recognize the 14 different kinds of policy sub-system (most of which, it turns out, have been identified and given names by policy scientists: see Ney, 2009) that populate the "excluded middle" between Dahl's two extremes. And, for good measure, this refurbishment of the classic theory of pluralist democracy makes clear that it is *discourse* – contending voices, narratives, storylines and so on – that is key.

So the conventional methods – the aforementioned single-metric-and-optimization calculus – is valid only for tame problems. When faced with wicked problems, we need to discard these time-honoured methods (cost-benefit analysis, for instance, general equilibrium modelling, probabilistic risk assessment, and all those approaches that assume that uncertainty is merely the absence of certainty) and reach for something very different: the new tool-kit for policy in a wicked and plurally perceived world (see Thompson et al., 1998; Thompson and Beck, 2015; Gyawali et al., 2016).

## Learning resources

Grundmann, R. (2016) 'Climate change as a wicked problem', *Nature Geoscience*, 9, August: 562–563.
This short, readable piece usefully compares tame and wicked problems to establish what qualifies as "wicked", using climate change as an example. As you read, reflect on what kinds of knowledge and expertise are required to manage wicked problems. So far, much of the public discourse about anthropogenic climate change has been led by geoscience.

## Bibliography

Bloor, D. (1982) "Polyhedra and the abominations of Leviticus". In M. Douglas (ed.) *Essays in the Sociology of Perception* (pp. 191–218). London: Routledge and Kegan Paul.

Dahl, R. (1971) *Polyarchy: Participation and Opposition*. New Haven, CT: Yale University Press.

Douglas, M. (1970) *Natural Symbols: Explorations in Cosmology*. London: Cresset Press.

Douglas, M., Thompson, M. & Verweij, M. (2003) "Is time running out? The case of global warming", *Daedalus*, Spring: 98–107.

Gyawali, D., Thompson, M. & Verweij, M. (eds) (2016) *Aid, Technology and Development: The Lessons from Nepal*. London: Earthscan.

Lakatos, I. (1976) *Proofs and Refutations: The Logic of Mathematical Discovery*. Cambridge: Cambridge University Press.

Ney, S. (2009) *Resolving Messy Policy Issues*. London: Earthscan.

Rittel, H.W.J. & Webber, M.M. (1973) "Dilemmas in a general theory of planning", *Policy Sciences*, 4(2): 155–169.

Rockström, J., Steffen, W., Noone, K., Persson, Å., Chapin, III, F.S., Lambin, E., Lenton, T.M., Scheffer, M., Folke, C., Schellnhuber, H., Nykvist, B., de Wit, C.A., Hughes, T., van der Leeuw, S.,

Rodhe, H., Sörlin, S., Snyder, P.K., Costanza, R., Svedin, U., Falkenmark, M., Karlberg, L., Corell, R.W., Fabry, V.J., Hansen, J., Walker, B., Liverman, D., Richardson, K., Crutzen, P. & Foley, J. (2009) "A safe operating space for humanity", *Nature*, 461: 472–475.

Shapiro, M. (1988) "Judicial selection and the design of clumsy institutions", *Southern California Law Review*, 61: 1555–1569.

Thompson, M. & Beck, M.B. (2015) *Coping with Change: Urban Resilience, Sustainability, Adaptability and Path Dependence*. UK Government Foresight Future of Cities Project. Online. Available at www.gov.uk/government/publications/future-of-cities-coping-with-change.

Thompson, M., Ellis, R. & Wildavsky, A. (1990) *Cultural Theory*. Boulder, CO: Westview Press.

Thompson, M., Rayner, S. & Ney, S. (1998) "Risk and governance, part 2: policy in a complex and plurally perceived world", *Government and Opposition*, 33(3): 330–354.

Verweij, M. (2011) *Clumsy Solutions for a Wicked World*. Basingstoke: Palgrave Macmillan.

Verweij, M. & Thompson, M. (eds) (2011) *Clumsy Solutions for a Complex World* (revised paperback edition). Basingstoke: Palgrave Macmillan.

# Part 3

## Classic approaches

# Part 3 Introduction

Classic environmental studies of the 1960s and 1970s drew upon a number of existing fields of research and teaching in order to more carefully examine environmental issues. Additionally, new subfields were created, sometimes drawing upon parallel fields outside of environmental studies (such as the branches of physical geography). In many ways, the hitherto unexamined question of environment motivated scholars of the classic era to utilise the tools and approaches they already possessed to offer clarification and, given the very applied nature of much environmental discourse, possible solutions. In this nascent stage of environmental studies, however, these existing approaches sometimes conveyed relatively circumscribed scholarly insights and prospective solutions to the issues – such as pollution and population growth – that arose to such prominence in the classic era.

Many classic approaches were environmental subfields of other existing disciplines, such as law, philosophy, economics, sociology, or natural science fields such as chemistry and physics. Thus, environmental law considered classic questions such as whether trees should have legal standing; environmental sociology considered whether emerging social attitudes to nature comprised a 'new ecological paradigm'; and environmental chemistry addressed the source, transport, and fate of air pollutants such as sulfur dioxide responsible for acid precipitation, a long recognised problem, but now increasingly understood and managed.

The tools available from these existing disciplines offered significant building blocks toward a fuller understanding of the environmental predicament, but there were gaps, some filled via new subfields. Thus, for instance, eco-theology was developed to address the challenge of Lynn White Junior's controversial thesis of the late 1960s (White, 1967), that environmental problems were in large part a result of technologies unleashed by a Judeo-Christian ethos of mastery of nature – and that solutions to these problems may similarly require drawing from religious traditions. Or, ecofeminism arose as a confluence of two emerging realms of scholarly interest and popular concern in the classic era: gender and environment. Could it be that the human domination of nature was not only analogous to, but actually a product of, the patriarchal domination of women by men? Might it be that women are better stewards of the earth given their more intimate biological connection with nature? Classic ecofeminist assertions such as these have been modified, and essentialist assumptions challenged, by more recent ecofeminist approaches, but, like classic environmental concepts, some remain firm in the popular realm today.

Whether drawn from related fields or created anew as subfields under the broad, cross-disciplinary field of environmental studies, this suite of classic approaches can, from the vantage point of the 21st century, be more critically assessed – both in terms of its insights and its profound lacunae. For instance, viewed from today's perspective, the 'new ecological paradigm' of the 1970s may be seen to have encoded specific features of middle-class white (albeit 'alternative') cultural practice onto the supposed whole of environmentalism. And then there are the silences. For example, issues of power and difference were only haphazardly incorporated into classic environmental approaches, many of which were firmly First World and universalist in their unwritten assumptions about humans and environment. Similarly, little attention was paid to the legacies of colonialism in the 'two thirds world', which shaped so many of the human and ecological landscapes of the newly independent nations of the 1960s and 1970s.

These limitations are true of earlier approaches of any field of scholarly study, but what is intriguing in environmental studies is the lag between more sophisticated contemporary approaches and the popular imagination, which in some respects remains heavily influenced by the approaches of the classic era – evidenced, for example, as one scans recent environmental titles in the local bookstore. Classic approaches are thus important to consider in environmental studies not only to appreciate the history of the field, but to critically survey its current landscape, as environmental studies continues to engage with the concerns, movements, and practices of a much broader circle of people.

## Bibliography

White, L. (1967) 'The historical roots of our ecological crisis', *Science*, 155(3767): 1203–1207.

# 3.1 Cultural theory

Åsa Boholm

The key message of cultural theory launched by social anthropologist Mary Douglas (1985, 1992) is that concern about environmental problems and technological risks is socially and culturally constructed. If, why and over what people are concerned all vary between groups according to social relationships, institutions, values and belief systems. Cultural theory has been influential in several areas of environmental studies: for example, in understandings of risk, in understanding public attitudes towards controversial social and environmental problems and lifestyle choices related to taste and consumption.

## The topic of risk

Cultural theory specifically addresses risk as a topic of investigation. It argues that the features, meanings, and moral implications of unwanted potential events derive from collectively shared representations. Since there is no single over-arching rationality underpinning knowledge of risk, or the ways in which it should be managed, understanding of risk in society is intrinsically multifarious. Criticizing psychological work on 'risk perception' Douglas (1985, 1992) argues that risk is culturally biased. Cultural theory breaks with methodological individualism in that it is not based on assumptions about individuals acting out of self-interest and utility maximization (Tansey & O'Riordan, 1999). It is proposed that awareness of risk, choices of risk taking and avoidance are made selectively on the basis of different collective 'ways of life', cosmologies or world views.

An underlying assumption of cultural theory is that universal experiences of misfortune demand a social and moral explanation to establish responsibility. Explanations of misfortune have a characteristic social dimension since individuals or groups are separated into victim(s) and culprit(s), the latter to be blamed. Various forms of practices such as taboo in traditional societies (inviolable rules of conduct to avoid causing harmful events) or risk assessment in modern societies (probabilistic calculations of cause and effect to avoid or mitigate harmful events) are understood to support and maintain social order – gender divisions, social hierarchies, labour divisions and power relations – by suppressing certain actions that potentially threaten the social order (Douglas, 1996). According to cultural theory, risk, similar to taboo, has a social function. It serves as a mechanism for social categorization and attribution of blame in cases of misfortune (Tansey & O'Riordan, 1999: 75).

Since 1982, when Mary Douglas and Aaron Wildavsky first published *Risk and culture: an essay on the selection of technological and environmental dangers*, a considerable amount of literature has addressed comparative dimensions of environmental concern and risk awareness from cultural theory framework. Central to this work is questions about how social groups differ regarding attitudes to environmental problems and risk issues and how such differences can be explained. Apart from Mary Douglas's own work (1978, 1985, 1992, 1996) cultural theory has been outlined by Michael Thompson, Richard Ellis and Aaron Wildavsky (1990) in the book *Cultural Theory*. It has also been explored in a number of other works by, among others, Dake (1991); Rayner (1992); Rayner and Cantor (1987) and Schwartz and Thompson (1990).

## The grid-group model

Cultural theory addresses a classical sociological problem, namely the structural relationship between individuals and society. The basis of cultural theory is grid-group analysis, a development within so-called 'structural-functionalism' in British social anthropology (Douglas, 1978). The grid-group model serves as the basis for a theory of 'cultural bias' which explains social and cultural differences based on a typology of group formation and coexisting cosmology (or world view). Cultural theory is constructed by way of deduction from a limited number of basic axioms underlying the grid-group model that Douglas (1978, 1992) developed as an analytical tool for comparison of societies. The dimensions grid and group are taken to be universal in human society and according to cultural theory these dimensions serve as coordinates that make it possible to measure "possible social structures" (Douglas 1978: 7).

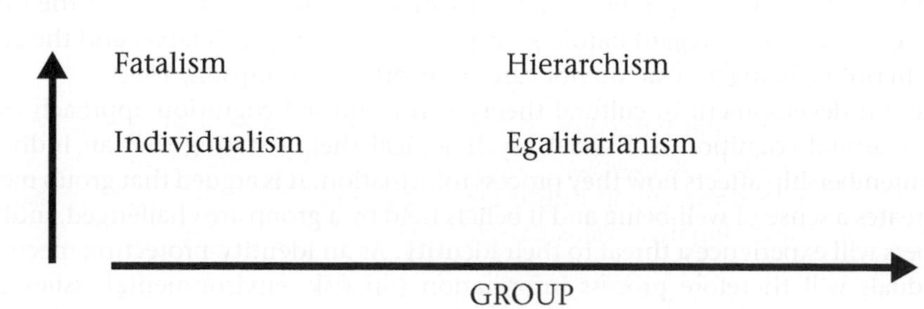


**Figure 3.1.1** The four cosmological types, or world-views according to cultural theory.

*Source*: Adapted from Schwartz and Thompson (1990).

Basically, the grid-group model postulates that all societies are configurations of two general parameters: social differentiation (grid) and social cohesion (group). If the extreme poles of the grid and group scales are combined with each other so that they form a diagram, four universal 'cosmological types' (also denoted 'ways of life') emerge, from which decisive cultural orientations can be deduced (Douglas, 1978). It should be noted that although grid and group might be understood as gradual scales, when combined to form a typology, they serve as binary pairs.

## Cosmologies, ways of life and myths of nature

Grid-group theory identifies several types which, in principle at least, are taken to coexist in every society: *fatalism* (high grid/low group), *individualism* (low grid/low group), *hierarchism* (high grid/high group) and *egalitarianism* (low grid/high group). A fifth possibility is considered occasionally in the literature, that of the autonomous state of *hermit-hood* holding a position detached from the grid-group classification (Thompson et al., 1990: 3).

According to cultural theory, these orientations include notions concerning fundamental and ever-present human matters such as nature, foreign people and places, animals, death, medicine, cookery, moral judgment, taste, aesthetics, justice and retribution. Cultural theory makes a basic distinction between on the one hand, 'cultural bias' defined as shared values and beliefs and, on the other, 'social relations' defined as relationships between individuals. The crucial explanatory concept, 'way of life' (or 'cosmology') is defined as a "combination of social relations and cultural bias" (Thompson et al., 1990: 1). A core assumption is that shared values and beliefs, and social relationships are closely knit together. Cultural theory proposes that each way of life implies a characteristic stance towards the environment, other people and the self and that they offer different solutions to problems encountered in life (Douglas, 1996).

Way of life has also been understood to include specific understandings of 'nature'. Hence, an innovation of the schema was made by combining each way of life with one of four (or even five) basic 'myths of nature' (Dake, 1992). By juxtaposing these myths with 'ways of life', culture theory predicts that *egalitarians* are afraid of technology and hold that it has little benefit, they think that nature is ephemeral, fragile and vulnerable; *individualists* believe that nature is benign, affluent and full of promise ('cornucopian'); *hierarchists* are afraid of social deviance, they regard nature as being 'perverse/tolerant', bound by certain rules: it will be benign if treated properly or dangerous if the rules are violated; while *fatalists* regard nature as 'capricious' and unpredictable; and the autonomists (hermits), being asocial, do not care (Schwartz & Thompson, 1990).

A recent development in cultural theory is the cultural cognition approach (Kahan, 2012). Cultural cognition is a social psychological theory stating that an individual's group membership affects how they process information. It is argued that group membership creates a sense of well-being and if beliefs held by a group are challenged, individual members will experience a threat to their identity. As an identity-protecting mechanism, individuals will therefore process information (on risk, environmental issues or any other matter) in a biased manner. Empirical findings about how American individuals perceive various risk issues such as climate change, nanotechnology, guns, vaccines and abortion, arguably demonstrate that 'cultural worldviews' explain risk perception.

## Critique of cultural theory

There are several strands of critique against cultural theory. A critique raised early on was that a typology made up of a limited number of archetypical world views is too simplistic. It reasonably cannot account for the contextual dimensions of the social dynamics of contested environmental problems or risk issues. Cultural theory has also been criticized for conceptual inconsistency and lack of clarity, for vagueness and circularity of reasoning (van der Linden, 2015). Another critique is that it fails to distinguish risk as a culturally specific epistemology in modern society, from pre-modern ways of understanding potential, unwanted events (Boholm, 2015).

Lack of empirical support for the proposed interlinkages between world views and risk perceptions is another area of critique (Marris et al., 1998); and it is argued that attempts to test the theory empirically have been unsuccessful (Sjöberg, 1997). Lack of empirical substantiation of cultural theory has raised concern about the validity of the measurement scales and more broadly, about whether or not the theory can actually be tested at all on an individual level (Rippi, 2002). The relationships between environmental values and beliefs and their connection to pro- or anti-environmental behavior have also been shown to be considerably more complex than culture theory assumes. Studies of environmental attitudes suggest that there are good reasons to expect that cultural biases are dynamic and malleable and that the typology of four (five) cosmologies do not in fact cover actual cultural biases regarding the environment (Price et al., 2014).

## Learning resources

This is a short film introducing and explaining the anthropological concept of culture:

'What is Culture? The Anthropological Concept of Culture (David Alexander Pamer)' https://www.youtube.com/watch?v=VpiIlemQK8o

This is a conversation between Michael Thompson and Matthew Taylor about how the grid-group model and the four cosmological types could be used to understand the behavior of the market:

'Michael Thompson and Matthew Taylor discuss cultural theory as a framework for explaining the faith in the market as a solution to societal problems' https://www.youtube.com/watch?v=Rmz_t_V9sJg

This is an interview with Mary Douglas conducted in 2006 where she looks back on her long career as an anthropologist and where she reflects on her different research interests. The topics of risk and environmental concern are addressed in Part 2 of the interview.

'Interview with Mary Douglas – February 2006 – Part 1' https://www.youtube.com/watch?v=xl3oMdIRFDs

'Interview with Mary Douglas – February 2006 – Part 2' https://www.youtube.com/watch?v=yeoYjMekgZY

## Bibliography


Boholm, Å. (2015) *Anthropology and risk*. London: Routledge.

Dake, K. (1991) 'Orientating dispositions in the perceptions of risk: an analysis of contemporary world views and cultural biases', *Journal of Cross-Cultural Psychology*, 22(1): 61–82.

Dake, K. (1992) 'Myths of nature: culture and the social construction of risk', *Journal of Social Issues*, 48(4): 21–38.

Douglas, M. (1978) 'Cultural bias. Royal Anthropological Institute of Great Britain and Ireland', *Occasional Papers*, 35.

Douglas, M. (1985) *Risk acceptability according to the social sciences*. New York: Russell Sage Foundation.

Douglas, M. (1992) *Risk and blame: essays in cultural theory*. London: Routledge.

Douglas, M. (1996) *Thought styles. Critical essays on good taste*. London: Sage Publications.

Douglas, M. & Wildavsky, A. (1982) *Risk and culture: an essay on the selection of technological and environmental dangers*. Berkeley: University of California Press.

Kahan, D.M. (2012) 'Cultural cognition as a conception of the cultural theory of risk', in, Roeser, S., Hillebrand, R., Sandin, P. & Peterson, M. (eds) *Handbook of risk theory* (pp. 726–759). Berlin: Springer.

Marris, C., Langford, I.H. & O'Riordan, T. (1998) 'A quantitative test of the cultural theory of risk perceptions: comparison with the psychometric paradigm', *Risk Analysis*, 18(5): 635–647.

Price, J.C., Walker, I.A. & Boschetti, F. (2014) 'Measuring cultural values and beliefs about environment to identify their role in climate change responses', *Journal of Environmental Psychology*, 37: 8–20.

Rayner, S. (1992) 'Cultural theory and risk analysis', in Krimsky, S. & Golding, D. (eds) *Social theories of risk* (pp. 83–115). Westport, CT: Praeger.

Rayner, S. & Cantor, R. (1987) 'How fair is safe enough? The cultural approach to societal technology choice', *Risk Analysis*, 7(1): 3–9.

Rippi, S. (2002) 'Cultural theory and risk perception: a proposal for a better measurement', *Journal of Risk Research*, 5(2): 147–165.

Schwartz, M. & Thompson, M. (1990). *Divided we stand: Redefining politics, technology and social choice*. New York: Harvester Wheatsheaf.

Sjöberg, L. (1997) 'Explaining risk perception: an empirical evaluation of cultural theory', *Risk Decision and Policy*, 2(2): 113–130.

Tansey, J. & O'Riordan, T. (1999) 'Cultural theory and risk: a review', *Health, Risk & Society*, 1(1): 71–90.

Thompson, M., Ellis. R. & Wildavsky, A. (1990) *Cultural theory*. Boulder, San Francisco: Westview Press.

van der Linden, S. (2015) 'A conceptual critique of the cultural cognition thesis', *Science Communication*, 38(1): 128–138.

# Ecotheology



Anne Marie Dalton

Ecotheology refers to the construction of religious doctrines and teachings to emphasize the human relationship to nature. Ecotheology is a so-called 'contextual theology'. A common context for all ecotheology is the environmental crisis. The question for ecotheologians is what can religion contribute to moral concern for the state of the natural environment. Strictly speaking, ecotheology is a Christian term having arisen within that religious tradition in the late twentieth century. Theology itself as a term to describe a discipline dates to fourteenth-century Christianity. However, there are analogous terms to theology in some other religions such as Judaism and Islam. Hence ecotheology also has a wider usage.

The religious response to the environmental crises of the twentieth century includes virtually all religions as well as groups that consider themselves spiritual but not affiliated with a specific religious institution. Other terms often used are religious environmentalism, religion and ecology, ecology and spirituality. While these terms are often used interchangeably, sometimes the preference for either ecology or environment is intentional. Ecology is used to convey the intimate relationship of humans to the rest of the natural world; environment, on the other hand, carries the notion of the human against nature as a background. Many ecotheologians want to emphasize that humans are part of nature and participate in its ecology.

## History of ecotheology

Virtually all religions contain ancient teaching about the relationships among God/gods, humans and the rest of nature. It was not until awareness of environmental crises became prevalent in the mid-twentieth century, however, that scholars of religion began to consider the importance of these teachings. Treatment of non-human nature became a moral concern. Religious leaders and scholars looked to their traditional teachings both to critique those that may be complicit in human disregard for and destruction of nature, as well as to find support for a responsible environmentalism. So ecotheology brought together the wisdom of ancient traditions and the call to respond to the ecological crisis.

Many scholars trace the beginning of ecotheology to an article by Lynn White, Jr., first published in 1967. In the article 'The Historical Roots of our Ecologic (sic) Crisis', White accused biblical religions, Christianity in particular, of being a main cause of the ecological crisis. White, along with other scholars such as the historian Arnold Toynbee, argued that the biblical injunction to humans to dominate the Earth, to populate and subdue it for their own benefit was largely to blame for the ecological crisis. He also claimed that the biblical tradition focused human aspirations and endeavours on another world and on a deity transcendent to it. Furthermore, he insisted these teachings were the source of an instrumentalist view of the rest of nature; nature existed only for the benefit of humans. Thus, the Bible contained the historic roots of our ecological crisis. Such accusations provoked an apologetic response from Christians. Many concurred with White that his interpretations were both popular and detrimental but most considered White's thesis too simplistic; it ignored teachings such as the biblical emphases on asceticism and simplicity of life. Titles, such as H. Paul Santmire's *The Travail of Nature: The Ambiguous Ecological Promise of Christian Theology* (1985) capture the general tenor of responses to White's accusations.

Ecotheology also had roots among theologians already engaged in the reconstruction of theology to meet contemporary concerns of the mid-twentieth century. They addressed human alienation, war, racial equality, gender equality as well as the challenge by new scientific discoveries to traditional religious understandings of the universe. In 1963, Joseph Sittler, the American Lutheran pastor, published *The Care of the Earth*. In it, as well as in prior sermons, he raised some of the issues treated by White later and admonished Christians that such a reading of the Bible was not justified; nature did not exist exclusively for human use. Such a view, he claimed, contributed to the sense of alienation of humans from their earthly home.

Other early responders to the ecological crisis include John Cobb, Thomas Berry and Rosemary Radford Ruether. All three were already engaged in a reconstruction of theology to meet the context of the twentieth century. For all of them, however, the ecological crisis became the primary lens through which they called for and contributed to an emerging ecotheology. Cobb described the long-term task of ecotheology as a conversion, turning religious believers toward regard for the earth. Berry claimed that the ecological crisis was primarily a religious crisis caused by the fragmentation of any comprehensive story from which humans found meaning. Ruether emphasized the relationship of ecological devastation to the oppressive exercise of power especially of patriarchy. She linked ecotheology with ecofeminism.

## Ecotheology and world religions

Ecotheology benefitted from the increased interest in interreligious dialogue of the mid-twentieth century. A 1996 conference sponsored by Harvard University brought together scholars from many traditions to study and write about the relationship of their traditions to the ecological crisis. This was a watershed event for religions and ecology. The conference resulted in a series of volumes with subjects such as Christianity and Ecology, Taoism and Ecology, Islam and Ecology, Indigenous religions and Ecology and so on. Each one featured articles from experts both critiquing and reconstructing teachings from one of the mainline religious traditions. The goal was to initiate or further deepen a re-thinking of how humans relate to the rest of the natural world as well as to encourage environmental activism.

## Strands of ecotheology

The Harvard conference revealed a growing awareness that religions provided not merely a source of piety and personal meaning, but also a "broad orientation to the cosmos and the human role in it" as Mary Evelyn Tucker pointed out in "Religion and Ecology: Survey of the Field." A large strand of ecotheology attempts to integrate the latest scientific cosmology with either specific religious traditions or broadly cast spiritual understandings. These strands of ecotheology challenge the hierarchy that places humans above the rest of the natural world. Rather the human species is understood to have its own special role as do all other species. The human is often described as the universe becoming conscious of itself. Some form of consciousness is usually attributed to the entire created order. The internationally popular film *Journey of the Universe*, written and directed by M. E. Tucker, B. Swimme and J. Grimm in 2011, and its companion teaching materials promotes this strand of ecotheology.

Besides cosmologically based ecotheology, there are more conservative or fundamentalist views that attempt to maintain a strict separation between the physical world and the spiritual one. In those traditions, human stewardship of nature is emphasized by some Jewish and Christian traditions and trusteeship of nature by Islam. Some in this strand deny the theory of evolution and accept a literal translation of the scriptural creation stories, but at the same time teach that believers are commanded to care for creation.

While most contemporary ecotheology acknowledges the relationship of the ecological crisis to political and economic systems of oppression and runaway consumerism, some are more clearly focused on this context than others. In *Cry of the Earth, Cry of the Poor* (1997) the liberation theologian, Leonardo Boff, wrote about the connection between the oppression of the poor and oppression of the earth. His ecotheology drew on social analysis of poverty, new turn to ecology in science, and on biblical teachings about justice and respect for creation.

Another critical movement that has influenced ecotheology is eco-racism. Similar to Boff's theology, this ecotheology has a strong claim that the earth is entitled to justice. Justice for the poor and powerless is closely tied to justice for the earth. Much of the impetus for this strand of ecotheology originates with grassroots campaigns in the United States by African American and minority ethnic groups. A 1987 study by the United Church of Christ in the USA found that African American neighbourhoods were more likely than all others to have polluting industries and garbage dumps. With this awareness, religious traditions have called upon resources in their historical traditions to address the effects of the ecological crisis on minorities and the poor.

Post-colonial ecotheology analyses the effects of colonialism on attitudes and practices of humans towards the natural world. Post-colonial ecotheologians often highlight pre-colonial religious beliefs and claim that these older religious cultures advocated a more responsible relationship between humans and the rest of nature. While many of these theologians may now be Christian, Muslim, or Buddhist, for example, they integrate ideas from the more ancient religions into present religious teachings. Chung Hyun Kyung brings specifically ancient Korean beliefs to her ecotheology, indicating how these same beliefs are manifest within Christianity. Likewise, Ivone Gebara brings specifically Latin American contexts to her ecotheology, while also acknowledging the debt to European and American theologies.

Ecotheology is a vibrant and dynamic development within religious traditions. It continues to develop and broaden its scope as scholars from various religions and disciplines interact and as the problems of the environmental crisis become more apparent. It continues, however, to be tied to a moral responsibility to respond actively to the ecological crisis.

## Learning resources

The following two websites provide continually updated information on developments in ecotheology across various cultures and religions as well as access to relevant films, interviews, podcasts and other learning aids:

ARC: Alliance of Religion and Conservation. www.arcworld.org
Forum on Religion and Ecology. www.fore.yale.edu

For a list of books, including specialized topics within specific religions as well as inter-religious and interdisciplinary topics in ecotheology see:

Eco-theology Book list. http://www.cep.unt.edu/ecotheo.html

## Bibliography

Berry, T. (1988) *Dream of the Earth*. San Francisco, CA: Sierra Club Books.
Boff, L. (1997) *Cry of the Earth, Cry of the Poor*. Maryknoll, NY: Orbis Books.
Cobb, J.B., Jr. (1972) *Is it Too Late? A Theology of Ecology*. Beverley Hills, CA: Bruce Publishing.
Deane-Drummond, C. (2008) *Eco-theology*. London: Darton, Longman and Todd, Ltd.
Gebara, I. (1999) *Longing for Running Water: Ecofeminism and Liberation*. Minneapolis, MN: Augsburg Fortress Press.
Kyung, C.H. (1990) *The Struggle to Be Sun Again*. Maryknoll, NY: Orbis Books.
Ruether, R.R. (1994) *Gaia and God: An Ecofeminist Theology of Earth Healing*. New York: HarperCollins.
Sittler, J. (2004) *The Care of the Earth*. Minneapolis, MN: (Facets) Fortress Press.
Tucker, M.E. (2006) 'Religion and Ecology: Survey of the Field', in R.S. Gottlieb (ed.) *The Oxford Handbook of Religion and Ecology* (pp. 398–418). New York: Oxford University Press.

# Environmental anthropology



## Laura Rival

'Environmental anthropology' has been practiced under different names at different times and in different locations. Known variously as ecological anthropology, cultural ecology, human ecology, cultural materialism, anthropology of landscape, historical ecology, and anthropology of nature, the name changes reflect theoretical or methodological disagreements among practitioners. It is therefore no surprise that a historical approach to anthropological conceptualisations of the natural environment reveals some of the discipline's most entrenched epistemological tensions. As illustrated below, these tensions are often linked to theoretical difficulties encountered when attempting to undo conceptual categories linked to Victorian geographical determinism and evolutionism, or to dichotomies predicated on the dual opposition of nature and culture, or environment and society. The period between the early 1970s and the late 1990s was particularly rich in studies that questioned the thinking habits characteristic of modern Western societies. The renewed attention paid during this period to ecology and environment, materialist themes that 1980s and 1990s social and cultural theory had relegated to the margins, paved the way for contemporary ontological shifts, including post-human approaches and multi-species ethnographies.

I focus below on an important author whose work stands at a crucial juncture between old and new ecological anthropology, Roy Rappaport. I show how his rich empirical data and unconventional ideas have influenced the revival of environmental approaches within the discipline. The chapter then explores areas of both heated debate and continued concern (adaptation, meaning and religion, and traditional environmental knowledge), before concluding on dualist framings of environment/society coupling in anthropology and in geography.

## Roy Rappaport's ecological anthropology

The 1984 enlarged edition of Rappaport's seminal *Pigs for the ancestors* contains a 200-page long epilogue in which the author responds to his critiques and forcefully defends his approach, thereby rekindling the debate about cultural adaptation to environmental constraints. First published in 1967, the monograph controversially used ecological theory and method to describe the Tsembaga Maring, a Melanesian community of horticulturists and pig herders, whose contact with modern civilisation and colonial rule is dated to no earlier than 1954. Autonomous from external market and state institutions and inseparable from kinship and religious institutions, the Tsembaga's natural economy is reproduced through ritual regulation of ecological relations. Through detailed empirical measurements based on the modelling of Tsembaga society as a population in the ecological sense (that is, in Rappaport's words, "as one of the components of a system of trophic exchanges taking place within a bounded area"), the author calculated that a human population of 200 can maintain perpetually a population of 140 to 240 pigs averaging 100 to 150 pounds each. The book also contains detailed information about the intricate linkages between hunting, horticulture and pig husbandry. Men hunt and capture wild boars that are brought to mate with female pigs in the village compound. Women do the bulk of the work of raising crops for the pigs (these receive 54 per cent of sweet potatoes and 82 per cent of manioc harvested each year). Pigs are ritually killed during periodic year-long festivals (*kaiko*) that reduce the pig population to one-sixth of its former size. For Rappaport, these rituals are best analysed as conventionalised acts that seek to involve non-empirical (supernatural) agencies in human affairs. Moreover, ritual cycles serve to prevent environmental degradation while regulating tribal warfare. Being the first monographic study of the total social-ecological system of a people, this book quickly became a classic; it remains the most widely cited ecological anthropology book. The methodology by which the interrelations of social, cultural and environmental elements in a wider system are described; the focus on homeostatic flows of nutrients in a biotic community and on negative feedbacks; and the role of ritual in local and regional resource management are some of the themes most commented upon.

The book's application of systems theory to an anthropological population interacting with other components of the landscape through trophic exchanges has often been misunderstood. Whereas Rappaport views culture as a cybernetic system that regulates the integration of social and biophysical factors, many have criticised him for ignoring social causation and historical processes that alone explain the particularities of Tsembaga prestige economy of feasting and exchange. Critiques of the type of quantitative and qualitative field data collected on a wide range of topics (such as diet, energy expenditure, myths or folk taxonomies) also reject Rappaport's ecological approach on the grounds that it goes against the premise that social phenomena are autonomous from physical reality. In defence of his empirical research and analytical work Rappaport reminded his critics that he sought to highlight the relevance of biophysical facts as they are, for their social impact cannot be reduced to the ways in which they are socially understood and culturally constructed.

The reception of the book's second edition made it particularly evident that there was more to the success of *Pigs for the ancestors* than the challenge it poses to anthropology's professional social bias. What made the book theoretically ambitious is not simply Rappaport's highly original application of ecological knowledge and evolutionary biology, but, more significantly, the ways in which ecological evidence is used to engage anew older theorisations of environment/society interactions. By sustaining a parallel and combined attention to ecological facts and cultural meanings ("The human species is a species that lives in terms of meanings in a physical world devoid of intrinsic meaning but subject to causal laws"), Rappaport was able to transcend the opposition between particularism, historicism and possibilism (Franz Boas) on the one hand, and determinism, materialism and evolutionism (Leslie White) on the other. Like Julian Steward before him, he focused on very specific and localised environment/society interactions, which enabled him to consider the mediating impact of tools and technological know-how in his analysis of the ecological determinants of Tsembaga society. Unlike his predecessor, however, he was not primarily interested in change, nor did he consider religion as secondary or peripheral to the organisation of subsistence activities. On the contrary, his thesis was that rituals perform an ecological function by creating order and regularity. Even if it is not the reason why people perform them, rituals regulate relationships between humans and non-humans (as well as among humans). Unlike Marvin Harris whose reductionist materialism entirely by-passes people's understandings of the world and of their actions in it, Rappaport paid equal attention to 'operational' and 'cognised' models of the environment. He explained that although his study is primarily concerned with the role ritual plays in the material relations of the Tsembaga, the fact that these people understand their ritual performance in terms of the need they feel to rearrange their relations with ancestral spirits should not be ignored: "It would be possible in an analysis of the empirical consequences of ritual acts to ignore such rationalisations, but anthropology is concerned with elucidating causes, as well as consequences, of behaviour, and proximate causes are often to be found in the understandings of the actors. It seems to me, therefore, that in ecological studies of human groups we must take these understandings into account" (Rappaport, 1984: 237).

The evolutionary and political implications of Rappaport's unique approach to natural systems and cultural constructions became understood more fully after his premature death in 1997. In his view (deeply influenced by Gregory Bateson and Leslie White), it is through the special features of language as a system of symbolic communication that culture has come to play a fundamental role not only in human evolution, but also in the evolution of life in general. While language enables humans to communicate information about the world, it can also be used to create entirely new content, thus greatly enhancing human adaptability and flexibility. As people can only live in terms of meanings they themselves must construct (Rappaport, 1994: 156), the ability to create reality without a material basis through language may result in maladaptation; wrong values may be over-sanctified, and proximate causes confused with ultimate causes. This occurs in modern societies dominated by market exchanges, where sacred values are degraded to the status of tastes or preferences. In the United States, for example, the cult of money and of individualism is endangering human survival and the evolution of life as a whole. Somewhat surprisingly, a comparative analysis of the relationship between naturally constituted physical laws and cultural values among Tsembaga and North-American people leads Rappaport to reach a Durkheimian conclusion: that religion is necessary to create social order. Whereas ritual performance ensures that ecology and economy are attuned in Tsembaga society, such attuning in industrialised societies will require the reinvention of moral theory, religion and science on the basis of ecological knowledge so that humanity can take its place within nature, that is, "inside rather than outside life" (Rappaport, 1994: 163–166).

## Person, environment, and the reinvention of environmental anthropology

Scientific and public interest in environment-society interrelations grew exponentially in the last decade of Rappaport's life, which was marked by a surge in anthropological publications on the topic. Cross-disciplinary research such as political ecology also soared, often with the active participation of anthropologists. Rappaport's work may have been increasingly overshadowed in the process, yet the ghostly presence of his ideas dwells in most research initiatives and ethnographic studies of the period, to say nothing of the theoretical elaborations of the time on the ontological status of nature in Western and non-Western cultures.

The specialised sub-field of environmental anthropology became increasingly focused on what Rappaport had analysed under the label 'ecology.' Much work was dedicated to keeping up with scientific developments in both ecology and evolutionary biology, and assessing the relative merit of an ecosystem approach in anthropology and archaeology. Environmental anthropologists also dedicated much of their work to redefine adaptation, a concept marred with passive connotations (what nature does to people), soon to be replaced by the more dynamic idea of human adaptability (how humans influence the environment). Throughout the period under discussion, environmental anthropologists actively engaged with the diversification of ecology as a scientific project into systems ecology, behaviour ecology and non-linear systems theory. Systems ecology appealed to environmental anthropologists ready to grant causal priority to groups and social wholes, while behaviour ecology strengthened the conclusions of those who located explanatory power at the level of individual rationality. A new approach, historical ecology, took on board many of the insights of nonlinear biology in analyses of the dynamics of gradual change in the Amazon biome and other tropical rainforests. Although often perceived to be frontiers of wilderness, such ecosystems could be shown to result from past human habitation. The application of historical ecology insights to other regions of the world popularised the notion of anthropogenic landscape.

Constructions of nature in ritual and religious beliefs became a central interest of mainstream social and cultural anthropologists looking for better ways of conceptualising objective and subjective dimensions of reality through notions such as practice, agency, consciousness, personhood, and so forth. Comparative analyses of Western and non-Western notions of nature, wilderness and the environment confirmed Rappaport's discernment regarding the unique features of ritual communication. If cosmovision (or 'super-nature') is wrought out of ritual performance, what relationship does it bear to nature? As ethnographers grew dissatisfied with dual oppositions such as 'symbolic' versus 'material' dimensions of life, they looked for more satisfactory ways of theorising the role played by language in mediating between minds and the world. Cognitivists offered new explanations of the connection between individual and public mental representations, while phenomenologists proposed direct perception as an alternative to representational approaches.

Rooted in ethnoecology and earlier 'ethnomethodology' approaches to indigenous science, studies of traditional environmental knowledge flourished in the 1990s, when biodiversity conservation and adaptive resource management became important areas of global policy. These studies applied Rappaport's holistic combination of scientific ecology modelling with ecological models found in tribal cultures to demonstrate that the worldviews, technologies and environmental management practices found in indigenous societies have scientific value. Many of these studies also sought to demonstrate the relevance of traditional environmental knowledge to sustainable development, thus challenging hierarchies of knowledge while overtly supporting indigenous political and economic autonomy. The contemporary relevance of indigenous science, as well as the mutual benefits achieved by coupling traditional environmental knowledge and scientific ecology, was further argued by some on the basis of a perceived convergence between indigenous worldviews and alternative scientific paradigms. The latter, characterised by a shift away from mechanistic understandings of the world, focus on natural laws of interdependence. These more organic and relational models are still explored today as sharing much in common with indigenous cosmologies and traditional value systems. Such studies are thus in direct line with Bateson and Rappaport, for whom ecology, the study of interrelationships among living organisms and between living organisms and the biophysical environment necessarily calls into question the autonomy of science and religion from society. In Rappaport's words, "the ecosystem concept and actions informed by it are part of the world's means for maintaining, if not indeed constructing, ecosystems."

## Nature, Western modernity, and historical particularisms

As this brief history of environmental anthropology shows, the process of breaking down the nature/culture dualism that lies at the core of Western modernity has been slow. It may have been even slower without Rappaport's intervention. In any case, none of the influential theoretical alternatives put forward so far (Tim Ingold's radically anti-evolutionist phenomenology, Marilyn Strathern's highly abstract mereology, or Philippe Descola's neo-Durkheimian theorisation of the world's diversity) can be fully comprehended without proper attention to his anthropology. Now that the widespread acceptance of anthropogenic climate change renders the notion of constructed nature redundant, who will take up the challenge of asking with Rappaport 'How natural is nature?'

## Learning resources

The most up to date – though high level – guide to environmental anthropology is *The Routledge Handbook of Environmental Anthropology* (2016), edited by Helen Kopnina and E. Shoreman-Quimet.

## Bibliography

Descola, P. & Palsson, G. (eds) (1996) *Nature and society. Anthropological perspectives*. London: Routledge.

Dudgeon, R. & Berkes, F. (2003) 'Local understandings of the land: traditional ecological knowledge and indigenous knowledge', in Helaine Selin (ed.) *Nature across cultures: views of nature and the environment in non-western cultures* (pp. 75–96). Dordrecht: Kluwer Academic Publishers.

Harris, M. (1980) *Cultural materialism. The struggle for a science of culture*. New York: Vintage Books.

Medin, D. & Atran, S (eds) (1999) *Folkbiology*. Cambridge, Mass: MIT Press.

Moran, E. (1979/2000) *Human adaptability: An introduction to ecological anthropology*, 2nd edition. Boulder, Col: Westview Press.

Moran, E. (ed.) (1990) *The ecosystem approach in anthropology. From concept to practice*. Ann Arbor: The University of Michigan Press.

Rappaport, R. (1967/1984) *Pigs for the ancestors. Ritual in the ecology of a New Guinea people*. New, enlarged edition. New Haven: Yale University Press.

Rappaport, R. (1994) *Humanity's evolution and anthropology's future*, in Robert Borofsky (ed.) *Assessing cultural anthropology* (pp. 153–167). New York: McGraw-Hill.

Steward, J. (1977) *Evolution and ecology. Essays on social transformation*. Edited by Jane C. Steward and Robert F. Murphy. Urbana: University of Illinois Press.

White, L. (1959/2016) *The evolution of culture*. London: Routledge.

# Environmental economics

## 3.4

Wim Carton

Environmental economics is a sub-discipline of economics that aims to understand, and influence, the economic causes of human impacts on the non-human world, such as atmospheric pollution. It seeks to apply the main concepts and methods of economic thought to environmental goods (i.e. various natural resources) and services (e.g. carbon sequestration) with the objective of managing those goods and services more efficiently. Underlying this approach is the conviction that economic concepts are useful tools for dealing with such problems as environmental degradation, resource depletion, and global environmental change. Even though this assumption is heavily contested, environmental economics has over the past decades proven highly influential in the design and implementation of environmental policies. The sub-discipline takes the neo-classical perspective that has dominated the economics discipline for about a century.

One of the foundations of environmental economics is the idea that environmental problems are a form of *market failure*. This is the inability of the market to account for the full environmental costs of economic production of goods and services. This concept goes back to the work of the English economist Arthur Pigou (1920) and essentially states that some (social and environmental) goods and services tend to be undervalued in market exchange, resulting in resource allocation and a supply/demand balance that is 'suboptimal'. Pigou elaborated this idea by developing British economist Alfred Marshall's notion of the economic *externality*, or the unintended positive or negative effects of economic activity on anyone who did not choose to experience these effects. From this perspective, air pollution, for example, is a market failure because polluting industries do not bear the costs of the unintended negative consequences that their activities have for wider society. Since air pollution is a 'free' factor in the production process, market mechanisms cannot give any incentives to limit its occurrence. Similarly, some environmental economists see global climate change as "the greatest example of market failure we have ever seen" (Stern, 2006, p. 1) because the social and environmental costs of greenhouse gas emissions are not considered in standard economic accounting.

Traditionally, environmental policy took the form of so-called *command and control* mechanisms, which involve direct government intervention in the market, for example through laws and regulations that set specific environmental standards or prohibit polluting practices. Conceptualizing environmental problems as market failures, however, allowed the design of policies that aim to *internalize* unaccounted-for environmental costs, by giving environmental externalities their 'proper' economic value. This has spurred debates not only about what the economic value of 'nature' is (Costanza et al., 1997), but also about how to account for this value through environmental policies. Over the years, environmental economists have proposed different ways of doing this. Pigou (1920) himself argued that market failures inevitably required government intervention in the form of subsidies or taxes. Presumably he would have argued that a problem like climate change should be dealt with by governments imposing a carbon tax on industries and energy producers, or conversely handing out subsidies to producers of renewable energy. In this scenario, it is the government that decides on the economic cost that externalities should have, and who then imposes that cost.

Over the years however, economists became more averse to the idea of environmental taxes and subsidies. In an influential article from 1960, Ronald Coase argued that there was no economic reason to suppose that government intervention in the market would be the preferred mechanism for internalizing costs. In order to evaluate solutions on the basis of their economic efficiency, he argued, one needed to account for the reciprocal character of the problem. Whereas Pigou's solution primarily recognized the rights of the victims of negative externalities, say, the right to clean air for people affected by air pollution, Coase also insisted on the economic value of polluting activities, and therefore the right to pollute. In other words, it was necessary also to take into account the economic benefits connected to negative externalities, hence the potential costs of foregoing these. Following this logic, the creation of scarce pollution rights should be at least as effective as taxes and subsidies, without the alleged negative effects connected to government interventions (Coase, 1960).

In the 1960s and 1970s a number of economists further developed this argument. Most notable among these is John H. Dales, whose treatise *Pollution, Property & Prices* (2002 [1968]) later became one of the cornerstones for the design of market-based environmental policies. Inspired by Coase, Dales (2002 [1968]) reworked Pigou's idea of market failure by defining environmental problems as a "failure to devise property rights" (p. 792) for the use of environmental goods and services. The logical solution to this then was the creation of a system of tradeable pollution rights. Using the example of water pollution, Dales noted how a government could set a limit on the amount of pollution (e.g. wastewater) it allows, and then sell the equivalent amount of pollution rights on the market. Anyone who wanted to discharge wastewater into the environment would then need to possess a corresponding amount of pollution rights or face a fine. Businesses that turn out to need less of these pollution rights than they had originally purchased could sell their excess rights on the market. Since, according to Dales, environmental objectives should ultimately be achieved in the least costly way, and because he expected a tradeable pollution scheme to have clear cost-saving benefits, this mechanism would be preferable to taxes and subsidies.

With respect to this focus on property rights, it is worth briefly mentioning another influential text published at that time, namely biologist Garrett Hardin's (1968) *Tragedy of the Commons*. Not unlike Dales, Hardin argued that environmental problems arise because individuals or companies strive to maximize their utility in a situation where the negative outcomes of their activities are commonly shared. Unlike Dales though, Hardin's primary concern was with population growth, which led him to the conclusion that the problem of the commons required coercive measures, not necessarily private property relations. Despite this, the 'Tragedy of the Commons' concept has in popular discourse become fused with Dales' arguments and now frequently serves as legitimation for market-based environmental regulation and the privatization of various environment assets (Harvey, 2010; Hawkshaw et al., 2012).

Post-1960s, environmental economists increasingly embraced the ideas of Coase and Dales, fine-tuning their arguments but leaving the main tenets untouched. Tradeable pollution rights (or 'permits') thereby gradually became the most popular approach for internalizing environmental costs (Tietenberg, 2010). Policy makers, however, initially proved reluctant to adopt the idea, holding on to their tried-and-tested combination of command and control, and taxation and subsidies. While the first local and regional attempts at implementing tradeable permit schemes occurred during the 1970s and 1980s, it was not until the beginning of the 1990s, propelled by a broader neoliberal trend in policy making, that the first large-scale experiment with tradeable permits was put in place. This was the United States $SO_2$ allowance trading system, a scheme designed to decrease emissions of sulphur dioxide ($SO_2$) and nitrogen oxides ($NO_x$) in the power sector. The scheme was seen as a success and quickly became a best-practice example justifying the adoption of similar mechanisms for other environmental problems (Stavins & Whitehead, 1997).

Since then, tradeable permit schemes have found their most widespread application in climate and energy policy. Under the guise of three 'flexibility mechanisms', they were a key component of the Kyoto Protocol, the world's first binding climate change mitigation treaty (UNFCCC, 1997). As with the US $SO_2$ scheme, these instruments aimed to internalize the environmental costs of climate change by creating tradeable emission rights. Regional and national adaptations of this approach have since proliferated, as with the EU Emission Trading Scheme (EU ETS), which was instituted in 2005 as one of the pillars of the European Union's climate policy (Carton, 2014, 2016). As such, recent trends in environmental policymaking clearly bear the hallmark of environmental economics. Reasons for the popularity of this approach are many, but can at least in part be attributed to the alleged (but contested) superiority of permit schemes in terms of their economic efficiency, the opportunities for 'flexibility' they provide to targeted businesses, and the opposition of politically influential businesses to regulatory instruments such as carbon taxes (Braun, 2009; Lane, 2012).

It is also in the design and implementation of these concrete market instruments that the ideas of environmental economists have met their most vocal critics, and that they have arguably faced their own limits and shortcomings. Instruments such as the EU ETS and the Kyoto mechanisms have struggled with serious and continuous problems over the years, exposing the enormous challenges involved in creating functional markets in pollution rights (Gilbertson & Reyes, 2009; Morris, 2012). A wide range of critics meanwhile opposes the underlying ideas of environmental economics on the grounds that fitting environmental goods and services into existing economic frameworks is both infeasible, undesirable, and insufficient for dealing with current environmental problems. For these critics, seeing environmental degradation as primarily a market failure is hardly self-evident, and certainly not inconsequential when taking a broader range of social, cultural and ecological perspectives into account (Hyams & Fawcett, 2013; Lohmann, 2009; Spash, 2010).

## Learning resources

For an introduction to the field of Environmental Economics, see:

Field, B.C. & Field, M.K. (2009) *Environmental Economics: An Introduction*. New York: McGraw-Hill.

For a critique of the assumptions of Environmental Economics, and an alternative take on the relationship between economy and environment, see:

Costanza, R., Cumberland, J.H., Daly, H., Goodland, R., Norgaard, R.B., Kubiszewski, I. & Franco, C. (2015) *An Introduction to Ecological Economics*. Broca Raton: CRC Press.

For an introduction to some of the problems that have arisen in the development of tradable permit schemes for climate change mitigation, see:

Gilbertson, T. & Reyes, O. (2009) Carbon trading: how it works and why it fails. *Critical Currents, Occassional Papers Series*, number 7. Available at http://www.dhf.uu.se/pdffiler/cc7/cc7_web.pdf

Lohmann, L. (2006) Carbon Trading: A critical conversation on climate change, privatisation and power. *Development Dialogue* (Vol. 36). Uddevalla, Sweden. Available at http://www.thecornerhouse.org.uk/sites/thecornerhouse.org.uk/files/carbonDDlow.pdf

## Bibliography


Braun, M. (2009) The evolution of emissions trading in the European Union: The role of policy networks, knowledge and policy entrepreneurs. *Accounting, Organizations and Society, 34*(3–4), 469–487.

Carton, W. (2014) Environmental protection as market pathology? Carbon trading and the dialectics of the "double movement." *Environment and Planning D: Society and Space, 32*(6), 1002–1018.

Carton, W. (2016) *Fictitious Carbon, Fictitious Change? Environmental Implications of the Commodification of Carbon*. Lund: Media-Tryck, Lund University.

Coase, R. (1960) The problem of social cost. *The Journal of Law and Economics, 3*, 1–44.

Costanza, R., D'Arge, R., de Groot, R., Farber, S., Grasso, M., Hannon, B., Limburg, K., Naeem, S., O'Neill, R.V., Paruelo, J., Raskin, R.G., Sutton, P. & van den Belt, M. (1997) The value of the world's ecosystem services and natural capital. *Nature*. 387, 253–260.

Dales, J. (2002) *Pollution, Property & Prices*. Cheltenham: Edward Elgar Publishing.

Gilbertson, T. & Reyes, O. (2009) Carbon trading: how it works and why it fails. *Critical Currents, Occassional Paper Series*, number 7.

Hardin, G. (1968) The tragedy of the commons. *Science, 162*(3859), 1243–1248.

Harvey, D. (2010) The future of the commons. *Radical History Review, 2011*(109), 101–107.

Hawkshaw, R.S., Hawkshaw, S. & Sumaila, U.R. (2012) The tragedy of the "tragedy of the commons": why coining too good a phrase can be dangerous. *Sustainability, 4*(11), 3141–3150.

Hyams, K. & Fawcett, T. (2013) The ethics of carbon offsetting. *Wiley Interdisciplinary Reviews: Climate Change, 4*(2), 91–98.

Lane, R. (2012) The promiscuous history of market efficiency: the development of early emissions trading systems. *Environmental Politics, 21*(4), 583–603.

Lohmann, L. (2009) Toward a different debate in environmental accounting: The cases of carbon and cost-benefit. *Accounting, Organizations and Society, 34*(3–4), 499–534.

Morris, D. (2012) *Losing the lead? Europe's flagging carbon market*. Retrieved from http://www.sandbag.org.uk/site_media/pdfs/reports/losing_the_lead.pdf

Pigou, A.C. (1920) *The Economics of Welfare*. London: Macmillan.

Spash, C.L. (2010) The brave new world of carbon trading. *New Political Economy, 15*(2), 169–195.

Stavins, R.N. & Whitehead, B. (1997) Market based environmental policies. In M.R. Chertow & D.C. Esty (Eds), *Thinking Ecologically, The Next Generation of Environmental Policy* (pp. 105–117). New Haven, CT: Yale University Press.

Stern, N. (2006) *Stern review report on the economics of climate change*. Retrieved from http://www.citeulike.org/group/342/article/919155.

Tietenberg, T.H. (2010) The evolution of emissions trading. In J.J. Siegfried (Ed.), *Better Living through Economics* (pp. 42–58). London: Harvard University Press.

United Nations Framework Convention on Climate Change (UNFCCC) (1997) *Kyoto Protocol to the United Nations Framework Convention on Climate Change*. Bonn: HeinOnline. Retrieved from http://unfccc.int/resource/docs/convkp/kpeng.pdf

# 3.5 Ecofeminism

## Greta Gaard

What does feminism have to do with climate change? Around the world, we see forests cut down to make room for cattle grazing, rivers dammed to create electricity for growing cities, waste dumps and e-wastes from industrialized nations exported to the least wealthy countries for women and children to sort. While feminists have a long-standing commitment to gender justice and to action toward ending oppression – from the anti-slavery movement to the suffrage movement, and from urban sanitation to sustainable agriculture and animal rights – ecofeminism emerged in the 1980s as a perspective and political movement that first brought together the gendered intersections of both human–human injustices and human–environment exploitations.

## Activist origins

Ecofeminist theories grew out of women's activism. In India, the ecofeminist physicist Vandana Shiva researched and organized around biotechnology, water, forestry, and oil for the ways these issues affect women's livelihoods and environments. In Kenya, Wangari Maathai observed the deforestation and desertification of her community where the kikuyu trees had once flourished, sheltering the land, retaining rainwater, and providing food. When she saw how the women struggled to grow food in this now barren land, Maathai recognized these crises of food scarcity, women's subsistence impoverishment, and lack of water were linked, and she created strength out of scarcity. By helping women to plant trees, Maathai's Green Belt Movement helped reforest Kenya and helped women feed their families.

In 1978, a young mother of two children in upstate New York, Lois Gibbs read the newspaper reports describing the toxic waste dump beneath her community, the chemical contaminants found in their air and water, and the dump's suspected links to her community's high rates of miscarriage, birth defects, and childhood illnesses. Armed with a high school education and skills in community networking, Gibbs discovered the cancer cluster atop Hooker Chemical's waste dump. Her work eventually forced a complete buyout of all homes in Love Canal at fair market value, a relocation of these families, and the launch of the Environmental Protection Agency's Superfund Program.

Women's peace activism also grounds ecofeminism. In Clayoquot Sound, British Columbia, feminists created a Peace Camp in the early 1990s to stop logging, and through their activism and conversations, they developed ecofeminist theory addressing forestry, gender and sexuality, peace and militarism (Moore, 2015). In England, the women's peace camp at Greenham Common (1981–2000) embodied their resistance to the possibility of a nuclear war in Europe. In the U.S.A. the links between feminism and ecology were forged at the "Women and Life on Earth: Ecofeminism in the 1980s" conference, and were quickly followed by the Women's Pentagon Actions of 1980 and 1981, where large numbers of women demonstrated against militarism and its violence against women, children, people of color, poor people and the earth. Launched in 1985, WomanEarth Feminist Peace Institute was a direct outgrowth of these anti-militarist actions, founded with the intention of creating an ecofeminist educational center producing theory, conducting research, and supporting political activities that would confront racism head-on (Sturgeon, 1997).

In northern California, Feminists for Animal Rights (FAR) was launched in 1982 to address the linked exploitation of women and animals. The sexualized objectification of women of color as desirous animals posing in animal skins and cages, along with the projection of human desire onto animals who claim they want to be eaten, raised their curiosity about Western culture's interlinked dehumanization of women and animals. FAR activists have advocated against industrialized animal agriculture, and for the rights of battered women, their children and companion animals: these activists built collaborations between battered women's shelters and animal shelters to ensure that women fleeing domestic violence could also protect the family pets.

Feminist interest in environmental issues took on a global focus through the Women's Environment and Development Organization (WEDO), formed in 1991 by former U.S. Congresswoman Bella Abzug and feminist activist and journalist Mim Kelber. Bringing together women from around the world to take action in the United Nations and other international policymaking forums, WEDO's primary events in the 1990s included the World Women's Congress for a Healthy Planet in 1991, and a number of Women's Caucuses at key UN conferences. In 2015 and beyond, WEDO has turned its focus on the critical issue of women and climate justice, participating in the United Nations' Council of the Parties discussions on climate change and advocating for gender justice.

## Ecofeminist philosophy

Are women like trees, toxic waste, or animals? Hardly. Ecofeminists soon rejected essentialist equations of women and nature for more robust analyses that utilize 'intersectionality' to explain the linkages between systems such as global economics, agriculture, energy production, interspecies justice, and waste disposal. Intersectionality is an analytical approach that explores how different factors combine to place many women (along with those marginalized by race, class, sexuality, species, age, and ability) in a disadvantaged position relative to most men.

At the same time that ecofeminist Karen Warren (1990) developed the boundary conditions for ecofeminist philosophy, and Val Plumwood (1993) explained the functioning of the so-called Master Model, Black feminists Kimberlé Crenshaw (1989) and Patricia Hill Collins (1990) developed the concept of intersectionality, explaining the ways that human–human oppression is simultaneously and inextricably gendered, raced, and classed. As Warren explained, a feminist issue can be any issue that contributes to understanding the oppression of women–from environmental degradation to food production. Building on the feminist notion of human self-identity as fundamentally relational, ecofeminism reconceives what it means to be human as emerging from our relations with networks of humans as well as other species and ecosystems. Deane Curtin's (1991) theory of contextual moral vegetarianism grew out of ecofeminism's foundation as a contextual ethic, and one that resists all forms of social and environmental domination. Like feminism, ecofeminism rejects claims of objectivity and universalism as factually impossible, and augments standpoint epistemology with the feminist values of contextual ethics and inclusiveness, valuing diversity in standpoints and including other species and environments interests as well.

Conceptual tools from ecofeminism have been used in developing other theories such as posthumanism, critical animal studies, material feminisms, ecocriticism, queer ecologies, as well as feminist perspectives on citizenship, democracy, economics, and climate change.

## Ecofeminism and climate justice

Like the anti-globalization movement that emerged at the Seattle World Trade Organization resistance in November 1999, the climate justice movement brings together environmental justice activists, feminists, labor activists and environmentalists. Initially articulated in the Bali Principles of Climate Justice (2002), its intersectional analysis can become more inclusive with an ecofeminist perspective (Gaard, 2015). Climate change disasters such as Hurricane Katrina in New Orleans illustrate these missing components.

### *Environmental sexism*

Women and children are more likely to die during and immediately after ecological disasters than men. In New Orleans, the fact of gendered wage inequities combined with race had already placed 41 per cent of female-headed households with children below the poverty line, and thus the majority of those left behind were women with children, the poor and the elderly. In the immediate aftermath of Katrina, domestic violence and sexual assaults spiked.

### *Environmental ageism*

The very elderly and the very young are more at risk in climate crises and in areas of ongoing toxicity, and the majority of elderly populations are women. During and immediately after Hurricane Katrina, the majority of deaths occurred disproportionately among the elderly.

## Environmental ableism

After Hurricane Katrina, horror stories emerged of people in hospitals and nursing homes being left to drown. While age and disability often co-occur, impairments of hearing, vision, cognition, speech, and mobility can affect people of all ages, making it difficult for them to seek protection in climate crises; for elderly people, these impairments are more likely and more challenging. Young people are at greater risk as well, given how the disproportionate racial impact of asthma among urban and lower-income children of color affects their ability to breathe while fleeing or surviving climate disasters.

## Environmental heterosexism

Climate change homophobia is evident in the media blackout of LGBTQ people in the wake of Hurricane Katrina, which occurred just days before the annual queer festival in New Orleans, "Southern Decadence," a celebration that drew 125,000 revelers in 2003. The religious right quickly declared Hurricane Katrina an example of God's wrath against homosexuals, waving signs with "Thank God for Katrina" and publishing detailed connections between the sin of homosexuality and the destruction of New Orleans. Environmental heterosexism also masks the presence of LGBTQI activists already present in the climate justice movement, yet their vulnerability as queers remains.

## Environmental speciesism

Speciesism obscures the ways that people's lives are lived in relationship with other animals as well as environments, and limits the scope of feminist analysis: nearly half of those who stayed behind during Katrina refused rescue helicopters and boats that offered safety only to humans, and stayed because of their companion animals – and many died together. Moreover, speciesism obscures climate change analyses of root causes, since industrial animal agriculture is one of the top three emitters of greenhouse gases as well as the cause of continued deforestation, water use and water pollution, and real material hunger for humans (Kemmerer, 2015).

# Conclusion

Around the world, women are on the frontlines of climate justice crises as well as climate justice solutions. International organizations bringing an ecofeminist or feminist environmental justice perspective to climate change now include not only the Women's Environment and Development Organization (WEDO) but the Global Gender and Climate Alliance, GenderCC: Women for Climate Justice, WoMin: an African ecofeminist organization, the Indigenous Environmental Network, Idle No More, and the Women's Earth and Climate Action Network (WECAN). With its long history of eco-activism, ecofeminism provides tools for understanding these climate crises and for developing an inclusive approach to climate justice.

## Learning resources

Adams, C.J. "About Ecofeminism." Available at http://caroljadams.com/about-ecofeminism/.

Anderson, K. and Kuhn, K. (2014). "Cowspiracy: The Sustainability Secret." Available at http://www.cowspiracy.com/.

Gaard, G. (1996). "Ecofeminism Now." Available on YouTube, at https://www.youtube.com/watch?v=BTbLZrwqZ2M

Hutner, H. (2015). "Eco-Grief and Ecofeminism." TEDx Talk. Available on YouTube, at https://www.youtube.com/watch?v=t6FuKhjfvK8.

Mellor, M. (2014). "Eco-Feminism and Handbag Economics." Available on Vimeo at https://vimeo.com/99156211.

Shiva, V. (2013). "Vandana Shiva Interview About Ecofeminism." Available on YouTube, at https://www.youtube.com/watch?v=fM8TLXjpWk4.

## Bibliography


Adams, C.J. & Gruen, L., (eds) (2014) *Ecofeminism: Feminist Intersections with Other Animals and the Earth*. New York: Bloomsbury.

Collins, P.H. (1990) *Black Feminist Thought: Knowledge, Consciousness, and the Politics of Empowerment*. New York: Routledge.

Crenshaw, K. (1989) 'Demarginalizing the Intersection of Race and Sex: A Black Feminist Critique of Antidiscrimination Doctrine, Feminist Theory and Antiracist Politics', *University of Chicago Legal Forum*, 140, 130–167.

Curtin, D. (1991) 'Toward an Ecological Ethic of Care', *Hypatia*, 6:1, 60–74.

Gaard, G. (2015) 'Ecofeminism and Climate Justice', *Women's Studies International Forum*, 49, 20–33.

Harper, A.B. (ed.) (2010) *Sistah Vegan: Black Female Vegans Speak on Food, Identity, Health, and Society*. New York: Lantern Books.

Kemmerer, L. (ed.) (2011) *Sister Species: Women, Animals, and Social Justice*. Urbana, IL: University of Illinois Press.

Kemmerer, L. (2015) *Eating Earth: Environmental Ethics & Dietary Choice*. London: Oxford University Press.

Moore, N. (2015) *The Changing Nature of Eco/Feminism: Telling Stories from Clayoquot Sound*. Vancouver: University of British Columbia Press.

Mortimer-Sandilands, C. & Erickson, B. (eds) (2010) *Queer Ecologies: Sex, Nature, Politics, Desire*. Bloomington, IN: Indiana University Press.

Plumwood, V. (1993) *Feminism and the Mastery of Nature*. New York: Routledge.

Shiva, V. (1988) *Staying Alive: Women, Ecology and Development*. London: Zed Books.

Shiva, V. (1997) *Biopiracy: The Plunder of Nature and Knowledge*. Boston, MA: South End Press.

Shiva, V. (2002) *Water Wars: Privatization, Pollution, and Profit*. Boston, MA: South End Press.

Shiva, V. (2008) *Soil Not Oil: Environmental Justice in an Age of Climate Crisis*. Boston, MA: South End Press.

Sturgeon, N. (1997) *Ecofeminist Natures: Race, Gender, Feminist Theory and Political Action*. New York: Routledge.

Warren, K.J. (1990) 'The Power and the Promise of Ecological Feminism', *Environmental Ethics*, 12, 125–146.

# Environmental ethics

3.6


## Ben A. Minteer


### The rise of environmental ethics

If the goal of ethical reasoning is the attempt to answer the question posed by Socrates – 'How should one live?' – the goal of *environmental* ethics is to determine how one should live in relation to plants, animals, and the natural world as a whole. A challenging question, it has motivated the writing and actions of philosophers, nature writers, conservationists, and environmental advocates for generations.

For early American Romantics like Henry David Thoreau, an ethical relationship to nature reflected a deep aesthetic and moral response to the beauty and transcendent qualities of the woods, waters, and meadows around him in mid-nineteenth-century New England. For the naturalist and wilderness advocate John Muir at the turn of the twentieth century, it entailed an attitude of reverence and humility toward the sublimity of the wild peaks and valleys of his beloved Yosemite. For the forester, conservationist, and amateur environmental philosopher Aldo Leopold in the 1940s, it meant adopting a 'land ethic', an ecologically grounded sense of moral obligation to and citizenship within the wider biotic community (see Figure 3.6.1). And for the marine biologist and nature writer Rachel Carson in the 1960s, it meant rejecting the morally chauvinistic idea that humans were superior to other species – and that plants and animals exist only for our own convenience.

In the 1970s, academic philosophers, drawing from these historical sources and spurred by the arrival of the wider environmental movement, began to develop a more formal response to the environmentalist version of Socrates' question. The field of environmental ethics emerged during this time alongside other branches of applied philosophy/ethics (e.g., biomedical ethics, engineering ethics, etc.) that brought philosophical scrutiny to pressing social, political, and scientific problems. First generation environmental ethicists from the beginning saw their challenge as developing an

**Figure 3.6.1** Aldo Leopold (1887–1948). Forester, conservationist, and amateur philosopher, Leopold's essay 'The Land Ethic' (collected in his 1949 book *A Sand County Almanac*) has inspired multiple generations of philosophers, ecologists, and conservationists.

*Photo source*: Aldo Leopold Foundation, permission granted.

alternative to the dominant environmental ethos, one in which nature was viewed merely as a set of commodities or resources for human use.

## Against anthropocentrism

This critique of the human-centered environmental ethic, an outlook referred to as 'anthropocentrism', would propel much of the work in academic environmental ethics in its first decades. A few key publications in the late 1960s and early 1970s would light the trail.

In the late 1960s, the historian Lynn White, Jr. (1967) published a controversial and much discussed essay in *Science* magazine arguing that this attitude of human superiority was the driver of contemporary environmental problems, from air and water pollution to extinction and overpopulation. White also put forward a provocative thesis about the roots of anthropocentrism in the Creation story of the Judeo-Christian tradition, a reading that appeared to place humans in a privileged and despotic position vis-à-vis nature. Although subsequent scholars would find more positive and less domineering environmental ethics in the Bible, the philosophical centerpiece of White's argument – the claim that a worldview driven by a profound human arrogance was the root of our environmental problems – proved persuasive to many early environmental ethicists.

Indeed, the idea that a viable environmental ethic had to reject anthropocentrism and embrace a view that recognized the independent moral status of nature would soon gain additional support in an influential paper by the New Zealand philosopher Richard Routley (1973). There, Routley argued that according to the standard anthropocentric accounts of Western moral philosophy the destruction of nature would only be unacceptable if it harmed humans. Because we could envision destroying species without any consequence for us and yet we had the intuition that doing so was somehow still wrong, Routley argued that a *non-anthropocentric* ethic was needed if we were serious about developing a defensible environmentalism.

Around this same time, the legal theorist Christopher Stone made a parallel legal case for the rights of trees and other 'natural objects' in a paper (and later a book) posing the question, 'Should Trees Have Standing?' (Stone, 1975). Hoping to influence the outcome of Supreme Court case involving a lawsuit filed by the Sierra Club to block the construction of a Ski area in California's Mineral King Valley, Stone offered an innovative and unconventional legal argument for granting nonhuman entities standing in the eyes of the law. Although the Sierra Club ultimately lost its appeal in the case, Stone's arguments would be cited approvingly in Supreme Court Justice William O. Douglas's dissenting opinion in the Mineral King case (see Figure 3.6.2).

**Figure 3.6.2** California's Mineral King Valley. The valley figured in a groundbreaking argument for recognizing the rights of natural objects. Legal and political arguments for nature's rights have been resurgent in recent years, as evidenced by their codification in Ecuador's national constitution.

*Photo source*: HikingMike/Wikimedia Commons.

The non-anthropocentric outlook quickly took hold in the young field of environmental ethics. By the 1980s most environmental ethicists, including prominent voices such as Paul Taylor (1986), Holmes Rolston (1988), and J. Baird Callicott (1989) advocated (each in different ways) for the widespread adoption of a worldview and moral framework able to respect all living organisms ('biocentrism'), and/or other species and ecosystems ('ecocentrism'). It's an ethical tradition that remains strong decades later, with many of these same thinkers refining and extending their earlier arguments for according nature intrinsic value (e.g., Rolston, 2012; Baird Callicott, 2013).

These writers typically make a pair of claims about the necessity of non-anthropocentrism and its defining emphasis on nature's intrinsic value: 1) it is the morally 'correct' position to hold given certain assumptions and arguments about the diversity and complexity of the natural world; and 2) it is also the most effective and powerful approach to arguing for the protection of wildlife, wilderness, and other environmental features. That is, non-anthropocentrists make both a *moral* and an *empirical* claim about why non-anthropocentrism is the most philosophically compelling and indispensible commitment of environmental ethics.

## Anthropocentrism rehabilitated

Not every environmental ethicist, however, agrees with these assertions. Bryan Norton, for example, has long questioned the logic, coherence, and pragmatic necessity of the non-anthropocentric approach. As early as 1984 he suggested that a broadened or enlightened form of anthropocentrism was a more philosophically sound and practically effective commitment in environmental ethics. Specifically, Norton had in mind a value system that took a more generous and expansive view of human values in nature to include a range of non-consumptive and non-economic 'uses' such as recreation, aesthetic enjoyment, and the therapeutic 'use' of experiences in nature to promote human flourishing (see Figure 3.6.3). This moral outlook of 'weak' anthropocentrism, he argued, was actually more likely to move people to conserve species and protect places than non-anthropocentric appeals to the intrinsic value of nature.

By the 1990s it became clear that this alternative understanding of environmental ethics, one focused less on the defense of intrinsic value in nature and more on understanding and promoting the many *instrumental* values of nature that support environmental protection, was growing into a wider movement in the field. Soon, this new approach would have a name: 'environmental pragmatism'. It was a moniker that captured the practical and policy emphasis of ethicists working in this vein while also acknowledging the influence of classic American pragmatist philosophers such as Charles Sanders Peirce, William James, and John Dewey (Menand, 2001). The new environmental pragmatists, including Anthony Weston, Andrew Light, and Ben Minteer, stressed the wide array and diversity of environmental values rather than the reduction of environmental ethics to a single kind of intrinsic value, emphasized decision-making processes and deliberation over the reliance on fixed moral principles, and generally argued for environmental ethicists to spend more time engaging real-world environmental policy and management problems (e.g., Norton, 1991, 2005, 2015; Light and Katz, 1996; Minteer, 2012).

The emergence of environmental pragmatism spurred a protracted and at times heated debate about the goals, methods, and commitments of environmental ethics,

**Figure 3.6.3** Weak anthropocentrism. Non-consumptive uses of nature, from hiking and bird watching to the seeking of spiritual fulfillment, can offer powerful reasons for environmental protection. Some environmental ethicists believe that these reasons are in fact more compelling to the public than arguments premised on the intrinsic value of nature.

*Photo source*: NPS photo by Michael Quinn/Wikimedia Commons.

one that continues to play out as the field evolves and matures as a form of applied philosophical practice. Pragmatists criticize the non-anthropocentrists for being too narrowly focused on intrinsic value and thus too disconnected from the many human motives and interests that shape everyday decision making, environmental and otherwise. Non-anthropocentrists respond by arguing that the pragmatist approach is far too compromising and too wedded to instrumentalist views of nature to offer much of a defense against accelerating environmental exploitation.

## Environmental ethics in the human age

The disagreement between pragmatists and non-anthropocentrists takes on even more significance in light of the recognition among scientists that we live today in the Anthropocene, an era in which the human presence and influence has become so profound that humans are effectively a geological force on Earth (Lewis and Maslin, 2015; Figure 3.6.4). A controversial and polarizing construct, the Anthropocene idea would seem to pose a significant challenge to the non-anthropocentric worldview. After all, how can we be truly 'nature-centered' on a planet that is increasingly human-driven?

Yet the acknowledgment that we may be living in the 'Age of Humans' given the extent and intensity of human impacts on Earth systems (from human-caused mass extinctions and rampant urbanization to global climate change) also poses a challenge

**Figure 3.6.4** The human planet? The Anthropocene raises serious questions for environmental responsibility in an era of human dominance.

Photo: Mark Klett, Trails of Weekend Explorers, near Hanksville. Reproduced with permission.

for more human-centered approaches in environmental ethics. The humility and self-possession traditionally counseled by more enlightened forms of anthropocentrism may well become lost in all the talk of intelligently shaping nature for our own purposes in the human age (Minteer and Pyne, 2015). An environmental ethics for the Anthropocene will therefore have to thread the moral needle, maintaining its long-standing respect for nature while at the same time acknowledging that the natural world is increasingly under our thumb.

## Learning resources

### *Websites*

Here are a few websites that offer additional materials on the field of environmental ethics, including its history and core concepts, as well as discussion of particular environmental problems and cases:

*Center for Environmental Philosophy at the University of North Texas*. The CEP is the home of the journal *Environmental Ethics*, the first and still the leading publication in the field. The site includes information about the development of the field as well as links to other programs and learning resources in environmental ethics: http://www.cep.unt.edu/.

*Nature Education Knowledge Project* series on Environmental Ethics. This is an open-access series of essays on environmental ethics topics, from introductory articles on the field's big questions to more advanced material on the ethical dimension of specific ecological and conservation challenges: http://www.nature.com/scitable/knowledge/environmental-ethics-96467512.

*Online Ethics Center*. The Online Ethics Center (OEC) is a web resource that carries case study materials and learning resources for professionals, educators, and students seeking to understand and grapple with the ethical questions raised by science and engineering. Maintained by the Center for Engineering Ethics and Society (CEES) at the National Academy of Engineering (NAE), the OEC materials emphasize engineering applications, though many of the resources on the site speak to an array of problems in applied environmental ethics: http://www.onlineethics.org/

## Bibliography


Baird Callicott, J. (1989) *In Defense of the Land Ethic: Essays in Environmental Philosophy*. Albany, NY: State University of New York Press.

Baird Callicott, J. (2013) *Thinking Like a Planet: The Land Ethic and the Earth Ethic*. New York: Oxford University Press.

Leopold, A. (1949) *A Sand County Almanac and Sketches Here and There*. New York: Oxford University Press.

Lewis, S.L. & Maslin, M.A. (2015) 'Defining the Anthropocene', *Nature*, 519: 171–180.

Light, A. & Katz, E. (eds) (1996) *Environmental Pragmatism*. London: Routledge.

Menand, L. (2001) *The Metaphysical Club: A Story of Ideas in America*. New York: Farrar, Straus and Giroux.

Minteer, B.A. (2012) *Refounding Environmental Ethics: Pragmatism, Principle, and Practice*. Philadelphia: Temple University Press.

Minteer, B.A. & Pyne, S.J. (eds) (2015) *After Preservation: Saving American Nature in the Age of Humans*. Chicago: University of Chicago Press.

Norton, B.G. (1984) 'Environmental Ethics and Weak Anthropocentrism', *Environmental Ethics*, 6: 131–148.

Norton, B.G. (1991) *Toward Unity Among Environmentalists*. New York: Oxford University Press.

Norton, B.G. (2005) *Sustainability: A Philosophy of Adaptive Ecosystem Management*. Chicago: University of Chicago Press.

Norton, B.G. (2015) *Sustainable Values, Sustainable Change: A Guide to Environmental Decision Making*. Chicago: University of Chicago Press.

Rolston, H., III (1988) *Environmental Ethics: Duties to and Values in the Natural World*. Philadelphia: Temple University Press.

Rolston, H., III (2012) *A New Environmental Ethics: The Next Millennium for Life on Earth*. New York: Routledge.

Routley, R. (1973) 'Is there a Need for a New, an Environmental Ethic?', *Proceedings of the XVth World Congress of Philosophy*, 1: 205–210.

Stone, C. (1975) *Should Trees Have Standing?* (rev. edn.) New York: Avon Books.

Taylor, P.W. (1986) *Respect for Nature: A Theory of Environmental Ethics*. Princeton, NJ: Princeton University Press.

White, L., Jr. (1967) 'The Historical Roots of our Ecological Crisis', *Science*, 155: 1203–1207.

# 3.7 Environmental geography

John G. Hintz

To write about environmental geography is, nearly, to write the history of geography itself. After all, the term *geography*, famously coined by the Greek scholar Eratosthenes over 2000 years ago, means to write about the Earth. And, to write about the Earth *is* to write about the *environment* – commonly understood as the living and non-living components of the world around us. But even as the case can be made that the discipline of Geography has been *environmental geography* from the start, fast-forward to the nineteenth century and – explicitly – environmental geography emerges, even as it would splinter into distinct forms by the end of the century.

## Nineteenth-century origins

Alexander von Humboldt's travels and writings serve as an appropriate starting point for nineteenth-century environmental geography (and, arguably, *modern* environmental geography as well). If the common ground of contemporary geography is a focus on space and spatial relations, Humboldt's geography was consistent with this emphasis, for sure, but was also always an environmental geography. As David Livingstone (1992) put it in *The Geographical Tradition*, for Humboldt, "the experiences that made him into a geographer [were those that] he learned first-hand about the spatial distributions of organic life and its umbilical ties with environment" (p. 135). Humboldt is most well-known for his epic multi-volume *Cosmos*, which can hardly be considered mere environmental geography (or even mere geography), as the sprawl of its ambitions and content blow apart disciplinary bounds and conventions. However, the sections of *Cosmos* that were drawn from his 5-plus year expeditions in Central and South America were all based on meticulous notes and field measurements of the natural world, *the environment*. The emphases on explaining spatial relations of environmental phenomena all the while recognizing and explaining the interconnections of local phenomena stand as lasting contributions to geography. By mid-century, the American George Perkins Marsh would match Humboldt in ambition and scope, but his environmental geography, in stark contrast to Humboldt's, firmly placed human agency as an essential component of geographical studies of the environment.

Marsh's seminal 1864 work, *Man and Nature: Or, Physical Geography as Modified by Human Action*, prefigures contemporary 'Anthropocene' perspectives by foregrounding the profound transformative power of human action on the Earth while simultaneously warning of dire potential futures. However, Marsh stressed that disastrous futures could be averted by appropriate, careful, technical management of land and resources. Passages from the preface echo (in content and spirit if not in style) much of the focus of contemporary environmental geography: "The object . . . is to indicate the character and . . . extent of the changes produced by human action in the physical conditions of the globe we inhabit; to point out the dangers of imprudence and the necessity of caution in all operations which . . . interfere with the spontaneous arrangements of the organic or the inorganic world" (p. 3). *Man and Nature* is, however, a work of its time. It reflects a nineteenth-century, pre-ecological, mechanistic worldview. The "or" in the passage above ("the organic *or* the inorganic world") implies a lack of, or inability to see, interdependence between biota and environment in Marsh's world. The man/environment (and organic/inorganic) dichotomy precludes a dialectical or nuanced analysis of humans living within a world of natural and social relations. Even as Marsh acknowledges that analyzing how the natural world, in all of its geographical variability, affects human societies is "perhaps the most interesting field of speculation," he chose to operationalize a methodology that seemed to just flip the causal arrow. Marsh's environmental-geographical analysis is always (often only) an analysis of how humans affect the earth. All that said, the book stands (like *Cosmos*) as a work of grand scope as well as one built from careful observation (although built less from direct observation than *Cosmos* and more from archival resources) and meticulous documentation (the length of the footnotes exceeds that of the text on many pages).

A distinct, yet equally dualistic, framework for assessing human–environment relations would gain prominence as the nineteenth century closed and European imperialism reached its apex. If Marsh gave humans supreme agency over nature, *environmental determinism* did just the opposite. For geographers such as Friedrich Ratzel of the University of Leipzig and his students Ellen Churchill Semple and Ellsworth Huntington (each of whom would go on to teach geography in the United States), local environmental conditions, particularly climate, were the dominant causal factors in determining the various types of societies and levels of development that occurred in certain areas. The geographers of this era were not the first to posit environmental-determinist explanations of societal development. Greek scholars, after all, hypothesized that great societies could only develop in temperate climates. In the nineteenth century, Darwinian evolution would add a new element to the broad brushes that the Greeks used to explain societal development. The implications for European imperialism were fairly obvious: European rule over colonized people and places could be rationalized as the natural order of things, predetermined by environment, or God, in whichever order one preferred. Contemporary scholars have argued for a revisiting of this scholarship, arguing that all geographical scholarship that argued along these explanatory lines should not be automatically implicated as racist and imperialist. However, it is hard not to wince when, for example, one reads Huntington's *Civilization and Climate*, as he posits that climate is the main reason that the development of the Bahamas has been "retarded" across time (Huntington, 1915, p. 29). According to Huntington, the Bahaman climate may not induce laziness to the degree of Africa, but it also certainly does not inspire the industriousness and innovation found in Maine or North Prussia.

## Twentieth-century developments

Perhaps not surprisingly, many geographers after World War II worked diligently to distance themselves from environmental determinism. As such, the 1950s and 1960s were a period when human–environment interactions faded into the background, as academic geography increasingly split into discrete human and physical sub-disciplines. Each was dominated by approaches that eschewed explanatory frameworks (*theory*) altogether, favoring descriptive, cataloging approaches to (human *or* physical) geographical analysis. Two trends that arose in the 1970s, however, would re-position the human–environment nexus front and center in academic geography. One was the rise of popular environmentalism; the other was the rise to prominence of Marxist and other 'critical' frameworks for explanation of geographical phenomena.

Various moments could be chosen as signal arcs that bridged popular environmentalism and academic environmental science (geography included) – Rachel Carson's *Silent Spring* and Aldo Leopold's *A Sand County Almanac* come to mind – but perhaps the most significant and enduring work is Donella Meadows and the Club of Rome's *The Limits to Growth* study. Published in 1972, this report was essentially a quantitative analysis (enabled by the then-new power of computer modeling) that assessed resource depletion, human population growth, and pollution over time and projected humans exceeding earth's carrying capacity, ushering in severe environmental crisis. This report inspired countless studies across a number of academic disciplines, many of which focus on notions of carrying capacity and environmental sustainability. By the 1990s such studies had somewhat coalesced under the interdisciplinary umbrella of 'environmental science', but no single discipline could claim proprietary ownership over the field. While some critical geographers voiced wariness toward the undifferentiated 'humans' that comprised the inputs of many environmental analyses (more on that below), others cited geography's long history of environmental analysis – coupled with new analytical tools such as geographic information systems (GIS) and remote sensing – as the potential base for a needed and explicitly *environmental* geography. Marsh and Grossa (2005), for example, situate geography as the natural home and framework for environmental science due to the long-standing attention to distance and scale in geography as well as its established emphasis on geographic holism and interconnectedness. Even with these claims and the scholarly accomplishments of many within the field, geography has failed to contain the growing field(s) of environmental science, which seem destined to remain interdisciplinary endeavors in higher education.

Marxist, post-Marxist and other 'critical' geographers also contributed to the renewal of an environmental focus in geography, sometimes in response to and highly critical of neo-Malthusian strands of environmental geography. Early critical environmental geography included Neil Smith's explicitly Marxist 'production of nature' framework, unapologetic in its critique of dualistic frameworks that posit pristine wilderness as a baseline for environmental analyses. By acknowledging the produced nature of *all* 'nature', geographers can better differentiate the wildly varying types, degrees, and agents of environmental modification in contemporary capitalist society (Smith, 1991). Poststructuralist and other post-Marxist scholars would expand Smith's produced nature to a broader socially constructed nature, one that is recognized as material reality (albeit one produced under capitalism) but also always a product of contested discourses and ideologies which govern people's actions in the world (see Castree and Braun, 1998). A third (overlapping) variant of 'critical' environmental geography is political ecology. Political ecology represents an incredibly diverse body of scholarship that, broadly, deploys a non-dualistic ecological understanding of human–environment interactions but is also informed by Marxist or post-Marxist social theory. Recent years have witnessed a spate of theoretically informed approaches appearing on the scene, including animal geographies, social nature, environmental pragmatism, and hybrid geographies. Geography may not have a monopoly on environmental science, but nature has certainly regained its centrality within the discipline and is the crucible of often heated debates about how, why and to what ends the non-human world matters.

## Learning resource

Castree, N. (2011) 'Nature and society.' In J.A. Agnew & D.N. Livingstone (eds), *The SAGE Handbook of Geographical Knowledge* (pp. 286–298). Los Angeles: Sage.

## Bibliography

Castree, N. & Braun, B. (1998) (eds) *Remaking Reality: Nature at the Millenium*. London: Routledge.
Huntington, E. (1915) *Civilization and Climate*. Cambridge, MA: Yale University Press.
Livingstone, D. (1992) *The Geographical Tradition*. Oxford: Blackwell.
Marsh, G.P. (1965) (1864). *Man and Nature: Or, Physical Geography as Modified by Human Action*. London: Oxford University Press.
Marsh, W. & Grossa, J. (2005) *Environmental Geography: Science, Land Use, and Earth Systems*. New York: John Wiley.
Smith, N. (1991) *Uneven Development: Nature, Capital, and the Production of Space*. Cambridge, MA: Blackwell.

# 3.8 Environmental health

## Brian King

Environmental health is the recognition that human interactions with the natural and built environment generate vulnerabilities to infectious disease. Additionally, environmental health considers how people are exposed to a number of factors that contribute to noncommunicable disease, such as chemicals or air pollutants that result in asthma, respiratory disease, and cancer. Obesity rates are blamed for increases in Type 2 Diabetes, and the rise of obesity is attributed to unhealthy lifestyles and the ways that the built environment reduces the opportunities for physical activity. Proximity to the conditions that contribute to human health and well-being, either in terms of access to clinics or hospitals, or green space for recreation, are unevenly experienced within societies. Social activism informed by environmental justice has helped demonstrate how a number of these dynamics produce disproportionate health outcomes for individuals, communities, and regions (Walker, 2012). Human health is directly connected to the interactions between social and environmental systems (King, 2017; King and Crews, 2013).

The World Health Organization (2014) defines environmental health as the "physical, chemical, and biological factors external to a person, and all the related factors impacting behaviours. It encompasses the assessment and control of those environmental factors that can potentially affect health. It is targeted towards preventing disease and creating health-supportive environments. This definition excludes behaviour not related to environment, as well as behaviour related to the social and cultural environment, and genetics." While this definition is helpful in articulating the relationships between human health and the natural environment, it is less useful in addressing the role of social and cultural factors that shape decision making while exposing people to unhealthy conditions. As will be discussed, the spread of infectious disease and exposure to the conditions that contribute to noncommunicable disease are generated by local processes that are contextual and dynamic.

The field of environmental health examines how human health is connected to the natural and physical environment. The disciplines of environmental epidemiology, toxicology, and exposure science are central to the study of environmental health though they provide different emphases. Environmental epidemiology studies the relationship between environmental exposures, such as chemicals, radiation, microbiological agents, and human health. Toxicology studies show how these exposures lead to specific health outcomes. Finally, exposure science examines human exposure to environmental contaminants by identifying and quantifying exposures. In addition to these research fields, several scientific disciplines, such as social epidemiology, medical and health geography, medical anthropology, and public health, have advanced understandings of the relationships between the natural environment and human health.

## Infectious disease and environmental change

Infectious disease patterns are connected to natural and built environments. Cholera, malaria, tuberculosis, and dengue fever are some examples of disease patterns that are produced through unsafe conditions that increase the possibility for exposure. The World Health Organization reports that within developing countries the main environmentally caused diseases are diarrheal disease, lower respiratory infections, unintentional injuries and accidents, and malaria. Within industrialized countries, cancer, cardiovascular disease, asthma, lower respiratory infections, and traffic injuries are major health hazards. According to the World Health Organization Public Health & Environment Global Strategy Overview (2011), environmental hazards are responsible for roughly a quarter of the total burden of global disease. In developing countries, the burden of environmental hazards is more strongly experienced by communicable diseases, such as malaria, dengue fever, and human immunodeficiency virus (HIV). In developed countries, environmental hazards are believed to have a larger impact upon noncommunicable (also known as non-infectious) diseases such as diabetes and cancer.

Biophysical processes that are responsive to global climate change generate vulnerabilities to infectious diseases. Increasing temperatures, variability in rainfall and flooding, and rising sea levels impact human health in a variety of ways. Shifts in ecosystem dynamics can allow disease vectors, which are the organisms that transmit disease from one entity to another, to enter new locations. One example of this is the *anopheles* mosquito that transmits malaria through the plasmodium parasite to human populations. Malaria causes the morbidity and mortality of millions of people each year. In 2015, an estimated 214 million cases of malaria occurred globally and 438,000 people died, of which most were children in sub-Saharan Africa (CDC, 2016). Public health campaigns to reduce the incidence of malaria can involve the spraying of pesticides to eliminate the *anopheles* mosquito.

The vulnerability that people have to infectious disease can change over time and is shaped by dynamic biophysical conditions. Research has shown that even small shifts in temperature can alter the habitats that allow the *anopheles* mosquito to survive, thereby demonstrating the tight coupling between infectious disease and the environment (Paaijmans et al., 2010).

Changing environmental dynamics can impact human health in other ways. The Intergovernmental Panel on Climate Change (IPCC) Fifth Assessment Report (Smith et al., 2014) identifies three specific pathways in which climate change impacts health. The first pathway includes direct impacts that are tied to changes in the frequency of extreme weather including heat, drought, and heavy rain. The second pathway considers effects produced through natural systems such as the role of disease vectors, waterborne diseases, and air pollution. Lastly, the third pathway addresses impacts from human systems, such as occupational impacts, undernutrition, and mental stress.

## Exposure and justice

Environmental health also considers exposure to noncommunicable disease and disproportionate effects for human populations. In an increasingly urbanized world, people face new challenges for health management. It is estimated that by 2050, 70 per cent of the world's population will be living in towns and cities (WHO, 2011). As the WHO report explains, road traffic injuries are the ninth leading cause of death around the world, and most road traffic deaths occur in low- and middle-income countries. Urban residents are exposed to air pollution that kills roughly 1.2 million people around the world each year. This air pollution is caused by transportation generated by motor vehicles along with industrial pollution, and the generation of energy. This means that the built environment, including infrastructure and strategies for energy generation, would need to be transformed to reduce rates of noncommunicable disease around the world.

Exposure to air pollution and other factors that contribute to poor health is unequally experienced within societies. The environmental justice movement that emerged in the United States in the early 1980s, which was paralleled by similar movements elsewhere around the world, was based on the recognition that people are differentially exposed to environmental health factors due to race, ethnicity, socio-economic class, and location. Environmental justice has also shown that people differentially experience conditions that contribute to health and well-being, such as proximity to clinics and hospitals or green space for recreation and contemplation.

The links between environmental justice and human health are often front page news stories. In April 2014, a public health crisis unfolded in the city of Flint, Michigan that demonstrates the links between human health and the built environment. Residents had been exposed to lead in drinking water because the Detroit Water and Sewerage Department had shifted its source to the Flint River. According to the Centers for Disease Control (CDC), the lead levels in children's blood increased shortly afterwards and did not decline until the discovery several months later. The CDC noted that children who drank water from this source had a 50 percent higher risk of dangerously elevated blood lead levels than before the change of supply (NBC News, 2016a).

In this case, exposure to lead in the drinking water coincided with a location that has historically struggled for economic reasons. Following the discovery, a massive public outcry put pressure upon state and national public officials, and resulted in at least nine officials facing criminal charges for not revealing a report several years before that suggested health issues from the switch in the source of the water supply. Michigan Attorney General Bill Schuette commented that the officials acted out of arrogance and by "viewing people in Flint as expendable" (NBC News, 2016b).

Environmental health helps show how exposure to infectious disease and the conditions that contribute to poor health are often the result of interactions with the natural and built environment. These conditions are unequally experienced within societies, which has generated attention to how human health is also about environmental sustainability and justice.

## Learning resources

The U.S. Centers for Disease Control and Prevention's National Environmental Public Health Tracking Network provides searchable information about environmental health conditions in your community: https://ephtracking.cdc.gov/showHome.action.

The Public Broadcasting System website showcases introductory video case studies that illustrate the connections between environment and public health: http://wpsu.pbslearningmedia.org/collection/enh.

For more information on the United Nations' international efforts to link public health solutions with sustainable development, see: http://www.un.org/sustainabledevelopment/health.

The World Health Organization and the U.S. Centers for Disease Control and Prevention are good ongoing sources for updates on infectious disease, current world events, and other health issues: http://www.who.int/en/ and https://www.cdc.gov.

## Bibliography


Centers for Disease Control and Prevention (CDC) (2016) Malaria. Accessed September 22, 2016 at: https://www.cdc.gov/malaria

King, B. (2017) *States of Disease: Political Environments and Human Health*. Berkeley, CA: University of California Press.

King, B. & Crews, K.A. (eds) (2013) *Ecologies and Politics of Health*. London: Routledge.

NBC News (2016a) CDC confirms lead levels shot up in Flint kids after water switch. Accessed September 26, 2016 at: http://www.nbcnews.com/storyline/flint-water-crisis/cdc-confirms-lead-levels-shot-flint-kids-after-water-switch-n598496.

NBC News (2016b) Six more officials charged in Flint water crisis for alleged cover-up. Accessed September 26, 2016 at: http://www.nbcnews.com/storyline/flint-water-crisis/six-more-offcials-charged-flint-water-crisis-alleged-coverup-n619811.

Paaijmans, K.P., Blanford, S., Bell, A.S., Blanford, J.I., Read, A.F., Thomas, M.B. & Denlinger, D.L. (2010) "Influence of climate on malaria transmission depends on daily temperature," *Proceedings of the National Academy of Sciences of the United States of America*. 107(34): 15135–15139.

Smith, K.R., Woodward, A., Campbell-Lendrum, D., Chadee, D.D., Honda, Y., Liu, Q., Olwoch, J., Revich, B. & Sauerborn, R. (2014) "Human health: impacts, adaptation, and co-benefits." In Field, C.B., Barros, V.R., Dokken, D.J., Mach, K.J., Mastrandrea, M.D., Bilir, T.E., Chatterjee, M., Ebi, K.L., Estrada, Y.O., Genova, R.C., Girma, B., Kissel, E.S., Levy, A.N., MacCracken, S., Mastrandrea, P.R. & White, L.L. (eds) *Climate Change 2014: Impacts, Adaptation, and Vulnerability. Part A: Global and Sectoral Aspects. Contribution of Working Group II to the Fifth Assessment Report of the Intergovernmental Panel on Climate Change* (pp. 709–754). Cambridge: Cambridge University Press.

Walker, G. (2012) *Environmental Justice*. London: Routledge Press.

World Health Organization (WHO) (2011) "WHO Public Health & Environment: Global Strategy Overview." *World Health Organization*. September 26. Accessed October 6, 2014 at: <http://www.who.int/phe/publications/PHE_2011_global_strategy_overview_2011.pdf>.

World Health Organization (WHO) (2014) Environmental Health. Accessed October 6, 2014 at: <http://www.who.int/topics/ environmental_health/en>.

# 3.9 Environmental history

## Stephen Mosley

### Origins and aims

Today, environmental history is one of the fastest growing and most innovative fields of scholarly research. Its origins can be found in the classic text *Man and Nature* (1864/2003) by the polymath-diplomat and conservationist George Perkins Marsh, which surveyed the destructive impacts of human civilisations on the natural world down the ages. Its early roots also extend across historical geography, archaeology, natural sciences and a range of other disciplines that deal with human-induced environmental change. But environmental history as a distinct field of study did not emerge until the 1970s, inspired in no small part by the rise of environmentalism as a mass social movement. As environmental concerns, particularly about indiscriminate pesticide use, urban-industrial sprawl, and nuclear threats, rose up the global political agenda, historians began to investigate human–nature interactions in earnest. One of its leading exponents, the American academic Donald Worster, defined environmental history concisely as the study of 'the role and place of nature in human life' (1988: 292).

Scholars in the United States were in the vanguard of efforts to better understand the coevolution of people and planet, founding the American Society for Environmental History in 1976. Its counterpart on the other side of the Atlantic, the European Society for Environmental History, was not established until 1999. Together they now coordinate H-Environment, an internet discussion forum that keeps environmental historians up-to-date with the latest developments in the field. There are also three academic journals dedicated to the discipline: *Environmental History*; *Environment and History*; and *Global Environment*. More recently, the Latin American and Caribbean Environmental History Association (2003), the Association for East Asian Environmental History (2009) and the International Consortium of Environmental History Organisations (2006) were formed, showing its increasing institutional strength.

Its initial connection with environmentalism saw a strong emphasis on the moral worth and political relevance of environmental history. Practitioners aimed to produce work that would speak forcefully to the green issues of the day, which for some critics conflicted with traditional historical caution and 'objectivity' (Mosley, 2006). Its narratives also tended to be declensionist in character, typically arguing that environments transformed by human action had either been seriously degraded or ruined completely (Hughes, 2015). However, overt advocacy for nature and the nonhuman world faded as the field matured, and more complex and optimistic stories started to be told about sustainable human–environment relationships over time. Environmental history now provides much-needed context for both policymakers and activists wrestling with current environmental problems (Coulter and Mauch, 2011).

## Approaches and methodology

Although there are numerous ways in which one might approach a subject as vast as environmental history, the four-pronged framework outlined below has proven to be most influential:

1. understanding the dynamics of ecosystems in time;
2. examining the interactions between environment, technology and the socioeconomic realm;
3. exploring cultural values and beliefs about the natural world;
4. analysing environmental policy and planning.

Based on Worster's ambitious model for 'doing' environmental history, it encourages scholars to make connections between the different levels of analysis (Worster, 1988). But rather than being a rigid schema, it is perhaps best viewed as a flexible programme for study. Pragmatically, most environmental historians have chosen to focus on just one or two levels, particularly the ecological impacts of technological and socioeconomic change since the industrial revolution. To date, there are still few works that link all four areas of investigation effectively.

Another challenge for practitioners is the inherently interdisciplinary nature of research in environmental history. To account for the roles that climate, forests, rivers and other non-human actors have played in shaping our histories, scholars in the field often have to work with both textual records and scientific data. Where traditional documentary sources are lacking or non-existent, modern scientific techniques such as ice-core analysis, dendrochronology and palynology can reveal important information stored in 'nature's archives', helping to reconstruct histories of climate change, forest clearance, human settlement, and much more besides (Mosley, 2006). Influenced from the outset by ecology's holistic principles, environmental historians are beginning to break down disciplinary boundaries between the humanities and the sciences.

The openness and inclusiveness of environmental history, welcoming contributors from a wide range of disciplinary backgrounds (from anthropologists to zoologists), has been seen as both a strength and a weakness. For some, the discipline is under-theorised and lacks coherence, limiting its ability to impact and influence debates in the historical mainstream (Sörlin and Warde, 2007). For many others, the 'big tent' approach – the diversity and 'messiness' involved in doing interdisciplinary research – has produced some of the most interesting and innovative historical work around. For example, the

History of Marine Animal Populations project (HMAP), part of the wider Census of Marine Life, brought together environmental historians and natural scientists to examine the long-term ecological impacts of overharvesting fish and other sea creatures to meet growing market demand. There is still a need to encourage closer cooperation and alliances across disciplines if we are to better understand the environmental problems that we have created.

## Scale and scope

As one of the youngest fields of historical study, environmental history has brought fresh thinking about, and approaches to, space and time. Given that political boundaries and ecological boundaries rarely coincide, the familiar organisational framework of the nation-state is not always an apt unit of analysis for exploring past human–nature interactions. Regional and local level research, centred on different types of ecosystems – coastal, rural and urban – can often produce more coherent case studies of how societies and environments shape and reshape each other over time. National histories remain important, however, as data are rarely gathered at a global level. Environmental history can be written on any scale, and the fast-growing literature in this new field – on themes such as biological exchanges, population growth, energy use, and the ecological impacts of agriculture, imperialism and industrialisation – has allowed scholars to weave the various threads of national, regional and local narratives together to produce some inventive work at the macro-scale (Crosby, 2004; McNeill, 2000; Richards, 2003; Mosley, 2010). And as 'everything connects' in nature, adopting a global approach often makes more sense from an analytical point of view.

Similarly, systems of periodisation in environmental history are often defined by long-term natural processes, rather than the conventional political markers of nation states such as the Tudor, Georgian and Victorian eras. For example, important works such as Anthony Penna's *The Human Footprint* (2010) begin with the origins of the earth itself, while Ian Simmons's *Global Environmental History* (2008) takes the end of the last ice age as its starting point. Timescales routinely encompass millennia, challenging historians to engage with chronologies that reduce the human lifespan to the blink of an eye.

Finally, environmental history, as well as encouraging practitioners to think back over much longer timescales, and across disciplinary and geographical borders, can also stretch time – moving beyond 'normal scales' of history. The ambitious Integrated History and Future of People on Earth project (IHOPE), which aims to combine human and Earth system histories, has adopted a chronology that stretches from the Holocene (11,600 YBP), through the present and into the future to better model and meet sustainability challenges. In the new geological epoch of the Anthropocene, environmental history is proving indispensable to our becoming better managers of the global environment.

## Learning resources

For those interested in environmental history, the Environment and Society portal offers online essays, virtual exhibitions, and a multimedia library: http://www.environmentandsociety.org.

Coordinated by both the American Society for Environmental History and the European Society for Environmental History, the H-Environment discussion network provides online book and film reviews, conference announcements, and discussion threads: https://networks.h-net.org/h-environment.

Currently containing over 45,000 annotated entries relating to books, articles, and dissertations on environmental history the Forest History Society's bibliography covers all aspects of the topic, and not just the history of forests: http://www.foresthistory.org/research/biblio.html.

## Bibliography

Coulter, K. & Mauch, C. (2011) *The Future of Environmental History: Needs and Opportunities*. Online. Available at: http://www.environmentandsociety.org/perspectives/2011/3/future-environmental-history-needs-and-opportunities, accessed on 16 October 2016.

Crosby, A.W. (2004) *Ecological Imperialism: The Biological Expansion of Europe, 900–1900*, 2nd edn. Cambridge: Cambridge University Press,

HMAP (History of Marine Animal Populations/Census of Marine Life) (2016) http://www.comlsecretariat.org/research-activities/history-of-marine-animal-populations-hmap/, accessed on 16 October 2016.

Hughes, J.D. (2015) *What is Environmental History?* Cambridge: Polity.

Integrated History and Future of People on Earth (IHOPE) (n.d.) Online. Available at: http://ihopenet.org, accessed on 16 October 2016.

Marsh, G.P. (1864/2003) *Man and Nature; or, Physical Geography as Modified by Human Action*. Seattle: Seattle University Press.

McNeill, J. (2000) *Something New Under the Sun: An Environmental History of the Twentieth Century*. London: Allen Lane.

Mosley, S. (2006) 'Common Ground: Integrating Social and Environmental History.' *Journal of Social History*, 39(3): 915–933.

Mosley, S. (2010) *The Environment in World History*. New York: Routledge.

Penna, A.N. (2010) *The Human Footprint: A Global Environmental History*. Chichester: Wiley-Blackwell.

Richards, J.F. (2003) *The Unending Frontier: An Environmental History of the Early Modern Period*. Berkeley, CA: University of California Press.

Simmons, I.G. (2008) *Global Environmental History: 10,000 BC to AD 2000*. Edinburgh: Edinburgh University Press.

Sörlin, S. & Warde, P. (2007) 'The Problem of the Problem of Environmental History: A Re-Reading of the Field', *Environmental History*, 12(1): 107–130.

Worster, D. (1988) *The Ends of the Earth: Perspectives on Modern Environmental History*. New York: Cambridge University Press.

# 3.10 Environmental modelling

## Mark Mulligan

### What modelling is and is not

A model is an abstraction from, or simplification of, reality suited to a particular purpose. Models can be conceptual, physical or mathematical and are increasingly important in environmental science. Most science today is highly specialised and reductionist, but increasingly pieces of knowledge must be brought together to understand whole environmental systems – down to the local scale. Models can help do this by providing a common framework (simulation) and a language (mathematics) for integration. Most physical and social scientists use models, at least conceptual models of the systems that they conduct research into. Moreover, since most measurement techniques are indirect, those scientists who monitor the environment are effectively deploying hardware models (for example, the height of a column of mercury or the voltage generated by a thermocouple, in response to temperature). Many scientists use statistical and other abstractions of their systems of study.

### Types of models

The three broad types of model are: conceptual (maps, diagrams, flowcharts), physical (scaled down hardware models) and mathematical (formulae, statistics, computer code). Most environmental models are inherently spatial (they include location or aspects of location as a variable) and dynamic (incorporating variation over time) as is the world they simulate. Spatial and temporal discretisation is often necessary. This means that reality is broken up into discrete spatial grid-cells (for example every hectare) and temporal time-steps (for example every hour).

## The role of modelling in environmental science

Models help integrate data with knowledge and often simulate processes and events over space and time, thereby managing variability in these two domains. This is crucial for understanding our heterogeneous physical and social environments. Modelling is used for a variety of reasons:

1. *Completion (of data gaps).* Data gaps in time or space may need to be filled (interpolation) or data series extended into the past, future or into new spaces beyond where data are currently available (extrapolation, projection). Understanding trends in data and evaluating data that we can measure as surrogates for those we cannot, can enable these gaps to be filled through modelling.
2. *Experimentation (and understanding).* Often models are needed to experiment and understand the impact of 'driving variables' on 'response variables': for example, understanding the impact on environments of a changing climate (without changing it for real) or the impacts of deforestation on streamflow (without having to cut trees). Models create a laboratory of the environment in which experiments can be performed and predictions (or projections) made about real-world systems. These sometimes have policy value, although caution about over-confidence in model predictions is always advisable.
3. *Communication and collaboration.* Models help operationalise, test and communicate theory by confronting it with data. Models can help many scientists work together on a common problem, to which each contributes a small component. They can also help bridge the gap between science and the wider world by communicating the outcome of complex processes in simple ways: as scenarios, maps, time series and other visualisations.

However, modelling only does any of this well if the models are appropriate abstractions of the systems that they represent, if they are built on robust knowledge of those systems, and if sufficient high quality data are available to represent the systems accurately. If not, then models can end up being no more than expensive, time-consuming and misleading distractions from more useful scientific activities.

## How (and how not) to build and use models

As just intimated, the first and most important step to building models is *abstraction*. This is the process by which the modeller distils from reality the key rules, functions, components, or processes of the system and their linkages. Abstraction involves simplification, and simplification involves making assumptions about what is and is not important, about how properties are related and how processes work. These assumptions must be clear, well-documented and appropriate to the purpose to which the model is to be put. The best model is not always the simplest but rather the *simplest that achieves its objectives*. This means the best model may still be complex, but should be no more complex than necessary. Poor abstraction can lead to great expense in model building, data collection and integration and to poor simulation of the system represented. Model building includes the following steps: abstraction; set boundary conditions (what is in the model system, what is supplied from outside); set initial conditions (what are the values of the variables at time zero); understand processes; build equations; parameterise (what values should parameters and constants have?); conduct sensitivity analysis; verify and validate; and calibrate (if necessary). This process may be iterated several times, so that models improve through learning in use.

## The present and future of environmental modelling

Modelling has grown significantly as a research activity over the last few decades and will very likely continue to do so, as a result of increased demand and increased supply (capacity of models). A number of recent trends are now apparent that will certainly affect the models of the future.

### *Changing hardware*

Since Moore (1965) postulated the doubling of the number of components on an integrated circuit every 2 years – effectively a doubling of computer power – this has, indeed, generally held true because of improved design and miniaturisation of Central Processing Units (CPUs), memory and storage. However in recent years, limits to this doubling have become more apparent such that the rate of growth of computing power per dollar has declined. This means it is no longer possible to rely on increasingly powerful processors to handle more and more sophisticated models. In many cases the need is now to adapt model codes so that they can run effectively across multiple processor cores and multiple processors, or indeed computers. Growth in processing power now often involves using more and more *computers* to do the processing (rather than a more and more powerful *computer*). Even high performance computing (HPC) can easily be hired via 'the cloud'. Services like Amazon Web Services (AWS), Microsoft Azure, and Google Earth Engine (GEE) make significant compute and storage capacity available for hire through the launch of instances of virtual machines that can run any of a range of operating systems and software on a variety of hardware within data centres around the world.

### *Changing software*

Non-Microsoft Windows operating systems like UNIX and Linux have always been important in development of environmental models. Many flavours of Linux are now available, supported by a powerful command line and simple GUI, supporting model development across a range of computers from desktops to servers and HPCs. The open source nature of Linux and associated software, means that many of these models have also been developed as open source codes. This means that modellers and users alike can see, share and adapt the model code under a range of creative commons licenses. Linux has made modelling cheaper, more powerful and more open. Programming languages have also changed. There has been a move away from low-level languages requiring significant coding for even simple tasks, to much higher-level languages capable of performing complex tasks with much simpler coding. This has been achieved through the use of in-built libraries that provide powerful spatial, display, numerical or analytical functions. Many of these model codes are integrated with powerful development environments like Eclipse for Python or RStudio for R. Modelling and associated data mining are also benefiting from increased capacity for the application of machine learning through, for example, deep learning approaches. Deep learning uses artificial neural networks (ANN, algorithms based on the structure and function of the brain). Deep learning ANN are much larger and deeper than early ANN, with more layers and more data. They are run with much greater computing power by coupling CPUs and Graphical Processing Units (GPUs), with the compute intensive work going through the GPUs to improve performance. CPUs typically have 4–24 cores, whereas GPUs can have thousands, thereby improving parallel processing.

## *Changing data*

Recent trends in data capture and availability for modelling indicate a trend towards reduced financing and support of national and international *in situ* data collecting and monitoring networks. Consequently there has been an increased focus on remotely sensed and crowdsourced data. Crowdsourced data are collected actively or passively by a crowd of (voluntary) contributors. Active examples include: *FreshWaterWatch* from EarthWatch, which engages the general public in citizen science collection of water quality data in collaboration with scientists. Passive examples include the *CleanSpace* air quality tag that continuously collects georeferenced air quality data when carried by users. This trend has both increased the number of data points and their geographic distribution, but provides significant challenges with respect to quality control and potential impact on important nationally funded environmental data collection services. We have also seen much recent growth in the democratisation of data (making data available and accessible through opendata agreements).

Developments in sensors, computing power, storage and networks have enabled much more easily integrated and thus much larger datasets to be produced: so-called *big data*. The Internet of Things (IoT) is the growing global network of connected devices. Some of these have sensing capability and can thus contribute to model parameterisation and validation. Low-cost and open source microprocessors like Raspberry Pi and Arduino open up the opportunity to design, build and deploy sensor networks widely and cheaply. These networks can connect directly into modelling platforms and blend with other forms of data, like the FreeStation project connected to the WaterWorld model. Since models are highly dependent on data these are very positive developments, but also ones that require careful thought and caution in use.

# Learning resources

## *Books*

The following books are all relatively accessible:

Beven, K. (2010) *Environmental Modelling: An Uncertain Future?* Boca Raton, FL: CRC Press.

Christie, M., Cliffe, A., Dawid, P. and Senn, S.S. (2011) *Simplicity, Complexity and Modelling*. New York: John Wiley & Sons.

Smith, J. and Smith, P. (2007) *Environmental Modelling: An Introduction*. Oxford: Oxford University Press.

Wainwright, J. and Mulligan, M. (2013) *Environmental Modelling: Finding the Simplicity in Complexity*, 2nd edition. New York: John Wiley & Sons.

## *Journals*

These journals contain new papers about different aspects of environmental modelling:

*Ecological Modelling*. Online. Available at: https://www.journals.elsevier.com/ecological-modelling.

*Environmental Modelling and Software*. Online. Available at: http://www.sciencedirect.com/science/journal/13648152.

*Modelling Earth Systems and Environment*. Online. Available at: http://www.springer.com/earth+sciences+and+geography/earth+system+sciences/journal/40808.

## Bibliography

Moore, G. (1965) *The Future of Integrated Electronics*. Sunnyvale, CA: Fairchild Semiconductor internal publication.

# Environmental sociology

3.11

Riley E. Dunlap

Environmental sociology emerged in response to the rapid societal awareness of environmental degradation that began in the 1960s in the USA, cresting with the 1970 Earth Day, and quickly spreading internationally as evidenced by the 1972 United Nations Conference on the Human Environment in Stockholm. Not surprisingly, much of the early research represented an effort to understand the sources, nature and consequences of environmental awareness, including studies of public opinion and varying environmental attitudes of differing sectors of society, environmental activists and environmental organizations, and governmental agencies and policies. Such research involved the application of traditional sociological approaches to a new domain, and was termed the "sociology of environmental issues" (Dunlap & Catton, 1979).

Growing discussion of 'limits' to economic and population growth was reinforced by the 1973–74 energy crisis in the USA, leading to a rapid increase in sociological analyses of the vital role of energy in modern societies and the relationship between such societies and their natural resource base more generally. This involved analyses of the relationships between modern societies and their biophysical environments, and was termed 'environmental sociology' to delineate the unique nature of the field: a focus on environmental phenomena in empirical, sociological research (Dunlap & Catton, 1979). As noted below, this was a major departure from mainstream sociology in the 1970s, which ignored the biophysical environment to focus squarely on 'society'.

Both strands of sociological research on environmental problems, along with a range of complementary work on outdoor recreation, social impact assessment of things like dams and natural hazards, were represented in the formation of a Section on Environmental Sociology within the American Sociological Association in 1976, widely regarded as the 'birth' of the field. Over time the key foci have evolved in response to emerging ecological problems and societal reactions to them, and nowadays all sociological scholarship focused on the biophysical environment is considered environmental sociology. The field has also spread internationally, and by the 1990s university-based environmental sociology groups were established in several nations as well as within the International Sociological Association.

## Departure from disciplinary norms

Although founding figures in sociology such as Emile Durkheim, Max Weber and Karl Marx gave some attention to environmental factors, Durkheim justified the new discipline of sociology (as it was in the late nineteenth century) by emphasizing that it explained 'social facts' with reference to other social—rather than physical, biological or psychological phenomena. Over time Durkheim's 'anti-reductionism' dictum, strengthened by sociologists' understandable reluctance to attribute racial, cultural and gender differences to factors like heredity and climate, led the discipline to adopt a socio-cultural determinism that viewed the bio-physical environment as irrelevant to sociological analyses. This tendency was reinforced by rapid scientific and technological advances and resulting faith that modern, industrial societies were increasingly exempt from ecological constraints. The result was that by the mid-twentieth century sociology shared Western society's Promethean worldview or 'human exemptionalism paradigm' that assumed a future of endless progress, an assumption challenged by environmentalism and especially the idea of ecological constraints (Catton & Dunlap, 1980).

In this context, environmental sociology's call to adopt an ecological worldview and examine societal–environmental relationships was a departure from mainstream sociology (Catton & Dunlap, 1980). However, the continual emergence of newer and often more serious ecological problems, ranging from toxic contamination at local levels to global climate change, has given the call increased credibility, and nowadays environmental sociologists routinely analyze bio-physical factors in empirical analyses (York & Dunlap, 2012) and develop ecologically grounded theoretical analyses (York & Mancus, 2009).

## Major emphases

Early on, sociologists interested in environmental issues recognized that environmental (nowadays 'ecological') problems are 'social problems', caused by human actions, seen as problematic primarily because of their threat to human societies, and requiring collective action for their solution. In addition, conditions must be successfully defined and accepted as problematic before being treated as 'ecological problems' (Hannigan, 2014). These characteristics yield major emphases in the field: analyzing the causes of ecological problems, their societal impacts, their solutions, and how they come to be defined as problems (York & Dunlap, 2012).

### *Causes*

An early contribution of environmental sociology was to go beyond overly simplistic analyses of the causes of ecological degradation offered by natural scientists, who tended to emphasize the primary role of population numbers or technology while downplaying that of social forces such as the quest for endless economic growth—which Schnaiberg (1980) conceptualized as a 'treadmill of production'. A range of political economy perspectives, all highlighting the link between capitalism's need for growth and environmental degradation, have greatly extended the treadmill model, especially to the international level (Rudel et al., 2011). An alternate human ecology perspective, involving a major advance over earlier natural science analyses, suggests that population and affluence as well as economic growth, urbanization and other institutional characteristics contribute to the ecological degradation (York et al., 2003). This has led to growing consensus among environmental sociologists on the causal processes producing degradation, especially the link between economic growth and ecological impacts (Jorgenson & Clark, 2012), and also yielded valuable analyses of the driving forces of global climate change (Dunlap & Brulle, 2015: Chap. 2).

## *Impacts*

The discovery of severe ground contamination at Love Canal in New York State in the late 1970s put toxic wastes in the limelight, and led sociologists to study the impacts of environmental hazards on local communities. Evidence of such hazards quickly emerged across the USA, and it became apparent that they were not randomly distributed but were disproportionately located in ethnic minority and/or economically disadvantaged communities. This led to 'environmental justice' research becoming a major emphasis within environmental sociology, and a rich body of literature has emerged on the sources and impacts of environmental 'injustice' (Brulle & Pellow, 2006). As this research has become increasingly sophisticated methodologically, progress has been made in understanding the relative roles of ethno-racial identity, socioeconomic status and the housing market in generating inequitable exposure to environmental hazards, long the subject of debate and controversy (Mohai & Saha, 2015). Growing attention is also being given to the inequitable impacts of environmental hazards internationally, and the growth of environmental justice movements—from the local to the global level—to fight for justice (Dunlap & Brulle, 2015: Chap. 5). While environmental sociologists study a wide range of impacts, environmental justice is a core focus, reflecting the central role of social inequality with the larger discipline (Pellow & Brehm, 2013).

## *Solutions*

Traditionally environmental sociologists have paid less attention to the solution of ecological problems than have political scientists, economists and others interested in environmental policy making. They often criticize popular approaches employed in the USA, especially those premised on promoting behavioral change at the individual level (Heberlein, 2012), but are increasingly analyzing international policy making such as efforts by governments to develop effective mitigation schemes to limit global warming (Dunlap & Brulle, 2015: Chap. 7).

The topic of solutions did gain more prominence by becoming the subject of a major debate in the field starting in the 1990s. Environmental sociologists from Northern Europe, where considerable progress in environmental protection was being made, argued that the region was experiencing a process of 'ecological modernization' in which market forces, technological innovations and enlightened government guidance reduces the ecological impacts of industrialization by decoupling economic growth from resource use and pollution. They further suggested that this 'ecological rationality' was spreading internationally (Mol et al., 2009). American sociologists have been highly critical of ecological modernization theory—methodologically, for inadequately measuring environmental impacts (York et al., 2010), theoretically for resurrecting outmoded theories of modernization and human exemptionalism (Foster, 2012), and empirically for evidence of strong decoupling not being found (Jorgenson & Clark, 2012; York et al., 2003). A quarter century of debate may be coming to a close, as ecological modernization's optimistic projections are being undermined by the dismantling of environmental protection programs by a growing number of conservative governments.

## *Defining ecological problems and subsequent debates*

A fundamental sociological insight is that conditions (e.g. toxic contamination) do not become treated as 'problems' until they are recognized and successfully defined as problematic in the public arena, typically by coalitions of activists, scientists, and policymakers with help from the media. Environmental sociologists have shed great light on the social construction of environmental problems, including the fact that there is often considerable reliance on scientific evidence—particularly for problems that are not easily detectable by human senses such as nuclear radiation and global warming (Hannigan, 2014). This reliance creates opportunities for anti-environmental interests to dispute the evidence, often manufacturing scientific uncertainty as fossil fuels corporations and their allies have done to dispute the seriousness of human-caused climate change (Dunlap & Brulle, 2015: Chap. 10).

The reliance on evidence from natural scientists about global warming and other problems also created tension within environmental sociology, with some 'constructivists' adopting a post-modernist stance of skepticism toward such evidence that provoked criticism from those in the 'realist' camp. The constructivist-realist debate tended to subside by the early 2000s, with the former camp emphasizing they were not questioning the reality of ecological problems and the latter acknowledging the socially contingent nature of scientific evidence. Yet, there remains two rather distinct stances: an 'agnostic' stance toward science, holding that sociologists should be cautious about grounding their analyses on natural scientists' findings; and a 'pragmatic' stance that employs the best-available evidence (on deforestation, $CO_2$ emissions, ecological footprints, etc.) to analyze the causes and impacts of ecological problems (York & Dunlap, 2012).

More recently broader, derivative debates have emerged over appropriate metatheoretical stances for the field, with one side promoting 'hybridity' and related concepts such as 'co-construction' to overcome a dualistic separation of humans and nature (White et al., 2016), and the other arguing that doing so undermines our ability to analyze societal-environmental relationships—the very core of environmental sociology (Murphy, 2016). While these debates will no doubt continue, they have modest impact on empirical research in the field, work premised on the materiality of ecological problems.

## Conclusion

In its four decades of existence, environmental sociology has developed an impressive body of evidence, methodological tools and theoretical insights, as exemplified by its wide-ranging analyses of climate change (Dunlap & Brulle, 2015). This has not only enhanced its status within sociology (Scott & Johnson, 2016) but also strengthened its ability to contribute to interdisciplinary programs and research and to the policy arena (Pellow & Brehm, 2013).

## Learning resources

The online journal *Environmental Sociology*, published by Taylor and Francis, is the leading disciplinary journal. It is sponsored by the International Sociological Association's Research Committee on Environment and Society (RC24) and publishes a diverse range of work that represents the current field of environmental sociology.

Dunlap, R.E. (2015) 'Environmental Sociology' in J.D. Wright (ed.) *International Encyclopedia of the Social and Behavioral Sciences*, 2nd edn Vol. 7 (pp. 796–803). Oxford: Elsevier. A concise overview that expands on the coverage presented in the current essay.

Harper, C. & Snowden, M. (2017) *Environment and Society: Human Perspectives on Environmental Issues*, 6th edn. New York: Routledge. This basic text offers a good overview of environmental sociology.

## Bibliography

Brulle, R.J. & Pellow, D.N. (2006) 'Environmental Justice: Human Health and Environmental Inequalities', *Annual Review of Public Health*, 27: 103–124.

Catton, W.R. Jr. & Dunlap, R.E. (1980) 'A New Ecological Paradigm for Post-Exuberant Sociology', *American Behavioral Scientist*, 24: 15–47.

Dunlap, R.E. & Brulle, R.J. (eds) (2015) *Climate Change and Society: Sociological Perspectives*. New York: Oxford University Press.

Dunlap, R.E. & Catton, W.R. Jr. (1979) 'Environmental Sociology', *Annual Review of Sociology*, 5: 243–273.

Foster, J.B. (2012) 'The Planetary Rift and the New Human Exemptionalism: A Political-Economic Critique of Ecological Modernization Theory', *Organization & Environment*, 25: 211–237.

Hannigan, J.A. (2014) *Environmental Sociology: A Social Constructionist Perspective*, 3rd ed. Abingdon: Routledge.

Heberlein, T.A. (2012) *Navigating Environmental Attitudes*. New York: Oxford University Press.

Jorgenson, A.K. & Clark, B. (2012) 'Are the Economy and the Environment Decoupling: A Comparative International Study', *American Journal of Sociology*, 118: 1–44.

Mohai, P. & Saha, R. (2015) 'Which Came First, People or Pollution? Assessing the Disparate Siting and Post-Siting Demographic Change Hypotheses of Environmental Injustice', *Environmental Research Letters*, 10(11): 115008.

Mol, A.P.J., Sonnenfeld, D.A. & Spaargaren, G. (eds) (2009) *The Ecological Modernization Reader: Environmental Reform in Theory and Practice*. New York: Routledge.

Murphy, R. (2016) 'Conceptual Lenses to Bring into Focus the Blurred and Unpack the Entangled', *Environmental Sociology*, 2: 333–345.

Pellow, D.N. & Brehm, H.N. (2013) 'An Environmental Sociology for the Twenty-First Century', *Annual Review of Sociology*, 39: 229–250.

Rudel, T.K., Timmons Roberts, J. & Carmin, J.A. (2011) 'Political Economy of the Environment', *Annual Review of Sociology*, 37: 221–238.

Schnaiberg, A. (1980) *The Environment: From Surplus to Scarcity*. New York: Oxford University Press.

Scott, L.N. & Johnson, E.W. (2016) 'From Fringe to Core? The Integration of Environmental Sociology', *Environmental Sociology*, 3: 17–29.

White, D.F., Rudy, A.P. & Gareau, B.J. (2016) *Environments, Natures and Social Theory*. London: Palgrave.

York, R. & Dunlap, R.E. (2012) 'Environmental Sociology' in G. Ritzer (ed.) *The Wiley-Blackwell Companion to Sociology* (pp. 504–521). Oxford: Wiley-Blackwell.

York, R. & Mancus, P. (2009) 'Critical Human Ecology: Historical Materialism and Natural Laws', *Sociological Theory*, 27: 122–149.

York, R., Rosa, E.A. & Dietz, T. (2003) 'Footprints on the Earth: The Environmental Consequences of Modernity', *American Sociological Review*, 68: 279–300.

York, R., Rosa, E.A. & Dietz, T. (2010) 'Ecological Modernization Theory: Theoretical and Empirical Challenges' in M.R. Redclift & G. Woodgate (eds) *The International Handbook of Environmental Sociology*, 2nd ed. (pp. 77–90). Cheltenham: Edward Elgar.

# Environmental politics

3.12

Shannon O'Lear

As explored in the "Classic concepts" section, "environment" can mean many things and refer to a variety of features and systems. There is also a growing body of work on hybridity which looks at "the environment" as not necessarily distinct from social systems. Arguably, there is not "the environment" but many different environments at different scales. The notion of "politics", however, often evokes an understanding of established interests and entrenched power. There is a sense of politics as being fixed or stuck in familiar and predictable tracks. A more useful way to think of politics, environmental or otherwise, is as a process to manage a variety of interests among groups of people, to compromise, and to resolve conflict (Agnew, 2011). That understanding allows us to ask more interesting questions about environmental politics. We can see how an environmental issue may shape a political process, or how a political decision may raise particular environmental issues. Regardless of the starting or end points of environmental politics, it is helpful to consider key elements of environmental politics: scale, actors, and the role of science.

## Why is scale important in environmental politics?

Scale may be spatial, in terms of area or place, or it may be temporal, in terms of time. Paying attention to spatial and temporal scale can be useful towards understanding more clearly how and why different priorities come into conflict – or resolution – in political processes involving environmental features.

## Spatial scale

First, spatial scale refers to the spatial scope of an environmental issue. That could mean the area of a forest, a watershed involving all of the tributaries of a main river, the distribution of a particular pollutant, or the locations of nuclear waste storage facilities. Spatial scale is not merely two-dimensional. It may have volume such as depth of the ocean, layers of the atmosphere, or an expanding plume of pollution moving through an underground aquifer. A particular environmental feature may reflect many spatial scales all at the same time. For instance, a single field of corn might be part of an international network of industrial agricultural practices, an issue for national policy on farm subsidies and ethanol production, a concern for local economic well-being in a small, Iowa township, and a context for studying the resilience of a genetically modified organism. The spatial scale at which any one person might focus their view will depend on what that person's interests are.

Just as we can discuss or map the spatial scale of environmental features, we can also think in terms of multiple spatial scales of politics. First, there are legal, jurisdictional scales which operate like nested containers where laws and policies are enforced. There is the sovereign state or country, and within that there are often counties or districts, and within those are townships or smaller areas where decision making is conducted. Much of environmental politics has to do with determining which political scale or jurisdiction is appropriate for taking the lead in decision making or policy implementation. For instance, when hazardous pollution was found to be leaking from the Rocky Mountain Arsenal, a World War II munitions plant east of Denver, a considerable dispute erupted over which political jurisdiction was responsible for the cleanup: The Federal government (since the arsenal had been commissioned by the Department of War), the state of Colorado, or Commerce City. However, when bald eagles were found nesting on the property, the National Fish and Wildlife Service suddenly took jurisdictional control since bald eagles were recognized as an endangered species (Wiley and Rhodes, 1998).

Environmental features and systems do not fit neatly within political jurisdictions. Likewise, political jurisdictions do not necessarily capture environmental concerns completely. Rivers and wildlife cross political boundaries, weather patterns and air pollution move freely, ecosystems and biomes from forests to wetlands to deserts in many cases existed before political boundaries were even established. So, the process of identifying and measuring problems and generating possible solutions – key aspects of environmental politics – likely involve not only nested political jurisdiction as in the case of the Rocky Mountain Arsenal, but also lateral or neighboring jurisdictions as well.

## Temporal scale

In addition to spatial scale, environmental politics often involve multiple temporal scales or windows of time. Again, time scales that are relevant for environmental features rarely coincide with time scales of politics. For instance, it may take 100 years to produce an inch of topsoil, several years for a forest to regenerate after a fire, and a few days for toxins at the headwaters of a river to cause massive fish kill miles downstream. Political time scales are often determined by election cycles. A politician who is looking to be re-elected in just a few years is more likely to pursue decisions that will show a quick result or avoid issues altogether that seem unresolvable in that limited time frame. As a process of negotiation that involves many actors and perspectives, politics can move forward very slowly. It may even serve some interests to stall the political process and delay decision making.

Any situation of environmental politics involves spatial scales and temporal scales. The process of politics necessarily also involves multiple types of actors and their interests. Often, each actor or group involved in an environmental politics situation takes a particular spatial and temporal focus.

## Who are actors in environmental politics?

Any given instance of environmental politics usually emerges when there are different views about how an environmental feature should be treated. These interests range widely from how we should extract and process resources from natural gas to strategic minerals, how we should produce our food, how and to what extent we should expand our urban areas, to what degree we should maintain pristine ecosystems and wildlife and other questions about human–environment relationships. Often, environmental politics is dominated by anthropocentric (human-centered), managerial attitudes that view "the environment" as something to be objectively studied, measured, and controlled. Other views are more biocentric and see inherent value in non-human features and systems.

The first and perhaps most obvious actors engaged in environmental politics tend to be states or countries. The participation of sovereign, territorial states is central to processes such as international negotiations over fishing rights or carbon emissions limits. We can also think of more local, political processes involving smaller scale political jurisdictions and citizen groups. Returning to the issue of spatial scale, actors in environmental political processes often prioritize a particular spatial scale. This is how a wide range of interests and values come into interaction in the process of environmental politics, and the challenge (and opportunity) comes in identifying how to bring about an end result that reflects the values of the interested parties. So, in addition to the representations of political jurisdictions discussed above, other actors represent an array of interests and priorities associated with the environmental concern at hand. Corporations with transnational operations may seek to streamline their productivity and maximize profits. Increasingly, corporations seek ways to do "green" business (sometimes these efforts are genuine, and in other cases it is a matter of "greenwashing" their image (Scanlan, 2013)). Nongovernmental organizations (NGOs) such as the Sierra Club or Greenpeace may operate through a network of members to promote a certain view of the environment. Not all NGOs share the same values, however. The Wise Use Movement in the U.S., for instance, has powerful members across the resource extraction industries and promotes private property rights and access to public lands by corporate interests. Citizen groups representing a certain place or sub-population such as outdoor enthusiasts or cancer survivors also participate in environmental politics as do indigenous groups, and consumer rights activists.

All actors and interests in any given environmental politics situation are not equally powerful. Some actors may be influential in the early stages of setting a political agenda or in determining how a particular issue is framed. Other, less powerful actors may only be able to engage in the formal environmental politics process by writing letters to their political representatives or in town hall type meetings. Radical activists may eschew the formal political process altogether and conduct acts, sometimes viewed as destructive, in line with their values. Not all views or relevant spatial scales may be represented in an instance of environmental politics. Who speaks for the trees?

## What is the role of science in environmental politics?

Finally, in the process of environmental politics, scientific evidence is often wielded in support of a particular view or agenda. Scientific studies or expertise may be used to promote different priorities or agendas. That is possible because there are different ways of studying and measuring environmental features. Returning to our corn field discussed earlier, scientific work focused on GMOs will certainly offer different insights and findings than scientific work on ecosystem processes or food security. Science cannot determine what we should prioritize in environmental politics. Instead, appropriate forms of science are best used in the environmental politics process as a guide to decision making only after priorities have been identified (Sarewitz, 2004).

## Learning resources

### *Books*

Death, C. (ed.). (2014) *Critical Environmental Politics*. New York: Routledge. Worth dipping into the many short chapters.

O'Lear, S. (2010) *Environmental Politics: Scale and Power*. New York: Cambridge University Press. This is an advanced introduction to how scale is a key dimension of environmental politics in a world of unequal capacity of people to act politically.

### *TV/Video*

*Years of Living Dangerously*. 2014. James Cameron, Jerry Weintraub & Arnold Schwarzenegger. http://yearsoflivingdangerously.com, Accessed 29 July 2016.

## Bibliography

Agnew, J. (2011) 'Presidential Address: Waterpower: Politics and the Geography of Water Provision', *Annals of the Association of American Geographers*, 101(3): 463–476.

Sarewitz, D. (2004) 'How Science Makes Environmental Controversies Worse', *Environmental Science & Policy*, 7(5): 385–403.

Scanlan, S.J. (2013) 'Feeding the Planet or Feeding Us a Line? Agribusiness, "Grainwashing", and Hunger in the World Food System', *International Journal of Sociology of Agriculture and Food*, 20(3): 357–382.

Wiley, K.B. & Rhodes, S.L. (1998) 'From Weapons to Wildlife: The Transformation of the Rocky Mountain Arsenal', *Environment*, 40(5): 4–35.

# 3.13

# Design, emotion, sustainability


## Jonathan Chapman


At its best, design is a powerful tool for cracking problems and leveraging opportunities for new products, services and systems that drive a more resource efficient economy and create value for policy makers, businesses and consumers. However, despite being an incredibly dynamic and vibrant cultural phenomenon, design is an extremely wasteful and destructive one too. This is largely due to its ephemeral nature, fuelled by the ceaseless consumer hunt for change, novelty and innovation. This chapter shows how *sustainable design* recalibrates the parameters of good design in an unsustainable age. It advances and broadens the agenda of the design system – with its established emphasis on economic sustainability and development, at all costs – such that it's fit for purpose in unraveling the Gordian knots of sustainability, through the design of more sustainable goods and services.

The Earth is finite, balanced, synergistic and reactive, and yet we design the world as though it were separable, mechanical and lasting, leading to what Bateson refers to as a fundamental epistemological error (Bateson, 1972) that shapes practically all that we do, and one that can be found at the very root of unsustainability. Indeed, human destruction of the natural world is a crisis of behaviour, and not one simply of energy and material alone, as is often assumed in design; the decisions we make as an industry, the values we share as a society and the dreams we pursue as individuals collectively drive all that we accomplish, while shaping the ecological impact of our development as a species.

The UK Government is one of several proposing an economy where resources are used sustainably through design for longer life, upgrading, re-use or repair. Product life extension strategies – like *emotionally durable design* (Chapman, 2005) – have a vital role to play here: combatting rising levels of e-waste and obsolescence; tackling the challenge of weaning people off their desire for the new; and, helping shape new sustainable business models; supporting users in keeping products, components and materials at their highest utility and value throughout their lifetime. Indeed, the success of a resource efficient, and circular, economy depends on new business models that are able to truly capitalise on longer product lifespans over time (Bakker et al., 2014).

Conventionally, industrial activity involves a linear production-consumption system with inbuilt environmental destruction at either end; sustainable product design activity over the past 45 years has made these wasteful and inefficient ends of the scale, marginally less wasteful and inefficient. However, we need to move away from this linearity in our design thinking, to reconnect with design on a more circular and systemic level, if we are to achieve the degrees of transformation our current situation demands. These new approaches require designers and manufacturers to take greater control over material flows; closing the loop through clear and systematised processes of product design, production, delivery and take back. A circular economy is one in which resources are kept in use for as long as possible. The maximum value is extracted from them, while materials and energy are recovered or recycled as much as possible at the end of any product's life. In the circular economy, materials and resources *flow through* products and into new ones, as opposed to being designed into products, locked into landfill.

Global businesses, supported by governments, are also beginning to look at product life extension as a viable route to waste reduction, and value creation. Electronic waste (e-waste) in particular is growing at three times the speed of any other form of waste in the EU. Today, practically everything is disposable – it is culturally permissible to throw away anything (Thackara, 2015) from a barely-used smartphone, television, or vacuum cleaner, to an entire three-piece suite or fitted bathroom. Given the huge quantities of precious resources (including gold and other rare metals) that find their way into our gadgets, it would surely be worth us taking more care of them, repairing them when broken, and keeping them for longer.

In fact, the opposite is happening: product lifespans are shortening as material culture becomes increasingly disposable. Hence, we live in a world drowning in objects (Sudjic, 2008): households with a television in each room; kitchen cupboards stuffed with waffle makers, blenders and cappuccino whisks; drawers filled to bursting with pocket-sized devices powered by batteries – batteries which themselves take a thousand times more energy to make than they will ever provide. A child's remote-control tank, for example, contains a thumbnail-sized microchip, containing over 65 per cent of the elements in the periodic table. There is more gold in a tonne of phones than a tonne of rock from a gold mine. Due to their design and manufacture, the rock-bound gold is more economically viable to extract than its phone-bound counterpart.

Of course, the notion of a throwaway society is nothing new. American economist Bernard London first introduced the term 'planned obsolescence' in 1932 as a means to stimulate spending among the few consumers who had disposable income during the depression. The concept was popularised by Vance Packard in his seminal book, *The Waste Makers* (1964). Though informed by the work of both Bernard London (1932) and Earnest Elmo Calkins (1932) on consumer engineering, Packard's dualistic theories of *functional obsolescence* and *psychological obsolescence* assert that the deliberate shortening of product lifespans was unethical, both in its profit-focused manipulating of consumer spending, and its devastating ecological impact through the nurturing of wasteful purchasing behaviours. In fact, the concept of disposability was a necessary condition for America's cultural rejection of tradition and acceptance of change (Slade, 2007).

Over the past decade, issues of sustainability have become well established within design; strategies like: design for recycling, disassembly, service and energy efficiency, for example, have become commonplace in today's process. Designing for *emotionally durable* products and user experiences helps reduce the consumption and waste of resources by building lasting relationships between users and the products they buy. The term 'emotional' is used here because wasteful patterns of consumption and waste are driven, in large part, by emotional and experiential factors. Here, longer-lasting products have the potential to build economic models around creating robust products, upgrade and repair services, and brand-loyal customers – all without excessive waste.

In product design terms, we can support greater levels of emotional longevity when we specify materials that age gracefully, and that develop quality over time. We can design products that are easier to repair, upgrade and maintain throughout their lifespan. These are effective product life extension strategies, and while they *can* come at an increased cost at point of purchase, they generate revenue downstream, through the introduction of service and upgrade packages. Furthermore, extending the life of a product has significant ecological benefits. For example, take a toaster that lasts about 12 months. Even if the toaster's life is extended to just 18 months through more durable design, the extra longevity would lead to a 50 per cent reduction in the waste consumption associated with manufacturing and distributing it. Scale this up to a national or international population of toaster-buyers, and it's clear how significant an impact this could be.

Designing products that can be kept for longer also nurtures a deeper relationship with both the product and the brand, which increases the likelihood of brand loyalty maturing. Therefore, such emotionally durable design doesn't just make sense from an environmental and resources perspective, but can be seen as a commercially viable business strategy in an increasingly competitive globalised world.

Simply having more *stuff* stopped making people in Britain, the USA and other wealthy countries happier decades ago; we need an economy of *better*, not *more*; one in which things last longer, age gracefully and can be repaired many times before being recycled.

## Learning resources

Sustainable design is supported by a number of excellent online resources. Of particular importance are:

- *The RSA Great Recovery Project* (http://www.greatrecovery.org.uk/) Helping designers make things better, turning waste into value and reducing environmental impacts through systems thinking.
- *The Living Principles for Design Framework* (http://www.aiga.org/the-living-principles-for-design/) A roadmap for sustainable design that is understandable, integrated and actionable.

## Bibliography


Bakker, C., Hollander, M. & van Hinte, E. (2014) *Products That Last: Product Design for Circular Business Models*, The Netherlands: TU Delft Library.

Bateson, G. (1972) *Steps to an Ecology of Mind*, Chicago: University of Chicago Press.

Calkins, E.E. (1932) 'What Consumer Engineering Really Is', in Sheldon, R. and Arens, E., *Consumer Engineering: A New Technique for Prosperity* (pp. 1–14), New York: Harper & Brothers.

Chapman, J. (2005) *Emotionally Durable Design: Objects, Experiences and Empathy*, London: Earthscan.

Chapman, J. (2014) in Webb, F. (Ed), 'Can Emotionally Durable Design Prolong Use-life and therefore Reduce Waste?', *Making It*, UN, Geneva, April.

London, B. (1932) *Ending the Depression Through Planned Obsolescence*, Pamphlet, US.

Packard, V. (1964) *The Waste Makers*, Middlesex: Penguin.

Slade, G. (2007) *Made to Break: Technology and Obsolescence in America*, Cambridge, MA: Harvard University Press.

Sudjic, D. (2008) *The Language of Things*, London: Allen Lane.

Thackara, J. (2015) *How to Thrive in the Next Economy: Designing Tomorrow's World Today*, London: Thames & Hudson.

# Environmental law

3.14

## David Delaney

Environmental law is conventionally understood as the set of formal legal directives (statutes, regulations, doctrines, judicial decisions and the like) that are explicitly oriented toward protecting, conserving or restoring components of 'the environment.' They are reflective of the idea that law provides a rational tool-kit with which humans govern themselves and solve problems of common concern. These directives include those promulgated by each of the 200 or so sovereign states (such as the National Environmental Protection Act and the Clean Water Act in the USA) as well as those issued by countless sub-sovereign provinces, states and municipalities (such as local bans or taxes on plastic bags). Included as well are those rules contained in international treaties and conventions, for example, the Convention on Trade in Endangered Species of Wild Flora and Fauna. But environmental law is also a field of practice structured and navigated by countless institutions, organizations and actors including ministries, agencies, activists, NGOs, international organizations, businesses, specialist law firms, academics and, of course, those subject to its prescriptions. Although there are antecedents dating back to the nineteenth century it's fair to say that in its present, elaborated global form, environmental law is a product of the last 50 years. That is to say, it is an institutional response to increased scientific awareness of the pervasiveness and severity of 'environmental problems' and to the related emergence of environmentalisms as political (and legal) projects that became prominent in the 1960s. The *amount* of 'law' that has, since then, been created to forestall environmental degradation is enormous. And, of course, it has not been inconsequential. This is especially so in comparison with what *would likely* be the case in its absence. For example, there is little doubt that, in some parts of the world, water and air quality are significantly better than they otherwise would have been; that some species of plants and animals that would have become extinct remain viable; that the depletion of the ozone has been arrested and so much more.

However, given the acceleration of global deforestation, the collapse of marine life, diminishing biodiversity, and, of course, the super-wicked problem of climate change, among many other urgent matters, there exists a strong sense among many that these massive legal interventions have been far from sufficient. Many scientists, activists, scholars and policy makers believe that environmental law, as presently imagined and practiced, has failed, or, indeed, has proven to be counterproductive. But because virtually every proposed route toward sustained environmental integrity acknowledges that law and legal change are necessary means to this end (if not law, what?) then answering the 'what's-the-matter-with-law' question is also a pre-requisite to any path forward. For some environmental scholars, lawyers and activists attending to this question requires re-imagining what law needs to become in the Anthropocene.

## The insufficiency of law

There are any number of explanations for legal failure. Many of these raise fundamental questions about law, its possibilities and limitations. One important factor is the spatial mismatch between the entities and processes associated with the environment – which are dynamic, ecological, and, ultimately, planetary – and the fragmented, often competitive, self-interested political nature of sovereignty as the ultimate foundation of modern law. Then too, at a relatively simplistic level of analysis, one must acknowledge the existence of a multi-faceted, contentious politics of environmentalism and anti-environmentalism within legal regimes. This acknowledgement highlights the social forces that both oppose genuine remediation and have the resources to block or 'zombify' legal avenues of intervention. Other explanations note that 'the environment' is, at best, a tertiary priority for most law makers/enforcers to the extent that such remedial measures as are acceptable – or, as is often put, 'feasible' – are constrained by other higher priorities such as those identified with 'the economy' and, more specifically what is called 'growth.' In another register, meaningful legal action on behalf of the environment must be compatible with the perceived imperatives and constraints of capital accumulation. These constraints are often understood as making the kinds of social-material transformations that are required virtually impossible to achieve. Such explanations implicate deeper 'structural' arguments concerning the relative autonomy of the state vis-à-vis capitalist firms and, derivatively, of law vis-à-vis politics. Analysis of the actual formulation, implementation and enforcement of 'environmental law' leaves little doubt that such political economic factors are indispensable to any minimal understanding of the designed insufficiency of environmental law (Layzer, 2012).

However, though it often goes without saying, we are not talking about 'law' as such but, most often, versions of distinctively *liberal* law as these may be manifest in a wide range of historical and geographical (state specific) forms. That is to say, there may be something *intrinsic* to the liberal legal imaginary – and to the ways in which this conditions how 'the environment' is itself imagined – that accounts for much of the failure of law to respond appropriately to environmental degradation. Liberalism, in this sense, refers to a set of normative and practical metaphysical commitments that became a pronounced ideological force in sixteenth-century Europe and have animated much of political life on the planet since that time. Across the variety of forms, liberal legal regimes, by definition, entail some (strong) commitment to property rights. These include the presumptive 'right' to use fragments of the material world according to the will or desires of those designated as 'owners.' Corollative to this are strong limitations on the capacity of state actors to 'interfere' with such rights. This is simply what it means to 'have a right.' That is to say, much environmental destruction is founded on the recognized and protected 'right' of owners to destroy. Other key elements of the present legal foundation of environmental degradation include the assimilation of enormous economic organizations (corporations and the state itself) to the abstract positions of legal personhood and 'ownership'; and the often severe constraints (no-rights) on third parties to pursue legal remedies against these rights bearing, state protected owners.

Other impediments to the efficacy of environmental law can be attributed to how the entity 'the environment' is itself conceptualized by the conventional discourses – and so, institutions – of law. The environment (water, air, wildlife and so on) is a legal specialty, like admiralty, family law, constitutional law or the law of war. It is conceptualized as a discrete area of governance and, indeed, often as a 'special interest' (Bosselmann, 2008). This accounts, in part, for the ease with which environmental issues are subordinated to other legal priorities such as corporate law, banking law, trade, militarism, development and so on. Proponents of more robust, effective legal programs insist that this is a profound misconception of the focal topic. *Ecological*, in contrast with *environmental*, legal thinking asserts the unity and complex, dynamic interdependencies of the physical world and all aspects of the human social world. To the extent that the (liberal, more generally, modern) legal imagination compartmentalizes and fragments 'the environment', then law simply gets the world wrong. And in getting the world wrong it incapacitates itself. These lines of critique suggest that instead of imagining environmental law as after-the-fact, ad hoc regulatory responses to discrete 'environmental problems' more attention should be paid to the ways in which the distinctively legal is always already *constitutive* of both 'the environment' and what counts and doesn't count as a 'problem.' It is important to acknowledge, however, that some 'other-than-liberal' set of legal presumptions would likely entail significant sacrificial trade-offs that many would consider to be intolerable. Moreover, there is little reason to believe that less than liberal or authoritarian legal regimes would accomplish what liberal legal regimes have not.

## Re-imagining law in the Anthropocene

Millions of human Earthlings are working diligently and in good faith within the constraints of conventional law to avoid the catastrophes subsumed under the sign of the Anthropocene. Law, again, appears to be a necessary tool that humans must use in mitigating, if not arresting, the urgent problems associated with this concept. But law's foundations in sixteenth-century European metaphysics may render it unsuitable for what is required in the twenty-first century and beyond. There are a number of contemporary theoretical projects that aim to create the intellectual conditions for thinking *about* law differently and thinking about the world *through* law differently so that it might be up to these unprecedented tasks. Among these are programs to 'green' sovereignty (Eckersley, 2004), international law, and global constitutionalism; comprehensive re-imaginings of 'earth governance systems' (Biermann, 2014); advocacy of a robust 'green criminology' (South and Brisman, 2013); the articulation of post-modern premises of 'earth jurisprudence' (Burdon, 2012) and even 'wild law' (Maloney and Burdon, 2014). Less speculative are projects such as Newell's 'political economy of global environmental governance' (2008). This project analytically suspends the liberal public/private distinction, thereby enlarging what counts as 'law' in the first place. But again, the paradox is that law would first need to be reformed such that it is capable of reforming itself, such that it is, then, capable of doing what is required.

## Learning resources

There are countless organizations that are devoted to reshaping the environment through the means of re-working law. Many law schools have environmental law centers, many environmental NGOs have their associated legal defense programs. Environmental law organizations may be local, regional, national or international in scope. Some are broad, some issue specific. Most are liberal, some less so. Good places to begin learning more include:

Earthjustice (earthjustice.org), formerly Sierra Club Legal Defense Fund;
Wild Law (wildlaw.org);
Environmental Law Institute (eli.org); and
Centre for International Environmental Law (ceil.org).

## Bibliography

Biermann, F. (2014) *Earth System Governance: World Politics in the Anthropocene*. Cambridge, MA: MIT Press.

Bosselmann, K. (2008) *The Principle of Sustainability: Transforming Law and Governance*. Burlington, VT: Ashgate.

Burdon, P. (2012) 'A Theory of Earth Jurisprudence', *Australian Journal of Legal Philosophy*, 37: 28–60.

Eckersley, R. (2004) *The Green State: Rethinking Democracy and Sovereignty*. Cambridge, MA: MIT Press.

Layzer, J. (2012) *Open for Business: Conservatives' Opposition to Environmental Regulation*. Cambridge, MA: MIT Press.

Maloney, M. and Burdon, P. (2014) *Wild Law – in Practice*. New York: Routledge.

Newell, P. (2008) 'The Political Economy of Global Environmental Governance', *Review of International Studies*, 34: 507–529.

South, N. and Brisman, A. (eds) (2013) *Routledge International Handbook of Green Criminology*. London: Routledge.

3.15

# Environmental management

Chris Barrow

The environment needs to be managed: humans are increasingly transforming nature and the intended and unplanned outcomes of this are the domain of environmental managers. Environmental management (EM) has developed during the last 30 years and is still evolving. It seeks to dovetail development with environmental and social awareness, to take advantage of opportunities, to avoid hazards, to mitigate problems, and to prepare people for difficulties that cannot be avoided by improving flexibility, adaptability and resilience (Barrow, 2006; Thompson, 2002).

Before the 1980s little effort was made to integrate environmental and natural resources exploitation with social and economic development. Today EM is implicated in virtually every aspect of human affairs. It has the following characteristics (four things are especially important and are in italics):

- it supports *sustainable development* as a key goal;
- it tries to be *proactive*;
- it often embraces the *precautionary principle* (this shifts the burden of proving a development proposal is safe from the potential 'victim' to the 'developer');
- it usually stresses stewardship and prudence, rather than exploitation and seeks to make the *'polluter pay' (i.e. those damaging the environment).*
- it deals with a world greatly affected by humans;
- it demands a multidisciplinary or even 'holistic' approach;
- it tries to integrate and reconcile different development viewpoints;
- it seeks to co-ordinate science, social science, policymaking and planning;

- it recognises the desirability of meeting, and if possible sustaining, more than basic human needs;
- it works with a timescale well beyond the short term, and looks beyond the local to the global situation;
- it seeks opportunities as well as address threats and problems.

Businesses, governments and societies may have no clear idea of what is good EM and can be in conflict, one with another. Consequently EM operates in the face of challenges. People often want things that are damaging to themselves, others, and the environment. Also, demands and fashions tend to change over time, sometimes suddenly. There are usually a number of potential routes to an EM goal: one may be the best all-round solution, one the best practical solution, one the cheapest, one the best for the environment, one may be favoured by the government in question, another supported by a powerful body.

Consequently, it is necessary for EM to advise, educate, lobby and police stakeholders (various groups with an interest in development) to pursue what it identifies as the 'best EM' option. It has to cope with politicians, NGOs (non-governmental organisations), special-interest groups, aid agencies, company directors, and so on – all of whom may already have decided objectives and strategies. Economists and law-makers have increasingly become aware of environmental issues and are developing approaches that support EM. For example, the former now seek to value and account for functions of the environment – *natural capital* – (e.g. sinks for wastes and sources of raw materials and amenities).

EM often needs to deal with problems that reach beyond sovereignty limits (i.e. address global and transboundary issues). EM commonly a faces a dilemma – to reconcile a need to take time to adequately research versus demands for rapid, affordable, and clear-cut decisions. Delay and poor research can result in costly, possibly irreparable environmental problems or missed opportunities. It is easy to collect misleading data, particularly when 'polarised perception' influences researchers (i.e. they have conscious and/or unconscious views that influence their work). EM must distinguish between reliable and inaccurate data. That data must be objectively interpreted: the apparent causes of a problem may in reality be symptoms and faulty diagnosis and advocacy can lead to delays, misspending and even serious environmental damage (Lomborg, 2001). EM demands co-ordination, ability to devise trade-offs, negotiation, diplomacy, and foresight. Most EM must assume a 'business as usual scenario' – i.e. address problems with human attitudes, economics, and so on little changed to help. EM involves decision making under uncertainty.

A key goal of EM is sustainable development. This can be said to be the stretching of what nature provides to optimally satisfy human needs (using knowledge, effort and technology to stretch) while maintaining that output indefinitely without causing environmental breakdown (possibly in spite of social change, economic problems and natural disasters). It is still unclear whether sustainable development is mainly a guiding principle or if there can be workable strategies and techniques to pursue it and ensure it lasts; history suggests it is a challenge and is seldom maintained (Diamond, 2005). Sustainable development demands trade-offs between present enjoyment versus investment ensuring future environmental function. People and institutions find it difficult to be altruistic to benefit future generations and non-relatives. Currently, a widespread view is that humankind have only a few decades to set in motion EM that will sustain indefinitely as many people as can be given a satisfactory 'quality of life' while causing as little environmental damage as possible. Failure to effectively act could mean an environmental catastrophe with disastrous human consequences. En-route to a sustainable development it will probably be necessary to support too large a global population and to cope with excessive environmental demands (a sort of overdraft), perhaps for several decades.

A growing number of businesses and institutions employ EM (Welford, 2000). This 'Corporate EM' includes: education of employees; updating management on environmental regulation, laws and issues; selecting specialists; ensuring waste management is satisfactory; avoiding legal costs and reducing insurance premiums, risk and hazard assessment; correcting mistakes of the past; decommissioning and rehabilitating, and much more. What is often unclear is: does a business just seek to comply with environmental regulations and use EM to improve profits and avoid legal liability, taxation, or insurance claims – or, truly support EM alongside corporate social responsibility (i.e. have concern for government and wider society)?

Although there is no universal EM approach there are widely adopted systems employed to steer it, maintain more consistent standards and make it more comprehensive. An eco-audit gives a largely retrospective 'snapshot view', narrow in time and breadth of coverage, seeking to establish what the current situation is or was. It is better if eco-audits are part of a structured and proactive EM system (EMS). An EMS enables a business, activity or organisation to set goals and monitor performance and it shows when to take corrective action or make improvements; also, it supports the development of a reflective outlook. One widely used EMS is ISO 14001. Launched in 1996, this now provides a virtually 'world standard' accredited framework and guidelines (Tinsley, 2001).

EM standards allow meaningful evaluation, exchange and comparison of data, improve objectivity of judgement, aid recognition of crucial thresholds and limits, support negotiation, law making and comparison. EM standards may be divided into broad groups: those concerned with ensuring human health and safety; those concerned with maintaining environmental quality; those concerned with the quality of consumer items. EM indicators help ensure things can be relatively quickly and easily measured, and have specific, often crucial, meaning. EM indicators are widely used to judge whether environmental quality, stability and vulnerability are getting better or worse. Other EM tools and techniques include: environmental monitoring and environmental modelling. Of great value is proactive assessment (attempting to predict what may happen in future) and this is provided through environmental impact assessment (EIA), hazard and risk assessment and forecasting. Eco-footprinting is a tool which seeks to measure EM performance of an individual, group, company or an activity, service, transport network, supply of a commodity, region, country, city and so on. EM may make use of lifecycle assessment to examine what needs to be done at each of the stages in development of something, such as an industry, problem, infrastructure, etc.

## Learning resources

Institute of Environmental Management and Assessment (IEMA)(UK)—http://www.iema.net.

International Network for Environmental Management (INEM) – http://www.inem.org.

*Journal of Environmental Management* (Elsevier) – http://www.journals.elsevier.com/journal-of-environmental-management.

Owen, L. and Unwin, T. (1997) *Environmental Management: Readings and Case Studies*. Oxford: Blackwell.

# Bibliography


Barrow, C.J. (2006) *Environmental Management for Sustainable Development*. London: Routledge.

Diamond, J. (2005) *Collapse: How Societies Choose to Fail or Survive*. London: Penguin/Allen Lane.

Lomborg, B. (2001) *The Sceptical Environmentalist: Measuring the Real State of the World*. Cambridge: Cambridge University Press (published in Danish 1998).

Meadows, D.H., Randers, J. and Meadows, D.L. (2004) *The Limits to Growth: The 30-year Update*. London: Earthscan.

Thompson, D. (ed.) (2002) *Tools for Environmental Management: A Practical Introduction and Guide*. Gabriola Island, Canada: New Society Publishers.

Tinsley, S. (2001) *Environmental Management Demystified: A Guide to Implementing ISO 14001*. London: Spon Press.

Welford, R. (2000) *Corporate Environmental Management: Towards Sustainable Development*. London: Earthscan.

3.16

# Environmental philosophy

Ned Hettinger

Environmental philosophy explores the fundamental concepts and values underlying humans' relationships to nature. Its origins are as old as our species, for humans engage in environmental philosophy whenever they reflect on their relationship to their environment, human and nonhuman. Philosophers have been writing about this relationship from ancient times, as when Aristotle suggested that plants were made for animals and animals were made for people (Bk. 1, Pt. 8). Earlier still, the Old Testament has God making mankind in his image and telling humans to "be fruitful and increase in number; fill the earth and subdue it. Rule over the fish in the sea and the birds in the sky and over every living creature that moves on the ground" (*Genesis*, 1:28). The rejection of this idea that nature's value is solely its usefulness to humans has been a central focus of the thriving field of academic environmental philosophy that has developed over the last half century.

The recent explosion in environmental philosophizing coincides with growing environmental consciousness that followed the 1960s realization that humans faced an environmental crisis, an idea fueled by classics in environmental writing such as Rachel Carson's *Silent Spring* (1962) which documented the harmful effects of widespread pesticides use and Lynn White's essay, "The Historical Roots of our Ecological Crisis," that pinned the blame for the environmental crisis on the science and technology spawned by "orthodox Christian arrogance toward nature" (1967: 1207). There were important environmental thinkers a hundred years earlier as well, including philosopher John Stuart Mill, the transcendentalist Henry David Thoreau, and environmental writer and activist John Muir. Mill worried about human overpopulation and the impossibility of endless growth (1848, Bk. IV. Ch. 6.). In *Walden* (1854), Thoreau wrote about the life of simplicity and self-sufficiency and he later proclaimed "in Wildness is the preservation of the world" (1862). John Muir was a scientist, nature mystic, and founder of the Sierra Club and he believed that "None of Nature's landscapes are ugly so long as they are wild" (1901: 4).

**Figure 3.16.1** John Muir, 1907.

Photo by Francis M. Fritz. Retrieved from: https://upload.wikimedia.org/wikipedia/commons/2/2a/John_Muir_Cane.jpg

He also argued that "When we try to pick out anything by itself, we find it hitched to everything else in the Universe" (1911: 110), an idea codified by scientist-activist Barry Commoner in *The Closing Circle* as "the first law of ecology," namely, that "Everything is connected to everything else" (1971: 39).

A major focus of environmental philosophy has been exploring whether, and to what extent, moral concern should be aimed at beings other than humans. While some contemporary environmental philosophers believe that only humans merit direct moral concern (Norton, 1991), most reject such "anthropocentrism" and argue that some nonhumans have value that transcends usefulness to others (that is, "intrinsic value"). Animal activists such as Peter Singer (*Animal Liberation*, 1975) and Tom Regan (*The Case for Animal Rights*, 1983) consider our lack of concern for animals' rights to be "speciesism," a type of arbitrary discrimination analogous to racism and sexism.

While extending moral concern to include nonhuman animals is an important step away from the anthropocentrism that has dominated much ethical philosophy, it is a small step compared to an environmental ethic that respects all life. Such a "biocentrism" was articulated early on by Nobel laureate Albert Schweitzer in his "reverence for life" philosophy (*Civilization and Ethics*, 1923) and was systematically defended by Paul Taylor's *Respect for Nature* (1986). Taylor argued that we have duties to other individual living beings to fairly share the planet, to minimize causing harm to them, and when such harm is necessary (but only for the pursuit of valuable human goals), we owe restitution on their behalf. While some environmental philosophers are egalitarian, claiming that all animals or living beings have equal value, others insisted that morally important beings have different degrees of intrinsic value, typically claiming the humans have the most such value.

Many environmental philosophers argue that neither the animal activist philosophies nor biocentrism do justice to the moral and conceptual foundations of environmental thinking. They argue that a focus on the rights of individual animals conflicts with important environmental goals, such as protecting ecosystems by culling animal populations (Callicott, 1980). The focus on living individuals has also come under criticism by those who believe intrinsic value and moral concern should primarily be directed at natural systems and processes (Rolston, 1994). Such views are called "ecocentric" and their focus is to protect ecosystems, species, and biodiversity in general and not just the individuals that compose them. Capturing all the remaining California condors for captive breeding purposes harmed the individuals but promoted the welfare and continuation of the species. Aldo Leopold's (see Chapter 3.6 Figure 1) moral maxim – "A thing is right when it tends to preserve the integrity, stability and beauty of the biotic community. It is wrong when it tends otherwise" – epitomizes this holistic environmental ethic (1949/1966: 240).

The philosophy of "deep ecology" developed by Norwegian philosopher and climber Arne Naess (1973) rejected the anthropocentrism of the "shallow ecology movement" seen as "the fight against pollution and resource depletion" aimed at "the health and affluence of people in the developed countries." According to deep ecology, reducing the richness and diversity of nonhuman life can only be justified by vital human needs. Human influence on the Earth is excessive and the flourishing of human life and cultures requires fewer people and an emphasis on quality of life rather than quantity of material possessions. Deep ecology also rejected "atomistic individualism", arguing for an expanded conception of self by which harm to nature was felt to be harm to oneself. (See deep ecology platform: http://www.deepecology.org/platform.htm).

**Figure 3.16.2** Aldo Leopold.

Courtesy of the Aldo Leopold Foundation, www.aldoleopold.org

Many environmental philosophers (including Naess) have been inspired by wild nature and so "wildness" or "naturalness", understood as the extent to which nature has not been influenced by humans, has become a key value in many environmental philosophies. Ecosystems and other natural entities with causal continuity with deep natural history not significantly altered by humans are valued more highly in virtue of this lack of humanization. A geyser whose regular explosion and magnificent height was not manipulated by park personnel is of greater value in virtue of its naturalness than one regulated by humans. Defenders of naturalness value object to calls to manage the planet in the so-called "Anthropocene" – perhaps by geoengineering the climate as a response to climate change – whether to benefit humans or nonhumans, arguing instead for restoration and rewilding human-damaged natural systems.

This focus on naturalness has elicited powerful criticisms. Many argue that humans are thoroughly a part of nature and object to the separation implied by the special valuing of things uninfluenced by humans. Some argue that, in the wake of long-standing human impact on earth and the more recent global pollution epitomized by climate change, there is no naturalness left to value (McKibben, 1989). Others argue that the focus on naturalness has led to the pernicious and elitist idea that the key environmental goal is to set aside and protect wild land (Cronon, 1995). Such a narrow-minded 'wilderness environmentalism' ignores the importance of improving environmental conditions in cities and eliminating the environmental racism whereby minority communities suffer a wildly disproportionate amount of pollution and resource degradation. It is also argued that the emphasis on setting aside wild lands, particularly in the developing world, amounts to cultural imperialism, for it often displaces people from lands on which they have long resided and forces relatively trivial environmental values onto cultures whose priority should be the economic development necessary to raise their populations out of poverty (Guha, 1989).

Questions of environmental justice have been a central focus, including justice between species, between human generations, and between the industrialized and developing countries. The problem of climate change highlights these injustices quite clearly for present humans get the benefits of a cheap, polluting, fossil-fuel energy source while pushing the costs of a destabilized climate onto our decedents who are powerless to object. Further, because the industrialized world's wealth has been achieved by exploiting and nearly exhausting the earth's resources and capacities for climate stabilization, the developing countries are being asked to constrain their opportunities for development for the sake of protecting what remains of those common resources (such as forests, biodiversity, and the atmosphere's ability to benignly absorb $CO_2$).

Environmental philosophy has also been exploring the various meanings of such key environmental concepts as nature, biodiversity, and sustainability. It has critically examined economic approaches to the environment, particularly the ideas that all environmental values can be monetized, that environmental decisions should be made based on cost-benefit analyses, and that the solution to the tragedy of the commons is to privatize public goods (Sagoff, 1988). Ecological feminism has argued that environmental degradation and the subordination of women are interlocking problems resulting from a common, masculine, and hierarchical logic of domination whereby some things are superior to others (male over female, rationality over emotion, mind over body, humans over nature) thereby justifying oppression of the inferior (Plumwood, 1993). Environmental virtue ethics has argued that it is important to address environmental problems with the concepts of virtue (humility, compassion, restraint) and vice (arrogance, insensitivity, greed) and to focus on questions of what sort of people we should be, rather than simply what we should do (van Wensveen, 1999). Much fruitful work has also been done in environmental aesthetics, both about the nature of natural beauty compared with artistic merit and about what role nature's aesthetic value should play in environmental protection (Carlson, 2000).

## Learning resources

The Environmental Ethics entry from Stanford Encyclopedia of Philosophy is most instructive: <http://plato.stanford.edu/entries/ethics-environmental>.

A useful collection of articles covering the diversity of the field: Jamieson, D. (1991) *A Companion to Environmental Philosophy*. New York: Wiley-Blackwell.

The two major societies of philosophers working in environmental philosophy:

International Association for Environmental Philosophy: https://environmentalphilosophy.org.

International Society for Environmental Ethics: https://enviroethics.org.

## Bibliography


Aristotle, *Politics*. Online. Available at: http://classics.mit.edu/Aristotle/politics.html.

*The Bible*

Callicott, J.B. (1980) 'Animal Liberation: A Triangular Affair', *Environmental Ethics*, 2(4): 311–338.

Carlson, A. (2000) *Aesthetics and Environment*. New York: Routledge.

Carson, R. (1962) *Silent Spring*. Boston: Houghton Mifflin.

Commoner, B. (1971) *The Closing Circle: Nature, Man, and Technology*. New York: Knopf.

Cronon, W. (1995) 'The Trouble with Wilderness; Or, Getting Back to the Wrong Nature', in *Uncommon Ground: Toward Reinventing Nature*. New York: W.W. Norton & Co.

Guha, R. (1989) 'Radical American Environmentalism and Wilderness Preservation: A Third World Critique', *Environmental Ethics*, 11(1): 71–83.

Leopold, A. (1949/1966) *A Sand County Almanac*. New York: Oxford University Press.

McKibben, B. (1989) *The End of Nature*. New York: Anchor Books.

Mill, J.S. (1848) *Principles of Political Economy*. Online. Available at: http://www.econlib.org/library/Mill/mlP.html.

Muir, J. (1901) *Our National Parks*. Boston: Houghton Mifflin.

Muir, J. (1911) *My First Summer in the Sierra*. Boston: Houghton Mifflin.

Naess, A. (1973) 'The Shallow and the Deep, Long Range Ecology Movements. A Summary', *Inquiry*, 16(1): 95–100.

Norton, B. (1991) *Toward Unity Among Environmentalists*. New York: Oxford University Press.

Plumwood, V. (1993) *Feminism and the Mastery of Nature*. New York: Routledge.

Regan, T. (1983) *The Case for Animal Rights*. Berkeley: University of California Press.

Rolston, H. (1994) *Conserving Natural Value*. New York: Columbia University Press.

Sagoff, M. (1988) *The Economy of the Earth*. New York: Cambridge University Press.

Schweitzer, A. (1923) *Civilization and Ethics*. London: A. & C. Black.

Singer, P. (1975) *Animal Liberation*. New York: Harper Collins.

Taylor, P. (1986) *Respect for Nature*. Princeton, NJ: Princeton University Press.

Thoreau, H.D. (1854) *Walden*. Online. Available at: http://thoreau.eserver.org/walden00.html.

Thoreau, H.D. (1862) 'Walking'. *The Atlantic*. Online. Available at: http://www.gutenberg.org/files/1022/1022-h/1022-h.htm.

van Wensveen, L. (1999) *Dirty Virtues: The Emergence of Ecological Virtue Ethics*. New York: Humanity Books.

White, L. (1967) 'The Historical Roots of Our Ecologic Crisis', *Science*, 155(3767): 1203–1207.

# Environmental planning

3.17

## Iain White

As knowledge concerning the effects of human activity has grown over time, so has the necessity to manage human impacts effectively. Environmental planning is a mechanism to influence decisions relating to land and resources in order to balance the freedom to develop against the need to consider the consequences for society as a whole. It has a wide role that encompasses elements such as permitting new buildings or regulating certain polluting activities, and deciding how future development should occur, like creating protected areas, or changing the use of land to allow for economic growth. It is therefore both reactive in determining development proposals and proactive in shaping how land should be used in future decades.

The easiest way to see the value and purpose of environmental planning is to take a moment to reflect upon your local neighbourhood or city and think about how land and resources are currently used – from where people live, work and play, to the quality and quantity of green space, to the ways that people or wildlife can move around – all of that was partly shaped by environmental planning. Now imagine a society without those policies, practices and processes. How may investments, resources or impacts be distributed differently? Would green spaces be built on, would the water or air quality be lower, would there be more neighbourly disputes due to insensitive building?

While there is no clear answer to these questions, the exercise is useful as it enables the discussion to turn to how environmental planning is much more than a series of statutory or non-statutory tools. It acts on behalf of society to mitigate some of the consequences of capitalism, such as those associated with the inherent desire to minimise costs or externalise pollution. As such, it can lay a strong claim to be a 'public good' that brings often unseen benefits that may last for generations.

This perspective gives the field a broad spatial and temporal remit that ties it closely to geographical thinking, while also linking it to related fields such as politics, economics, or law. Taking a step backward to see the bigger picture we can now see how the field is situated in a challenging interdisciplinary location that links land and resources, people and places, and politics and power. While there are many aspects to environmental planning, it may be simply defined as a mechanism to help societies manage how space and natural resources are used over time. White (2015: 12) sums it up in this way: "to live on this planet is to affect it: societies need to develop land and consume resources, and there will be impacts from this, but where, who or what is affected, and to what extent?" Environmental planners seek to answer this question in a reasoned and accountable way.

This definition also serves to illustrate the scope, connectivity and complexity of environmental planning, which in theory has significant power, but is tempered by the real world political realities that need to balance public and private interests, and deliver what are sometimes competing economic, environmental and social objectives.

## The development of environmental planning

The birth of what we would now recognise as environmental planning began during the midst of the Industrial Revolution. Population growth and industrial activity were producing both significant changes to the environment and providing a poor quality of life for people. In the United Kingdom, the initial response was on regulating water pollution and improving conditions in the urban slums, while in the United States conservation of wilderness areas was an early policy, leading to the formation of the world's first National Park, Yellowstone, in 1872.

During the twentieth century the scale and rationale of environmental planning continued to develop in scope and sophistication. Daniels (2009) handily categorises the major ideas that emerged in the United States during this time as being a series of sequential eras, starting with a primary focus on parks and conservation, being gradually influenced by the rise of science, ecological knowledge and assessments, to the more recent arrival of sustainability and global concerns.

Appreciating the historical development of the subject is important as it emphasises two aspects. First, that the mandate and methods of environmental planning are rooted in the environmental, social and political contexts of the time, as such it will continue to change in conjunction with aspects, such as the nature of the impacts, changes in science and technology, or societal expectations. Second, that the scale and complexity of both problems and responses has changed dramatically since its original inception. This issue is worth exploring a little further.

## Changing scale and complexity

In contrast to the visible and localised human impacts of a century ago, environmental planning in the current context provides a more substantial and complex managerial challenge. The spatiality of environmental concerns now ranges from the very local to encompass our entire biosphere, while the temporal remit is similarly wide-ranging and considers intergenerational equity and the possible effects on people who are not yet born.

This creates a friction between the nature of problems and the means to address them. Not only do global issues, such as climate change, mesh poorly onto the traditional boundaries and policies of national, regional and local governments, but they have opaque and dispersed causal chains when compared to the more identifiable smoke, pipes or deforestation of a century ago. As such, it can be frequently hard to link problems to reasons and potential solutions. For instance, just who is responsible for climate change and how, specifically, should they change their behaviour? Or precisely which farm in the catchment is causing diffuse pollution of a watercourse?

The governance of environmental planning is also much more fragmented than in the past. No longer does the national or local state simply solve problems by regulation, now sectors as wide ranging as forestry, energy, water or agriculture are all seen as playing a role in improving the environment, and partnership working with the private sector or community organisations is commonplace. In part this is a response to the nature of concerns, which may challenge traditional managerial boundaries, but it also relates to changes in politics more generally, which has witnessed a shift away from a centralised, top-down approach to environmental management. The state now plays more of an enabling role that can achieve objectives by working with stakeholders to engender behaviour change, design best practice examples, or supply incentives, rather than relying merely on the blunt laws or regulations that are more obvious.

These aspects emphasise that environmental planning is an intensely political activity. You may initially appreciate how a mechanism associated with determining development proposals or the protection of places will involve a degree of conflict between parties, but the discipline itself is subject to wider political debates concerning its power, approach or priorities. For example, cross border issues may demand negotiation between political actors and agencies, as without a consistent approach you may simply move an unsustainable practice from one area to another where rules are not so restrictive.

As such, the spatiality of environmental plans and policies also links to the spatiality of impacts and capital. Countries and cities are in economic competition and so the perceived 'red tape' in one place can serve to move capital elsewhere. While this may lead to fears of a 'race to the bottom', it is increasingly apparent that if the environment isn't attractive it can be difficult to attract skilled workers or mobile multi-nationals. Therefore, the notion of environmental planning mediating a conflict between economic and environmental interests is a little simplistic. In reality, good environments can attract capital and many citizens value the environment whether they happen to run a business or not.

## Conclusion

At first glance, environmental planning appears to be a very practical subject concerned with using designation, regulation or consents to manage resource use. But by delving a little deeper we can see how there are multiple geographies and spatial tensions at play, from the scale of problems, to the politics and administrative areas of the agencies concerned, to the boundaries of natural ecosystems (Haughton, 2017). As a consequence, environmental planning targets some problems more successfully than others. Regulating industrial activity at the site scale is much easier than tackling problems that may be rooted beyond national boundaries and demand international agreements. Indeed, the continued existence of many environmental issues despite scientific awareness and public concern signifies that they may fit poorly onto the means by which societies manage land and resources.

Above all, environmental planning is a state-led activity. As such it is influenced by wider trends in economics, politics and science that both reflect the extent to which the state should intervene in how markets or businesses operate, and what is of value to citizens. After all, the environment doesn't have a voice, it relies on society to do that, and environmental planning is one of the ways by which we collectively manage that shared resource.

## Learning resources

This report is a good source for learning more about the benefits of environmental planning and the various instruments and tools that can be used. The Cities Alliance (2007) *Liveable Cities: The Benefits of Urban Environmental Planning*. Retrieved from: http://www.unep.org/urban_environment/PDFs/LiveableCities.pdf.

This link provides a resource to a variety of articles that discuss planning and sustainability http://www.sustainablecitiescollective.com/all/11?ref=navbar.

The United Nations Urban Environmental Planning site provides a quick explanation of the benefits of environmental planning and has more links to other relevant policy and resources: http://www.unep.org/resourceefficiency/Policy/ResourceEfficientCities/FocusAreas/UrbanEnvironmentalPlanning/tabid/101663/Default.aspx.

## Bibliography

Daniels, T.L. (2009) 'A Trail Across Time: American Environmental Planning from City Beautiful to Sustainability', *Journal of the American Planning Association*, 75(2): 178–192.

Haughton, G. (2017) 'Environmental Planning,' in *The International Encyclopedia of Geography* (pp. 1–7). Washington, DC: American Association of Geographers.

White, I. (2015) *Environmental Planning in Context*. Basingstoke: Palgrave Macmillan.

3.18

# Environmental psychology

Patrick Devine-Wright

Environmental Psychology is the research, teaching and practice conducted by psychologists on both the built and natural environment. This chapter presents an overview of this academic sub-discipline since its origins in the 1960s. It is structured in terms of a brief introduction and history of the field, key theories and methods, topics of research and conclusions.

## Origins and place within psychology

The discipline of Psychology emerged in the late nineteenth century, but it was in the 1960s that a movement arose to direct psychological research attention towards the environment. The early field was dominated by an interest in the built environment, reflected in attempts to inform and collaborate with architects, and publications such as *Ecological Psychology* (Barker, 1968), *Psychology for Architects* (Canter, 1974) and 'The Theory of Affordances' (Gibson, 1977). Research was motivated by a concern that design had significant consequences for human well-being and could be usefully informed by psychological theories and research. Nearly half a century later, the built environment remains a prominent topic of research by environmental psychologists. However, alongside this, interest has grown in the contribution psychological research can play towards conservation of the natural environment (Gifford, 2014). The topic of pro-environmental behaviour – understanding its causes and informing the design of interventions to foster behavioural change – is now driving research by psychologists worldwide.

There are two principal journals in the field: the *Journal of Environmental Psychology* and *Environment and Behaviour*, several handbooks and reviews have been published (e.g. Stokols & Altman, 1987; Bechtel & Churchman, 2002; Clayton, 2012; Gifford, 2014) and umbrella organisations organise regular conferences (e.g. International Association of People-Environment Studies; International Association of Applied Psychology, Environmental Psychology Division). Although established and thriving, it remains a minority interest across the discipline of psychology as a whole (Gifford, 2014). Few university psychology departments have staff with specialised training in environmental psychology and few postgraduate training courses are available (see the learning resources below for further info).

## Theory and method

Environmental psychology is similar to many social science disciplines in its interest in human–environment relations, yet is distinguished by two fundamental attributes: a focus upon the individual as the unit of analysis and an emphasis upon a scientific approach to research.

A long-standing debate concerns how human–environment relations should be theorised. Much early research in environmental psychology was concerned with investigating the impacts of environments on human experience and behaviour, focusing upon issues such as stress and adaptation in noisy or overcrowded settings. Subsequent research has moved away from a deterministic view of human–environment relations. This orients the sub-discipline towards a more transactional approach that recognises human impacts on the environment as well as environmental impacts on people and proposes research on how these develop over time. More recently, informed by theory from social psychology, researchers have investigated attitudes towards the environment using questionnaire survey methods, seeking to better understand why behaviours occur, and how these can be changed.

The positioning of environmental psychology as a science reflects the broader emphasis in psychology upon the importance of robust measurement, researcher objectivity, hypothesis testing, replication of findings and the identification of cause-effect relations. Taken together, the focus on the individual and the adoption of a scientific approach makes environmental psychology distinctive from other approaches to human–environment relations, including environmental sociology, which focuses upon societies and social groups rather than individuals and approaches such as 'psychogeography' and 'ecopsychology' that are less scientific and empirical.

Environmental psychology has always been a field-based discipline of inquiry, investigating how certain environments impact human experience and behaviour, and in turn how people shape those environments over time. Research on these human–environment 'transactions' has been conducted in locations such as national parks, hospital wards, open-plan offices, school classrooms, railway stations and football grounds. The field also has a strong applied focus, seeking to inform better design of the built environment or strategies to change environmentally impactful behaviour. For these reasons, unlike the discipline of psychology more generally, relatively few environmental psychological studies involve experiments conducted in laboratory environments.

Environmental psychologists have devised ways to adapt experimental methods to field contexts and to address environmental problems. An example is research on the topic of pro-environmental behaviour, which has compared the impact of different interventions undertaken in 'matched' locations (e.g. the provision of information to residents, the provision of incentives, new facilities, attempts to change social norms) that are compared against a 'control' location where interventions did not take place. Research has investigated the impact of 'block leaders' (i.e. local residents who take on the active role of encouraging other residents to recycle waste products) on rates of domestic recycling has adopted this design. A landmark study (Hopper & Nielsen, 1991) showed that the actions of block leaders can lead to an increase in recycling, and this body of research has informed subsequent policies and municipal interventions to this day.

Although quasi-experimental methods of this kind reflect the predominant approach, it would be a mistake to assume that *all* Environmental Psychologists work in this way. Some researchers take a less individualistic approach to human–environment relations by focusing upon discourse – ways of talking about human–environment issues problems that are shared across a group or culture and accessible through communication (e.g. media articles, conversations within focus groups). Some adopt qualitative methods that aim to reveal how individuals experience environmental problems in their own words. Others use a mix of quantitative and qualitative methods. Therefore, considerable diversity has and continues to exist in how environmental psychologists approach research.

## Topics of research

The topics of environmental psychology research are highly diverse. Some researchers take fundamental psychological processes and apply them to environmental issues and problems, for example perception, values, personality traits, attachment, identity, learning, childhood experience, norms and habits (Gifford, 2014). Below, some of the ways that these processes have been applied to research on built and natural environments are presented.

### *Built environments*

Initial research focused on how people come to know and learn about the environment, researching topics such as environmental perception and cognition. This led to discoveries about differences between expert and lay understandings of the built environment and how formal training in architecture produces these differences (Wilson, 1996). A substantial body of literature has investigated wayfinding in cities, drawing on the concept of 'cognitive mapping' to reveal how individuals orient themselves around landmarks and routes and move through space (Kitchin, 1994). Research on tragedies in locations such as football grounds and railway stations revealed how individuals and groups move through space in stressful circumstances, challenged popular assumptions (e.g. about 'panic') and informed policy and design (Canter et al., 1989). A related theme is building appraisal, where research has developed standardised tools to investigate the quality of built environments from the experience of the users, including the impact of design features such as open-plan offices on employee satisfaction. At a broader level, researchers have investigated residents' satisfaction with their neighbourhood and what are the key dimensions underlying their evaluation. Related to this, studies have investigated the meaning of 'home' (Moore, 2000), how residents become emotionally attached to places that come to form an important part of their self or identity (Altman & Low, 1992) and the implications of changes to these places (Manzo & Devine-Wright, 2014).

## *Natural environments*

Environmental psychology has revealed how we make sense of the environment, identifying underlying dimensions including complexity, symmetry etc. Research has investigated preferences for landscape types, using photography to reveal the consistent effects of 'green' and 'blue' environment features on judgements of landscape quality. Researchers have identified and examined the impacts of restorative environments, revealing the qualities of places that restore depleted resources and promote psychological well-being (Kaplan & Kaplan, 1989). Concerned with the increasing degradation of the natural environment, psychologists have investigated the causes of environmentally significant behaviour, distinguishing between different forms of behaviour (Stern, 2000) in domains such as waste, energy conservation, transport and water consumption. Research has concluded that enduring behavioural change requires multi-dimensional interventions that combine strategies such as education, incentives, new facilities and regulations (Gardner & Stern, 2002). These studies have been developed in relation to climate change, focusing on both adaptation and mitigation. Adaptation research has investigated resilience and willingness to take preventative action to reduce vulnerabilities (Clayton et al., 2015). Mitigation research has investigated willingness to adopt behaviours that reduce greenhouse gas emissions such as choosing public transport over private car use (Steg, 2003), accepting renewable energy projects nearby (Devine-Wright & Howes, 2010) and using less energy in the home (Abrahamse et al., 2005).

# Conclusions

Environmental Psychology is a thriving area of research investigating many important environmental challenges, concerning both built and natural environments. Research is founded upon the adoption of rigorous, theory-driven empirical research methods and aims to inform policies and improve human well-being.

# Learning Resources

The American Psychologcal Association's website features more information about the field:

http://www.apa.org/action/science/environment.

For an example of an accredited training programme at the University of Surrey, see:

http://www.surrey.ac.uk/postgraduate/environmental-psychology-2017.

For a real-world example of the application of the Block Leader idea, see:

http://www.ci.minneapolis.mn.us/solid-waste/recycling/Recycling_Block_Leader.

# Bibliography


Abrahamse, W., Steg, L., Vlek, C. & Rothengatter, T. (2005) 'A review of intervention studies aimed at household energy conservation', *Journal of Environmental Psychology*, 25: 273–291.

Altman, I. & Low, S. (1992) (Eds) *Place Attachment*. New York: Plenum.

Barker, R. (1968) *Ecological Psychology: Concepts and Methods for Studying the Environment of Human Behaviour*. Palo Alto, CA: Stanford University Press.

Bechtel, B. & Churchman, A. (2002) *Handbook of Environmental Psychology*. New York: Wiley.

Canter, D. (1974) *Psychology for Architects*. London: Elsevier.

Canter, D., Comber, M. & Uzzell, D. (1989) *Football in its Place: An Environmental Psychology of Football Grounds*. London: Routledge.

Clayton, S. (2012) (Ed.) *The Oxford Handbook of Environment and Conservation Psychology*. Oxford: Oxford University Press.

Clayton, S., Devine-Wright, P., Stern, P.C., Whitmarsh, L., Carrico, A., Steg, L., Swim, J. & Bonnes, M. (2015) 'Psychological research and global climate change', *Nature Climate Change*, 5: 640–646.

Devine-Wright, P. & Howes, Y. (2010) 'Disruption to place attachment and the protection of restorative environments: a wind energy case study', *Journal of Environmental Psychology*, 30(3): 271–280.

Gardner, G. & Stern, P. (2002) *Environmental Problems and Human Behavior*. Boston, MA: Allyn & Bacon.

Gibson, J.J. (1977) 'The Theory of Affordances'. In R. Shaw and J. Bransford (Eds) *Perceiving, Acting and Knowing* (pp. 67–82). New Jersey: Lawrence Erlbaum Associates.

Gifford, R. (2014) 'Environmental psychology matters', *Annual Review of Psychology*, 65: 541–580.

Hopper, J.R. & Nielsen, J. (1991) 'Recycling as altruistic behavior: normative and behavioral strategies to expand participation in a community recycling program', *Environment and Behavior*, 23: 195–220.

Kaplan, R. & Kaplan, S. (1989) *The Experience of Nature: A Psychological Perspective*. New York: Cambridge University Press.

Kitchin, R.M. (1994) 'Cognitive maps: what are they and why study them?', *Journal of Environmental Psychology*, 14: 1–19.

Manzo, L. & Devine-Wright, P. (2014) *Place Attachment: Advances in Theory, Method and Application*. Abingdon: Routledge.

Moore, J. (2000) 'Placing home in context', *Journal of Environmental Psychology*, 20: 207–217.

Steg, L. (2003) 'Can public transport compete with the private car?', *IATSS Research*, 27(2): 27–35.

Stern, P.C. (2000) 'Toward a coherent theory of environmentally significant behavior', *Journal of Social Issues*, 56(3): 407–424.

Stokols, D. & Altman, I. (1987) (Eds) *Handbook of Environmental Psychology*. New York: John Wiley & Sons.

Wilson, M.A. (1996) 'The socialisation of architectural preference', *Journal of Environmental Psychology*, 16: 33–44.

# 3.19 Natural hazards research

Greg Bankoff

Since World War II, research on natural hazards has been dominated by two concepts: vulnerability or why certain groups of people are unequally exposed to risk, and resilience or how some people are better able to deal with that exposure. These are historical concepts born out of the intellectual debates about how to explain the new world order that emerged post-1945 and its tripartite division between First (industrialised), Second (Communist), and Third (newly emergent states) Worlds. As regards the non-Western world, during the Cold War, these regions were generally depicted as vulnerable to natural hazards, and, then, following the collapse of the Soviet Union in 1991, increasingly recast as resilient. Natural hazard research over these decades duly reflected this momentous political shift in world affairs.

## Vulnerability

Vulnerability as a concept emerged as the Cold War intensified during the 1970s. Previous research on natural hazards had largely attributed disasters to physical or meteorological causes with little regard to the state of the human populations affected. Vulnerability's chief proponents were motivated by concern for the plight of citizens in the newly denominated Third World and shared a growing suspicion of the development policies pursued by Western governments and transnational corporations in these new nations. By demonstrating that there was nothing "natural" about disasters and that people were put at risk as much by the political and social structures of the societies in which they lived as by any physical hazard or event, scholars began to question the hitherto unchallenged assumption that the growing incidence of disasters was due to a rising number of natural phenomena. In the process, they offered a critique of both the means and the intent behind development. Rather than lifting people out of poverty, these policies too often made of their life a "permanent emergency". Instead, the emphasis was on what rendered communities unsafe, a condition that depended primarily on a society's social order and the relative position that a particular group occupied within it. The term coined to assess the nature and extent of this risk was "vulnerability", where the latter is not only a gauge of people's exposure to hazard but also a measure of its capacity to recover from loss.

The Cold War origins of the term begin with its definition or, rather, the way vulnerability is applied in practice. Everybody is made vulnerable to some extent by a combination of variables such as class, gender, age, disability and the like that affects their entitlement to basic necessities and their empowerment to enjoy fundamental rights. However, certain people were more vulnerable than others and lived mainly in vulnerable places that lay in the developing world subjected to the modernisation projects of the post-World War II era. This meaning was made clear in the Pressure and Release Model (PAR) that explained not only how risk is directly attributable to the physical hazard but also the extent to which the social order disadvantages people over time. By offering a framework for linking the impact of hazards to a series of societal factors that generate vulnerability, the PAR model exposed the historical processes that transformed colonial territories into the states of the Third World. The critique was unequivocal: imperial heritage, development policies, and unequal power relationships rendered some communities less able to deal with disasters and left them more at risk.

Vulnerability offered a means of critiquing developmentalism and the untrammelled pursuit of material prosperity that became the dominant model of economic progress after 1945. Nations were increasingly assessed in terms of their development or lack of it and some societies began to be regarded as underdeveloped, a condition associated with backwardness, poverty and, implicitly, vulnerability. The Third World was not only disease-ridden and poverty-stricken but it was also increasingly disaster-prone, a zone where repeated hazards inflicted upon people sudden death and damaging losses that left communities physically weak, economically impoverished, socially dependent and psychologically harmed. Development was supposed to ameliorate the conditions that put people at risk but largely failed to do so because it was too much a part of the root causes that underlay societies' vulnerability in the first place. In this newly-constructed Third World, many people began to perceive development projects as disadvantageous rather than beneficial.

## Resilience

With the end of the Cold War in 1991, the emphasis in research on how societies should be viewed began to shift. Societies were seen as no longer simply vulnerable but people began to be considered as primarily resilient with capacities to organise, resist, learn, change and adapt. A change of discourse was politically advisable in the new international climate. The rationale behind Overseas Development Aid (ODA) and projects funded by the World Bank to contain the spread of Communism were no longer required. Foreign direct investment and private capital flows began to replace ODA as the favoured development paradigm. At the heart of the new approach was an emphasis on the importance of macroeconomic stability and integration into the international economy. Under this new financial regime, funding was made conditional on fiscal discipline, tax reform, trade liberalisation, privatisation, deregulation and a reduced role for the state. In this new political climate, it was expedient to stress what made people resilient rather than what made them vulnerable.

If research that emphasised vulnerability was a product of the Cold War, to what extent was the subsequent focus on resilience a result of neoliberalism? The uncomfortable truth, as scholars have recently pointed out, is that the two discourses have much in common and share many policy approaches even if for different reasons. The neoliberal agenda envisages a state whose primary responsibility is the creation of a fully functioning market. It effectively devolves public safety to civil society and expects the market and non-state actors to meet the social needs of the population. As regards disaster management, the state increasingly depends on NGOs to fulfil the public safety roles it wishes to divest. Reframing the state's responsibilities in this manner casts poverty largely as a voluntarily choice: the poor choose to be poor and only have themselves to blame for being poor. Likewise, those who are vulnerable choose to be vulnerable and have only themselves to blame for being vulnerable. The proponents of neoliberalism regard social responsibility as optional, and vulnerability as voluntary: "resilient" people do not have to look to the state to secure their well-being as they have already made themselves secure.

The commonalities in practice between a neoliberal agenda and the shift from vulnerability to resilience in research led to a new emphasis on disaster risk reduction (DRR) and a focus on community-based disaster risk management. DRR began to emerge in the 1980s and has gradually become the most dynamic discourse in the global policy field of disasters even if it still remains "marginal" to mainstream international development. DRR prioritises improving the quality and security of people's lives by improving livelihoods and increasing social mobilisation as identified in the 2005 Hyogo Framework for Action. Although for very different intentions, neoliberalism and resilience end up advocating much the same approach by much the same methods. They both emphasise an active citizenship whereby people take responsibility for their own social and economic well-being, and they both share a general distrust of centralised state systems and a desire to decentralise responsibilities. The emphasis is on local capacity, local decision making, local responsibility and, of course, local funding. To one, however, this championing of civil society is a way to disguise the imposition of market discipline; to the other, resilience is a continuing critique of existing international development and aid programmes.

## Conclusion

Both vulnerability and resilience are concepts in natural hazards research that originated at a particular moment in recent history. Their meaning was shaped by a particular historical perspective and their significance can only really be understood through a consideration of the way power operated at the time in the prevailing socio-environmental systems. If vulnerability expressed a profound unease with the developmental model that dominated the Cold War era and that depicted natural hazards as largely physical events for which there were mainly technical solutions, then the subsequent discourse of resilience fitted well with pre-established neoliberal ideas about competition and entrepreneurship that viewed disaster impacts after the collapse of Communism as largely the result of individual choice. The current emphasis in research on "adaptation" implies accepting a world in which disturbance and crisis are constant features whether caused by climate change and/or social upheaval. It is also one where there is a continual need for neoliberally-sanctioned discourses about resilience and change. It accepts disaster as an endemic condition in anticipation of which society must remain in a permanent state of high alert. It is also a profoundly conservative discourse that largely obscures questions about the role of power and culture in society, and about whose environments and livelihoods are to be protected and why.

## Learning resources

EM-DAT The International Disaster Database, Centre for Research on the Epidemiology of Disasters (CRED): the most complete data on historical and contemporary natural hazards http://www.emdat.be.

NASA Global Exchange Master Directory: largely US-focused but still a useful resource http://gcmd.gsfc.nasa.gov/learn/pointers/hazards.html.

UNISDR PreventionWeb: worldwide resources http://www.preventionweb.net.

## Bibliography

Bankoff, G. (2001) Rendering the World Unsafe: 'Vulnerability' as Western Discourse, *Disasters*, 25(1): 19–35.

Blaikie, P., Cannon, T., Davis, I. and Wisner, B. (1994) *At Risk: Natural Hazards, People's Vulnerability and Disasters*. London and New York: Routledge.

Hannigan, J. (2012) *Disasters without Borders*. Cambridge: Polity Press.

Hewitt, K. (1997) *Regions of Revolt: A Geographical Introduction to Disasters*. Edinburgh: Longman.

O'Keefe, P., Westgate, K. and Wisner, B. (1976) Taking the Naturalness out of Natural Disasters, *Nature*, 260: 566–567.

United Nations (2005) *Hyogo Framework for Action 2005–2015: Building the Resilience of Nations and Communities to Disasters, Extract from the Final Report of the World Conference on Disaster Reduction A/CONF.206/6*. Geneva: United Nations International Strategy for Disaster Reduction.

# 3.20 Science and technology studies

Sergio Sismondo

Science and Technology Studies (STS) is an interdisciplinary field that examines science, technology and closely connected areas as social and material activities. The central stance of STS is the idea that all aspects of science and technology can be studied and understood as *constructed* – that is, made by particular people with particular aims, operating in specific contexts using certain materials. In part, the development of STS has been a history of increasing the scope of this *anti-essentialism* about science and technology, starting with scientific knowledge, and expanding to artifacts, methods, observations, phenomena, classifications, institutions and cultures. What follows is a sketch of this expansion of STS's attention.

## Demystifying science

In the 1970s, several different lines of thinking arrived at the position that scientific knowledge could be explained historically and sociologically, as the outcome of social, material and contingent processes. Whereas popular and then-standard accounts tended to portray scientific developments as if they were processes of the rational uncovering – in some universal sense of 'rational' – of facts, early STS accounts understood scientific developments in causal and relative terms, where cultures, ideologies, interests and even idiosyncrasies combined to produce what were or are taken to be.

Studies of scientific controversies make the constructed nature of scientific knowledge particularly clear, because the parties during controversies interpret any or all of materials, data, methods and theories differently, often providing reasons for their interpretations, and reasons to disregard their opponents' interpretations. For example, in the 1970s and 1980s there was a debate about the efficacy of vitamin C as a treatment for cancer, a debate studied by STS researcher Evelleen Richards. The initial clinical trials, in the UK, found that high doses of vitamin C improved the health and well-being of patients; the hypothesis was that vitamin C helped the immune system to control the effects of the cancer, though it did not attack the cancer itself. Clinical trials in the US, performed to disprove this apparent 'quackery', found no positive effect. However, the original researchers argued that the US patients were significantly different from the UK ones, having received prior chemotherapy and thus having already weakened immune systems; the difference was in part a difference between norms and expectations of the two health systems. In addition, the proponents and opponents disagreed about the idea of 'treatment', with the former measuring quality of life and the latter measuring survival rates. The proponents argued that other aspects of the methodologies of the trials importantly differed, and the opponents argued that they were essentially the same. Although mainstream medicine eventually dismissed vitamin C as a treatment for cancer, the original researchers maintained, and had apparently strong arguments for their position, that their experiments were never properly replicated and their results were never refuted.

The natural world does not by itself, then, determine scientific knowledge. Case studies show that controversies are resolved by actions that define one position as the right and reasonable one for members of the expert community to hold. Thus pieces of accepted scientific knowledge contain ineliminable references to particular social configurations.

## Varieties of STS

STS is methodologically diverse, valuing a variety of anthropological, historical, philosophical and sociological approaches. However, the field has particularly embraced ethnographic and participant-observer studies of scientists and engineers at work. In the 1970s and early 1980s, a number of researchers – most prominently Harry Collins, Karin Knorr-Cetina, Bruno Latour, Michael Lynch, and Sharon Traweek – entered laboratories to watch and participate in the work of experimentation, the collection and analysis of data, and the refinement of claims. Their results were strikingly productive, drawing attention to the importance of mundane and specialized skills of manipulation and observation, they negotiated the nature of data and results that could be published, and the roles of local and broader cultures. Close empirical attention to scientific work challenged many entrenched assumptions about the nature of science.

Not only data are constructed in laboratories, but even phenomena themselves are: inputs are extracted, refined, or invented for the purpose, are shielded from outside influences and placed in innovative contexts. Laboratory phenomena, then, are not in themselves natural, but are made to stand in for nature. In their purity and artificiality they are typically seen as more fundamental and revealing of nature than the natural world itself can be. The general acceptance of laboratory phenomena is itself the outcome of contingent historical processes, starting in the seventeenth century.

In these observations we can see an attention to not just the constructed nature of ideas and institutions, but also of the material world. Several different traditions in STS have paid particular attention to this.

Trevor Pinch and Wiebe Bijker's transfer of concepts from the study of science to the study of technology, under the label the 'Social Construction of Technology' (SCOT), argued that the success of a technology depends upon the strength and size of the groups that take it up and promote it. Even a technology's definition is a result of its interpretation by that and other 'relevant social groups': artifacts may be interpreted flexibly, because what they do and how well they perform are the results of competing goals or competing senses of what they should do. Thus SCOT points to contingencies in the histories and meanings of technologies, contingencies on actions and interpretations by different social groups.

Actor-Network Theory (ANT), developed by Michel Callon, Bruno Latour, John Law and others, further broadens that picture by representing the work of technoscience as the attempted creation of larger and stronger networks. Technoscientific actors build networks that become machine-like when their components are made to act together to achieve a consistent effect, or that become facts when their components are made to act as if they are in agreement. Distinctive about ANT is that the networks are heterogeneous, including diverse components that span types from materials, equipment, components, people, and institutions. In ANT's networks, bacteria may rub shoulders with microscopes and public health agencies, and experimental batteries may be pulled apart by car drivers and oil companies. All of these components are actors, and are treated as simultaneously semiotic and material; ANT combines the anti-essentialism with which STS began with the materialism of laboratory studies – even though it is often difficult to talk about the actions and effects of non-humans while maintaining anti-essentialism about science and technology. ANT's step in the history of constructivist STS is to integrate human and non-human actors in analyses of the construction of knowledge and things, seeing technoscience and society as inseparable, each contributing to the constitution of the other.

A number of perspectives adopted within STS – for example, symbolic interactionism, ethnomethodology and some feminist positions – have some affinities with ANT, but approach issues differently. For example, Donna Haraway's perspectivalism also emphasizes the conjunctions of humans and non-humans, and the distinct situated knowledges that result from these. Karen Barad's 'agential realism' emphasizes how 'intra-action' produces objective entities, thus allowing understandings of the material and social worlds as parts of and resulting from the same sets of actions. Some scholars taking up Barad's work have contributed to a 'new materialism' popular in environmental studies, though this has had a strong tendency to undercut the anti-essentialism that has been so productive in STS: new materialism scholars often attribute quite definite properties to the natural world, thus replicating, rather than studying, accounts that scientists provide.

For scientific knowledges and technological artifacts to be successful, they must be made to fit their environments or their environments must be made to fit them, or both. The process of adjusting pieces of technoscience and their environments to each other, or of simultaneously creating both knowledge and institutions, is a process of the co-production, in Sheila Jasanoff's terms, of the natural, technical, and social orders. For example, classifications of diseases allow the diagnoses that reinforce those classifications, regimes of scientific knowledge validate industrial over traditional agriculture, and climate science has created institutions that help to establish and address climate knowledge. Part of the work of successful technoscience, then, is the construction not only of facts and artifacts, but of the societies that accept, use, and validate them.

There have been many more extensions of constructivist approaches. For example, Thomas Gieryn's study of 'boundary work' displays the construction and reconstruction of the edges of disciplines, methods, and other social divisions. Efforts to create public policy invariably require translations of relevant knowledge; 'boundary actors' and 'boundary organizations' between science and policy help re-construct the science and the policy. Recent research has examined the construction of public participation and engagement with technological and environmental issues, seeing the nature of engagement in non-essentialist terms. Researchers have even turned from studying the construction of scientific knowledge to studying the construction of ignorance – a project called 'agnotology' by Robert Proctor – such as Naomi Oreskes and Erik Conway's study of doubt around climate change. A challenge that STS researchers are increasingly taking on is to understand the construction of large structures in political economies and ecologies of knowledge, rather than particular developments within those structures. Thus the constructivist STS project continues to find new objects to analyze and new tools of analysis.

## Learning resources

Two helpful general accounts of STS, which can provide references for much of the work described above, are:

Sismondo, S. (2010) *An Introduction to Science and Technology Studies*, 2nd ed. Chichester: Wiley-Blackwell.

Yearley, S. (2005) *Making Sense of Science*. London: Sage Publications.

A select number of the central journals in the field are (in alphabetical order):

*BioSocieties*, http://www.springer.com/social+sciences/journal/41292.
*Catalyst: Feminism, Theory, Technoscience*, http://catalystjournal.org.
*Engaging Science, Technology and Society*, http://estsjournal.org.
*Isis*, http://www.journals.uchicago.edu/toc/isis/current.
*Science and Technology Studies*, http://ojs.tsv.fi/index.php/sts.
*Science as Culture*, http://www.tandfonline.com.
*Science, Technology, & Human Values*, http://sth.sagepub.com.
*Social Studies of Science*, http://sss.sagepub.com.
*Technology and Culture*, http://muse.jhu.edu/journal/194.

# 3.21 Social ecology

Stephen M. Wheeler

The idea that human societies are interwoven and evolving systems similar to those of the natural world has been widespread within Western society since at least the nineteenth century. At a general level this notion underlies academic disciplines such as sociology and political science, shown for example in Karl Marx's view of struggles between social classes and evolving modes of economic production in *Das Kapital* (1867). But within the twentieth and now the twenty-first centuries multiple schools of thought have arisen that specifically employ the term 'ecology' and related concepts to the study of human societies. Such analysis raises the hope that humans can understand and manage their own social systems in more sustainable ways.

## Roots of the concept

Understandings that human and natural systems are dynamic and interwoven have been common within pre-industrial cultures, and within religious belief systems such as Buddhism and Hinduism. Although in the Middle Ages European societies were relatively static, pre-industrial philosophers such as fourteenth-century Islamic historian Ibn Khaldun and sixteenth-century French writer Michel de Montaigne, argued that societies change over time due to cultural, political, social, and economic forces. In later centuries Enlightenment concepts of social progress based on reason, tolerance, and intellectual discourse gained ground. Thinkers such as the Marquis de Condorcet, Auguste Comte, and Henri de Saint-Simon emphasized the historical growth of social organization and humanism, setting the stage for a more ecological understanding of social systems (e.g. Condorcet, 1795).

Theories of evolution and ecology arose in the mid-nineteenth century, leading people to wonder if these concepts might be used to analyze human societies as well. Charles

Darwin, Herbert Spencer, Lewis H. Morgan, and Marx all considered how social systems might evolve in ways analogous to natural systems. The previous western view of 'man' as separate from 'nature' also began to break down. Early conservationist George Perkins Marsh analyzed the effect of human systems on natural ecologies in *Man and Nature* (1864), while early landscape architects such as Frederick Law Olmstead considered the reverse — how carefully designed semi-natural landscapes could positively affect human society, leading to creations such as New York's Central Park (designed in 1858). Several decades later, in 1907, Ellen Swallow Richards coined the term 'human ecology' for her analysis of household systems of provisioning. This phrase has appeared sporadically in academic contexts ever since. It is usually defined loosely and has similar connotations to 'social ecology'.

## The Chicago School

Sociologists based at the University of Chicago made early explorations of 'social ecology' beginning in the 1910s. Robert Park, Ernest Burgess, and Roderick McKenzie developed a spatial model of urban development and demographic change for the city of Chicago, in which they employed ecological concepts such as invasion, succession, and evolution (McKenzie et al., 1925). Their model described how the expansion of some population groups encroaches spatially on others, and how lower-income, ethnic communities ascend the economic ladder into the middle class. This view has been criticized as being overly deterministic and simplistic, and for focusing on the supposed disorganization of lower-income populations. But it was a pioneering attempt to apply social ecological concepts to an entire metropolitan area.

## Murray Bookchin

A very different take on social ecology was developed by social philosopher Murray Bookchin in the 1960s and 1970s. Bookchin argued that ecological problems arise as a result of particular political and economic contexts, an idea that presaged the later academic field of political ecology. Bookchin's concept of 'social ecology' did not so much apply ecological concepts to human societies as argue that radical social change was necessary in order to protect the natural environment. In books such as *Post-Scarcity Anarchism* (1971) and *The Ecology of Freedom* (1982), Bookchin was strongly critical of capitalism and advocated a decentralized, communalistic society.

## Recent work

Other thinkers beginning in the 1960s and 1970s applied ecological concepts to the analysis of human societies. These included anthropologist Gregory Bateson (1972, 1979), environmental historian and feminist Carolyn Merchant (1980), physicist Fritjof Capra (1983, 1996), and Buddhist-influenced thinker Joanna Macy (1991). Biologist Richard Dawkins's 1976 book *The Selfish Gene* proposed the idea of 'memes' — ideas or concepts that spread themselves using humans as vehicles — as units of cultural evolution analogous to genes within individuals. This perspective would suggest that within

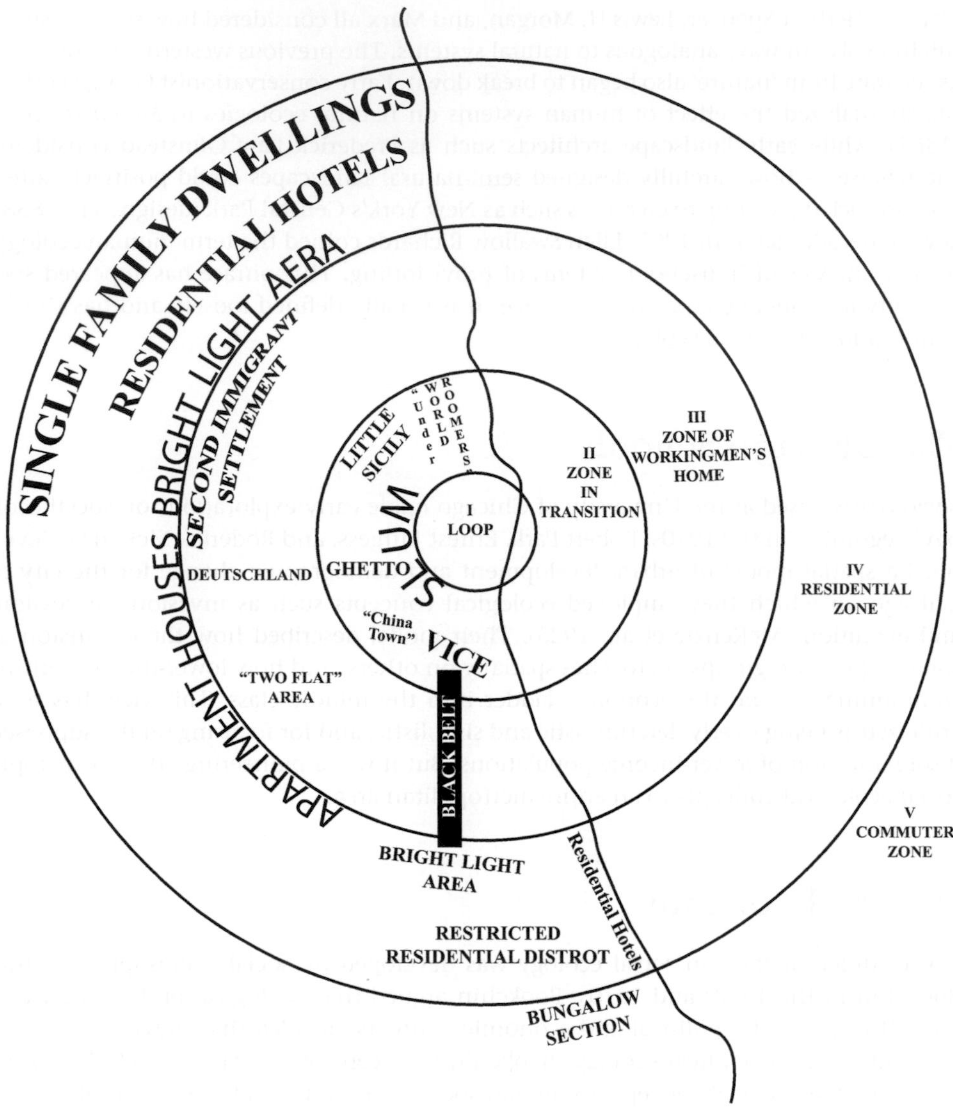


**Figure 3.21.1** Social ecology map of Chicago.

*Source*: McKenzie, Park and Burgess (1925).

social ecologies individuals may be less important than the trends, ideas, ideologies, or technologies they employ.

Through the *Coevolution Quarterly*, an alternative magazine published between 1974 and 1985, 'green' businessman Stewart Brand and others helped popularize the concept of coevolution. This term within ecological science refers to the joint evolution of multiple species. However, these visionaries used it more loosely to connote rapid social evolution due to the interplay of technology and social factors. In his 1996 book *Development Betrayed: The End of Progress and a Coevolutionary Revisioning of the Future*, ecological economist Richard Norgaard provided one illustration of how different types of forces might influence coevolution:

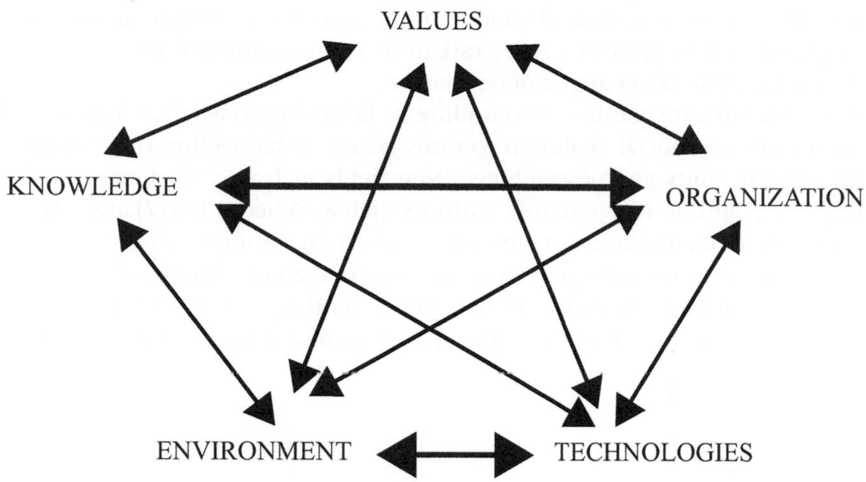


**Figure 3.21.2** The coevolutionary process.

*Source*: Norgaard (1996).

In recent years researchers interested in how human societies can sustainably manage natural resources such as fisheries and forests have sought to analyze 'social-ecological systems' (e.g. Ostrom, 2009). A main question has been whether human users of these resources such as fishermen can cooperate to manage the resource sustainably, or whether government must step in to impose a regulatory framework. The answer seems to be that both approaches are possible and necessary at times, but that particular conditions can increase the likelihood of human users self-organizing to protect resources. These conditions include the size of the territory, the type of resource, the cost, the presence of social capital (traditions of trust and cooperation), and leadership.

## Looking ahead

Using the metaphor of natural ecologies for human societies has a number of implications. First, natural ecosystems are best understood holistically, in terms of the interactions between many different species and environmental factors. A similar holistic outlook seems necessary in order to understand social ecologies. All system elements are potentially important. No single lens on society or nature is going to provide full understanding; rather, the challenge is to weave together multiple perspectives to understand as many dimensions of the system as possible.

Second, natural ecosystems don't necessarily change in a straightforward, linear fashion. Change arises from shifting interactions among many variables in addition to random factors. Natural evolution is 'punctuated', happening in fits and starts rather than smoothly. Probably the evolution of social ecologies will be similar. Small-scale dynamic changes are constantly underway, but larger-scale systems may be relatively stable until tipping points are reached and new social equilibria are found.

Third, interactions between ecosystem components are of many sorts. Cooperation and synergy may be at least as important as competition. In contrast to dark views of nature as 'the war of all against all', both natural ecosystems and human societies benefit from these gentler processes. A key task of social evolution is thus to encourage, in Lincoln's words, 'the better angels of our nature'.

Perhaps the ultimate question surrounding social ecology is whether human societies can direct their own social evolution towards greater sustainability (including toward improved social equity and peacefulness). New fields such as evolutionary psychology may help shed light on this potential. Authors such as Wheeler (2012) argue that more conscious human management of our own social ecology is imperative for dealing with problems such as climate change. One must be careful about extrapolating too directly from natural systems to human societies. But potentially social ecology offers a tool that can help us better guide our own social evolution so as to live more sustainably on a small planet.

## Bibliography


Bateson, G. (1972) *Steps to an Ecology of Mind: Collected Essays in Anthropology, Psychiatry, Evolution, and Epistemology*. Chicago: University of Chicago Press.

Bateson, G. (1979) *Mind and Nature: A Necessary Unity*. New York: Dutton.

Bookchin, M. (1971/1986) *Post-Scarcity Anarchism*. Montreal: Black Rose Press.

Bookchin, M. (1982) *The Ecology of Freedom*. Palo Alto, CA: Cheshire Books.

Capra, F. (1983) *The Turning Point: Science, Society, and the Rising Culture*. New York: Bantam Books.

Capra, F. (1996) *The Web of Life: A New Scientific Understanding of Living Systems*. New York: Anchor Books.

Condorcet, Marquis de. (1795) *Outlines of an Historical View of the Progress of the Human Mind*. Online. Available at: http://oll.libertyfund.org/index.php?option=com_staticxt&staticfile=show.php%3Ftitle=1669&Itemid=27.

Dawkins, R. (1976) *The Selfish Gene*. Oxford: Oxford University Press.

Macy, J. (1991) *Mutual Causality in Buddhism and General Systems Theory: The Dharma of Natural System* (Buddhist Studies Series). New York: State University of New York Press.

Marsh, G.P. (1864) *Man and Nature, or Physical Geography as Modified by Human Action*. New York: Charles Scribner.

Marx, K. (1867/2007) *Das Kapital*. New York: Synergy International of the Americas.

McKenzie, R., Park, R. and Burgess, E. (1925/1967) *The City*. Chicago: University of Chicago Press.

Merchant, C. (1980) *The Death of Nature: Women, Ecology, and the Scientific Revolution*. New York: Harper & Row.

Norgaard, R. (1996) *Development Betrayed: The End of Progress and a Coevolutionary Revisioning of the Future*. London and New York: Routledge.

Ostrom, E. (2009) "A general framework for analyzing sustainability of social-ecological systems", *Science*, 325 (5939): 419–422, DOI: 10.1126/science.1172133.

Wheeler, S. (2012) *Social Ecology and Climate Change*. London and New York: Routledge.

# Part 4

# Contemporary approaches

# Part 4 Introduction

If the approaches constituting the early life of environmental studies were in large part existing at the time issues of environment arose to global prominence in the 1960s and 1970s, more contemporary approaches arose from the maturation of the field, as well as important developments in ancillary fields and inclusion of entirely different fields. This has resulted in a greater diversity of approaches. The impact of these recent subfields and tools in environmental studies has been a more inclusive and plural, though possibly increasingly divergent, analysis of environment.

One important feature of classic approaches was the role of science: following the typical narrative of mid-20th century environmentalism, it was scientific discoveries of the sort featured in, for instance, Rachel Carson's *Silent Spring* that ushered in the environmental era and the field of environmental studies. Science, as a set of settled facts evidencing environmental pollution and disruption, was a cornerstone of many classic approaches to environmental scholarship. Contemporary approaches maintain some indebtedness to the physical, biological, and behavioral sciences, yet increasingly diverge along two axes: the first an invigoration of humanistic perspectives, and the second a more forthright consideration of politics.

The disciplinary span of contemporary approaches in environmental studies now includes the natural sciences, social sciences, and humanities. Indeed, some perspectives outside the sciences emerged in the classic era in subfields, such as environmental history or environmental economics. But there has been a much greater flourishing in recent times, with for instance animal studies extending a range of interpretive approaches to the nonhuman, ecocriticism engaging literature and environment, and queer ecology addressing sexuality and environment.

Contemporary approaches not only include those beyond classic science; they also extend and/or challenge science itself via a far more robust consideration of power and politics. Thus, for instance, the assumptions and implications of 'post-normal' science are studied; political ecology considers how political economy and power structure both environmental knowledge and practice; and approaches such as resilience proceed not from the assumptions of balance and equilibrium that characterized a great deal of classic environmental science, but rather of dynamism, disruption, and change – which understandably ring true to scholars of the contemporary political world.

If these contemporary approaches offer exciting, even inspiring new vehicles to navigate the emerging environmental questions we are asking, they do present their own challenges. One important challenge environmental studies has not yet overcome – indeed, may be increasingly unable to overcome – is the question of methodological coherence or integrity: how do you do environmental studies in some sort of identifiable manner, if the proliferation of new approaches simply multiplies the investigative options?

One integrative approach on the natural sciences end of the disciplinary spectrum is earth system science, an attempt to relate data, theory, and ultimately management of a number of interconnected earth sub-systems, like the hydrosphere and atmosphere (see Noel Castree's chapter, this volume). There have been attempts at integrative approaches to environmental studies emanating from the interpretive social sciences and humanities as well: one approach that deliberately levels the playing field across a wide range of disciplines is known as actor-network theory or ANT, in which the network of relations between manifold human and nonhuman actors is ultimately seen to determine our unfolding environmental (and other) realities. Even so, ANT has been viewed with suspicion from social scientists with more of a 'structural' bent, like Marxists, and in a broader sense the perennial lack of paradigmatic settlement across the social sciences and humanities may preclude more widespread adoption of ANT or any similar common method.

Examining contemporary approaches to environmental studies is unlikely to offer some unified way of doing environmental studies. Rather, it offers an appreciation that these diverse approaches represent the leading edge – in many ways, an unsettled, raw, plural edge – of the field as it moves beyond its classic roots. In doing so it engages with an unruly, changing reality and an unruly, changing intellectual landscape in ways that will perhaps forever preclude tidy settlement. Environmental studies as an interdiscipline thus offers many contemporary intellectual virtues – provided one is prepared to accept what may ultimately seem like a hodgepodge of scholarly tools.

# 4.1 Animal studies

## Jody Emel and Ilanah Taves

Animal studies is a burgeoning new field focusing on both contemporary and historical relations of human and non-human animals (hereafter 'animals'). Informed by cognitive ethology and conservation biology, by the humanities and the social sciences, its terrain includes cultural norms between humans and animals, human treatment of animals, consideration of cultural differences in relations and beliefs, representations of animals and animal difference, and animal subjectivity (consciousness, agency, and sense of self). Concerns about species decline and extinction, concentrated animal feeding operations, pet keeping, urban wildlife and other human–animal relations continue to push animal studies to the forefront of academia, attracting a wide range of diverse disciplines. Animal studies scholars use content from anthropology, geography, feminist and gender studies, science and technology studies, sociology, philosophy, literature, psychology and other fields to examine the spaces and standpoints occupied by nonhuman animals existing in an increasingly human-dominated world. They engage with post-humanist and post-anthropocentric thought, which challenges commonly held beliefs in which humans are valued over other animals. Decolonial, queer, and indigenous studies also provide perspectives for envisioning ways of rethinking human–animal relations in departure from mainstream utilitarian hierarchies. These new ways of perceiving nonhuman beings are at the center of animal studies, which also attempts to incorporate the animal experience into research, not only through cognitive ethology but also through multispecies ethnography. Animal studies presents material that is crucial to deconstructing the existing divide between humans and animals, and reconsiders the problematic interactions between the human and other animal species.

## Disciplinary contributions

In this section, we briefly review the fields contributing to animal studies. Cognitive ethology examines the mental state of nonhuman animals – specifically the effects of consciousness and other thought processes on animal behavior. The field received attention in the late 1970s through the work of Donald R. Griffin (1992), a zoologist publishing works concerning cognition of nonhuman animals. Frans de Waal (2016) is a primate ethnologist noted for researching altruism and emotions; his impact on informing the public about animal minds is significant, especially through his easily readable and popular books. The main objective of cognitive ethology is to understand natural animal behaviors and the roles they play in evolution and species survival. These ideas establish validity in nonhuman animal existence and experience, exposing insights into the complexity of animals' brains and similarities to humans. These notions disrupt the major divide drawn between humans and animals, disputing the concept that humans solely possess the ability to think and reason. This raises ethical questions about the rights of animals and the threat of human development to the existence of nonhuman species.

Anthropology is the study of human beings. Studies of human culture and experience are not complete without a consideration of animal actors in societies. Tim Ingold (1988) argued for a greater focus on nonhuman actors in anthropology. He discussed the validity of animal cultures, addressing how the term 'culture' itself is not peculiar to humans alone. A consideration of shared characteristics between humans and animals opens a doorway for inclusion of animals in a largely anthropocentric discipline. These considerations provide a platform for recent subfields of anthropology. Anthrozoology and human–animal studies are recent branches of anthropology accounting for nonhuman others in an effort to gain further insight into human lives. Many human–animal interactions, such as the domestication of animals, are key to human development as a whole. These new considerations necessitate revised approaches to the analysis of human–animal relations. Anthropologist Barbara Noske referred to a misguided view of animals in traditional anthropology 'as passive objects that are dealt with and thought and felt about' (1993, 185). Noske outlined the shortcomings of anthropology's disciplinary approach and its lack of animal perspective. Eben Kirksey's (2015) multispecies ethnographies constitute a new turn in human–animal studies, placing emphasis on ecological assemblages which might be creative enough to foster both conservation and social justice. Similarly, Timothy Pachirat's (2011) undercover work in an Omaha slaughterhouse illustrates the injustices and exploitation visited upon both workers and animals.

Philosophical questions concerning animals are at the center of animal studies as a discipline. Ethical issues with common practices involving animals and their treatment drive the increase of attention to nonhuman beings in academia. A major influence in this area was Peter Singer's book *Animal Liberation*, published originally in 1975. Singer began by addressing 'the basic principle of equality' and applies it to a revised consideration for nonhuman animals. He argued understanding animals as equals does not necessarily mean they be given the same rights, but that they are not thought of as less than. Singer was one of the first to introduce the term 'speciesism', referring to the superior attitude of humans in considering animals as inferior. He disavowed the commonly held idea that anything can be done to animals, in any manner. His words significantly influenced a widely accepted framework for philosophical reconsideration of animals. Bernard Rollin (1995) carries this work forward, particularly as it applies to industrially farmed animals. Using biopolitical and posthumanist theories, Carey Wolfe has advanced ethical consideration of animals in several excellent books (see Wolfe 2013 in particular).

As with numerous other disciplines, in the 1990s scholars from the field of geography began to incorporate animal-centered ideas into research and literature. The subfield of animal geography grew as a direct response to growing interest in how nonhuman animals engage with space; human–animal interactions are of particular interest. Recent changes in animal populations and distribution call for more attention to animal spaces and how those spaces are often ignored in a human-dominated world (Wolch and Emel 1998). These themes encouraged further critiques of the 'animal condition' (Gillespie and Collard 2015), where the commodification and exploitation of animals has manifested in the form of auction grounds and large-scale industrialized farming systems. Geography provides an important lens for animal studies because it explicitly engages with particular spaces and places of animal lives, not only describing these locational interactions but actively pursuing intersectional justice in living together.

## Theoretical and political threads

Besides disciplinary contributions, there are multiple, if overlapping, ways to distinguish the threads of animal studies. One of the principal divisions is between those authors who are critical of existing human–animal relations and those who are describing and interrogating such relations without overt judgement. Even within this 'critical' vein, there are shades of difference. The Institute for Critical Animal Studies group advocates scholarship plus direct action and aims for total liberation. These scholar/activists are adamant about animal emancipation and harbor no compromise (Best and Nocella 2004). Critical animal studies is overtly negative about mainstream factory farming, zoos, laboratory experiments using animals, the failures of wildlife protection, and human–animal relations that are judged speciesist.

Others who are considered critical scholars are less sanguine about the possibilities for total emancipation. Donna Haraway examines the ways in which politics and money inform science (and vice versa), and she interrogates human–animal relationships, notably those involving primates, dogs and genetically bred mice (1997, 2008). She ultimately decides on the notion of respect (or 'respecere') as the guiding principle for such relationships – not emancipation or some sort of equality. For Haraway, a professed Marxist-feminist, the importance of animal worlds is significant, as is the connectedness of humans and non-humans. But while she sees the connections between the mice used for developing cancer treatment ('Oncomouse') and her human sisters (subject to breast and other cancers), she is not willing to endorse emancipation of such mice.

Another thread that distinguishes parts of the animal studies *oeuvre*, is the use of social theory. Derrida's 2008 published work, *The Animal That Therefore I Am*, contributed significantly to the gravitas of animal studies within the academy. Many feminist theorists, particularly ecofeminists, have problematized the human–animal divide using sophisticated theoretical arguments. Val Plumwood (1993) wrote of the ways in which woman, animal, nature, indigenous, and others were marginalized and 'othered' by a hierarchical duality which placed man, culture and rationality on the privileged side of

**Figure 4.1.1** A Mongolian Fall Camp 2013.

*Source*: Image by Jody Emel

the dualism. Within the past few decades, several new ideas about animals have been promoted by academic and activist writers – ideas that go beyond the older controversy between 'rights' and 'welfare'. Posthumanism and postanthropocentrism are two of the most important (see Braidotti 2013). In many ways, these theories resonate with the values and stories of many decolonial and indigenous writers (e.g. Leslie Marmon Silko 1977). Diverging from the liberal philosophies that claim something akin to human rights for animals, these ideas emphasize instead the value of animals for themselves, their vitality and unpredictability, the relatedness of beings and their material worlds, and a non-hubristic and non-individualist view of humans relative to other animals and nature in general.

# Learning resources

## *Websites*

Journal for Critical Animal Studies website: http://journalforcriticalanimalstudies.org.
Animals and Society Institute: https://www.animalsandsociety.org/human-animal-studies.
Minding Animals International: http://www.mindinganimals.com.
Journals Pertaining to Animal Studies: <http://guides.main.library.emory.edu/c.php?g=50800&p=326053>.

## *Books*

*Sistah Vegan: Black Female Vegans Speak on Food, Identity, Health and Society* edited by A. Breeze Harper.
*Placing Animals: An Introduction to the Geography of Human-Animal Relations* by Julie Urbanik.
*Animals and Society: An Introduction to Human-Animal Studies* by Margo DeMello.
*Queer Ecologies: Sex, Nature, Politics, Desire* edited by Catriona Mortimer-Sandilands and Bruce Erickson.

## *Films*

*Earthlings*
*Cowspiracy*

## Bibliography

Best, S. & Nocella, A. (2004) *Terrorists or Freedom Fighters: Reflections on the Liberation of Animals*. New York: Lantern Books.
Braidotti, R. (2013) *The Posthuman*. Cambridge and Malden: Polity Press.
de Waal, F. (2016) *Are We Smart Enough to Know How Smart Animals Are?* New York: W.W. Norton Publishing.
Derrida, J. (2008) *The Animal That Therefore I Am*. New York: Fordham University Press.
Gillespie, K. and Collard, R.C. (2015) 'Introduction.' In Critical Animal Geographies: Politics, Intersections and Hierarchies in a Multispecies World, edited by Kathryn Gillespie and Rosemary-Claire Collard, 1–16. London: Routledge.
Griffin, D. (1992) *Animal Minds: Beyond Cognition to Consciousness*. Chicago: University of Chicago Press.
Haraway, D. (1997) *Second Millennium: FemaleMan Meets Oncomouse*. London and New York: Routledge.
Haraway, D. (2008) *When Species Meet*. Minneapolis and London: University of Minnesota Press.
Ingold, T. (1988) *What is an Animal?* London: Unwin Hyman.
Kirksey, E. (2015) *Emergent Ecologies*. Durham and London: Duke University Press.
Noske, B. (1993) 'The Animal Question in Anthropology: A Commentary', *Society and Animals*, 1(2): 185–190.
Pachirat, T. (2011) *Every Twelve Seconds: Industrialized Slaughter and the Politics of Sight*. New Haven and London: Yale University Press.
Plumwood, V. (1993) *Feminism and the Mastery of Nature*. London and New York: Routledge.
Rollin, B. (1995) *Farm Animal Welfare*. Ames: Iowa State University Press.
Silko, L.M. (1977) *Ceremony*. New York: Penguin.
Singer, P. (2001) *Animal Liberation*. New York: ECCO Press.
Wolch, J. & Emel, J. (1998) *Animal Geographies: Place, Politics, and Identity in the Nature-Culture Borderlands*. London and New York: Verso.
Wolfe, C. (2013) *Before the Law: Humans and Other Animals in a Biopolitical Frame*. Chicago and London: University of Chicago Press.

# Business studies and the environment

4.2

## Martina K. Linnenluecke and Tom Smith

At the 2015 Paris "climate summit", 195 countries adopted the first ever universally, legally binding climate deal to come into force in 2020; with a focus on mitigation, adaptation, loss and damage and support for developing countries. The Paris Accord follows scientific findings that have generated significant concern about the level of human induced global environmental change. The case for urgent and immediate business action is thereby not just based on the harmful effects that business and human endeavors are having on the Earth system, but also on the recognition that the ongoing degradation of these systems will have impacts on corporate and industrial activities (Winn et al., 2011). Modern business scholarship has started to critically reflect on the way modern firms utilize the environment relationship and argues that companies need to manage their relationship with the natural environment to not destroy the very life-supporting bases provided by nature (e.g., Winn and Pogutz, 2013). There is now a high likelihood that environmental changes will significantly impair the assets, profits and human capital of businesses (Linnenluecke, Birt et al., 2015).

In this chapter, we summarize studies at the intersection of business and the environment, set out the case for action to confront environmental threats, and also point out the many opportunities that arise. We believe that many future opportunities will come about as a result of the next technological breakthrough in clean technology. Hong et al. (2008) identify eight examples of technological breakthroughs in modern markets that have driven growth and wealth: namely, railways, electricity, automobiles, radio, microelectronics, personal computers, biotechnology, and the internet.

Because of the Paris Accord, firms worldwide have started to commercialize their clean technology patents. A new technological breakthrough in clean technology is, arguably, imminent. China, Europe and the US hold the majority of the over half a million clean tech patents and are leading the way to a technological breakthrough in clean tech which will drive wealth and growth for decades to come.

## Business and the environment

The recognition that corporate and industrial activities cause significant adverse impacts on the environment is now new, and became fully visible through a number of environmental disasters that led to the environmental movement of the 1970s and 1980s (Hart, 1997). At the time, growing public pressure and stricter government regulations required companies to change the way that they do business, leading to a significant stream of research that sought to reconcile strategies for business growth with improvements in environmental performance (or "corporate greening"), often through pollution prevention and environmental management efforts (for a summary, see Schot et al., 1997; Sharma, 2003). Many researchers argued that well-designed resource-efficiency or energy-savings programs could bring significant financial benefits to firms through adopting environmental technologies and achieving cost savings (e.g., Shrivastava, 1995).

While these early studies predominantly focused on how companies can minimize their impact on the environment and minimize severe – typically local or regional – environmental problems, the late 1980s and early 1990s saw the emergence of a new type of concern about global degradation. This was prompted by a growing body of evidence suggesting that climate change, population growth and ongoing levels of resource depletion pose real, long-term threats (Linnenluecke et al., 2016). For example, the US Global Change Research Act of 1990 stated that: "'Global change' means changes in the global environment (including alterations in climate, land productivity, oceans or other water resources, atmospheric chemistry, and ecological systems) that may alter the capacity of the Earth to sustain life". On political levels, the increasing concerns about climate change prompted the establishment of the Intergovernmental Panel on Climate Change (IPCC), as well as the United Framework Convention on Climate Change (UNFCCC) which significantly impacted the global agenda on mitigating climate change, with implications for high polluting industries and companies through the implementation of carbon reduction measures in the form of emissions trading schemes or carbon taxes (Busch and Hoffmann, 2011).

## Business action on global environmental change

The case for business to act on global environmental change revolves around several dimensions (Linnenluecke et al., 2016). The moral case is summarized by the Catholic Pope's 2015 encyclical which makes the case that it's the poorest of the poor that will ultimately have to pay the price for environmental change unless effective and immediate action is taken, prompting policy-makers and private-sector leaders to take immediate action. The legal case is best exemplified by the recent decision in the

Netherlands (*Urgenda Foundation vs The State of the Netherlands, Ministry of Infrastructure and the Environment*), where nine hundred citizens sued the Netherlands government forcing the government to take more effective action on climate change. The Netherlands government had agreed to cut greenhouse emissions by 17 per cent by 2020 but the court ordered them to cut by 25 per cent. The science said that cuts had to be in the range of 25–40 per cent. The decision was based on the tort of negligence, a duty of care was owed by the government to its citizens and its future citizens. As one might expect this case has been followed by similar lawsuits in countries worldwide. In other legal actions, a farmer in Peru is suing a German utility, RWE AG, for its proportional contribution to global climate change. Action is to be heard in a regional court in Essen, Germany. According to a 2014 study (Heede, 2014), RWE AG is responsible for half a per cent of the total emissions that were released into the atmosphere since the beginning of industrialization.

The business case takes the view that businesses exist to create value and that any business that does not add value will either be forced to change or cease to exist. This raises fundamental questions such as "what is value and how is it measured?" and "how does a well-run business create or add value?". The natural environment is now undergoing new changes that alter the foundation of value creation. These changes create threats through adverse impacts and the needs to adapt (e.g., Noble et al., 2014) – but also creating new opportunities in the form of business solutions to some of the world's most pressing issues, including clean tech development (e.g., Alstone et al., 2015; Bertram et al., 2015). Research is now increasingly focusing on ways to value such opportunities, especially as the traditional discounted cash flow methods (i.e., assessing investments by discounting expected future free cash flows) do not work under conditions of future uncertainty. Opportunities are best valued using a real options approach, whether in the form of analytical models (Black and Scholes, 1973; Merton 1973), simulated models (Schwartz and Moon, 2000) or strategic decision trees (Kellogg and Charnes, 2000).

## The state of the field and future opportunities

Studies on business and the environment are now increasingly drawing on interdisciplinary findings to develop powerful insights for future business opportunities (Winn and Pogutz, 2013). Research now combines viewpoints from economics, finance, organizational strategy, and the science disciplines to tackle the most pressing environmental and social issues. One example is the potential impairment of assets due to environmental changes because of extreme weather events, regulatory restrictions on the extraction of fossil fuels (Linnenluecke, Birt et al., 2015). Climate change has prompted a worldwide divestment movement (Linnenluecke, Meath et al., 2015). Proposed limits on the exploration and use of fossil fuels (e.g., several countries announcing plans to ban petrol cars) has promoted many to be concerned that fossil fuel companies might become "stranded" assets. The science says that if proven reserves are converted to energy then the 2°C target would be greatly exceeded – to remain within the target of 2°C less than half of these reserves can be burned in the future (Allen et al., 2009; Meinshausen et al., 2009). Dietz et al. (2016) estimate the Value at Risk (VaR) for global assets as US$2.5 trillion if we continue on a business as usual trajectory for emissions which has significant implications for corporations and industries with investments in such assets.

Another important area of interdisciplinary research is the question of how companies should best adapt to climate change. Weather extremes are already causing large losses. Environmental volatility is likely to give rise to risk transfer solutions such as geographical diversification (Tang and Jang, 2011), insurance measures (Phelan, 2011), weather derivatives (Bank and Wiesner, 2011) and catastrophe bonds (Johnson, 2014). The international architecture for adaptation funding has also become a key research priority area. The Special Climate Change Fund, the Least Developed Countries Fund, the Green Climate Fund, and the Kyoto Protocol Adaptation Fund are all designed to support action on climate change. Research on the optimal governance of funds, fund allocation, and the integration of adaptation and mitigation (Locatelli et al., 2016; Dzebo and Stripple, 2015) are very important future research directions.

Lastly, an increasing number of studies are focusing on the important area of how the technological breakthrough in clean-tech will unfold over time. Gunderson and Holling (2002) outline four phases of transformation that can be applied to technological breakthroughs such as the automobile (see also Hong et. al., 2008). The four phases are (i) rapid growth, (ii) a period of stability, during which time the industry becomes less resilient, (iii) creative destruction, otherwise known as a post-bubble market crash, and (iv) a period of renewal. The coming technological breakthrough has given rise to studies on innovation challenges, funding streams and venture capital activity (Cumming et al., 2016), as well as policy regimes supporting clean tech opportunities and uptake, also in developing countries (Gosens et al., 2015).

## Conclusion

While business and corporate activity has often been viewed as the culprit behind many of the world's pressing environmental issues, emerging research prompts opportunities for business to be involved in creating future solutions. This chapter has provided a short overview of studies on businesses and the environment, has outlined a case for action and is pointing to many future opportunities.

## Learning resources

Linnenluecke, M.K. and A. Griffiths. (2015) *The Climate Resilient Organization.* Cheltenham: Edward Elgar.

The reader is directed to the website of the Organizations and the Natural Environment Division of the Academy of Management for additional teaching and learning resources. Available at: <http://one.aom.org/index.php/online-resources>.

## Bibliography

Allen, M.R., D.J. Frame, C. Huntingford, C.D. Jones, J.A. Lowe, M. Meinshausen and N. Meinshausen (2009) "Warming caused by cumulative carbon emissions towards the trillionth tonne." *Nature*, 458(7242): 1163–1166.

Alstone, P., D. Gershenson and D.M. Kammen (2015) "Decentralized energy systems for clean electricity access." *Nature Climate Change*, 5(4): 305–314.

Bank, M. and R. Wiesner (2011) "Determinants of weather derivatives usage in the Austrian winter tourism industry." *Tourism Management*, 32(1): 62–68.

Bertram, C., G. Luderer, R.C. Pietzcker, E. Schmid, E. Kriegler and O. Edenhofer (2015) "Complementing carbon prices with technology policies to keep climate targets within reach." *Nature Climate Change*, 5: 235–239.

Black, F. and M. Scholes (1973) "The pricing of options and corporate liabilities." *The Journal of Political Economy*, 81(3): 637–654.

Busch, T. and V.H. Hoffmann (2011) "How hot is your bottom line? Linking carbon and financial performance." *Business & Society*, 50(2): 233–265.

Cumming, D., I. Henriques and P. Sadorsky (2016) "Cleantech venture capital around the world." *International Review of Financial Analysis*, 44: 86–97.

Dietz, S., A. Bowen, C. Dixon and P. Gradwell (2016) "'Climate value at risk' of global financial assets." *Nature Climate Change*, 6: 676–679.

Dzebo, A. and J. Stripple (2015) "Transnational adaptation governance: an emerging fourth era of adaptation." *Global Environmental Change-Human and Policy Dimensions*, 35: 423–435.

Gosens, J., Y. Lu and L. Coenen (2015) "The role of transnational dimensions in emerging economy 'Technological Innovation Systems' for clean-tech." *Journal of Cleaner Production*, 86: 378–388.

Gunderson, L.H. and C.S. Holling (2002) *Panarchy: Understanding Transformations in Human and Natural Systems*. Washington, DC: Island Press.

Hart, S.L. (1997) "Beyond greening: strategies for a sustainable world." *Harvard Business Review*, 75(1): 67–76.

Heede, R. (2014) "Tracing anthropogenic carbon dioxide and methane emissions to fossil fuel and cement producers, 1854–2010." *Climatic Change*, 122(1–2): 229–241.

Hong, H., J. Scheinkman, and W. Xiong, (2008) Advisors and asset prices: a model of the origins of bubbles. *Journal of Financial Economics*, 89(2): 268–287.

Johnson, L. (2014) "Geographies of securitized catastrophe risk and the implications of climate change." *Economic Geography*, 90(2): 155–185.

Kellogg, D. and J.M. Charnes (2000) "Real-options valuation for a biotechnology company." *Financial Analysts Journal*, 56(3): 76–84.

Linnenluecke, M.K., T. Smith and B. McKnight (2016) "Environmental finance: a research agenda for interdisciplinary finance research." *Economic Modelling*, 59: 124–130.

Linnenluecke, M.K., J. Birt, J. Lyon and B.K. Sidhu (2015) "Planetary boundaries: implications for asset impairment." *Accounting & Finance*, 55(4): 911–929.

Linnenluecke, M.K., C. Meath, S. Rekker, B.K. Sidhu and T. Smith (2015) "Divestment from fossil fuel companies: confluence between policy and strategic viewpoints." *Australian Journal of Management*, 40(3): 478–487.

Locatelli, B., G. Fedele, V. Fayolle and A. Baglee (2016) "Synergies between adaptation and mitigation in climate change finance." *International Journal of Climate Change Strategies and Management*, 8(1): 112–128.

Meinshausen, M., N. Meinshausen, W. Hare, S.C.B. Raper, K. Frieler, R. Knutti, D.J. Frame and M.R. Allen (2009) "Greenhouse-gas emission targets for limiting global warming to 2°C." *Nature*, 458(7242): 1158–1162.

Merton, R.C. (1973) "Theory of rational option pricing." *The Bell Journal of Economics and Management Science*, 4(1): 141–183.

Noble, I., S. Huq, Y. Anokhin, J. Carmin, D. Goudou, F. Lansigan, B. Osman-Elasha and A. Villamizar (2014) "Adaptation Needs and Options." In *Climate Change 2014: Impacts, Adaptation, and Vulnerability: Contribution of Working Group II to the Fifth Assessment Report of the Intergovernmental Panel on Climate Change*. Online. Available at: <http://ipcc-wg2.gov/AR5/report>, edited by IPCC.

Phelan, L. (2011) "Managing climate risk: extreme weather events and the future of insurance in a climate-changed world." *Australasian Journal of Environmental Management*, 18(4): 223–232.

Schot, J., E. Brand and K. Fischer (1997) "The greening of industry for a sustainable future: building an international research agenda." *Business Strategy and the Environment*, 6(3): 153–162.

Schwartz, E.S. and M. Moon (2000) "Rational pricing of internet companies." *Financial Analysts Journal*, 56(3): 62–75.

Sharma, S. (2003) "Research in corporate sustainability: what really matters?" In *Research in Corporate Sustainability: The Evolving Theory and Practice of Organizations in the Natural Environment*, edited by S. Sharma and M. Starik. Cheltenham: Edward Elgar.

Shrivastava, P. (1995) "Environmental technologies and competitive advantage." *Strategic Management Journal*, 16(S1): 183–200.

Tang, C.H. and S. Jang (2011) "Weather risk management in ski resorts: financial hedging and geographical diversification." *International Journal of Hospitality Management*, 30(2): 301–311.

Winn, M.I. and S. Pogutz (2013) "Business, ecosystems, and biodiversity: new horizons for management research." *Organization & Environment*, 26(2): 203–229.

Winn, M.I., M. Kirchgeorg, A. Griffiths, M.K. Linnenluecke and E. Gunther (2011) "Impacts from climate change on organizations: a conceptual foundation." *Business Strategy and the Environment*, 20(3): 157–173.

# Environmentalism and creative writing

4.3

Richard Kerridge

'Creative writing': the term names two things. There is the art itself, also known as literature, and there is the academic discipline, long established in the USA and now offered by most universities in the UK, and emerging in many other countries besides. In this essay, I will look at how Creative Writing in the second sense (perhaps indicated by the capital letters) interacts with environmental concern – but, of course, the university discipline that is Creative Writing is an attempt to make a place for the art in the academic world, as a combination of mentored instinct and taught technique connected with the student's personal vocation, and as a form of practical research. So I cannot talk about the second sense without also discussing the first.

In universities, the subject is commonly taught by means of 'workshops'. These are classes in which students present short pieces of work for detailed discussion by the group – parts of novels or non-fiction works in progress, or single poems or short stories. Usually the work has been circulated in advance. Discussion concerns itself with particular words, lines, sentences, paragraphs, stanzas, voices and rhythms, and also with larger aspects of the writing, such as the overall conception and aim, the style and technique, and the appropriateness of form and structure to subject matter. The aim of the class is to help each student make decisions and explore the practical implications of those decisions by editing, re-writing and re-shaping. Students write instinctively, write self-consciously, think critically, read observantly, articulate arguments, listen, change the writing in response to the arguments, and change it again if the arguments do not work out in practice, or sometimes change it back, after all, to something close to the first version. All these activities are testing each other. Critics introducing, for example, feminist or ecocritical priorities, will question the writing and then have to explore, with the writers, the practical consequences of the questions. Practice speaks back to theory, and both are honed by the process. Classes of this kind are called 'workshops' because something is on the bench being made. Through a series of steps in which critical perspectives come up against practical testing, the novel, poem, story or work of non-fiction begins to emerge, and so does the distinctive evolving identity of the writer – the combination of literary form, recurrent subject matter and personal voice and history.

So, what intervention can environmental concern make in Creative Writing, in both senses of the term? What can the workshop do for environmentalism?

The most obvious answer is that environmental problems call for adaptations of literary form, and that these adaptations, and the difficulties writers encounter in attempting to make them, are of more than purely literary importance. Our environmental crisis consists of multiple dangers, all developing on a large scale. These dangers include global warming, the imminent wave of species extinctions, the general dwindling of wildlife numbers, ocean acidification, topsoil erosion, overconsumption of water and the spread of chemical and radioactive pollution. Human overpopulation is implicated in most of these problems. Dangers of this kind almost defy representation. They present writers with specific technical challenges, but these are specialized literary versions of the general problems of representation, imagination and action that such dangers pose. Any successful innovations in literary form that emerge in response will be sources of hope, not only because the books may influence people's behaviour directly, but also because they may be a sign, early or late, of a more general shift in how we can imagine ourselves living. What we need to do, in rich consumer societies, is reasonably well known. It is the strong will to make the changes that is so elusive. We lack the capacity to imagine ourselves making them, individually or collectively.

Narrative 'point of view' is one of the most common topics of discussion in the Creative Writing workshop. Often it is the starting-point. From whose viewpoint is the story being told? Who is the dramatic speaker of the lyric poem? If the point of view belongs to one of the characters in the story, are the cadences and vocabulary consistent with this character? Has the narrative remained within the limitations of that character's knowledge? Is the character speaking from the midst of the action, perhaps with the full immediacy of the present tense? In that case, has the narrative kept to the character's sightlines? Or is the character looking back at the novel's events from a later moment, comparing what she knows now with what she knew then? In that case the trick, or trade-off, is to combine the dramatic intensity that comes from not knowing what is about to happen with the deeper understanding provided by the longer view – or the different understanding, anyway, since though the older narrator sees a wider context there may be things he has forgotten. Or perhaps the writer, unfashionably, is using an 'omniscient voice', in which any knowledge can be expressed that the writer possesses or gains through research. Such knowledge can be deployed in explanation of the story, regardless of whether it is shared by the characters involved. Here there will be another trade-off, or alternation, between immediate intensity and more distant perspectives.

When the story is about climate change, these questions are perplexing, since that subject demands strange combinations of perspective. There is a disjuncture between the immediate manifestations and the large-scale continuing systemic change. It is a gulf that challenges efforts to represent both together or move smoothly from one to the other. Representation of the large-scale change requires unwieldy amounts of scientific information, while the immediate manifestations, such as floods, storms and crop failures, are hard to distinguish from similar events in the past. Characters in stories may struggle dramatically to escape these events, but the danger is that such immediate struggles will become self-contained stories that do not clearly represent large-scale collective efforts to avert or mitigate climate change. A novel that set out to engage with the large-scale subject may thus fall back into a story of a more familiar kind.

Disjunctures of this kind attend all the environmental problems listed above. These problems are dispersed across large distances of space and time in a way that challenges narrative 'point of view'. This is why the eco-critic Timothy Morton calls these problems 'hyperobjects' (see Morton, 2013). They can be manifest in the form of figures, diagrams and general descriptions, but if these forms of representation cannot, without loss of the large-scale outlines, be translated into particularities – into dramatic events or symbolic or metonymic entities that fit into the frame of 'point of view' – then a gulf appears that may be a barrier to action. Effective action is difficult to envisage if the distance between the individual experience and the large-scale result to which it contributes is so great that the two cannot be held in view together, or held in the same frame. Another eco-critic, Timothy Clark, describes this impasse as a 'derangement of scales' (Clark, 2011: 136). The gulf is between large-scale and small-scale perceptions. Microcosm does not simply convert into macrocosm. Instead, there is a dangerous conflict between the two, which thwarts such attempts at conversion.

As a result of scale effects, what is self-evident or rational at one scale may well be destructive or unjust at another. Hence, progressive arguments designed to affirm individual rights and help disseminate Western levels of prosperity may even resemble, on another scale, an insane plan to destroy the biosphere. Yet, for, say, any individual household or motorist a scale effect in their actions is invisible. It is not present in any phenomenon in itself, but only in the contingency of how many other such phenomena there are, have been and will be, at even large distances in space or time.

> Can the Leviathan of humanity *en masse*, as a geological force, be represented? No, at least not in the realist mode still dominant in the novel. Its effects are global and non-localizable.
>
> (Clark, 2015: 73)

Clark sees literary realism as unable to accommodate the shift of perspective advocated by New Materialist eco-critics: a shift of emphasis in the way we imagine the self, from the self as an atomised individual with hard-boundaries to a self through which the world flows. That is, a self that is as conceptually inseparable as it is materially inseparable from the larger ecosystem that sustains its physical body. Ecocritic Hannes Bergthaller sums up the New Materialist ecocritical project as:

> a re-description of the world that dissolves the singular figure of the human subject, distinguished by unique properties (soul, reason, mind, free will or intentionality), into the dense web of material relations in which all things are enmeshed.
>
> (Bergthaller, 2014: 37)

Ecological perception dissolves unifying notions of selfhood and strong dualistic separations between culture and nature, subject and object or human and nonhuman. Instead of these hard selves and boundaries, we have assemblages, shared ancestry, coevolution, system, process, energy flow, hybridity, actor-networks, post-humanism, symbiosis, biosemiotics and the continuous mutual constitution of self and world: the system of relationships that Timothy Morton calls 'the mesh', Stacy Alaimo calls 'trans-corporeality' and New Materialist theorists call 'distributed agency'.

Is that shift as incompatible with literary realism as Clark suggests? This is an urgent question for creative writers wanting to engage with environmental concerns. Is there promise in some of the combinations of local and global achieved by recent nature writing? Jean Sprackland's *Strands*, for example, is a book about walks on a beach. Each chapter starts with something Sprackland finds there. The object prompts her to research, which carries her far distances in space and time. She zooms, for example, from plastic toys and other familiar items littering the beach to the immense accumulation of such objects in the Great Pacific Garbage Patch, and then back in time to the astonishingly recent introduction of mass-produced plastic (see Sprackland, 2012: 103–118). In such a short period – still within living memory – the world has been filled with this stuff. Such insignificant items have such threatening implication. It is a Clarkian derangement of scale, the more so because the elasticity is contained within a dramatized point of view and a short chapter recounting an afternoon's leisurely walk. Or is that containment – that ease of turning away and rejoining normality – the most ominous feature, reproducing the gulf after all? Such questions arise from close consideration of the book's form and technique, but they are large and essential questions. Since the gulf between theory and literary practice is related to Clark's other derangements of scale, the Creative Writing workshop is a good place to attempt to build bridges.

## Bibliography


Bergthaller, H. (2014) 'Limits of Agency', in Iovino and Opperman (eds), *Material Ecocriticism* (pp. 37–50). Bloomington: Indiana University Press.

Clark, T. (2011) *The Cambridge Introduction to Literature and the Environment.* Cambridge: Cambridge University Press.

Clark, T. (2015) *Ecocriticism on the Edge: The Anthropocene as a Threshold Concept.* London: Bloomsbury Academic.

Morton, T. (2013) *Hyperobjects: Philosophy and Ecology After the End of the World.* Minneapolis: University of Minnesota Press.

Sprackland, J. (2012) *Strands: A Year of Discoveries on the Beach.* London: Jonathan Cape.

# Ecocriticism

4.4

Greg Garrard

## Romantic nature

The origins of modern Western environmentalism are literary as much as they are ethical or scientific. Romantic writers in eighteenth- and nineteenth-century Europe reacted against the environmental changes caused by industrialization, and the 'disenchanting' effect of Enlightenment science, by idealizing Nature. In his 1798 poem 'The Tables Turned', William Wordsworth wrote:

> Sweet is the lore which Nature brings;
> Our meddling intellect
> Mis-shapes the beauteous forms of things:
> We murder to dissect.

For the Romantics, Nature was encountered in wild and pastoral landscapes, far from the sooty, burgeoning cities of Western Europe. It was either uninhabited or else populated by 'noble savages': supposedly primitive people like 'red Indians' who were thought to live lightly on the earth. The myth of the noble savage morphed in the twentieth century into the stereotype of the Ecological Indian, a figure of largely European invention (Krech, 1999).

The Romantic preference for dramatic landscapes, such as Wordsworth's English Lake District, directed the first efforts to preserve Nature in the late nineteenth and early twentieth century. Enthusiastic nature writers, notably John Muir in the USA, popularized the Romantic aesthetic, leading to the designation of national parks: Yellowstone (established 1872) and Yosemite (1890) were the models for numerous others around the world. In the absence of Romantic landscapes, parks like Kruger in South Africa (1926) were founded to preserve populations of wild animals for the benefit of white big game hunters. Whereas British National Parks, first established in the 1950s, allowed for continuing human inhabitation, parks in colonized nations, including Kruger and Yosemite, often required the forced removal of indigenous residents and colonial squatters to enhance their wildness. When scholars began to explore the role of literature in environmentalism in the 1990s, they initially celebrated its contribution to nature conservation movements (Bate, 1991), alongside a determined effort to revalue nonfictional nature writing. As 'ecocriticism' developed, though, its practitioners increasingly questioned the landscape bias and human impact of the idea of Nature.

## Resisting scientification

Rachel Carson's *Silent Spring*, the founding text of modern Western environmentalism, is clearly indebted to literary forebears: its evocative title alludes to John Keats's poem 'La Belle Dame Sans Merci', which is the source of one of Carson's epigraphs: 'The sedge is wither'd from the lake, / And no birds sing.' The first chapter of her book, 'A Fable for Tomorrow', depicts the world disrupted by synthetic pesticides in pastoral language, and portrays the impact of these 'elixirs of death' using apocalyptic imagery of blight and toxic miasma (Carson, 1999, pp. 21–22). Eco-critical readings of Carson treat *Silent Spring* as the epitome of what Lawrence Buell has dubbed 'toxic discourse', which constructs concern about poisoning by providing an 'interlocked set of [cultural conventions] whose force derives partly from the anxieties of late industrial culture, partly from deeper-rooted habits of thought and expression' (Buell, 2001, p. 31). For example, Carson's 'images of a world without refuge from toxic penetration' (p. 38) influenced the representation of insidious chemical pollution in Hollywood movies like *Erin Brockovich* (2000).

Regardless of *Silent Spring*'s literary sophistication, and its seminal impact on later depictions of pollution, Carson was clear that her authority derived primarily from the scientific evidence she presented. The role of literature is clearly subsidiary: ecocritics might illuminate the *forms* chosen by Carson to convey her opposition to indiscriminate use of synthetic pesticides, but the *substance* of her objections was scientific. To this day, 'environmental studies' is typically constructed with a core of scientific subjects, a periphery of social sciences, and an outer rim of the arts and humanities. While Carson herself sought greater democratic engagement in environmental governance, environmental issues are inevitably 'scientificated' into universalizing claims made by experts (Yearley, 1996, p. 140). Literary specialists might be called upon for public relations, but not analysis.

Scientification has some impressive victories to its credit. The discovery of the effects of CFCs on stratospheric ozone, and the subsequent negotiations that led to the Montreal Protocol, though facilitated by public awareness (especially the compelling metaphor of the 'hole in the sky'), exemplify the linear science–policy relationship that is the ideal form of scientification. The United Nations response to the threat of climate change attempted to use a similar model – the Intergovernmental Panel on Climate Change supplied the science that would inform the Conference of the Parties' policy implementation – with rather less success. As Mike Hulme, a former IPCC scientist, acknowledges in his book *Why We Disagree about Climate Change*, "humanity has a very long cultural history of understanding the idea of climate and of experiencing its sensual intimacy [whereas] There is no such cultural or sensual history of relating to stratospheric ozone" (Hulme, 2009, p. 292). 'Climate' was, in retrospect, a poor candidate for scientification. Despite the publication of five IPCC Assessment Reports, cultural differences continue to be conspicuously influential in public discourse about climate change.

When eco-critics began to consider climate change from a literary perspective, they identified the gap between (scientific) knowledge and (personal or political) action as a key concern. The roots of ecocriticism are in environmental activism, and it continues to attract scholars who wish to bring about social change. Thus individual novels, such as Barbara Kingsolver's *Flight Behavior* and Ian McEwan's *Solar*, and literary genres, from science fiction to modernist poetry, were assessed for their potential contributions (Kerridge, 2014). More recently, though, eco-critics have adopted a more diagnostic approach. Ursula Heise has shown that our sense of environmental risk is inherently enculturated – how else can we explain the different place of nuclear power in Germany and France, for example? – and pointed out that "Risk theorists have paid relatively little attention to the role that particular metaphors, narrative patterns, or visual representations might play in the formation of risk judgments" (Heise, 2008, p. 137). She has concluded that "literary critics' detailed analyses of cultural practices stand to enrich and expand the body of data that an interdisciplinary risk theory can build on" (Heise, 2008, p. 136). In addition to criticizing scientification on principled and pragmatic grounds, then, eco-critics can make positive contributions to understanding and resolving environmental problems.

## Eco-criticism after nature

According to Bill McKibben, a leading environmental activist and writer, climate change has brought about 'the end of Nature':

> We have changed the atmosphere, and thus we are changing the weather. By changing the weather, we make every spot on earth man-made and artificial. We have deprived nature of its independence, and that is fatal to its meaning. Nature's independence *is* its meaning; without it there is nothing but us.
>
> (McKibben, 1990, p. 54)

Given that the climate crisis has only worsened in the decades since he wrote these words, it is hard not to share McKibben's elegiac mood. At the same time, though, eco-critics are also aware that 'Nature' has historically functioned to divide, stigmatize and oppress social groups: gays have been condemned for their 'unnatural' sexuality, for instance, while women and black people have been dehumanized by depiction as 'closer to nature' than white males. As Paul Outka observes,

> This legacy – in which whites viewed black people as part of the natural world, and then proceeded to treat them with the same mixture of contempt, false reverence, and real exploitation that also marks American environmental history – inevitably makes the possibility of an *uncomplicated* union with the natural world less readily available to African Americans than it has been to whites.
>
> (Outka, 2008, p. 3)

Outka's study shows how immersion in the natural world, which a white writer such as John Muir might represent as sublime, is experienced by African American writers as traumatic.

Nature is not only a dubious ally for marginalized social groups; it also, increasingly, seems an obstacle, not an asset, for environmentalists. As literary theorist Timothy Morton puts it:

> 'Nature' fails to serve ecology well. I . . . use a capital *N* to highlight its 'unnatural' qualities, namely (but not limited to), hierarchy, authority, harmony, purity, neutrality, and mystery. Ecology can do without a concept of a something, a thing of some kind, 'over yonder', called Nature.
>
> (Morton, 2010, p. 3)

Morton praises artworks that seem to anticipate life 'after' Nature, a future condition of 'interconnectedness in the fullest and deepest sense' (p. 7) that he calls 'the ecological thought'.

Ecocriticism has arrived in a wholly unexpected place. It started out with the intention of reconnecting readers and students to Nature in a critical, reflective mode that would not have been unfamiliar to William Wordsworth. It is now, to the distress of some of its early proponents, more 'nature-skeptical' than 'nature-endorsing' (Soper, 1995, p. 34). It has, moreover, been transformed in the course of constructive debates with other literary theories, notably feminism, post-colonialism, and queer theory. Appropriately for a subfield of one of the disciplines still known as 'humanities', ecocriticism has shifted focus from 'books with trees in them' – Romantic poetry and nature writing – to art that registers the strange predicament of being human after Nature: the loneliness of life as a dominant species; the unexpected intimacies of urban ecology; the pleasures of literature that characterizes our symbionts, not merely a cast of *Homo sapiens*; the unbearable lightness of being just one of seven, eight or nine billion people.

## Learning resources

The principal academic organization for ecocritics is the Association for the Study of Literature and the Environment (ASLE), which started in the USA and now has branches and affiliated organizations around the world: www.asle.org. The leading journals for ecocritical research are *Interdisciplinary Studies in Literature and Environment* (ISLE), *Green Letters: Studies in Ecocriticism*, and *Ecozon@*.

Accessible book-length introductions to ecocriticism include Lawrence Buell, *The Future of Environmental Criticism* (Oxford: Blackwell, 2005); Timothy Clark, *A Cambridge Introduction to Literature and the Environment* (Cambridge: Cambridge University Press, 2011); Greg Garrard, *Ecocriticism* (London: Routledge, 2011).

## Bibliography

Bate, J. (1991) *Romantic ecology: Wordsworth and the environmental tradition*: London: Routledge.

Buell, L. (2001) *Writing for an endangered world: literature, culture, and environment in the U.S. and beyond*. Cambridge, MA: Belknap.

Carson, R. (1999) *Silent spring*. London: Penguin.
Heise, U.K. (2008) *Sense of place and sense of planet: the environmental imagination of the global* (Kindle edn.). Oxford: Oxford University Press.
Hulme, M. (2009) *Why we disagree about climate change: understanding controversy, inaction and opportunity*. Cambridge: Cambridge University Press.
Kerridge, R. (2014) Ecocritical approaches to literary form and genre: urgency, depth, provisionality, temporality. In G. Garrard (Ed.), *The Oxford handbook of ecocriticism* (pp. 361–376). New York: Oxford University Press.
Krech, S. (1999) *The ecological Indian: myth and history*. New York: W.W. Norton & Company.
McKibben, B. (1990) *The end of nature*. New York: Viking.
Morton, T. (2010) *The ecological thought*. Cambridge, MA: Harvard University Press.
Outka, P. (2008) *Race and nature from transcendentalism to the Harlem Renaissance*. Basingstoke: Palgrave Macmillan.
Soper, K. (1995) *What is nature? Culture, politics and the non-human*. Oxford: Blackwell.
Yearley, S. (1996) *Sociology, environmentalism, globalization: reinventing the globe*: Thousand Oaks, CA: SAGE Publications.

# 4.5 Ecological Marxism

Alf Hornborg

The theoretical framework developed by Karl Marx and Friedrich Engels in the mid-nineteenth century to explain the underlying logic of the Industrial Revolution was founded on the insight that the economy can only be understood by juxtaposing the social sphere of exchange values and the natural sphere of biophysical metabolism. Indeed, this is the essence of 'historical materialism' as a scientific theory of capital accumulation. From a modern perspective it appears that Marx's main challenge was to account for the conversion, in production, of physical energy into exchange value. As a century and a half of debate illustrates, however, his conceptualization of this relation was far from final. The modern concept of energy was not established in physics until around 1860, and to the extent that early economists aspired to model their science on physics, they were unable to draw on the Second Law of Thermodynamics, the so-called entropy law (Mirowski, 1988; Georgescu-Roegen, 1971). A long line of theorists have, over the course of a century and a half, aspired to reinterpret or modify Marx's framework. While united by the conviction that his critical understanding of industrial capitalism (as based on compulsive accumulation, exploitation, and alienation) is fundamentally correct, their debates have highlighted many of the inconsistencies and ambiguities in his texts. As for Marx himself, the recurrent problem for these theorists has been how to handle the relation between the economic and the biophysical – the social and the natural. Over the past 30 years, augmented by the expansion of environmental concerns, this interdisciplinary dilemma has become the focus of debates within the field known as *ecological* Marxism.

## The ambiguous relation between energy and value

In the 1980s, after two decades of environmentalist discourse, several socialist theorists criticized mainstream Marxism for its neglect of ecological issues. Joan Martinez-Alier and José M. Naredo (1982) observed that the response, in the early 1880s, of Friedrich Engels to the ideas of Sergei Podolinsky reflected the difficulties of integrating Marxian value theory and thermodynamics. The observation is repeated in Martinez-Alier's (1987) book *Ecological Economics*. Podolinsky had recognized that Marx was aiming to articulate a natural science of society by explaining economic value with reference to the investment of physical labor-power in production. Engels nevertheless dismissed his proposal that Marx's concept of 'surplus value' could be understood in terms of energy. In hindsight, this failure of communication illuminates how Marx's aspiration to analytically merge nature and society was blocked by the incommensurable ontologies of physicists and economists, and by the conceptual confusion accompanying the transition from organic to inorganic energy in production. John Bellamy Foster's and Paul Burkett's efforts (Foster & Burkett, 2004, 2008; Burkett & Foster, 2006) to justify Engels' dismissal of Podolinsky's energy theory of value as reductionist are valid, but unfortunately they make no effort to understand what drove Podolinsky to contact Marx in the first place, or how his proposal ultimately highlights some fundamental flaws in the Marxian labor theory of value. Paradoxically, Foster and Burkett themselves repeatedly assert that Marx's theory of surplus value is a matter of net energy flows from workers to capitalists (e.g., Burkett & Foster 2006: 120, 126; Foster & Burkett, 2008: 26, 29) and confirm the Marxian axiom that the increase of economic value in production can be analytically derived from expenditures of labor energy. A step forward would have been to conclude that *any* energy theory of economic value is reductionist (Georgescu-Roegen, 1979), regardless of whether that energy is provided by fuels, draft animals, or human labor. Energy and economic value should be kept analytically distinct. Quite to the contrary, Foster has recently endorsed (Foster & Holleman 2014) the explicit energy theory of value of Podolinsky's intellectual heir Howard T. Odum (1971, 1988).

## Critiques of classical Marxist 'productivism'

In 1989, Ted Benton (1989) identified a general technological optimism in Marxist theory which underplayed natural limits to production and thus seemed to make it ill-equipped to deal with environmental problems. He traced this 'Promethean' approach to the discourse of classical political economy in which Marx was involved. This historical context contributed, for instance, to his labor theory of value and his Enlightenment belief in the progress of technology. Similar, ecological critiques of Marxism were published by André Gorz (1994), Alain Lipietz (2000), and Joel Kovel (2002). Several ecological feminists have also questioned mainstream Marxian productivism and the labor theory of value (Soper, 1991; Salleh, 1994; Brennan, 1997).

In 1988, James O'Connor launched the first issue of the eco-Marxist journal *Capitalism Nature Socialism* with a theoretical introduction (O'Connor, 1988) outlining his proposition that capitalism suffers from not one but two crisis-generating contradictions. In addition to the tendency of capital to keep wage-based demand for its commodities below what is necessary to realize their surplus value, the second contradiction of capitalism denotes the tendency of capital to degrade its 'conditions of production', including the natural environment. Ten years later, the text was reprinted in O'Connor's (1998) essay collection *Natural Causes*, aimed to complement the classical Marxist framework by developing its potential for theorizing natural constraints and identifying the actors of socio-environmental conflicts, importantly demonstrating that the global protagonists of 'environmental justice' are generally external to the industrial proletariat. The two contradictions of capitalism, in this view, thus oppose different categories of social actors.

## Defense of Marxism as inherently ecological

In a review of O'Connor's book, Paul Burkett (1999a) criticizes him for accusing Marx and Engels of having omitted ecological concerns, rather than acknowledging the ecological insights of the nineteenth-century founders of Marxism. In a book published the same year, Burkett (1999b) uses exegesis of nineteenth-century texts to demonstrate that Marx's theory of value was indeed intended to accommodate the natural, material aspects of the capitalist economy by conceptualizing them as 'use values' obscured by the exchange values of the market. The following year, John Bellamy Foster published his book *Marx's Ecology* (Foster, 2000), in which he similarly showed that there is plenty of textual evidence that Marx himself was highly concerned about ecological issues, particularly what Foster calls the 'metabolic rift' in nutrient flows between rural and urban areas. In his view, the abandonment of Marx's ecological concerns in twentieth-century Western Marxism was a consequence of the critique of technological rationality and Enlightenment by the Frankfurt School (cf. Foster & Clark, 2016). In several more recent publications, Burkett and Foster have continued to defend Marx and Engels against charges of Prometheanism and ecological ignorance (Burkett, 2005; Foster et al., 2010; Foster & Burkett, 2016). This defensive posture has puzzled many socialists, who believe that more would be gained from updating and modifying Marx's theoretical framework than from denying any indication of inconsistency or ambiguity in it (cf. Rudy, 2001).

## Distinguishing energy and value

To pursue a genuinely interdisciplinary ecological Marxism should mean unpacking the concept of underpaid "use values." In conflating a commodity's material properties and its anthropocentric (and inevitably cultural) *value*, it misleadingly suggests that there is a common metric for assessing the discrepancy between its natural and social features. When Foster sets out to reframe the theory of ecologically unequal exchange in classical Marxist terms, he and Hannah Holleman (Foster & Holleman, 2014) thus suggest that asymmetric transfers of biophysical resources on the world market are to be understood as unequal exchanges of *values*. This is to confuse a cultural and a material perspective on the economy. The significance of asymmetric transfers of material resources such as energy is not that they represent underpaid values, but that they contribute to the physical expansion of productive infrastructure at the receiving end. The accumulation of such technological infrastructure may yield an expanding output of economic value, but this is not equivalent to saying that the resources that are embodied in infrastructure have an objective value in excess of their price.

In recent decades, the essential role of fossil energy in defining the operation and environmental history of industrial capitalism has been reconceptualized by several researchers who would no doubt identify with the label of ecological Marxism (Deléage, 1989; Altvater, 1994, 2007; Hornborg, 2001, 2013; Huber, 2008, 2013; Mitchell, 2011; Barca, 2011; Malm, 2016). Rethinking the relation between energy and exchange value, and its transformation by the turn to fossil fuels, makes it possible to pose new questions regarding the intuitions underlying Marx's conceptual framework.

## Learning resources

The following journals all contain many articles about Marxism and the environment:

*Capitalism Nature Socialism*: http://www.tandfonline.com/toc/rcns20/current.
*Climate and Capitalism*: http://climateandcapitalism.com.
*Historical Materialism*: http://www.historicalmaterialism.org.
*Monthly Review*: http://monthlyreview.org.

## Bibliography


Altvater, E. (1994) 'Ecological and Economic Modalities of Time and Space', in O'Connor, M. (ed.) *Is Capitalism Sustainable? Political Economy and the Politics of Ecology* (pp. 76–90), New York: Guilford.

Altvater, E. (2007) 'The Social and Natural Environment of Fossil Capitalism', *Socialist Register*, 43: 37–59.

Barca, S. (2011) 'Energy, Property, and the Industrial Revolution Narrative', *Ecological Economics*, 70: 1309–1315.

Benton, T. (1989) 'Marxism and Natural Limits: An Ecological Critique and Reconstruction', *New Left Review*, 178: 51–86.

Brennan, T. (1997) 'Economy for the Earth: The Labour Theory of Value without the Subject/Object Distinction', *Ecological Economics*, 20: 175–185.

Burkett, P. (1999a) 'Fusing Red and Green', *Monthly Review*, 50(9): 47–56.

Burkett, P. (1999b) *Marx and Nature: A Red and Green Perspective*, New York: St. Martin's Press.

Burkett, P. (2005) *Marxism and Ecological Economics: Toward a Red and Green Political Economy*, Leiden: Brill.

Burkett, P. & Foster, J.B. (2006) 'Metabolism, Energy, and Entropy in Marx's Critique of Political Economy: Beyond the Podolinsky Myth', *Theory and Society*, 35: 109–156.

Deléage, J.P. (1989) 'Eco-Marxist Critique of Political Economy', *Capitalism Nature Socialism*, 1(3): 15–31.

Foster, J.B. (2000) *Marx's Ecology: Materialism and Nature*, New York: Monthly Review Press.

Foster, J.B. & Burkett, P. (2004) 'Ecological Economics and Classical Marxism: The "Podolinsky Business" Reconsidered', *Organization & Environment*, 17(1): 32–60.

Foster, J.B. & Burkett, P. (2008) 'Classical Marxism and the Second Law of Thermodynamics: Marx/Engels, the Heat Death of the Universe Hypothesis, and the Origins of Ecological Economics', *Organization & Environment*, 21(1): 3–37.

Foster, J.B. & Burkett, P. (2016) *Marx and the Earth: An Anti-Critique*, Leiden: Brill.
Foster, J.B. & Clark, B. (2016) 'Marx's Ecology and the Left', *Monthly Review*, 68(2): 1–25.
Foster, J.B. & Holleman, H. (2014) 'The Theory of Unequal Ecological Exchange: A Marx-Odum Dialectic', *The Journal of Peasant Studies*, 41(2): 199–233.
Foster, J.B., Clark, B. & York, R. (2010) *The Ecological Rift: Capitalism's War on the Earth*, New York: Monthly Review Press.
Georgescu-Roegen, N. (1971) *The Entropy Law and the Economic Process*, Cambridge, MA: Harvard University Press.
Georgescu-Roegen, N. (1979) 'Energy Analysis and Economic Valuation', *Southern Economic Journal*, 45: 1023–1058.
Gorz, A. (1994) *Capitalism, Socialism, Ecology*, London: Verso.
Hornborg, A. (2001) *The Power of the Machine: Global Inequalities of Economy, Technology, and Environment*, Walnut Creek: AltaMira.
Hornborg, A. (2013) 'The Fossil Interlude: Euro-American Power and the Return of the Physiocrats', in Strauss, S., Rupp, S. & Love, T. (eds), *Cultures of Energy: Power, Practices, Technologies* (pp. 41–59), Walnut Creek: Left Coast Press.
Huber, M.T. (2008) 'Energizing Historical Materialism: Fossil Fuels, Space and the Capitalist Mode of Production', *Geoforum*, 40: 105–115.
Huber, M.T. (2013) *Lifeblood: Oil, Freedom, and the Forces of Capital*, Minneapolis: University of Minnesota Press.
Kovel, J. (2002) *The Enemy of Nature: The End of Capitalism or the End of the World?*, London: Zed Books.
Lipietz, A. (2000) 'Political Ecology and the Future of Marxism', *Capitalism Nature Socialism*, 11: 69–85.
Malm, A. (2016) *Fossil Capital: The Rise of Steam Power and the Roots of Global Warming*, London: Verso.
Martinez-Alier, J. (1987) *Ecological Economics: Energy, Environment and Society*, Oxford: Blackwell.
Martinez-Alier, J. & Naredo, J.M. (1982) 'A Marxist Precursor of Energy Economics: Podolinsky', *The Journal of Peasant Studies*, 9(2): 207–224.
Mirowski, P. (1988) 'Energy and Energetics in Economic Theory: A Review Essay', *Journal of Economic Issues*, 22(3): 811–830.
Mitchell, T. (2011) *Carbon Democracy: Political Power in the Age of Oil*, London: Verso.
O'Connor, J. (1988), 'Capitalism, Nature, Socialism: A Theoretical Introduction', *Capitalism Nature Socialism*, 1(1): 11–38.
O'Connor, J. (1998) *Natural Causes: Essays in Ecological Marxism*, New York: Guilford.
Odum, H.T. (1971) *Environment, Power, and Society*, New York: Wiley-Interscience.
Odum, H.T. (1988) 'Self-Organization, Transformity, and Information', *Science*, 242: 1132–1139.
Rudy, A. (2001) '*Marx's Ecology* and Rift Analysis', *Capitalism Nature Socialism*, 12(2): 56–63.
Salleh, A. (1994) 'Nature, Woman, Labor, Capital: Living the Deepest Contradiction', in O'Connor, M. (ed.), *Is Capitalism Sustainable? Political Economy and the Politics of Ecology* (pp. 106–124), New York: Guilford.
Soper, K. (1991) 'Greening Prometheus', in Osborne, P. (ed.), *Socialism and the Limits of Liberalism* (pp. 271–293), London: Verso.

# Ecopoetry

4.6

Samantha Walton

*Ecopoetry* can be defined as poetry that addresses, or can be read in ways that address, the current conditions of our environmental crisis. *Ecopoetics* refers to its theorisation, and *ecopoets*, of course, to the writers themselves—although few writers adopt the label without some qualification. These terms emerged in the late 1990s. Since then, they have become increasingly recognised, and their meanings debated and honed by scholars and writers. Magazines dedicated to new ecopoetry have been established, and the publication of ecopoetry anthologies has contributed to the formation of a new canon of ecologically-oriented verse. Scholarly work has furthered this project of retrospective canon-formation, as ecocritics have sought to trace and name alternative environmental traditions in Western and non-Western literary canons.

Ecopoetry can, then, be divided into two categories: that which is consciously written as 'ecopoetry', and that which has been claimed or reclaimed as such. For living poets writing environmental or nature-oriented poetry, the label is one that they might choose to adopt, or from which they might separate themselves. By adopting the label, poets make a conscious statement about the intentions and orientations of their work. For example, authors and critics often propose that ecopoetry might lead humanity back into conscious awareness of the ecological entanglement and stimulate care and concern. Seen in this way, ecopoetry can be framed as an active and activist form of writing and reading, contributing to the task of repairing divisions between humanity and the ecosystems that constitute and support us. There is no one form or style definitive of ecopoetry. In terms of content and theme, ecopoetry might focus attention on the language used to describe nature, be it poetic, scientific, technical or commonplace. It might expose tropes and traditions of nature representation, and challenge dominant discourses such as landscape aesthetics, cartography, or environmental economics. Ecopoetry may draw attention to specific places in order to deepen understanding of natural processes and cultural histories, or reflect on the kinds of attachments and feelings people experience in relation to the more-than-human world.

Since its inception, ecopoetry has been entangled with environmentally-oriented literary criticism. In attempting to expose the cultural-roots of environmental crisis, ecocritics interrogated constructions of nature in Western intellectual history: in particular, they examined the pastoral visions of harmony, retreat, and nostalgia derived from Virgil's *Eclogues* (see Gifford, 2010), and the Romantic construction of Nature as "a dynamic, living, self-transforming whole" (Rigby, 2004, p. 24), of which we are part. The first theorisations of ecopoetry came from critics, not from poets, and arose in ecocritical reappraisals of Romanticism. Debates about whether Romanticism really did *romanticise* nature persist, and they influence ecopoetry in decisive ways. Some critics insist that Romantic poetry established an attitude to the natural world that is of value to contemporary writers who try to address the environmental crisis through literature that sparks the imagination and the senses. For others, such as Timothy Morton, Romantic Nature-worship and yearning for restored unity only deepens the separation: "Nature fails to serve ecology well" (2010, p. 3).

Lawrence Buell's definition of 'environmental literature' provides another key starting point for theories of ecopoetics. According to Buell (1995), not all writing about nature is 'environmental', but that which is may a) show how human and natural history are interconnected; b) represent nature as a process, not a constant given or static; c) express ethical concern that extends beyond the human; and/or d) acknowledge responsibility for anthropogenic environmental damage (pp. 7–8). This 'it-is-or-it-isn't' approach was pursued by Jonathan Bate in *The Song of the Earth* (2000). Here, he defines ecopoetry as "not a description of dwelling with the earth, not a disengaged thinking about it, but an experiencing of it" (p. 42). For Bate, Wordsworth comes out as an (or rather, *the*) ecopoet, because of his attention to the affective, emotional, ethical and intellectual impact that nature has on the lyric 'I'. Ecopoetry is, in his definition, more phenomenological than it is political. It is concerned less with the kinds of representational politics essential to feminist and critical race theory, and more with the evocation or stimulation of a mood, tone, or force of attachment capable of remaking the lost connection with nature: what he terms an 'ecotone'. Language organised as poetry need not separate us from the earth (as Bate believed was the prevailing theory amongst postmodern literary critics at the time), but could be a means of "answering nature's own rhythms, and echoing of the song of the earth itself" (p. 76).

Unsurprisingly—given Bate's depoliticisation of the moment of 'nature-experience'—the ecopoets he names are white, Western men, whose major obstacles to unmediated contact with nature are the conditions of modernity, knowledge of science, the 'meddling' intellect, and language itself. Ecopoetry was first established, then, as the preserve of "a rather exclusive club of neo-romantic, male poets", as Harriet Tarlo (2007) puts it. This selective bias is reproduced in many studies; for example, in *Sustainable Poetry* (1999), Leonard Scigal selects four white men, including Gary Snyder and Wendell Berry, as America's foremost ecopoets. Their poetry is able to ponder the referential capacity of language, and the question of whether it brings humanity closer to the natural world or pushes us further away, without any of the messy complications presented by race, gender, or sexuality.

Many writers now consciously reject Romanticism as a source of ecopoetics. Romanticism, after all, addressed historically specific conditions of urban expansion and agricultural and industrial revolutions, responding with National Parks and nature reserves.

Such thinking and organising is arguably incapable of addressing the globalised conditions of our environmental crisis. It also excludes non-white, non-Western environmental understandings and spiritualties, and instead centres the experiences of able-bodied men trained in the intellectual traditions of the European philosophy. The existential and emotional dramas of those writers have proved to be far from universal. Camille Dungy (2009) explains: "Many black writers simply do not look at their environment from the same perspective as Anglo-American writers . . . The pastoral as diversion, a construction of a culture that dreams, through landscape and animal life, of a certain luxury or innocence, is less prevalent." Instead, poems are "written from the perspective of the workers of the field"; these poems are undeniably pastorals, which "describe moss, rivers, trees, dirt, caves, dogs, fields: elements of an environment steeped in a legacy of violence, forced labor, torture, and death" (2009, p. xxi). Bearing this out, Lucille Clifton's 'Surely I am able to write poems' reveals a conflicting and deeply ambivalent perspective on nature, informed by African American history, Western Romanticism and personal sense-experiences of nature-contact. The poem begins by celebrating the natural world, but ends with the question: "why / is there under that poem always / an other poem?" (Dungy, p. vii). LaTasha N. Nevada Diggs's 'My First Black Nature Poem™' (2013) deals with Nature™ as an artificial construct which excludes black bodies and colludes in the erasure of legacies of slavery and racist violence in the supposedly pristine American wilderness: "that water got too much memory." Attention to black nature writing does not just reveal the political dimension of cultural histories of nature: it is fundamental to understanding the ways in which perception of nature is never free from social conditions. The affective quality of the moment of nature-contact which ecopoetry, according to Bate, is meant to capture, will be qualitatively different depending on the extent to which one is both vulnerable to and traumatised by bigotry and oppression.

Poets and critics have begun to chart alternative histories of ecopoetry, broadening out the initial emphasis on the Romantic heritage, recognising the contributions of women and poets of colour, incorporating more intersectional approaches to nature and ecology, and exploring the ecopoetical significance of experimental and avant-garde traditions. Ann Fisher-Wirth and Laura-Grey Street's *The Ecopoetry Anthology* (2013) includes Harlem Renaissance writers Langston Hughes and Jean Toomer, whose poem 'Reapers' addresses experiences of dislocation, diaspora and oppressive land-relations under conditions of slavery and rural poverty (p. 59). The catastrophic social and environmental injustices of colonisation and its aftermath are reflected across diverse postcolonial ecopoetries. In 'Genocide, Again' Kwame Dawes connects a land overgrown and depopulated of human inhabitants through brutality and slaughter, while Craig Santos Perez's 'All with ocean views' uses found-text to critique the ways his native Guam has become an idyllic tourist destination for its former and current military occupiers at the expense of displaced indigenous Chamorro people (Fisher-Wirth & Street, 2013, p. 225; p. 461).

The critique of Romanticism and—perhaps more importantly—its legacies, is particularly potent in these works. This is not least because of the modes of dwelling and inhabitation favoured by early theorists of ecopoetry are either not relevant to indigenous environmental understandings, or have so utterly colonised the imaginations of Western tourists and indigenous people alike that many native understandings have been lost along with land rights, languages, and mutually sustaining ecological relations. Decolonising ecopoetics might take the form of reclaiming lost and forcibly suppressed indigenous language. Such an approach is found in diverse postcolonial literatures. For example, the poet Hugh MacDiarmid found that Scots Vernacular words 'watergaw' and 'yow-trummle' described 'natural occurrences and phenomena of all kinds which have apparently never been noted by the English mind. No words for them exist in English' (Grieve, 1923, p. 28). Another example is the Anishinaabe word 'puhpowee' rediscovered by Robin Wall-Kimmerer, an ethnobiologist and member of the Potawatomi tribe. Puhpowee is "the force which causes mushrooms to push up from the earth overnight" (2015). This word reveals an understanding of natural processes outside the grasp of Western science, delimited as it is by an imprecise technical vocabulary. Although she is not a poet, Wall-Kimmerer's work has been picked up by poets interested in the ways in which language influences socio-cultural perception of the natural world, and behaviour towards it (see Keller, 2015).

Critical gender theory has also contributed to ecopoetics. Feminists and queer theorists have focused on exposing and critiquing binary understandings which underpin human–nature relations, and which revolve around constructions of man/woman, mind/body, intellect/emotion and, of course, culture/nature. While much early ecofeminism was deeply essentialist in its understanding of gender, more recent ecofeminist and queer scholarship has focused on revealing the ways in which hierarchical dualisms have reinforced the exploitation of women, queer people and nature. In ecopoetry, this might mean addressing both the ways in which nature-appreciation and its expression has been gendered, and the obstacles women writers face in a patriarchal society. For example, a poem by Lila Matsumoto responds to Dorothy Wordsworth's writing, including the gendered division of labour on walking excursions with her brother William: "(the men write their poems) . . . she looks for cottages where they might take refreshments and pass the night" (Matsumoto, 2016).

As with all feminist scholarship, feminist contributions to ecopoetics have involved retrospective recuperation of women's writing. Anglo-American modernists H.D., Muriel Rukeyser and Marianne Moore have been reclaimed as ecopoets. The attempt to inhabit non-human subject positions in H.D.'s 'Oread', or Moore's attention to the movement and points of contact between bodies and matter in 'The Fish', each avoid altogether the experiences of the lyric 'I' (Fisher-Wirth & Street, p. 40; p. 48). For the poet and editor Harriet Tarlo, the rejection of the 'dominant' and 'domineering' lyric 'I' is an essential move in escaping from legacies of Romanticism (Fisher-Wirth, qtd. in Tarlo, 2007). In the work collected in her *How2 Journal* special edition, poems resist the lyric 'I' through multi-voice translation pieces or the use of found text as a form of ecological practice modelled on recycling. Ecopoetry moves, then, from being a late-Romanticist movement to a late-Modernist one. Formal innovations and conceptual practices are used to denaturalise 'nature' language; to reveal socio-environmental interdependencies; and to model ecological relations. According to Lynn Keller, "experimental poetics . . . might be helping shift our sense of human–nonhuman relations away from the anthropocentric and might enhance our sense of kinship and interdependence with other life forms" (Keller, 2015).

Experimental poetry shares ground and techniques both with disability ecopoetics and queer ecopoetics when it crosses imaginative boundaries, and challenges humanity's supposed separateness from the 'natural world' and from other bodies. Through multimedia works, collaboration and performances, poets stage encounters between bodies which transgress normative forms of intimacy (see Kuppers & Leto, 2012). As Angela Hume states: "*non*normative intimacies are the stuff of disability ecopoetics—a poetics of interrelation between humans and other-than-humans on a shared path" (2013). Ecopoetics here means an orientation towards bodies and matter that can play out in a range of ways, often unexpected, in individual poems or creative projects.

New terms are being coined to better distinguish between—or further worry at—different kinds of ecologically-oriented work, including biopoetics; lithopoetics (stone poetics, see Weishaus, 2010); hydropoetics (influenced by the connective capacities of water); Mestizo poetics (created through cultural clash and hybridity in Latin America context, see Vicuña, 2009); and what Tarlo terms 'radical landscape poetry', concerned with new approaches to landscape representation. It is likely that ecopoetics will remain an overarching term for referring to consciously environmental writing and practice, albeit one which writers, critics, editors and readers will continue to question, redefine and transform.

## Learning resources

The following are all useful and interesting resources:

### *Online*

*Ecopoetics*. Publishing 2001–2009. https://ecopoetics.wordpress.com.
*Ecotone*. Publishing since 2005. https://ecotonemagazine.org.
*Epizootics Zine: Online Literary Journal for the Contemporary Animal*. Publishing since 2016. https://epizooticszine.wordpress.com.
Jonathan Skinner has contributed a number of posts on ecopoetry to *Jacket2*: http://jacket2.org/commentary/jonathan-skinner.

### *Print*

Abs, P. (ed.) (2002). *Earth Songs: A Resurgence Anthology of Contemporary Eco-poetry*. Dartington: Green Books.
Dungy, C. (ed.) (2009) *Black Nature Poetry: Four Centuries of African American Nature Poetry*. Athens: University of Georgia.
Tarlo, H. (ed.) (2011). *The Ground Aslant: Radical Landscape Poetry*. Bristol: Shearsman Books.

## Bibliography


Bate, J. (2000) *The Song of the Earth*, London: Macmillan.
Bryson, S.J. (2002) *Ecopoetry: A Critical Introduction*, Salt Lake City: University of Utah Press.
Buell, L. (1995) *The Environmental Imagination: Thoreau, Nature Writing and the Formation of American Culture*, Cambridge and London: Harvard University Press.
Diggs, L.N.N. (2013) 'My First Black Nature Poem™' *Poetry Foundation*. Online. Available at: https://www.poetryfoundation.org/poems-and-poets/poems/detail/56492 [Accessed 2.12.16].
Dungy, C. (ed.) (2009) *Black Nature Poetry: Four Centuries of African American Nature Poetry*, Athens: University of Georgia.
Fisher-Wirth, A. & Street, L.G. (eds) (2013) *The Ecopoetry Anthology*, San Antonio: Trinity University Press.

Gifford, T. (2010) *Pastoral*, Oxford: Routledge.

Grieve, C.M. [Hugh MacDiarmid] (1923/2004) 'Causerie: A theory of Scots letters' in M. McMulloch (ed.) *Modernism and Nationalism: Literature and Society in Scotland* (pp. 26–28), Glasgow: ASLS.

Hume, A. (2013) 'Queering ecopoetics: Nonnormativity, (anti)futurity, precarity' in *Jacket2*. Online. Available at: http://jacket2.org/commentary/queering-ecopoetics-nonnormativity-antifuturity-precarity [Accessed 2.12.16].

Keller, L. (2015) 'a. rawlings: Ecopoetic Intersubjectivity' in *Jacket2*. Online. Available at: http://jacket2.org/article/arawlings-ecopoetic-intersubjectivity [Accessed 2.12.16].

Kuppers, P. & Leto, D. (2012) *A Radiant Approaching* (film poem). Online. Available at: https://www.youtube.com/watch?v=WihPJuFaBNg [Accessed 2.12.16].

Matsumoto, L. (2016) from 'She points out features of the landscape'. *Wordsworth and Bashō; Walking Poets; Encounters with Nature*. Kakimori Bunko, Itami, Japan. 17 September–3rd November 2016. Exhibition.

Morton, T. (2010) *The Ecological Thought*. Cambridge: Harvard.

Rigby, K. (2004) *Topographies of the Sacred: The Poetics of Place in European Romanticism*, Charlottesville: University of Virginia Press.

Scigal, L. (1999) *Sustainable Poetry: Four American Ecopoets*, Lexington: University of Kentucky Press.

Tarlo, H. (2007) 'Women and ecopoetics: an introduction' in Women and Ecopoetics, *How2 Journal*. 3(2). Online. Available at: <https://www.asu.edu/pipercwcenter/how2journal/vol_3_no_2/ecopoetics/introstatements/tarlo_intro.html> [Accessed 2.12.16].

Vicuña, C. (2009) 'An Introduction to Mestizo Poetics', in E. Livon-Grosman and C. Vicuña (eds) *The Oxford Book of Latin American Poetry: A Bilingual Anthology* (pp. xix–xxxii). Oxford: Oxford University Press.

Wall-Kimmerer, R. (2013) *Braiding Sweetgrass: Indigenous Wisdom, Scientific Knowledge and the Teachings of Plants*. Minneapolis: Milkweed Editions.

Wall-Kimmerer, R. (2015) 'Learning the grammar of animacy', *The Moon Magazine*. Online. Available at: <http://moonmagazine.org/robin-wall-kimmerer-learning-grammar-animacy-2015–01–04> [Accessed 2.12.16].

Weishaus, J. (2010) 'Toward a Lithopoetics', *Splitting the Stone*. Online. Available at: <http://www.cddc.vt.edu/host/weishaus/Splitting/Page.htm> [Accessed 2.12.16].

# Earth System Science[1]

Noel Castree

4.7

A system is any natural or humanly created entity comprised of functionally interdependent parts connected by flows of energy. Earth System Science (ESS) goes back 30 years and today has considerable momentum. As one commentator puts it, ESS "embraces chemistry, physics, biology, mathematics and applied sciences in transcending disciplinary boundaries to treat the Earth as an integrated system and seeks a deeper understanding of the physical, chemical, biological and human interactions that determine the past, current and future states of the Earth" (Ruzek, 2013: n.p.). ESS thus has grand intellectual ambitions. As a science it uses evidence, logic and physical laws to identify patterns, trends and explanations that aspire to accurately represent the interactions between the components of the Earth surface (namely, the atmosphere, hydrosphere, lithosphere, cryosphere and biosphere). However, unlike the specialist sciences (such as marine biology) ESS aims to 'join the dots' by identifying relations and feedbacks between Earth surface phenomena often studied in relative isolation. Because the Earth surface comprises a set of interlinked 'open systems' existing at a range of spatio-temporal scales, ESS's quest to be interdisciplinary has faced some formidable intellectual challenges, as this chapter will explain. Geologist Richard Alley (2000: 7) likened it to creating the world's most detailed ever 'operator's manual' for a 'machine' of unequalled complexity.

## The origins of ESS

The precursor to Earth System Science was the idea of an 'operator's manual for spaceship Earth' formulated in 1968 by the remarkable American architect-inventor Richard Buckminster Fuller. Fuller criticized academic specialization and challenged his readers to aspire to 'total knowledge' of a world they were fast-changing. He wrote at a time when humans, courtesy of the American Apollo space missions, had seen their planet from space for the first time. The term Earth System Science originated 17 years later with a British mathematician and modeller of ocean and atmosphere dynamics, Francis Bretherton (1985). A University of Wisconsin professor, in 1983 he was appointed by NASA (America's National Aeronautics and Space Administration) to chair a committee convened to consider how the Administration could most effectively fulfil its mission to observe the Earth surface. The committee comprised 16 members covering meteorology, atmospheric chemistry, marine biology, plant ecology, soils and vegetation interactions, agronomy, and geophysics (among other fields).

After a series of meetings in 1984 and 1985, the committee produced a report entitled *Earth System Science Overview* (NASA, 1986). The report included a diagrammatic presentation of the Earth system that would subsequently circulate widely in the research communities devoted to understanding global environmental change (see Figure 4.7.1). The diagram expressed the holistic view recommended 12 years earlier in a (now quite famous) paper. That paper proposed that we see Earth as a self-regulating system and was authored by James Lovelock and Lynn Margulis (1974). Lovelock, a former NASA employee, went on to promote the so-called 'Gaia hypothesis'. One aspect of this, which few took seriously in the early 1970s, was that humans' combined activities might be sufficiently powerful to set in train alterations to the Earth system as a whole, not just one or other component of it.

As this diagram made clear 30 years ago, a number of important ingredients would be required if ESS was to become more than a grand, possibly idealistic, research idea. First, mechanisms for fostering cross-disciplinary exchange among the various geosciences would be essential. Second, high resolution data about Earth surface patterns and processes at a range of spatio-temporal scales would be required. Third, there would be a need to ensure international academic cooperation among researchers in all parts of the world. Finally, there would be a need for computational and analytical techniques able to portray (and predict) complex Earth system dynamics across geographical space and through time. Recognizing this, NASA quickly used the Bretherton committee report to involve the USA's National Science Foundation (NSF) and the National Oceanic and Atmospheric Administration (NOAA) in its aspiration to make ESS a reality.

One result of this outreach was the creation of the federal Global Change Research Program (GCRP) in 1989, underpinned by the 1990 Global Change Research Act. This program, involving multiple federal government agencies, committed considerable financial resources to promoting integrated environmental research. Its original strategic vision was three-fold: "1. Establish an Integrated, Comprehensive Monitoring Program for Earth System Measurements on a Global Scale; 2. Conduct a Program of Focused Studies to Improve Our Understanding of the Physical, Chemical, and Biological Processes that Influence Earth System Changes and Trends on Global and Regional Scales; and 3. Develop Integrated Conceptual and Predictive Earth System Models" (Committee on Earth Sciences, 1989: 11–12). This trio of aims was framed by the overarching desire to "establish the scientific basis for national and international policymaking related to natural and human-induced changes in the Earth System" (p. 9). Because of America's leading role in academic and applied research globally, the creation of the GCRP had an almost immediate impact on research programmes into anthropogenic environmental change elsewhere. For instance, the UK's Natural Environment Research Council (NERC) announced an ESS initiative of its own from 2001. Around the same

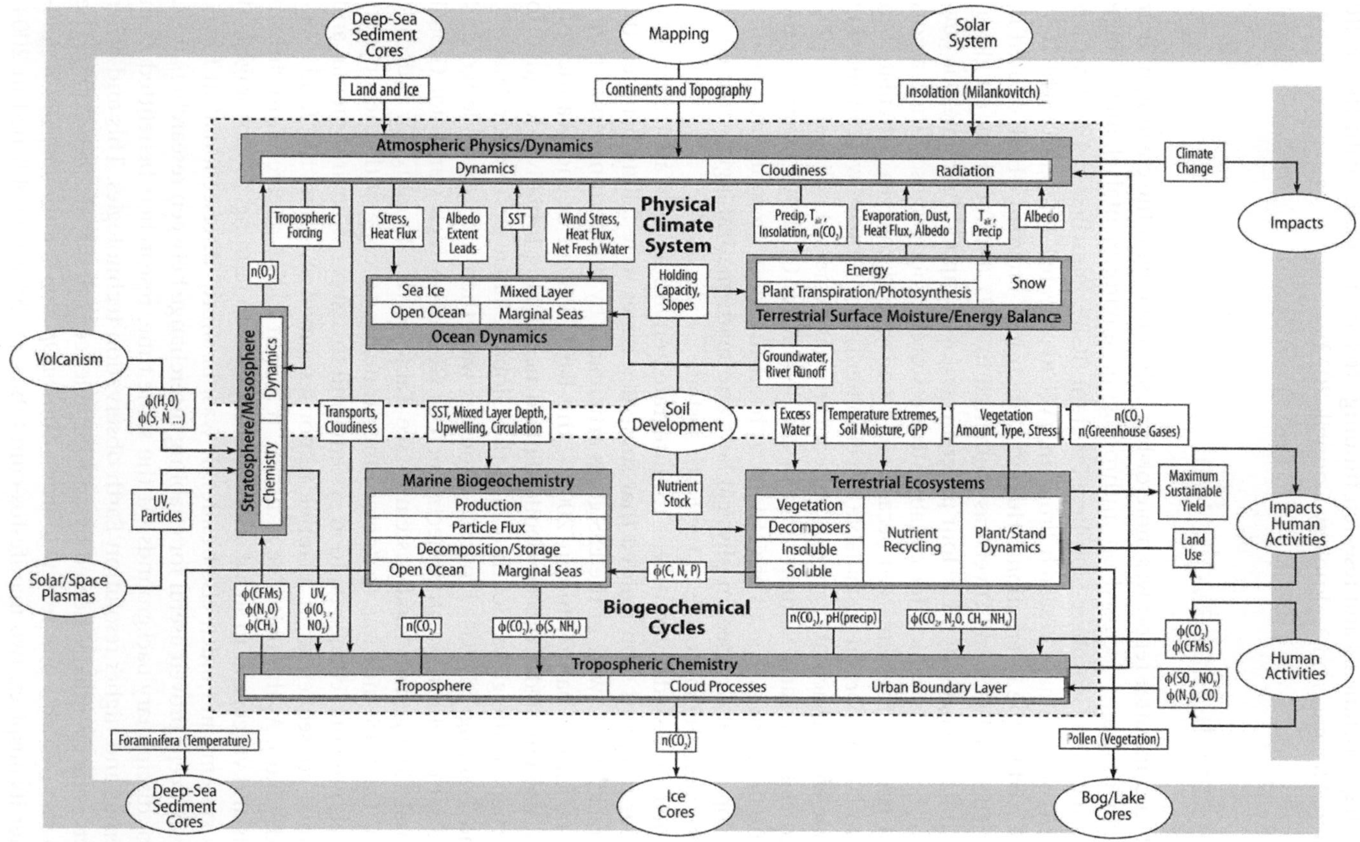


$\phi$ = flux, n = concentration

**Figure 4.7.1** Conceptual mode of Earth system processes and interrelations.

*Source*: Reproduced from NASA (1986: 24–25).

time, leading German physicist Hans Joachim Schellnhuber (1999) made 'Earth system analysis' central to the Potsdam Institute for Climate Impact Research (founded 1992 and today a 'hotspot' of ESS). Meanwhile, the term 'Earth system' is routinely found in the latest strategy documents and research funding streams associated with the GCRP (e.g. see National Science and Technology Council, 2012).

## ESS's international evolution

The Bretherton committee report was more-or-less coincident with the creation of four global research programs focussed on anthropogenic environmental change. These were the International Geosphere-Biosphere Program (IGBP, launched in 1987), which followed the World Climate Research Program, (WCRP, created 1980), and which in turn was followed by the International Human Dimensions Program (IHDP created in 1990, re-launched in 1996) and Diversitas (launched in 1991 and focussing on global biodiversity and biogeography). The four programs enjoyed support by numerous governments and became a vehicle for collaborative research among geoscientists across international borders. They emerged because of a widespread concern, articulated in the 1987 Bruntland Report on Sustainable Development, that humans were instigating planetary-scale environmental changes. That concern was expressed politically in 1992 at the United Nations 'Earth Summit' held in Rio de Janeiro. Of all the programs, the IGBP took the Bretherton report's aspirations to heart the most. From its inception, it tried to develop models – based on physical laws and mathematical logic – that could represent accurately the intricacies of the Earth system.

Despite their professed commitment to 'integrated environmental analysis' members of all four programs soon realised that some key issues were falling between the proverbial cracks. Well aware of the ESS goals articulated by Bretherton, senior members agreed to form a partnership in 2001 – the Earth System Science Partnership (ESSP). They also enrolled a capacity building and networking initiative designed to reach researchers in the global South (called the Global Change SysTem for Analysis, Research and Training: START). Program resources were then used to co-design and implement a set of joint research projects, such as GECAFS (which stands for Global Environmental Change and Food Systems – see Table 4.7.1). Running for a decade (2001–2011), GECAFS' aim was to "determine strategies to cope with the impacts of global environmental change on food systems and to assess the environmental and socio-economic consequences of adaptive responses aimed at improving food security" (GECAFS, n.d.). As this objective indicates, it sought to link research into climate change, soils and vegetation to social scientific inquiry into farming, food storage and distribution systems and much else besides. It was, in short, 'intersectional' in focus. The systems concept proved useful for enabling interchange between researchers from a diversity of disciplinary backgrounds. At the same time, researchers benefitted from more numerous and higher resolution Earth observation technologies. This and other joint projects were eventually steered by an ESSP science committee (from 2007). The Partnership also created a peer review journal, *Current Opinion in Environmental Sustainability*. Under its auspices, two manifestos-cum-bibles of ESS were published in 2004: one was the jointly authored *Global Change and the Earth System: A Planet Under Pressure* (Steffen et al., 2004); the other was an edited book *Earth Systems Analysis for Sustainability* (Shellnhuber et al., 2004).

**Table 4.7.1** The lexicon of global environmental change research organizations and initiatives

| | *Definitions* | *Organizations* |
|---|---|---|
| Global environmental change programs | Programs are legally recognized scientific organizations that coordinate GEC research. They are co-sponsored by major agencies, such as the International Council for Science, the United Nations Educational, Scientific and Cultural Organization and the World Meteorological Organization. | DIVERSITAS, IHDP, IGBP, WCRP |
| Partnerships | Partnerships are informal arrangements established by the GEC research programs to exchange ideas, synthesize and communicate integrative GEC research findings and conduct interdisciplinary research. | ESSP |
| ESSP joint projects | Joint Projects are sponsored by at least three GEC research programs, promoting interdisciplinary research across disciplinary boundaries (natural and social science). The ESSP Joint Projects are designed to directly address the two-way interaction between GEC and global and regional sustainability issues. The Joint Projects also benefit from the expertise and synthesized knowledge of the Core Projects and the GEC research community. | GCP, GECAFS, GWSP & GECHH |
| Core projects | Core projects are disciplinary enterprises sponsored by one GEC research program, designed to research one specific field/scientific challenge. | For example, bioGENESIS (DIVERSITAS); Integrated Land Ecosystem-Atmosphere Processes Study (IGBP); Urbanization and Global Environmental Change (IHDP); Stratospheric Processes And their Role in Climate (WCRP). |
| Regional networks | Regional networks provide opportunities to enhance GEC research and networking capacity, particularly in developing countries. | Asia-Pacific Network for Global Change Research (APN), Inter-American Institute for Global Change Research (IAI), and global change SysTem for Analysis, Research, and Training (START). |

*Source*: Adapted from Ignacuik et al. (2012: 149) with permission.

Despite progress being made in GECAFS, specific separate projects within the IGBP, WCRP, IHDP and Diversitas have also contributed to the ESS aspiration for holistic analysis of global environmental change and its regional components. For instance, the IGBP's AIMES project (standing for Analysis, Integration and Modeling of the Earth System) made great strides in representing computationally the dynamic interrelations between atmosphere, water and land. Unlike the traditional image of a system as closed and self-regulating (unless externally forced beyond operational boundaries), AIMES has made use of wider scientific thinking about complexity, non-linear behaviour and irreversible tipping points. The new incarnation of AIMES aspires to incorporate human behaviour and responses, and to help decision makers anticipate serious socio-environmental problems before they appear (see van der Leeuw, 2013).

This aspiration returns us to the original inspiration for the creation of the ESSP. Many projects within the four global change research programs have arguably advanced 'narrow' interdisciplinarity rather than 'wide' interdisciplinarity. For instance, computer models of the sort used in many WCRP and IBGP projects have thus far left out many social scientists whose research into 'human dimensions' of environmental change focuses on politics, social power, societal resilience and community capacity building as they relate to avoiding or adapting to biophysical hazards. This is, perhaps, consistent with the externalization of those dimensions evident in the 1985 Bretherton diagram (placed off to the right: see Figure 4.7.1). Even the joint projects conducted under ESSP auspices have had their limits. Consider the Global Carbon Project (2001– ). Its professed aim is to offer a "complete picture of the global carbon cycle, including both its biophysical and human dimensions together with the interactions and feedbacks between them" (Global Carbon Project, n.d.).

However, it has been dominated by geoscientists and, among the social scientists involved, economists have loomed large. This means that other disciplines focussed on other human dimensions of carbon acquisition and use have been rather sidelined. Meanwhile, some nationally funded ESS projects have remained unapologetically focussed on physical dimensions – such as the QUEST project funded by the UK's NERC (QUEST stands for Quantifying and Understanding the Earth System). What is more, some have wanted to take ESS in an applied direction by treating couple human–environment systems as objects for a new generation of engineering and management approaches. For instance, in his manifesto for ESEM (Earth systems engineering and management), Brad Allenby argued that "the anthropogenic Earth is a difficult, highly complex, tightly integrated system that challenges society to rapidly develop tools, methods, and understandings that enable reasoned responses. Engineers in general, and civil and environmental engineers in particular, must be a critical part of any such response" (2007: 7961). This chimes with Richard Alley's earlier mentioned notion of ESS as providing an 'operator's manual' if it achieves sufficient analytical precision and predictive accuracy.

Why the limited focus on human dimensions, going back to Bretherton – notwithstanding the creation of the IHDP? Arguably, it reflected the wider, long-standing 'divide' between the natural sciences and the disciplines studying people – a divide British novelist and scientist C.P. Snow (1959) famously described as academia's 'two cultures problem'. In essence, the divide reflected the ontological differences between the biophysical world and the world of people. Unlike rocks and rivers, people are creatures of both reason and emotion, able to be proactive and reactive in relation to nature and other humans. Consequently, 'scientific' approaches are not wholly appropriate to understanding human thought and behaviour. As far back as the 1960s, the German social theorist Jürgen Habermas (1972) argued that 'hermeneutic' and 'critical-emancipatory' approaches were equally important. The former elucidates various human values, beliefs, feelings and desires using methods like ethnography, interviews and focus groups. These methods do not aim for objectivity and precision in the same sense as most natural science research techniques. The latter highlights the inequalities rife in various societies and reveals the values and goals associated with marginalized or oppressed social groups (with a view to achieving them in practice). Again, where natural science aims for objectivity, critical-emancipatory research is overtly political and researchers may align their own activities with those of the disadvantaged people they are researching.

The limitations in applying a scientific approach to understand human dimensions help us understand why ESS, on the international stage, is now morphing into something less dominated than heretofore by climatologists, oceanographers, and other geoscientists. This mirrors changes in the way research into climate–society interactions has changed since the Intergovernmental Panel on Climate Change published its fourth global assessment in 2007. The most recent report (published in 2014 and 2015) makes a considerable effort to focus on human responses to climate change and policy options for better mitigation and adaptation measures.

However, the intellectual and practical barriers to success in broadening ESS's intellectual reach remain extremely high. Despite (i) improvements in the quantity and quality of available Earth surface data, (ii) increased computational power and sophistication, and (iii) success in international cross-disciplinary team working, problems remain. First, most geoscientists are interested in identifying what *has happened* and *will happen* to the Earth's environment. However, many social scientists (and humanists) are concerned with normative issues – that is, with debates over what *could and should happen*, depending on the values and institutions that might influence human decision making looking ahead. Normative questions can be logically debated; but they are not amenable to measurement and modelling. They speak to humans' capacity to change the way they think and act, perhaps dramatically so in the face of perceived existential threats. Large – perhaps unprecedented – changes would radically alter the future course of the Earth system.

Secondly, practical issues arise in trying to synthesize knowledge among researchers who have different 'worldviews'. For instance, where environmental economists might be keen to put a price on carbon dioxide emissions, an anthropologist might want to focus on the ensemble of institutions, social relations and cultural norms that dispose whole societies to burning fossil fuels on a mass scale. Finally, practical issues also arise in assembling data on current patterns of human behaviour. Such behaviour cannot be monitored by satellites in the way Amazonian forest cover can. Instead, different sorts of data from a myriad of sources (governmental and academic) needs somehow to be synthesized to better understand patterns, variations and transitions in how people influence and are influenced by global environmental change. One complication here is that past and present human behaviour is not governed by a universal rationality; instead, it is profoundly conditioned by the norms, relationships and power dynamics of different societies. Added to this, predicting such behaviour is very difficult when compared to the challenging, but more tractable, problem of predicting future environmental change in light of current biophysical knowledge.

## ESS today: the *Future Earth* framework for global change research

As noted above, ESS is now transitioning into a more transdisciplinary endeavour. A recent definition by some of those involved captures an intellectual broadening that is central to this. According to Leemans et al. (2009: 5), ESS today aims to "observe, understand and predict global environmental changes involving interactions between land, atmosphere, water, ice, biosphere, *societies, technologies and economies*" (emphasis added). The ESSP has come to an end, along with the IHDP. With the three other global change research programs and START, the ESSP has fed into a new initiative called Future Earth (see http://www.futureearth.org). Future Earth (FE) was launched at the 2012 Earth Summit (otherwise known as Rio+20). It has high level support globally. It is sponsored by the Science and Technology Alliance for Global Sustainability comprising the International Council for Science (ICSU), the International Social Science Council (ISSC), the Belmont Forum of national funding agencies, the new Sustainable Development Solutions Network (SDSN), the United Nations Educational, Scientific, and Cultural Organization (UNESCO), the United Nations Environment Program (UNEP), the United Nations University (UNU), and the World Meteorological Organization (WMO).

Several joint projects and separate projects from the ESSP and the four global change research programs are rolling forward under FE auspices. However, more than in previous research, FE aims to foster 'wide-angle and relevant' inquiry that speaks more to (i) human dimensions (including development in the global South) and (ii) the needs of decision makers (in politics, commerce and civil society) and stakeholders. To quote from its strategy document, FE places "a strong emphasis on full integration among scientific disciplines, on engagement with societal partners in co-designing and co-producing knowledge, on international collaboration, on producing knowledge that is valuable to decision-makers, and on generating the solutions that society needs" (Future Earth, n.d.). Indicative of this, only the first of its three overarching themes speaks to the 'pure research' interests of many geoscientists (the themes are, respectively, Dynamic Planet, Global Sustainable Development and Transformations Towards Sustainability). This said, it will be several years before projects more focused on the human side of the human–environment relationship come to fruition. In turn, such fruition will only be possible if sufficient funds are made available to pay for the time and travel needs of myriad researchers from all parts of the globe. Additionally, the intellectual and practical challenges of achieving 'wide interdisciplinarity' – noted above – will remain. In the meantime, various environmental scientists aspire to deepen the complexity and predictive power of their Earth system concepts and models (see, for instance, Liu et al., 2015).

## Conclusion

To summarise, ESS is not so much a single coherent research approach as a set of approaches to research, supported by various funding streams and institutions. These approaches, sometimes overlapping, sometimes sharing a family resemblance, together aim to provide a holistic analysis of Earth surface change (including prediction of future states in some cases). For many years dominated by geoscientists favouring large-scale Earth observation technologies and computational models, we have seen that ESS is now broadening out. In the process, the notion of an Earth system is likely to serve as little more than a metaphor for many social scientists (and humanists) who are less disposed than some geoscientists to see humans as one (albeit important) 'component' within a wider set of parts and connections.

This is evidenced by the recent World Social Science Report (WSSR) sponsored by the International Social Science Council (2013). Entitled *Changing Global Environments*, it does not foreground the Earth system concept, even as it accepts the need for 'joined up' analysis across the natural sciences and the 'people disciplines'. Instead, it accents the diversity of human values and ambitions, the need for radical change to our fossil-fuel driven way of life, and the role of social science in such change through fostering democratic debate about alternative future pathways for humanity. Where the 'Earth system' idea invites the image of an omniscient engineer 'fixing' the machine, the WSSR promotes the idea that any 'fix' will need to be socially negotiated between diverse cultures, governments, NGOs, researchers and so on. In this context, geoscientists will be servants of diverse political agendas not 'value free' experts presenting objectively relevant facts and neutral technological solutions. The seeds of this alternative approach were already sown in some projects sponsored by the IHDP, notably the Global Environmental Change & Human Security project (1999–2010: see O'Brien & Barnett, 2013).

## Note

1 A longer version of this chapter appears under the same title in D. Richardson et al. (eds) *The International Encyclopaedia of Geography* (Wiley-Blackwell: Malden). The text is here reproduced, in abbreviated form, with permission.

## Learning resources

The following offer rich insight into the history, evolution and substance of ESS:

Cornell, S., Prentice, I.C., House, J.I. & Downy, C.J. (eds) (2012) *Understanding the Earth System*. Cambridge: Cambridge University Press.

Ignacuik, A., Rice, M., Bogardi, J., Canadell, J.G., Dhakal, S., Ingram, J., Leemans, R. & Rosenberg, M. (2012) 'Responding to complex societal challenges: A decade of Earth System Science Partnership (ESSP) interdisciplinary research', *Current Opinion in Environmental Sustainability*, 4: 147–158.

Mooney, H.A., Duraiappah, A. & Larigauderie, A. (2013) 'Evolution of natural and social science interactions in global change research programs', *Proceedings of the National Academy of Science*, 110, supplement 1: 3665–3672.

Wainwright, J. (2009) 'Earth system science', in N. Castree, D. Demeritt, D. Liverman & B. Rhoads (eds) *A Companion to Environmental Geography* (pp. 145–167). Oxford: Wiley Blackwell.

## Bibliography

Allenby, B. (2007) 'ESEM: a manifesto', *Environmental Science & Technology*, December 1st: 7960–7965.

Alley, R. (2000) *Two Mile Time Machine*. Princeton: Princeton University Press.

Bretherton, F.P. (1985) 'Earth System Science and remote sensing', *Proceedings of the IEEE*, 73, 1118–1127.

Committee on Earth Sciences (1989) *Our Changing Planet*. Washington, DC: Office of Science & Technology Policy.

Ernst, W.G. (2000) 'Synthesis of Earth systems and global change', in W.G. Ernst (ed.) *Earth Systems: Processes and Issues* (pp. 519–532). Cambridge: Cambridge University Press.

Fuller, R.B. (1968) *Operating Manual for Spaceship Earth*. Illinois: Southern Illinois University Press.

Future Earth (n.d.) Home page. Online. Available at: http://www.futureearth.org/media/strategic-research-agenda-2014. Accessed May 18th 2016.

Global Carbon Project (n.d.) Home page. Online. Available at: http://www.globalcarbonproject.org. Accessed May 18th 2016.

Global Environmental Change and Food Systems (GECAFS) (n.d.) Home page. Online. Available at: http://www.gecafs.org. Accessed May 20th 2016.

Habermas, J. (1972) *Knowledge and Human Interests*. London: Heinemann.

Ingram, J., Ericksen, P. & Liverman, D. (eds) (2010) *Food Security and Global Environmental Change*. London: Earthscan.

International Social Science Council (ISSC) (2013) *World Social Science Report 2013*. Paris: UNESCO.

Kates, R., Clark, W.C., Corell, R., Hall, J.M., Jaeger, C.C., Lowe, I., McCarthy, J.J., Schellnhube, H.J., Bolin, B., Dickson, N.M. et al. (2001) 'Sustainability science', *Science*, 292(5517): 641–642.

Lambin, E.F. & Geist, H.J. (eds) (2006) *Land-use and Land-cover Change: Local Processes and Global Impacts*. Berlin: Springer.

Leemans, R., Asrar, G., Busalacchi, A. & Young, O. (2009) 'Developing a common strategy for integrative global change research and outreach', *Current Opinion in Environmental Sustainability*, 1: 4–13.

Liu, J., Mooney, H., Hull, V., Davis, S.J., Gaskell, J., Hertel, T., Lubchenco, J., Seto, K.C., Gleick, P., Kremen, C. et al. (2015) 'Systems integration for global sustainability', *Science*, 347(6225): 963.

Lovelock, J.E. & Margulis, L. (1974) 'Atmospheric homeostasis by and for the biosphere – the Gaia hypothesis', *Tellus*, 26: 2–10.

NASA (1986) *Earth System Science Overview*. Washington, DC: NASA.

National Science and Technology Council (2012) *The National Global Change Research Plan, 2012–22*. Washington, DC: NSTC.

O'Brien, K. (2011) 'Responding to environmental change: a new age for human geography?', *Progress in Human Geography*, 35(4): 542–549.

O'Brien, K. & Barnett, J. (2013) 'Global environmental change and human security', *Annual Review of Environment and Resources*, 38: 373–391.

Richardson, D., Castree, N., Goodchild, M.F., Kobayashi, A., Liu, W. & Marston, R.A. (eds) (2017) *The International Encyclopaedia of Geography*. Malden: Wiley-Blackwell.

Ruzek, M. (2013) 'ESS in a nutshell'. Online. Available at: http://serc.carleton.edu/introgeo/earthsystem/nutshell/index.html. Accessed May 16th 2015.

Schellnhuber, H.J. (1999) 'Earth system analysis and the second Copernican revolution', *Nature*, 402: C19–C23.

Schellnhuber, H J., Crutzen, P.J., Clark, W.C., Claussen, M. & Held, H. (eds) (2004) *Earth Systems Analysis for Sustainability*. Cambridge, MA: MIT Press.

Snow, C.P. (1959) *The Two Cultures*. Cambridge: Cambridge University Press.

Steffen, W., Sanderson, R.A., Tyson, P.D., Jäger, J., Matson, P.A., Moore III, B., Oldfield, F., Richardson, K., Schellnhuber, H J., Turner, B.L. et al. (2004) *Global Change and the Earth System*. Berlin: Springer.

van der Leeuw, S. (2013) 'AIMES 2: towards a global earth systems science', *Global Change Magazine*, 81: 10–13.

# Energy studies



## Benjamin K. Sovacool and Michael Jefferson

"Energy studies" is an inherently interdisciplinary field looking at the intersection of energy conversion, production, and consumption in society. It encompasses a range of topics revolving around the intersection of energy technologies, fuels, and resources on the supply side, as well as social processes and influences – including communities of energy users, people affected by energy production, social institutions, customs, traditions, behaviors, and policies – on the demand side. It therefore involves a mosaic of different academic disciplines, including a core of physics, engineering, environmental science, economics, and public policy supported by various other disciplines across the natural sciences, social sciences, and arts and humanities (Sovacool, 2014).

Despite being relatively new, the field has a rich history, and in this chapter we explore three distinct dimensions: energy technologies and services, phases of energy research, and future directions.

## Energy technologies and services

Given that modern energy use involves resources and fuels, prime movers, and delivery mechanisms, a significant amount of energy studies research has focused on these topics. The plight of those without access to modern energy services has also received some attention.

Analysts often divide resources and fuels into renewable and non-renewable categories, further differentiating them along primary and secondary sources, according to the typology presented in Table 4.8.1 (Smil, 2010). Renewable resources include solar radiation and all of its biospheric transformations, such as plant mass, wind, and

**Table 4.8.1** Energy resources and fuels

| | *Renewable* | *Nonrenewable* |
|---|---|---|
| Primary | Solar radiation, plant mass, wind, moving water | Coals, crude oils, natural gases, uranium, other minerals |
| Secondary | Biodiesel, ethanol, refuse derived fuel, processed wood pellets, electricity | Charcoal, coke, coal gas, refined crude oils, nuclear fuel rods, electricity |

moving water, temperature differences between the surface and depths of water, and geothermal heat. They are generally considered to share one common characteristic: natural processes are considered to regenerate faster than they are likely to be depleted by any probable system of usage. This common assumption has only been challenged in the context of biomass/biofuels in some specific contexts where whole trees and forests are cut down. Non-renewable resources include fossil fuels – coals and hydrocarbons – deposits that are the product of transformations of ancient biomass, buried in the earth, processed by high pressures and temperatures for millions to hundreds of millions of years. They all share the dominant presence of carbon, whose content can be close to 100 percent for the best anthracite coals or below 60 percent for natural gas and methane.

As Table 4.8.1 indicates, primary types of resources refer literally to the stores of chemical energy inherent in harvested wood or crop residues, or extracted from the earth, such as fossil fuels. Their combustion provides heat (thermal energy) or light (electromagnetic and radiant energy). Secondary fuels have to be processed, which changes their physical state, such as making solid briquettes by compressing coal dust, or harnessing charcoal from trees (which, by the way, can have an energy density as good as the best coals, at 30 MJ/kg). Electricity is unusual, in that it can be both primary and secondary: primary when converted from renewable flows, secondary when released from the combustion of fossil fuels or second-order renewable fuels such as refined biofuels, compacted wood pellets, or refuse derived fuel (which comes from treated, sorted, and processed municipal solid waste).

Prime movers are the technologies that convert primary and secondary fuels into useful and usable energy services. Without prime movers, all of the dazzling advances human civilization has made over the past millennia would remain nothing more than unrealized concepts. Human muscles are the classic prime movers; those muscles enabled us to hunt, gather, and farm. The first mechanical prime movers were simple sails, water wheels, and windmills; the industrial revolution had its steam engines and turbines; the modern era has internal combustion engines, jet turbines, compact florescent light bulbs, and household electric appliances (Jefferson, 2015).

Energy resources and prime movers need delivery infrastructure to connect them, and while such transportation and distribution systems are breathtakingly variegated, the three most prominent are pipelines, tankers, and electric transmission and distribution lines. Taken together, this infrastructure occupies a substantial chunk of land, with one assessment estimating that roughly 30,000 square kilometers – the size of Belgium – are currently dedicated exclusively to supporting the oil, gas, coal, and electricity industries (Smil, 2010). This seems a very large figure, yet placed in a broader context is relatively modest. Land use is a key factor in estimating the power densities of different sources of energy (power per unit of area, usually expressed in Watts per square meter). These calculations are difficult, and consideration needs to be given to area and quality of land used, above ground disturbance (which may be very limited for wind turbines but often result in visual intrusion and the effects of aerodynamic modulation over a wide area), longevity of structures, and, of course, whether they are on land or in water. Vaclav Smil (2015: 21) has concluded: "The power densities realized by harnessing these renewable energy flows are appreciably lower than the power densities of fossil fuel-based systems."

## Phases of energy research

Drawing from Rosa et al. (1988), there have been four distinct modes of energy analysis looking primarily at energy resources, prime movers, or delivery systems to date.

From about the 1940s to the early 1970s, energy discussions were primarily dominated by economists emphasizing the importance of energy to economic performance. Studies frequently measured economic performance and growth – using indicators such as GNP (Gross National Product) – and compared these to the amount of energy a given country consumed. These assessments found strong parallelism between energy use and economic growth, and perpetuated the widely uncontested idea that increases in energy consumption were essential to the continued growth of industrialized economies. Energy policy analysis, then, largely consisted of calibrating economic performance against energy consumption, and of devising strategies to ensure adequate supply to guarantee economic expansion.

A more refined type of energy analysis emerged with the energy crises of the 1970s, challenging the assumption that energy consumption and economic well-being were destined to grow in a lock-step relationship. New studies suggested that advanced societies differed greatly in their per capita energy consumption. Longitudinal studies of energy use patterns and cross-national surveys comparing countries with similar standards of living all seemed to point in the same direction: a threshold level of high energy consumption had to be met for a society to achieve industrialization, but after that threshold had been crossed a wide latitude in the amount of energy needed to sustain standards of living existed. Energy analysis became a means of finding out how much efficiency could be achieved and of exploring alternatives for those countries that had already crossed the consumption threshold.

A third type of analysis started to take hold in the 1970s and continues to the present. Such studies are predominately concerned with producing forecasts of the future and, given that the future is essentially unknowable, developing scenarios of alternative possible – and hopefully likely – futures. Reports from the US Energy Information Administration (EIA), Environmental Protection Agency (EPA) and International Energy Agency (IEA) typically focus on estimating generation capacities, projecting fuel costs and predicting the environmental impacts of particular energy technologies. For example, the paragon of excellence among these types of reports, the EIA's 'Annual Energy Outlook', predicts the current and future technical potential for energy technologies, but does not anticipate expected policy changes or provide policy recommendations. This type of analysis focuses on different technological options, and provides insight into how supply and demand should be managed. Analysis frequently extrapolates current trends, creates a picture of a future world, and informs policymakers of the different options for accomplishing such a vision. More substantive scenario work acknowledges a range of uncertainty across a very broad front, efforts which can be traced back to oil company Shell's scenario work from 1970 (Jefferson, 2012).

A fourth type of analysis concerns technology assessment, or focuses on how to diffuse a particular technology into the marketplace. The Edison Electric Institute (EEI) and Electric Power Research Institute (EPRI) tend to center purely on the economics of electricity supply and demand, while reports from groups like the Pew Center on Global Climate Change and the Natural Resources Defense Council emphasize the environmental dimensions of energy consumption. The US National Academies of Science and Union of Concerned Scientists have produced insightful analysis of the security and infrastructure challenges facing the energy sector, while groups like the Alliance to Save Energy and the American Council for an Energy Efficient Economy remain principally concerned with conservation and energy efficiency. Other groups – including the Nuclear Energy Institute, Global Wind Energy Association, the International Renewable Energy Agency, the International Energy Agency – focus on particular types of technologies. Energy analysis, in this light, involves picking optimal technologies and then attempting to integrate them into society.

## Future research

Notwithstanding the utility of such phases, a few gaps exist. One is that to date most analysis has focused on the technical side to energy production and consumption, with less emphasis on social dimensions. Yet the human causes and consequences of energy-related activities and processes as well as social structures shape how people interact with energy systems. Energy analysis therefore needs to look beyond the dimensions of technology and economics to include these social and human elements (Stern et al., 2016).

A second gap is moving beyond models and scenarios to include other types of data-driven analysis (Jefferson, 2014). Here, the community needs to embrace more human-centered forms of data collection such as interviews with users, focus groups, and qualitative household surveys (Sovacool et al., 2015).

A third gap is the widespread failure, apart from a few specialists, to recognize some of the most profound challenges to more sustainable energy provision and use. Three obvious challenges are: (1) the relatively low power densities of renewable forms of energy when compared to the fossil fuels; (2) the relatively low return on energy invested in renewable forms of energy; and (3) the avoidance of the "rebound effect" whereby those who do save on their energy costs or find themselves materially better off then spend their money on goods and services which means greater energy production and use. One result of failure to recognize these challenges is the plethora of unrealistic, over-optimistic, targets and policies which exist. This fosters further public distrust of those politicians, bureaucrats, and "special interests" involved.

A final challenge is more collaboration among the disparate disciplines that compose the energy studies communities. Cross-disciplinary teams are rare – limiting the ability to comparatively test theories and also blunting the exploratory power of most studies, which remain wedded to single conceptual frameworks or isolated case studies. The greater the recognition that energy studies emerged out of human aspirations and behavior, not readily amenable to mathematics and model-building, the better the chances of enlightenment.

Such efforts – more social analysis, inclusion of broader forms of data, a recognition of challenges, and cross-disciplinary collaboration – will only serve to enhance a dynamic and rapidly growing field of inquiry.

## Learning resources

This book is a great introduction to the world of energy studies as well as one of its particular themes, renewable energy:

MacKay, D.J.C. (2008) *Sustainable Energy – Without the Hot Air*. Cambridge: UIT. Download it for free at https://www.withouthotair.com.

*Energy Policy* is an international peer-reviewed journal addressing the policy implications of energy supply and use from their economic, social, planning and environmental aspects. See http://www.journals.elsevier.com/energy-policy.

*Energy Research & Social Science (ERSS)* is a peer-reviewed international journal that publishes original research and review articles examining the relationship between energy systems and society. See http://www.journals.elsevier.com/energy-research-and-social-science.

## Bibliography

Jefferson, M. (2012) "Shell scenarios: what really happened in the 1970s and what may be learned for current world prospects", *Technological Forecasting & Social Change*, 79: 186–197.

Jefferson, M. (2014) "Closing the gap between energy research and modelling, the social sciences, and modern realities", *Energy Research & Social Science*, 4: 42–52.

Jefferson, M. (2015) "There's nothing much new under the sun: the challenges of exploiting and using energy and other resources through history", *Energy Policy*, 86: 804–811.

Rosa, E.A., G.E. Machlis and K.M. Keating (1988) "Energy and society", *Annual Review of Sociology*, 14, 149–172.

Smil, V. (2010) *Energy Transitions: History, Requirements, Prospects*. Santa Barbara: Praeger.

Smil, V. (2015) *Power Density: A Key to Understanding Energy Resources and Uses*. Massachusetts: MIT Press.

Sovacool, B.K. (2014) "What are we doing here? Analyzing fifteen years of energy scholarship and proposing a social science research agenda", *Energy Research & Social Science*, 1, 1–29.

Sovacool, B.K., S.E. Ryan, P.C. Stern, K. Janda, G. Rochlin, D. Spreng, M.J. Pasqualetti, H. Wilhite and L. Lutzenhiser (2015) "Integrating social science in energy research", *Energy Research & Social Science*, 6: 95–99.

Stern, P.C., B.K. Sovacool and T. Dietz (2016) "Towards a science of climate and energy choices," *Nature Climate Change*, 6: 547–555.

# 4.9 Environmental discourse analysis

## Adrian Peace

### Why discourse?

Environmental discourse comprises the linguistic and other symbolic resources which people draw upon to express and elaborate arguments about the relations between themselves and their natural environments. In the past these discourses have seen nature as self-regulating and the balance between nature and culture as self-perpetuating. This is no longer the case. Contemporary environmental discourses acknowledge that relations between people and their environments are distinguished by conflict, contradiction and even catastrophe. We are not even confident we have the knowledge, organization or will to turn around the disruptive and destructive forces already underway.

A 'realist' view of discourses presumes that they accurately reflect the realities of the biophysical world. Though not exactly incorrect, this underplays the extent to which people culturally constitute discourses to reflect their own values and goals where natural resources are concerned. The specific ways in which these cultural understandings are brought into the social world is the major analytic concern of a 'constructionist' approach to environmental discourse. In practice, it also aims to establish why certain modes of discoursing about the environment acquire influence and authority when others fail to do so. Whether the conflict is over chemicals, coffee cultivation or the climate, it is assumed that command over language and other communicative resources results in one line of argument becoming dominant. Such conflicts can be thought of as sites of struggle between rival discourses, but these are often by no means reliant on the same sources of authority. Natural resources can be valued in religious, economic, or scientific terms; combatants over the significance of even one such resource are unlikely to begin from the same epistemological position. The divide has to be addressed somewhere down the track, however, which is why all discourses are marked by modification and refinement. Charting these changes is integral to environmental discourse analysis.

The concerns of this kind of analysis are often recognized by those amongst whom we work. Environmental actors describe themselves as being engaged in 'a war with words', leaders are explicitly chosen because they have 'a way with words' as they seek the attention of elites, the media or the public at large. Equally important, academic disciplines go about their interrogation of discourses in different ways. Here I take the case of social anthropology which has had considerable success unravelling the discursive threads woven into sites of struggle. Anthropologists spend lengthy periods in the field, work hard at building rapport, and, most relevant here, become familiar with the natural discourses local people draw upon to describe environments of greatest significance to them. As a result, anthropologists have been able to describe the major contours of discursive conflicts and detail the semantic finessing on which effective advocacy turns.

## Loaded language

An early but seminal analysis (Lee, 1989) focussed on the loaded language of protagonists over the annual hunt for the harp seal in Canada. One of the terms introduced by the environmentalist organization that spearheaded the seal's defence was that it was an 'endangered species'. Greenpeace thus aimed to provoke widespread anger in Canadian society. However, scientists employed by the government which sought an extensive cull demonstrated that the species wasn't even threatened, still less endangered. In response, the environmentalists changed tack and argued that Canadians were about to witness 'the largest slaughter of marine animals in the world' – all for the sake of the luxury goods market.

This put national pride at the centre of the conflict whilst other terms in Greenpeace's linguistic armoury aimed to anthropomorphize the seal. By describing the whalers' target as a 'baby seal', it was linked to a human infant and this was reinforced by pictures of wide-eyed, cuddly creatures. The most contested issue was what the looming encounter out on the ice was to be called. The environmentalists argued that the government's options of 'cull' and 'hunt' were too suggestive of a rational conservation strategy. Their preference was for 'a huge slaughter', even 'species extermination', either of which would bring global opprobrium down on Canadian society.

## Analyzing emotions

Evidently focussing on key words and select terminologies is integral to the anthropological contribution to environmental discourse analysis. Likewise with the proliferation of striking images without which no environmental argument would get past first base. But in certain respects these concerns are preliminary to anthropologists asking why key terms and other symbolic forms stir the emotions of large numbers of people in the way they evidently do. It is often acknowledged that the emotional intensity provoked by environmental issues is among their most distinctive qualities, but the specific reasons for this are relatively unexamined. Milton argues that it is because we are in direct and continuous involvement with the environment: 'Throughout our lives we learn from our whole environment. . . . What we learn about the world depends on how we, as individual organisms, engage with it' (2002: 32). But even this is debatable: is it not the case that many environmental issues which generate powerful sentiments involve subjects selectively filtered through magazines, television, and social media? As far as most people are concerned, climate change and greenhouse warming would fall into this category.

What is indisputable is that these discursive conflicts often prove long-lasting, and even after protracted dispute turn out to be inconclusive. Is this precisely because of the emotional intensity that environmental issues generate? This question has been posed by Kalland whose ethnography *Exploring the Whale: Discourses on Whales and Whaling* (2009) explores the most enduring of environmental conflicts. His analysis is exclusively about rival cultural interpretations of whales and whaling, the still-unresolved conflict between a hegemonic anti-whaling discourse and its rival pro-whaling discourse.

## Symbolic power

Kalland argues that the outstanding discursive innovation by the anti-whaling lobby occurred in the early 1990s. Its leading proponents drew together several human attributes that different whale species were said to display and so constructed a fully anthropomorphized 'super-whale' which could be promoted as our closest marine relative. The super-whale was not only the planet's largest animal: it was especially intelligent, capable of communication through song, devoted to family life, distinctly sociable, and so on. The super-whale – quite inappropriately labelled 'an endangered species' – thus exhibited valorized qualities that we humans had possessed until relatively recently. This powerful totem harked back to a cultural era and a material life which we in the West had lost in our headlong rush into modernity.

Kalland argues that construction of the super-whale was a discursive masterstroke: it was the symbolic potency of this entirely mythic creature that stirred the emotions of the Western middle classes. Most significant, it became integral to political developments inside the International Whaling Commission (IWC), which was established post-World War II to regulate the commercial hunting of whales, because henceforth the IWC became an arena of intense opposition between anti-whaling and pro-whaling discourses. Proponents of the former discourse proved adept at recruiting new members and rallying external support that could be relied upon to oppose continuation of commercial whaling in any form; the IWC was transformed from a 'whalers' association' to a 'protectors' club'.

The main theme of Kalland's study is of broader relevance than his immediate subject of rival representations of whales and whaling. For he argues that, from the 1990s, the war of words and the battle for control inside the IWC and across the international community became more polarized, more vitriolic, and increasingly remote from any prospect of resolution. Such was the intensity of emotions generated on both sides over this iconic natural resource that any prospect of definitive conclusion disappeared from view. Kalland describes this as a state of 'cultural schismogenesis'. This technical term refers to structured conflicts in which rival camps become wholly concerned with consolidating their control over those who support their goals rather than trying to negotiate some kind of compromise with their opponents. The gap between the major protagonists can only deepen, any prospect of compromise remains as distant as ever. Anthropologists have persuasively taken this argument to other environmental disputes, from the introduction of wolves to Norwegian forests to the development of salmon fisheries in Alaska. The term cultural schismogenesis may be off-putting, but the analytic thrust does seem relevant to an increasing range of regional and global environmental conflicts.

## Next steps

As we look to the future, we can ask: whose environmental discourses will come to matter most? Who will have the power to enforce on others their definitions of environmental problems – and indeed possible solutions? Swyngedouw (2010) argues that we live in a 'post-political age' where oppositional discourses that fundamentally challenge the present socio-environmental order are either silenced or heard without being at all effective. Evidently, this is a different conclusion to Kalland's insight concerning ever-deepening polarization where natural resources are at stake. Finally, those of us researching environmental discourses might ask ourselves: what we can contribute to the intense, sometimes remorseless, rivalries which we study. Should we strive for professional neutrality or join the political and discursive war of words?

## Learning resources

To read further into the ways anthropologists and others explore the role of language and other symbolic resources in environmental conflicts, consider the following source materials. Three broad-ranging reviews of the field are followed by three case studies of specific environmental disputes:

Larsen, B. (2011) *Metaphors for Environmental Sustainability: Redefining Our Relationship with Nature.* New Haven, CT: Yale University Press.

Milton, K. (1996) *Environmentalism and Cultural Theory: Exploring the Role of Anthropology in Environmental Discourse.* London: Routledge.

Mühlhäusler, P. & Adrian, P. (2006) 'Environmental discourses' *Annual Review of Anthropology*, 35: 457–479.

Peace, A. (1997) *A Time of Reckoning: The Politics of Discourse in Rural Ireland.* St John's, Newfoundland: Institute of Social and Economic Research.

Richardson, M., Sherman, J. & Gismondi, M. (1993) *Winning Back the Words: Confronting Experts in an Environmental Public Hearing.* Toronto: Garamondi Press.

Strang, V. (2004) *The Meaning of Water.* Oxford, New York: Berg.

## Bibliography

Kalland (2009) *Exploring the Whale: Discourses on Whales and Whaling.* New York: Berghahn Books.

Lee, J.A. (1989) Waging the seal war in the media, *Canadian Journal of Communication*, 14(1): 43–61.

Milton, K. (2002) *Loving Nature: Towards an Ecology of Emotion.* London: Routledge.

Swyngedouw. E. (2010) Apocalypse forever? Post-political populism and the spectre of climate change, *Theory, Culture & Society*, 27(2–3): 213–232.

# 4.10 Environmental humanities

## Thom van Dooren

The environmental humanities began to take shape in the early 2000s. This area of enquiry brings together a range of disciplines within the humanities, the arts and the interpretive social sciences. Environmental sub-disciplines in history, philosophy and anthropology, for example, had been gaining traction in some parts of the world from the 1960s or 1970s; eco-criticism since perhaps the 1990s; and the environment had always been, in different ways, a key concern in fields like science and technology studies, cultural studies, political ecology, and human geography. Other fields of scholarship and activism, perhaps especially those emerging out of the eco-feminist, indigenous and environmental justice movements, also formed key parts of this foundation. Drawing these diverse threads together—many of them from outside the traditional humanities—scholarship in the environmental humanities centres on the thick inter-weavings of human cultures, histories, values, imaginaries and ways of life with a dynamic more-than-human-world. More than an umbrella term, the environmental humanities has become a gathering ground upon which new interdisciplinary questions, collaborations and approaches are being imagined and crafted, often in dialogue with the sciences and with broad publics beyond the academy. These conversations have begun to establish a set of shared ideas and commitments which are making clear precisely what it is that the humanities (broadly construed) have to offer in thinking through environmental issues and our responses to them, while also lending critical mass to disciplinary approaches, and marginalised peoples, that have often been largely excluded from formal environmental fora (Palsson et al., 2013; Holm et al., 2015; Castree, 2014).

There are two fundamental propositions at the heart of the environmental humanities. The first is that the humanities—in particular an engaged, interdisciplinary, humanities—have a great deal to offer in developing better understandings of, and approaches to addressing, pressing contemporary environmental challenges. The second proposition, centred on a movement in the opposite direction, is that taking the more-than-human world seriously might enrich the humanities as fields of inquiry in important ways.

## Thick and entangled notions of human life

Scholars in the environmental humanities begin with a principled refusal of the "compartmentalization of 'the environment' from other spheres of concern" (Neimanis et al., 2015: 67). From this perspective, nature and culture, facts and values, scientific and human dimensions, cannot be neatly separated out from each other. The always culturally and historically specific ways in which societies—including our own—understand and relate to their environments, matter profoundly. Taking but one, albeit centrally important, example, we might note that now dominant modes of understanding and concern are bound up with, and profoundly shaped by, the emergence of "the environment" as an empirical object around the 1950s, as well as its subsequent enrolment within very specific discourses of expertise and knowledge (Robin et al., 2013; Sörlin, 2013). But other ways of relating to, of *enacting*, worlds are, of course, possible and indeed persist. Appreciating these kinds of processes requires us to accept that "the ecological crisis is not only a crisis of the physical environment but also a crisis of the . . . systems of representation and of the institutional structures through which contemporary society understands and responds to environmental change" (Bergthaller et al., 2012: 262).

To some extent this claim is an old one. But an appreciation of this basic fact does not guarantee an adequate response to it. Scholars in the environmental humanities also insist that the ways in which human lives have been included within environmental policy and decision making—approaches that have drawn in large part on economics and cognitive psychology—have often wrongly depicted people as simply rational decision makers whose attitudes, behaviours and choices might be moderated effectively through education campaigns and market mechanisms (like green taxes). These approaches "do not capture the full range of commitments, assumptions, imaginaries, and belief systems" that guide and inform diverse ways of life (Holm et al., 2015: 978). This "thicker" notion of human life (Geertz, 1973) is the central concern of the humanities. Alongside providing a better picture of who people are and how they live, such an approach also works against the problematic homogenisation of humanity, paying attention to the ways in which cultural and historical difference, as well as the specificity of the social, economic and infrastructural networks we inhabit, produce very different forms of both accountability and vulnerability, perhaps especially in the face of environmental change. "We may all be in the Anthropocene but we're not all in it in the same way" (Nixon, 2014). How, for example, do histories and ongoing realities of colonialism and gross inequality, shape both the landscapes people inhabit and their understandings of the environment and efforts to conserve it?

## Towards which worlds?

Rethinking human life in this way might contribute to the development of better ways of engaging individuals and communities, of understanding and perhaps even "transforming human preferences, practices and actions" (Holm et al., 2015). But environmental humanities scholars have tended to resist straightforward "operationalization" towards the achievement of pre-given "pro-environmental" outcomes. Rather, what will count as an environmental "solution", or indeed as a "problem", are highly contestable questions. While any effective global response to climate change might be a relief at this stage, can we imagine solutions that are not only technically and economically feasible, but also democratic, creative and just—might this be a part of what it means to be truly "effective"? Approaches that instead of exacerbating existing inequalities, or even just leaving them unchanged (if such a thing were possible given the world remaking responses required), might begin to actively redress them? Or, are there some other desirable outcomes—alongside or instead of democracy and justice—that ought to be at the core of our efforts to address environmental problems? What cultural and intellectual resources—from rich ethical and religious traditions, to poetry and sci-fi—do we have for taking up these questions? From this perspective, a thicker notion of human life is also understood to be a vital component of the cultivation of culturally appropriate, life-enriching, futures.

## Rethinking the humanities

Bringing the humanities into dialogue with the more-than-human-world is transforming key approaches, ideas and assumptions. While the humanities have developed rich accounts of human life, they have also usually cordoned it off from the world beyond. In thinking through the environment, however, the human is drawn out of itself; we are required to take seriously all of the ways in which our lives are shaped and made possible by nonhuman others. The human is positioned as a being suspended not only within the "webs of significance" and meaning that anthropologist Clifford Geertz famously described (Geertz, 1973: 5), but within webs of material nourishment and exposure that constitute our "ecological embodiment" (Plumwood, 2003). Both "sets" of webs are, of course, tangled up with each other, and both draw us into connection with the many others—human and not—that make possible both our lives and our specific ways of life. As anthropologist Anna Tsing has succinctly put it "human nature is an interspecies relation" (Tsing, 2012). But "the human" is just the tip of the iceberg. Bursting the anthropocentric bubble in this way, scholars in the environmental humanities are asking how key concepts that have structured many of our understandings about the world—from freedom (Chakrabarty, 2009) to rationality (Plumwood, 2002)—might themselves be key parts of our environmental problems, in desperate need of critical rethinking in more-than-human terms.

## Knowing-with others

Approaching the environment in this way is increasingly leading humanities scholars to conduct ethnographic, participatory and collaborative research. While these research methods have long histories in some disciplines, one of the most interesting aspects of the environmental humanities is the way in which they are being taken up and put to use in new contexts and developed in new ways. Much of this work has sought to engage communities whose needs and perspectives are often marginalised, to develop new forms of participatory or "citizen humanities" research that might unsettle conventional notions of expertise and assumptions about what constitutes a legitimate form of knowing. In its more ambitious formulations, it is hoped that these kinds of collaborations might "initiate a renovation of the university with far-reaching consequences for how knowledge is organized, produced, disseminated, and understood" (Bergthaller et al., 2012: 263). Beyond purely human publics, other scholars are asking similar questions about how various nonhumans might be brought into our research, perhaps through the diverse approaches collecting under labels like "multispecies studies" (van Dooren et al., 2016) and "more-than-human participatory research" (Bastian, 2016).

Each of these broad insights on its own is important. Each has the potential to enrich scholarship in different ways. Taken together, however, they seems to demand a profound rethinking, a reorientation of the humanities. Perhaps most excitingly of all, the environmental humanities is a fundamentally "experimental" field (Bergthaller et al., 2012)—with all of the promises and dangers this implies—committed not to a single outcome, but to an ongoing process that "self-reflexively acknowledges and even nurtures its own contradictions, variances, and necessary open-endedness" (Neimanis et al., 2015: 69).

## Learning resources

The following explore the emergence, significance and key dimensions of the environmental humanities in an accessible style:

Bergthaller, H., R. Emmett, A. Johns-Putra, A. Kneitz, S. Lidström, S. McCorristine, I. Pérez Ramos, D. Phillips, K. Rigby and L. Robin (2012) "Mapping common ground: ecocriticism, environmental history, and the environmental humanities", *Environmental Humanities*, 5: 261–276.

Holm, P., J. Adamson, H. Huang, L. Kirdan, S. Kitch, I. McCalman, J. Ogude, M. Ronan, D. Scott, K.O. Thompson et al. (2015) "Humanities for the environment: a manifesto for research and action", *Humanities*, 4(4): 977–992.

Rose, D., T. van Dooren, M. Chrulew, S. Cooke, M. Kearnes and E. O'Gorman (2012)_ "Thinking through the environment, unsettling the humanities", *Environmental Humanities*, 1: 1–5.

Tsing, A.L. (2012) "Unruly edges: mushrooms as companion species", *Environmental Humanities*, 1: 141–154.

## Bibliography

Bastian, M. (2016) "Towards a more-than-human participatory research." In M. Bastian, O. Jones, N. Moore and E. Roe (eds) *Participatory Research in More-than-Human Worlds*. London & New York: Routledge.

Bergthaller, H., R. Emmett, A. Johns-Putra, A. Kneitz, S. Lidström, S. McCorristine, I. Pérez Ramos, D. Phillips, K. Rigby and L. Robin (2012) "Mapping common ground: ecocriticism, environmental history, and the environmental humanities", *Environmental Humanities*, 5: 261–276.

Castree, N. (2014) "The Anthropocene and the environmental humanities: extending the conversation", *Environmental Humanities*, 5: 233–260.

Chakrabarty, D. (2009) "The climate of history: four theses", *Critical Inquiry*, 35: 197–222.

Geertz, C. (1973) "Thick description: toward an interpretive theory of culture." In *The Interpretation of Cultures* (pp. 3–30). New York: Basic Books.

Haraway, D. (2008) *When Species Meet*. Minneapolis: University of Minnesota Press.

Holm, P., J. Adamson, H. Huang, L. Kirdan, S. Kitch, I. McCalman, J. Ogude, M. Ronan, D. Scott, K.O. Thompson et al. (2015) "Humanities for the environment: a manifesto for research and action", *Humanities*, 4(4): 977–992.

Neimanis, A., C. Åsberg and J. Hedrén (2015) "Four problems, four directions for environmental humanities: toward critical posthumanities for the Anthropocene", *Ethics & the Environment*, 20(1): 67–97.

Nixon, R. (2014) "The Anthropocene: the promise and pitfalls of an epochal idea", *Edge Effects*. Online. Available at: http://edgeeffects.net/anthropocene-promise-and-pitfalls.

Palsson, G., B. Szerszynski, S. Sörlin, J. Marks, B. Avril, C. Crumley, H. Hackmann, P. Holm, J.S.I. Ingram, A. Kirman, et al. (2013) "Reconceptualizing the 'Anthropos' in the Anthropocene: integrating the social sciences and humanities in global environmental change research", *Environmental Science & Policy*, 28: 3–13.

Plumwood, V. (2002) *Environmental Culture: The Ecological Crisis of Reason*. London & New York: Routledge.

Plumwood, V. (2003) "Animals and ecology: towards a better integration", unpublished article. Online. Available at: http://hdl.handle.net/1885/41767.

Robin, L., S. Sörlin and P. Warde (2013) *The Future of Nature: Documents of Global Change*. London: Yale University Press.

Sörlin, S. (2013) "Reconfiguring environmental expertise", *Environmental Science and Policy*, 28: 14–24.

Tsing, A.L. (2012) "Unruly edges: mushrooms as companion species", *Environmental Humanities*, 1: 141–154.

van Dooren, T., E. Kirksey and U. Münster (2016) "Multispecies studies: cultivating arts of attentiveness", *Environmental Humanities*, 8(1): 1–23.

# Environmental image analysis

4.11

## Sidney I. Dobrin

In their 1991 landmark book *Ecospeak: Rhetoric and Environmental Politics in America*, M. Jimmie Killingsworth and Jacqueline S. Palmer explore how rhetoric is used by those seeking to shape public understandings of environmental issues. For them 'ecospeak' referred to the discourses used by different actors to align their own identities and political agendas with environmentally friendly values and actions. These discourses, they showed, were rhetorical because they leaned heavily on persuasive metaphors and statements that often served to mask the reality of environmentally harmful practices (e.g. by big oil companies like British Petroleum). The rhetorical analysis that Killingsworth and Palmer show-cased—and the methodologies derived from their analyses—became fundamental to the evolving discipline of ecocriticism and eventually to eco-composition. More recently, eco-rhetoricians and ecocritics have begun to employ their methodologies to analyze a range of texts beyond word-rich ones, including visual texts and images. Located at the intersection of ecocriticism, ecocomposition, visual rhetoric, visual communication, art and design, and digital media studies, research about the role of the visual in constructing and circulating information about environmental and ecological thought has become central to environmental studies. We might say that environmental image analysis—or environmental visual rhetoric—asks to how we read environmental images or how we picture nature.

In the 2009 collection *Ecosee: Image, Rhetoric, Nature*, influenced by *Ecospeak*, Sidney I. Dobrin and Sean Morey moved beyond Killingsworth and Palmer's attempt to understand "the relationships among language, thought, and action in environmental politics" (p. xi) to take up the visual facet of environmental rhetoric. For Dobrin and Morey, 'ecosee' is the "study and production of visual representations of space, environment, ecology, and nature in photographs, paintings, television, film, video games, computer media, and other forms of image-based media. Ecosee considers the role of visual rhetoric, picture theory, semiotics, and other image-based studies in understanding the construction and contestation of space, place, nature, environment, and ecology" (p. 2). According to Dobrin and Morey, "Ecosee is not (only) an analysis of existing images, it is a work toward making theories that put forward ways of thinking about the relationship between image and environment, nature, and ecology, as well as a theory (or more accurately, a number of theories) of visual design for those who make images" (p. 2). Thus, the production and interpretation of images has become central to how we read and write concepts such as the environmental, the natural, the ecological, the animal, and so on.

The image is central to how we come to know the environment. How we learn to interpret the image is as important to environmental education/environmental studies as is any other form of content delivery (and some might argue more important than in contemporary hyper-circulatory, digital communications systems). Since the emergence of environmental movements of the 1960s and the work of environmental studies to examine complex relationships between humans and 'the environment,' the image has played a central role in how we have analyzed such relationships and how we have provided evidence about those relationships. For example, many working in environmental fields have identified the importance of William Anders' 1968 Apollo 8 photograph of the Earth and part of the Moon's surface known popularly as 'Earthrise' as altering the human perception of our relationship with the Earth. This photograph, often identified as one of the most influential environmental photos taken, is exemplary of how image affects environmental understanding. While simply 'seeing' an image might have tactical influence, it is the ability to critically read/analyze images or present rhetorically-charged images that becomes central to arguments from all positions within environmental conversations. For example, many environmental organizations long ago turned to using before/after images as evidence of environmental change. For example, the two pictures included here depict Lake Erie, Ohio, USA before an algal bloom and after a bloom. The lake, which is susceptible to such blooms because of its shallow depth, is a primary source of tap water for the region. Comparative images like these are often used when issuing advisory warnings about possible algal contamination in order to visually convey a clean/contaminated dichotomy. Such a comparison is intended to persuade viewers not only about the accuracy of the claims of contamination, but to move them to react in particular ways, such as avoiding contact with tap water during periods of contamination. Such comparative images are only effective when viewers know how to interpret them (that is, know how to make visual comparison) and image writers construct them to highlight their environmental message. In such cases, as well, the perceived authenticity and accuracy of the image also drives its effectiveness. Visual analysis for environmental studies requires substantial concern about the accuracy/authenticity of the representation. For instance, in these images, the photographer likely framed the images to include enough visual imagery to convey evidence of location, evidence of contamination, and evidence of timeliness, while also considering similarity in point of view between the images in order to convey that these photographs are of the same location (note, for example the curvature of the shoreline).

**Figure 4.11.1** "Earthrise."

William Anders (1968); image is in the public domain.

Engagement with images—environmental or otherwise—requires understanding the image not as inherently authentic or accurate in its portrayal, but as always already encumbered in the politics and rhetorics of representation. From the moment that the first cave paintings were produced, human imaging of nature/environment has been inextricably bound up in the politics of representation. The cave painter, for example, choosing to depict prey or predator, human or animal or interaction between the two affects the meaning one reads from the painted image. Such politics and rhetorics allow us to recognize that visual representation of environment is always presented with an agenda and never able to be represented truly objectively. Consider the intention of the Pioneer Plaque, a pictorial message placed aboard the 1972 Pioneer 10 and the 1973 Pioneer 11 space crafts. The plaque, designed by Carl Sagan, was intended to convey messages of welcome and information about Earth and humans to any extra-terrestrial being that might find the space crafts. However, every symbol in the image is reliant on a strictly anthropocentric visual literacy (not to mention the assumption that an alien audience would communicate by way of visual stimulus).

Visual rhetoric teaches us that images communicate, but that in order for one to understand what is communicated, one must be visually literate. That is, one must have learned the conventions of visual communication—just as one learns to be linguistically literate, culturally literate, computer literate, and so on—before one can interpret the meaning of an image. Visual interpretations are influenced by numerous visual theories derived from fields as diverse as semiotics, biosemiotics, art history, photography, linguistics, classical and contemporary rhetorics, cultural studies, digital design, graphic design, and technical/scientific design. Each of these (and other) methods contributes to how one sees, analyzes, interprets, and understands what an image represents about the world. Visual analysis of any image may occur with only

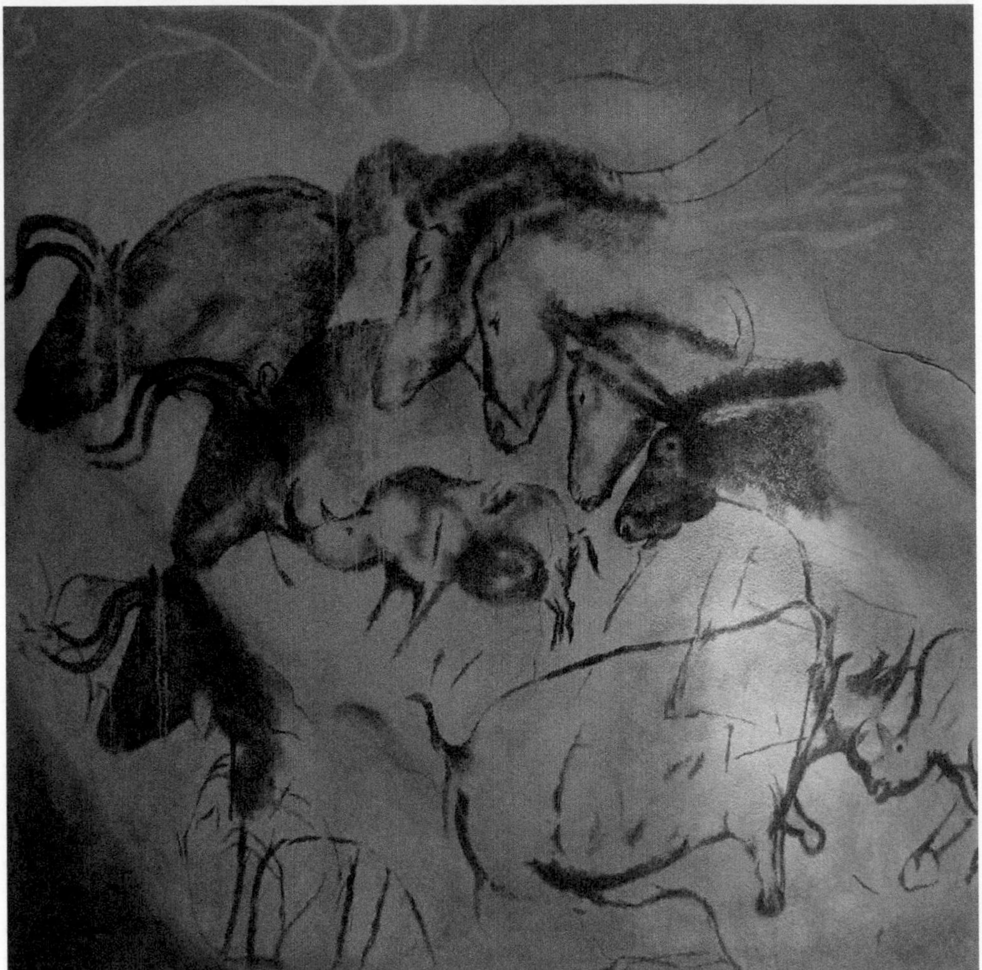

**Figure 4.11.2** Cave painting from Chauvet.

*Source*: Image is in the public domain.

**Figure 4.11.3** Cave painting from Chauvet.

*Source*: Image is in the public domain.

tacit visual literacy, but critical analysis requires more developed analytical methodologies in order to interpret the nuances of an image's meaning. For example, reading a digital marine chart requires a specific visual literacy that is distinct from reading a seventeenth-century maritime map, or, likewise, reading an environmental modelling schematic requires specific disciplinary as well as visual literacies in order to fully understand the image's information.

The study of visual representation in environmental studies is particularly important given that a large part of an individual's environmental experience and awareness is explicitly visual.[1] For example, much of the language used to describe places and spaces relies on visual language and the formation of mental images: grand vistas, crystal clear waters, vast oceans. That is, linguistic and literary imagery has been paramount in environmental discourse. Thus, the individual learns not only to see the image, but to see the image through a variety of interpretive lenses: an environmentalist's lens, an ecologist's lens, a naturalist's lens, and so on. Each visual experience the individual encounters can be interpreted and written from any given lens.

As Sean Morey has explained, "Images—whether appearing as words, pictures, movies, or any visual medium—are not just hermeneutic stabs at showing nature; instead, images compose nature" (Dobrin & Morey, 2009, p. 24). Thus, visual analysis in environmental studies also takes up questions of how images are made and how the choices in making images affect how we read and interpret those images. Images, that is, mediate between the individual and the environment and how that mediation is enacted alters how the individual "sees" the environment. For example, visual analysis might ask as to why a writer opted to include a photograph in order to convey a degree of realism and authenticity instead of a schematic to suggest a more scientific or technical ethos.

Part of the question of visual representation, then, also turns to questions of mediation and the technologies used to produce images that are assumed authentic, accurate representations. For example, technologies like sonar/sonograms and deep space imaging rely on sound and other wave forms to reach objects inaccessible to the human eye (note, too, the role sonogram imaging has played in the politics of women's bodies). These wave forms are then "translated" by way of algorithms and digital devices to produce images that are more easily understood through visual response than would be the pre-translated data. Environmental visual analysis must also consider the ways in which such technological translations/mediations affect how we see the world. Such mediated sight should be addressed from sight augmenting technologies like simple eyeglasses/sunglasses to the highly-complex technologies used to relay pictures of deep space entities. Thus, environmental visual analysis also accounts for the technologies used to make images and the role such technologies play in meaning-making, questioning the ways in which image-mediating technologies are frequently naturalized (think, for example, about eyeglasses or cameras) as always providing accurate representations.

Likewise, environmental visual analysis takes into account the material impact of the technologies used to make images, concerning itself with the environmental consequences of the forging and disposal of image-making technologies. For example, environmental visual analysis considers the role of mineral mining and e-waste associated with the construction and destruction of digital cameras, phones, and computer devices used to make and circulate images. This is to say, environmental visual analysis concerns itself not only with the analysis of the image itself, but the mediating devices used to make those images in order to more thoroughly understand the relationship between the image and environment.

## Note

1 Though not taken up in this chapter, any address of the visual must account for visual impairment, as well. To fail to do so privileges vision and the sighted and excludes a significant portion of the world's population from participation in visual culture. Failure to account for visual impairment in addressing visual analysis in environmental studies (or any other area) is a serious ethical failure.

## Learning resources

http://visual-memory.co.uk/daniel/Documents/S4B
http://twp.duke.edu/uploads/assets/photography.pdf
https://web.archive.org/web/20050412074521/
http://www.arthist.lu.se/kultsem/sonesson/pict_sem_1.html
Aloi, G. (2011) *Art and Animals*. New York: I.B. Taurus.
Barthes, R. (1978) *Image-Music-Text*. New York: Hill and Wang.

Baker, S. (2001) *Picturing the Beast: Animals, Identity, and Representation*. Chicago: University of Illinois Press.
Berger, A. (2011) *Seeing is Believing*. New York: McGraw Hill.
Berger, J. (1992) *About Looking*. New York: Vintage.
Berger, J. (1995) *Another Way of Telling*. New York: Vintage.
Berger, J. (2016) *Landscapes: John Berger on Art*. London: Verso.
Brereton, P. (2015) *Environmental Ethics and Film*. Abingdon: Routledge.
Crow, D. (2011) *Visible Signs*. Worthing: AVA Publishing.
Cubitt, S. (2005) *Ecomedia*. Amsterdam: Rodopi.
Dobrin, S.I. (2015) (ed.) *Ecology, Writing Theory, and New Media: Writing Ecology*. New York: Routledge.
Dobrin, S.I. (2017) (ed.) *Ecocomix*. Jefferson, NC: McFarland.
Dobrin, S.I. & Morey, S. (2009) (eds) *Ecosee: Image, Rhetoric, Nature*. Albany, NY: State University of New York Press.
Downs, R., Stea, M.D. & Boulding, K. (2005) (eds) *Image and Environment: Cognitive Mapping and Spatial Behavior*. Chicago: Aldine Transaction.
Eco, U. (1986) *Semiotics and the Philosophy of Language*. Bloomington, IN: Indiana University Press.
Handa, C. (2004) (ed.) *Visual Rhetoric: A Critical Sourcebook*. Boston: Bedford/St. Martin's.
Mitchell, W.J.T. (1995) *Picture Theory: Essays on Verbal and Visual Representation*. Chicago: University of Chicago Press.
Mitchell, W.J.T. (2013) *Iconology: Image, Text, Ideology*. Chicago: University of Chicago Press.
Mitchell, W.J.T. (2015) *Image Science: Iconology, Visual Culture, and Media Aesthetics*. Chicago: University of Chicago Press.
Rothfels, N. (2002) *Representing Animals*. Bloomington, IN: Indiana University Press.
Rust, S. & Monani, S. (2015) (eds) *Ecomedia: Key Issues*. Abingdon: Earthscan/Routledge.
Rust, S., Monani, S. & Cubitt, S. (2012) (eds) *Ecocinema Theory and Practice*. New York: Routledge.

## Bibliography

Dobrin, S.I. & Morey, S. (eds) (2009) *Ecosee: Image, Rhetoric, Nature*. Albany, NY: State University of New York Press.
Killingsworth, M. & Palmer, J.S. (1991) *Ecospeak: Rhetoric and Environmental Politics in America*. Urbana, IL: Southern Illinois University Press.

# 4.12 Environmental political economy

Jennifer Clapp

The complex interplay between the economy and the environment is often hotly debated and has varying effects on different actors. The field of environmental political economy (EPE) seeks to understand these dynamics, although there are multiple approaches within this broad field. Political economists study the relationships between economic activity and the decisions and policies of governments. They show that what we call 'the economy' is political through and through, both containing and affecting relations of authority and power, while serving (by design or default) certain social interests and values over others. Environmental political economists have extended the focus of political economy. Some scholars focus their analysis on trying to understand the ways in which economic activity affects environmental outcomes, while others concentrate their efforts on the political challenges of designing and implementing economic policy tools to address environmental problems. The uneven exercise of power and the norms guiding different political institutions also shape EPE dynamics.

## Problems at the intersection of economy and environment

The natural environment and the economy interact in myriad ways at various points along economic supply chains—spanning local, national, and global scales. The production of economic goods relies to a great extent on natural resources as inputs, and production processes utilize energy and often emit pollutants. For example, the meat and diary industries rely on livestock operations that often occupy large tracts of land cleared of forests, consume enormous amounts of water and feed crops, and result in the release of environmental pollutants that can contaminate water sources (Weis, 2013). The marketing and global trade of products also consumes energy, including for the transportation of goods and the operation of retail outlets, in addition to resource use and waste associated with product packaging. Financial relationships, although they may appear at first glance to be largely separate from physical production and marketing activities (and their associated environmental impacts), are important to consider because they fund activities along supply chains from production to consumption. Recent years have also seen an increased financialization of nature, including financial trading in carbon, biodiversity, water, land, and agricultural commodities that can have a profound effect on the prices and availability of these natural resources.

## Policy contexts, key actors, and the exercise of power

States, markets, and local governance arenas are key institutional sites of environmental and economic policymaking, and as such are important policy contexts for the study of EPE. The political dynamics of environmental policymaking are unique for each issue, as policy unfolds at different jurisdictional scales. Some issues are clearly global in scope, such as climate change, making political contests in international environmental agreements important to examine in those instances (Newell & Paterson, 2010). At the same time, climate change is also governed by policies that are more national or local in scale, and thus require also examining those policy contexts (Betsill & Bulkeley, 2006).

There are a number of actors at the intersection of environment and economy issues. Corporations are key economic actors, being the main producers of goods and services, and thus they have a primary concern in their economic bottom line (Levy & Newell, 2004). States also play a central role as overseers of policies and regulations for both economic activity and environmental protection (Lemos & Agarwal, 2006). Civil society actors, such as environmental NGOs and social movements, frequently advocate for improved environmental conditions through more stringent policy and regulation (Wapner, 1995). Consumers, meanwhile, have increased their demand for environmentally sound products, while at the same time enjoying high levels of consumption that takes a toll on the environment (Dauvergne, 2008).

These various actors have access to different kinds of power and influence in policy contexts (Fuchs, 2005). States and corporate actors can exercise instrumental power (the ability to influence other actors directly) as well as structural power (the ability to set the rules of the game that conditions the behaviour of other actors). NGOs have little access to these forms of power, but they do have access, along with state and corporate actors, to what is known as discursive power (the ability to frame issues with ideas that condition the likely policy outcomes advocated by other actors). Consumers are often credited with purchasing power (the ability to demand products that are environmentally sound). EPE dynamics are profoundly shaped by the ability of some actors to exercise certain kinds of power that are not as easily available to other actors.

## Competing perspectives in EPE

Given the complexities outlined above, the issues that arise at the intersection of environment and economy are necessarily questions of political economy, because not all actors are affected equally, and their interests often clash. Political institutions are not always agile in responding to new problems that emerge at this interface, and which are unique to their specific jurisdictional scale and context. A variety of perspectives have emerged within the field of EPE to help explain these dynamics, each taking a very different view of the roles of both state and nonstate actors in both encouraging problems and in implementing policies to address them (see Table 4.12.1). Each of these perspectives is outlined in more detail below.

*Market liberals* focus on the power of economic incentives to address environmental problems. Influenced by neoclassical economic ideas, the market liberal perspective considers market distortions, especially those caused by state intervention, as a major cause of inefficiencies that result in environmentally damaging outcomes (Hanley et al., 2013). These problems include insufficient economic growth, which they argue has the twin effect of fueling poverty that can lead to the over-exploitation of natural resources as well as reducing resources available to devote to the development of more environmentally sound technologies. Market distortions, such as subsidies for the energy or forestry sectors, are also seen by market liberals to skew incentives in ways that can lead to environmental harm. Market liberals tend to promote environmental solutions that are market-friendly and promote economic growth, such as voluntary market-based sustainability initiatives.

*Institutionalists* endorse strong economic and environmental governance frameworks that include provisions to protect the environment. Drawing on theories of cooperation, the institutionalist perspective considers weak institutional and governance frameworks for the economy and environment as key culprits that hinder environmental protection. The primary solutions promoted by institutionalists are a stronger role for governance frameworks, including those established by the state, in setting economic and environmental regulatory frameworks that are legitimate and accountable (Biermann et al., 2012). These thinkers do not necessarily dismiss a

**Table 4.12.1** Perspectives on environmental political economy

| *Perspective* | *Market Liberal* | *Institutionalist* | *Critical Approaches* |
|---|---|---|---|
| *Key concept(s)* | Economic incentives; efficiency | Cooperation; governance | Inequality; environmental limits to growth; environmental justice |
| *Perceived drivers of environmental harm* | Market intervention; market distortions; lack of economic growth | Weak institutional and governance frameworks; market failures | Global industrial capitalism; overconsumption; marginalization |
| *Proposed solution(s)* | Liberalized markets; economic growth; voluntary market-based governance | Stronger role for both public and private governance institutions | Degrowth; localism; empowerment of marginalized people to govern themselves |

*Source*: draws on Clapp and Dauvergne (2011).

potential role for markets in policy solutions, but view the state as having a legitimate role in addressing market failures by using a range of tools from environmental taxation to regulation.

*Critical approaches* are not a single school of thought, but rather encompass several strands of thinking that draw on Marxism (O'Connor, 1998), green political theory (Cato, 2009) and ecological economics (Daly & Farley, 2010). Although each strand has a distinct theoretical approach to EPE issues, what these critical thinkers share is their concern about the ways in which the rise of global industrial capitalism and economic growth has fuelled both overconsumption and injustice. These dynamics are problematic for these thinkers because they cause enormous environmental damage that falls disproportionately on marginalized people, while simultaneously propping up the very system that creates these problems. Many critical scholars are also deeply skeptical of the state and its ability to insulate itself from powerful capitalist interests, and thus call for more transformational solutions. Critical scholars advocate for anti-capitalist strategies of degrowth, justice and empowerment for marginalized populations, and more localized economies that reject large-scale industrial capitalism (Martinez-Alier et al., 2010).

## Conclusion

The field of EPE is diverse and complex. It is characterized by competing ideas on the ways in which the environment–economy relationship plays out, with different perspectives offering diverging interpretations of the causes of environmental degradation as well as on the most promising solutions. The debate over policy responses is made more complex by the varying access to power and influence available to different actors, as well as the wide range of policy contexts, each with their own unique political norms and structures.

## Learning resources

The following publications are accessible syntheses of key debates in the field of environmental political economy:

Cato, M.S. (2011) *Environment and Economy*. London: Routledge.

Clapp, J. & Dauvergne, P. (2011) *Paths to a Green World: The Political Economy of the Global Environment*. Cambridge, MA: MIT Press.

Newell, P. (2012) *Globalization and the Environment: Capitalism, Ecology and Power*. Cambridge: Polity.

## Bibliography

Betsill, M. & Bulkeley, H. (2006) 'Cities and the multilevel governance of global climate change', *Global Governance: A Review of Multilateralism and International Organizations*, 12(2): 141–159.

Biermann, F., Abbott, K., Andresen, S., Bäckstrand, K., Bernstein, S., Betsill, M.M., Bulkeley, H., Cashore, B., Clapp, J., Folke, C. et al. (2012) 'Navigating the Anthropocene: improving earth system governance', *Science*, 335(6074): 1306–1307.

Cato, M.S. (2009) *Green Economics: An Introduction to Theory, Policy and Practice*. London: Earthscan.

Daly, H and Farley, J. (2011) *Ecological Economics: Principles and Applications*. Washington: Island Press.

Dauvergne, P. (2008) *The Shadows of Consumption: Consequences for the Global Environment*. Cambridge, MA: MIT Press.

Fuchs, D. (2005) 'Commanding heights? The strength and fragility of business power in global politics', *Millennium-Journal of International Studies*, 33(3): 771–801.

Hanley, N., Shogren, J. & White, B. (2013) *Introduction to Environmental Economics*. Oxford: Oxford University Press.

Lemos, M. & Agrawal, A. (2006) 'Environmental governance', *Annual Review of Environment and Resources*, 31(1): 297.

Levy, D. & Newell, P. (2004) *The Business of Global Environmental Governance*. Cambridge, MA: MIT Press.

Martínez-Alier, J., Pascual, U., Vivien, F-D. & Zaccai, E. (2010) 'Sustainable de-growth: Mapping the context, criticisms and future prospects of an emergent paradigm', *Ecological Economics*, 69(9): 1741–1747.

Newell, P. & Paterson, M. (2010). *Climate Capitalism: Global Warming and the Transformation of the Global Economy*. Cambridge: Cambridge University Press.

O'Connor, J. (1998) *Natural Causes: Essays in Ecological Marxism*. London: Guilford Press.

Wapner, P. (1995) 'Politics beyond the state environmental activism and world civic politics', *World Politics*, 47(3): 311–340.

Weis, A. (2013) *The Ecological Hoofprint: The Global Burden of Industrial Livestock*. London: Zed Books.

# Environmental political theory

John M. Meyer

Environmental political theory (EPT) – sometimes called green political theory – is an interdisciplinary field that contributes to our understanding of environmental challenges, and strategies designed to address these challenges, in several interconnected ways. First, EPT is rooted in the tradition of political theorizing, which closely and critically examines concepts including democracy, justice, freedom, and representation. EPT extends its critique to concepts including sustainability, environment, and nature. Doing so requires us to recognize the ongoing significance of these concepts to public discourse. It also requires us to recognize that disagreement about these concepts often reflects underlying differences in meaning and values, rather than simply erroneous definitions that can be corrected with more information. It is in this sense that these concepts are often described as 'essentially contested' (Gallie, 1955). Second, EPT is committed to the critical analysis of structures of power. Third, EPT moves beyond critique to advance normative arguments for alternative arrangements of social and political ideas and institutions.

By combining these elements, EPT helps us identify the profound limitations of rationalist assumptions about environmental action and change. From such a rationalist viewpoint, problems including climate change require us to ensure that conclusive scientific evidence gets recognized, that experts formulate policy to address the problem, and that both the general public and policy-makers then be persuaded to adopt these policies and implement them (Dryzek et al., 2013: 7–9). It is easy to recognize that this rationalist model fails to correspond with actual experience. Yet the explanation for this lack of correspondence remains. One important line of argument highlights the economic self-interest of polluters and the 'greenwashing' of consumer products, for example, to explain the deviations from this rationalist model. There is much to be said for this and political theorists often join with others in saying it. This work can demonstrate how arguments for change are refracted through systems of power and privilege in political decision-making. Yet political theorists can also help us to understand why such widespread rationalist assumptions or models are *inherently* flawed, by drawing out underlying differences in our conceptions of the role of democracy and expertise, about the meaning of justice and freedom, or about what it is that 'sustainable development' is understood to sustain or develop. In these cases, the differences are not simply a reflection of errors or lies perpetrated by those who have an interest in derailing rational approaches, but are an inherent characteristic of politics.

## Role of democracy and politics

As one might expect, then, EPT highlights the importance of politics to addressing environmental concerns. But 'politics', here, is understood more broadly than is often the case. In this sense, it is far more than just election campaigns or the policymaking process. It is a label for all forms of a messy dialogue and struggle to define what society should value and how it should be organized. It can therefore be understood as the struggle to define – at least for a time – essentially contested concepts including those noted above. Confronting the contested character of our political concepts and cultural contexts can help us to see how even aspiring to rationalist models sets us up for frustration and failure.

Understood in this expansive way, political theorists often look for politics in places where it is left unstated or neglected. To ask 'what the politics of that proposal is' is to seek to unpack implicit assumptions about what sort of changes are needed in order to realize a proposal, who will be most affected by these changes, and what assumptions it makes about the desired contours of social and political institutions.

All this tends to position environmental political theorists as 'small d' democrats in environmental discourse. They worry that when others characterize the problem as getting an ignorant public to understand climate change science properly, or when economists focus on calculating the proper discount rate in a policy proposal, they often assume – and sometimes assert – that a top-down, expert-driven or managerial process is the most efficient and so best way to implement change. By contrast, many environmental political theorists argue that while scientists and other experts play a vital role in addressing environmental challenges, this role is necessarily limited given the differences over which values to promote and which meanings to sustain. These differences can be addressed through a variety of formal and informal processes – voting, legislative debate, organized rallies, or collective actions to block dumping, construction, and other environmentally destructive projects – that are all encompassed within a broad understanding of democratic citizenship and politics.

## Theory and practice

While EPT is by definition a form of theorizing, it is only coherent if understood in relation to action or practice. There are at least two broad approaches to the theory-practice relationship. First, some environmental political theorists write with the goal of enabling their audiences (whether students, fellow academics, policymakers, activists, or other citizens) to be more receptive or attentive to important elements of political practice. they might highlight the dilemmas of representation or of over-reliance upon centralized structures as a basis for positive change. Here, the theorist might be thought of in the role of advisor or strategist, providing insight or warnings for practitioners. But environmental political theorists also often reverse this theory-practice relationship, by looking first to the actual practices of activists, organizers, and social movements for inspiration and insight. For instance, study of environmental justice activism has enriched our understanding of the diverse conceptions of justice itself. On these occasions, the role of the theorist is to tease out the implicit or perhaps underdeveloped vision of justice or democracy or sustainability that is at the heart of practice.

## Where it is located and where it comes from

Despite its interdisciplinarity, and its overlapping concerns with others working in the environmental humanities and social sciences, EPT remains strongly influenced by the disciplinary location from which it emerges. Most scholars who identify their work as EPT have been trained as political theorists within the academic discipline of political science. This position can be understood as both a strength and a limitation.

Among the strengths cultivated by their disciplinary location, environmental political theorists are typically trained to read texts closely and critically, often focused upon influential thinkers from earlier eras who shape contemporary understandings both for better and for worse. They look not just for what is said, but what is left unsaid or unexamined. Their attention to contested concepts can also expand our understanding of the sources of political conflict, avoiding dogmatic agendas or ideological rigidities.

The disciplinary location is – perhaps especially in the United States – also distinctive in that it is typically understood as *in*, but not *of*, the broader discipline of political science: whereas the latter generally understands itself as an empirical social science (often emphasizing quantitative data analysis and rational choice modeling), political theory is largely a normative and critical project that straddles the boundary between the social sciences and humanities. A valuable consequence is that political theorists can be especially well situated to bridge the divide between these divergent ways of knowing, which often plague interdisciplinary environmental inquiry. Finally, the position on the margins of their own discipline has often allowed political theorists an unusual degree of freedom in their work – to draw ecumenically from diverse fields and methodological approaches – something political theorist Michael Walzer has described as a 'political theory license' (2013).

A limitation is that like the broader field of political theory, EPT has been largely an Anglo-American and European discourse, both in its practitioners and in its grounding in a distinctly Western tradition of political philosophizing. While the boundaries of both have been challenged and expanded in recent years, discourses deeply informed by non-Western theoretical approaches remain a minority. Another limitation is that while a consideration of the non-human in political theorizing can be traced back at least as far as Aristotle in Ancient Greece, study of ecological- and earth-systems is not integral to the training or referents common among most political theorists today. In this sense, EPT represents a departure from the discipline most are trained in, and requires its practitioners to immerse themselves in fields of study at some distance from their own. For all the reasons above and more, this immersion is valuable – both for those who pursue it and for their ability to contribute a vital perspective to the conversation in environmental studies and environmental discourse itself.

## Learning resources

The most comprehensive introduction to the field is T. Gabrielson, C. Hall, John M. Meyer and D. Schlosberg (eds) (2016) *Oxford Handbook of Environmental Political Theory*. Oxford: Oxford University Press. This handbook is also available via many university libraries in the electronic database 'Oxford Handbooks Online'.

The interdisciplinary blog 'Inhabiting the Anthropocene' includes at least two series of short posts by environmental political theorists. These can be read at: https://inhabitingtheanthropocene.com

The environmental political theory website contains listings of books, articles, journals, and other resources, as well as announcements of upcoming conferences and other events relevant to those with interests in this field: http://www.cddc.vt.edu/ept

## Bibliography

Dryzek, J.S., Norgaard, R.B. & Schlosberg, D. (2013) *Climate-Challenged Society*. Oxford: Oxford University Press.

Gallie, W.B. (1955) 'Essentially Contested Concepts', *Proceedings of the Aristotelian Society*, 56: 167–198.

Walzer, M. (2013) 'The Political Theory License', *Annual Review of Political Science*, 16(1): 1–9.

# Political ecology

4.14

## Simon Batterbury

Political ecology (PE) is concerned with how humans relate to the biophysical world. Political ecologists have investigated the many environmental challenges that vulnerable communities across the world must face. These include the unequal impacts of global warming on different societies, the health effects of environmental toxics in food, air and water, the environmental crimes of corporations and syndicates, tropical deforestation, wars over control of natural resources, land grabbing, and urban environmental injustices. Political ecology has been important in explaining such phenomena, and particularly the social and political inequities both causing them and mediating their impacts (Bryant, 2015).

PE is interdisciplinary and most closely associated with the disciplines of geography, anthropology and development studies. PE's distinguishing feature is tracing environmental problems and human vulnerabilities to inequalities in power, although many political ecologists also carry out analysis of environmental processes. In general, political ecologists believe that the human struggle for resources and healthy environments is strongly influenced by how much power societies, and individuals hold, and how they use it.

## Access to resources

Inequalities in 'access to natural resources' is a central theme of political ecology. Consider the widely reported acquisition of African land by investors, foreign governments and corporations wanting to grow biofuels and conduct agribusiness in the 2000s. This saga of 'accumulation by dispossession' (Harvey, 2004) peaked in 2011–2012, but has major effects on access to land and water. In SW Cameroon in West Africa, deals were made with government that permitted vast tracts of land to be occupied legally by off-shore industrial corporations for palm oil plantations, including the US-based Herakles Farms. Tens of thousands of hectares of gallery forests and farmland has been cleared and replanted for palm and other monocrops. Political ecologists identify the process as political, since although SW Cameroon has suitable environmental conditions for palm, that is not why companies have gained a foothold in this particular place – rather, there is weak and sometimes corrupt governance, outdated and undemocratic land tenure rules, as well as divided communities who are poor and vulnerable to demands for land. Foreign investors have exploited these conditions at different scales. Worse affected are rural women, who lack a political mandate to speak out collectively. A further inequality is that the loss of land is not compensated by the few agro-plantation jobs and minor financial returns from palm oil production. And, because this region is geopolitically marginalised and in conflict with the state, legal challenges to land grabbing are going unanswered. The political ecology of land access remains central to the region and its people (Batterbury & Ndi, 2017).

## Hatchet and seed

So, a broad-ranging political ecology investigation is required to understand such complex environmental issues, and particularly to reveal inequalities and injustices. For geographer Paul Robbins, political ecology can be used as 'hatchet' and 'seed' (Robbins, 2004). In this case, the first is an exposé of power inequities, and the effects of deforestation and palm cultivation. Empirical research may be useful in defending the rights of local residents. Indeed in other locations, the 'green economy' of conservation reserves, biofuel plantations or renewable energy installations also closes down access to resources. Sometimes this pits political ecologists against conservation biologists and large environmental NGOs (today, sometimes in partnership with business) anxious to preserve habitat, at the expense of local livelihoods (Adams, 2017).

'Seeding' involves generating fresh and useful ideas that may find a home in direct advocacy, legal challenges, or activism. PE has supported better environmental governance, and efforts to 'fight back' against injustice. Most directly, PE practitioners have not just researched the origins of injustice, but tackled it, with constituencies including local scholars and community organisations. There is an 'environmentalism of the poor' involving political coalition-building, an Environmental Justice Atlas, and concerted action against tough opponents, (Martinez-Alier et al., 2014). While some struggles are international and combative, others are more localised community efforts to capture or sustain access to resources.

Political ecologists note that persuasive narratives skew our judgements about human–environment relationships. The 'power to convince' is important in explaining how land was ceded in Cameroon for example; local chiefs were persuaded that benefits would result from palm oil plantations. This is a common story; narratives embody power, and can have tangible effects on environments and people (Adger et al., 2001; Escobar, 2008). The 'post-truth', pro-business, anti-environmentalist narratives rolled out by the 2017 Donald Trump administration in the United States are not supported by scientific analysis of climate change and other environmental phenomena. This is a struggle for truth, which for political ecologists and their ilk involves "resisting truth claims that lead systematically to un-freedoms and objectionable practices" (Sullivan, 2017: 234). There are potential alliances here between the objectives of climate science, political ecology and global coalitions of activists and affected communities – and of course, many questions to be resolved about the values and justice claims that are adopted.

## Methods and theories

Methodologically, political ecologists use a wide range of approaches, to illuminate exactly how political and economic activities influence the fate of ecosystems and local cultures, and how institutional arrangements and organisations are (or have been) responsible for these outcomes. Geographer Piers Blaikie and colleagues were the originators of a distinctive 'regional PE' approach which analyses processes that operate at different scales, but are interlinked. Blaikie identified a "chain of explanation", or a cascade of effects, linking soil erosion to changes in land use practices, caused not by poor local land management but by poverty and denial of access, even arguing that "soil erosion in lesser developed countries will not be substantially reduced unless it seriously threatens the accumulation possibilities of the dominant classes" (Blaikie, 1985: 147).

This type of explanation has been applied in other contexts, including lifting the gaze to global issues like carbon emissions and the new 'green economy' (Peet et al., 2011). PE draws upon a range of theories to explain human–environment relationships (Bixler et al., 2015). Theorists including Karl Marx and Michel Foucault, show how inequalities are embedded in capitalism, and in the subjugation and control of entire populations through monitoring and categorising. There are rational underpinnings to environmental struggles that create winners and losers (Bixler et al., 2015; Hornborg, 2017). Bruno Latour's work is rather different, used by some political ecologists to look closely at relational networks or 'assemblages' involving objects (like trees, and genetically modified seeds) as well as people (Latour, 1991). Feminist political ecology sees power as gendered, focusing on the marginalisation of women but also their vital role in maintaining livelihoods and in struggling for access to resources (Rocheleau et al., 1996). Anthropologist Arturo Escobar's work, based on the Pacific Coast of Colombia, develops unique theories of resistance to modernity and development (Escobar, 2008).

## Conclusion

Political ecology addresses political and economic agendas that have real effects on resources, environments, and people. Its practitioners dig deep, exposing these agendas but also the practices of those who survive in an unequal world. This work is important in the world of 'post-truth', in which 'facts' seem negotiable if they are unwelcome to powerful interests. 'Received wisdoms' can too easily direct bad policy. In the current, desperate context where unscientific narratives are imperilling everybody's environmental future, political ecology has come of age as a necessary dimension of environmental studies.

## Learning resources

The following offer rich resources written by political ecologists:

*Journal of Political Ecology* (free access) http://jpe.libary.arizona.edu
POLLEN (political ecology network) https://politicalecologynetwork.com

ENTITLE – collaborative writing on PE https://entitleblog.org
Environmental Justice Atlas https://ejatlas.org

## Bibliography

Adams, W.M. (2017) 'Sleeping with the enemy? Biodiversity conservation, corporations and the green economy', *Journal of Political Ecology*, 24: 243–257.

Adger, N., Benjaminsen, T.A., Brown, K. & Svarstad. H. (2001) 'Advancing a political ecology of global environmental discourses', *Development and Change* 32, 4: 681–715.

Batterbury, S.P.J. & Ndi, F. (2017) 'Land grabbing in Africa'. In Binns J.A., K. Lynch and E. Nel (eds) *Handbook of African Development*. New York: Routledge.

Bixler, R.P., Dell'Angelo, J., Mfune, O. & Roba, H. (2015) 'The political ecology of participatory conservation: institutions and discourse', *Journal of Political Ecology*, 22: 164–182.

Blaikie P.M. (1985) *The Political Economy of Soil Erosion in Developing Countries*. London: Longman.

Bryant, R. (ed.) (2015) *The International Handbook of Political Ecology*. Cheltenham: Edward Elgar.

Escobar, A. (2008) *Territories of Difference: Place, Movements, Life, Redes*. Durham, NC: Duke University Press.

Harvey, D. (2004) 'The "new" imperialism: accumulation by dispossession', *Socialist Register*, 40, 63–87.

Hornborg, A. (2017) 'Artifacts have consequences, not agency: toward a critical theory of global environmental history', *European Journal of Social Theory*, 20(1): 95–110.

Latour, B. (1991) *We Have Never Been Modern*. Hemel Hempstead: Harvester Wheatsheaf.

Martinez-Alier, J., Anguelovski, I., Bond, P., Del Bene, D., Demaria, F., Gerber, J-F., Greyl, L., Haas, W., Healy, H., Marín-Burgos, V. et al. (2014) 'Between activism and science: grassroots concepts for sustainability coined by environmental justice organizations', *Journal of Political Ecology*, 21: 19–60.

Peet, R., Robbins, P. & Watts, M.J. (2011) *Global Political Ecology*. London: Routledge.

Robbins, P. (2004) *Political Ecology: A Critical Introduction*. Oxford: Blackwell.

Rocheleau D, Thomas-Slayter, B. & Wangari, E. (eds) (1996) *Feminist Political Ecology: Global Issues and Local Experiences*. London: Routledge.

Sullivan, S. (2017). 'What's ontology got to do with it? On nature and knowledge in a political ecology of the "green economy"', *Journal of Political Ecology*, 24: 217–242.

# Post-normal science

4.15

## Silvio Funtowicz and Jerome Ravetz

The term 'post-normal science' (PNS) was coined by science and technology philosophers Silvio Funtowicz and Jerome Ravetz in the years around 1990. It was developed in numerous publications in subsequent years, mainly in the journal *Futures* (Funtowicz & Ravetz, 1993). It's now become an accepted, even mainstream, term in many areas of discourse about science and technology, in particular in relation to scientific advice for policy. It both defines the context in which science and technology now exist, and recommends ways of making science and technology responsive to that context.

### What is post-normal science?

PNS can be expressed in a diagram, a 'quadrant-rainbow' with three zones. The two axes are, respectively, Systems Uncertainties and Decision Stakes. The three quarter-circular zones are, respectively, Applied Science, Professional Consultancy, and Post-Normal Science. There's a clear similarity of this diagram with those once used to characterise technological and environmental risks faced by societies (Funtowicz & Ravetz, 1985), but in the previous context the axes referred to (supposedly) quantifiable attributes like Probability and Harm. The PNS diagram (Figure 4.15.1) accomplishes a subversion of the implicit message of the 'standard approach' to such risk problems, which is that all relevant variables can be quantified.

If normal science typically occurs in university laboratories and is curiosity-driven, other forms of science connect it to the wider world in different ways. In the diagram above, first devised by the authors, (Funtowicz & Ravetz, 1993), different forms of science correspond to different degrees of uncertainty in the behaviour of that being studied and

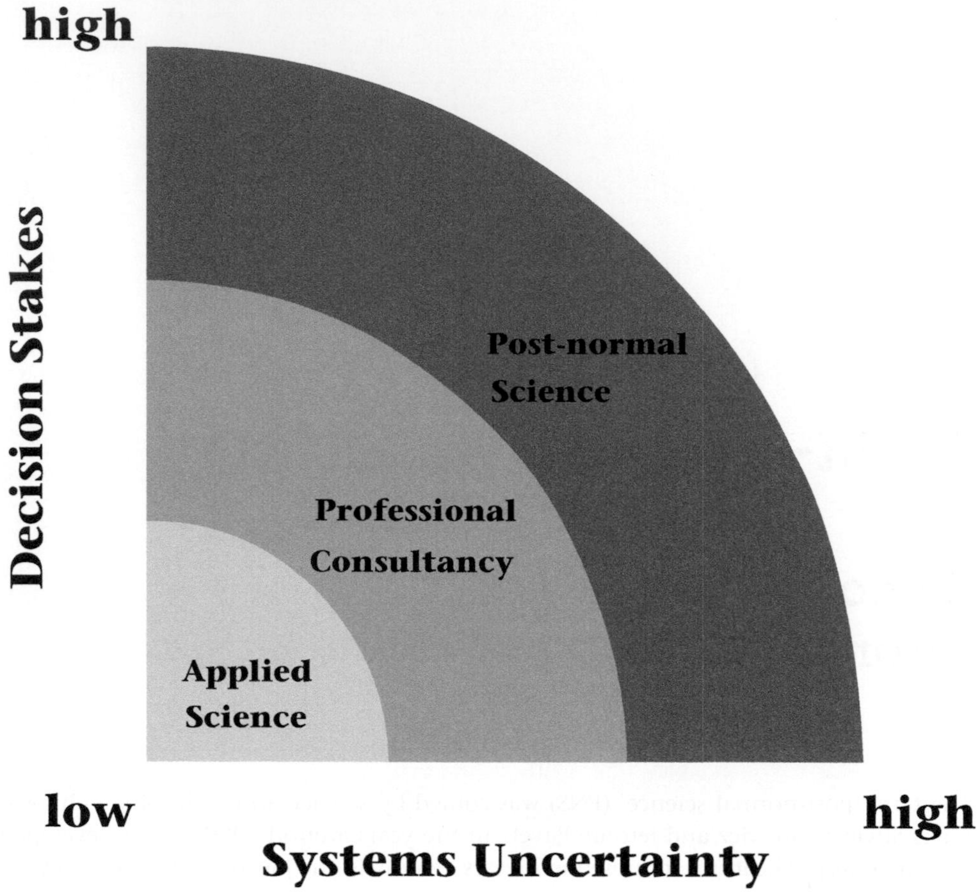


**Figure 4.15.1** Beyond 'normal science'.

*Source*: Reproduced from Funtowicz, S. and Ravetz, J. (1993).

different societal and environmental stakes attending decisions made on the basis of scientific knowledge.

The other defining feature of the PNS idea is captured in a mantra, with a fourfold challenge: 'facts uncertain, values in dispute, stakes high and decisions urgent'. This challenges the common assumptions about scientific knowledge, even when employed in the policy process: scientific facts are by definition certain; scientific knowledge is essentially value-free; and decision stakes and urgencies are irrelevant to scientific knowledge, which is simply true. These assumptons define 'normal science'. In response to the challenge, some have denied that PNS is a special sort of science, relegating it to the subjective world of policy formulation and implementation.

## The origins of PNS

PNS has many roots. The immediate precursor is NUSAP, a notational system for the management and communication of uncertainty in science for policy, created by

Funtowicz and Ravetz (1990). It was an attempt to bring clarity and discipline to the expression of uncertainty and quality in policy-relevant science (see, for instance, an application at the Netherlands Environmental Assessment Agency, discussed by Petersen et al., 2011). Attempts to quantify risk analysis, such as those seriously challenged in the Three Mile Island nuclear power plant accident of 1979, provided a backdrop to both of their endeavours.

Further back there is Ravetz's somewhat visionary discussion of 'critical science' in his 1971 foundational work *Scientific Knowledge and its Social Problems* (Ravetz, 1996). Early examples of anti-establishment science involving 'housewife epidemiology', as at Love Canal and Woburn, MA (in the 1970s in the USA), were an inspiration.

The name post-normal carries an obvious reference to the seminal work of Thomas Kuhn, with his contrast between the limited, puzzle-solving 'normal science' and the 'revolutionary science' in which qualitative progress is made. On occasion this name has caused confusion, since for many scientists 'normal' simply means 'accepted', and the critique that is implied in the name PNS is lost, misinterpreted or confused, leading to an identification of post-normal with 'post-modern'.

A precursor to PNS was Alvin Weinberg's 'trans-science' (Weinberg, 1972), where the inability of 'normal science' to solve policy-relevant problems was seen as a quantitative, rather than a qualitative deficiency. Comparisons have been made to 'mode 2' science (Gibbons et al., 1994) but that is better conceived as analogous to professional consultancy, for in that theory there is no hint of an extension of the peer community.

## The growth of PNS

PNS has become far more common in contemporary societies. This reflects the fact that science and technology are ever more intimately entwined in daily life, affecting both people and the environment more profoundly as new inventions like synthetic biology are applied on a large scale. There exists a growing class of policy issues involving science in which the normal assumptions of certainty, objectivity, value-neutrality, and (implicitly) expertise do not hold.

For resolving such issues there needs to be an extension of the peer community who are responsible for quality control and governance in science and technology. It can no longer be restricted to scientific experts or even accredited professionals, but must include a multiplicity of actors, including what are now called grassroots 'citizen scientists'. The full participation of an 'extended peer community' in the resolution of a growing class of policy issues is necessary not only because of ethical or political reasons, in its creation of an open space of democratic deliberation, but also because it enhances the quality of the relevant scientific input. At the time of first announcement, this extension of the peer community was indeed heretical, and many commentators have substituted 'extended peer *review*' for *community*. This response was unsurprising because it's part of a tradition that is committed to the reduction of all political/practical issues to techno-scientific problems.

Yet, contrary to this tradition, these 'extended peers' may use as evidence 'extended-facts', obtained through sources and methods other than the traditional peer-reviewed literature authored by credentialed experts. PNS has legitimised the use of facts that are, in many occasions, based on experiential, practical or ancestral knowledge. Thanks to the emergence of an educated citizenry and of powerful and ubiquitous Information and Communication Technologies (ICT) citizens (and among them, 'citizen scientists') create useful and relevant knowledge on a massive scale. And as they engage in scientific and policy debates, they show more systematic attention to issues of quality than are frequently observed among 'normal' scientists.

## PNS today

PNS has been maturing constantly in several respects. There is now a small but flourishing presence in academe (see, for instance, the special issue in *Science, Technology and Human Values*, edited by Turnpenny et al. in 2011), with one vigorous group, in the Centre for the Study of the Sciences and the Humanities at the University of Bergen in Norway, teaching and researching along PNS lines. Another sign of vitality and relevance is an effort by the PNS community to engage with the present crisis in science's governance and quality control (Benessia et al., 2016). This activity involves two other clusters of PNS scholarship and practice: the Joint Research Centre of the European Commission (EC-JRC) and the Integrated Assessment of Sociology, Technology and the Environment group at the Universitat Autonoma de Barcelona (ICTA-UAB).

PNS continues to function as a path to a modest enlightenment for scientists in policy-relevant fields whose professional training inhibits or even prohibits attention to significant uncertainties or the fundamental role of values and context. To discover that it's really natural and all right for a science to lack genuine three-digit precision in many of its findings and predictions can be a liberating insight. And in their way, activists and grassroots citizen science groups derive validation from their recognition and identification as extended peer communities.

A most significant development has been the adoption of PNS by some progressive science policy advisors, notably Professor Sir Peter Gluckman in New Zealand. From his experience, he has found that PNS can help to explain why science advice is really *not* a case of simple 'speaking truth to power'. In his discussions, he places the extended peer community in the context of the multiplicity of actors and channels by which science policy is now influenced (Gluckman, 2014). The politics of science is being transformed in ways that were not imagined in the original formulation of PNS. But if the core insight of PNS remains sufficiently illuminating and empowering, these enrichments can be integrated today into the conduct of science.

## Bibliography

Benessia, A., Funtowicz, S., Giampietro, M., Guimarães Pereira, A., Ravetz, J., Saltelli, A., Strand, R. & van der Sluijs, J. (2016) *The Rightful Place of Science: Science on the Verge*, Tempe, AZ: The Consortium for Science, Policy and Outcomes, Arizona State University.

Funtowicz, S. & Ravetz, J. (1985) 'Three Types of Risk Assessment: A Methodological Analysis in Environmental Impact Assessment', in V.T. Covello, J.L. Mumpower, P.J. M. Stallen & V.R.R. Uppuluri (eds) *Technology Assessment, and Risk Analysis* (pp. 831–848). Berlin: Springer Berlin Heidelberg.

Funtowicz, S. & Ravetz, J. (1990) *Uncertainty and Quality in Science for Policy*, Alphen aan den Rijn: Kluwer.

Funtowicz, S. & Ravetz, J. (1991) 'A new Scientific Methodology for Global Environmental Issues', in Costanza, R. *Ecological Economics. The Science and Management of Sustainability* (pp. 137–152). New York, Oxford: Columbia Press.

Funtowicz, S. & Ravetz, J. (1993) 'Science for the Post-normal Age', *Futures*, 25(7): 739–755.
Funtowicz, S. & Ravetz, J. (1994) 'The Worth of a Songbird: Ecological Economics as a Post-normal Science', *Ecological Economics*, 10(3): 197–207.
Gibbons, M., Limoges, C., Nowotny, H., Schwartzman, S., Scott, P. & Trow, M. (1994) *The New Production of Knowledge: The Dynamics of Science and Research in Contemporary Societies*. London: SAGE.
Gluckman, P. (2014) 'Policy: The Art of Science Advice to Government', *Nature*, 507: 163–165.
Petersen, A.C, Cath, A., Hage, M., Kunseler, & van der Sluijs, J.P. (2011) 'Post-normal Science in Practice at the Netherlands Environmental Assessment Agency', *Science Technology & Human Values*, 36(3): 362–388.
Ravetz J. (1996) *Scientific Knowledge and its Social Problems*, New Brunswick: Transaction Publishers. (Originally published by Oxford University Press, 1971).
Turnpenny J., Jones, M., & Lorenzoni, L. (2011) 'Special Issue on Post Normal Science', *Science, Technology & Human Values*, 38(3).
Weinberg, A. (1972) 'Science and Trans-Science', *Minerva*, 10(2): 209–222.

# 4.16 Queer ecology

## Nicole Seymour

'Queer ecology' refers to a conceptual framework that considers how sexual and environmental issues intersect – in cultural representations, scientific research, and other realms.[1] This framework casts a particular eye on how human norms such as heterosexuality influence our appraisal of the natural world, and vice versa.

Activism and art have been crucial to the development of queer ecology, but this framework has been articulated most explicitly within academia. It draws from a wide variety of disciplines and traditions – from ecocriticism to queer theory, feminism to evolutionary biology, posthumanism to performance studies, landscape ecology to cultural studies, architecture to farming practices, philosophy to geography. (Due to particularly strong roots in second-wave ecocriticism, some scholars use the more specific term 'queer ecocriticism'.) Queer ecology's objects of study are diverse; scholars have analyzed the sexual politics of, to offer some examples, literary genres such as the pastoral, policies such as United Nations environmental protocols, and physical spaces such as national parks.

This framework emerged in the last two decades, owing largely to a 1995 special issue on 'Queer/Nature' from *UnderCurrents: Journal of Critical Environmental Studies* and to Greta Gaard's 1997 *Hypatia* essay, 'Toward a Queer Ecofeminism'. This confluence of sexual and environmental inquiries in the late twentieth century is not coincidental: as Gavin Brown has pointed out, "while modern sexual identities (including, later, the queer challenge to them) came into being contemporaneously with the ascendency of neoliberal capitalism, they also coincide with height of high-carbon economies" (Anderson et al., 2012: 86). Among other things, then, queer ecologists examine such coincidences.

## Major themes and insights

Much queer ecology work proceeds from the insight that 'the queer' and 'the natural' have been opposed in cultural, scientific, and political discourses. More specifically, 'the natural' is typically associated with reproductivity, health, futurity – and, of course, heterosexuality – whereas 'the queer' is associated with urbanity, disease, and death. Queer ecology both challenges the discriminatory implications of these associations and argues that queers/sexuality and nature/the environment/animals actually have much in common. To wit: both sets of entities have been subjected to biopolitical control and surveillance; both have been objects of scientific scrutiny; both have historically been oppressed or exploited; both have been positioned on the low end of sociocultural hierarchies; both are understood to exceed standards of human civility and decency; and both have been feared, pathologized, fetishized, and commodified.

This framework has a significant, and complicated, relationship with science, particularly the life sciences such as biology and zoology. Some queer ecology scholars critique these fields – for example, pointing out how heteronormative bias has led scientists to downplay evidence of homosexual behaviors among nonhuman animals. Others embrace scientific findings, using them to dethrone both heterosexuality and humanity as the yardsticks of what is 'normal' in nature – and, hence, to establish the queerness of biotic life. Some scholars, for example, revel in the fact that sexual reproduction (not to mention *hetero*sexual reproduction) is much less common than one might assume, when all living organisms are taken into account. Others have highlighted how species classifications – much like sex and gender classifications – are sometimes arbitrary, occluding messy overlaps and ambiguities. And still others have noted that the concept of 'deviation' is central to both queer sexualities and evolutionary biology; they argue that evolution can itself be considered queer because it depends on deviation from the norm.

Considering the historic opposition between 'the queer' and 'the natural', it is not surprising that the specific fields of queer theory and ecocriticism have also been opposed. For example, queer theorists have complained that the heterosexual, reproductive family serves as the emblem of the future in the dominant cultural imagination. Meanwhile, in their concern with the future impacts of current human actions, ecocritics have regularly deployed this emblem. Queer theory has also centered almost exclusively on human issues. Queer ecology acknowledges the gap between these fields as well as the need to bring them together. For example, queer theory's attention to social construction allows queer ecologists to trouble idealized notions of 'pure' Nature. And queer theory's emphases on fluidity and intimacy allows queer ecologists to study how human and nonhuman lives are deeply entangled and sometimes even indistinguishable – as with, for example, the microbes that live in and on our bodies.

Queer ecology also makes use of the commonalities that queer theory and ecocriticism do have – such as their mutual interests in questioning notions of 'progress' and breaking down conceptual binaries. Among others, queer ecology takes aim at binaries such as Culture/Nature, Constructed/Natural, Human/Nonhuman, Male/Female, Reason/Eroticism, Straight/Gay, Normal/Abnormal, Self/Other, and Life/Nonlife.

Some queer ecology scholars focus on particular histories, practices, and sites. For example, much work has been published on LGBTQ farms and rural communities, 'cruising' or public sex in parks, and how physical spaces such as public campgrounds are designed to promote and/or prohibit particular sexual practices. Other queer ecology work is more abstract and theoretical in nature. For example, drawing on some of the scientific findings referenced above, many queer ecology scholars have noted that biotic life is always interdependent and dynamic – and thus, along with other contemporary scholars, they ask us to focus on 'becoming' as opposed to 'being', relationships over individuals.

## Debates and criticisms

The concept of 'nature', as it has appeared in queer ecology scholarship, has been a source of controversy and contradiction. Many scholars attempt to naturalize phenomena such as human homosexuality, by, for example, pointing to the aforementioned evidence of same-sex behavior among nonhuman animals. However, others note that such moves run counter to the anti-essentialism of queer theory – not to mention potentially engaging in anthropomorphism.

Some critics have lamented that much queer ecology work, in drawing attention to that (supposed) queerness of all life, empties the term 'queer' of meaning, decontextualizing it and stripping it of its political implications. Indeed, while some queer ecology scholars connote 'queer' positively, and thus seek to advocate specifically for LGBTQ communities through their criticism, others generalize that term and, arguably, neutralize it.

Some critics also worry that, in seizing upon the queer implications of environmental crises – such as how toxins apparently trigger androgyny and sterility in many animals – some queer ecologists actually celebrate such crises or, at best, fail to condemn them.

## New directions

Though much queer ecology work addresses the whiteness of mainstream environmentalism and the co-implication of homophobia, colonialism, and racism, the framework lacks racial and national diversity; it has been developed primarily by white scholars and has tended to focus on American, British, and Canadian texts. However, recent scholarship from places such as Brazil and Japan indicates potential for expansion.

In the future, queer ecology might also diversify in terms of its scope of inquiry. For example, while literature and film have served as primary texts for many queer ecology scholars, other arts such as dance, music, and painting have been largely ignored. And at least one scholar has observed that "a queer ecological politics of gentrification is lacking" (Patrick, 2014: 940); thus, queer ecology scholars might study, for example, how public greening projects go hand-in-hand with the gentrification of neighborhoods by white, wealthy gay males. Such work might complicate the notion that 'the queer' and 'the natural' are always oppressed.

Queer ecology scholars might also grapple with the paradigm of 'ecosexuality', recently introduced by U.S. activist/artist Annie Sprinkle and U.S. academic/artist Beth Stephens. With the mantra, "the Earth is our lover [not our mother]," so-called ecosexuals recognize that eroticism, pleasure, and desire often inhere in relations between humans and nonhumans. Ecosexuality resonates with the playful affect found in much recent queer ecology work – another new direction for this framework. However, queer ecology scholars might ask how ecosexuality differs from older sexualized discourses – in which, for example, male colonizers conquer feminized, 'virgin' land.

Queer ecology might also move beyond its focus on queer *sexuality* to consider queer *gender*. Thus, another possible direction – or interlocutor – might be the nascent framework of trans ecology, which considers the intersection of gender identity and environment. Specifically, queer ecology scholars might engage directly with recent transgender studies scholarship on animality.

Moving forward, queer ecology scholars might pointedly explore their relationship to the term 'queer' – perhaps seeing it as neither positive nor neutral, but deeply ambivalent. Indeed, as some scholars have recently indicated, the original definition of 'queer' as 'odd' or 'perplexing' may help us reflect critically on the strange new world of our present environmental era. Relatedly, queer ecology scholars might continue to take a pragmatic approach – refraining from both naïve optimism and gloom and doom, and instead grappling with our messy reality, in which environmental destruction is already here and, in many cases, irreversible. As Heather Davis and Sarah Ensor have argued, respectively, "we need to find ways of *living with toxicity*" (Davis, 2015: 244) and "environmentalism needs . . . a model for ethically relating to the terminally ill" (Ensor, 2016: 43). Queer ecology is poised to offer those things.

## Note

1 Some of the work surveyed here does not identify itself explicitly as 'queer ecology'; however, because it shares with explicitly-self-identified queer ecology work a concern with the intersection of the sexual and the environmental, we can reasonably include it.

## Learning resources

Annie Sprinkle and Beth Stephens' documentary *Goodbye Gauley Mountain*, which depicts queer activists' playful protests against mountaintop removal, can be purchased as a digital copy here: http://gauleymountain.com.

Writer and conservationist Alex Carr Johnson offers suggestions on how to queer our ecological thinking in his *Orion* magazine essay, "How to Queer Ecology: One Goose at a Time": https://orionmagazine.org/article/how-to-queer-ecology-once-goose-at-a-time.

Out for Sustainability is a U.S.-based organization that "mobilizes the LGBTQ community for environmental and social action": http://out4s.org.

The author's sample syllabus on queer ecology can be found here: https://www.academia.edu/29132350/Queer_Ecologies_Seminar_Syllabus.

## Bibliography

Anderson, J., Azzarello, R., Brown, G., Hogan, K., Brent Ingram, G., Morris, M.J. & Stephens, J. (2012) 'Queer Ecology: A Roundtable Discussion', *European Journal of Ecopsychology*, 3: 82–103.

Azzarello, R. (2012) *Queer Environmentality: Ecology, Evolution, and Sexuality in American Literature*. Burlington, VT: Ashgate.

Bagemihl, B. (1999) *Biological Exuberance: Animal Homosexuality and Natural Diversity*. Boston: St. Martin's Press.

Barton, D.L. (2014) 'The Extinction Project', Rhinofest. Chicago, IL. Prop Theatre. Directed by Anna C. Bahow.

Chen, M.Y. (2012) *Animacies: Biopolitics, Racial Mattering, and Queer Affect*. Durham, NC: Duke University Press.

Chen, M. & Luciano, D. (eds) (2015) *'Queer Inhumanisms'*, Spec. issue of *GLQ: A Journal of Lesbian and Gay Studies*, 21(2–3): 183–458.

Clare, E. (2015) *Exile and Pride: Disability, Queerness, and Liberation*. (1999.) Durham, NC: Duke University Press.

Davis, H. (2015) 'Toxic Progeny: The Plastisphere and Other Queer Futures', *philoSOPHIA*, 5(2): 231–250.

Ensor, S. (2012) 'Spinster Ecology: Rachel Carson, Sarah Orne Jewett, and Nonreproductive Futurity', *American Literature*, 84(2): 409–435.

Ensor, S. (2016) 'Terminal Regions: Queer Ecocriticism at the End', in Hunt, A. & Youngblood, S. (eds) *Against Life* (pp. 41–62). Evanston, IL: Northwestern University Press.

Estok, S.C. (2009) 'Theorizing in a Space of Ambivalent Openness: Ecocriticism and Ecophobia', *ISLE: Interdisciplinary Studies in Literature and Environment* 16(2): 203–225.

Gaard, G. (1997) 'Toward a Queer Ecofeminism', *Hypatia*, 12(1): 114–137.

Gaard, G., Estok, S.C. & Oppermann S. (eds) (2013) *International Perspectives in Feminist Ecocriticism*. New York: Routledge.

Gabriel, A. (2011) 'Ecofeminismo e Ecologias *Queer*: Uma Apresentação', *Revista Estudios Feministas*, 19(1): 167–173.

Gandy, M. (2012) 'Queer Ecology: Nature, Sexuality, and Heterotopic Alliances', *Environment and Planning D*, 30(4): 727–747.

Garrard, G. (2010) 'How Queer is Green?', *Configurations*, 18(1–2): 73–96.

Giffney, N. & Hird, M.J. (eds) (2008) *Queering the Non/Human*. Farnham: Ashgate.

Gough, N., Gough, A., Appelbaum, P., Appelbaum, S., Aswell Doll, M. & Sellers, W. (2003) 'Tales from Camp Wilde: Queer(y)ing Environmental Education Research', *Canadian Journal of Environmental Education*, 8(1): 44–66.

Griffiths, T. (2015) 'O'er Pathless Rocks: Wordsworth, Landscape Aesthetics, and Queer Ecology', *ISLE: Interdisciplinary Studies in Literature and Environment*, 22(2): 284–302.

Grover, J.Z. (1997) *North Enough: AIDS and Other Clearcuts*. Minneapolis: Graywolf Press.

Hannah, D. (2013) 'Invitations and Withdrawals: Queer Romantic Ecologies in William Blake's *The Book of Thel* and John Clare's *The Nightingale's Nest*', *Essays in Romanticism*, 20(1): 1–18.

Herring, S. (2010) *Another Country: Queer Anti-Urbanism*. New York: New York University Press.

Krupar, S. (2013) 'Transnatural Review: Irreverent Counterspectacles of Mutant Drag and Nuclear Waste Sculpture', in *Hot Spotter's Report: Military Fables of Toxic Waste*. Minneapolis: Minnesota University Press.

Morita, K. (2013) 'A Queer Ecofeminist Reading of 'Matsuri' [Festival] by Hiromi Ito', in Simon Estok & Won-Chung Kim (eds) *East Asian Ecocriticisms: A Critical Reader*. London: Palgrave.

Morton, T. (2010) 'Queer Ecology', *PMLA*, 125(2): 273–282.

Nyong'o, T. (2012) 'Back to the Garden: Queer Ecology in Samuel Delany's *Heavenly Breakfast*', *American Literary History*, 24(4): 747–767.

Patrick, D. (2014) 'The Matter of Displacement: A Queer Urban Ecology of New York City's High Line', *Social and Cultural Geography*, 15(8): 920–941.

Queer Cultural Center (n.d.) 'Queer Ecologies: A Gallery Event and Conversation Featuring Kim Anno and Adrian Parr'. Online. Available at: http://qcc2.org/queer-ecologies.

Queer/Nature (1994) *Undercurrents: Journal of Critical Environmental Studies*, 6 Spec. issue. Online. Available at: http://currents.journals.yorku.ca/index.php/currents/issue/view/2157/showToc.

Queer/Nature to Queer Ecologies: Celebrating 20 Years of Scholarship and Creativity (2015) *Undercurrents: Journal of Critical Environmental Studies* 19 Spec. issue. Online. Available at: http://currents.journals.yorku.ca/index.php/currents/issue/view/2202/showToc.

Rosenberg, J. (2014) 'The Molecularization of Sexuality: On Some Primitivisms of the Present', *Theory and Event*, 17(2). n.p.

Roughgarden, J. (2009) *Evolution's Rainbow: Diversity, Gender, and Sexuality in Nature and People*. Berkeley: University of California Press.

Sandilands, C. (2005) 'Unnatural Passions? Notes toward a Queer Ecology', *Invisible Culture: An Electronic Journal for Visual Culture*, 9. Online. Available at: https://www.rochester.edu/in_visible_culture/Issue_9/issue9_sandilands.pdf.
Sandilands, C. & Erickson, B. (eds) (2010) *Queer Ecologies: Sex, Nature, Politics, Desire*. Bloomington: Indiana University Press.
Seymour, N. (2013) *Strange Natures: Futurity, Empathy, and the Queer Ecological Imagination*. Champaign: University of Illinois Press.
Stein, R. (ed.) (2004) *New Perspectives on Environmental Justice: Gender, Sexuality, and Activism*. New Brunswick, NJ: Rutgers University Press.
Sturgeon, N. (2008) *Environmentalism in Popular Culture: Gender, Race, Sexuality, and the Politics of the Natural*. Tucson: University of Arizona Press.

# 4.17 Resilience science

## Manjana Milkoreit

### Definition and history

Resilience science is concerned with the capacity and tendencies of social-ecological systems to respond to change. The definition that established the field emphasized "the persistence of systems and their ability to absorb change and disturbance" (Holling, 1973: 17). However, resilience is not about resisting change or recovering from a disturbance back to a previous state. More important is the system's ability to adapt to changing circumstances, while maintaining its core characteristics, i.e., its functions and identity (Walker, 2004). Preserving a stable identity while being internally flexible requires renewal, innovation and reorganization (Folke, 2006). More recently, scholars have expanded this definition of resilience, adding two new components: development and human well-being as desirable system goals and the ability to transform (i.e., change the system's identity) as an important dimension of managing (conserving) resilience. Folke writes that resilience is the "capacity of people, communities, societies, cultures to adapt *or* even transform into new development pathways in the face of dynamic change" (2016: 4).

Resilience science has conceptual origins in multiple disciplines, including engineering and mathematics, but it is most firmly rooted in ecology. Building on observations of ecosystem behavior, it recognizes multifaceted linkages in social-ecological systems (SES)—interacting systems of people and nature. The insight that human systems at all scales are embedded in and affect the biosphere is fundamental to resilience science.

Transferring knowledge about ecosystems to social-ecological systems or even entirely social phenomena is often problematic. Resilience science has—with few exceptions (e.g., Adger, 2000)—not yet grappled sufficiently with the question of whether and how social and ecological processes differ. A resilience approach is not naturally attuned to the conditions that characterize social relationships, which are at the root of all social and political contention (Brown, 2014). This has opened the field up to multiple critiques from the social sciences, including that charge that resilience science ignores questions of power, equity, or resource access.

Resilience science emerged in the context of growing concern about global environmental change. Recognizing humans as agents of ecosystem change from the local to the global scale, resilience science is concerned with environmental degradation and sustainable resource governance. Over the last decade, it has become increasingly embedded in sustainability science and global governance efforts for sustainable development and climate change.

## Key concepts and debates

Resilience science is grounded in a systems perspective, focusing on patterned interactions between different system components. Most scholars explore the dynamics of complex-adaptive systems (CAS), which exhibit complex behaviors, such as non-linearity, feedbacks, tipping points and emergence. CAS have the ability to change in a self-organizing fashion in response to an external event.

Two core concepts of resilience science are the adaptive cycle and Panarchy. The adaptive cycle conceptualizes the change dynamics of an SES, distinguishing four distinct phases: growth and exploitation (r), conservation (K), release (Ω) and reorganization (α).

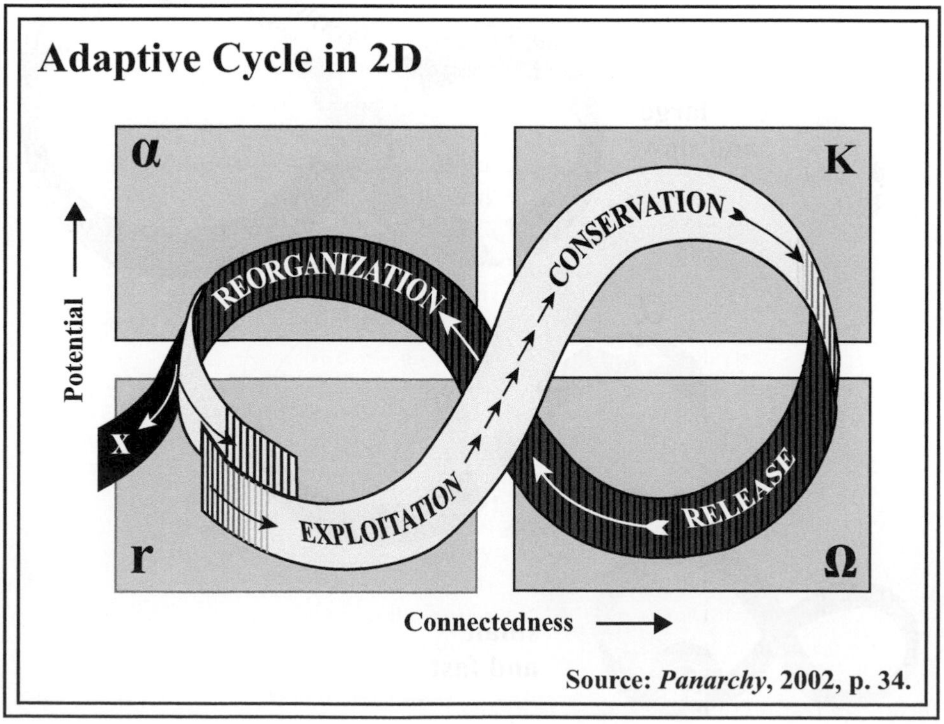


**Figure 4.17.1** The adaptive cycle.

Credit: http://www.peopleandplace.net/2008/11/14/adaptive-cycle/

The first two phases ("the front loop") represent slow and incremental processes of growth and accumulation. The shorter and faster back loop is a rapid process of renewal.

Any SES is embedded in a multi-scale, nested hierarchy of SES. Adaptive cycles can take place at any scale within that hierarchy. Higher-scale cycles tend to be slower than lower-scale cycles. Important cross-scale effects draw attention to the timing of adaptive dynamics. This multi-scale perspective is captured in the Panarchy framework (Gunderson & Holling, 2002).

Resilience science seeks to establish the conditions for developing high adaptive capacity to foster the ability of an SES to deal with future change. It also promotes adaptive resource management, a flexible governance approach that recognizes the inherent uncertainty in complex systems and allows for learning (trial-and-error) and adjustments over time.

Resilience science pays particular attention to the possibility of so-called regime shifts—large, persistent changes in the structure and function of SES. For example, lakes can shift between a clear and a turbid state; the Arctic is currently moving from a regime of perennial summer sea ice to one that is summer ice-free. Moving between different regimes involves passing a tipping point—a moment at which gradual change becomes rapid, internally-driven and unstoppable.

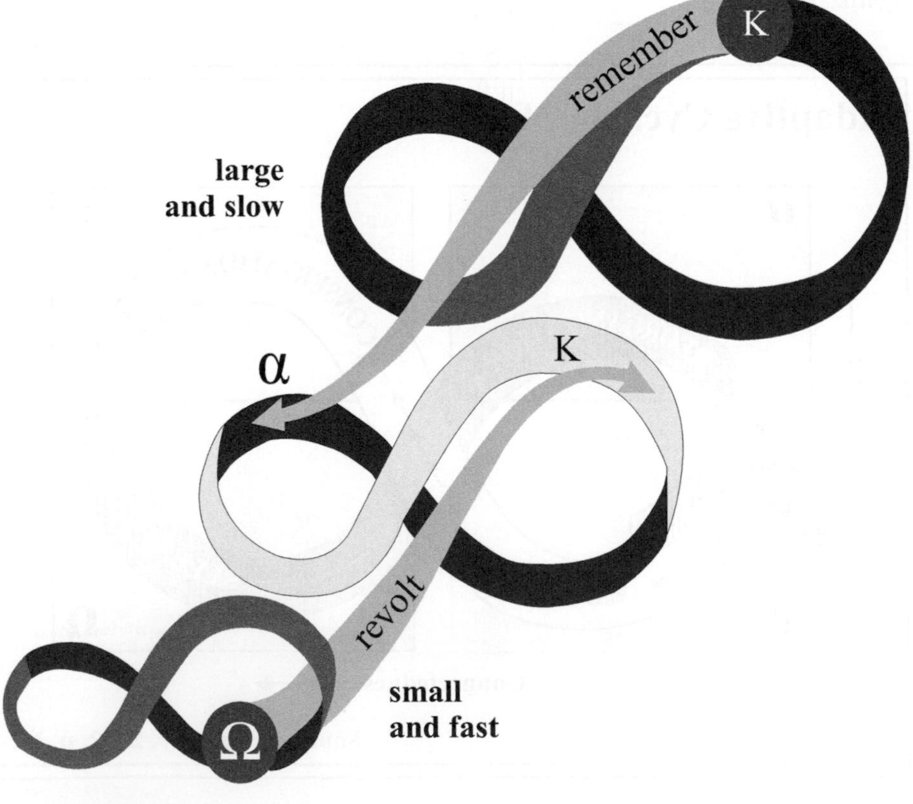


**Figure 4.17.2** Panarchy

Credit: Resilience Alliance website, http://www.resalliance.org/panarchy

Regime shifts are important examples of system transformations, affecting the identity of an SES. They are usually considered undesirable, especially in SES that are important for human well-being. But some resilience scholars study the possibility of 'proactive tipping'—intentionally pushing an SES out of its current, undesirable but resilient state into a more desirable one (Westley et al., 2011).

One of the major debates in the field concerns the relationship between adaptation and transformation—two distinct types of systemic change. Generally, adaptation refers to identity-maintaining changes that take place while the SES remains in a particular regime. Identity-altering changes, including regime shifts, are transformational. Transformations usually result from a loss of resilience in the old regime and involve (re-)establishing resilience in a new one. Folke adds a development context to this distinction: "Adaptation refers to human actions that sustain development on current pathways, while transformation is about shifting development into other emergent pathways and even creating new ones" (2016: 4).

Ultimately, resilience science is interested in fostering resilient SES that provide ecosystem services and allow for human flourishing, which often requires social change. Consequently, questions of sustainable resource governance, institutional design and policy making have grown in importance over time. These topics raise fundamental questions about values, social relations, agency and power, knowledge, learning and decision making. Resilience science tends to favor participatory, democratic and polycentric approaches to governance, and emphasizes the value of diverse kinds of knowledge and their integration.

## Applying resilience thinking

Based on these conceptual foundations, resilience scientists have developed seven basic principles of resilience practice (Biggs et al., 2015).

1) *Maintain diversity and redundancy*: More diverse systems with many different kinds of components tend to be more resilient than less diverse systems. Not only do they provide a larger set of response options to unexpected changes, redundancy also serves as a back-up system, or 'insurance', in case of failure.
2) *Manage connectivity*: The connectivity of a system—the number and structure of links between its various components—determines how easily both good and bad things can move within the system. A well-connected system can both spread and recover from a disturbance more easily than a poorly connected one. Since connectivity can both enhance and reduce resilience, it needs to be managed depending on the context.
3) *Manage slow variables and feedbacks*: Slow variables are those that do not change much over time. They are connected by feedback mechanisms that can either reinforce or dampen changes, e.g. triggered by a disturbance. This cooperation between slow variables and feedbacks keeps a system stable, ensuring it remains in a certain stability domain.
4) *Foster CAS thinking*: Adopting a systems-perspective and acknowledging the uncertainty and to some extent unpredictability of CAS is a key first step towards building resilience in SES.
5) *Encourage learning*: SES are constantly changing and developing, rendering decision-makers' knowledge incomplete. Managing SES hence requires continuous learning

and updating of one's understanding of the system, including the change of social norms and values. Often, experimentation and knowledge integration across multiple domains are the most successful learning strategies.

6) *Broaden participation*: Engaging a broad range of stakeholders in SES management leverages a large and diverse base of knowledge and experience. It also builds the necessary trust and legitimacy for collective action.
7) *Promote polycentric governance*: The distributed and nested nature of polycentric governance systems promotes resilience because it enables fast, fitting and diverse responses to change.

## Resilience thinking across disciplines

Resilience has become a very popular concept outside of resilience and sustainability science. Scholars in many social science fields discuss resilient communities (Norris et al., 2008), economies (Simmie & Martin, 2010), or societies (Redman & Kinzig, 2003). Political scientists apply resilience to questions of national and international security or development policy. Psychologists raise questions of mental resilience, for example in times of crisis or depression. Lately, questions of planetary resilience and global tipping points have emerged among Earth system and climate scientists (Rockström et al., 2009). This process of conceptual borrowing across disciplinary boundaries often involves the bending and changing of definitions, sometimes losing the initial conceptual meanings imbued by resilience science.

## Resilience science beyond science

Political actors, public policy-makers as well as corporate and civil society actors have appropriated resilience concepts for various purposes far beyond academia, for example, for natural resource or disaster risk management. There is a rich political discourse on resilience at the global scale, linked to discussions about Planetary Boundaries, the Sustainable Development Goals, and, more broadly, questions of international development in a time of rapid global environmental change (Brown, 2015). For example, a 2012-report of United Nation's *High-level Panel on Global Sustainability* was entitled "Resilient People, Resilient Planet – A Future Worth Choosing". The latest example for the spread of resilience ideas in global governance is the 2015 Paris Agreement on Climate Change. Article 2 of the agreement established the global goals of climate governance, which include fostering "climate resilience and low greenhouse-gas emissions development" as well as "making finance flows consistent with . . . climate-resilient development."

## Learning resources

The website of the Resilience Alliance provides a useful introduction to the core concepts of resilience science as well as other resources (e.g., publications). The Resilience Alliance is the network of scholars that pioneered resilience science starting in the 1990s: http://www.resalliance.org/key-concepts.

The Stockholm Resilience Center (SRC) is a major hub of resilience research and education. The Center's website is another excellent resource for questions about resilience science and current research topics: http://www.stockholmresilience.org/research/our-research-focus.html.

The SRC also offers Massive Open Online Courses on major resilience science topics: http://www.stockholmresilience.org/education/independent-courses.html.

## Bibliography


Adger, W.N. (2000) 'Social and Ecological Resilience: Are They Related?', *Progress in Human Geography*, 24(3): 347–364.

Biggs, R., Schlüter, M. & Schoon, M.L. (eds) (2015) *Principles for Building Resilience: Sustaining Ecosystem Services in Social-Ecological Systems*. 1st edition. Cambridge: Cambridge University Press.

Brown, K. (2014) 'Global Environmental Change I A Social Turn for Resilience?', *Progress in Human Geography*, 38(1): 107–117.

Brown, K. (2015) *Resilience, Development and Global Change*. Abingdon: Routledge.

Chandler, D. (2012) 'Resilience and Human Security: The Post-Interventionist Paradigm', *Security Dialogue*, 43(3): 213–229.

Folke, C. (2006) 'Resilience: The Emergence of a Perspective for Social-Ecological Systems Analyses', *Global Environmental Change*, 16(3): 253–267.

Folke, C. (2016) 'Resilience', *Oxford Research Encyclopedia of Environmental Science*. Oxford Research Encyclopedias. New York: Oxford University Press.

Gunderson, L.H. & Holling, C.S. (2002) *Panarchy: Understanding Transformations in Human and Natural Systems*. Washington: Island Press.

Holling, C.S. (1973) 'Resilience and Stability of Ecological Systems', *Annual Review of Ecology and Systematics*, 4(1): 1–23.

Masten, A.S., Best, K.M. & Garmezy, N. (1990) 'Resilience and Development: Contributions from the Study of Children Who Overcome Adversity', *Development and Psychopathology*, 2(4): 425–444.

Norris, F.H., Stevens, S.P., Pfefferbaum, B., Wyche, K.F. & Pfefferbaum, R.L. (2008) 'Community Resilience as a Metaphor, Theory, Set of Capacities, and Strategy for Disaster Readiness', *American Journal of Community Psychology*, 41(1–2): 127–150.

Redman, C. & Kinzig, A. (2003) 'Resilience of Past Landscapes', *Ecology and Society*, 7(1). Online. Available at: https://asu.pure.elsevier.com/en/publications/resilience-of-past-landscapes-resilience-theory-society-and-the-l.

Rockström, J., Steffen, W., Noone, K., Persson, Å., Chapin, F.S., Lambin, E.F., Lenton, T.M., Scheffer, M., Folke, C., Schellnhuber, H.J. et al. (2009) 'A Safe Operating Space for Humanity', *Nature*, 461(7263): 472–475.

Simmie, J. & Martin, R. (2010) 'The Economic Resilience of Regions: Towards an Evolutionary Approach', *Cambridge Journal of Regions, Economy and Society*, 3(1): 27–43.

Walker, B., Holling, C.S., Carpenter, S.R. & Kinzig, A. (2004) 'Resilience, Adaptability and Transformability in Social-Ecological Systems', *Ecology and Society*, 9(2): 5.

Westley, F., Olsson, P., Folke, C., Homer-Dixon, T., Vredenburg, H., Loorbach, D. Thompson, J., Nilsson, M., Lambin, E., Sendzimir, J. et al. (2011) 'Tipping Toward Sustainability: Emerging Pathways of Transformation', *AMBIO*, 40(7): 762–780.

# 4.18 Sustainability science

Thaddeus R. Miller

## Science for sustainability

It has been 30 years since sustainable development emerged on the global stage with the publication of the World Commission on Economic Development's seminal report *Our Common Future* (or, as it is more commonly known, the Brundtland Report [WCED, 1987]). The Report provided the oft-cited definition of sustainable development as "development that meets the needs of the present without compromising the ability of future generations to meet their own needs" (WCED, 1987: 43). In the long wake of the Report, sustainability has gone from a radical idea to a sometimes vilified buzzword. Along the way, the values and ideals behind the term – ecological health, social justice, intergenerational equity – have influenced business practices (Corporate Social Responsibility, Triple Bottom Line), urban planning and development, international development (e.g., United Nations Sustainable Development Goals), infrastructure and buildings (e.g., US Green Building Council Leadership in Energy and Environmental Design [LEED] standards), education, and scientific research. This chapter examines the emergence of sustainability science and discusses the common barriers – intellectual and institutional – such a transdisciplinary field encounters.

Sustainability science is a diverse, transdisciplinary field of inquiry focusing on the dynamics of complex, coupled social, ecological and technological systems. A core aim of the field is to link research to social action and impact (Miller, 2015). At its most expansive interpretation, sustainability science is research on coupled human, natural and technological systems that attempts to link knowledge to action. Sustainability science, at least in the European and North American contexts, emerged in the late 1990s in response to a growing call from scientists (Lubchenco, 1998) to conduct research on pressing social and environmental challenges, including climate change and biodiversity loss (Kates et al., 2001). Led by Pamela Matson, Robert Kates, and William Clark, the

US National Research Council's (NRCs) Board on Sustainable Development 1999 report, *Our Common Journey*, proposed the development of sustainability science that would bring "significant advances in basic knowledge, in social capacity and technological capabilities to use it, and political will to turn this . . . into action" (NRC, 1999: 7) to foster a sustainability transition that "should be able to meet the needs of a much larger but stabilizing human population, to sustain the life support systems of the planet, and to substantially reduce hunger and poverty" (NRC, 1999: 31).

Since the publication of the NRC report in 1999, sustainability science research, education programs, and funding has rapidly expanded around the world. The concept began to gain significant traction in academic circles with the publication of the short article 'Sustainability Science' in *Science* (Kates et al., 2001). Kates and his coauthors (2001: 641) defined sustainability science as a new field that seeks "to understand the fundamental character of interactions between nature and society" and enhance "society's capacity to guide those interactions along more sustainable trajectories." In 2006, Arizona State University established the first School of Sustainability. ASU's initial efforts coincided with a proliferation of research and education programs and centers, including the Lund University Centre for Sustainability Studies, the University of Tokyo Graduate Program in Sustainability Science, broad based work in sustainability at Leuphaha University, and the Institute for Sustainable Solutions at Portland State University, to name but a few. Sustainability science has also shaped science policy with large investments from national science funding bodies, including the US National Science Foundation in Science, Engineering and Education for Sustainability (SEES) and Sustainability Research Networks (SRNs); Australia's Commonwealth Scientific and Industrial Research Organization (CSIRO) in Sustainable Futures and Sustainable Ecosystems; and, the European Research Council (ERC). Sustainability science has also emerged as a central piece of scientific governing and advising bodies, including the American Association for the Advancement of Science (AAAS) Forum on Science and Technology for Sustainability, the Roundtable on Science and Technology for Sustainability Program at the National Academy of Sciences, and the Initiative on Science and Technology for Sustainability sponsored by the International Council of Science. More recently, in 2015, Future Earth was established to advance global sustainability science over the next 10 years. A centerpiece of Future Earth and allied efforts is a focus on linking knowledge to action.

## Linking knowledge to action

At the core of sustainability science and similar efforts is a critical question: How can science and technology most effectively inform and foster social action for sustainability? Implicit in this question is an acknowledgement that the so-called 'linear model' of science and policy is broken. Enshrined in US science policy and scientific research, more broadly, by engineer Vannevar Bush's efforts to formalize science policy and funding post-World War II, the linear model holds that the best way to ensure that science generates valuable social, technological, and economic outcomes is to invest in basic science. In other words, let scientists pursue their own intellectual interests independent of any potential outcome and those outcomes will eventually come. Over the last several decades, however, the technological disasters, environmental concerns, decreased trust in science, the increasingly urgent need for science and technology to be applied to pressing needs, have brought the traditional linear model into question.

Many scientists and policymakers have called for a new social contract for science (Nowotny et al., 2001) – one that directly engages in addressing society's most urgent needs and works to explicitly link to social action and decision making. Sustainability science is perhaps the most direct and concentrated effort to embody this new social contract. Many sustainability scientists, building on science policy scholar Donald Stokes' (1997) work, have called for sustainability science to conduct use-inspired basic research (see Figure 4.18.1). Yet, how different communities in the field are navigating this link between knowledge and action varies with different implications for the future of sustainability science.

Research agendas for sustainability are broadly forming into two broad camps: (1) the coupled systems approach, and (2) the social change approach (Miller, 2013; 2015). The couple systems approach, as Carpenter et al. (2009: 1305) note, is "motivated by fundamental questions about interactions of nature and society as well as compelling and urgent social needs." Bridging ecology and human and physical geography, this vision of sustainability science positions scientific research on human–environmental interactions as central to the pursuit of sustainability. In addition, these scientists call for the co-production of knowledge with stakeholders to help ensure its impact (Clark, 2010; Matson, 2009).

| *Research inspired by...* | | Considerations of use: No | Considerations of use: Yes |
|---|---|---|---|
| Fundamental understanding | Yes | Pure basic research (Bohr) | Use-inspired basic research (Pasteur, Sustainability science) |
| | No | | Applied research (Edison) |

**Figure 4.18.1** Quadrant model of scientific research. This model, adapted from Stokes (1997) and Clark (2007), depicts how sustainability scientists view the position of their field – pursuing questions that contribute to the fundamental understanding of human–environment interactions while also contributing to an ability to make decisions that contribute to more sustainable outcomes.

*Source*: From Miller (2015).

The social change approach holds that sustainability science should "drive societal learning and change processes" and focus "on the design and running of processes linking knowledge with action to deal with persistent problems of unsustainability and to foster transitions to sustainability" (Jäger, 2009: 3). Emerging out of work on socio-ecological transitions and transitions management (Loorbach & Rotmans, 2009), the social change approach is largely (but not exclusively) populated by social scientists with a more explicit focus on engagement with communities to develop clear normative goals for what sustainability means in context and what social, policy or technological pathways are required to meet them (Robinson & Tansey, 2006; Wiek & Lang, 2016).

## The future of sustainability science

Despite significant advances in the institutional capacity of sustainability science – peer-reviewed journals, academic schools and departments, degree programs, research centers, funding programs, etc. – there remains a dearth of data and analyses on how science and technology are able to advance more sustainable social and ecological outcomes. Much of the work in sustainability science has advanced the understanding of coupled systems in important ways. But, how have these and other advances translated into outcomes? Future work in sustainability must engage this question. After nearly 20 years, sustainability scientists have generated a wealth of qualitative and quantitative data on couple systems and sustainability transitions. Yet, both conceptually and methodologically, the core approaches in sustainability science remain relatively isolated. If sustainability science is to link knowledge to action for sustainability, researchers must begin to collaborate and compare across efforts to understand the perils, pitfalls and opportunities – and to realize the limits of scientific knowledge in an era of political turmoil and rapid social and technological change.

## Learning resources

The USA's National Academies of Sciences, Engineering and Medicine's Science and Technology for Sustainability Program (STS) curates an impressive library of sustainability related material at http://sites.nationalacademies.org/pga/sustainabilitytopics/index.htm. Learning for Sustainability provides a wealth of material about sustainability and transdisciplinary research: http://learningforsustainability.net/.

## Bibliography

Carpenter, S.R., Mooney, H.A., Agard J., Capistrano, D., DeFries, R.S., Diaz, S., Dietz, T., Duraiappah, A.K., Oteng-Yeboah, A., Pereira, H.M., Perrings C., Reid, W.V., Sarukhan, J., Scholes, R.J. & Whyte, A. (2009) 'Science for managing ecosystem services: beyond the Millennium Ecosystem Assessment', *Proceedings of the National Academy of Sciences*, 106(5): 1305–1312.

Clark, W.C. (2007) 'Sustainability science: a room of its own', *Proceedings of the National Academy of Sciences*, 104(6): 1737–1738.

Clark, W.C. (2010) 'Sustainable development and sustainability science'. In report from *Toward a Science of Sustainability Conference*, Airlie Center, Warrenton, VA (2009).

Jäger, J. (2009) 'Sustainability science in Europe'. (Background paper prepared for European Commission's DG for Research). Online. Available from: http://ec.europa.eu/research/sd/pdf/workshop2009/background_paper_sust_scienCe_workshop_october_2009.pdf.

Kates, R.W., Clark, C.C., Corell, R., Hall, J.M., Jaeger, C.C., Lowe, I., McCarthy, J.J., Schellnhuber, H.J., Bolin, B., Dickson, N.M. et al. (2001) 'Sustainability science', *Science*, 292(5517): 641–642.

Loorbach, D. & Rotmans, J. (2009) 'The practice of transitions management: examples and lessons from four distinct cases', *Futures*, 42(3): 237–246.

Lubchenco, J. (1998) 'Entering the century of the environment: a new social contract for science', *Science*, 279(5350): 491.

Matson, P. (2009) 'The sustainability transition', *Issues in Science and Technology*, Summer 2009: 39–42.

Miller, T.R. (2013) 'Constructing sustainability science: emerging perspectives and research trajectories', *Sustainability Science*, 8(2): 279–293.

Miller, T.R. (2015) *Reconstructing Sustainability Science: Knowledge and Action for a Sustainable Future*. Science in Society Series. Abingdon: Routledge/Earthscan.

National Research Council (NRC) (1999) *Our Common Journey: A Transition Toward Sustainability*. Washington, DC: National Academy Press.

Nowotny, H., Scott, P. & Gibbons, M. (2001) *Re-thinking Science: Knowledge and the Public in an Age of Uncertainty*. Malden, MA: Polity.

Robinson, J. & Tansey, J. (2006) 'Co-production, emergent properties and strong interactive social research: the Georgia Basic Futures Project', *Science and Public Policy*, 33(2): 151–160.

Stokes, D. (1997) *Pasteur's Quadrant: Basic Science and Technological Innovation*. Washington, DC: Brookings Institution Press.

World Commission on Environment and Development (WCED) (1987) *Our Common Future*. New York: Oxford University Press.

Wiek, A. & Lang, D. (2016) 'Transformational sustainability research methodology'. In H. Heinrichs, P. Martens, G. Michelsen & A. Wiek, (eds) *Sustainability Science: An Introduction*. London: Springer.

# Vulnerability science

4.19

## Thomas Webler

*Vulnerability* helps explain why not all people respond to a hazard the same way. Hazards (also called "stressors") are anything that can be damaging. For example, ultraviolet radiation (UV) in sunlight can damage skin. But some people develop sunburn more quickly than others. Tornados are another kind of hazard. But a tornado does not affect every house in its path the same way. One house can be completely obliterated while the one right next door is barely damaged. We use the concept of vulnerability to account for these differences.

There are three dimensions to understanding why people or things are affected differently by a hazard. The first has to do with *exposure* to the hazard. A person can be in the sun and avoid exposure to the hazard by using sunscreen, which blocks UV light. A house in the path of a tornado may be sheltered by the worst winds by a small hill or valley. Second, people or things are not equally *sensitive* to being damaged by a hazard. People's bodies react differently to sunlight. Houses are built with different degrees of strength. Third, it is possible to *adapt* in a way that reduces exposure, decreases sensitivity, or helps cope with and recover from harm. For example, one person might not get a sunburn because they built up a protective tan by going to the beach all summer long. They were able to adapt to the environment. These three elements – exposure, sensitivity, and adaptive capacity – make up the notion of vulnerability (Adger, 2006; Turner et al., 2003).

## Vulnerability science

Vulnerability science is the application of the scientific process to characterize the state of vulnerability and to learn how it could be reduced (Dow, 1992). Vulnerability science has much in common with sustainability science (Turner et al., 2003). Both refer to the application of a wide range of scientific approaches and tools to achieve a general social purpose: to reduce harm (vulnerability science) or to create a sustainable society (sustainability science).

Vulnerability scientists are interested in mitigating harm to a wide variety of things including people's health, nature and ecosystems, property, social relations, cultural practices, and more. Consequently, vulnerability science needs to draw on knowledge from virtually every discipline. Historically, vulnerability science developed to understand how climate change would affect human communities (Clark et al., 1998). It usually begins with a focus on a specific kind of hazard and the first step is to characterize the risks presented by that hazard. For example, the science of climate vulnerability estimates future climatic conditions and the harm they are likely to cause (Stern et al., 2013).

Climate scientists project that hazards such as sea level rise, storms, wildfires, droughts, and flooding will increase as the planet warms. Vulnerability scientists seek to understand: 1. What places are likely to be affected and when? 2. What is at risk of harm from these threats? 3. What can be done to reduce the expected harms? The first question is the domain of climate scientists and meteorologists, who develop models that represent planetary or regional weather systems.

## Social indicators of vulnerability

Geographers have focused on characterizing the spatial variation in exposure, sensitivity, and adaptive capacity to hazards associated with climate change. Many collections of indicators to measure vulnerability have been developed (see Cutter et al., 2009 for a review). The Social Vulnerability Indicators (SoVI) approach uses existing data to measure vulnerability (Cutter et al., 2003). Data can have many sources. The first SoVI data were gathered at the unit of a census block in the United States. These were combined to produce vulnerability scores. Indicators of vulnerability have been calculated for many places and for many types of hazards. For instance, Suk et al. (2014) recently used a variety of indicators to estimate vulnerability of European Union Member States to infectious disease associated with climate change.

There are several variables that can measure vulnerability. To take an example from America, race – particularly being African American – is often associated with a reduced access to the resources needed to avoid, cope, or adapt to hazards. To be more specific, it is known that African Americans do not have access to the bank loans that White Americans do, even when individuals make the same income. The inability to acquire loans prevents people from upgrading their homes or businesses to be more resilient to wind or storm surge.

The place-based data are overlaid onto meteorological projections for hazards, such as sea level rise, storm surge, or infectious disease, to arrive at a general depiction of who is most at risk of harm. Vulnerability datasets and maps allow emergency response, hazard mitigation, and public health agencies to focus on people and places who are most vulnerable. However, certainly not everyone in a data-collection unit is equally vulnerable. When using spatial datasets for policy-making purposes, one needs to keep in mind that vulnerability is a characteristic of the individual as much as a characteristic of place.

## Vulnerability scoping diagram

Given the wide variety of hazards and exposures, and given the wide range of things that could affect the sensitivity and adaptive capacity of the entities at risk, vulnerability scientists need some way to focus in on what matters most. Inspired by the scoping process used in environmental impact assessment, a group of scientists developed a generalized approach to scope out the relevant vulnerability issues for a specific context (Schröter et al., 2005). The idea behind the vulnerability scoping diagram (VSD) is that it is possible to start from any given hazard and then identify who or what is exposed, what determines sensitivity, and what are the options and resources needed to adapt (Polsky et al., 2007).

A VSD is a target-type diagram, circular with two concentric rings that is 'cut' into three large slices. Information relevant to characterizing vulnerability is written onto the diagram. One-third of the diagram is for information about exposure, another third is for sensitivity, and the last third is for information about adaptive capacity. In the bullseye is written the hazard of interest. In the first ring is written words that summarize

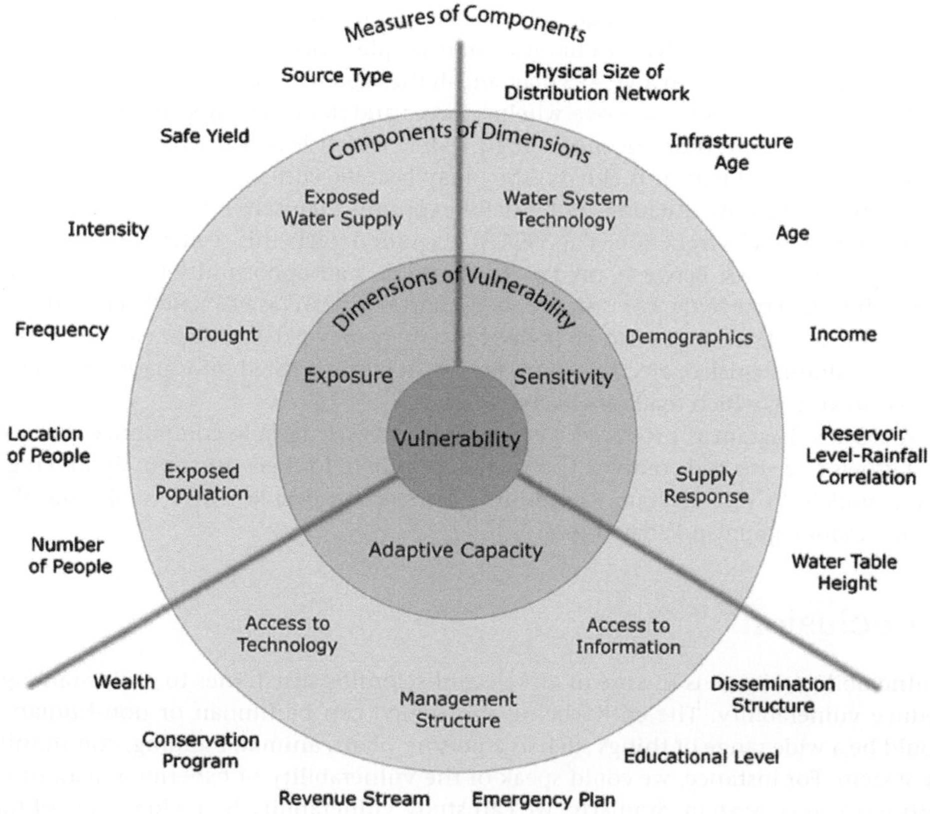


**Figure 4.19.1** Hypothetical VSD based on HERO research. Hazard = drought; Exposure unit = generic community water system.

*Source*: Polsky et al. (2007).

relevant components of the system. In the second ring is written the names of variables that can be used to measure the state of each component. The diagram does not specify what should change first. It merely summarizes what is normally a quite complicated system. For instance, in a study of the vulnerability communities to drought, researchers had community members come together in workshops to create a VSD (Polsky et al., 2007). Figure 1 summarizes the main elements of this VSD.

## Vulnerability consequences and adaptation planning scenarios (VCAPS)

VSDs are one approach vulnerability scientists use to summarize the broad range of issues that need to be taken into consideration when estimating vulnerability and planning how to reduce it. The VCAPS approach is similar in that it is an attempt to assist communities to prepare for climate change hazards by depicting the system that is affected (Kettle et al., 2014; Webler et al., 2016). The unique contribution of VCAPS is that it shows how elements in the system affect one another. This is an important contribution because other approaches like SoVI or VSD are snapshots in time, whereas VCAPS attempts to characterize the dynamics in the affected system.

A VCAPS diagram is simply a graphical depiction of how a sequence of events unfolds as a community is subjected to a hazard. For example, storm surge can affect a community in numerous ways. In a VCAPS diagram all these pathways are charted. An example path is as follows: storm surge overwhelms levees and floods neighborhoods. As people attempt to evacuate they encounter high water; which leaves them trapped in their cars. If the trapped person is elderly, they may become confused and attempt to walk out of the dangerous situation, potentially exposing themselves to risk of drowning, hypothermia, or electrocution. The VCAPS diagram depicts this scenario as a sequence of boxes and arrows. Between any two boxes there is an opportunity to intervene and break the chain of events. For instance, neighborhoods can be evacuated before the levees are breached; roadways can be policed by emergency personnel; or electricity shutoffs can eliminate risk of electrification. Below each box is listed information specific to the context (e.g. which roads are likely to flood).

A VCAPS diagram is produced by a group of knowledgeable community members and grows to a size and complexity that is appropriate to their situation. By looking at the complete VCAPS diagram, community leaders can decide where to allocate their resources for coping and adaptation.

## Conclusion

Vulnerability science is the use of all relevant scientific disciplines to understand and reduce vulnerability. The entity being threatened can be human or non-human. It could be a wide range of things such as a person, plant, animal, building, community, or system. For instance, we could speak of the vulnerability of fishermen, fish, or the entire ocean ecosystem. Similarly, we can study vulnerability to a wide range of hazards or stressors including chemicals, storms, accidents, disease, or even something as broad as climate change.

While people in different fields use the concept of *vulnerability* in slightly different ways, there are commonalities. *Exposure* generally refers to the amount of a hazard (or stressor) that an entity experiences or absorbs. *Sensitivity* is understood as the degree of reaction the entity has to the exposure. *Adaptation* refers to things an entity does reduce, manage, or recover from harm. This is similar to *resilience* and the terms are sometimes used interchangeably. In addition, while adaptation refers to what has happened, *adaptive capacity* refers to the ability to adapt in the future. Adaptive capacity includes things such as skills, knowledge, social networks, finances, and other resources an entity – be it an individual, group, or system – needs to reduce exposure, reduce sensitivity, cope, recover, or change in anticipation of harm. Together, the three concepts of exposure, sensitivity, and adaptive capacity provide insight into differences in susceptibility to harm.

## Learning resources

### Videos

Centers for Disease Control (CDC)/Agency for Toxic Substances and Disease Registry (ATSDR). (2016) Introduction to CDC's social vulnerability index. 4 min. Accessed at: https://www.youtube.com/watch?v=u5m0Lb3B4UY.

Jones, L. (2015) Imagine America without Los Angeles: Applying science to understand the vulnerability of modern society to natural disasters. University of California Television. 1 hour. Accessed at: https://www.youtube.com/watch?v=ONqqEUb7Njg.

IFRC (2009) ABC of VCA. 5 min. https://www.youtube.com/watch?v=wS719VN-HfU

Van Zandt, S. (2014) In harm's way: Measuring and mapping social vulnerability. Webinar given to Virginia Sea Grant, College of William and Mary on 30 January 2014. 1 hour, 15 min. Accessed at: https://www.youtube.com/watch?v=uYFRJ75V6pg.

### Websites

CDC/ATSDR on social vulnerability indicators: https://svi.cdc.gov.

International Federation of Red Cross and Red Crescent Societies (IFCR) method for vulnerability and capacity assessment (VCA): http://www.ifrc.org/vca.

Vulnerability Consequences and Adaptation Planning Scenarios (VCAPS): http://www.vcapsforplanning.org.

## Bibliography

Adger, W.N. (2006) Vulnerability, *Global Environmental Change, 16*(3), 268–281.

Clark, G.E., Moser, S.C., Ratick, S.J., Dow, K., Meyer, W.B., Emani, S., Jin, W., Kasperson, J.X., Kasperson, R.E. & Schwarz, H.E. (1998) Assessing the vulnerability of coastal communities to extreme storms: the case of Revere, MA., USA, *Mitigation and Adaptation Strategies for Global Change, 3*(1): 59–82.

Cutter, S.L., Boruff, B.J. & Shirley, W.L. (2003) Social vulnerability to environmental hazards, *Social Science Quarterly, 84*(2): 242–261.

Cutter, S.L., Emrich, C.T., Webb, J.J. & Morath, D. (2009) *Social Vulnerability to Climate Variability Hazards: A Review of the Literature. Final Report to Oxfam America.* Charleston: Hazards and vulnerability research institute, University of South Carolina.

Dow, K. (1992) Exploring differences in our common future(s): the meaning of vulnerability to global environmental change, *Geoforum, 23*(3): 417–436.

Kettle, N.P., Dow, K., Tuler, S., Webler, T., Whitehead, J. & Miller, K.M. (2014) Integrating scientific and local knowledge to inform risk-based management approaches for climate adaptation, *Climate Risk Management, 4*: 17–31.

Polsky, C., Neff, R. & Yarnal, B. (2007) Building comparable global change vulnerability assessments: The vulnerability scoping diagram, *Global Environmental Change, 17*(3): 472–485.

Schröter, D., Polsky, C. & Patt, A.G. (2005) Assessing vulnerabilities to the effects of global change: an eight step approach, *Mitigation and Adaptation Strategies for Global Change, 10*(4): 573–595.

Stern, P.C., Ebi, K.L., Leichenko, R., Olson, R.S., Steinbruner, J.D. & Lempert, R. (2013) Managing risk with climate vulnerability science, *Nature Climate Change, 3*(7): 607–609.

Suk, J.E., Ebi, K.L., Vose, D., Wint, W., Alexander, N., Mintiens, K. & Semenza, J.C. (2014) Indicators for tracking European vulnerabilities to the risks of infectious disease transmission due to climate change, *International Journal of Environmental Research and Public Health, 11*(2): 2218–2235.

Tuler, S., Agyeman, J., da Silva, P.P., LoRusso, K.R. & Kay, R. (2008) Assessing vulnerabilities: integrating information about driving forces that affect risks and resilience in fishing communities, *Human Ecology Review, 15*(2): 171.

Tuler, S.P., Webler, T. & Polsky, C. (2013) A rapid impact and vulnerability assessment approach for commercial fisheries management, *Ocean & Coastal Management, 71*: 131–140.

Turner, B.L., Kasperson, R.E., Matson, P.A., McCarthy, J.J., Corell, R.W., Christensen, L., Eckley, N., Kasperson, J.X., Luers, A., Martello, M.L. et al. (2003) A framework for vulnerability analysis in sustainability science, *Proceedings of the National Academy of Sciences, 100*(14): 8074–8079.

Webler, T., Tuler, S., Dow, K., Whitehead, J. & Kettle, N. (2016) Design and evaluation of a local analytic-deliberative process for climate adaptation planning, *Local Environment, 21*(2): 166–188.

# Urban ecology

4.20

Robert A. Francis

Ecologists primarily study the relationships and interactions between organisms and their environment. Classical ecology has generally focused on ecosystems that are more 'natural' than anthropogenic, such as tropical forests, though there is a healthy tradition of investigation into agricultural ecosystems. The advent of urban ecology – the study of ecological patterns and processes in towns and cities – from the 1970s is a contemporary response to the ubiquitous processes of urbanisation that are occurring globally. It is increasingly apparent that understanding the ecology of urban ecosystems is important not just for determining the potential impacts of urbanisation on environmental attributes such as biodiversity and ecosystem services, but also for the design and management of more ecologically supportive and sustainable cities. Such efforts are by necessity interdisciplinary, as urban ecosystems are characterised by dramatic human influence, arguably more so than any other type of ecosystem in the twenty-first century.

This chapter briefly discusses (1) what particularly characterises urban ecology as a contemporary approach to environmental studies; (2) geographical trends in the study of urban ecology; and (3) how urban ecology as a way of understanding the urban environment can influence the planning and management of cities.

## What characterises urban ecology?

Any ecosystem comprises the physical environment and the organisms that inhabit it. Uniquely, urban ecosystems are designed primarily as living and working spaces for people, combining the built environment (buildings, transport infrastructure) with vegetated areas (often termed 'green space(s)') that may remain from pre-urban land use, be regenerating after disturbance, or may be planted (Figure 1). Outside of the ecological

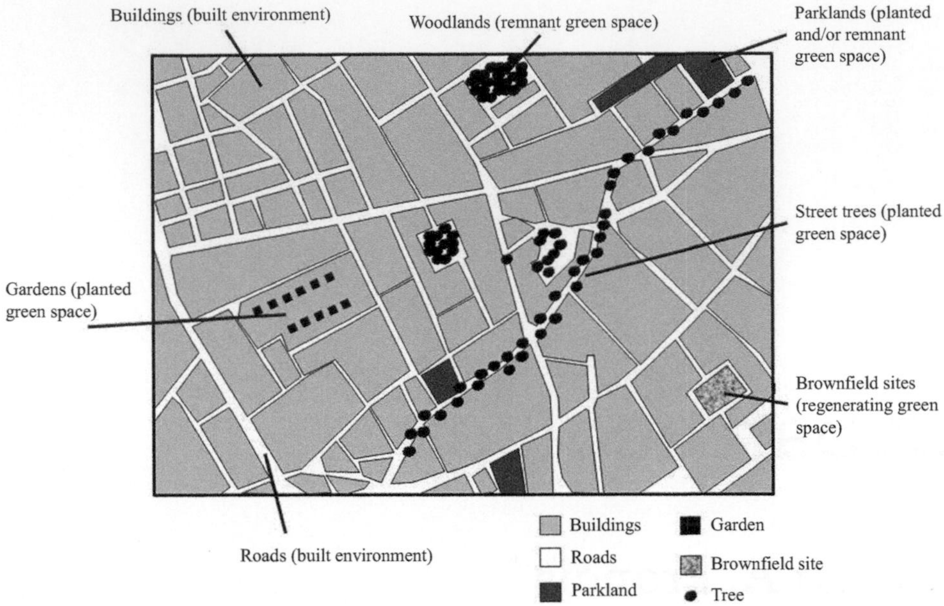


**Figure 4.20.1** Urban ecosystems are essentially mosaics of the built environment and different types of vegetated 'green space(s)'.

*Source*: Modified from Francis & Chadwick (2013).

sciences, nonhuman organisms in cities are usually considered only in relation to their use or desirability (or indeed undesirability) to citizens; for example garden plants and pets, or pests such as rats. As a result, societal drivers of ecological patterns and processes are particularly pronounced in urban ecosystems and operate over a range of scales; from individuals deciding how to manage their garden and what species to plant, to local government deciding on patterns of urban development or green space planning. Urban ecology therefore incorporates social sciences, geography and urban design and planning, and is inherently interdisciplinary.

Urban ecology is also highly spatial. The increasing sophistication of remote sensing technology from the 1980s onwards has allowed the quantification of patterns of land use types in spatially heterogeneous cities, which has often been linked to field measurements to enable relationships between landscape patterns and ecological conditions to be determined. These have often been along an urban-rural gradient, wherein spatial differences in the dominance of the built environment, for example from the city centre to the suburban fringe, are used to infer the influence of the urban environment on ecology. However, clear patterns are hard to determine as cities are highly heterogeneous at multiple scales, can have rapid rates of change (e.g. patterns of development and redevelopment) and green spaces vary substantially in their area and connectivity. Combined with differential responses of organisms to their varying environmental conditions, generalisations can be hard to make. For example, intermediate levels of urbanisation may increase the number of plant species resident compared to more natural areas, but this is certainly not the case for most animals (McKinney, 2008). Much more work needs to be done to both standardise what urban characteristics are being

**Table 4.20.1** Concepts that have emerged or that are particularly relevant to the study of urban ecology.

| *Concept* | *Description* | *Example reference* |
|---|---|---|
| Recombinant communities | Urban ecosystems contain many habitats in close proximity, and receive large numbers of species from regional and global transportation. This results in plant and animal communities that are dissimilar to those found elsewhere and which may contain species from forests, grasslands, wetlands, mountains and so on, all living together. | Meurk (2010) |
| Ecological engineering | The construction or modification of ecosystems for the mutual benefit of humans and nature. In an urban context, this might include the construction of urban wetlands or living roofs and walls that provide ecosystem services to citizens whilst also acting as habitat for species. | Francis & Lorimer (2011) |
| Reconciliation ecology | The usual methods of species conservation include the setting aside of nature reserves and the restoration of degraded land. Both are problematic and there is limited land available for these options. A third option is the design of human landscapes such as cities to allow other species to share the space and live alongside us. This may sometimes be achieved using ecological engineering technologies such as living roofs, or by behavioural changes such as gardening for wildlife. | Francis & Lorimer (2011) |
| Novel ecosystems | Cities contain ecosystems or habitats that exist in conditions that would not be found in nature, for example heavily engineered rivers, soils polluted with heavy metals or oils, or walls of concrete and glass. These present particular environmental conditions that might select for particular species or communities, and thereby shape patterns of urban ecology. | Kowarik (2011) |

measured and how this may take into account aspects like changing land use. Understanding these elements of urban systems becomes important when ecosystem function and services are being managed.

Urban ecology is also a frontier discipline and as such is not bound by the norms of more traditional ecological thought. The challenges of understanding and managing the ecology of such complex systems has resulted in some original concepts and perspectives to be developed or exapted from other contexts. Examples include work on recombinant ecological communities, ecological engineering, reconciliation ecology and novel ecosystems (see Table 4.20.1). Such developments have occurred in cities in particular due to the large amount of environmental novelty that can be found in urban ecosystems (Kowarik, 2011).

## The geography of urban ecology

Much of the early work on urban ecology came from Europe, with local studies of habitat and communities from old cities such as London and Berlin. Most of the recent work has emerged from the Global North, in particular from the USA. The USA has two Long-Term Ecological Research (LTER) sites that are urban: Baltimore and Central Arizona-Phoenix. These two locations alone have produced a substantial amount of information on urban ecology. A healthy amount of research is also coming from some newly urbanising countries, in particular China, though studies have tended to focus more on urban landscape patterns than ecology (e.g. Francis et al., 2016). Asia in particular has seen the development of ecocities, wherein the principles of urban ecology have been applied to urban design to ensure a more ecologically-focused environment. Most urbanisation is predicted to occur in Asia and Africa (United Nations, 2014), though (with the notable exception of China) these regions account for a relatively small proportion of studies published in the international literature. Such areas will be particularly important for the future of urban ecology, and are the most likely to benefit from the lessons that will be learnt over the coming decades.

Most studies performed have focused on relatively discrete aspects of urban ecology in individual cities, for example the diversity of urban woodlands or the ecological quality of an urban river. Apart from a few notable exceptions where extensive meta-analyses have been performed to determine common trends across multiple cities or regions (e.g. Beninde et al., 2015), there remains limited work on the processes of becoming urban, pre-post urban comparisons and comparative urban ecology. As the discipline of urban ecology develops, there will be greater focus on multi-city studies, analysis of multiple patterns and processes, linkage between disciplines and more extensive research in newly urbanised and urbanising countries.

## Urban ecology and the planning and management of cities

Urban ecology remains to a large part an academic pursuit, but recently there has been an increase in the incorporation of ecological principles into urban planning, particularly in some major global cities. This has mainly focused on spatial planning of green space to increase coverage and to encourage accessibility for citizens; and to a lesser extent for ecological aspects such as species conservation. For example, some global cities have developed plans for green grids or networks, based around the principles of increasing area and connectivity of green space (e.g. Greater London Assembly, 2012). This can be controversial however, as often these networks of green infrastructure are conceived as solving multiple problems (such as air quality, flooding and access to nature for citizens), though this is often not possible as multi-functional green space is challenging to create and uses may conflict (such as recreation and species conservation). Increasingly the ecosystem services framework has gained traction with urban planners, and is factored into considerations of the planning and management of green space to solve both environmental and social issues within cities.

An emerging idea in urban ecology is the integration of nature more firmly into the built fabric of the city, so that rather than portioning the environment into discrete patches and corridors of green space, the city becomes permeated with habitat that may support a range of species. Examples of this include the installation of living roofs and walls and wildlife gardening, as well as the potential for urban agriculture and the cultivation of public or abandoned spaces ('guerrilla gardening'). These actions are along the lines of reconciliation ecology and may utilise ecological engineering technology (Table 4.20.1). However, this does present tensions in determining how habitat may best be created and which species are being encouraged. The species that colonise urban habitat may not always be welcome in proximity to people, and there are important questions to be asked about the types of exposure to nature that are welcome, unwelcome or indeed essential within an urban context. Developing new ecological engineering technologies and their socio-ecological implications will remain at the forefront of urban ecology in the foreseeable future.

## Learning resources

The Long-Term Ecological Research sites for Central Arizona-Phoenix and Baltimore show the latest findings from ongoing research in these areas:

https://lternet.edu/sites/cap
https://lternet.edu/sites/bes

The Society for Urban Ecology (SURE) helps to encourage and develop the practice of urban ecology worldwide:

http://www.society-urban-ecology.org

An interesting and approachable site on urban wildlife in the UK can be found at the BBC:

http://www.bbc.co.uk/nature/habitats/Urban_ecosystem

The Nature of Cities collective blog has some interesting and high-profile discussions:

http://www.thenatureofcities.com/

## Bibliography


Beninde, J., Veith, M. & Hochkirch, A. (2015) 'Biodiversity in cities needs space: a meta-analysis of factors determining intra-urban biodiversity variation', *Ecology Letters*, 18: 581–592.

Francis, R.A. & Chadwick, M.A. (2013) *Urban Ecosystems: Understanding the Human Environment*, London: Routledge.

Francis, R.A. & Lorimer, J. (2011) 'Urban reconciliation ecology: the potential of living roofs and walls', *Journal of Environmental Management*, 92: 1429–1437.

Francis, R.A., Millington, J.D. & Chadwick, M.A. (2016) 'Introduction: an overview of landscape ecology in cities', in Francis, R.A., Millington, D.A. & Chadwick, M.A. (Eds) *Urban Landscape Ecology: Science, Policy and Practice* (pp. 1–18). London: Routledge.

Greater London Assembly (2012) *Green Infrastructure and Open Environments: The All London Green Grid: Supplementary Planning Guidance*. London: GLA.

Kowarik, I. (2011) 'Novel urban ecosystems, biodiversity, and conservation', *Environmental Pollution*, 159: 1974–1983.

McKinney, M.L. (2008) 'Effects of urbanization on species richness: a review of plants and animals', *Urban Ecosystems*, 11: 161–176.

Meurk, C.D. (2010) Recombinant ecology of urban areas, in Douglas, I., Goode, D., Houck, M.C. & Wang, R. (Eds) *The Routledge Handbook of Urban Ecology* (pp. 198–220). London: Routledge.

United Nations (2014) World Urbanization Prospects, the 2014 Revision. Online. Available at: https://esa.un.org/unpd/wup [accessed 6th Sep 2016]

# Part 5

## Key topics: environmental challenges and changes

# Part 5 Introduction

Present day humans are transforming all aspects of the biophysical world to an unprecedented degree. From molecular genetics to civil engineering, from farm fields to rapidly expanding cities, people are together altering the planet at a speed and scale that is unprecedented. Though changes in one place reverberate elsewhere, as does change to any one element of the biophysical world (e.g. atmospheric temperature), to understand human impacts in a manageable way it is useful to proceed topic by topic. This part of the book presents snapshots of trends in human uses of everything from water resources to forests. It identifies some of the challenges resulting from over-use or mismanagement. It also considers those aspects of nature that some people are subject to because they exceed human capacities for control – such as earthquakes. Though by no means comprehensive, the section says a lot about how reliant upon, and vulnerable to changes in, the material world contemporary humans are. Though anthropogenic climate change commands a lot of media, political and academic attention, this section reminds readers that people–environment relations are highly multidimensional in both scope and scale. Through its 'false' topical divisions, it also reminds us that even the most local or small-scale environmental challenges are multidimensional – they require joined-up approaches to analysis and decision making. This has long been recognized in what is called 'watershed management', where hydrological resources and threats are regarded as interlocking elements within large-scale landscapes. But, today, it needs to be recognized at *every* spatial and temporal scale for *all* elements of the biophysical world. This poses a challenge for both understanding and for the attempt to manage human–environment relations. For instance, we know very little about the ecology of the oceans, even after decades of research, and such is the scale of marine ecosystems that 'management' may be something of a pipe-dream. However, such ignorance is generative. Environmental blind-spots, challenges and changes spur human ingenuity and creativity, and foster cooperation and mutual support. When challenges are not met this often incites more intensive efforts to overcome barriers and achieve success.

# Anthropogenic climate change

5.1

Alice Larkin

The Earth's global climate has always changed, principally in response to (i) shifts in the Earth's position relative to the Sun, (ii) changes in radiation coming from the Sun, (iii) changes in the Earth's atmospheric composition. Variations in incoming solar radiation contribute to the Earth's natural climate variability, as does a shift in atmospheric composition due to a volcanic explosion. Both alter the balance of incoming and outgoing radiation, which changes the Earth's temperature, as it adjusts to bring the system into equilibrium.

'Anthropogenic climate change' is relatively new. The term describes changes in the Earth's temperature, and connected climate parameters, affected by human actions, largely through changes to the Earth's atmospheric composition. In recent decades it has become widely accepted that changes already set in train, and anticipated, have the potential to impact profoundly on the Earth's physical, social and biological systems. Two primary responses are: (1) enact policies and measures to prevent further changes to atmospheric composition – mitigation; (2) establish support measures to adapt to change – adaptation. This chapter describes the causes and effects of anthropogenic climate change, outlining trends and responses.

## The causes

The Earth's atmospheric composition influences its temperature through interactions with incoming 'solar', and outgoing 'terrestrial' radiation. Atmospheric greenhouse gases raise the global mean surface temperature from –18°C to around 15°C, by temporarily trapping the 'terrestrial' radiation. Greenhouse gases are therefore essential to support life as we know it. Anthropogenic climate change is caused by humans increasing atmospheric concentrations of greenhouse gases and other emissions.

## The energy system

Carbon dioxide ($CO_2$) is the dominant anthropogenic greenhouse gas, with the majority produced by combusting fossil fuels to deliver energy services such as mobility, nourishment, comfort and entertainment. By 2015, the largest share of global annual $CO_2$ was produced by combusting coal for electricity and heat, followed by oil for transport, then gas.

## Industrial and agricultural emissions

One of the biggest contributors to industrial process emissions is the cement sector, which also underpins global infrastructure. In 2014, cement production contributed

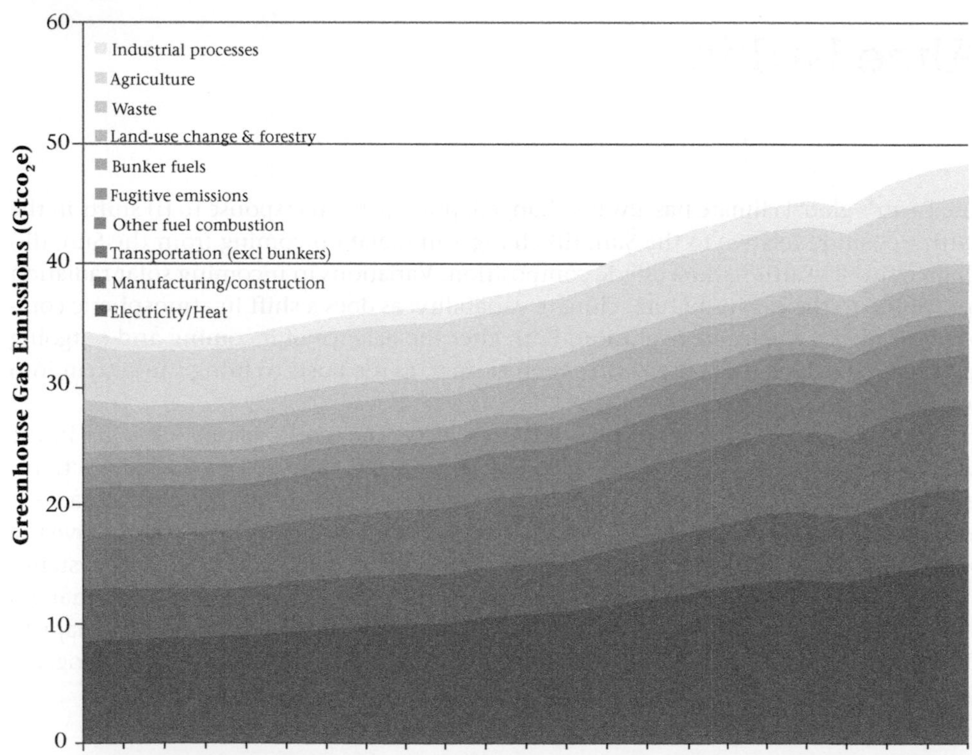


**Figure 5.1.1** Time-series of global greenhouse gas emissions by source measured in $GtCO_2$ equivalent, from 1990 to 2012.

Data from CAIT Climate Data Explorer (CAIT, 2015; FAO, 2014; International Energy Agency, 2014).

nearly 6 per cent of annual $CO_2$. Other industrial emissions include sulphate aerosol from the combustion of fossil fuels, Chlorofluorocarbons from refrigeration and air conditioning systems, and methane emissions released by landfill. The agricultural sector produces the largest share of non-$CO_2$ greenhouse gases, principally through nitrous oxide from fertilizer use, and methane released by manure and enteric fermentation (farting cows!).

## *Land-use change and forestry*

Land-use change can produce $CO_2$ through deforestation. Although some parts of Asia and North America are reforesting agricultural land, deforestation accounted for 8–10 per cent of annual $CO_2$ from anthropogenic sources in 2015.

## The effects

Evidence connecting humans with climate change relies on long-run data records, often aggregated over a large spatial scale. Such data underpins statements from the Summary for Policymakers in the Intergovernmental Panel on Climate Change's IPCC (2013) report including, 'Global atmospheric concentrations of carbon dioxide, methane and nitrous oxide have increased to levels unprecedented in the last 800,000 years'; 'The rate of global sea level rise has been larger since the mid-nineteenth century, than the mean over the previous two millennia'; 'Warming of the climate system is unequivocal, and since the 1950s, many of the changes are unprecedented over decades to millennia'. However, attribution of observed changes remains a significant challenge.

## *The language of uncertainty*

There is a large amount of natural decadal and inter-annual variability within the climate system, challenging the detection of human-induced change. Some evidence uses indirect ice-core records, dating back millions of years, but only since around 1950 has there been a diverse suite of instrumental, direct global scale observations. Even then, establishing definitive statements tends to be avoided, as uncertainty exists when measuring any variable. Instead, probabilities are used to convey confidence. Within the IPCC (2013) report, probabilities are translated into common terms. 'Likely' means a 66–100 per cent chance, 'Very Unlikely' a 0–10 per cent chance. For instance, 'it is extremely likely that more than half of the observed increase in global average surface temperature from 1951 to 2010 was caused by the anthropogenic increase in greenhouse gas concentrations . . .'.

## *Annual temperature variations*

The parameter drawing most attention, and also used by climate deniers, is the global temperature record. Annual temperatures are influenced by natural variability, as well as internal circulation patterns, such as the El Niño Southern Oscillation. The anthropogenic part of the temperature rise can therefore be somewhat hidden. While scientists do not expect temperatures to rise chronologically, much was made of a so-called hiatus between 1998 and 2012. Specifically, 1998 had a strong El Niño event, elevating global temperature, and so a series of subsequent cooler years fuelled a view that warming had paused. Since then, data analysis shows that warming has continued, albeit at a slightly slower rate between 2001 and 2010 than previous recent decades. 1998 now ranks as the sixth warmest year on record, with 2014, and then 2015, also an El Niño year, breaking previous records (NOAA, 2016). The 2015 annual temperature was 0.9°C above the twentieth-century average, strengthening the anthropogenic climate case to the extent that some climate deniers turned their attention to critiquing solutions, rather than the existence of the cause.

## *Impacts*

There is considerable evidence for a myriad of anthropogenic climate impacts. These include ice-sheet and glacial melting; sea level rise; more frequent hot days and nights over land; a greater number and intensity of extreme rainfall events; increased duration of droughts, to name a few. All impact on humans and ecosystems. There are other physical changes caused by rising greenhouse gases too, such as ocean acidification, already affecting ocean ecosystems. Whilst attributing particular weather-related events, or localised shifts, to climate change will continue to challenge scientists, the consensus on human-induced global change is probably as strong as any scientific community will reach.

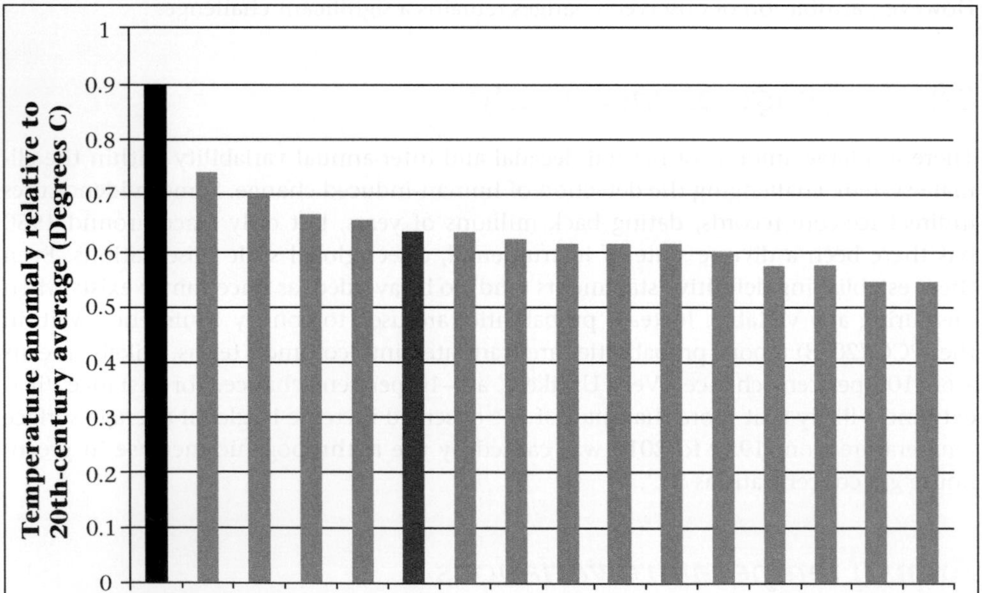


**Figure 5.1.2** The 16 warmest years on record, noting 1998 as a particularly warm year due to El Niño, as well as 2015 being the warmest year to date.

*Source*: NOAA (2016).

## Responses and policies

December 2015 marked an important historical milestone for addressing climate change. All Parties to the United Nations Framework Convention on Climate Change (UNFCCC) agreed to strengthen the global response to climate change through the Paris Agreement (UNFCCC, 2015), in which it specifies:

> holding the increase in global average temperature to well below 2°C above pre-industrial levels and pursue efforts to limit the temperature increase to 1.5C.
> and to implement the Agreement "to reflect equity".

Critiques of the Agreement highlight, for instance, an absence of the term 'decarbonisation', reference to international transport emissions or indeed a global carbon budget constraint. Nevertheless, the Agreement is sending signals to a much broader community than those engaged in the science, regarding the ultimate move away from fossil fuels.

### *Target setting*

Before Paris, nations submitted 'Intended Nationally Determined Contributions (INDCs)' outlining feasible mitigation plans. The INDCs varied, with some referring to absolute levels of decarbonisation, and others focusing on relative emission reductions, compared with 'Business As Usual', or an intensity-type target, relative to the size of the economy. Yet what became clear was that these INDCs did not match up with the 2°C goal that has been an anchor point for climate negotiations (Jordan et al., 2013), and is now enshrined within the Paris Agreement.

### *Mitigation solutions*

The accumulation of greenhouse gases over a century timescale – 'cumulative emissions' – is linearly related to global temperature rise (Allen et al., 2009). This relationship allows policymakers to think in terms of a diminishing 'carbon budget', drawing attention to the limited carbon 'spend' associated with a 2°C goal. With little room to manoeuvre, but armed with complex models coupling physics with economics, some conclude that breaching the 2°C warming threshold can only be avoided with a good chance, if 'negative emissions technologies', are assumed to work at scale, in addition to global decarbonisation. This requires, for instance, widespread use of biomass for energy combined with capturing the $CO_2$, and ensuring the $CO_2$ is locked in for centuries. Critiques argue that outputs from sophisticated modelling can mislead policymakers, with assumptions behind the models hiding high levels of uncertainty around, for example, future energy technology prices and feasibility to work at scale, (Anderson & Peters, 2016). Moreover, the models struggle to account for sociological aspects, so 2°C 'solutions' focusing on rapid social and technical change, or considering equity, are few and far between. Yet, the later mitigation happens, the more challenging it will be to avoid a 2°C rise, or greater levels of adaptation to higher temperatures will be necessary.

## Summary

The Earth's climate has always changed, and only recently have humans had a discernable impact on it. With our combustion of fossil fuels for energy services, large-scale landscape change through forest removal and industrial development, humans have contributed to very large increases in greenhouse gas concentrations. Attributing the rise in these concentrations to human activities is almost universally accepted, as is the resulting impact on recent temperatures. Spurred on by the scientific evidence, global leaders had little choice but to devise a landmark Agreement to tackle climate change, but for many this Agreement has come too late.

Most global nations are now pursuing mitigation policies, yet years of delay means a 2°C rise may be too difficult to avoid – as forthcoming solutions become increasingly challenging. Reframing the debate around what can be achieved through rapid socio-technical change on the demand-side offers some opportunities and is important not to overlook, given that every year's delay increases the chance that the 2°C threshold will be exceeded.

## Learning resources

*Getting to grips with climate policy*: United Nations Framework Convention on Climate Change (UNFCCC) accessible information giving the big picture of the climate agreements and protocols to date: http://bigpicture.unfccc.int/#content-the-paris-agreement.

*Mapping climate change*: this website illustrates how carbon data and climate risk can be interpreted in different ways, using a world map that expands and contracts countries depending on the climate indicator chosen: http://www.carbonmap.org.

*Carbon data and trends*: this site allows a range of different types of user to see the latest carbon dioxide emissions data, by country, over time, and under different accounting regimes http://www.globalcarbonatlas.org.

## Bibliography

Allen, M.R., Frame, D.J., Huntingford, C., Jones, C.D., Lowe, J.A., Meinshausen, M. & Meinshausen, N. (2009) 'Warming caused by cumulative carbon emissions towards the trillionth tonne', *Nature*, 458: 1163–1166.

Anderson, K. & Peters, G. (2016) 'The trouble with negative emissions: reliance on negative emission concepts locks in humankind's carbon addiction', *Science*, 354(6309): 182–183.

CAIT (2015) *CAIT Climate Data Explorer*. Washington, DC: World Resources Institute.

Food and Agriculture Organisation (FAO) (2014) *FAOSTAT Emissions Database*. Rome: Food and Agriculture Organisation of the United Nations.

Intergovernmental Panel on Climate Change (IPCC) (2013) *Climate Change 2013: The Physical Science Basis. Contribution of Working Group I to the Fifth Assessment Report of the Intergovernmental Panel on Climate Change*, Cambridge: Cambridge University Press.

International Energy Agency (2014) *$CO_2$ emissions from fuel combustion* (2014 edition). International Energy Agency (ed.) Paris: OECD/IEA.

Jordan, A., Rayner, T., Schroeder, H., Adger, N., Anderson, K., Bows, A., Quéré, C.L., Joshi, M., Mander, S., Vaughan, N. & Whitmarsh, L. (2013) 'Going beyond two degrees? The risks and opportunities of alternative options', *Climate Policy*, 13: 751–769.
National Oceanic and Atmospheric Administration (NOAA) (2016) Global Analysis – Annual 2015. Online. Available at: http://www.ncdc.noaa.gov/sotc/global/201513 [Accessed 28 Sept 2016].
United Nations Framework Convention on Climate Change (UNFCCC) (2015) Adoption of the Paris Agreement. Switzerland: http://unfccc.int/resource/docs/2015/cop21/eng/l09r01.pdf.

# 5.2 Agro-food systems

## Colin Sage

The agro-food system comprises all those activities related to the production, processing, distribution, sale, preparation and consumption of food. The prefix 'agro-' (or 'agri-') to the term 'food systems', however, invites us to place somewhat greater importance upon the farm sector and the production of *primary* foods than to subsequent stages where these materials are refined, manufactured into *final* foods for ultimate consumption. Moreover, as Robinson (2004) reminds us, agriculture is distinct from many other economic activities as it deals with living organisms, that is the plants and animals that possess particular biological characteristics. The degree to which these are adapted to the prevailing environmental conditions will largely determine their productivity. Consequently, agro-food systems encourage us to pay attention to their particular geographical context. Yet while climate, moisture, soils and other ecological services provide the key physical parameters of production potential, farmers influence this environment through limiting biological (and genetic) diversity and managing inputs to create a farmed *agroecosystem*. The nature and extent of farmer intervention is itself a consequence of the social and economic circumstances in which they find themselves, such that agricultural systems can range from extensive, low-input cultivation to highly intensive and industrialised operations. An understanding of agro-food systems must consequently appreciate the array of environmental and socio-economic factors that influence the ways in which these systems develop.

## Global or local agro-food systems?

We have come a very long way since the time when subsistence farming to meet household needs prevailed around the world. While there may be some isolated pockets of self-provisioning remaining, most farmers are in the business of producing surpluses for sale into markets. While these were, and for many remain, oriented to the supply of local towns and cities, elsewhere we have witnessed a growing sophistication and distance covered by food supply chains. Indeed, accompanying the process of trade liberalisation from the 1980s onward has been a growing enmeshment of farmers in international markets such that we now commonly refer to a singular 'global agro-food system'. This term reflects a complex and dynamic intensification of worldwide relations involving food and agriculture underpinned by a raft of global agreements in trade, intellectual property rights, financial investments and food safety regulations. The conclusion of the Uruguay round of trade talks in 1994 that led to the creation of the World Trade Organisation was especially significant in forcing the removal of protective measures (tariffs) that defended domestic farmers from large volumes of low-cost surplus food staples. Together with the imposition of structural adjustment programmes that strongly encouraged the pursuit of export revenues, such policy measures had the effect of realigning the agricultural sectors of many countries away from domestic staples and toward high value export crops.

Invariably this enabled some of the largest companies with a presence on international markets and recognised brands and services to establish themselves in many more countries worldwide. This has become a feature of every stage of the entire agro-food system: from agri-technology suppliers of machinery, seeds and chemical inputs; through the 'merchants of grain' such as Cargill and ConAgra; food processors and product manufacturers with their globally branded soft drinks, snacks and other convenience products; fast food franchises and, ultimately, supermarkets catering to the rising numbers of middle-income consumers in cities worldwide. It is no exaggeration to state that the parameters and modus operandi of each of these stages has been set by a relatively small number of large corporations that now exercise a global reach all the way to the farm sector in distant countries. Each of these stages are represented as discrete boxes in Figure 5.2.1 which also indicates the complex array of financial and regulatory arrangements that underpin and have enabled the development of the global agro-food system.

Yet such globalisation is not simply measured by rising volumes of internationally traded agricultural commodities but reflects more profound changes within the realms of food production and consumption. The expanding application of advanced agronomic science and technology – such as plant and animal genetic engineering and precision farming involving big data analytics – demonstrates how the locus of power is moving off the land. Developed for the purpose of maximising agricultural output, rather than meeting local food and livelihood needs in sustainable ways, such technologies easily displace the empirically-derived and context-specific knowledge of farmers that has evolved over generations and which has served to sustain their livelihoods in risk-prone environments. Yet, the challenges faced by farmers in many very different regions around the world are becoming more complex, intractable and reflect dynamic interactions arising from a spectrum of environmental and socio-economic factors. In particular, climate change, freshwater depletion, loss of topsoil, biodiversity loss and rising levels of food-related waste demonstrate that, in contrast to the imposition of inappropriate technologies, knowledge-based innovations responding to *local conditions with local resources* are likely to provide for more sustainable, equitable and capacity-enhancing solutions in an era of global environmental change (Thompson & Scoones, 2009).

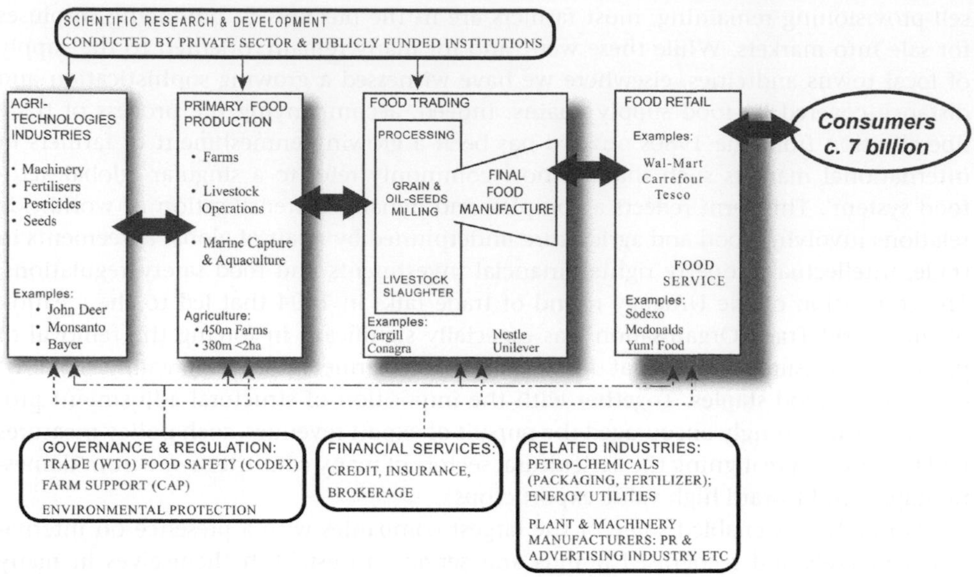


**Figure 5.2.1** The global agro-food system.

## Widening the frame of analysis

The proponents of the existing global agro-food system are keen to point out that it has delivered more calories for a lower proportion of consumer spending than ever before and has diversified the diets of a majority of the world's population too, a process generally known as the nutrition transition. But overall the evidence suggests an unsatisfactory performance given the environmental cost, for 800 million people in the world are suffering from hunger, 2 billion more are affected by micronutrient deficiencies, while a further 1.9 billion are classed as obese or overweight and susceptible to a range of non-communicable diseases (WHO, 2016). More people die today from diet-related ill-health than any other cause of premature death: a consequence of a food system busy promoting foodstuffs that do not optimise human nutrition and well-being.

For this reason, it is becoming increasingly necessary to adopt a much wider – indeed a holistic – perspective on the agro-food system, one that understands that its problems are deeply interconnected and mutually reinforcing. In my local supermarket in Ireland I can find green beans from Kenya and asparagus from Peru, two countries that appear to have successfully embraced the opportunities offered by non-traditional, high-value fruit, flower and vegetable contract production for the major supermarket chains in the global North. Yet while both countries' export earnings have grown significantly over the past two decades, this commodity 'boom' rests upon extremely fragile water resource endowments. While the rapid depletion of the Ica aquifer may result in asparagus production relocating to China from where Western consumers can still be fed, Peruvian farmers without irrigation in this desert region will be left with little option but to abandon farming altogether. Meanwhile, Irish agriculture continues to intensify dairy production in the hope of selling more infant formula feed to Chinese parents. Yet Ireland's greenhouse gas emissions per capita are now amongst the highest in the European Union.

These complex interconnections reveal the critical importance of widening our frame of analysis when considering the agro-food system, for it is not only a matter of *how* and *where* we are producing food, but *what* and *why* we are eating the way we do. Technology-focused innovations preoccupied with raising output not only fail to address the complexity of environmental and socio-economic drivers facing agriculture, they simply do not connect agriculture with the realm of consumption. Yet if we are to devise solutions to rectify the unsustainable nature of the current agro-food system, we need to begin by challenging prevailing assumptions about consumers' rights to 'cheap' food and the maintenance of current dietary practices that are so wasteful of resources. For example, it is well established that the low price paid by consumers for their food conceals huge externalities along the supply chain (Pretty et al., 2005). Moreover, the continuing upward trajectory in global demand for meat presupposes increased production of intensively reared livestock fed by grains and oil seeds that could otherwise support 4 billion people directly (Carolan, 2011). As meat and dairy are the most greenhouse-gas-intensive foods, the climate change implications are consequently extremely significant.

## Conclusions

While agro-food systems are highly complex arrangements, there is growing concern that their interconnected structure and scale – as well as the locus of power that drives them – has left them increasingly vulnerable to a variety of short-term episodic shocks, long-term stresses and lacking the resilience to cope effectively with unexpected events. A growing number of international expert panels have called for a fundamental revision of industrialised agriculture arguing for greater diversification and a move toward agro-ecological farming methods that would work with nature rather than against it (IPES, 2016; IAASTD, 2009). The challenge, however, should not be regarded simply as a way of maintaining output in a more environmentally-friendly manner; rather it is about addressing production and consumption of food across a complex landscape involving environmental sustainability, social justice and nutritional security. Arguably, eating has the potential to move wealthier societies in a more sustainable direction than many other activities because it entangles every one of us as consumers in webs of relations with producers and connects us to the ecological processes and services that underpin agro-food systems, many of which are under threat (Sage, 2012). For these reasons informed citizens have a responsibility to acquire a better understanding of the agro-food systems that feed us and, in turn, make dietary choices that sustain those systems.

## Learning resources

An excellent introduction to the competing interests that shape the food system and the role of public policy to improve human health is provided by:

Lang, T. & Heasman, M. (2015) *Food Wars: The Global Battle for Mouths, Minds and Markets*. 2nd edition. Abingdon: Earthscan/Routledge.

The multiple interconnections that tie the agro-food system to the use of environmental resources and the challenges this presents for our food choices in the future is explored in:

Sage, C. (2012) *Environment and Food*. Abingdon: Routledge.

## Bibliography

Carolan, M. (2011) *The Real Cost of Cheap Food*. Abingdon: Earthscan.

International Assessment of Agricultural Knowledge, Science and Technology for Development (IAASTD) (2009) *Synthesis Report: A Synthesis of the Global and Sub-global IAASTD Reports*. Washington, DC: Island Press.

International Panel of Experts on Sustainable Food Systems (IPES) (2016) *From Uniformity to Diversity: A Paradigm Shift from Industrial Agriculture to Diversified Agroecological Systems*. Online. Available at: www.ipes-food.org.

Pretty, J., Ball, A., Lang, T. & Morison, J. (2005) Farm costs and food miles: an assessment of the full cost of the UK weekly food basket, *Food Policy*, 30: 1–19.

Robinson, G. (2004) *Geographies of Agriculture: Globalisation, Restructuring and Sustainability*. Harlow: Pearson.

Sage, C. (2012) *Environment and Food*. Abingdon: Routledge.

Thompson, J. & Scoones, I. (2009) Addressing the dynamics of agri-food systems: an emerging agenda for social science research, *Environmental Science and Policy*, 12: 386–397.

World Health Organization (WHO) (2016) *Obesity and Overweight* factsheet. Updated June 2016. Online. Available at: http://www.who.int/mediacentre/factsheets/fs311/en. [Accessed 20 December 2016].

# Biofuels

5.3


## Vaclav Smil


Biofuels are one type of energy resource. They include traditional phytomass fuels, waste from wood processing, and fuels produced commercially by modern conversion methods. Traditional phytomass fuels include woody matter (fuelwood or firewood, ranging from dry litter fall and broken branches to cut stem wood), charcoal (produced by pyrolysis of wood), dry crop residues (mostly cereal straws, corn and sorghum stalks and leaves, sugar cane bagasse, legume and potato vines and cotton stalks) and dried dung. Waste from pulp and paper plants and sawmills is used by those industries to generate heat and electricity. Production of modern biofuels, by conversion, is dominated by ethanol and biodiesel. Gasification of phytomass yields gas of low energy density, biogas is produced by fermentation of organic wastes, and charcoal is made in modern retorts for industrial and household use.

Environmental challenges and changes arising from the harvesting, cultivation and conversion of biofuels extend to more than half of the categories treated in this segment. Inefficient combustion of traditional biofuels is a major source of local outdoor, and even more dangerously, indoor air pollution. Excessive harvesting of natural growth and expansion of monocultural plantations contributes to deforestation, loss of biodiversity and land degradation, affects water-retention capacity of soils, and increases soil erosion and the risk of flooding. Large-scale cultivation of energy crops (corn, soybeans, oil palm) alters agricultural systems, changes their carbon budgets and water demand, contributes to the global climate change and affects food prices. Biofuels are hardly the paragons of desirable renewable energy supply and inevitable expansion of their production and use in the world gradually moving away from fossil fuels is a major cause for concern.

## Traditional biofuels

In 1850 traditional biofuels supplied more than 90 per cent of the world's primary energy. By 1900 their share fell to about half, by 1950 to roughly a third. By the century's end they delivered about 12 per cent and by 2015 their share declined to about 8 per cent of the total. But during the entire twentieth century this relative decline was accompanied by doubling of their combustion, followed by a slight decline in the new century. Woody biomass (often harvested not from forests but from groves, roadside and backyard trees or from small fuelwood lots), crop residues (in deforested regions) and (in parts of Asia) also dried cattle dung remain principal sources of household energy for more than one billion people. This reliance is easily supported in forested regions (where deforestation for grazing or crop production has far larger impacts), but not in arid environments where the removal of shrubs and heavy pruning and cutting of the last remaining trees is a leading cause of devegetation. Indoor burning of biomass fuels in unvented or poorly vented stoves creates high levels of air pollution that have been implicated in higher incidence of respiratory illnesses.

While the ascent of fossil fuels eliminated wood from industrial uses in affluent countries, wood-based industries continue to use waste biomass, and many households in wood-rich countries (Russia, Finland, Sweden, Canada, US) still burn wood (in modern, efficient stoves) for heating and cooking. As a result, wood never disappeared from energy balances of even the richest countries, and at the beginning of the twentieth century it supplied about 7 per cent of the US and about 20 per cent of Finland's primary energy, with the EU mean at about 5 per cent. Efficient use of woody phytomass in affluent countries creates minimal environmental impacts; perhaps the most common undesirable effect is air pollution from fireplaces (emissions of polycyclic organic matters) in valleys where common inversion layers can cause high concentrations of those potential carcinogens.

## Ethanol

An entirely new set of environmental concerns has been introduced with using crops as energy sources. This practice began on a large scale in Brazil in the 1975 (with the ProÁlcool program to produce ethanol from sugar cane) and in the US in 1980 (with the fermentation of ethanol from corn). By 2015 ethanol output reached 55 billion liters in the US (where up to 10 per cent of ethanol by volume is blended with gasoline or the fuel is used in flexible-fuel vehicles burning up to 85 per cent ethanol) and billion liters GL in Brazil (as a blend of 25 per cent ethanol and 75 per cent gasoline or for a growing fleet of vehicles running on pure ethanol). These two countries now produce about 85 per cent of the global crop-based ethanol and the EU, China and Canada supply nearly all of the rest. This limited diffusion must be expected: inherently inefficient photosynthesis means that even with such high-yielding cops as corn and sugar cane the best fermentation techniques annual ethanol yields in Iowa are equivalent to less than 2 tonnes of crude oil per hectare (t/ha) and the Brazilian mean is about 3 t/ha.

Obviously, only countries with abundant farmland can contemplate large-scale crop-based ethanol production but they, too, face clear limits. In recent years the US has been diverting roughly 40 per cent of its corn crop into ethanol, and yet in overall energy terms that fuel has displaced less than 10 per cent of the country's annual motor gasoline demand. Consequences of expanding corn monoculture have been well documented, the most worrisome being accelerated erosion from row-crop cultivation, nitrate leaching and formation of dead zones in the Gulf of Mexico from heavy fertilization and depletion of the Ogallala aquifer from supplementary irrigation. Moreover, land-use changes triggered by the replacement of gasoline by corn-based ethanol may actually increase overall carbon emissions, and the net energy return of the entire process is only mildly positive. Brazilian ethanol has higher energy returns and dominant cultivars (with endophytic nitrogen-fixing bacteria) do not require fertilization.

## Biodiesel

Biodiesel is made by transesterification of plant oils, most commonly of rapeseed or soybean oil. Because these crops have much lower yields than corn or cane, productivity per hectare is commensurately lower, typically an annual equivalent of less than 1 t of crude oil per hectare. Again, the US and Brazil are the leading producers (using soybeans), followed by Germany and France (using rapeseed) and, again, obvious competition with food production (edible oils are in high demand) will limit biodiesel's expansion. The best alternative is to use oil from high-yielding oil palm, but expanded cultivation of that crop would cause further tropical deforestation.

Small-scale biogas production (using animal and human wastes and crop residues) has been common for decades in rural China and India, and its large-scale, optimized versions are useful primarily as an effective way to treat manures produced by large animal feedlots. Among the affluent nations only Germany, with its subsidized quest for higher shares of non-fossil energies, has a national biogas program. Feedstocks for more than 7,000 plants whose feedstock is about equally divided between energy crops (mainly corn) and livestock excrements. The biogas is used to generate electricity and, not surprisingly, the entire process has very low productivity of useful energy in terms of crude oil equivalents, typically well below 1 t/ha.

In 2015 the output of two leading modern liquid biofuels, bioethanol and biodiesel, was equivalent to only about 75 million tonnes of crude oil or roughly 3 per cent of the world's demand for gasoline, kerosene and diesel fuel used by land, air and water transportation. Raising this share to a significant level of future liquid fuel demand (20–30 per cent) would be an enormous scaling challenge even if there were only negligible environmental impacts. Obviously, it would be preferable to make ethanol by enzymatic hydrolysis of phytomass whose production does not use additional farmland or does not divert food crops. Production of ligno-cellulosic ethanol can use any woody matter, crop residues (mostly cereal straws) and cultivated grasses (switchgrass, reed canary grass, miscanthus) but its commercial deployment has a long way to go. The first two commercial plants (in Iowa converting corn stover and in Brazil using sugar cane bagasse) have aggregate capacity equal to just 0.005 per cent of current global requirement for liquid transportation fuels.

Moreover, productivities of this conversion would be broadly comparable to crop-based ethanol and large-scale production could not use only wood waste and it would have to rely on expanded planting of fast-growing trees or high-yielding grasses, raising, again, the concerns about land use changes, biodiversity loss, soil erosion, fertilization and water use. Intensive cultivation of genetically engineered algae would combine the lowest land claim and highest phytomass productivity but large-scale commercial operations remain elusive. Synthetic Genomics was hired by Exxon Mobil to develop the process but after spending more than $100 million between 2009 and 2013 Exxon abandoned the quest as uneconomical.

As the world moves gradually away from fossil fuels to renewable energies we can think of a new global energy system powered solely by inexpensive electricity that might be used to produce hydrogen as the main transportation fuel. But that is not an imminent solution, and we will need alternatives to fuels refined from crude oil. Biofuels produced from waste and from lignocellulosic phytomass is the most sensible way to pursue. Plantations of fast-growing leguminous trees (with symbiotic nitrogen fixation) on currently unused marginal land are also a desirable component of new energy strategy – but the production of biofuels derived from crops grown on farmland should be minimized.

## Learning resources

Berndes, G., Hoogwijk, M. & Van den Broek, R. (2003) 'The contribution of biomass in the future global energy supply: a review of 17 studies', *Biomass and Bioenergy*, 25: 1–28.

Giampietro, M. & Mayumi, K. (2009) *The Biofuel Delusion: The Fallacy of Large-Scale Agro-Biofuel Production*. London: Earthscan.

Howarth, R.W. & Bringezu, S. (eds) (2009) *Biofuels: Environmental Consequences and Interactions with Changing Land Use*. Ithaca, NY: Cornell University.

Kongsager, R. & Reenberg, A. (2012) *Contemporary Land-use Transitions: The Global Oil Palm Expansion*. Copenhagen: Global Land Project.

Smil, V. (2013) *Harvesting the Biosphere: What We Have Taken from Nature*. Cambridge, MA: MIT Press.

Smil, V. (2014) *Power Density: A Key to Understanding Energy Sources and Uses*. Cambridge, MA: MIT Press.

# Carbon budgets

5.4

Róisín Moriarty

The release of carbon in the form of carbon dioxide ($CO_2$) to the atmosphere is one consequence of human, or anthropogenic, activities including burning fossil fuels for energy, industrial processes such as cement, iron and steel production, and forest clearance for agriculture and other large-scale land-use changes. The $CO_2$ released from these anthropogenic processes disrupts the dynamic equilibrium of the carbon that cycles between land, ocean and atmosphere, creating an imbalance, resulting in the accumulation of $CO_2$ in the atmosphere (Figure 5.4.1). Natural sinks (in the sense that they take up $CO_2$) of $CO_2$ such as the oceans, forests and soils, are currently increasing in response to $CO_2$ emissions from human activities and taking up just under half of the $CO_2$ that is released to the atmosphere. The concentrations of $CO_2$ in the atmosphere has increased from 277 parts per million (ppm) since 1750 (Joos & Spahni, 2008) to around 400 ppm today (Dlugokencky & Tans, 2017). Cumulative or total anthropogenic $CO_2$ is a good indicator of atmospheric $CO_2$ concentration and of global temperature increase.

## The concept of a carbon budget

Carbon budgets are a key concept from the Fifth Assessment Report (AR5; Ciais et al., 2013) of the Intergovernmental Panel on Climate Change (IPCC). In this instance the budget is not monetary, but like money it is restricted and based on a fixed amount of carbon or $CO_2$[1] which may be released to the atmosphere. A carbon budget is based on the cumulative or total anthropogenic $CO_2$ emissions that can be released to the atmosphere while remaining below a given climate target. It is the extent to which climate change is constrained by limiting total or cumulative anthropogenic $CO_2$ emissions.

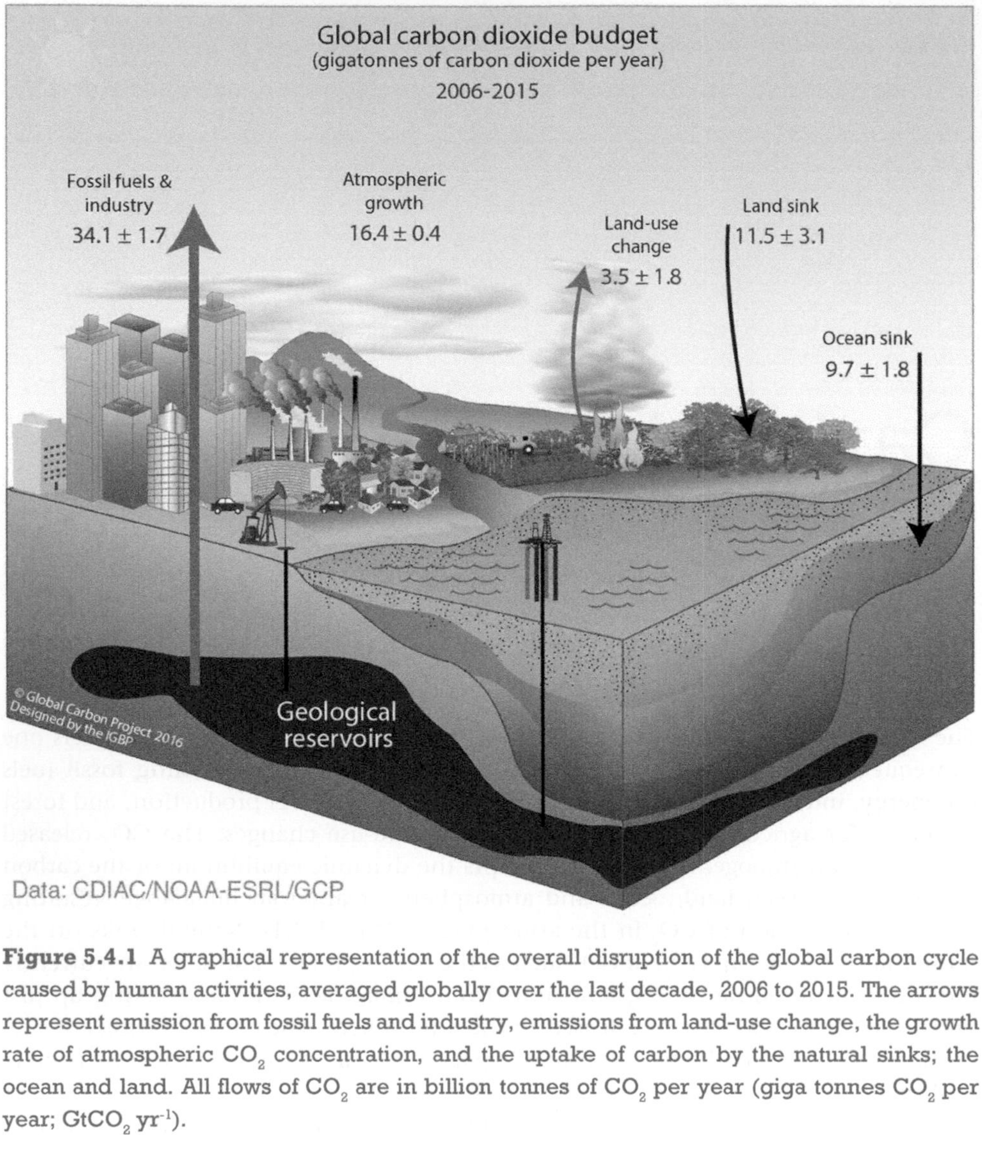


**Figure 5.4.1** A graphical representation of the overall disruption of the global carbon cycle caused by human activities, averaged globally over the last decade, 2006 to 2015. The arrows represent emission from fossil fuels and industry, emissions from land-use change, the growth rate of atmospheric $CO_2$ concentration, and the uptake of carbon by the natural sinks; the ocean and land. All flows of $CO_2$ are in billion tonnes of $CO_2$ per year (giga tonnes $CO_2$ per year; $GtCO_2\ yr^{-1}$).

*Source*: Le Quéré et al. (2016); Boden & Andres (2013); Dlugokencky & Tans (2017).

## The relationship between cumulative carbon, temperature and carbon budgets

Cumulative anthropogenic $CO_2$ emissions since 1870 released to the atmosphere amount to 2000 billion tonnes of $CO_2$ (giga tonnes $CO_2$; $GtCO_2$; Le Quéré et al., 2016). This is the amount of the carbon budget that has been used to date since 1870. Carbon budgets bring together findings from across the complex range of interactions between the carbon cycle, the climate system and a diverse set of feedback mechanisms, many of which are not yet fully understood.

Before the IPCC's Fifth Assessment in 2013, the manner in which the carbon cycle and the climate system responded to increasing concentrations of $CO_2$ in the atmosphere was not clear. While there is still uncertainty related to elements of the carbon cycle and its interactions with climate, there is a robust near-linear relationship between cumulative emissions of $CO_2$ and medium-to-long-term temperature change (Collins et al., 2013; see Figure 5.4.3).

In preparation for AR5, a new set of scenarios, known as Representation Concentration Pathways (RCPs; see Figure 5.4.2) were developed as an intermediate between levels of greenhouse gas emissions and projected climate impacts. Many different emissions scenarios were developed, based on various assumptions about the deployment of $CO_2$ mitigation technologies, carbon taxes based upon 'polluter pays' to encourage investment in cleaner and more efficient energy, as well as different policies and social change. Independently, climate modelers analyzed a large number of different scenarios of the future to assess potential climate impacts.

Through modeling the carbon cycle and climate interactions these scientists demonstrated a positive relationship that has been found to be consistent across all coupled carbon cycle–climate models (Collins et al., 2013). This relationship has allowed the definition of a new attribute, the transient climate response to cumulative carbon emissions (TCRE), which is the ratio of global temperature change to cumulative anthropogenic

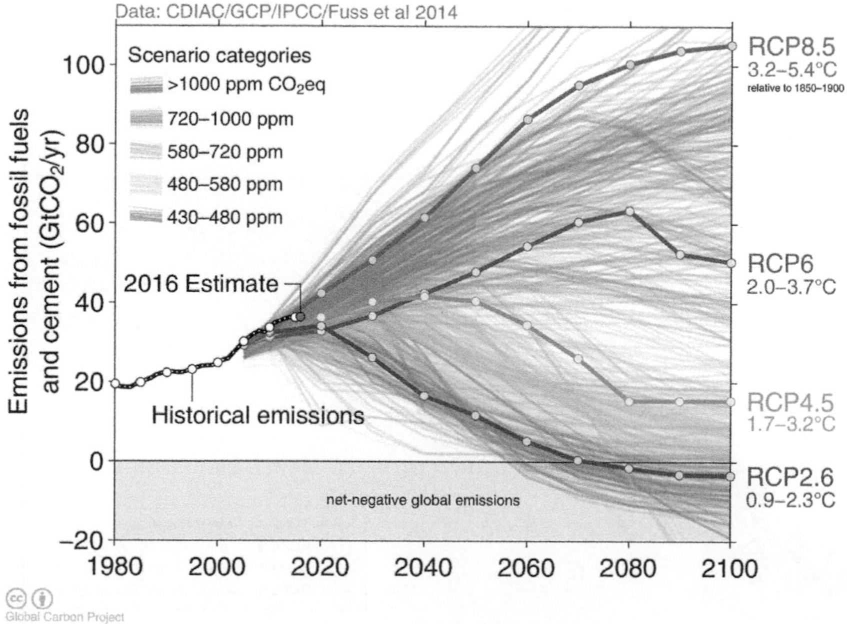


**Figure 5.4.2** Emission pathways for $CO_2$ until 2100 and the extent of net negative $CO_2$ emissions in 2100. Past emissions from fossil fuel combustion and industry (black) are compared with the IPCC AR5 Working Group 3 emissions scenarios (pale colours) and to the RCPs used to project climate change in the IPCC Working Group 1 contribution to AR5 (dark colours).

*Source*: Fuss et al. (2014); Boden & Andres (2013); Clarke et al. (2014).

$CO_2$ emissions (Collins et al., 2013) and provides the scientific basis for linking temperature change to carbon budgets (Figure 5.4.3).

## One problem and two simple concepts

The ultimate goal of the United Nations Framework Convention on Climate Change (UNFCCC) is the 'stabilization of greenhouse gas concentrations in the atmosphere at a level that would prevent dangerous anthropogenic interference with the climate system' (Article 2, UNFCCC, 1992). This definition is scientific and abstract to most people. Establishing more readily understandable climate targets aligned with the goal of stabilizing atmospheric greenhouse gases has been an area of significant and active discussion between scientists and policy makers for 40 years (see Jaeger & Jaeger, 2011; Randalls, 2010).

Climate change requires global cooperation and concepts that are universally understood. The well-below 2°C target, now included in in the UN Framework on Climate Change's Paris Agreement of 2015, potentially acts as a 'guard rail' (Jaeger & Jaeger, 2011; WBGU, 2009) to keep climate within a 'safe operating space' for humans (Rockström et al., 2009; Steffen et al., 2015; Schellnhuber et al., 2016). Temperature limits are shorthand to describe the potential impacts of climate change that will occur within that range

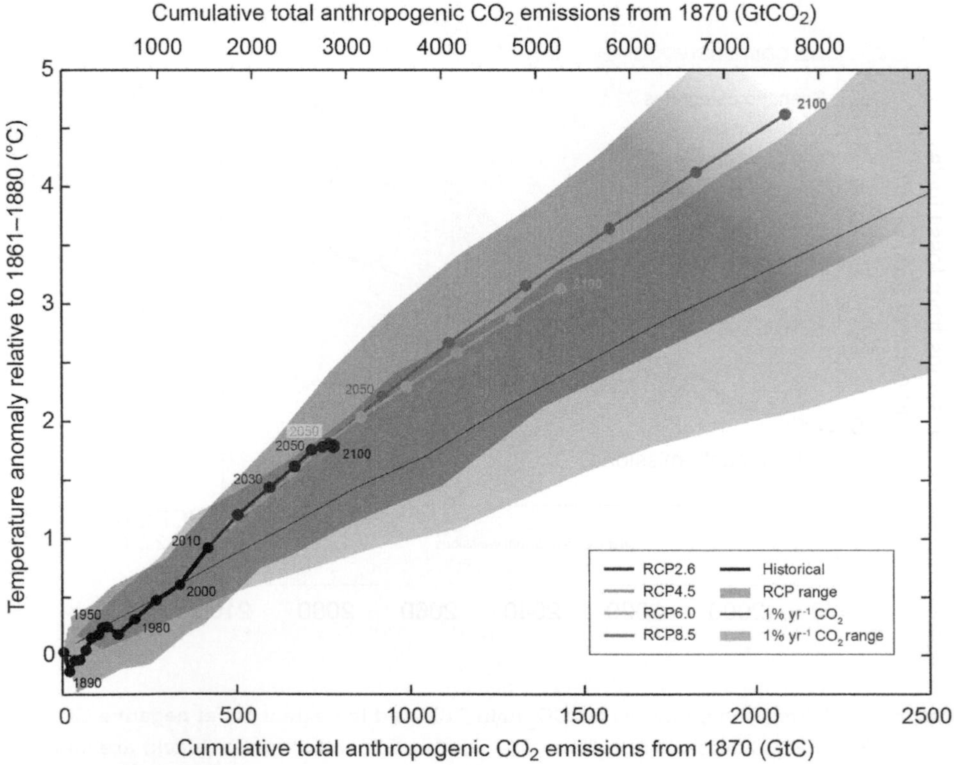


**Figure 5.4.3**

*Source*: IPCC, 2013; *WGI AR5 SPM.10*. Cambridge: Cambridge University Press.

of temperature limits. Staying within the 2°C limit should prevent extreme and more frequent weather conditions such as heat waves and storms and subsequent impacts such as severe economic loss, reducing risk of crop failure, the collapse of coral reefs through ocean acidification, species extinction as habitats change and sea level rise as the oceans expand and land-ice melts. Note that 2°C is average global warming, meaning that average temperature in specific regions, in particular polar latitudes, are where warming is most amplified.

The difficulty of linking temperature targets used in climate policy to the stabilization of greenhouse gas concentrations in the atmosphere has long been recognized. It was only the emergence of the near-linear relationship between global temperature change and cumulative anthropogenic $CO_2$ emissions that allowed temperature increase and carbon budgets to be linked. Temperature limits now come with an implicit carbon budget; a maximum amount of $CO_2$ that can be emitted to the atmosphere before that temperature limit is reached. For a 2°C limit to global average temperature increase, with a likely (>66 per cent) chance of achieving this target, the carbon budget amounts to 2900 $GtCO_2$ budget (IPCC., 2013).

The 2°C temperature target demonstrates how concepts familiar and understood by policy makers and the public can be easily remembered and communicated. Like the 2°C target, carbon budgets are also simple and relatively easy to communicate. They are based on the rigorous scientific analysis that informed AR5 and became linked directly to temperature limits (Figure 5.4.3), making the near-linear relationship between temperature increase and carbon budgets easier to understand. The available carbon budget has recently been adjusted to include other greenhouse gases such as methane ($CH_4$) and nitrous oxide ($N_2O$) but expressed in carbon, $CO_2$ or $CO_2$ equivalent for relativity. The carbon budget can also be expressed as 'equivalent emissions-years', that is, the time remaining until the global carbon budget is exceeded (Friedlingstein et al., 2014). This is a less complex way of communicating carbon budgets and is illustrated below.

For a 2°C or RCP2.6 pathway (RCP2.6 has a radiative forcing of 2.6 Watts per meter squared; $W/m^2$; see Figure 5.4.2), a low emission future, anthropogenic $CO_2$ emissions are required to fall to zero in the latter half of the twenty-first century and become net negative thereafter (see Figure 5.4.2).

## A carbon budget for a 2°C climate target and our current status

Cumulative anthropogenic $CO_2$ emissions between 1870 and 2016 are around 2000 billion tonnes of $CO_2$ (giga tonnes carbon; $GtCO_2$; Le Quéré et al., 2016) leaving around 900 $GtCO_2$, or around 30 per cent, of a 2900 $GtCO_2$ budget (IPCC, 2013) remaining if temperature increase is to be limited to 2°C with a likely (>66 per cent) chance of achieving this target. The current rate of anthropogenic emissions of $CO_2$ is around 30 $GtCO_2$ a year (Le Quéré et al., 2016). Based on this rate there are around 30 years until the carbon budget for the 2°C limit is exhausted. If emissions rates increase faster than current rates, for example further coal burning in rapidly developing countries, construction and continued tropical and other forest clearances for domestic agriculture, then the number of years remaining decreases.

As part of the Paris Agreement, countries declare national determined contributions (NDCs; UNFCCC, 2015) where they decide on the level of effort to decarbonize that will be made to limit their $CO_2$ pollution as their contribution towards the 2°C target. So far 187 countries have submitted 160 NDCs (Rogelj et al., 2016). They add up to a global average temperature rise of around 3.2°C (Rogelj et al., 2016), exceeding the 2°C target. Any near-term overspend of emissions, as with any budget, makes it far harder to balance the budget in subsequent years, requiring even more effort in decarbonization.

Scientists, policy makers and others involved in climate and environmental protection should continue working to rigorously quantify, assess and communicate the carbon budget in order to identify the progress – or lack of progress – in keeping climate change within safe operating limits.

## Learning resources

To see a very neat graphical representation of how the 2°C carbon budget, atmospheric concentrations of $CO_2$ and global average temperature increase are related visit openclimatedata.net at: http://openclimatedata.net/climate-spirals/from-emissions-to-global-warming-line-chart.

To learn more about the carbon cycle and different carbon futures please visit the Global Carbon Atlas at: http://www.globalcarbonatlas.org/en/outreach.

## Note

1 1 tonne of $CO_2$ equals 44/12 × 1 tonne of carbon (see Table 1 in Le Quéré et al., 2016). This conversion coefficient is not arbitrary or even uncertain, but a matter of elementary chemistry.

## Bibliography

Boden, T.A. & R.J. Andres (2013) Global, Regional, and National Fossil Fuel $CO_2$ Emissions. Oak Ridge, Tenn., USA: Oak Ridge National Laboratory, US Department of Energy. Online. Available at: http://cdiac.ornl.gov/ trends/emis/overview_2013.html, last access: October 2013.

Ciais, P., C. Sabine, G. Bala, L. Bopp, V. Brovkin, J. Canadell, A. Chhabra, R. DeFries, J. Galloway, M. Heimann, et al. (2013) Carbon and Other Biogeochemical Cycles. In Stocker, T.F., D. Qin, G-K. Plattner, M. Tignor, S.K. Allen, J. Boschung, A. Nauels, Y. Xia, V. Bex & P.M. Midgley (eds) *Climate Change 2013: The Physical Science Basis. Contribution of Working Group I to the Fifth Assessment Report of the Intergovernmental Panel on Climate Change.* Cambridge: Cambridge University Press.

Clarke L., K. Jiang, K. Akimoto, M. Babiker, G. Blanford, K. Fisher-Vanden, J.-C. Hourcade, V. Krey, E. Kriegler, A. Löschel, D. McCollum, S. Paltsev, S. Rose, P.R. Shukla, M. Tavoni, B.C.C. van der Zwaan and D.P. van Vuuren (2014) Assessing Transformation Pathways. In Edenhofer, O., R. Pichs-Madruga, Y. Sokona, E. Farahani, S. Kadner, K. Seyboth, A. Adler, I. Baum, S. Brunner, P. Eickemeier, B. Kriemann, J. Savolainen, S. Schlömer, C. von Stechow, T. Zwickel and J.C. Minx (eds) *Climate Change 2014: Mitigation of Climate Change. Contribution of Working Group III to the Fifth Assessment Report of the Intergovernmental Panel on Climate Change.* Cambridge: Cambridge University Press.

Collins, M., R. Knutti, J. Arblaster, J-L. Dufresne, T. Fichefet, P. Friedlingstein, X. Gao, W.J. Gutowski, T. Johns, G. Krinner, et al. (2013) Long-term Climate Change: Projections, Commitments and Irreversibility. In Stocker, T.F., D. Qin, G-K. Plattner, M. Tignor, S.K. Allen, J. Boschung, A. Nauels, Y. Xia, V. Bex, P.M. Midgley (eds) *Climate Change 2013: The Physical Science*

*Basis. Contribution of Working Group I to the Fifth Assessment Report of the Intergovernmental Panel on Climate Change*. Cambridge: Cambridge University Press.

Dlugokencky, E. & P. Tans (2017) *Trends in Atmospheric Carbon Dioxide*, National Oceanic & Atmospheric Administration, Earth System Research Laboratory (NOAA/ESRL) Online. Available at: http://www.esrl.noaa.gov/gmd/ccgg/trends/global.html. Last access: 31 January 2017.

Friedlingstein, P., R.M. Andrew, J. Rogelj, G.P. Peters, J.G. Canadell, R. Knutti, G. Luderer, M.R. Raupach, M. Schaeffer, D.P. Van Vuuren & C. Le Quéré (2014) Persistent growth of $CO_2$ emissions and implications for reaching climate targets, *Nature Geoscience*, 7(10): 709–715.

Fuss, S., J.G. Canadell, G.P. Peters, M. Tavoni, R.M. Andrew, P. Ciais, R.B. Jackson, C.D. Jones, F. Kraxner, N. Nakicenovic & C. Le Quéré. (2014). Betting on negative emissions, *Nature Climate Change*, 4(10): 850–853.

German Advisory Council on Global Change (WBGU) (2009) *Special Report 2009. Solving the Climate Dilemma: The Budget Approach*. Berlin: WBGU.

Intergovernmental Panel on Climate Change (IPCC) (2013) Summary for Policymakers. In Stocker, T.F., D. Qin, G-K. Plattner, M. Tignor, S.K. Allen, J. Boschung, A. Nauels, Y. Xia, V. Bex & P.M. Midgley (eds) *Climate Change 2013: The Physical Science Basis. Contribution of Working Group I to the Fifth Assessment Report of the Intergovernmental Panel on Climate Change*. Cambridge: Cambridge University Press.

Jaeger, C.C. & J. Jaeger (2011) Three views of two degrees, *Regional Environmental Change*, 11(1): 15–26.

Joos, F. & R. Spahni (2008) Rates of change in natural and anthropogenic radiative forcing over the past 20,000 years, *Proceedings of the National Academy of Sciences USA*, 105: 1425–1430.

Le Quéré, C., R.M. Andrew, J.G. Canadell, S. Sitch, J.I. Korsbakken, G.P. Peters, A.C. Manning, T.A. Boden, P.P. Tans, R.A. Houghton, et al. (2016) Global Carbon Budget 2016, *Earth System Science Data*, 8: 605–649.

Randalls, S. (2010) History of the 2°C climate target, *Wiley Interdisciplinary Reviews: Climate Change*, 1(4): 598–605.

Rockström, J., W. Steffen, K. Noone, Å. Persson, F.S. Chapin, E.F. Lambin, T.M. Lenton, M. Scheffer, C. Folke, H.J. Schellnhuber, B. Nykvist, et al. (2009) A safe operating space for humanity, *Nature*, 461(7263): 472–475.

Rogelj, J., M. Den Elzen, N. Höhne, T. Fransen, H. Fekete, H. Winkler, R. Schaeffer, F. Sha, K. Riahi, & M. Meinshausen (2016) Paris Agreement climate proposals need a boost to keep warming well below 2°C, *Nature*, 534(7609): 631–639.

Schellnhuber, H.J., S. Rahmstorf & R. Winkelmann (2016) Why the right climate target was agreed in Paris, *Nature Climate Change*, 6(7): 649–653.

Steffen, W., K. Richardson, J. Rockström, S.E. Cornell, I. Fetzer, E.M. Bennett, R. Biggs, S.R. Carpenter, W. de Vries, C.A. de Wit, C. Folke, et al. (2015) Planetary boundaries: guiding human development on a changing planet, *Science*, 347(6223): 1259855.

United Nations Framework Convention on Climate Change (UNFCCC) (1992) Secretariat. UNFCCC.

United Nations Framework Convention on Climate Change (UNFCCC) (2015) *Adoption of the Paris Agreement* FCCC/CP/2015/L.9/Rev.1. UNFCCC.

# 5.5 Fire


## Joshua Whittaker


The ability to use fire sets humans apart from other species. Early humans were attracted to fires by opportunities to forage resources such as bird eggs, rodents, lizards and other small animals. The ability to light and tend fire saw its use for small-scale domestic purposes, such as cooking and heating, and to modify nearby habitats. Aboriginal Australians, for example, used fire to clear vegetation, promote grass and plant growth, control insects and vermin, and hunt. Their use of fire for productive purposes was famously termed 'fire-stick farming' by the archaeologist Rhys Jones (1969). Humans also mastered fire for use in technological processes, such as firing pottery and metalworking. Fire regimes were altered by the increased number of ignitions and changes in their timing, but also by changes to fuel structure and quantity through burning and clearing vegetation. While it is indisputable that human fire has altered environments across the globe, scientists and anthropologists disagree over the extent of human modification, whether changes were intentional, and the degree of impact on 'natural' fire regimes. Indeed, as Bowman et al. (2011, p. 2225) point out, "The intertwined relationships between humans, landscapes and fire throughout human history argue against a clear distinction between natural and anthropogenic fires".

In more recent times, fire has been used for larger scale environmental transformations, such as rainforest burning in the Amazon to make way for agriculture. Anthropogenic climate change is expected to increase the frequency and severity of wildfires in many parts of the world (Field et al., 2014), further highlighting the inseparability of humans and fire regimes. Humans, then, are inherently embedded in global fire processes. Fire can be considered both a source of opportunity and hazard for people, but is also critical to the survival and well-being of non-human species and communities.

## Fire in nature

Fire plays a crucial role in the functioning of many ecosystems, where it is required to maintain a diverse range of plant and animal species and landscape-scale biodiversity (e.g. Keeley et al., 2012). Much of Australia, for example, has a long history of exposure to fire, with large parts of its vegetation assuming its 'modern character' within the last 60 million years or so. Fires gradually increased in frequency as the continent became drier, accelerating the transition from rainforest to more open savannah-type woodland by the elimination of fire-sensitive rainforest species (Kemp, 1981). Most sclerophyllous species, which make up the majority of the Australian flora, not only tolerate but require fire to regenerate. However, different species and ecosystems may require fire at varying frequencies, times of year and intensities. Some species are fire-sensitive or fire-intolerant and can be made locally extinct by fires. For example, fires that burnt stands of Tasmania's endemic King Billy and Pencil pines in 1960–61 and 2016 have been considered 'the most disastrous fire in Australian ecological history' (Gill, 1996) and 'a global tragedy' (Rickards, 2016) due to the loss of 1000-year-old trees. Fire can also cause direct mortality of animals, increase predation, and promote vegetation growth that favours certain species.

Fire plays important ecological roles on many other continents. In North America, for example, the giant sequoia-mixed conifer forests of the southern Sierra Nevada have evolved with and adapted to frequent, low-intensity fires (approx. every 5–15 years). Sequoias rely on fire to release seeds from their cones, expose soil for seedlings to take root, recycle nutrients to the soil, and open the forest canopy so that sunlight can reach seedlings. Human-induced 'fire exclusion' led to the accumulation of forest litter and the encroachment of white fire and incense cedar, which are shade tolerant and able to recruit seedlings in heavy litter. Fire also plays important roles in the functioning of African savanna ecosystems, where there is a much longer history of human involvement in fire regimes.

## Fire as opportunity

Human influences on fire regimes originated with the opportunities presented by fire. Fire continues to be used by people around the world for a range of purposes including foraging, hunting, cooking, heating, and modifying vegetation, albeit on greater scales and in new ways. Kull (1994), for instance, documents the intensification of fire use on the island of Madagascar through slash and burn farming known locally as *Tavy*. Fire is used to clear slashed vegetation and fertilise the soil, which is then sown with rice or other crops. Despite the important role of *Tavy* in local livelihoods, increased use of this practice and shorter fallow times are contributing to deforestation. Fire is also used widely throughout the world to prepare land for agriculture, and to produce a nutrient rich 'green pick' for grazing livestock. In Australia, some farmers use fire to protect their livelihoods by burning vegetation in mild conditions to prevent intense bushfires during hot, dry and windy weather.

## Fire as hazard

For all the opportunities it presents, human fire use also creates hazards. Fires in Indonesian peatlands and peat-swamp forests, for instance, are mostly lit to clear land for agriculture (particularly palm oil production) but have significant human health impacts. Major fires in 1997 are thought to have affected the respiratory health of 20 million Indonesians, with estimates suggesting between 20,000 and 50,000 premature mortalities. Fires in informal settlements – sometimes called 'slum fires' – are another problem in many parts of the world. Because construction of informal settlements is not guided by urban planning, there are rarely street grids, named streets or house numbers. Settlements often lack running water, sewage systems, electricity and garbage disposal. The high density of buildings and flammability of materials used in their construction creates potential for devastating fires, which are often ignited by cooking fuels or poorly wired electrical connections. Formal firefighting services are limited, if they exist at all, and are often hampered by poor road access. For example, in 2011 a fire tore through the 'shanty town' of Sinai in Nairobi, Kenya, killing approximately 100 people and injuring many more. The fire was caused by a leaking pipeline, operated by the state owned Kenya Pipeline Company, from which Sinai residents had started to collect fuel for their own use. A discarded cigarette is thought to have ignited the blaze.

Wildfires are a threat to people and property in many parts of the world including Australia, Canada, France, Greece, Portugal, Spain, the USA and Russia. A wildfire is an unplanned fire that burns through forests, grasslands, shrublands or other vegetation types. Many people choose to live and work in wildfire prone locations to enjoy the landscape and amenity values of regions such as the Sierra Nevada in California or the Blue Mountains in New South Wales, Australia. Some inhabit wildfire risk areas to sustain their livelihoods, as is the case for many agriculturalists, or because costs of living are lower. In South-eastern Australia, for instance, population growth in high bushfire risk areas has occurred as people move from metropolitan areas for the amenity and lifestyle of coastal and rural settings (the 'sea change' and 'tree change' phenomenon) and as housing in Melbourne and Sydney has become unaffordable (Whittaker et al., 2012). Countries like the USA and Australia have sophisticated systems in place to manage wildfires, including fire and emergency response, land management practices (such as forest-thinning and hazard reduction burning), land use planning and building codes, and community awareness and education strategies. Nevertheless, wildfires are inevitable, particularly in the summer months, and often impact on people and property.

The 2003 Cedar Fire is the largest wildfire in Californian history. Driven by strong Santa Ana winds, the fire burned over 280,000 acres of land in San Diego County, killed 15 people and destroyed more than 2200 homes. Fire investigators concluded that the fire was caused by a novice hunter who became lost and started a small fire to signify his whereabouts to rescuers. The fire escaped due to the hot, windy weather and low moisture content of the surrounding vegetation. Subsequent reviews identified numerous issues with the firefighting response, including a government policy that prohibited deployment of firefighting aircraft less than 30 minutes before sunset, which may have prevented earlier containment.

Australia's worst bushfire disaster occurred on 7 February 2009 in the south-eastern State of Victoria. The 'Black Saturday' bushfires burned under the most severe fire weather conditions on record in Victoria, with a record-high maximum temperature of 46.4°C (115°F) in Melbourne, record low relative humidity and strong winds throughout the State. The day saw more than 400 fires across Victoria, with most of the major fires started by fallen powerlines or arson. Fires burned out of control as communities

**Figure 5.5.1** Wildfires are a threat to human life and property in many parts of the world.

Photo by David Bruce, Bushfire and Natural Hazards CRC.

came under threat with little or no official warning. The speed, intensity and extent of the fires meant that firefighting capacities were stretched and most residents responded without direct assistance from fire services. By the time the fires were contained, 173 people had lost their lives, 414 were physically injured and more than 2000 houses were destroyed. A subsequent Royal Commission identified problems with fire service and community preparedness and responses to the fires and recommended significant changes to the way bushfires are managed.

## Managing fire in human landscapes

Humans have irreversibly changed fire regimes across the globe. Fire presents humans with innumerable opportunities, but can also threaten human life, health and property, as well as non-human species and communities. A key challenge is to manage human use of fire to maximise these opportunities while minimising hazards. This is complex, however, precisely because fire is so profitable. Short-term economic benefits tend to prevail over longer-term environmental costs, particularly in developing countries where environmental protections are often limited and people may be economically precarious. In terms of wildfire, management strategies may involve trade-offs. For example, in parts of Australia where regular, low intensity prescribed burning is undertaken to reduce bushfire hazard there may be adverse impacts on biodiversity. In areas of ecological value, less frequent burning may be required to conserve biodiversity at the cost of hazard reduction. These challenges are likely to be exacerbated by population growth and urban development in wildfire prone areas and by climate change, which is expected to increase the frequency and severity of landscape-scale fires into the future.

## Learning resources

Bankoff, G., Lübken, U. & Sand, J. (eds) (2012) *Flammable Cities: Urban Conflagrations and the Making of the Modern World.* Madison: The University of Wisconsin Press.

Bradstock, R.A., Gill, A.M. & Williams, R.J. (eds) (2012) *Flammable Australia: Fire Regimes, Biodiversity and Ecosystems in a Changing World.* Collingwood: CSIRO Publishing.

Pyne, S.J. (2001) *Fire: A Brief History.* Seattle: University of Washington Press.

NEO – NASA Earth Obervations 'Active fires': http://neo.sci.gsfc.nasa.gov.

## Bibliography

Bowman, D.M.J.S., Balch, J., Artaxo, P., Bond, W.J., Cochrane, M.A., D'Antonio, C.M., DeFries, R., Johnston, F.H., Keeley, J.E., Krawchuk, et al. (2011) 'The human dimensions of fire regimes on Earth', *Journal of Biogeography*, 38(12): 2223–2236.

Field, C., Barros, V.R., Mastrandrea, M.D., Mach, K.J., Abdrabo, M.A-K., Adger, W.N., Anokhin, Y.A., Anisimov, O.A., Arent, D.J., Barnett, J. et al. (2014) *Climate Change 2014: Impacts, Adaptation, and Vulnerability.* Summary for policymakers. Intergovernmental Panel on Climate Change. Cambridge: Cambridge University Press.

Gill, A.M. (1996) *How Fires Affect Biodiversity*. Biodiversity Series, Paper No.8. Canberra: Biodiversity Unit, Department of the Environment, Sport and Territories.

Jones, R. (1969) 'Fire-stick farming', *Australian Natural History*, 16: 224–228.

Keeley, J.E., Bond, W.J., Bradstock, R.A., Pausas, J.G. & Rundel, P.W. (eds) (2012) *Fire in Mediterranean Ecosystems: Ecology, Evolution and Management.* Cambridge: Cambridge University Press.

Kemp, E.M. (1981) 'Pre-quaternary fire in Australia'. In Gill, A.M., Groves, R.H. & Noble, I.R. (eds) *Fire and the Australian Biota* (pp. 3–21). Canberra: Australian Academy of Science.

Kull, C.A. (1994) *Isle of Fire: The Political Ecology of Landscape Burning in Madagascar.* Chicago: University of Chicago Press.

Rickards, L. (2016) 'Goodbye Gondwana? Questioning disaster triage and fire resilience in Australia', *Australian Geographer*, 47(2): 127–137.

Whittaker, J., Handmer, J. & Mercer, D. (2012) 'Vulnerability to bushfires in rural Australia: a case study from East Gippsland, Victoria', *Journal of Rural Studies*, 28: 161–173.

# Fisheries



Charles Mather

Since the mid-1990s the Food and Agriculture Organization (FAO) has published biennial reports on fisheries and aquaculture (FAO, 2016). Titled the *State of the World's Fisheries and Aquaculture* (SOFIA), these reports provide key insights into fish production and consumption trends, as well as a selection of detailed studies on key issues facing fisheries and aquaculture globally. Over more than two decades the SOFIA reports have also traced the most important changes in global fisheries including the rise of aquaculture, the changing and dynamic role of fish as food, and new developments in best practice for the management and governance of fisheries. In this chapter, I use the SOFIA reports as a way of framing three key issues related to human–environment issues in fisheries: the changing dynamics of fish capture and production, the complex and geographically diverse patterns of fish consumption, and the ongoing efforts to improve the management and governance of fisheries in a context of rapid environmental change.

## Producing fish: global shifts and the rise of aquaculture

There have been several key changes in the production of fish over the last 40 years. The first is the very significant geographical shift in fish production from developed to developing countries. Fish production in developed countries increased rapidly up until the late 1980s driven largely by technological innovations and increases in harvesting capacity. Since then, however, the production of fish in developed countries has slowed and, most recently, declined by several million tonnes. In developing countries, in stark contrast, fish production has increased dramatically: since the early 1970s, developing countries have more than doubled the volume of fish production for food (Delgado et al., 2003). By the late 1990s developing countries were producing more than double the volume of fish produced by developed countries. Not surprisingly, fish exports have become a significant source of trade income for developing countries. Revenues from fish now exceed the combined value of exports of traditional agro-commodities such as sugar, coffee and fruits and vegetables (Campling et al., 2012).

A second significant change has been the apparent limit that has been reached in the production of fish from wild capture fisheries. The volume of fish produced through capture fisheries based on FAO data reached an upper limit of around 90 million tonnes in the early 1990s and has remained around that level ever since. The stabilization of capture fisheries production should not, however, be interpreted as an indication that this sector has reached equilibrium between harvesting and natural production. On the contrary, global reports continue to suggest that pressure on marine and freshwater fisheries continues to be unsustainably high (Pauly & Zeller, 2016). The impact of a rapidly changing climate is expected to place additional pressure on the sustainability of wild capture fisheries.

The third change in fish production has been the rise in aquaculture, arguably the most important development in global fisheries in the last 40 years. The production of fish from aquaculture has grown at a blistering pace. Between 1980 and 2012 production increased at an average of 8.6 percent a year. China has played a key role in the growth of aquaculture and currently produces more than 60 percent of total fish production from aquaculture. The role of aquaculture in sustaining the global growth of fish production in the context of stagnant production from wild capture fisheries has been celebrated, especially given rapid increases in global population levels, and the growing dependence of people on fish as a source of protein (FAO, 2016). Yet aquaculture faces significant challenges including the management of disease, the problem of waste and environmental degradation, the issue of finding sustainable sources of feed, as well as a range of long identified social issues associated with aquaculture especially in the developing world.

## Consuming fish: between food security and risky food

Fish consumption at a global scale has increased significantly in the last 50 years. Between the 1960s and 2015, fish consumption has increased from just 10kg to over 20kg per capita. Current estimates suggest that as many as 3 billion people worldwide depend on fish as a key source of animal protein (FAO, 2014).

Fish consumption is promoted by national governments and by multilateral institutions because this form of protein contains vitamins and micronutrients that are important to the development of infants and young children. In industrialized countries, fish consumption is encouraged as it is a source of essential fatty acids, vitamins and minerals which may play a role in mitigating heart disease and mental illnesses such as dementia and depression. In a context of a rapidly growing global population, fish has the potential to play an even greater role in addressing food insecurity and global health challenges given that the production of fish has far outstripped increases in global population.

The problem of promoting fish consumption is that some farmed and wild caught fish have high levels of environmental toxins, mercury and dioxins. Farmed fish may also contain dyes and antibiotics that can be harmful to human health. The risk of contaminants is especially high when fish are harvested from polluted waters, or in the case of long lived predator fish, where the bio-accumulation of contaminants may be a problem. The challenge for those promoting fish consumption is how to highlight the benefits of fish consumption while at the same time addressing the risks associated with eating fish that is harmful to human health (Mansfield, 2011). There are no straightforward solutions to this challenge because of the wide variety of fish species consumed around the world, and because of the different vulnerability of the population to contaminants.

Concerns over the health benefits and risks of eating fish tend not to feature in the rapidly developing private eco-labeling systems in fisheries (examples include the Marine Stewardship Council and Friends of the Sea). Fish consumption advisories, like the Monterrey Bay's Seafood Watch, also tend to focus on the sustainability of fish production practices rather than the health of seafood. This is mainly because these eco-labeling systems emerged in response to the concerns over the sustainability of global fish stocks. While eco-labeling systems represent a new and dynamic feature of fish consumption, they are more prominent in the United States and the European Union. Only a relatively small volume of total fish consumption carries an eco-label, and the overall volume of fish that is certified as sustainable is very small.

## Managing fish: governance and policy for sustainable fisheries

Fisheries governance has for many years focused on the importance of rights-based management systems, which aim to provide fishers with secure and long-term rights to fish resources. Secure rights, according to some policy makers, leads to the sustainable and efficient use of fish resources. There is a very substantial literature on the effect of rights-based management mechanisms – such as individual transferable quotas – in fisheries (Olson, 2011). The literature remains divided on the effect of rights-based management: while many researchers have pointed to the negative social and environmental effects of formal property rights in fisheries, government policy makers around the world continue to promote these systems as a solution to overfishing.

Approaches to fisheries governance continue to focus on rights-based management. Yet in many parts of the world there has been an important shift in fisheries governance towards 'blue growth' approaches. Blue growth is justified on the basis that the sustainable use of coastal and marine resources, including but not restricted to fish, can play an important role in supporting livelihoods and sustaining coastal communities around the world (FAO, 2016). A central assumption of this new approach is that the unsustainable use of fish and other coastal resources represents an enormous lost opportunity to those directly and indirectly involved in fisheries and aquaculture.

Blue growth policies stress the importance of rehabilitating coastal and inland waters and wetlands and mangroves. They also emphasize the importance of improved governance in order to limit the incidence of illegal, unreported and unregulated fishing practices, and they focus on improving the social and environmental sustainability of aquaculture. There is a strong emphasis within blue growth approaches to sustainable livelihoods and small-scale fisheries in the recognition of the key role they play in addressing food security. The shift to blue growth represents an attempt to mobilize the potential of marine resources in contrast to previous policies that have focused on reducing fishing capacity and securing rights to fish resources.

## Conclusion

Fish production through aquaculture and wild capture fisheries is frequently presented as a solution to a growing global population that is consuming increasing amounts of animal protein. The role of fish production in meeting the demands of a hungry planet depends on sustainable production in aquaculture and wild capture harvesting. A more sustainable fisheries sector has the potential to not only satisfy the consumption demands of a growing global population, it also has the potential to sustain livelihoods and fish dependent coastal communities in a rapidly changing environment.

## Learning resources

*Journal of Agrarian Change* published a special issue on capture fisheries in 2012 with a range of interesting and relevant papers.

Friends of the Sea (http://www.friendofthesea.org) and the Monterrey Bay Seafood Watch (http://www.seafoodwatch.org) are two non-governmental organizations committed to sustainable fisheries and aquaculture. They are involved in certification and they also provide consumers with the knowledge they need to make sustainable seafood choices.

FAO State of the World's Fisheries (http://www.fao.org/fishery/sofia/en) releases biennial reports on the state of capture fisheries and aquaculture. The reports provide regular updates on volumes of production and consumption for both capture fisheries and aquaculture. They also provide thematic studies of key issues relating to global fisheries.

The Sea Around Us is a project based at the University of British Columbia that is concerned with the impact of fisheries on marine ecosystems. Their website (http://www.seaaroundus.org/) is a good source for new cutting edge research on fisheries.

## Bibliography

Campling, L., Havice, E. & McCall Howard, P. (2012) The political economy and ecology of capture fisheries: market dynamics, resource access and relations of exploitation and resistance, *Journal of Agrarian Change*, 12(2–3): 177–203.

Delgado, C.L., Wada, N., Rosegrant, M.W., Meijer, S. & Ahmed, M. (2003) *Fish to 2020: Supply and Demand in Global Markets*, Washington, DC: International Food Policy Research Institute.

Food and Agriculture Organization (FAO) (2014) *The State of World Fisheries and Aquaculture: Opportunities and Challenges*, Rome: Food and Agriculture Organization of the United Nations.

Food and Agriculture Organization (FAO) (2016) *The State of World Fisheries and Aquaculture: Contributing to Food Security and Nutrition for All*, Rome: Food and Agriculture Organization of the United Nations.

Mansfield, B. (2011) Is fish health food or poison? Farmed fish and the material production of un/healthy nature, *Antipode*, 43(2): 413–434.

Olson, J. (2011) Understanding and contextualizing social impacts from the privatization of fisheries: an overview, *Ocean & Coastal Management*, 54(5): 353–363.

Pauly, D. & Zeller, D. (2016) Catch reconstructions reveal that global marine fisheries catches are higher than reported and declining, *Nature Communications*, 7: 10244.

# Forest resources

5.7

## William Nikolakis and Harry W. Nelson

Forests provide fuel, fibre, and food to human communities. Forests are also a source of aesthetic, cultural, and spiritual values to people. Across the globe there are diverse forest-types, including tropical, sub-tropical, temperate, and boreal forests. Some countries have multiple climate zones and diverse forest types, like Australia, Canada, China, and the US. Each forest type provides different kinds of forest resources. There are also different ownership patterns and governance regimes that influence access and use of those forests. While this creates great institutional diversity, there is commonality in forest resource management, where approaches have focused either on timber values (production) or preservation (conservation and restoration). However, with a shift towards Sustainable Forest Management (SFM) the importance and value of forest ecosystem services are increasingly being integrated and oftentimes quantified into forest resource management decisions.

## Natural and plantation forests

Forests are the most widespread terrestrial ecosystem on Earth. In 2015, natural forests were reported to account for 93 per cent of global forest area, or 3.7 billion hectares (FAO, 2016). But increasingly plantation forests, which make up the remainder of global forest area, are an important source of forest products. Plantations account for one-third of all industrial harvests, and genetic engineering has enhanced the survival and yield from plantation timbers, making these an efficient source of forest resources like lumber, pulp and paper. There are criticisms around the ecological impacts from plantation monocultures, including conversion from native forests and associated reductions in water quality and biodiversity.

## Global forest resources

The Food and Agricultural Organization of the United Nations (FAO) Global Forest Resources Assessment (FAO, 2016) produced every 5 years since 1948, covers a 25-year period from 1990 to 2015. Drawing on data generated at the national level (from 234 countries), the assessment documents the anticipated forest area and projected change in forest area across the globe (to 2030), global timber production, and total conservation forest area (to 2050). Forests are defined in terms of tree cover, but the FAO assessment can include areas devoid of tree cover as forest land (MacDicken, 2015).

In 1990, the global assessment revealed there were 4.128 billion hectares of forest (or 31.6 per cent of total land area), and by 2015, this had been reduced to 3.999 billion hectares (or 30.6 per cent of total land area). A net loss of total forest area is described as deforestation, while forest gain is described as forest expansion. Understanding the change in total forest area provides insight into how land use impacts forest resources, but deforestation and forest gain are dynamic and difficult to monitor across the globe. Forest degradation is also drawing concern, defined as a reduction in the capacity of a forest to provide ecosystem services.

A revealing statistic from the global assessment is the net loss of natural forest – a reduction of 129 million hectares between 1990 and 2015, an annual net loss of 0.13 per cent of natural forest land. This deforestation has largely been in the tropics, South America and Africa, where forests have been cleared and converted for agriculture uses, resulting in habitat loss and carbon emissions. The annual net loss of forests has slowed from a height of 0.18 per cent of total forests from 1990 to 2000 (or 10.6 million hectares) to 0.08 per cent (or 6.5 million hectares) during 2010 to 2015. Reductions in tropical forest loss can be attributed to interventions by NGOs and governments to mitigate deforestation (MacDicken, 2015). From 1990 to 2015, average per capita forest has declined from 0.8 hectares to 0.6 hectares, consistent across all climatic zones (except temperate zones).

Degradation is particularly difficult to measure. From 1990 to 2015, the world's forests have changed in various ways including stocking density, species composition and diameter distribution. Climate change is likely to have significant impacts on forest health and species distribution. Hence, new instruments are required to effectively manage forest resources to support efficiency and equity goals.

Only 26 per cent of natural forests are classed as primary forest (i.e. old-growth or ancient woodland). From 1990, some 31 million hectares of primary forest has been modified or cleared; modified primary forest is then defined as "other naturally regenerated forest", which accounts for the remaining 74 per cent of natural forests.

Approximately 31 per cent of global forests are designated as production forest, and in 2011, about 3 billion cubic meters of wood was removed from global forests, and 49 per cent of this was for wood fuel. From 1990 to 2015, planted forests have increased by over 105 million hectares.

Almost 28 per cent of global forest area is classed as multiple use that provides a wide range of goods and services. Another 13 per cent of global forest area is dedicated to conservation, with 150 million hectares added from 1990.

## Forest ecosystem services

Forests are being valued for a broader range of products and services, for which historically market prices have not existed. These goods and services are now being recognized, valued, managed and traded. These include carbon sequestration, biodiversity, soil quality, and downstream water quality. Forests designated as protecting soil and water now represent 31 per cent of total forest area in the 2015 global assessment (FAO, 2016).

## Forest resources management and governance

According to the FAO (2016), most of the world's forests are publicly owned (about 3 billion hectares). But the amount of privately owned forest land has increased from 15 per cent to 18 per cent of forest land (or 774 million hectares), mostly in the developed world. Private management rights in public forests held by companies has increased from 6 per cent in 1990 to 14 per cent of forest land in 2010. Half of the world's forests are managed under a forest management plan (FAO, 2016), and this is split equally between conservation and production activity.

Industrial logging can crowd out local values and livelihoods, causing tensions between outside forest companies and local peoples. One particular approach to support these local values is the development of community forests.

Some 96 per cent of the world's forests are managed according to SFM principles, which involves a recognition of stakeholder values, forest ecosystem services, local livelihoods, and non-timber forest products (NTFPs) in management decisions (FAO, 2016). SFM also includes more sustainable harvesting practices like selective rather than clearcut logging, which can sustain aesthetic, recreational and ecological values.

The amount of forest area managed under a voluntary forest management certification scheme has increased from 14 million hectares in 2000, to 438 million hectares in 2014 (FAO, 2016). These schemes include the Forest Stewardship Council (FSC) (covering 42 per cent of forest land under certification), and the largest by forest area, the Programme for the Endorsement of Forest Certification (PEFC) (covering the remaining 58 per cent). These standards have emerged to support forest practices that protect diverse forest resources and encourage forest companies to internalize the costs (or externalities) from their activities. This includes standards to protect forest ecosystem values and for the FSC, achieving free prior and informed consent (FPIC) where harvesting impacts local indigenous peoples. Forest companies can apply these certification standards to their products which can facilitate market access through procurement policies. While such schemes have been widely adopted in temperate forests, there has been far less uptake in tropical forests.

## Forest resources trade

For most of the last century, public forests have been managed intensively for income and employment values and traditional forest resources like timber, and by-products from milling lumber, like woodchips for pulp and paper, and particleboard. Over time, modernization has made all aspects of the industry more capital intensive, and employment in the forest sector has decreased globally. Globalization through increased trade has expanded markets but also increased competitive pressures.

Globalization and modernization has facilitated the expansion of industrial forestry models where large-scale logging tenures encourage clear-cut harvesting in primary forests and intensively managed timber crops managed on financially-driven rotations. These models are focused on economies of scale, highly automated and optimized manufacturing facilities, and global distribution networks, to compete on world markets. On public forest lands, income from industrial forestry can finance public infrastructure such as roads and interventions into forest management, but forestry has declined in importance to more high value products from mining and energy. In tropical jurisdictions, weak governance can lead to unsustainable rates of logging and degradation. Elsewhere, government policies promoting higher value land uses (agriculture, infrastructure for other development) can also contribute to deforestation and degradation.

## The bio-economy

The digital era and growth in the use of plastics has seen significant reductions in paper use, which has facilitated a restructure among forest product companies because of declining demand (for more information see Roberts & Nikolakis, 2014). The bio-economy has emerged from this restructure, where wood-based feedstock produces renewable resources that replace products from non-renewable sources including bio-plastics, bio-fuels and bio-chemicals. These products are less toxic and are less greenhouse gas-intensive than fossil fuel based substitutes.

## Learning resources

There are some useful overviews of global forest resources and trends in forest resource management. The text below offers a broad perspective of forest resources from across the globe:

Nikolakis, W. & Innes, J. (Eds) (2014) *Forests and Globalization: Opportunities and Challenges for Sustainable Development.* Abingdon: Earthscan.

The FAO provides the most comprehensive analysis of forest resources at national and global scales. The following link is to the FAO Global Resources Assessment, 2015. This document presents data on forest resources at national and global levels: http://www.fao.org/3/a-i4808e.pdf.

The Rights and Resources Initiative, based in Washington, DC, documents rights to forest resources held by local communities and Indigenous Peoples across the globe, available at: http://rightsandresources.org/en/#.V-cszfArKM8.

The World Resources Institute provides information on sustainable forest management from across the globe: http://www.wri.org/our-work/topics/forests.

The World Resources Institute's 'Global Forest Watch' initiative allows users to observe and monitor deforestation through an interactive mapping tool, available at: http://www.globalforestwatch.org.

# Visual aids for this chapter

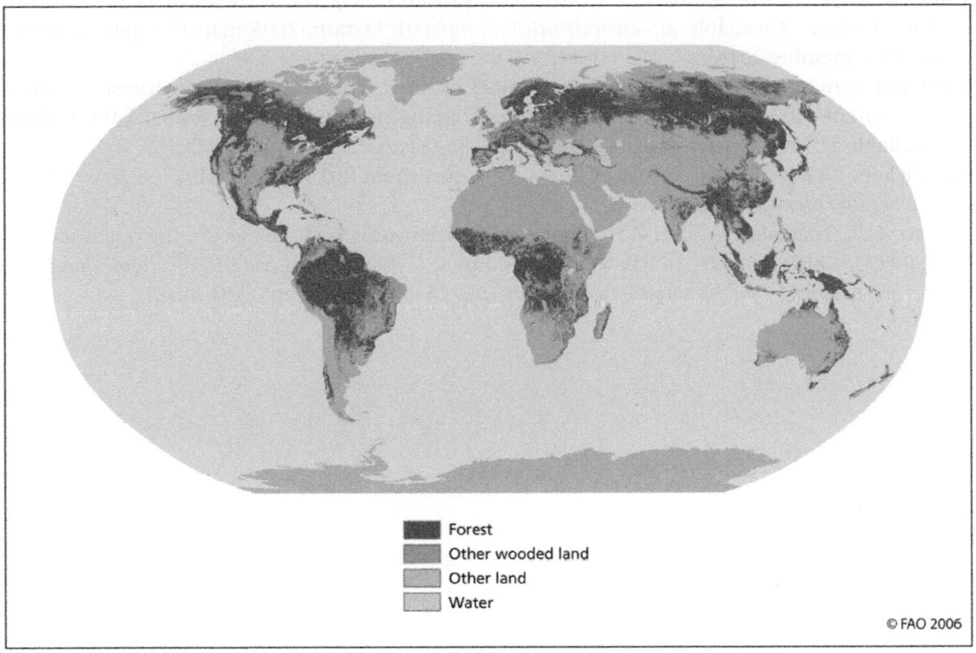


**Figure 5.7.1** Global forests.

*Source*: FAO (2006).

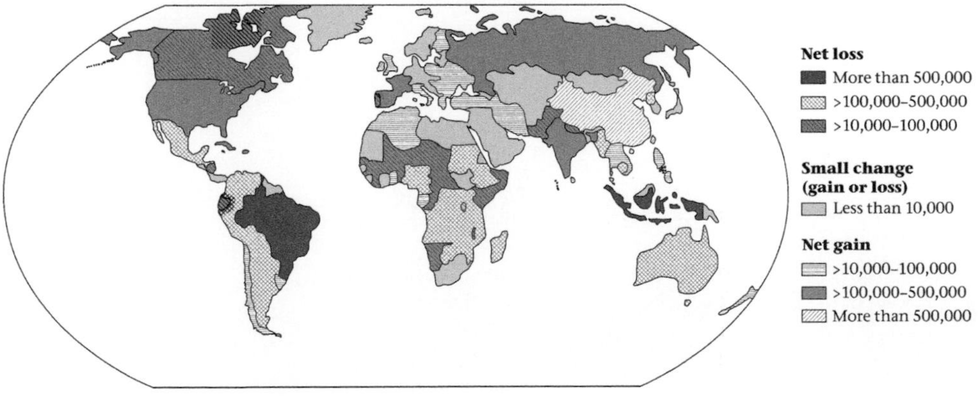


**Figure 5.7.2** Net forest gain/loss.

*Source*: FAO (2016).

## Bibliography

Food and Agricultural Organization of the United Nations (FAO) (2006) The world's forests. Rome: FAO. Online. Abvailable at: <http://foris.fao.org/static/data/fra2005/maps/2.2.jpg> (accessed on 24 September 2016).

Food and Agricultural Organization of the United Nations (FAO) (2016) Global Forest Resources Assessment 2015: How are the world's forests changing? 2nd Edition. Rome: FAO. Online. Available at: <http://www.fao.org/3/a-i4793e.pdf> (accessed on 25 August 2016).

MacDicken, K.G. (2015) 'Global Forest Resources Assessment 2015: What, why and how?', *Forest Ecology and Management*, 352: 3–8.

Roberts, D. & Nikolakis, W. (2014) 'Thoughts on transforming the forest sector: the potential (and reality) for bio-products', in Nikolakis, W. & Innes, J. (Eds) *Forests and Globalization: Opportunities and Challenges for Sustainable Development* (pp. 25–35). Abingdon: Earthscan.

# Floods

5.8

## Gemma Carr, Alberto Viglione and Magdalena Rogger

People benefit from living close to rivers, lakes and the sea. For millennia, water bodies have been used as trade and communication corridors, offered a defence measure against invaders, their fisheries provide an abundant high protein food supply that supports social and economic development, they supply energy and support industry, and they are often considered to be culturally and emotionally valuable. Flood plains also provide highly fertile agricultural land and their proximity to water offers an irrigation source. However, these societal benefits gained from settling close to the water come at the price of periodic flood damage.

Floods occur when land that is normally above water becomes temporarily submerged (see Figure 5.8.1). They can result from heavy rainfall, when water overflows a river's levees or embankments, or from high tides and/or storms in coastal areas. Floods occur over many different spatial and temporal scales. Some floods take place very rapidly, such as flash floods (where high rainfall leads to a rapid rise in the river level), dam-break floods, glacial lake floods or ice jam floods (where a natural or man-made dam breaks suddenly leading to a torrent of water flowing down the river) or tsunamis. Other floods lead to a slower rate of water level rise, such as large river floods due to high rainfall or snow melt, or groundwater floods resulting from long periods of rainfall leading to a rise in the water table.

## Flood risk

Floods that occur in areas inhabited or utilised by people are problematic because they can lead to damages resulting in human or economic losses. Flood risk deals with the

**Figure 5.8.1** Left: Large-scale flooding in New Orleans, USA following levee breach after Hurricane Katrina in 2005; Right: Small-scale flooding in a mountain valley in Tirol, Austria.

Left image, from: http://en.wikipedia.org/wiki/File:KatrinaNewOrleansFlooded_edit2.jpg; right image, from alpinesicherheit.com. Both images non-proprietary.

probability of negative consequences from flooding. It is determined by the size (magnitude) and regularity (frequency) of flooding (called the flood hazard) combined with the vulnerability of the people and their properties. Vulnerability is a somewhat complex concept. It relates to the exposure of people and goods to flooding and the capacity of the population to cope with them (see Merz et al., 2010).

People's actions may result in an increase in flood risk in several ways by changing flood hazard and vulnerability (Figure 5.8.2). Vulnerability and exposure to floods increases as human populations expand and people occupy and use more land in previously unpopulated flood plains and coastal areas. Flood risk increases when a greater number of people, property and possessions are exposed to flooding which raises the potential for loss of life and economic loss, and as the GDP of a region or population increases the size of the damages also increases. Structural flood protection measures and river regulations give a false impression of safe settlement areas. However, in the event that the measures fail, such as for very large flood events, damages are very high, because of the so called 'levee effect', i.e., the fact that vulnerability has increased (more exposure and less preparedness) due to the lack of experience of flooding (see e.g., Viglione et al., 2014). Structural flood protection measures may also lead to a change in the hazard component of the flood risk. In many regions of the world, river regulations have been widely installed over the last centuries to protect people from floods, but at the same time they increase the flood risk downstream by transferring and concentrating the flow water to other areas. In addition, population growth has led to significant land use changes that include urbanisation and changes in agricultural practices that increase floods at the local scale. By replacing permeable soil with impermeable concrete and drainage systems, and modifying the land surface for agriculture, surface water can move faster to the river. Human induced climate change may also lead to an increase in flood risk by causing more frequent and larger storms or driving changes in seasonal snowmelt patterns due to temperature changes (IPCC, 2014).

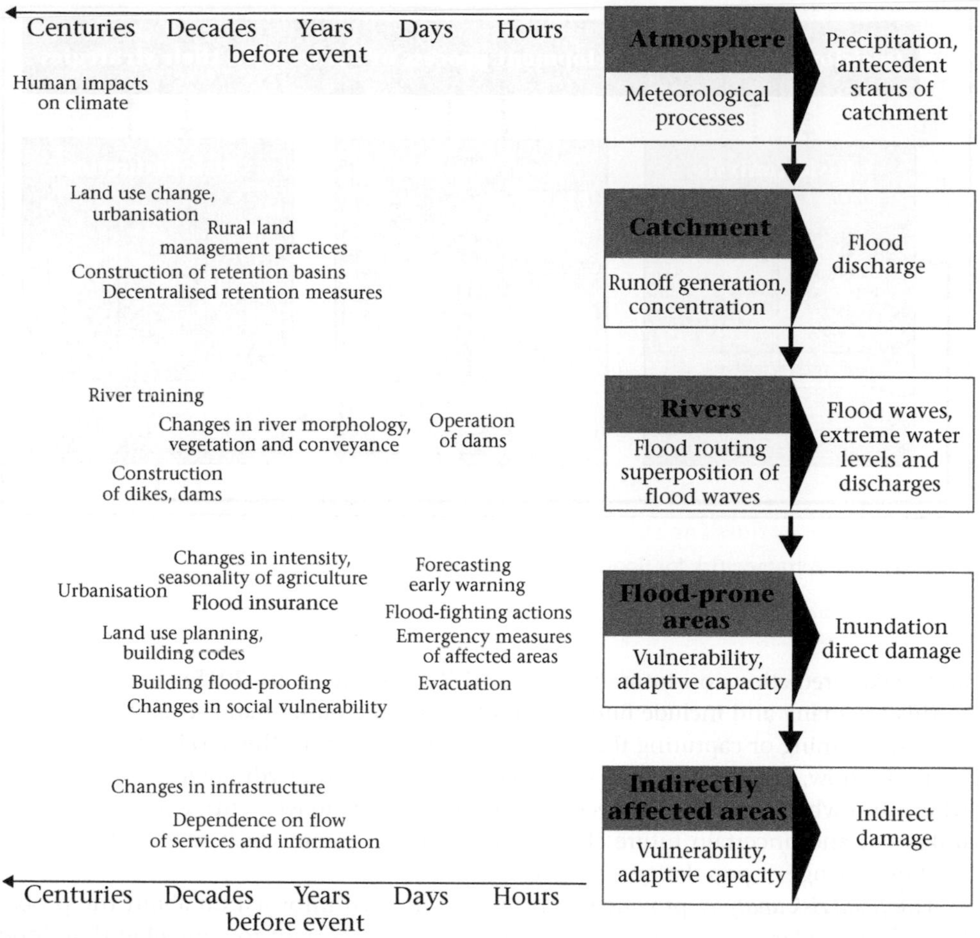


**Figure 5.8.2** Human actions and interventions that impact on flood risk and their timescales (left side of figure) and drivers of flood risk (right side of figure).

Reproduced from Merz et al. (2010).

## Flood risk management

Schanze (2006) categories flood risk management according to three aspects: *risk analysis, risk assessment and risk reduction* (Figure 5.8.3). *Flood risk analysis* is concerned with determining the past, present and future risks of flooding. It aims to comprehensively capture the flood hazard (natural system) and the vulnerability (human system) and the interactions between the two systems. Knowledge, understanding and information from meteorology, hydrology, hydraulics, geology, ecology, economics, sociology, law, political sciences and psychology have to be brought together to fully understand the flood hazard and the vulnerability to achieve the most complete analysis possible of flood risk. Modelling methods that can integrate understandings and data from both natural and social sciences, and capture the feedbacks between the two are, in many ways, in their infancy. Their development is essential to fully understand flood risk

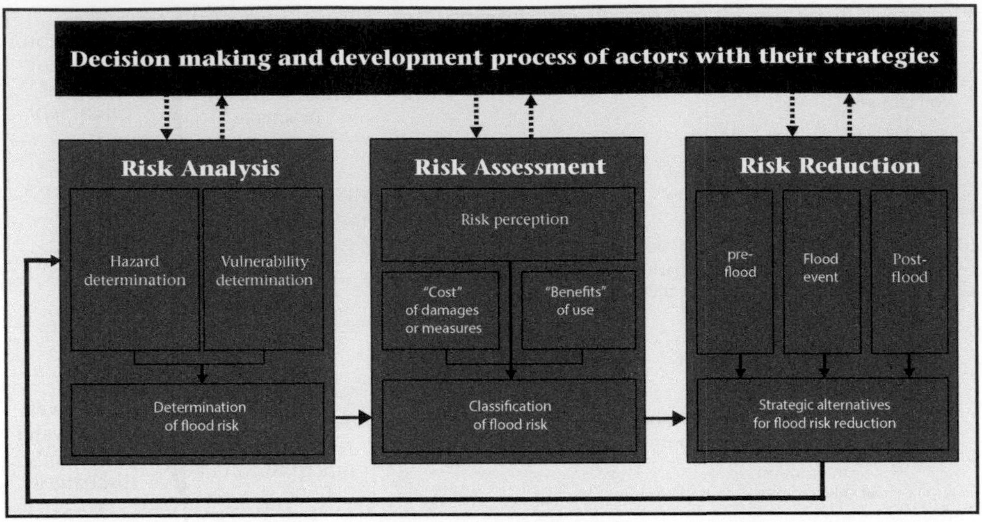


**Figure 5.8.3** A framework for flood risk management.

Reproduced from Schanze (2006).

and make predictions on future risks (see Di Baldassarre et al., 2015). The challenges to this are many and include finding ways to combine natural and social science data sets, overcoming or capturing the current day uncertainty in the flood risk system (for example, how, where and when water moves through the hydrological system; how, where and when people make decisions on flood risk management), and dealing with unknown and uncertain future changes in the river system (such as climate change, land use change or population change).

The *flood risk analyses* provide (scientifically) derived information about the probability of a flood event and the related damages. It is important to remember that flood risk can never be completely eliminated. During *flood risk assessment* society makes a decision as to the level of flood risk that they are willing to accept. This decision will not only be shaped by the flood risk analyses provided by scientists, but also by the risk perception perceived by the communities or populations and their decision makers combined with their financial and institutional capacity to implement desirable flood risk reduction measures. For example, the Netherlands has a very low acceptance of flood risk and choose (and have the capacity) to spend on expensive flood management strategies that aim to protect against large but highly unlikely flood events. It is also important to note that the collective risk perception, and therefore risk assessment, is highly fluid, and can be changed or influenced by the experiences of the population. As such, experiencing a flood event may lead to a decrease in risk acceptance even though the actual risk of flooding (as determined by the flood risk analysis) remains the same.

There are a great range of measures that can be implemented for *flood risk reduction* (see Figures 5.8.2 and 5.8.4). Preventative measures include dikes, dams and retention basins in the river basin and land use zoning and regulations to prohibit, for example, high flood risk areas being used for residential properties. Precautionary measures encompass building codes to ensure properties are flood proofed, insurance requirements that ensure land owners recognise flood risks and contribute to their

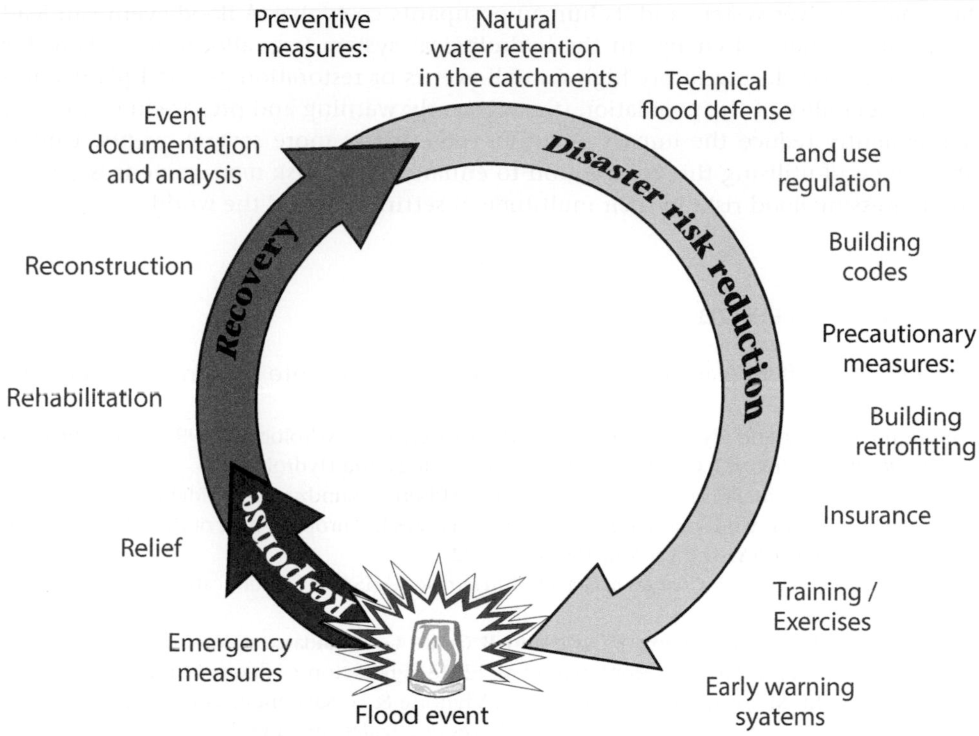


**Figure 5.8.4** Flood risk management cycle.

Reproduced from DKKV (2003).

costs, risk communication and awareness raising, preparation and training, and early warning systems.

Flood risk management therefore takes places in different forms at different times in the management process and can be termed the Flood Risk Management Cycle (Figure 5.8.4) (Thieken et al., 2016). Pre-flood management involves understanding and predicting where and when a flood might occur using mapping, modelling and scenario development. At this stage, structural and non-structural flood protection measures are explored and implemented. Unknown future changes, for example climate changes that may lead to heavier and more frequent rainfall events, or social change such as mass displacement and relocation of people due to war, need to be considered. Ways to do this include identifying robust and "no-regret" strategies, those that perform well under a variety of possible scenarios and give benefits regardless of future changes and that do not limit future options (Blöschl et al, 2013). For example, early warning systems and risk awareness raising. Employing a diverse range of approaches also helps to ensure that in the worst case scenario possible, if one flood management approach fails (such as dikes being overtopped) others are present to reduce the impact of the event (such as early warning systems and evacuation plans). Post-flood management involves recovery and taking measures to prevent further negative impacts such as outbreak of disease. Post-flood management should be followed by pre-flood management in order to enhance preparations for future floods. Embedded within this process of continual learning and refinement of flood risk management is the fact that the river system and its human occupants co-evolve. A flood event can lead to a human induced change in the hydrological system (e.g. allocation of land for water retention during future high rainfall events or restoration of flood plains) and the vulnerability of the population (through early warning and preparedness), which subsequently reduce the impact of future (potentially more severe) events. Understanding and utilising this coevolution to enhance flood risk management is critical for addressing flood risks in their multitude of settings around the world.

## Learning resources

The following offer detailed, practitioner focussed insights into flood risk estimation:

Centre for Ecology and Hydrology (formerly the Institute of Hydrology) (1999) *Flood Estimation Handbook*. Wallingford, Oxfordshire: Centre for Ecology and Hydrology.

Hall, J., B. Arheimer, M. Borga, P. Claps, A. Kiss, T.R. Kjeldsen, Z. Kundzewicz, N. McIntyre, B. Merz, R. Merz, et al. (2014) Understanding flood regime changes in Europe: a state of the art assessment, *Hydrology and Earth System Sciences*, 18: 2735–2772.

Kundzewicz, Z (ed.) (2012) *Changes in Flood Risk in Europe*. IAHS Special Publication 10. Boca Raton, FL: CRC Press.

Rosbjerg, D., G. Blöschl, D.H. Burn, A. Castellarin, B. Croke, G. Di Baldassarre, V. Iacobellis, T.R. Kjeldsen, G. Kuczera, R. Merz, A. Montanari et al. (2013) Prediction of floods in ungauged basins, in: G. Blöschl, M. Sivapalan, T. Wagener, A. Viglione & H. Savenije (eds.) *Runoff Prediction in Ungauged Basins – Synthesis across Processes, Places and Scales* (pp. 135–162). Cambridge: Cambridge University Press.

World Resources Institute (2015) *Aqueduct: Global Flood Risk Analyzer*. Online. Available at: http://www.wri.org/resources/maps/aqueduct-global-flood-analyzer.

## Bibliography

Blöschl, G., Ardoin-Bardin, S., Bonell, M., Dorninger, M., Goodrich, D., Gutknecht, D., Matamoros, D., Merz, B., Shand, P. & Szolgay, J. (2007) 'At what scales do climate variability and land cover change impact on flooding and low flows?', *Hydrological Processes*, 21: 1241–1247.

Blöschl, G., Viglione, A. & Montanari, A. (2013) 'Emerging approaches to hydrological risk management in a changing world', in: Pielke, R. (ed), *Climate Vulnerability: Understanding and Addressing Threats to Essential Resources* (pp. 3–10). Cambridge, MA: Elsevier Inc., Academic Press.

Deutsches Komitee Katastrophenvorsorge (DKKV) (eds) (2003) *Flood Risk Reduction in Germany. Lessons Learned from the 2002 Disaster in the Elbe Region*. Bonn: DKKV 29.

Di Baldassarre, G., Viglione, A., Carr, G., Kuil, L., Yan, K., Brandimarte, L. & Blöschl, G. (2015) 'Changes in flood risk: modelling feedbacks between physical and social processes', *Water Resources Research*, 51(6): 4770–4781.

International Panel on Climate Change (IPCC) (2014) *Impacts, Adaptation, and Vulnerability. Fifth Assessment Report of the Intergovernmental Panel on Climate Change*. Cambridge: Cambridge University Press.

Merz, B., Hall, J., Disse, M. & Schumann, A. (2010) 'Fluvial flood risk management in a changing world', *Natural Hazards and Earth Systems Science*, 10: 509–527.

Schanze, J. (2006) 'Flood risk management: hazards, vulnerability and mitigation measures', 1–20, in Schanze, J., Zeman, E. & Marsalek, J. (eds) Volume 67 of the NATO Science Series. New York: Springer.

Thieken, A.H., Kienzler, S., Kreibich, H., Kuhlicke, C., Kunz, M., Mühr, B., Müller, M., Otto, A., Petrow, T., Pisi, S. & Schröter, K. (2016) 'Review of the flood risk management system in Germany after the major flood in 2013', *Ecology and Society*, 21(2): 51.

Viglione, A. & Rogger, M. (2015) 'Flood processes and hazards', in Paron, P. & Di Baldassarre, G. (eds) *Hydro-meteorological Hazards, Risks and Disasters*. Netherlands: Elsevier.

Viglione, A., Di Baldassarre, G., Brandimarte, L., Kuil, L., Carr, G., Salinas, J.L., Scolobig, A. & Blöschl, G. (2014) 'Insights from socio-hydrology modelling on dealing with flood risk – roles of collective memory, risk-taking attitude and trust', *Journal of Hydrology*, 518: 71–82.

# 5.9 Glaciers

Jorge Daniel Taillant[1]

Only 2 per cent of the world's water is freshwater and, remarkably, three fourths of this water is in glacier ice in the large ice fields of Greenland and the Antarctic but also in smaller mountain glaciers widely distributed around the world.

While glaciers exist at sea level near the poles (thanks to prevailing cold temperature), as we move towards the Equator, glaciers can only survive at much higher elevations (where again, temperatures are very cold). If the elevation is high enough (and the temperature low enough) glaciers can be found in tropical or in Mediterranean areas of the planet, including the Rocky Mountains, the Swiss Alps or in the South American Andes, where glaciers exist somewhere between 2,000–5,000 meters. Near the Equator, glaciers can only survive at extremely high altitudes usually only above 4,000 meters. There are glaciers in countries most might not associate with glaciers, such as Mexico, Colombia, Spain, China, Japan, Kyrgyzstan, Kenya, and even in Indonesia.

## Glacier typology

There are various "types" of glaciers, distinguished by size, location and physical appearance. Usually glaciers have to be at least $1km^2$ to technically be considered a glacier, although by other definitions, including some legal definitions, glaciers can be of *any* size and shape. Some of the more common types of glaciers include: Ice Fields, Valley Glaciers, Mountain Cirque Glaciers, as well as some smaller glaciers the size of a football field or even smaller, called glacierets.

**Figure 5.9.1a** The Aletsch Glacier, Switzerland is a valley glacier.

*Source*: NASA.

**Figure 5.9.1b** Cirque glaciers in Alaska.

*Source*: USGS.

**Figure 5.9.1c** The Pircas Negra Mountain Glacier, San Juan, Argentina.

*Source*: Barberis.

**Figure 5.9.1d** The Harding Ice Field, Alaska.

*Source*: National Park Service, USA.

**Figure 5.9.1e** The Snezhnika glacieret in the Pirin Mountains of Bulgaria.

*Source*: Todor Bozhinov.

## Definition and function

A key element of our cryosphere (our frozen world), a glacier is defined as a body of perennial ice formed by the accumulation of snow that turns to ice over time. As snow accumulates it condenses and, over the course of a few weeks, the soft powdery snow turns into ice, and as the weeks turn into months and progress into years, harder and denser ice will form—this snow to ice process is called the *cryogenic process*. If, through the years, more new snow falls onto the ice during cold months, than melts away during warm months, a glacier will be formed. Generally, a glacier is on an incline and moving slowly downhill but in some cases, it may be trapped by the surrounding geology, for example, in the crater of a dormant volcano. The higher portion of the glacier is receiving new snow (the *accumulation zone*) and converting it into new ice, while the lower portion is melting away older ice (the *ablation zone*). During the summer months, when the seasonal surface snow has melted away, and glacier surfaces are visible, glaciers will appear whiter towards their fresher newer tops and darker towards their older melting bottoms. An *equilibrium line* will be visible where the newer portion of the glacier meets the older melting portion.

Characteristically, glaciers are found in ecosystems favorable to the formation of perennial ice, which is ice that has survived one or more summers. These can be considered *glaciosystems*. The *glaciosystem* is the glacier and its surrounding ecosystem that influences its constitution and composition, with respect to its water and ice accumulation

**Figure 5.9.2** Equilibrium line of a glacier divides the snow accumulation zone from melting zone.

*Source*: Google Earth.

and ablation, determining its biological process, its natural evolution during its periods of charge and discharge and which, if affected, could impact or cause the alteration of the glacier and/or impact the ecosystem in which it exists.

In the Southern Hemisphere, shady South Pole-facing mountain slopes in freezing environments are especially conducive to glacier formation. In the Northern Hemisphere this glacial characteristic is reversed, with glaciers tending to face towards the North Pole.

Most glaciers move due to their own critical weight and size. This is because ice begins to deform and collapse upon itself after it reaches critical mass and weight, combined with factors such as Earth's surface inclination and base humidity as well as temperature changes occurring in the local atmosphere.

## Glacier eras – the ice ages

Glaciers come and go in cycles of tens of thousands of years occurring approximately every 50,000 to 100,000 years, with warmer climates occurring in between. We are currently in an inter-glacial era and in a period of *de-glaciation*. That is, most of our glaciers are currently melting away after having occupied extensive areas of land. Because glaciers melt mostly into the sea, *de-glaciation* causes sea levels to rise significantly, even by hundreds of meters from glaciated to de-glaciated eras.

Occasionally, brief eras of several decades or centuries can bring about mini glacial eras, which can replenish glacier volume for already existing glaciers, or which can produce temporary glaciers that can feed ecosystems for several centuries. One such "mini" glaciation period was the recent *Little Ice Age* that occurred in several regions of the world during various periods ranging from 1300AD to 1900AD. During the Little Ice Age many smaller glaciers were formed in mountain ranges around the world, including: the Sierra Nevada in California, the Central Andes, the Rocky Mountains, the Swiss Alps, and ranges in Mauritania, Ethiopia, Kenya and other African countries.

Much like we count rings on a tree to determine its age, we can calculate the age of a glacier by observing atmospheric contamination (natural and anthropogenic) deposited on the surface of a glacier. During the winter glaciers accumulate clean bright white

**Figure 5.9.3** Contamination particles deposited in yearly cycles on the glacier permit calculating a glacier's age and other contaminating moments.

*Source*: Geostudios.

snow that gets compacted into fresh new ice. During the summer, airborne particles such as black carbon soil the glacier's surface leaving thin lines of contamination. By counting these lines one can determine the glacier's age. Glaciers can also tell us a lot about past climate. By taking out ice cores from millenary glaciers, we can isolate air particles and other pollutants such as black carbon, in the core relating them to the age of the glacier, and study the Earth's climate centuries, even millennia ago. We can also identify precisely when there might have been a significant volcanic eruption nearby or in another part of the world or when other sources of contamination might have affected the atmosphere.

## The role of glaciers in our ecosystems

Glaciers are important to our ecosystems for many reasons. They store most of our planet's freshwater, feeding ecosystems in a regulated manner (as opposed to quick melting of snow occurring immediately after the winter). The vast white surfaces of glaciers reflect sunlight back into space, helping cool the Earth's climate. When glaciers melt, they expose darker terrain that absorbs heat, while also exposing previously decayed organic material trapped under the glacier, potentially releasing vast amounts of methane into the atmosphere. Glacier melt is thus a key source of climate change.

Mountain glaciers, which only make up about 1 per cent of the glacier ice on Earth actually account for much of the freshwater consumed by our human environment around the world. These glaciers, commonly found in high-mountain environments located upriver from human settlements, are especially important to our ecosystems and to human and animal populations. In places like Western Europe, or the Central Andes, and even California, glaciers provide a little understood but nonetheless significant portion of water supply to local populations and agriculture especially during dry seasons or during extended drought periods.

## Vulnerability

Glaciers are especially vulnerable to the Earth's current warming climate, which is accelerating their melt at an alarmingly rapid rate. Human-induced climate change is rapidly accelerating natural de-glaciation. A heating atmosphere, as well as black carbon emissions from industry and transport or for example, from wood burning stoves or fireplaces, all contribute to soil and darken glaciers, increasing their heat absorption (because of their darker color) and causing glaciers to melt around the world. The recent dismemberment of the Thwaites Glacier (an ice sheet larger than much of Europe) is now floating off of Antarctica. When this single piece of floating ice finally melts off into the ocean over the next few decades, it will raise global sea level by four feet.

In select parts of the world, including areas in Chile, Argentina, Kyrgyzstan and others, mining operations have ventured into glacier terrain. These sometimes remove parts of glaciers to get at mineral deposits, or soil them due to massive earth removal to make room for roads or project infrastructure, explosions and stripping away of mountain surfaces, and other activity related to standard mining operations. Public works such as roads have also been known to affect glaciers, either by cutting through delicate

**Figure 5.9.4** Glacier at a mining project site completely covered due to debris from mining operations.

*Source*: Anonymous.

glaciosystems or by actually altering periglacial areas that also affect a glacier system's sustainability.

Melting glaciers can also be deadly, particularly in high mountain environments. Cirque glaciers especially, as they melt, commonly retract, leaving large depressions in the ground surrounded by glacier induced rock walls called *moraines*. Moraines act as natural dams accumulating water from glacier melt above which can in turn form large and very deep glacier lakes. These lakes are usually hundreds of feet deep and are oftentimes perched more than 3,000 or 4,000 meters above human populations residing below. Colossal portions of the glacier remain either partially submersed into the glacier's lake, or hanging high above the surface of the lake. As climate change continues to melt and destabilize the remaining glacier, large portions of ice as big as city blocks can come crashing down into the glacier lake, causing glacier waves that induce glacier tsunamis called *glacier lake outburst floods*, which can come racing down mountain gorges and valleys picking up rocks, trees, mud and other debris which can cause deadly tragedies, as occurred in the valleys of the Cordillera Blanca Mountains of the Andes in Peru several decades ago.

## Laws and regulations

Despite the importance of glaciers to our freshwater supply, and despite their growing vulnerability, societies and governments in particular, have largely failed to consider the need to protect this critical natural resource from its increasingly precarious situation. Perhaps this is due to their remoteness, and to the little existent knowledge of their importance and vulnerability. Until 2008, not a single country in the world had a specific law or regulation to protect glaciers.

A handful of local governments adopted ordinances in Western Europe in the early twentieth century establishing glacier ownership, mostly concerned with the commercial benefits of glacier use (such as for ski resorts—in Europe, a lot of skiing is actually on glaciers). A local government in Austria passed a regional environmental code, protecting glacier scenery. In 2008 Argentina became the first country to adopt a national law to protect glaciers but it was vetoed soon after because it was in conflict with proposed commercial activity.[2] The world's first specific glacier protection law was a provincial law in Santa Cruz province of Argentina.[3] This was followed in 2010 with the world's first federal law adopted in 2010 by Argentina. Argentina's new national glacier law protects both glacier and periglacial areas (areas with permanently frozen grounds).[4] Argentina's glacier law protects all types of glaciers, irrespective of shape or size and prohibits any activity that would destroy glacier resources. Chile and Kyrgyzstan, also concerned with mining impacts to glaciers, attempted unsuccessfully to introduce similar glacier legislation.[5] Like in Argentina, these efforts were thwarted due to conflicts with mining interests. The former passed a glacier protection policy in 2008 and 2009 to establish federal due diligence for glacier protection and is still attempting to get a glacier law passed in Congress.

## Periglacial environments and invisible glaciers

Somewhere below the glaciated environment and above the timberline, large swaths of frozen terrain cyclically freeze and thaw providing further cryospheric water contribution to our ecosystems. The freezing and thawing occurring in these strips of land produce a natural phenomenon called "frost heave", pushing ice and water down into crevasses of the surface of the Earth and between rocks, while pushing rock debris upwards to the surface. Over time, this generates large masses of completely or partially buried *invisible* ice that display similar active properties to uncovered visible glaciers.

In these "periglacial" areas, debris-covered glaciers and rock glaciers as large as visible surface glaciers contain colossal amounts of ice.

## Making and protecting glaciers

The protection and even the fabrication of ice for human and agricultural use is not very well known or documented, but has existed for centuries. Local tribes in countries such as Pakistan learned to conserve ice by taking ice blocks to higher environments and grafting them to cold rocks in caves or mountain surfaces, so that they could survive for longer times in the warmer months.

**Figure 5.9.5** Debris covered glacier in Nepal shows thick mantle of rock and solid ice inner core.

*Source*: Jorge Garcia-Dihinx.

In Chile and India, experiments are underway today to create small glaciers by manipulating snowdrifts or packing snow in artificially created glaciosystems. In Peru, other efforts have utilized sawdust to cover glaciers and provide thermal insulation thereby protecting winter snow and ice from heat.[6] As climate change threatens the survival of some European ski resorts, resort owners are experimenting with large white tarps to protect their commercially valuable ski ice. Others are experimenting with painting mountainsides white to generate more solar reflectivity thereby lowering temperature to help create ecosystems more apt for glacier creation. These experiments are showing that glacier ice *can* be artificially created and protected to provide ecosystems with much needed ice and water during dryer and warmer periods.

Environmental groups concerned with glacier vulnerability have given rise to a new field of environmental activism which some have labeled "cryoactivism", or environmental advocacy to protect the world's cryosphere (the frozen environment).

## Notes

1 The author would like to thank Peter Collins for his assistance in editing this chapter.

2 This was a bill proposed by Congresswoman Martha Maffei, but it was subsequently vetoed by then president Cristina Kirchner due to pressure from the mining sector: http://wp.cedha.net/wp-content/uploads/2013/05/Proyecto-Maffei-Ley-de-Glaciares.pdf.

3 Law 3123, Glacier Protection Law of Santa Cruz: http://wp.cedha.net/wp-content/uploads/2012/02/LEY-N per centC2 per centBA-3123-Glaciares-Santa-Cruz1.pdf.

4 Law 26.639, National Glacier Act (Unofficial Translation): http://wp.cedha.net/wp-content/uploads/2012/10/Argentine-National-Glacier-Act-Traducci per centC3 per centB3n-de-CEDHA-no-oficial.pdf.

5 Ortúzar, Florencia. 03/2015. 'Toward a law to protect glaciers and water in Chile' http://www.aida-americas.org/blog/toward-law-protect-glaciers-and-water-chile.

6 Grossman, Daniel. 09/2012. 'With sawdust and paint, locals fight to save Peru's glaciers' http://www.pri.org/stories/2012–09–25/sawdust-and-paint-locals-fight-save-perus-glaciers

# Land degradation and restoration

5.10

Ilan Chabay

Humans are creatures of the land and live largely from the benefits that land provides. Land is at the nexus of the necessities of life – food, shelter, energy – for all humans and the source of many valuable intangible and tangible assets, including recreation, minerals and gems, construction materials, and carbon storage. Humanity is dependent on land and soil in so many ways. Not only is it literally the foundation on which humanity builds its structures and from which it extracts resources, but it provides for the needs of humanity in the form of soil in which most of our food is produced, water is filtered and purified, and on which trees and plants provide essential forms of shelter, energy, and other ecosystem services.

However, despite humanity's fundamental dependence on them, vast areas of the Earth land and soil for productive use has been lost or its capacity to provide essential ecosystem services severely degraded. Especially in the face of a growing population with its burgeoning demands, this is a tremendous problem across the globe with diverse scales, contexts, causes, and consequences. Of the approximately 36 per cent of land across the planet that is either arable or permanent cropland, about one-third of that is now moderately to severely degraded. The area of land on which food can be produced is being increasingly limited not only by degradation, but also by the multiple demands for space on which to build the infrastructure (e.g., housing, businesses, recreational facilities, transportation corridors) to house and support a growing population. Land degradation in some areas has cascading impacts as people move to urban metropolises when they are no longer able to survive on their decreasingly productive land or when the land has been taken over by agribusinesses or resource extraction industries. The urban and peri-urban spaces then spread further, covering land and removing it from food production or other essential eco-services, including those that provide for the expanding urban and peri-urban areas themselves. Productive land and soil is being lost rapidly just as a growing population requires more food production and eco-system services leading to escalation of the conflicts and migrations triggered or exacerbated by land degradation and incompatible uses.

## Causes and consequences of land degradation

What are the causes and consequences of land degradation? Who benefits from the process of degradation and who is adversely affected? While some may derive some benefit from land-degrading practices either through intensive commercial agriculture that depends on heavy use of fertilizers and irrigation or out of the necessity for poor rural populations to meet their daily needs by subsistence farming, the overwhelming effect is negative. On a global as well as local and regional scale, we are all affected in multiple ways.

Degraded land decreases the capacity of the soil to capture carbon from the atmosphere. Soil is a massive global carbon sink and healthy soil plays a major role in mitigating and slowing the rate of rise of $CO_2$ in the atmosphere, thereby slowing the upward trend in the trajectory of global warming and climate instability. Loss of soil quality and thereby its ability to store $CO_2$ leads to a damaging positive feedback loop. Land degradation and the decrease in stores of carbon in soil and vegetation are significantly, although indirectly, affected in terms of the increasing frequency of extreme weather events (floods, landslides, drought, wildfires) exacerbated by climate change. Not only is soil eroded by intense rainfall or hail, but these events, as well as prolonged drought, result in soil becoming more compacted and less productive for crop growth. Irrigation, though necessary in arid conditions, also can contribute to soil degradation by conveying mineral salts to the root and surface levels, thereby making the soil inhospitable to plants that are not salt tolerant.

Available land for productive use must also be reclaimed in some cases from toxic spills, mining and timber industry wastes, and military munitions. Sea level rise due to climate change is also diminishing land area in some regions. All told, an area roughly equivalent to three times the size of Switzerland of fertile soil is lost every year due to the combination of these negative effects.

But the challenges of land degradation and restoration are not only matters of nature and technology. Profound issues of culture, gender, economics, and politics are operative in many societies at multiple levels in ways that affect the condition and treatment of the land and its resources. The impact of land degradation affects all of society, but falls disproportionately on the poorest agroforestry populations who subsist on the land. When they can no longer survive from their crop yields due to decreased soil productivity or when the forests have been denuded, accessible firewood for fuel diminished, and erosion of fertile soil accelerated, pressure increases on them to abandon the land and migrate to cities or other lands that may be more productive. In some cases, migration may result in conflict over land tenure and mode of use, as has occurred when pastoralists and agriculturalists occupy the same landscapes. Thus, contextually appropriate forms of governance of land and its uses are essential, as are ways to manage conflicts in accord with the competing needs of the stakeholders. Conflicts over land and its ecosystem services can act as security threat amplifiers and may spill over into active warfare or insurgency, if the root causes in the land are not addressed.

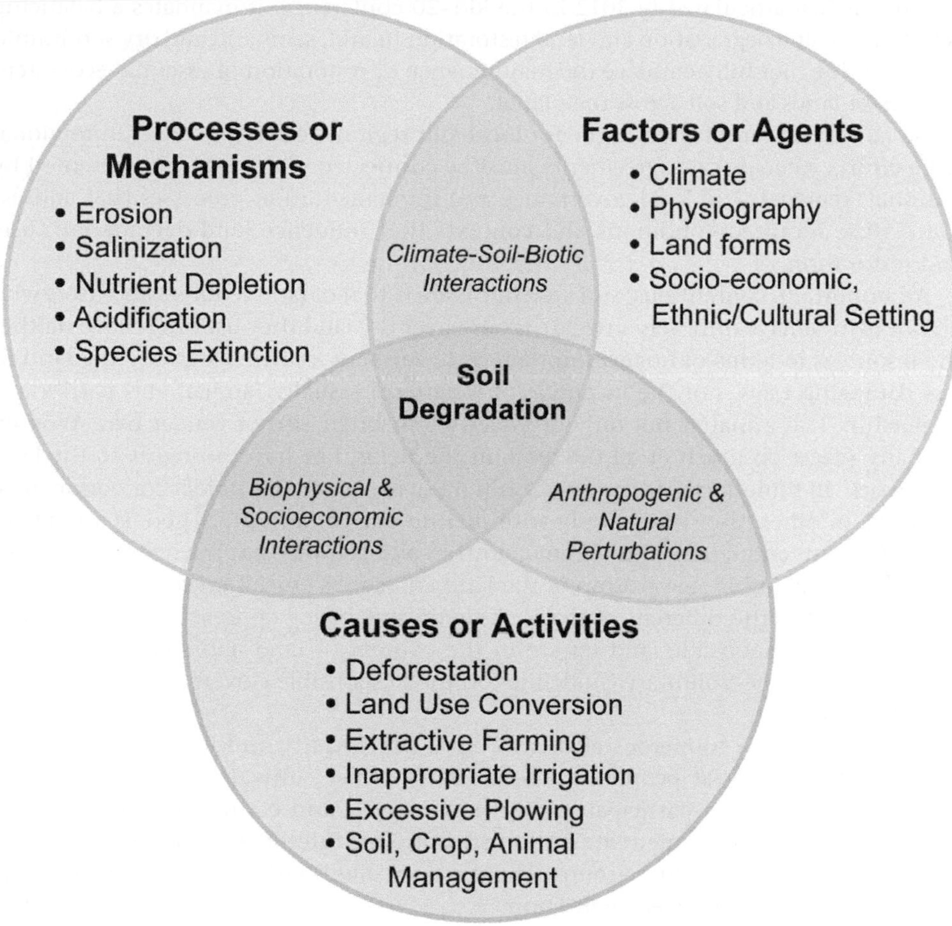


**Figure 5.10.1** The process-factor-cause nexus as a driver of soil degradation.

From Lal, R. (2015), Figure 3. Used under Creative Commons license.

## Governance of land and land restoration

Halting or reversing degradation and dealing with the multiple consequences require forms of governance of the land and land use practices at different spatial scales and governance levels.

Land governance on a global scale, though difficult to achieve, has been developing over the past decades. One of the first initiatives to address land degradation in the context of drylands was the United Nations Conference on Desertification (UNCOD) in 1977, which was limited mainly to the Sahel region. Degradation other than from drought was not addressed. A later agreement in 1994, the United Nations Convention to Combat Desertification (UNCCD), acknowledged the widespread seriousness of land degradation of social factors and ecosystems in dry land regions, which are a major concern, but the global issues are far broader and more comprehensive than are seen in dry lands alone. A much broader, significant international step toward slowing, if not reversing the loss of land and soil is the notion of "land degradation neutral world," which was first articulated in 2012 in the Rio+20 conference. It mandates a balancing act between land degradation and land restoration in arid, semi-arid and dry sub-humid land, but does not fully embrace the maintenance or restoration of essential ecosystem services of lands and soil across the planet.

Social and cultural factors argue for local and regional efforts in addition to global scale efforts. Global scale agreements must be connected with and complemented by national, regional, and local governance and implementation processes that address the diverse localized conditions and contexts that influence land degradation, use, and restoration.

An important, contentious, and systemic issue is that of land tenure. Land users who do not own or have only very insecure tenure over the land they use often have neither the resources in terms of finances nor access to relevant knowledge to use the land in less damaging ways, nor the incentive of long-term (usually familial and patrilineal) ownership. It is a matter not only of poverty, but often also of gender bias. Women in many places do much or all the work in the fields, but have no rights to the land they work. In some societies the ownership lies with large land-holders or commercial interests; in other cases it rests only with men in the community. There is a range of property rights and tenure in different countries and cultures, but the need to establish a basis for responsible stewardship of the land and soil is crucial however that may be accomplished in the different contexts. A significant policy approach to rights-based, equitable and responsible land tenure in the context of land and soil restoration is encapsulated in the Voluntary Guidelines on the Responsible Governance of Tenure of Land (VGGT).

There have been numerous efforts to build both equity and long-term responsibility for the land that people in a community either cultivate or use for pasture. Processes that involve participatory decision making and conflict management for sustainable land management are noteworthy and are intended to address power inequalities among stakeholder groups. These are usually highly charged processes and demand skilled, experienced facilitation.

Initiatives for Land, Lives and Peace (ILLP) hosts the annual Caux Dialogues on Land and Security in Switzerland in which the participants engage not only in issues of land and soil degradation directly, but also building the trust and dialogue to address the conflicts that are triggered or exacerbated by deteriorated conditions of the land.

## Restoration strategies and methods

The rate and extent of land degradation make it evident that land restoration is urgently needed through specific interventions. Simply allowing for natural restoration requires a period of 25 to 50 years, which highlights the urgent need for active human interventions that can increase the health of the soil and productive uses of lands to meet the present and near future needs of society.

A wide range of methods is being employed for restoring damaged or depleted soil. These affect not only the soil itself, but it also may affect the rate of recharging aquifers and the quality of the water in them. A valuable resource called The World Overview of Conservation Approaches and Technologies (WOCAT) contains an extensive library of case studies on and methods for land restoration.

The methods used vary greatly depending on local climate impacts, culture and traditional use patterns, economic circumstances, and available resources. Methods

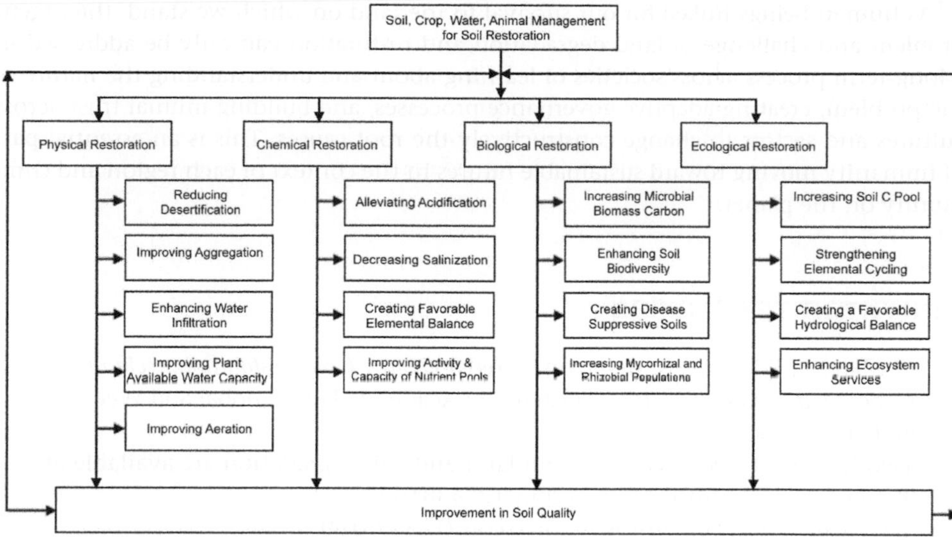


**Figure 5.10.2** Strategies of restoring soil quality.

From Lal, R. (2015), Figure 6 – used under Creative Commons license.

commonly in use include limiting soil erosion by physical structures, such as trees for wind breaks and rock emplacements to channel heavy rainfall runoff into storage areas; improving soil quality through low tillage and crop cover (to limit erosion, drying, and compacting); better matching of irrigation to plant requirements in order to avoid increasing salinity and mineralization of the soil; establishing protected and nutrient restorative areas on a permanent or rotating basis; large scale and very costly removal of heavily polluted toxic soil (e.g., from resource extraction processes or radioactive contamination from accidents or depleted uranium dust from munitions or land mines and other munitions).

## Conclusions: actions needed at multiple levels and scales

The multitude of devastating effects on our planet and our societies from past and continuing processes that degrade the quality of the soil and land are evident. They provide ample and compelling reasons for remediation of the conditions that lead to degradation, as well as steps to restore the land and soil.

Among the initiatives and programs that are responding to the urgent need for land restoration is an important step on a global scale that was taken starting in 2012 with the establishment of the Global Soil Week through the initiative of Professor Klaus Töpfer, then scientific director of the Institute for Advanced Sustainability Studies in Potsdam. It has continued on an annual basis and with some regional offshoots. The Global Soil Week brings together an array of scientific experts, policy makers, practitioners at multiple levels, and NGOs to explore and exchange approaches to scientific and technical issues, information, and governance.

As human beings linked for our survival to the land on which we stand, the urgent problem and challenge of land degradation and restoration can only be addressed in a long-term process across societies of learning about and understanding the nature of the problem, creating adaptive governance processes, and building mutual trust across cultures and sectors to change constructively the root causes. This is an essential part of humanity moving toward sustainable futures in the context of each region and community on the planet.

## Learning resources

- Chabay, I., Frick, M. & Helgeson, J. (eds) (2016) *Land Restoration: reclaiming landscapes for a sustainable future*. Amsterdam: Academic Press. Global Soil Week: http://globalsoilweek.org/
- Two free short animated videos on land and soil degradation are available at:
  'Better Save Soil': https://vimeo.com/125438160.
  'Let's Talk About Soil': https://vimeo.com/125438160.
- Initiatives for Land, Lives and Peace (ILLP): http://www.iofc.org/initiatives-for-land-lives-and-peace.

## Bibliography

Lal, R. (2015) 'Restoring soil quality to mitigate soil degradation" *Sustainability*, 7(5): 5875–5895.
UNCCD: http://www.unccd.int/en/Pages/default.aspx.
UNCOD: http://www.unccd.int/en/about-the-convention/history/Pages/default.aspx.
Voluntary Guidelines on the Responsible Governance of Tenure of Land www.fao.org/nr/tenure/voluntary-guidelines/en.
WOCAT (The World Overview of Conservation Approaches and Technologies): www.wocat.net

# Mining and the environment

5.11

R. Anthony Hodge

Mining dates from the dawn of human society. Mined metals gave rise to the names of the Bronze and Iron Ages. Early mythology – reflected in JRR Tolkien's *The Lord of the Rings* – describes dwarves that worked deep within the earth gathering metals with which they craft magical swords and chalices.

Through millennia, empires and civilizations have come and gone. As populations grew, lifestyles changed and urbanization set in. With this evolution, material demands of society exploded and more mined materials were sought.

These materials now touch every walk of life – our homes, buildings, farms, computers, the internet, phones, vehicles, roads, medical equipment, energy systems, water and sewage systems, pollution control, sporting equipment, musical instruments, food, planes, ships, industrial machinery, jewellery, batteries, and the list goes on. In short, mined materials are essential to contemporary society and their production is a significant component of economies all over the world (see ICMM, 2014).

But these materials come at a significant cost to both people and the environment, a cost that has grown with human society. Along the way, the early mystique of magical mined materials has given way to one of concern in today's world about mining's social and environmental footprint. This evolving relationship between mining, people and the enveloping ecosystem, is the focus of this chapter.

# Mining and the mine project life cycle

## *Mining and sustainability*

Mined materials are considered non-renewable resources – they are fixed in absolute quantity. This characteristic was a central focus of debate in the 1960s through the 1980s. It contributed greatly to the push for efficiency through the three 'Rs': reduce, reuse, and recycle. The importance of seeking efficiencies is no less today than it was then. However, today in mining, the focus is not so much on the limits of mined materials but on the implications of mining activities to host communities and ecosystems.

Mining itself is generally associated with large open pits and underground workings and associated infrastructure that produce a range of metals (such as gold, copper, iron, nickel and zinc), bauxite for aluminium, coal for energy and steel-making, and fertilizers such as potash amongst many others. However, in addition, mining encompasses artisanal and small-scale mining (ASM) and the many quarries and pits where sand, gravel and stone are taken for construction purposes.

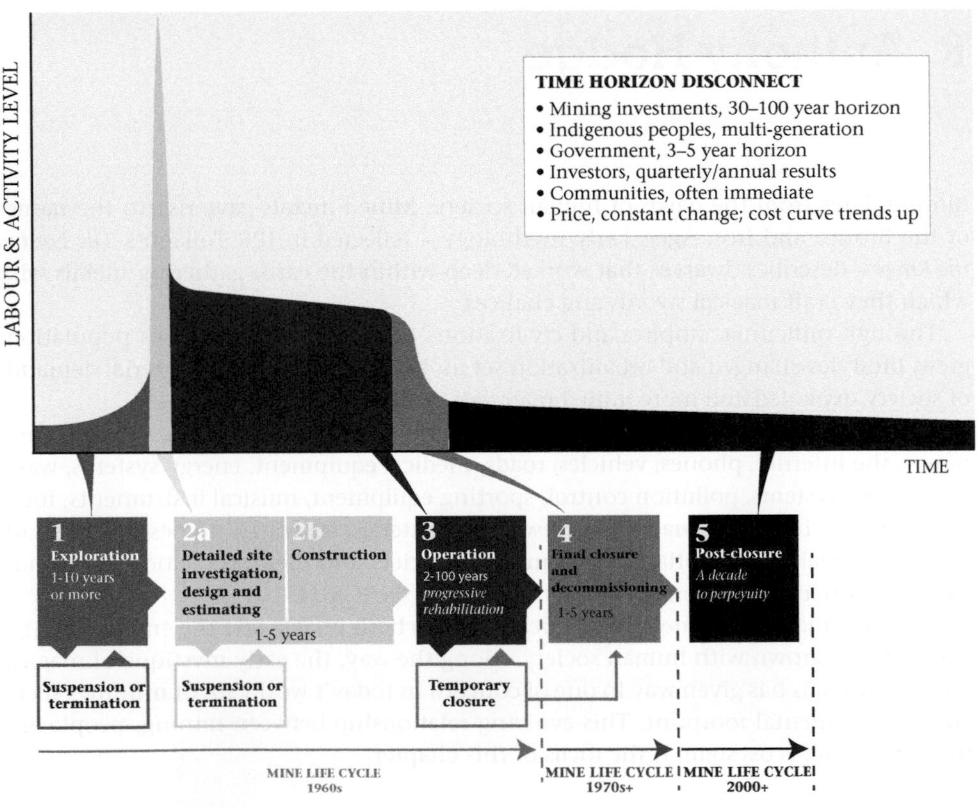


**Figure 5.11.1** The Mine Project Life Cycle with indicative levels of labour and activity and time horizon of each phase.

*Source*: modified after Hodge (2011) and ICMM (2014).

Any mining operation has an end and in that sense, is not 'sustainable'. However, the implications to people and the ecosystem of mining through the activity and the products produced go on indefinitely. To be consistent with the application of sustainability concepts, mining must be implemented in a way that ensures that these implications are net positive over the long term for host citizens, communities and countries, and the enveloping ecosystem (Hodge, 2011).

## *Mine project life cycle and time horizon*

Figure 5.11.1 shows the five phases of every mine's life cycle with indicative levels of labour and activity. Even a small, short-term mining project involves at least several decades. Long-term mines span centuries. Sweden's Falun mine operated for a thousand years from the tenth century until 1992. Sometimes, the post-closure phase must include provision for water treatment in perpetuity. Such long-term thinking brings out disconnects in the perspectives of different interests (Figure 5.11.1).

## *Implications of location*

A mine's location is pre-determined by where the ore body is. Often that location is remote, in areas that are socially and economically depressed, and close to locations that are special to Indigenous people and their rich cultures. Critically, this means that mine project implementation is not simply a 'technical' challenge addressing rocks, minerals, water and investment planning. Rather, it is as much a social and cultural challenge as it is technical. This is particularly the case for artisanal mining.

## Mining-environment implications by life-cycle phase

### *Exploration*

Only about one out of 1,000 exploration projects ever evolve towards a mine (MMSD, 2002, p. 15); a much smaller proportion actually become a mine. Exploration geologists move quickly, are highly competitive and are therefore secretive by nature and design. In a world moving towards increasing transparency, this can be challenging. Environmental implications of exploration are generally less than in subsequent phases. However, they can be significant. Examples include the biodiversity effects of creating access to areas that are previously untouched, and in generating noise and contamination from drilling activities. In response, the Prospectors and Developers Association of Canada has created an environmental and social performance guide called E3-plus (PDAC, 2016).

### *Investigation, design, estimating, construction*

This phase involves further testing for feasibility through additional drilling, excavation and sometimes tunnelling and development of preliminary underground workings.

When feasibility is established and government approvals are obtained, construction begins. Construction involves a short pulse of high activity (Figure 5.11.1) during which the workforce expands dramatically. Control systems and inspections are often early in their development and untested. It is a learning period and unintended environmental consequences can occur. Examples include mismanaged stream crossings and sediment and contaminant discharges to water and air.

## Operation

The main environmental footprint is created in this phase. Underground workings and/or large surface pits (open pit or open cast mines) are built to remove cover and extract the ores (rock material that contains enough of the sought minerals to make is economically retrievable). The ore is then crushed to enable processing through mechanical-chemical-biological processes to extract the sought minerals. Large quantities of waste products are generated including waste rock from the excavations and 'tailings' – the residue from the milling processes. The resulting physical, chemical and biological changes imposed on the environment are significant.

Waste rock and tailings are typically stored in impoundments and if the stability of these impoundments is compromised for some reason (weak design, inappropriate operation, unexpected natural event) they can fail with catastrophic environmental and human consequences. The design criteria for such impoundments have been progressively strengthened over the years in response to such failures and most jurisdictions (but not all) now insist that they are built to withstand maximum possible rainfall and earthquake events.

In the past, operations often discharged a large volume of processing water into adjacent water bodies. Most modern mineral processing facilities use closed circuit water systems that recycle water to limit contaminant discharges.

The hydrologic setting of mines varies from wet to dry depending on the annual rainfall. Rainfall enters the earth and feeds the groundwater. Groundwater movement is slow and controlled by the local and regional geology and topography. Mining operations can trigger contaminant migration in groundwater and once started, this problem is very difficult to control. Sometimes the host rock – often enhanced by bacterial action – triggers an acidification process that leads to mobilization of contaminants that can be deleterious to people, cattle, and the broader ecosystem. This is the Acid Rock Drainage (ARD) issue.

## Closure and post-closure

It is only since about 2000 that the post-closure phase has received significant attention (Figure 5.11.1). This is because of the growing recognition of historic mining legacies, the massive social and environmental costs and liabilities that can be incurred, and the very long time-horizon through which implications can be felt. The 2015 Annual General Meeting of the Intergovernmental Forum on Mining, Minerals, Metals and Sustainable Development focussed entirely on closure and post-closure issues (see http://igfmining.org/meeting-reports/#2015). Figure 5.11.2 provides a summary of mining–environment interactions.

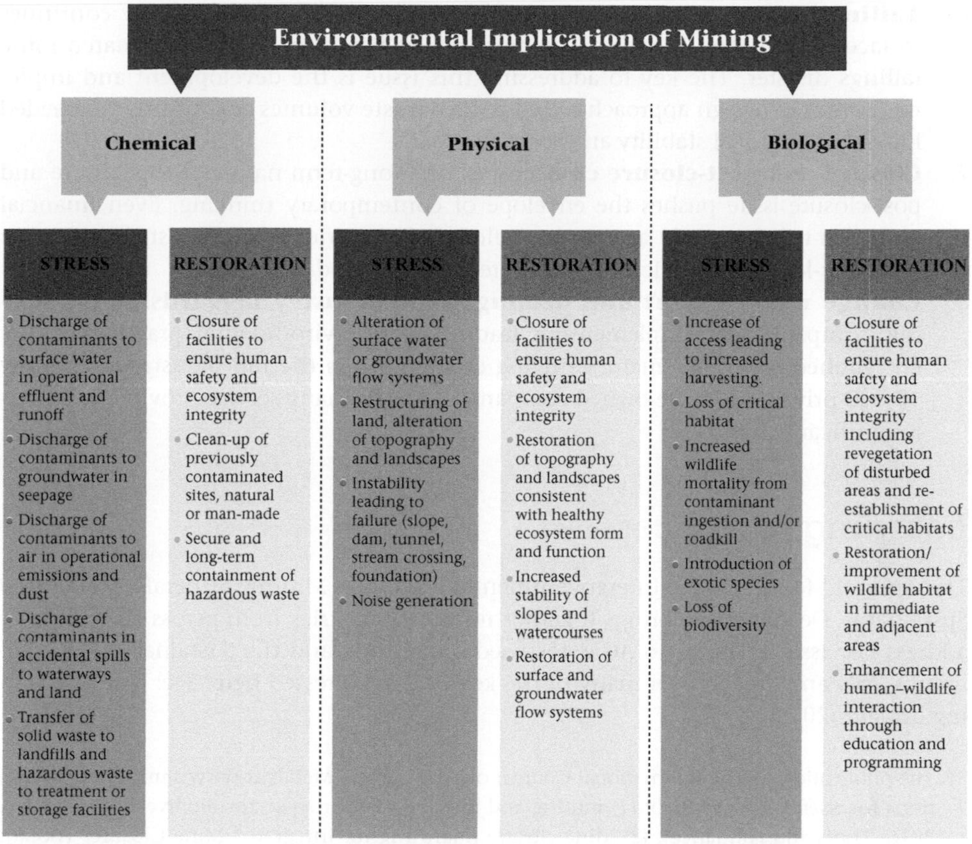


**Figure 5.11.2** Chemical, physical, and biological aspects of mining–environment interactions.

## Eight key mining–environment issues

1. **The distribution issue**. Today, establishing 'fairness' in the distribution of environmental, social, cultural and economic benefits, costs and risks, responsibilities, and accountabilities between mining proponent and host community/country, has become the most important issue facing the industry. This issue plays directly into implementation of the Sustainable Development Goals.
2. **Artisanal and Small-Scale Mining (ASM)**. Tens of millions are involved including many women and children, mostly under dire environmental and social conditions. Mercury releases poison the environment; regulation is extremely difficult.
3. **The legacy of old operations**. Much remains to be addressed related to the environmental legacies of past mining operations.
4. **Biodiversity preservation and enhancement**. The issue of biodiversity preservation is perhaps the greatest environmental issue facing humankind. Mining has a role to play in addressing this challenge – both in limiting its own immediate affects and in contributing to strengthening ecosystem functions.
5. **Water and geochemical management**. Water management – both from a quantity and quality perspective – remains a key issue. See leading edge work at https://www.icmm.com/publications/pdfs/8329.pdf.

6. **Tailings dam and impoundment design criteria**. The industry continues to face tailings dam breaches, the most recent being the Brazilian Samarco mine tailings disaster. The key to addressing this issue is the development and implementation of design approaches that reduce waste volumes and ensure the needed long-term physical stability and security.
7. **Closure and post-closure challenges**. The long-term nature of the closure and post-closure issue pushes the envelope of contemporary thinking. Even financial methods to adequately analyse cash flows of centuries do not exist. Decades- or centuries-long geochemical processes are ill-understood.
8. **Change management and dealing with industry laggards**. While leading companies are implementing leading-edge environmental practices, many entrenched laggards remain. Bringing change across the full industry of roughly 10,000 private and state-owned companies and the many service providers remains a significant challenge.

## Learning resources

The Annual Meeting of the Intergovernmental Forum on Mining, Minerals, Metals and Sustainable Development brings together mining ministries from across the world to address key issues. The 2016 AGM focussed on Mining and the Sustainable Development Goals and ranged over many of the key issues addressed here. See: http://igfmining.org/agm2016.

See the publications of the International Council on Mining and Metals at www.icmm.com/publications for useful treatments of (1) mining and biodiversity (Cross-sector Biodiversity Initiative, 2016. Tools and Guidance); (2) Mine closure (Planning for Integrated Mine Closure: Toolkit, 2008); and (3) Mining and water (A practical guide to a catchment-based water management for the mining and metals industry, 2015).

For an overview of Artisanal and Small-Scale Mining (ASM) and related issues see http://www.worldbank.org/en/topic/extractiveindustries/brief/artisanal-and-small-scale-mining. For an insightful review, see also http://ehp.niehs.nih.gov/120-a424/

For a useful overview of mining and the Sustainable Development Goals, see UNDP, WEF, and CCSI, 2016. Mapping Mining to the Sustainable Development Goals – An Atlas. New York and Geneva: United Nations Development Program, World Economic Forum, and the Columbia Centre for Sustainable Investment. http://unsdsn.org/wp-content/uploads/2016/01/160115-Atlas_full.pdf.

For a useful exploration of mining and indigenous people's issues see http://www.iisd.org/sites/default/files/publications/natres_out_of_respect.pdf

For background on the uses of materials in contemporary society, see: (1) Aluminium: https://en.wikipedia.org/wiki/Aluminium#General_use last visited 11 Sept 2016; (2) Coal: https://www.worldcoal.org/coal/uses-coal last visited 6 Sept 2016; (3) Copper: http://geology.com/usgs/uses-of-copper/ last visited 11 Sept 2016; (4) Diamonds: https://www.reference.com/beauty-fashion/primary-uses-diamonds-7ca4361125e8cb9e last visited 11 Sept 2016. See also, http://www.heartsonfire.com/history-of-diamonds/unusual-diamond-uses.aspx last visited 11 Sept 2016; (5) Iron and steel: http://www.australianminesatlas.gov.au/education/down_under/iron/used.html last visited 11 Sept 2016; (6) Gold: http://geology.com/minerals/gold/uses-of-gold.shtml last visited 11 Sept 2016; (7) Lead: http://geology.com/usgs/lead/ last visited 11 Sept 2016; (8) Nickel: https://www.nickelinstitute.org/NickelUseInSociety/AboutNickel/WhereWhyNickelIsUsed.aspx last visited 11 Sept 2016; (9) Platinum group minerals: http://www.miningweekly.com/article/the-uses-of-platinumgroup-metals-2006–11–10 last visited 11 Sept

2016; (10) Zinc: http://geology.com/usgs/uses-of-zinc/ last visited 11 Sept 2016; (11) Industrial minerals: http://www.ima-na.org and http://www.ima-europe.eu/; and (12) Quarrying: https://en.wikipedia.org/wiki/Quarry

## Bibliography

Hodge, R.A. (2011) Mining and Sustainability, in Peter Darling (ed.) *SME Mining Engineering Handbook*, 3rd edition. Englewood, Colorado: Society for Mining Metallurgy and Exploration Inc.

International Council on Mining and Metals (ICMM) (2014) *The Role of Mining in National Economies*, 2nd edition. London: International Council on Mining and Metals.

Mining Minerals and Sustainable Development (MMSD) (2002) *Breaking New Ground: Mining, Minerals and Sustainable Development*. London: International Institute for Environment and Development. Online. Available from: http://www.iied.org/mmsd-final-report. Last visited 11 September 2016.

Mining Minerals and Sustainable Development (MMSD) North America (2002) *Seven Questions to Sustainability – Assessing How a Mine/Mineral Project or Operation Contributes to Sustainability*. Winnipeg: International Institute for Sustainable Development. Online. Available from: http://www.iisd.org/pdf/2002/mmsd_sevenquestions.pdf last visited 11 Sept 2016.

Prospectors and Developers Association of Canada (PDAC) (2016) *E3 Plus*. Toronto: Prospectors and Developers Association of Canada. Online. Available from http://www.pdac.ca/programs/e3-plus.

# 5.12 Oceans: climate change, marine ecosystems and fisheries

Yoshitaka Ota and William W. L. Cheung

This chapter focuses on the interactions between $CO_2$ emissions, oceans and the world's fisheries. There is a growing concern among marine scientists over the impacts of global environmental changes, specifically climate change, ocean acidification, and the loss of biodiversity, on the state of our future oceans and their capacity to produce seafood. Thus far, scientists evaluate the emergence of $CO_2$ emission-driven patterns in four key properties: warming, acidification, deoxygenation and disturbance to biological productivity (Pörtner et al., 2014).

## Warming

As the global climate becomes warmer due to an increase in the concentration of greenhouse gases (GHGs) in the atmosphere (such as $CO_2$), the ocean's temperature is also expected to rise. The average sea surface temperature is projected to increase by a range of 2.0–3.5°C by the end of the twenty-first century if the current GHG emissions rate continues. Furthermore, because of water's natural capacity to absorb heat, combined with global ocean circulation patterns that transfer warm surface water across the entire ocean basin, the projected warming is likely to continue even if $CO_2$ emissions are brought under control. Changes in sea temperature disrupt movements of water, distributions of marine organisms, and structures of ecosystems.

## Acidification

As the chemistry of the Earth's atmosphere is altered through continued $CO_2$ emissions, so too will there be demonstrable changes in the chemistry of the oceans. In addition to serving as a heat sink, the oceans also act as a carbon sink by removing $CO_2$ from the atmosphere. Through this, the oceans can play a vital role in lessening the magnitude of temperature change in the atmosphere. However, this service comes at a cost as increased $CO_2$ in the water acidifies the ocean (ocean acidification). As the water becomes more acidic, it causes problems for some marine life, particularly those that form calcium-based shells (such as corals, crustaceans and shellfish). Under high $CO_2$ emissions scenarios, the acidity of the world's oceans, as measured by pH, is likely to increase by 0.33 units by the end of the century, a rate that is unprecedented over the past millions of years.

## Deoxygenation

With the melting of ice caps in sub-polar regions and the dilution of the surface water, stratification (formation of water layers with different properties, such as salinity, that act as barriers to water mixing) between warm, light surface water and cooler, denser deep ocean water is expected to intensify. Moreover, this stratification of ocean water is expected to limit the transfer of nutrients critical for marine life from the deep ocean to the surface, as well as the diffusion of oxygen from the surface to the deep ocean.

## Disturbance to biological productivity

The impact of climate change on marine biological productivity will vary geographically, due to the physiological condition, the habitat structure, and the composition of the species that occur in these areas. The projected changes in phytoplankton production will ultimately affect the growth and survival of animals that feed on the phytoplankton, as well as the fisheries that depend on the production of a variety of fishes and invertebrates. On a regional level, the impacts of climate change on marine biological productivity could be more distinct and influenced by local conditions. For example, in some areas, the heating patterns between land mass and sea surface may be more pronounced and result in stronger wind currents, intensified up-welling of deep sea water to the surface, and greater phytoplankton production.

In terms of impact on marine ecosystems, there is a pressing concern in the tropics where coral reefs are susceptible to warming and acidification – the two conditions that are projected to emerge earliest in the area (Frieler et al., 2013; Frölicher et al., 2016). More generally, in regions where these changes in ocean properties emerge most strongly, marine ecosystems may be pushed beyond their natural capability to cope. Given the observed and projected changes in ocean conditions, specifically the water temperature, acidity and oxygen concentration, marine species are becoming increasingly exposed to conditions beyond their tolerance level. On an individual level, some organisms may show decreased growth and body size (Cheung et al., 2013a). On a species level, some species may change in distribution by moving to areas more favourable to their survival (Poloczanska et al., 2013). All of this results in shifts in community composition of marine species and trophic interactions within ecosystems (Ainsworth et al., 2011; Jones and Cheung, 2015). In addition, the losses of ecologically important habitats (such as coral reef systems) are also exacerbating the changes in ecosystem structure and species distribution (Sunday et al., 2016).

In addition to food web interactions between phytoplankton, fish and other marine animals, climate change is expected to affect fisheries industry and food security of coastal communities globally (Lam et al., 2016; Golden et al., 2016). Fish and invertebrates (e.g. crustaceans and shellfish) are at the stake of their survival when they are exposed to ecological conditions outside the tolerance ranges for their body functions, such as growth and reproduction. This biological and ecological response of marine species to climate change and ocean acidification has already been observed in the fisheries through changes in diversity, composition, distribution, and the quality and quantity of the catch (Cheung et al., 2013). This type of change is expected to continue and will have major implications for future fisheries yields and coastal food security (Cheung et al., 2016).

As global fishing capacity – both in terms of the number of vessels and technology available – expands and an increasing number of fish stocks become fully or overexploited, the prospect for further growth in the world's marine fisheries production is diminished (Pauly & Zeller, 2016). Meanwhile, the intensification of other ocean and coastal activities, including offshore mining, coastal development and renewable energy generation, is likely to continue and may result in potential conflicts with marine fisheries (Merrie et al., 2014). These developments may also place coastal ecosystems and habitats under greater strain.

Seafood is now one of the most widely traded food commodities in the world, with the seafood industry increasingly becoming highly integrated and globalized. The consequences of large-scale ecological change due to climate change may, therefore, have significant implications throughout the global fisheries economy (Lam et al., 2014; Lam et al., 2016). As climate change impacts on ocean ecosystems, by reducing their productivity and biodiversity, will intensify the effects of these fisheries-driven pressures, the trade-offs between the economic objectives of commercial fisheries and conservation targets, particularly with regards to protection of vulnerable species, will become severe.

With improvements in the projections of future ocean changes and marine productivity, the fisheries status, seafood availability and security, there is a demand for better understanding of how climate change can influence the future outcomes of local marine systems and related socio-economic impacts (Österblom et al., 2013). The oceans in the past will not be the same as the oceans in the future. They are expected to change at a rate and magnitude that are unprecedented in human history, and empirical relationships between oceans, fish and fisheries observed in the past may differ from future dynamics. We need to improve our ability to anticipate and respond to future ocean changes by exploring the evolving nature in both ecological and socio-economic systems (Pershing et al., 2015).

Finally, as an attempt to apply our knowledge to responsible, ecosystem-based and precautionary management, we have identified six strategies that can be used to address immediate challenges to the sustainability of marine living resources, with a particular emphasis on fisheries. These six strategies are also topics that require further research to fully appreciate their importance and influence in contributing to the sustainability of the future oceans:

1. **Bringing $CO_2$ emissions under control**. Reduce the rate and magnitude of climate change, ocean acidification and other related changes in ocean properties.
2. **Maintaining biodiversity, habitat and ecosystem structure**. Protect the capacity of marine ecosystems to adapt to impacts from all human-induced stressors and enhance ecosystem services to human societies.
3. **Diversifying the "tool-kit" for fisheries management**. Ensure that fisheries management has the capacity to implement a diverse range of strategies to address the increased uncertainties in the marine ecosystems arising from multiple human-induced stressors and the impacts of climate change.
4. **Adopting economic systems that support sustainable practice**. Create a market model that is capable of recognizing and rapidly responding to the effects of all human-induced stressors on the oceans.
5. **Enhancing cooperation and coordination between international fisheries regulation and regulation of other maritime activities**. Address challenges to managing and responding to global environmental changes by increasing the coordination between existing global and regional regulatory frameworks.
6. **Ensuring equitable distribution and access for fishing in vulnerable communities**. Safeguard the rights of coastal communities that are vulnerable to the impacts of ocean changes driven by human-induced stressors. This includes more equitable distribution of access to the world's fish stocks.

## Learning resources

The following is a great introduction to ocean and climate change as well as one of its particular themes, fisheries:

Pörtner, H-O., Karl, D.M., Boyd, P.W., Cheung, W.W.L., Lluch-Cota, S.E., Nojiri, Y., Schmidt, D.N. & Zavialov, P.O. (2014) Ocean systems in *Climate Change 2014: Impacts, Adaptation, and Vulnerability. Part A: Global and Sectoral Aspects. Contribution of Working Group II to the Fifth Assessment Report of the Intergovernmental Panel on Climate Change* (pp. 411–484). Cambridge: Cambridge University Press.

## Bibliography


Ainsworth, C.H., Samhouri, J.F., Busch, D.S., Cheung, W.W.L., Dunne, J. & Okey, T.A. (2011) Potential impacts of climate change on northeast pacific marine foodwebs and fisheries. *ICES Journal of Marine Science*, 68: 1217–1229.

Cheung, W.W.L., Reygondeau, G. & Frölicher, T.L. (2016) Large benefits to marine fisheries of meeting the 1.5°C global warming target. *Science*, 354: 1591–1594.

Cheung, W.W.L., Watson, R., & Pauly, D. (2013) Signature of ocean warming in global fisheries catch, *Nature*, 497(7449): 365–368.

Frieler, K., Meinshausen, M., Golly, A., Mengel, M., Lebek, K., Donner, S.D. & Hoegh-Guldberg, O. (2013) Limiting global warming to 2°C is unlikely to save most coral reefs. *Nature Climate. Change*, 3: 165–170.

Frölicher, T.L., Rodgers, K.B., Stock, C.A. & Cheung, W.W.L. (2016) Sources of uncertainties in 21st Century projections of potential ocean ecosystem stressors. *Global Biogeochemical Cycles*, 30: 1224–1243.

Golden, C.D., Allison, E.H., Cheung, W.W.L., Dey, M.M., Halpern, B.S., McCauley, D.J., Smith, M., Vaitla, B., Zeller, D., Myers, S.S. et al. (2016) Fall in fish catch threatens human health. *Nature*, 534: 317–320.

Jones, M.C. & Cheung, W.W.L. (2015) Multi-model ensemble projections of climate change effects on global marine biodiversity. *ICES Journal of Marine Science*, 72: 741–752.

Lam, V.W.Y., Cheung, W.W.L., & Sumaila, U.R. (2014) Marine capture fisheries in the Arctic: winners or losers under climate change and ocean acidification? *Fish and Fisheries*, 17(2): 1467–2979.

Lam, V.W.Y., Cheung, W.W.L., Reygondeau, G. & Sumaila, U.R. (2016) Projected change in global fisheries revenues under climate change. *Scientific Reports*, 6: 32607.

Merrie, A., Metian, M., Boustany, A.M., Takei, Y., Ota, Y., Christensen, V., Halpin, P.N. & Österblom, H. (2014) Human use trends and potential surprise in the global marine common. *Global Environmental Change*, 27: 19–31.

Österblom, H., Merrie, A., Metian, M., Boonstra, W.J., Blenckner, T., Watson, J.R., Rykaczewski, R.R., Ota, Y., Sarmiento, J.L., Christensen, V. et al. (2013) Modeling social–ecological scenarios in marine systems. *BioScience*, 63(9): 735–744.

Pauly, D. & Zeller, D. (2016) Catch reconstructions reveal that global marine fisheries catches are higher than reported and declining. *Nature Communications*, 7: 10244.

Pershing, A.J., Alexander, M.A., Hernandez, C.M., Kerr, L.A., Le Bris, A., Mills, K.E., Nye, J.A., Record, N.R., Scannell, H.A., Scott, J.D. et al. (2015) Slow adaptation in the face of rapid warming leads to collapse of the Gulf of Maine cod fishery. *Science*, 350: 809–812.

Poloczanska, E.S., Brown, C.J., Sydeman, W.J., Kiessling, W., Schoeman, D.S., Moore, P.J., Brander, K., Bruno, J.F., Buckley, L.B., Burrows, M.T. et al. (2013) Global imprint of climate change on marine life. *Nature Climate Change*, 3: 919–925.

Pörtner, H.O., Karl, D., Boyd, P.W., Cheung, W.W.L., Lluch-Cota, S.E., Nojiri, Y., Schmidt, D.N. & Zavialov, P. (2014) Ocean systems. In Field, C.B., Barros, P., Dokken, D.J., Mach, K.J., Mastrandrea, M.D., Bilir, T.E., Chatterjee, M., Ebi, K.L., Estrada, Y.O., Genova, R.C. et al. (eds). *Climate Change 2014: Impacts, Adaptation, and Vulnerability. Part A: Global and Sectoral Aspects. Contribution of Working Group II to the Fish Assessment Report of the Intergovernmental Panel on Climate Change* (pp. 411–484). Cambridge: Cambridge University Press.

Sunday, J.M., Fabricius, K.E., Kroeker, K.J., Anderson, K.M., Brown, N.E., Barry, J.P., Connell, S.D., Dupont, S., Gaylord, B., Hall-Spencer, J.M. et al. (2016) Ocean acidification can mediate biodiversity shifts by changing biogenic habitat. *Nature Climate Change*, 7: 81–85.

# The commons

5.13

Fikret Berkes

Many centuries ago Aristotle identified the fundamental commons problem: "What is common to the greatest number gets the least amount of care . . . Men pay most attention to what is their own; they care less for what is common." Many of our shared environmental resources such as fisheries, forests, ground water, as well as large parts of our planetary environment, the oceans and the atmosphere, are held in common and are thus called *commons.* Sustainable use and stewardship of the commons is therefore one of the greatest imperatives of the Anthropocene. But our economically "rational" short-term self-interest often trumps long-term societal and global interest. This clash between the individual interest and collective interest is one of the great dilemmas of sustainability. Biologist Garrett Hardin (1968) described this paradox as "the tragedy of the commons" (see the chapter on this tragedy elsewhere in this book).

## Solving the tragedy: multiple paths

Commons or common pool resources share two characteristics. First, the exclusion of beneficiaries is costly. That is, controlling the access of potential users to the resource in question is difficult and/or expensive. How can you prevent someone from fishing at your favorite fishing spot, or the atmospheric pollution from additional cars on the road? Second, exploitation by one user reduces resource availability for others, what the economists call the sub-tractability problem. As one user pumps water from a groundwater aquifer, there is less water left for all other users. This is shown in Figure 5.13.1, where an open pasture becomes over-grazed through individually "rational" behaviour that ends-up being "irrational" for farmers as a whole.

To solve or avoid the "tragedy," commons have been typically managed in one of three ways: as state property, as private property, or through community management.

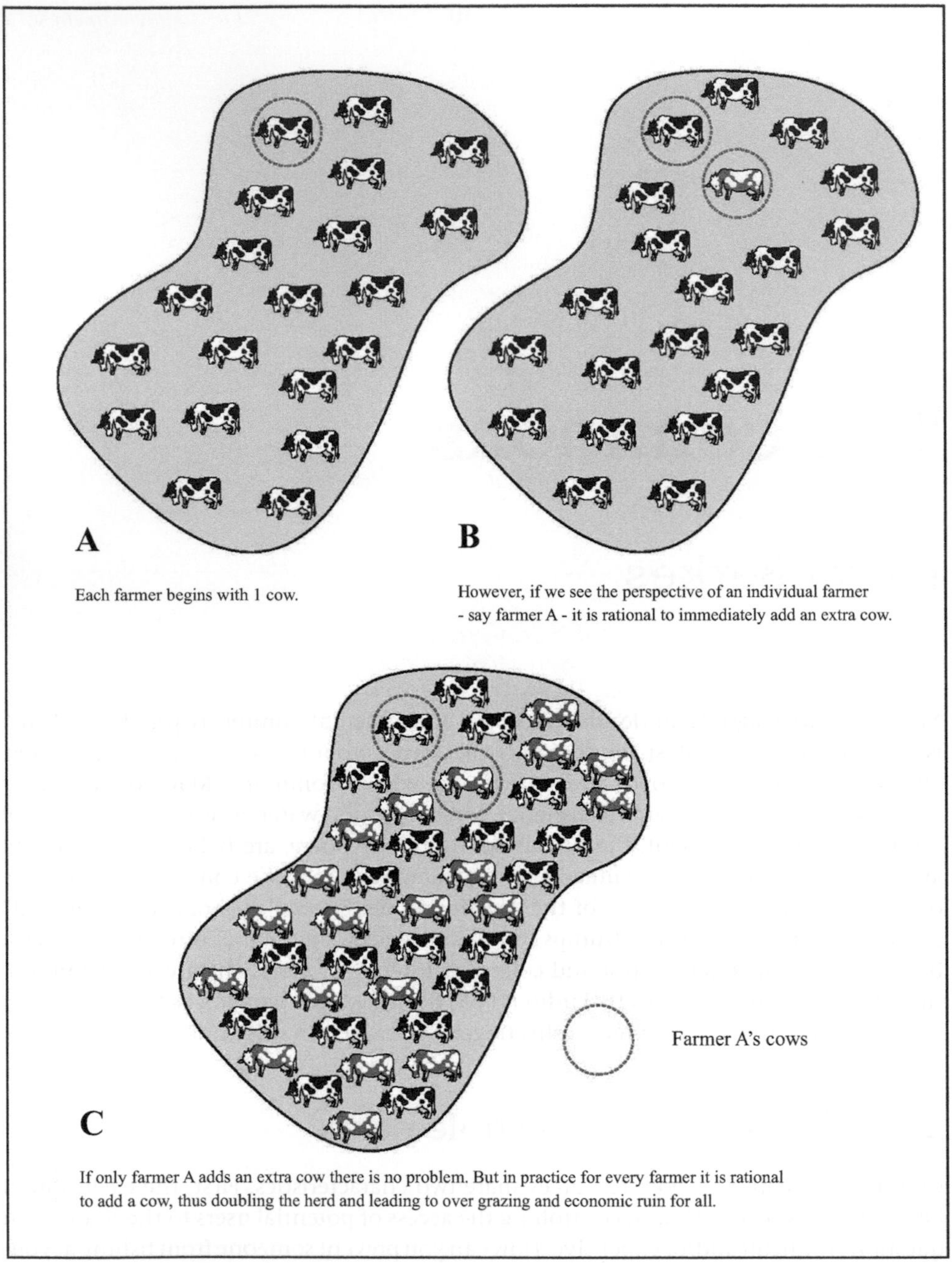

A
Each farmer begins with 1 cow.
B
However, if we see the perspective of an individual farmer
- say farmer A - it is rational to immediately add an extra cow.
Farmer A's cows
C
If only farmer A adds an extra cow there is no problem. But in practice for every farmer it is rational
to add a cow, thus doubling the herd and leading to over grazing and economic ruin for all.


**Figure 5.13.1**



Each has certain strengths and shortcomings. Hence management through a mixture of these three approaches may work better than any one alone. Until recent decades, the state property approach based on top-down, centralized, regulatory control was the norm throughout much of the world. Essentially, this is a *leave it to the government* solution. Obviously, many things can and should be regulated, but there are a number of shortcomings of this approach. For example, it stifles local stewardship and the ability of people to solve their own problems, it is often insensitive to local needs, it may result in narrow interests dominating decision making and it can be very expensive.

If government management has shortcomings, how about leaving the "invisible hand" of the market to work its magic? The logic of this solution is that, if an environmental resource can be privatized, then both the costs and benefits would accrue to the same owner, assuring sustainability. Again, the solution works in some cases but not in others. Some resources such as agricultural land can be privatized. Others, like oceans, cannot because they are simply not ownable. However, economists have devised various ways to get around this. For example, rights to some ocean resources, such as harvest quotas for commercial fish, can be bought and sold. Similarly, markets can be created for rights to pollute the atmosphere. However, there are some shortcomings: market solutions create winners and losers, creating an equity problem. Markets do not necessarily deliver the best solution for the public interest, nor for sustainability. Biodiversity conservation, for example, cannot be achieved through the market because conservation values are largely non-market values.

The community management approach offers a third model. It involves the collective management of commons under locally designed rules. Such community-based solutions have been successful for local commons such as village forests around the world. Community-based solutions have also been successfully scaled-up and applied regionally (for example with irrigation systems), nationally and internationally. The major problem of the community management approach is how to get the users of the commons to engage in collective action for everyone's long-term benefit, rather than for individual short-term self-interest. Can it be done by creating the conditions under which collective action become more (rather than less) feasible?

## Elinor Ostrom and principles for collective action

American political scientist and economist Elinor Ostrom demonstrated that the community management approach was workable under three general conditions: good communication, trust, and reciprocity. She developed principles for collective action, detailed in her 1990 book, *Governing the Commons*, and won the 2009 Nobel Prize in Economic Sciences. These principles have since been tested against the findings of a large number of empirical studies, verified and updated by Cox et al. However, Ostrom's principles do not add up to a solution of the commons problem once and for all. She notes that there is no panacea or simple blueprint solution that can be applied to all cases of commons management. The approach most likely to succeed, she says, is hard work on a case-by-case basis, involving a mix of the three kinds of solutions as appropriate to the case. Such an approach stimulates social learning and knowledge-building, leading to collective action.

Lessons from commons governance indicate that we can have government or market or community failure. Judicious combinations of the three approaches make it more likely to overcome the shortcomings of each, safeguarding sustainability and addressing equity concerns. The combination is more likely to accommodate the public interest and allow the representation of a broader diversity of specific interests. Perhaps most important, such an approach fosters local stewardship, solutions starting closer to home, with a tighter feedback loop between action and outcome.

These approaches to managing the local commons have implications for managing the global commons, even though local solutions cannot simply be scaled up to the global level without major modification. As the climate change and greenhouse emissions example illustrates, global-level problems require some additional considerations.

## A global commons dilemma: climate change

In recent years, much scientific attention and public concern have focused on one overriding environmental problem, climate change. The atmosphere is a global commons that is heading, by most accounts, toward a tragedy. Atmospheric pollution by greenhouse gases is a colossal problem, but it is neither unique nor isolated as commons problems go. Much can be learned from the management of local and regional commons to gain insight into managing global commons.

What can be done to reduce greenhouse gases currently emitted by all nations? At the global scale, the individual players are the nation states, and the main mechanism has been voluntary international agreements, which can be considered a kind of top-down, regulatory arrangement. We do have examples of international agreements that have worked. For example, the 1982 United Nations Convention on the Law of the Sea resulted among others in a drastic reduction of global marine oil pollution from the routine operation of tankers. It did not completely eliminate oil pollution, as there are still accidents that contribute to oil pollution, but it did curtail major sources of pollution.

Another example is ozone depletion. The Montreal Protocol of 1987 banned ozone-depleting chemicals from industrial use. This international ban, plus changing technology for refrigeration, resulted in a long-term solution to atmospheric ozone depletion. Acid rain was a more complex problem, brought under control through a mix of international agreements, emission markets (capping total emissions and trading polluting rights), technology change, improved national standards and enforcement. In each of these cases, good research and pressure from scientists and citizens, provided the push initially for national-level action, followed by international agreements.

In the case of greenhouse gases, the situation is yet more complex, mainly because energy systems throughout the world are tied largely to petroleum hydrocarbons. Past international agreements such as the Kyoto Protocol of 1997 were ineffective to even slow down the increase of atmospheric carbon dioxide. This was largely because many countries did not sign it, or did not pass and enforce national standards. International agreements become binding on sovereign nation states only when they pass their own legislation to make it so. If major players are unwilling to do that, the international agreement remains on paper only. Collective action theory holds that the vast majority of players have to be "on board" for collective action to succeed. In the case of greenhouse gases, the promising development is that major players such as the United States and China have joined Western European countries for policy change towards renewable energy and away from fossil fuels.

In conclusion, commons can be managed in one of three ways: as state property, as private property, or through community management. Each has certain strengths but each can also fail. Hence the pathway most likely to succeed is a combination of the three. As the climate change example shows, this is because emission limits cannot come from the top down, in the absence of policy and planning that makes it feasible to develop and adopt green technologies and reduce greenhouse gases. The impetus for change has to start at the community level, driven by sustainable cities, energy efficient buildings, public transportation networks, and sustainable agriculture and forestry. Starting at the local level, community management, government regulation and market forces together can be scaled up to create sustainable commons at the level of regions, nations and the biosphere.

## Learning resources

*International Journal of the Commons* is an interdisciplinary peer-reviewed open-access journal, dedicated to furthering the understanding of institutions for use and management of commons: https://www.thecommonsjournal.org.

*The Commons Digest* is the publication of the International Association for the Study of the Commons (IASC). It includes articles, news, conferences and publications regarding commons: http://www.iasc-commons.org/commons-digest.

Anderies, J. and M. Janssen (2016) *Sustaining the Commons*, second edition Open-access e-book. https://sustainingthecommons.asu.edu/

Digital Library of the Commons, Indiana University, provides a free and open-access gateway to the international literature on the commons: http://dlc.dlib.indiana.edu/dlc.

Ostrom, E. (1990) *Governing the Commons: The Evolution of Institutions for Collective Action*. Cambridge: Cambridge University Press.

# 5.14 Transportation systems

Tim Schwanen

The transport of people and goods and oil consumption are intimately connected: transport is the most oil dependent of all economic sectors, while the oil sector relies on transport as its biggest customer (Hickman & Banister, 2014; Figure 5.14.1). It is this mutual dependency rather than the continued growth of worldwide emissions of greenhouse gases (GHGs) and air pollution from transport that defines the global transport system. As there are few sectors where path dependencies in infrastructure, governance, finance and investment, technology, cultural values and lifestyles are as profound as in transport, the transport system arguably offers a litmus test for humanity's collective ability to transition to low-carbon, socially just futures more generally.

A pessimist can easily argue that the prospects of comprehensive transformation towards a low-carbon, non-polluting transport system catering to the mobility needs of all social groups are bleak. Genuinely clean technologies remain sometimes unavailable (aviation) and more often are not viable or seen as undesirable by potential users (road and maritime transport); for decades their diffusion has lagged behind expectations. In road transport both battery electric and hydrogen propulsion have gone through "hype-disappointment cycles" during which societal expectations surged and then deflated in respectively the 1990s and 2000s (e.g., Geels, 2012). Expectations around electric vehicles have recently increased again, in conjunction with the hype around autonomous vehicles and recent improvements in battery energy density and costs. Market shares nonetheless remain small in all but two countries (Figure 5.14.2), and much of the used electricity still comes from fossil fuel sources.

There are, secondly, vast economic – and political – interests to oil-dependent transport (Urry, 2004; Paterson, 2007). Transport is an important source of employment and profits, directly via the manufacturing of vehicles, ships and planes, road works, etc., and indirectly via activities premised on the use of vehicles, ships and planes (e.g., much of tourism, hospitality and trade). Overcoming oil dependence in transport has large

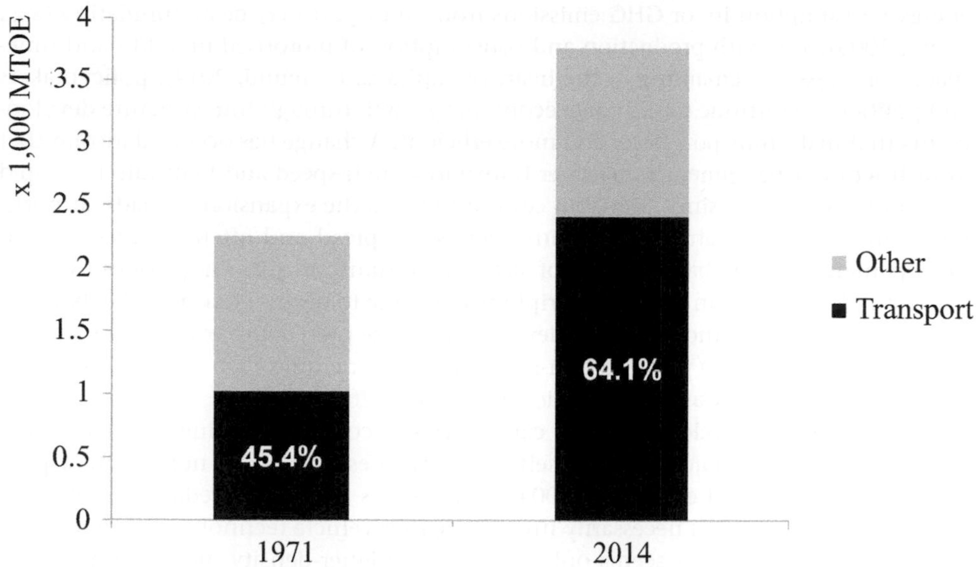


**Figure 5.14.1** Transport's role in global oil consumption.

*Source*: IEA (2016a).

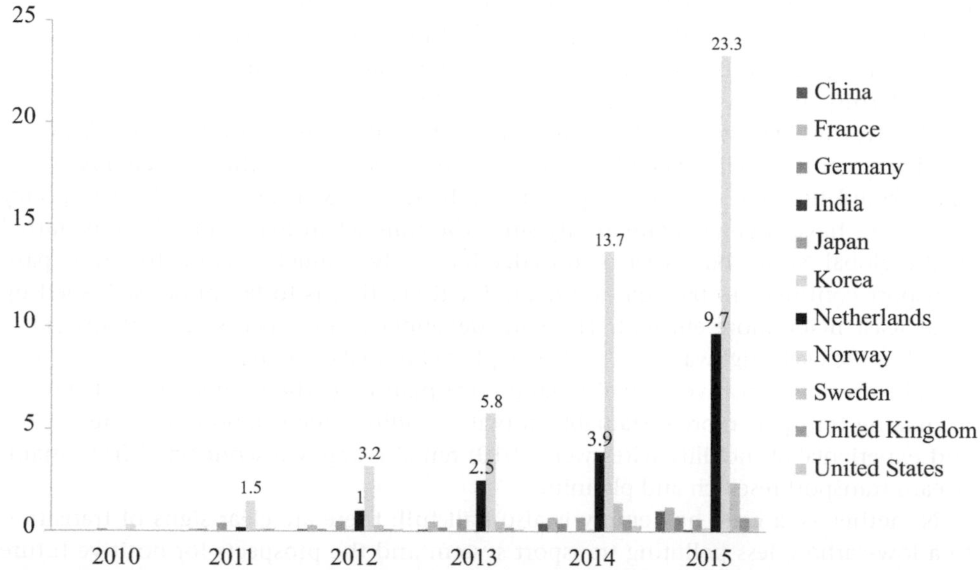


**Figure 5.14.2** Market shares of battery electric and plug-in hybrid vehicles in vehicle sales, by year and country.

*Source*: IEA (2016b).

economic implications and is resisted by powerful actors, many of whom have close ties to governments. Moreover, economic growth and increasing mobility are widely seen as intimately linked. Evidence of relative decoupling of GDP growth from increased energy consumption in, or GHG emissions from, transport may be accumulating (e.g., Tapio, 2005). Yet, with production and consumption of motorised mobility and time-space compression remaining at the heart of capitalism (Freund, 2014), policymakers and politicians continue to stimulate economic growth through infrastructure developments that make transport faster and more efficient. A change has occurred insofar that infrastructure development for cleaner transport – high-speed and light rail, bus rapid transport – has surged since 1980, but commitment to the expansion of roads, airports, ports, and container terminals remains deep, widespread and often dominant. Even in Copenhagen – a global exemplar of sustainable transport planning – does the construction of new roads in the city's periphery continue to be undertaken and debated as congestion reduction and economic development take precedence over other concerns (Driscoll, 2014). Once in place, carbon-intensive infrastructures such as roads, ports and airports shape mobility and urbanisation for decades, if not longer.

Thirdly, in the developed world, car use has become deeply ingrained into the everyday life of the majority of households, with lifestyles, social norms and aspirations revolving around cars (Urry, 2004). This means that major reductions in emissions from transport will necessarily involve cleaner vehicle technologies; a wholesale shift to non-car transport seems only possible in higher-density, urban locations. It is true that car ownership and use are unaffordable for many poorer households and that both have "peaked" (Goodwin & Van Dender, 2013): particularly among younger generations and in cities across the global North, driving licensing, car ownership and car use have stabilised or decreased since about 1990. However, this trend has proven insufficient to radically reduce GHG emissions and air pollution from transport, and its effects are dwarfed by the rapid increase in driving licensing, car ownership and car use in emerging and developing economies where car use remains a powerful symbol of modernity, progress and upward social mobility.

Finally, the expertise regarding transport held by planners, consultants and also academics remains heavily dependent on concepts, logics and methods originally developed to aid the expansion of oil-powered vehicle use (Schwanen, 2016). The policy challenges have changed dramatically since the time when mass motorisation started in the global North, but expert knowledge has evolved much less: for the most part, transport continues to be seen as a neutral activity that is to be minimised, sped up or at least made more efficient. The path dependency in expertise is perhaps epitomised in the cycle highway, now widely deployed to make cycling more attractive but a straightforward derivative of traditional pro-car planning. Their potential contribution to making transport more sustainable notwithstanding, understandings of the politics and experience of mobility (Cresswell, 2010) remain largely disconnected from mainstream transport research and planning.

Nonetheless, a glass half empty is also half full: there are clear signs of transition to a low-carbon, less polluting transport system and the prospects for positive future change are perhaps better than ever before. Transitions take place first and foremost in the central parts of global North cities where "peak car" is accompanied by a renaissance of urban rail, bus transport and cycling and the rise of car sharing. In London, for instance, the share of car trips in total trips has declined from 50 per cent in 1990 to 37 per cent in 2013 with public transport usage increasing correspondingly (Metz, 2015). Comparative research in UK cities suggests place-specific transition trajectories towards low-carbon personal transport systems (Schwanen, 2015), suggesting that transition dynamics are multiple and national-level policy incentives generate spatially differentiated outcomes. Enacted by a wide diversity of actors, including end users, employers, policy-makers, politicians and private sector organisations, these transition processes are linked to, and reinforced by, such developments as densification, gentrification, and community-led initiatives around community energy and localised food production/consumption.

Yet, the optimism about cities as sites of low-carbon, cleaner transport should be qualified, for various reasons. Not only are initiatives and changes outside cities at risk of being overlooked, there is also the danger that cities become seen as "islands of eco-rationality that are largely delinked from the broader territorial formations in which they are . . . embedded" (Brenner & Schmid, 2015: 157), and transport is key to the functioning and evolution of those formations. For instance, aeromobility between cities and connecting cities with popular tourist destinations and production sites of high-order consumption goods (e.g., miniaturised electronics, flowers) is increasing rapidly. Such long-distance mobilities cancel out a substantial share of the environmental benefits of shorter car trips and increased walking, cycling and public transport use to access everyday activities (e.g., Ottelin et al., 2014). Celebratory accounts of how intra-urban transport is becoming more environmentally sustainable are only one side of the story.

Discourses on urban mobility also tend to privilege personal over freight transport. Inter- and intra-urban freight movements are growing fast for many reasons, including increasing wealth and globalised production and logistics systems in the global South and increasing popularity of e-commerce in the North (e.g., Figueroa et al., 2014). Spatial co-location of production and consumption is often proposed as a means to reduce (the growth of) freight mobility, but this may be neither beneficial nor easily achievable. In agriculture, for instance, "food miles" may be reduced but total GHG emissions can increase if more heating/cooling or fertiliser use is required (Kreidenweis et al., 2016), and when global production is optimised on the basis of just-in-time principles, as in the automotive industry, localisation strategies require disruptive innovations in business models and organisational routines.

Moreover, transition trajectories in personal mobility within cities cannot be reduced to a logic of eco-rationality because they are replete with conflict and contestation. A nascent research literature on cycling and urban rail demonstrates that environmentally beneficial transport can be socially unjust (e.g., Revington, 2015; Stehlin, 2015). This happens when infrastructure development caters disproportionally to the mobility needs of middle-class and gentrifying households, creates mobility systems that are unaffordable for low-income groups, and in the longer run displace the latter because housing around stops and stations has become too expensive. In addition, initiatives to reduce private car use in cities often meet significant resistance. As the 2010 election of Toronto's infamous ex-mayor Rob Ford on a "stop the war on the car" ticket (Walks, 2015) suggests, such initiatives can trigger a public backlash that perpetuates the environmental and social costs imposed by automobility and associated politico-economic interests.

Transport is one of the most challenging domains for the realisation of just transitions to futures of minimal environmental impact. Addressing the challenges will require huge determination, a lot of good luck and probably a good deal of social struggle.

## Acknowledgement

Work on this chapter has been made possible by RCUK grant EP/KO11790/1.

## Learning resources

A useful resource for the latest social science perspectives on transport and mobility is offered by the Mobile Lives Forum website: http://en.forumviesmobiles.org. This website contains short articles, interviews, videos, etc. on a wide range of different themes, albeit mostly with a focus on Western Europe.

Also useful is the website of the U.S. Transportation Research Board, http://www.trb.org, which offers a useful portal to the main research topics, debates and policy issues in the North American context.

Various UN-related organisations, including the IPCC and UN Habitat, regularly publish authoritative publications on transport with an explicitly global perspective. One of the most comprehensive discussions is offered by Sims et al. (2014) as part of the IPCC's *Fifth Assessment Report.*

## Bibliography


Brenner, N. & Schmid, C. (2015) Towards a new epistemology of the urban?, *City*, 19(2–3): 151–182.

Cresswell, T. (2010) Towards a politics of mobility, *Environment and Planning D: Society and Space*, 28(1): 17–31.

Driscoll, P.A. (2014) Breaking carbon lock-in: path dependencies in large-scale transportation infrastructure projects, *Planning Practice & Research*, 29(3): 317–330.

Figueroa, M., Lah, O., Lewis M., Fulton, L.M., McKinnon, A. & Tiwari, G. (2014) Energy for transport, *Annual Review of Environment and Resources*, 39: 295–325.

Freund, P. (2014) The revolution will not be motorized: moving toward nonmotorized spatiality, *Capitalism Nature Socialism*, 25(4): 7–18.

Geels, F.W. (2012) A socio-technical analysis of low-carbon transitions: introducing the multi-level perspective into transport studies. *Journal of Transport Geography*, 24: 471–482.

Goodwin, P. & Van Dender, K. (2013) Peak car – themes and issues, *Transport Reviews*, 33(3): 243–254.

Hickman, R. & Banister, D. (2014) *Transport, Climate Change and the City*, Abingdon: Routledge.

International Energy Agency (IEA) (2016a) *Key World Energy Statistics*, Paris: IEA.

International Energy Agency (IEA) (2016b) *Global EV Outlook 2016: Beyond One Million Electric Cars*, Paris: IEA.

Kreidenweis, U., Lautenbach, S. & Koellner, T. (2016) Regional or global? The question of low-emission food sourcing addressed with spatial optimization modelling, *Environmental Modelling & Software*, 82: 128–141.

Metz, D. (2015) Peak car in the big city: reducing London's transport greenhouse gas emissions, *Case Studies on Transport Policy*, 3(4): 367–371.

Ottelin, J., Heinonen, J. & Junnila, S. (2014) Greenhouse gas emissions from flying can offset the gain from reduced driving in dense urban areas, *Journal of Transport Geography*, 41: 1–9.

Paterson, M. (2007) *Automobile Politics*, Cambridge: Cambridge University Press.

Revington, N. (2015) Gentrification, transit, and land use: moving beyond neoclassical theory, *Geography Compass*, 9(3): 152–163.

Stehlin, J. (2015) Cycles of investment: bicycle infrastructure, gentrification, and the restructuring of the San Francisco Bay area. *Environment and Planning A*, 47(1): 121–137.

Schwanen, T. (2015) The bumpy road toward low-energy mobility: case studies from two UK cities, *Sustainability*, 7(6): 7086–7111.

Schwanen, T. (2016) Rethinking resilience as capacity to endure: automobility and the city, *City*, 20(1): 152–160.

Sims R., Shaeffer, R., Corfee-Morlot, J., Creutzig, F., Cruz-Núñez, X., Dimitriu, D., D'Agosto, M., Figueroa Meza, M.J., Fulton, L., Kobayashi, S. et al. (2014) Transport. In Edenhofer, R. et al. (eds) *Climate Change 2014: Mitigation of Climate Change. Contribution of Working Group III to the Fifth Assessment Report of the Intergovernmental Panel on Climate Change* (pp. 590–670), Cambridge: Cambridge University Press.

Tapio, P. (2005) Towards a theory of decoupling: degrees of decoupling in the EU and the case of road traffic in Finland between 1970 and 2001, *Transport Policy*, 12: 137–151.

Urry, J. (2004) The 'system' of automobility, *Theory, Culture and Society*, 21(4–5): 25–39.

Walks, A. (2015) Stopping the 'war on the car': Neoliberalism, Fordism, and the politics of automobility in Toronto, *Mobilities*, 10(3): 402–422.

# 5.15 Volcanoes

## Susanna F. Jenkins and Sarah K. Brown

Around 10 per cent of the world's population (~800 million people) live within 100 km of a potentially active volcano (Brown et al., 2015). Dense populations, agriculture and industrial developments in many volcanically active regions are testament to the varied benefits of volcanic eruptions: fertile land for agriculture and aggregate for construction; favourable climate; geothermal energy; and tourist attractions. As volcanic areas become ever more populated and urbanised, the potential for volcanic disasters increases. Historically, human and economic losses from volcanic eruptions have been much smaller than those from other natural hazards. However, many volcanoes are capable of eruptions much larger than those witnessed in historic times. Volcanic eruptions are the only natural hazard aside from a large asteroid or comet impact that can threaten the global environment and civilisation as we know it.

This chapter provides an overview of volcanoes and volcanic eruptions, with a particular focus on the hazards and risks they pose for society.

## Volcano basics

Volcanoes are commonly classified as active (erupting now), dormant (asleep, but expected to erupt in the future) or extinct (not expected to erupt again). Volcanoes can erupt after thousands of years of dormancy, making the distinction between dormant and extinct volcanoes difficult.

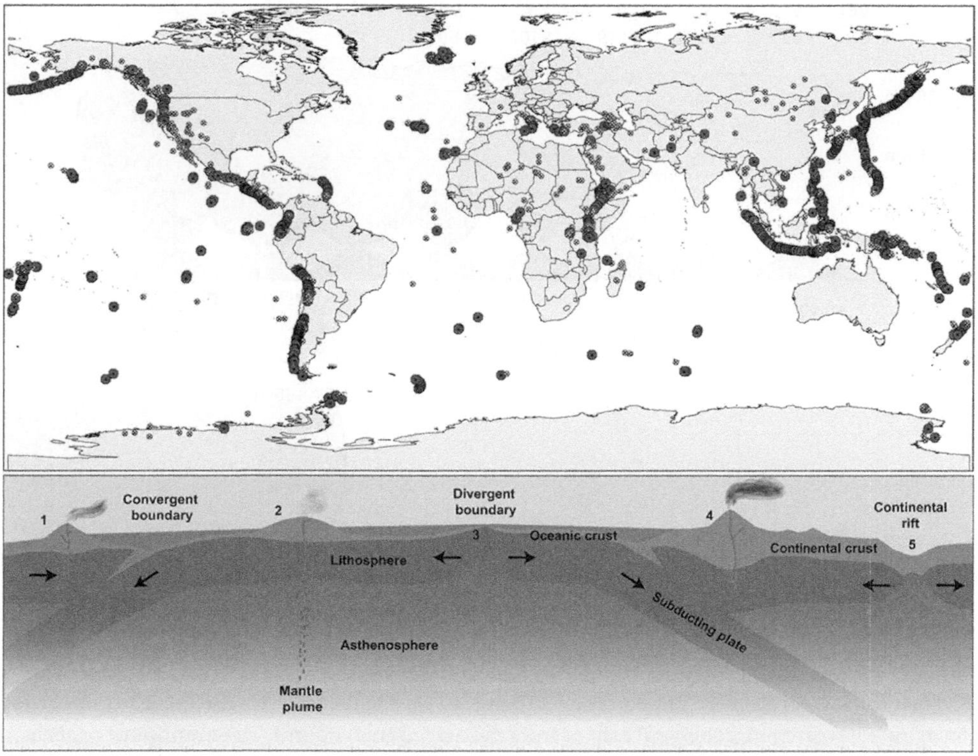


**Figure 5.15.1** (Top) Global distribution of active volcanoes. Dark grey circles indicate volcanoes that have erupted since 1500 AD; small crossed circles are Holocene volcanoes active before 1500 AD. (Bottom) Plate tectonics sketch showing convergent plate boundaries resulting in (1) island arc and (4) continental stratovolcanoes; intraplate volcanism from a mantle plume producing (2) shield volcanoes; divergent plate boundary with (3) sea-floor volcanism and an oceanic spreading ridge; and a continental rift (5) with rift volcanism.

Credit: Figures created by Sarah Brown.

Most volcanoes are distributed along plate boundaries (Figure 5.15.1) that are either divergent (where tectonic plates move apart forming a rift zone) or convergent (where tectonic plates collide, subducting one below another). Intraplate volcanoes, such as in Hawaii and Yellowstone, USA, form away from plate boundaries as a consequence of mantle plumes. Intraplate and divergent plate volcanoes produce the largest amount of magma, but are typically non-explosive. Yellowstone is a notable exception, being capable of extremely large explosive super-eruptions. Convergent plate volcanoes produce just 10% of the world's magma production but represent more than 85% of reported eruptions (Simkin & Siebert, 2000) being commonly explosive and subaerial, increasing the probability of eyewitnesses.

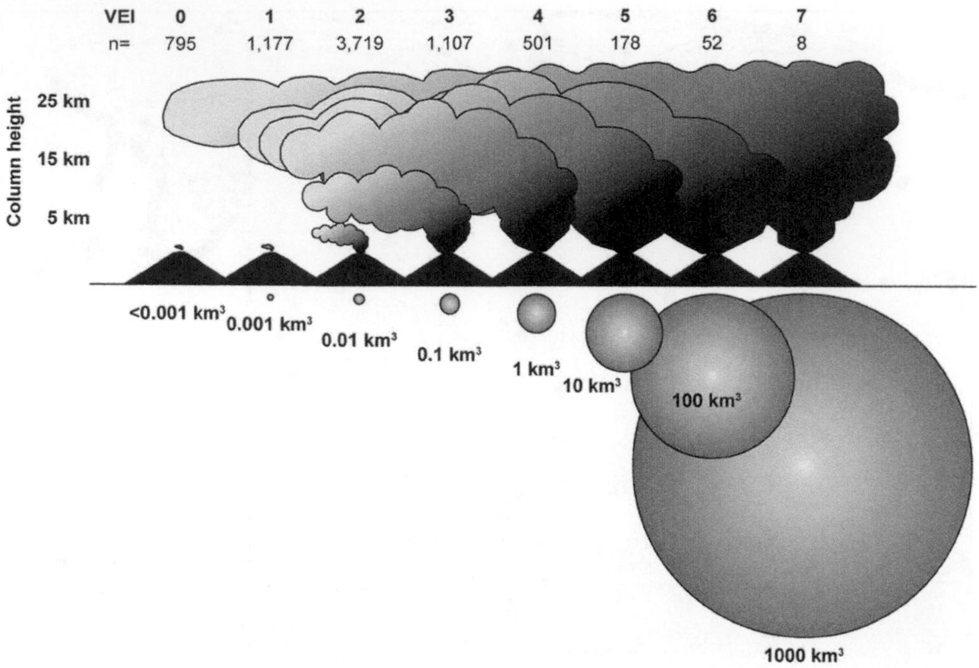


**Figure 5.15.2** A representation of the VEI scale of Newhall and Self (1982). VEI is estimated from the total erupted volume of ash, or from the ash column height. The number of confirmed eruptions in the last ~10,000 years (n) with attributed VEIs are shown.

*Eruption data source*: www.volcano.si.edu, VOTW4.22. Figure created by Sarah Brown.

Eruption size is measured either on a Magnitude (M) scale derived from the total erupted mass (Pyle, 2000), or as a Volcanic Explosivity Index (VEI), based largely upon the volume of material erupted and the eruption column height (Newhall and Self, 1982). Both are logarithmic scales (Figure 5.15.2): i.e. a VEI 5 eruption has ten times the volume of a VEI 4 eruption. Eruption frequency is inversely proportional to the size of the eruption. Globally, eruptions of VEI or M ≤3 occur many times a year, while larger eruptions of VEI or M ≥4 occur every few years (Simkin & Siebert, 2000). Very large eruptions (VEI or M ≥7) may only occur around once every 1000 years.

## Volcanic hazards

A hazard is a physical process or product that adversely affects society or the environment. Volcanoes are unusual amongst natural hazards in that they produce multiple direct and indirect hazards that can occur repeatedly or in combination throughout an eruption, with varying duration, spatial impact and intensity. Some hazards, e.g. volcanic mudflows or gas emissions, can take place even when there is no eruption. Hazardous volcanic processes have been described in detail by a number of authors (e.g. Blong, 1984; Sigurdsson et al., 2015).

Explosive hazards are often rapid-onset and include pyroclastic density currents, ash falls, lahars (volcanic mudflows), blasts and volcanic earthquakes. Effusive eruptions are

**Figure 5.15.3** (A) Lava flows on Etna, Italy, with the city of Catania in the distance[1]; (B) Lava flows destroying a house at Fogo, Cape Verde[2]; (C) Pyroclastic flows at Sinabung, Indonesia. Note the church for scale bottom mid-left[1]; (D) Destroyed town at Merapi, Indonesia, following a pyroclastic flow[2]; (E) Ash fall at Sinabung, with collapsed and damaged houses[1]; (F) House buried in lahar deposits at Chaiten, Chile[1].

*Source*: 1 = Martin Rietze, www.mrietze.com, reproduced with permission; 2 = Susanna Jenkins.

characterised by lava flows and large gas emissions (Figure 5.15.3). Additional volcanic hazards include ground deformation, lightning, acid rain and, in very large explosive or effusive eruptions, significant climatic variations. While effusive eruptions are more frequent than explosive eruptions, explosive eruptions typically have a more significant and widespread impact.

Hazardous extent varies widely: ballistics (volcanic bombs) are commonly confined to within 5 km of the vent, but pyroclastic density currents, lava flows and lahars can affect distances from 1 km to beyond 100 km. Volcanogenic tsunami or ash falls have the potential to travel thousands of kilometres (Blong, 1984). Eruptions can generate hazards for many years, or even decades. Quiescent intervals between hazards may be on the order of months, allowing little respite for affected communities. Conversely, quiescent periods between eruptions can be hundreds or thousands of years. Eruptions are often preceded by days or months of unrest phenomena, such as small earthquakes, ground deformation or changes in gas emissions. Unfortunately, only 35 per cent of the world's volcanoes are continuously monitored (Brown et al., 2015). Unmonitored volcanoes that haven't erupted for many years may not be considered hazardous (or even recognised as volcanoes) and an eruption may therefore catch people by surprise. In such cases, monitoring and emergency management planning are critical. The VEI 6 eruption of Pinatubo, Philippines, in 1991 after more than 500 years of quiescence, is a good example of successful monitoring and eruption forecasting. Mass evacuations and protective measures taken before the eruption saved thousands of lives and hundreds of millions of dollars in property losses (Newhall et al., 1997).

## Volcanic risk

Risk is distinct from hazard in that it considers both the probability of a hazard *and* its negative consequences. Nearly 300,000 people have been killed through volcanic activity since 1600 AD (Auker et al., 2013; Figure 5.15.4). Volcanic risk is a function of the volcanic hazard/s, the exposure, and the vulnerability of society and its assets to the hazard/s.

The exposure and vulnerability of a population, and therefore the likely risks associated with a future volcanic eruption, will vary in space and time, as they do for hazards. Exposure is simply a count of the number of people or assets that may be affected by a hazard, and therefore requires an estimation of the likely hazard 'footprint'. The vulnerability of a community is more complex. There are many aspects that can make a population and its assets vulnerable, including physical, social, systemic, environmental and economic factors. For example, poorly constructed buildings can make a population physically vulnerable to ash falls, while poorly communicated or absent evacuation plans and low public awareness can make a population socially vulnerable to an eruption.

The majority of volcanic disasters have socio-economic and political root causes. These include factors that draw people to volcanic areas, as well as factors that affect how well prepared a society is to respond and recover from an eruption. For example, in 2002, fast-moving lava flows from Nyiragongo, Democratic Republic of Congo, buried farms, villages and 13 per cent of the city of Goma. Over 245 people were killed, 250,000 were evacuated and thousands lost their livelihoods and faced soaring prices of basic goods (Jenkins & Haynes, 2011). While the lavas were devastating, regional politics were key in increasing the exposure and vulnerability of the population. Refugees from the conflict with neighbouring Rwanda had doubled the population of Goma in the 10 years leading up to the eruption, and no volcanic evacuation plan existed (Jenkins & Haynes, 2011).

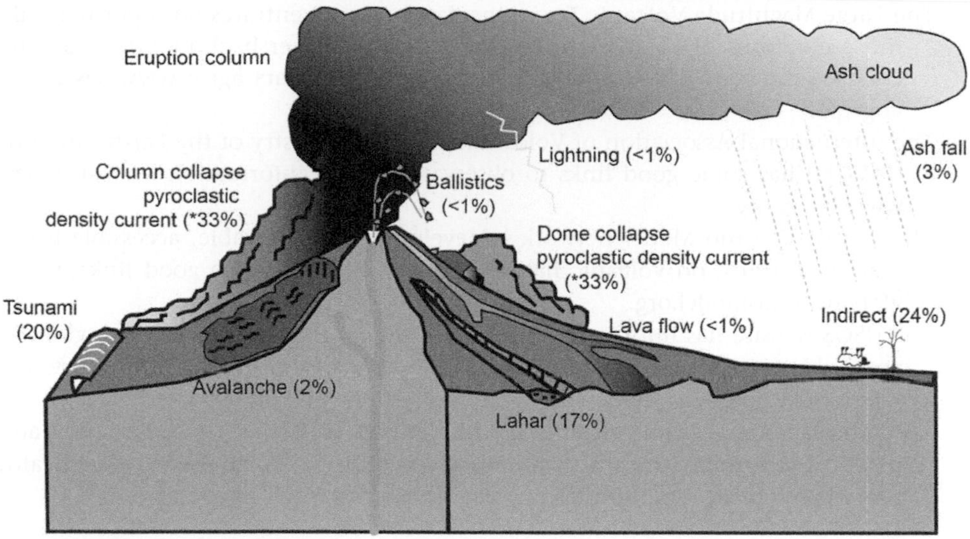


**Figure 5.15.4** The main volcanic hazards and the percentage of fatalities they were responsible for since 1600 AD (data from Auker et al., 2013). *33% of fatalities due to pyroclastic density currents of any origin.

Diagram modified after Myers and Driedger (2008).

# Conclusion

The history of volcano losses is dominated by a few major events that caused devastating mass casualties or losses. This demonstrates the potential for devastating future events with mass casualties. With increasing urbanisation in volcanic areas, the best route to risk reduction is continuous monitoring of potentially active volcanoes, and the assessment of their volcanic hazards and risks. However, volcanic hazards, exposure and vulnerability are diverse in space, intensity and time, complicating the assessment of future hazard and risk. Alongside improved monitoring, a better understanding and quantification of the potential impacts, and the varied factors contributing to increased vulnerability to volcanic hazards is key. Each volcanic eruption teaches us something new about how physical processes interact with society. Quantitative post-eruption field assessments that catalogue the nature of the hazards, exposure and vulnerability are imperative for forecasting future risk, or impacts. Outreach activities and experimental testing in the quiescent periods can help to limit exposure or vulnerability, and prepare a community for future eruptions.

# Learning resources

There are some very useful online resources for volcanic hazards and risk:

The Smithsonian Institution Global Volcanism Program (GVP) eruption database has information and updates on Holocene (~last 10,000 years) active volcanoes and their eruptions: www.volcano.si.edu.

The Large Magnitude Volcanic Eruptions database concentrates on reporting only those eruptions of M 4 or greater, but extends further back in time than the GVP (to the start of the Quaternary, ~2.5 million years ago): www.bgs.ac.uk/vogripa.

The International Association of Volcanology and Chemistry of the Earth's Interior (IAVCEI) has some good links to observatories and information on volcanoes: www.iavcei.org.

The Global Volcano Model network is developing a sustainable, accessible information platform on volcanic hazard and risk and has some good links: www.globalvolcanomodel.org.

The USGS website has information on US volcanoes, monitoring and also globally applicable material on volcanic hazards, impacts and mitigation options: www.volcanoes.usgs.gov.

The International Volcanic Health Hazard Network (IVHHN) has some useful and practical pamphlets and information (in a number of languages) on the health hazards and mitigation options of volcanic ash: www.ivhhn.org.

The following books are excellent overview resources on everything you could ever want to know about volcanoes:

Loughlin, S.C., Sparks, R.S.J., Brown, S.K., Jenkins, S.F. & Vye-Brown, C. (eds) (2015) *Global Volcanic Hazards and Risk*. Cambridge: Cambridge University Press. Open-source: www.cambridge.org/volcano

Sigurdsson, H., Houghton, B.F., McNutt, S.R., Rymer, H. & Stix, J. (eds) (2015) *Encyclopedia of Volcanoes* (2nd edition). Netherlands: Elsevier.

# Bibliography

Auker, M., Sparks, R., Siebert, L., Crosweller, H. & Ewert, J. (2013) A statistical analysis of the global historical volcanic fatalities record, *Journal of Applied Volcanology*, 2(2):1–24.

Blong, R.J. (1984) *Volcanic Hazards: A Sourcebook on the Effects of Eruptions*. Cambridge, MA: Academic Press, Elsevier.

Brown, S.K., Auker, M.R. & Sparks, R.S.J. (2015) Populations around Holocene volcanoes and development of a Population Exposure Index. In Loughlin, S.C., Sparks, R.S.J., Brown, S.K., Jenkins, S.F. & Vye-Brown, C. (eds), *Global Volcanic Hazards and Risk* (pp. 173). Cambridge: Cambridge University Press. Open Access: cambridge.org/volcano.

International Air Transport Association (IATA) (2010) Volcano crisis cost airlines $1.7 billion in revenue – IATA urges measures to mitigate impact, Press Release No. 15, 21 April 2010. Montreal: IATA.

Jenkins S.F. & Haynes, K. (2011) Volcanic eruption. In Wisner, B., Gaillard, J.C. & Kelman, I. (eds) *Handbook of Natural Hazards and Disaster Risk Reduction and Management* (p. 880). Abingdon: Routledge.

Myers, B. & Driedger, C. (2008) *Geologic Hazards at Volcanoes: U.S. Geological Survey General Information Product 64*, 1 sheet. Online. Available at: http://pubs.usgs.gov/gip/64.

Newhall, C.G. & Self, S. (1982) The volcanic explosivity index (VEI) – an estimate of explosive magnitude for historical volcanism, *Journal of Geophysical Research*, 87: 1231–1238.

Newhall, C., Hendley II, J.W. & Stauffer, P.H. (1997) *Benefits of volcano monitoring far outweigh costs – the case of Mount Pinatubo*, U.S. Geological Survey Fact Sheet 115–97.

Pyle, D. (2000) Sizes of volcanic eruptions. In Sigurdsson, H., Houghton, B.F., McNutt, S.R., Rymer, H. & Stix, J. (eds) *Encyclopedia of Volcanoes* (2nd edition) (pp. 263–269). Netherlands: Elsevier.

Sigurdsson, H., Houghton, B.F., McNutt, S.R., Rymer, H. & Stix, J. (eds) (2015) *Encyclopedia of Volcanoes* (2nd edition). Netherlands: Elsevier.

Simkin, T. & Siebert, L. (2000) Earth's volcanoes and eruptions: An overview. In Sigurdsson, H., Houghton, B.F., McNutt, S.R., Rymer, H. & Stix, J. (eds) *Encyclopedia of Volcanoes* (2nd edition). Netherlands: Elsevier.

# 5.16 Water resources

## Joseph Holden

Water is vital for life. Humans need water not only for drinking but in the provision of food and sanitation and also use large quantities of water for industrial purposes, such as energy production or manufacturing. In some developed countries the supply of safe, clean drinking water is taken almost for granted, while in other areas of the world access to safe water is severely lacking. However, even in prosperous nations, droughts can occur, and when combined with inappropriate use of water this can lead to major water shortages with significant economic impacts (e.g. Californian drought of 2011 to 2016).

### Water sources

Of all of the water on the planet, only 3 per cent is freshwater. Of this freshwater around 29.8 per cent is held as groundwater (stored mainly inside the cavities of rock), 0.8 per cent is in lakes, 0.1 per cent is in the atmosphere, and only 0.01 per cent is within rivers. Most of the rest of global freshwater is locked up in ice sheets.

Some groundwater is inaccessible due to its depth or the fact that it is held in rocks that permit almost no flow within them (i.e. the water is held in disconnected pore spaces). The available groundwater resource, however, is very important, with 22 per cent of the world's population reliant solely on groundwater. The USA obtains around half its water use from groundwater and a quarter of all drinking water. Some countries such as Hungary and Denmark obtain almost all of their drinking water from groundwater. Abstraction in these countries is normally (but not always) done in a sustainable manner, using groundwater sources replenished by rainfall during winter.

Unfortunately, much groundwater can be considered non-renewable as it may have taken millennia to build up and replenishment may be very slow. For example, the

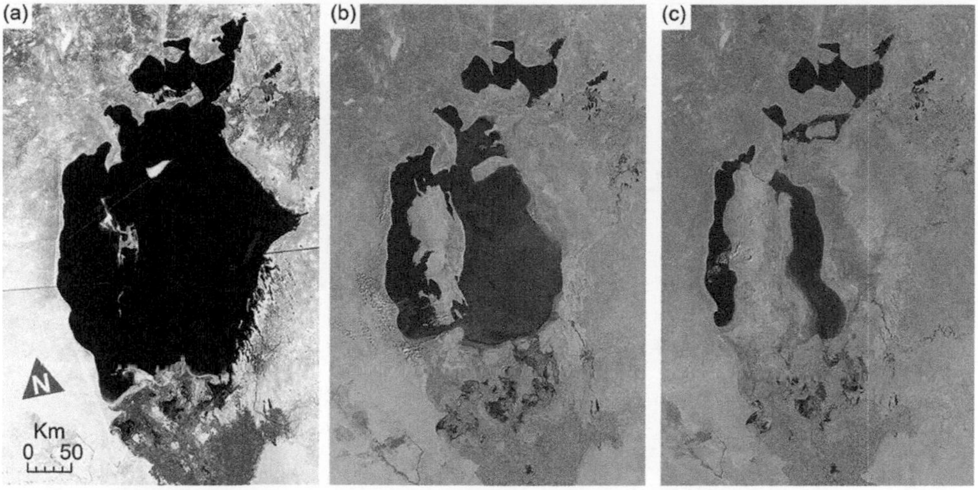


**Figure 5.16.1** The shrinking Aral Sea (Kazakhstan and Uzbekistan) caused by diversion of feeder rivers into agriculture: a) 1972, b) 1998, c) 2010.

*Source*: USGS EROS Data Center.

Nubian Sandstone Aquifer System under the north eastern Sahara contains around 50,000 km$^3$ of water, mostly older than 20,000 years with some of the water deposited a million years ago (Patterson et al., 2005). It is not replenished today due to arid desert conditions in North Africa. However, its use is ever increasing with pipelines northwards into Libya and Egypt supporting agriculture and urban growth. Over-abstraction can deplete groundwater resources and have knock-on consequences including subsidence which has occurred in parts of Mexico City or Tucson, Arizona. Around 10 per cent of groundwater consumption is thought to be extracted at a greater rate than it is being replenished. Of further concern is that around 10 per cent of global food supply is based on unsustainable groundwater use.

Abstraction of water from rivers and lakes is common and many dams have been constructed on major rivers to provide more reliable water resources and for energy production. However, the result of this is that many major rivers that once continuously flowed into the oceans no longer do so (e.g. the Indus and Colorado Rivers) or the river flows have become so highly regulated that biodiversity has declined (Poff & Zimmerman, 2010). Such changes can destabilise coastal areas particularly on deltas because coastal erosion and saltwater incursions can occur at an accelerated rate as the system is diminished of freshwater and sediment inputs from land. Coastal retreat on the Indus delta is ~50 m per year. Many major natural lakes have also shrunk to a tiny proportion of their former size (e.g. Aral Sea, Dead Sea, Lake Chad) due to abstraction (Figure 5.16.1).

## Distribution of water resources

Regions with little water may not be a cause of concern unless they are also areas with high populations. The OECD suggest that water scarcity occurs when there is less than 1000 m$^3$ of water per person per year. Note this figure is not just the resource for drinking but for all other water uses too, such as sanitation or food production. There are

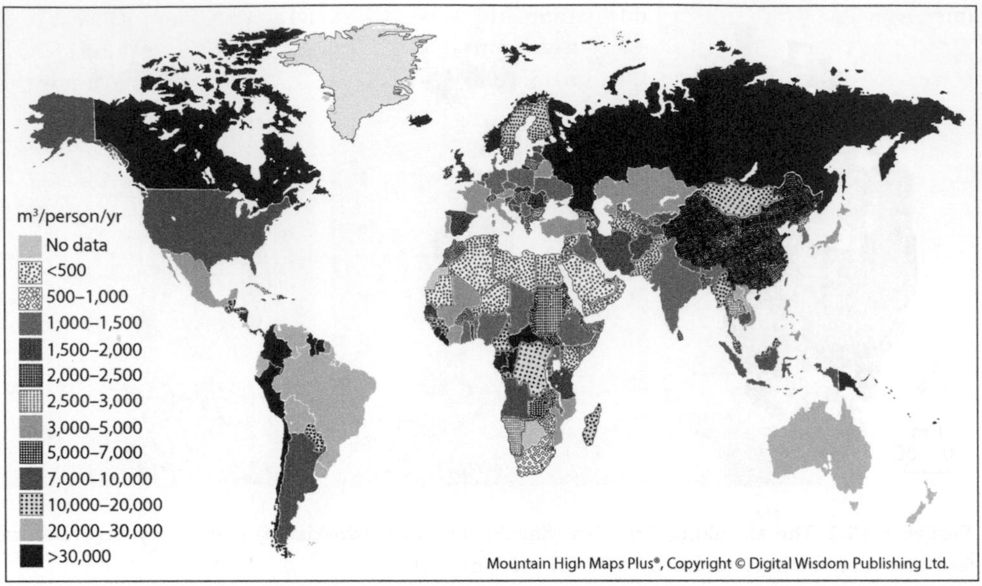


**Figure 5.16.2** Global water resource availability per person per year.

*Source*: Food and Agriculture Organization, AQUASTAT database.

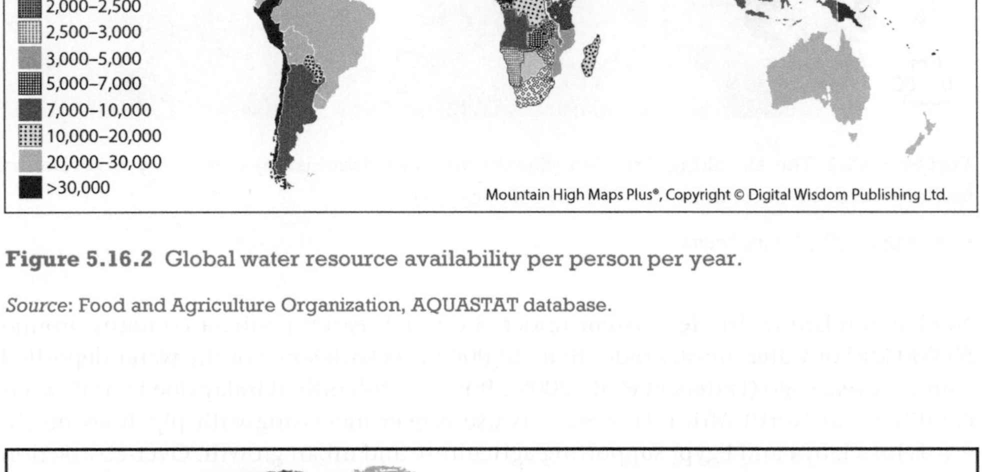

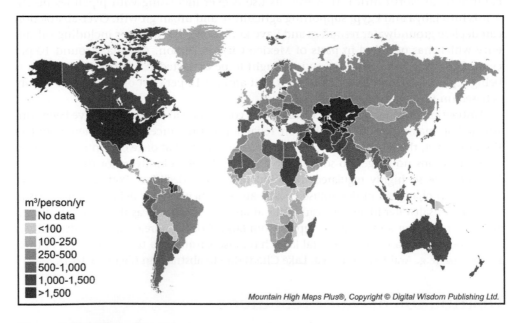


**Figure 5.16.3** Water demand per person per year.

*Source*: Food and Agriculture Organization, AQUASTAT database.

currently 28 countries with less than 500 $m^3$ of water resource per person per year (e.g. Singapore, Saudi Arabia, Niger, Pakistan). Broadly speaking, the Middle East and North Africa are water scarce regions (Figure 5.16.2). Large parts of Asia and Europe are water stressed or water vulnerable.

Water demand varies enormously between countries and depends on population, lifestyle and economy (Figure 3). Therefore true water scarcity may not simply be related to a threshold value of water resource, but will depend on the balance between availability and demand.

## Changing water resources

Agriculture accounts for 86 per cent of global water use, while industry accounts for 10 per cent and the remaining 4 per cent is for domestic use. There is increasing demand for water by a rate of about 2 per cent compound per year (McDonald & Mitchell, 2014). As more water is used for agriculture to feed the world's population, less becomes available for other uses. The remaining water resources may also become poorer quality as pollution from agriculture can impact both surface and groundwaters. As countries become more developed, industrial and domestic uses of water increase with enhanced lifestyle expectations. For example, it takes 1 $m^3$ of water to produce 1 kg of food in the form of pulses, roots and tubers. However, meat consumption often tends to increase with development and it takes 6 $m^3$ of water to produce 1 kg of poultry and 15 $m^3$ of water to produce 1 kg of beef. Countries undergoing rapid industrialisation combined with population growth (e.g. India) may find they quickly enter a water stressed state. By 2030 there is likely to be a huge gap between demand and availability unless demand is reduced through more efficient processes (e.g. better technology for irrigation). Around 2.25 billion of the 2.5 billion people who are to be added to the world's population by 2050 will be in developing countries, many in those areas that do not have access to safe drinking water and adequate sanitation.

Climate model predictions indicate a future with both more intense rainfall and more droughts (IPCC, 2013). The predictions indicate that wet regions may get wetter and dry regions become drier. Changes in snow and ice cover can also have dramatic influences on global water resources. For example, it is expected that spring and summer water supplies from winter snowpacks and glaciers in South Asia will decrease (because more precipitation will fall as rain, or because glaciers will disappear) so that there will be less discharge within the upper Indus, Ganges and Brahmaputra by the 2050s (Immerzeel et al., 2010). This will have major impacts on agricultural production in the region. However, there may be enhanced flows in the Yellow River because it depends less on meltwater and rainfall may increase under climate change.

## Conflict and trade

There are 44 countries that depend on other countries for more than half of their water resources. This means that international political negotiations over water resources are required. Conflict commonly occurs where there is perceived over-abstraction in an upstream area impacting the availability of water resources downstream (e.g. Ganges), or where pollution upstream affects downstream water quality (e.g. Rhine). As water resource use intensifies in the coming decades it will be necessary to intensify political negotiations to avoid serious conflict over water.

When global trade is considered it is possible to show how globally interconnected water resource use is. For each product that is manufactured or each foodstuff provided, the water used in that process can be calculated. This is sometimes called the water footprint of a product. However, that product may be purchased and consumed in another region of the world. Thus the water use or footprint of that product might be ascribed to the consumer. USA citizens have twice the water footprint of the global average while people in China have around half the global average water footprint. Many European and Middle Eastern countries are highly dependent on other nations for the water used in products that they consume. In fact trade between countries means that products consumed in rich countries may directly affect water resources in poor countries. Allan (2011) considered these exports and imports of products to be associated with the virtual movement of water between countries. Many African countries are net exporters of 'virtual water' (Figure 5.16.4) and yet suffer from water scarcity. Of course, through such trade there are instances when global water resource use is more efficient. For example, virtual water may flow from regions that take little water to grow a particular crop towards countries where it would have taken a lot of water to grow that same crop. The result in such a case is a net benefit to overall global water resource use. It is estimated that around 300 km³ of water is saved annually in this way (Fader et al., 2011). Such global analysis can be important for national planners and big business who seek to ensure their own food security by assessing risks of future water scarcity for a distant supply region due to population, political or climate change.

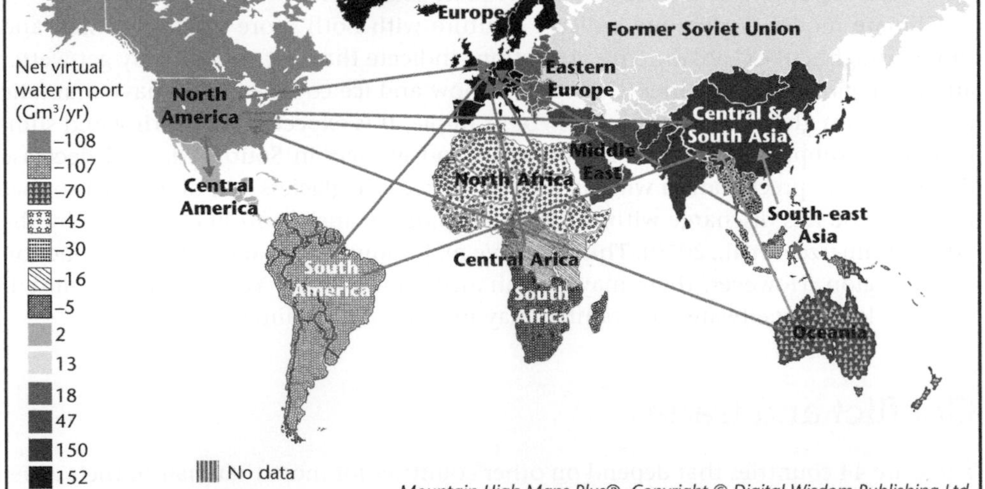




**Figure 5.16.4** Net virtual water balance for major global regions for the water flows associated with agricultural trade. The arrows show some of the largest net movements.

*Source*: waterfootprint.org, reproduced under Creative Commons CC-BY-SA 3.0 licence.

# Learning resources

A useful textbook written by 23 contributors that covers water resources from many different perspectives including natural and socio-economic processes:

Holden, J. (ed) (2014) *Water Resources: An Integrated Approach.* Abingdon: Routledge.

A short overview of some key frameworks for conceptualising the global interconnectedness of water resource use:

Konar, M., Evans, T.P., Levy, M., Scott, C.A., Troy, T.J., Vörösmarty, C.J. & Sivapalan, M. (2016) Water resources sustainability in a globalizing world: who uses the water?, *Hydrological Processes*, 30: 3330–3336.

Dozens of excellent water resource reports, with different applications and areas of interest are available at this website:

www.iwmi.cgiar.org

The latest findings from a water resources Earth observation programme are presented at this website:

www.earth2observe.eu

# Bibliography

Allan, J.A. (2011) *Virtual Water: Tackling the Threat to Our Planet's Most Precious Resource.* London: I.B. Tauris.

Fader, M., Gerten, D., Thammer, M., Heinke, J., Lotze-Campen, H., Lucht, W. & Cramer, W. (2011) Internal and external green-blue agricultural water footprints of nations, and related water and land savings through trade, *Hydrology and Earth System Sciences*, 15: 1641–1660.

Immerzeel, W.W., van Beek, L.P.H. & Bierkens, M.F.P. (2010) Climate change will affect the Asian water towers. *Science*, 328, 1382–1385.

Intergovernmental Panel on Climate Change (IPCC) (2013) *Climate Change 2013: The Physical Science Basis. Contribution of Working Group I to the Fifth Assessment Report of the Intergovernmental Panel on Climate Change*, Stocker, T.F., D. Qin, G.-K. Plattner, M., Tignor, S.K. Allen, J. Boschung, A. Nauels, Y. Xia, V. Bex & P.M. Midgley (eds). Cambridge: Cambridge University Press.

McDonald, A.T. & Mitchell, G. (2014) Water demand planning and management. In Holden, J. (ed) *Water Resources: An Integrated Approach* (pp. 203–221). Abingdon: Routledge.

Patterson L.J., Sturchio, N.C., Kennedy, B.M., Van Soest, M.C., Sultan, M., Tian Lu, Z., Lehmann, B., Purtschert, R., El Alfy, Z., El Kaliouby, B. et al. (2005) Cosmogenic, radiogenic and stable isotope constraints on groundwater residence time in the Nubian Aquifer, Western Desert of Egypt, *Geochemistry Geophysics Geosystems*, 6: 1–19.

Poff, N.L. & Zimmerman, J.K.H. (2010) Ecological responses to altered flow regimes: a literature review to inform the science and management of environmental flows, *Freshwater Biology*, 55: 194–205.

# Part 6

## Key topics: human responses to environmental change

# Part 6 Introduction

Natural and human-caused environmental events, both small and large, inspire a remarkably wide range of imaginative and practical responses. The same can be said of longer term and large-scale environmental changes, like the melting of the polar ice caps or the conversion of forests to farmland. From 'cli-fi' (fictional books about anthropogenic climate change and its future effects) to carbon capture and storage (CCS) to the specialist field of environmental ethics, societies register their response to the non-human world in a variety of ways, and act accordingly. Why such diversity? There are at least two reasons. First, in our world of cultural diversity, people look at the physical environment through different 'templates', meaning they see and value elements of the world in more than one way. Indigenous cultures in Australia, for instance, see Country, whereas Australians descended from European settlers often see 'resources' to be exploited or managed. Second, such is the diversity of the non-human world that it inevitably provokes different cognitive and affective responses, depending on which people are affected in which situations. For instance, low income people in areas that are proximate to landfill sites have been key players in advancing the idea of 'environmental injustice'. Others are more or less forced to leave their home-places as 'environmental refugees' displaced by chronic flooding or by drought, exacerbated by political unrest and oppression. By contrast, some well-paid and famous Hollywood stars like Leo DiCaprio have seen fit to use their public profiles to act as advocates for 'greener' ways of life. Still others, recognizing that considerable power lies in the hands of politicians, long ago sought to 'speak for nature' in the political arena, forming Green political parties in Germany and elsewhere. These parties remain active today, though still have yet to form a single party government in any major country.

This part explores the plethora of ways in which different people have sought to represent, react to and influence human–environment interactions – visually, linguistically, economically, technically and so on. It takes readers beyond much discussed responses, like carbon trading schemes, in order to examine numerous forms of response, many of which are opposed. For instance, some argue for an 'ecocentric' (or nature first) approach to the world, regarding institutions like carbon trading as deeply 'anthropocentric' (or people-focused). This opposition reflects the politically and morally charged nature of human–environment relations.

# Corporate environmental responsibility

6.1

Tomas Frederiksen

Debates over corporate environmental responsibility are largely between those that see businesses as the cause of environmental problems and those that see them as the source of environmental solutions. Governments increasingly regulate the environmental impacts of corporate activity and new societal expectations of 'responsible' behaviour by corporations has seen important changes in business activities. Companies have struggled to maintain legitimacy for their activities in the face of a series of environmental disasters and sought to manage the environmental footprint of their increasingly complex and globally disparate activities. Within companies, the last 50 years has seen a shift in attitudes and approaches to the idea of corporate environmental responsibility from 'heresy to dogma' (Hoffman, 2001). This chapter gives a broad chronology of these changes before discussing current key concepts in corporate environmental responsibility and finally looking at a range of critiques of corporate environmentalism.

## A changing history of corporate environmental responsibility

The current move towards corporate environmental responsibility began with the birth of the modern environmental movement in the 1960s. Many studies trace this to the publication of Rachel Carson's 1962 novel *Silent Spring* which explored the cumulative effects of pesticide use on wildlife. In response to the changing science which highlighted the ways in which corporate activities can harm the environment, governments increasingly sought to regulate their activities. Initially, businesses pushed back strongly against this 'unwelcome interference' which sought to curb their activities. Monsanto, for example, attacked Rachel Carson personally and released a parody of her book called *The Desolate Year* (Hoffman, 2001). Companies in this era saw themselves as autonomous entities not to be encumbered or interfered with by external actors, and some argued that companies should ignore environmental regulation that was more costly to implement than subsequent fines may be. In 1970 economist Milton Friedman famously argued that companies which did anything beyond that which was strictly required by the law were 'unwitting puppets', 'preaching pure unadulterated socialism' (Friedman, 2007: 173). Environmental offices were introduced to companies in this period as they sought to achieve regulatory compliance, though they were often peripheral to companies and had little internal influence.

In the 1970s and 1980s, awareness of environmental issues was rising and governments took an increasingly interventionist stance. In 1970 for example, the US Environmental Protection Agency was created and Environmental Impact Assessments (EIAs) became a widespread requirement for new operations. A series of high profile environmental disasters helped colour the public, and regulatory, mood. Most notably, in 1984, an accident at a Union Carbide pesticide plant in Bhopal, India released 45 tonnes of toxic chemicals killing 3500 and injuring 300,000. In response, the Government of India issued an arrest warrant for the CEO of Union Carbide before later accepting a settlement of USD $470 million. Union Carbide itself was subject to a hostile takeover bid after their share price dropped (Hoffman & Bansal, 2012). As large fines were meted out for lack of environmental compliance and insurers began to refuse to cover non-sudden environmental liabilities, companies began to change their approach to environmental issues. Moving to a more proactive stance and taking on notions such as 'corporate social responsibility', companies sought to define themselves as part of the solution, not just the cause of problems. In terms of organisational structure, this saw environmental departments growing in size, importance and influence within companies and environmental considerations expanding out of small teams.

In the 1990s, 'strategic environmentalism' emerged, framing environmental concerns as of strategic importance for entire firms. In this period, environmental management had reached the highest echelons of corporate structures and environmental considerations were pushed into line operations. As some companies begin to adopt 'beyond compliance' 'policies [that] specifically propose to exceed the requirements of extant laws' (Prakash, 2000: 3), approaches like waste minimisation, pollution prevention and product stewardship emerged. These approaches frame the environment as less an 'externality' to the production process and more as part of companies' 'social license to operate' (legitimacy) (Morrison, 2014). By the 2000s, corporate environmentalism had become mainstream and discourses of 'corporate sustainability' abounded. Debates about large systemic environmental change such as *climate change*, the *anthropocene and planetary boundaries* have increased pressure on companies to not only integrate environmental concerns into all levels of their decision making, but to proactively work to produce better and more sustainable environmental futures. What was once heresy – that companies should proactively reach beyond regulations and actively voluntarily work to curtail their environmental impact – is now a dogma within many corporate sectors and companies. Corporations now often show greater recognition of the ways environment and society are intertwined and how environmental performance is central to wider legitimacy.

## The current landscape of corporate environmental responsibility

The current landscape of corporate environmental responsibility paints a contrasting picture. On the one hand, the global economy continues to grow in size and scope and companies face constant competitive pressure to maximise profits and returns to shareholders. On the other, this same growth and human activity show only greater and more concerning environmental impacts, and societal expectations of corporate environmental responsibility are increasing, creating legitimacy problems for companies. The rise of the internet and social media has produced a new level of transparency around corporate actions where transgressions are quickly recorded and disseminated with companies finding themselves exposed in new ways. Since the 2008 economic crash, states have been increasingly willing to develop and enforce legislation which seeks to control the activities of companies both within and beyond national borders (Scherer et al., 2016). For example, the 2010 US Dodd-Frank act requires companies listed on US stock exchanges to report on their due diligence procedures around conflict materials globally. Courts and financial settlements have emerged as key mechanisms for regulating corporate behaviour with cases brought by both governments and civil society actors. In one recent example in Sanmarco, Brazil, 12 people were killed when a village was swamped by mining slurries after a tailings dam broke at a nearby mine operated by two of the world's largest mining companies, Vale and Anglo American. The companies settled with the Brazilian government for USD $4 billion. This pales in comparison to BP's liabilities for the 2010 Deepwater Horizon oil spill in the Gulf of Mexico which are estimated to be up to USD $37 billion.

Companies' responses have sought to continue to redefine their role in relation to the environment and find new ways of conceiving and managing their environmental performance. Sensing a steady rise in regulations over time, the search continues for win-win solutions which make it 'pay (profitable) to be green' and help define corporations as the source of environmental solutions and not problems. The rise of voluntary private regulation is one example of this, where companies choose to adopt environmental standards which are stricter than those legally required such as the Responsible Care initiative in the chemicals industry and the Cyanide Code in the gold mining industry. Seeking to embed environmental concerns deeper into management and ethical structures, companies have worked to improve how they account for, measure and report their environmental impact. The UN Global Compact and the Global Reporting Initiative (a 'triple bottom line' approach covering 50 separate indicators of environmental, social and economic performance) are high profile initiatives in this direction. Companies have seen their responsibilities extended to a wider range of stakeholders, including to suppliers through initiatives which call for 'supply chain stewardship'. The intertwined nature of companies' social and environmental impacts, are frequently integrated into decision making as strategic risks within a wider risk management framework.

## Criticisms

Despite decades of changing approaches to corporate environmental responsibility, environmental problems generated by corporations continue on a large scale. Events like Sanmarco and Deepwater Horizon show that corporate environmental *irresponsibility* is still very much present. A recent UN report found that in 2008 the top 3000 publicly traded companies produced over $2.15 trillion in environmental damages (UNPRI, 2011). Many of the efforts to produce more sustainable business practices refuse to compromise the profit motive, and do not acknowledge that limits on growth and consumption may exist (Banerjee, 2012). Environmental regulation which curtails companies' ability to generate profits is frequently seen as a burden to be resisted or evaded – as evidenced by companies relocating polluting industries to countries with less stringent environmental standards. Instead of changing behaviour, critics argue that companies instead engage in 'greenwashing' where they engage in some high-profile efforts to minimise waste while ignoring more problematic processes which impinge on their business models, i.e. the tourism industry promoting 'ecotourism' holidays at the other end of long haul flights, a large and growing global contributor to $CO_2$ emissions. Further, some see efforts to seek win-win solutions as the effective capture of environmental discourses for the benefit of corporations, a hollowing out and replacing of ecological priorities with economic ones. The notion of corporate environmental responsibility is seen thus as a contradiction in terms – how can, for example, companies which have funded organisations seeking to discredit science on anthropogenic climate change be trusted to help develop global mechanisms to govern carbon emissions? For these critics, what is required is a greater focus on corporate *accountability*, not responsibility; to replace woolly notions of responsibility and legitimacy with hard lines of regulation and clear consequences for companies that damage the environment (Levy & Newell, 2005).

## Learning resources

For some interesting and accessible insight into corporate responsibility for social and environmental harms see the following:

BBC (2004) 'One Night in Bhopal' (BBC)
BBC (2011) 'Deepwater Horizon—the real damage': http://www.bbc.co.uk/programmes/b0106rvq.
Bichlbaum, A & Bonnano, M. (2009) *The Yes Men Fix the World* (HBO)

## Bibliography

Banerjee, S.B. (2012) 'Critical Perspectives on Business and the Natural Environment'. In *The Oxford Handbook of Business and the Natural Environment* (pp. 572–590). Oxford: Oxford University Press.

Carson, R. (1962) *Silent Spring*. Boston, MA: Houghton Mifflin.

Friedman, M. (2007) 'The Social Responsibility of Business is to Increase its Profits'. In W.C. Zimmerli, M. Holzinger & K. Richter (eds) *Corporate Ethics and Corporate Governance* (pp. 173–178). Berlin, Heidelberg: Springer.

Hoffman, A.J. (2001) *From Heresy to Dogma*. Palo Alto, CA: Stanford University Press.

Hoffman, A.J. & Bansal, P. (2012) 'Retrospective, Perspective and Prospective: Introduction'. In *The Oxford Handbook of Business and the Natural Environment* (pp. 3–28). Oxford: Oxford University Press.

Levy, D. & Newell, P.J. (eds) (2005) *The Business of Global Environmental Governance*. Cambridge, MA: MIT Press.

Morrison, J. (2014) *The Social License*. London: Springer.
Prakash, A. (2000) *Greening the Firm*. Cambridge: Cambridge University Press.
Scherer, A.G., Rasche, A., Palazzo, G. & Spicer, A. (2016) Managing for Political Corporate Social Responsibility: New Challenges and Directions for PCSR 2.0, *Journal of Management Studies*, 53(3): 273–298.
United Nations Principles for Responsible Investment (UNPRI) (2011) *Universal Ownership: Why Environmental Externalities Matter to Institutional Investors* (pp. 1–69). Brussels: UNPRI.

# 6.2 Ecological modernisation

## Giorel Curran

Ecological modernisation (EM) is a prominent approach to achieving sustainable development in modern market societies. It seeks to do so in a minimally disruptive way and without diminishing economic growth. These objectives help explain EM's overarching appeal and why it is one of today's pre-eminent environmental management forms. Through a process of decoupling the negative impacts of development from economic development itself, it offers a co-benefits paradigm – a 'win-win' – that sees technological innovation protecting the environment at the same time as it generates economic buoyancy (Hajer, 1995: 64). By utilising a range of familiar economic and market tools, EM offers a relatively routine path out of the morass of environmental degradation. Its 'discourse of reassurance' (Dryzek, 2005: 172) is central to its widespread uptake by both governments and businesses across the globe. This reassurance stems from its focus on what is possible rather than ideally necessary in achieving sustainability.

## The idea

The idea of EM emerged in earnest in a largely western European context in the early 1980s (see Dryzek, 2005). It offered a middle course between the radical change demanded by some sections of the then environment movement and a business community wary of such change (see Buttel, 2000). To the former it sought to show that significant environmental progress could be won without transforming economic and social systems altogether, and to the latter that the change that was required would ultimately benefit them in any case. Ecological modernisation does not baulk from acknowledging the gravity of environmental problems. But it believes that the industrial capitalism that helped create these problems also offers the solutions to address them. The focus is hence on harnessing the efficiency of the market – particularly its capacity for innovation and entrepreneurialism – to not eliminate growth but to do it differently. But in order to ensure business' cooperation, it was first necessary to demonstrate that this new approach would be actually good for business – that is, that environmental protection could pay.

Ecological modernisation's theoretical legacy is nonetheless diverse, with a range of different foci: from one on social theory (Buttel, 2000); technological and industrial development (Huber, 1982; Weale, 1992); discursive strategy (Hajer, 1995); and political modernisation (Mol, 1996). Joseph Huber (1982) and Martin Jänicke (1985) are widely acknowledged as among the first to discuss the notion of EM and propose it as a way forward in environmental management. The key, and compelling, proposition of early EM theorists was that, with some creative re-thinking and re-tooling, a capitalist industrial economy could render industrial practices more environmentally benign. While this was not expected to be easy, Jänicke (2008: 563) nonetheless observes that the capacity of an EM approach to 'radically reduce the environmental burden of industrial growth' – to 'green' a capitalist economy rather than depose it – 'is without any alternative'.

Arthur Mol (1996) and Gert Spaargaren (Mol & Spaargaren, 1993) were among some of the key theorists that built on their colleagues' earlier legacy by developing the concept into a sophisticated and 'distinctive theoretical argument' and by establishing a 'core literature' that contributed significantly to the widespread dissemination of the idea (Buttel, 2000: 58). According to Mol (1996: 313–315), EM's promise resides in its core characteristics. First, rather than discarding the institutions that have admittedly contributed to environmental ruin, these very same institutions – the market, the state, science and technology – could instead be ecologically modernised and redeployed for environmental good. Second, Mol singles out the market, particularly through the technological innovation impetus that drives it, for its remediation potential. Civil society and the state are also important drivers, but without the market economy in tow, the modernisation project is set to fail. Third, to be an effective driver, statist practices require considerable transformation. Here a shift is envisaged from hierarchical decision-making processes to more inclusive, decentralised and participative forms that engage the market in more creative ways.

Overall, this modernisation strategy, would reconcile economic and ecological goals in such a way that 'pollution prevention [would] pay' (Hajer, 1995: 26). And both economy and ecology would be winners. For business, investment in technological innovation would create efficiencies and reduce waste – hence minimising costs and enhancing profits. Not only would the environment benefit, but this new modernised thinking could open up a wealth of further environmentally-oriented business opportunities. This is consistent with EM's intent to not stymie development but advance it. The co-benefits paradigm that underpins EM; its embrace of the prevailing, albeit 'greened', institutional infrastructure; and its allocation of technological entrepreneurism to the heart of the ecological enterprise, helps explain the central place it has won in environmental management.

## Versions and criticisms

EM is not a one-size-fits all, however. Nor has this rosy picture of EM's promise stalled its critics. The literature on EM is often conceptualised in terms of a reformist-transformative spectrum – even as there remains considerable overlap. Positions captured within this spectrum include Christoff's (1996) 'stronger or weaker', Hajer's (1995) 'techno-corporatist' or 'reflexive', and Buttel's (2000) mainstream or 'social constructivist' versions. These different forms are distinguished by the underpinning critique of prevailing political economic arrangements; the understanding of the environment within these scenarios, and the identification of the key actors and processes that drive socio-ecological change.

The overriding criticism is that the largely mainstream form of EM we have described above does not go far enough to generate genuine and substantive environmental reform. This is despite its narrative appeal. Mainstream EM is charged with having too limited an understanding of the complex relationship between ecology and society and with an over-reliance on technology and the cooperation of business. In overlooking the difficult power relations of environmental change, it also fails to address the significant equity and justice considerations of environmental problems.

As we saw, a mainstream EM places technological innovation and eco-efficiency gains at the centre of its model, and while arguing that reform of existing political economy arrangements is required, it makes clear that it does not reject the capitalist model, believing that environmental problems will themselves trigger the necessary changes to industrial capitalism. This is not to deny that hard choices and risky investments will need to be made. But in the end it foresees that the ensuing gains – for example, the burgeoning green consumerism markets – will drive change.

Those at the more critical transformative end remain circumspect. On their reading, mainstream EM pays limited attention to the power relations of the prevailing political economy and their link to socio-ecological degradation. While not advocating the revolutionary proposals of some of the 'first generation' radical ecologists, these critics nonetheless maintain that mainstream EM simply tinkers at the edge of necessary social change rather than directly confronting it. A stronger eco-modernisation would urge that institutional restructuring and technological innovation be directed to addressing the social justice and ecological considerations of environmental decline, rather than prioritising commercial gain. While mainstream EM's emphasis on innovation and the technological re-tooling of industry are important elements of ecological change, mainstream EM over-emphasises the production side of the sustainability equation vis-à-vis its consumption side. More efficient and better quality production may indeed reduce pollution loads, or the *quality* of production, but this does not necessarily address the *quantity* of production (York & Rosa, 2003). Moreover, EM is primarily directed to the renovation of industrial processes – environmental problems go beyond this: biodiversity disruption and equity considerations, for example.

Not all of EM's critics urge its dismantling altogether, however. Some may reject it entirely (see Blühdorn, 2007), but others subscribe to elements of its logic while seeking to revise and strengthen it. Christoff, for example, talks of the 'weak and strong features of EM' as not 'simply mutually exclusive binary opposites' since an 'enduring ecologically sustainable outcome . . . does not abandon technological change, economic instruments or instrumental reason' (1996: 491). Rather, the shortcomings of EM can be ameliorated with a stronger focus on more participative and inclusive processes and stronger political change. Indeed, thorough-going political modernisation is key to EM's success.

The central insight of political modernisation is that the governing institutions of the state need to undergo considerable change if the goals of sustainability are to be met. Political commitment to ecological reform, at all levels of government including the elite level, is key. Such commitment would be devoted to overseeing an extensive renovation of the political and institutional infrastructure of environmental decision making to make it more inclusive, integrated and robust. And considerable political will would need to be mounted to challenge the sectoral and structural resistance to ecological goals. This is clearly no easy task. Analysts highlight the difficulty of achieving such modernisation in the political context of neo-liberalism where the goal is to limit, rather than enhance, government's role. Nonetheless, EM's success relies on new ways to 'green' the (neo-liberal) state; hence the centrality of political modernisation to the ecological modernisation enterprise.

## Learning resources

The following offer rich introductions to EM:

Dryzek, J. & Schlosberg, D. (eds) (2005) *Debating the Earth: The Environmental Politics Reader*. Section VII Ecological Modernisation. Oxford: Oxford University Press.

Mol, A.P.J. & Sonnenfeld, D.A. (eds) (2013) *Ecological Modernisation around the World: Perspectives and Critical Debates*. New York: Routledge.

Mol, A.P.J., Sonnenfeld, D.A. & Spaargaren, G. (eds) (2009) *The Ecological Modernisation Reader: Environmental Reform in Theory and Practice*. New York: Routledge.

## Bibliography

Blühdorn, I. (2007) 'Sustaining the Unsustainable: Symbolic Politics and the Politics of Simulation', *Environmental Politics*, 16(2): 251–275.

Buttel F.H. (2000) 'Ecological Modernization as Social Theory', *Geoforum*, 31(1): 57–65.

Christoff, P. (1996) 'Ecological Modernisation, Ecological Modernities', *Environmental Politics*, 5(3): 476–500.

Dryzek, J. (2005) *The Politics of the Earth: Environmental Discourses*. 2nd edition. Oxford: Oxford University Press.

Hajer, M.A. (1995) *The Politics of Environmental Discourse: Ecological Modernisation and the Policy Process*. Laderley: Clarendon Press.

Huber, J. (1982) *The Lost Innocence of Ecology: New Technologies & Superindustrial Development*. Frankfurt: Fisher Publishing.

Jänicke, M. (1985) *Preventative Environmental Policy as Ecological Modernisation and Structural Policy*. Berlin: Wissenschaftszentrum.

Jänicke, M. (2008) 'Ecological Modernisation: New Perspectives', *Journal of Cleaner Production*, 16(5): 557–565.

Mol, A.P.J. (1996) 'Ecological Modernisation and Institutional Reflexivity', *Environmental Politics*, 5(2): 302–323.

Mol, A.P.J. & Spaargaren, G. (1993) 'Environment, Modernity and the Risk-Society: The Apocalyptic Horizon of Environmental Reform', *International Sociology*, 8(4): 431–459.

Weale, A. (1992) *The New Politics of Pollution*. Manchester: Manchester University Press.

York, R. & Rosa, E.A. (2003) 'Key Challenges to Ecological Modernization Theory: Institutional Efficacy, Case Study Evidence, Units of Analysis, and the Pace of Eco-Efficiency', *Environment and Organisation*, 16(3): 273–288.

# Ecotourism

6.3

## Robert Fletcher

Ecotourism is often considered one of the fastest growing segments of a global tourism market that now rivals oil production as the world's largest industry (UNWTO, 2016). Yet what is actually being promoted under this label remains a matter of some confusion. The most widely accepted definition of ecotourism, offered by The International Ecotourism Society (TIES), defines it as "Responsible travel to natural areas that conserves the environment and improves the well-being of local people" (cited in Honey, 2008: 6). Yet it must be recognized that this is not merely a factual statement, but rather a political argument intended to assert that these are the only activities that the concept *should* designate. In wider popular discourse, of course, the term is used to describe all manner of activities that generally endeavor to sell an encounter with 'nature' very broadly defined. This promiscuous use of the term has led to criticism that it has become an empty label – or worse, a cover for 'greenwashing', allowing operators to conceal their detrimental practices beneath a veneer of social and environmental responsibility (Mowforth & Munt, 2008). Efforts by TIES and others to define the term more narrowly, therefore, seek to focus attention on the ways in which the activities designated may or may not actually fulfill the social and environmental promises that they commonly make. Hence, advocates seek to distinguish 'nature-based' tourism from ecotourism specifically, where the former "is defined solely by the recreational activities of the tourist" while "ecotourism is defined as well by a set of principles that include its benefits to both conservation and people in the host country" (Honey, 2008: 7).

## A global panacea?

Understood in this way, ecotourism has, over the past several decades, been enthusiastically promoted by all manner of actors and organizations, including international financial institutions (IFIs) like the World Bank and the United Nations Environment Programme (UNEP), prominent non-governmental organizations (NGOs) like Conservation International (CI) and The Nature Conservancy (TNC), as well as countless private businesses worldwide, as a key component of sustainable development and environmental conservation. Martha Honey, former Executive Director of TIES and current co-director of the Center for Responsible Tourism (CREST), summarizes this enthusiasm quite nicely:

> Around the world, ecotourism has been hailed as a panacea: a way to fund conservation and scientific research, protect fragile and pristine ecosystems, benefit rural communities, promote development in poor countries, enhance ecological and cultural sensitivity, instill environmental awareness and a social conscience in the travel industry, satisfy and educate the discriminating tourist, and, some claim, build world peace.
>
> (Honey, 2008: 4)

As a result of such promotion, ecotourism has become a core component of the so-called Integrated Conservation and Development Programmes (ICDPs) that have been promoted throughout the world since the 1990s, intended to inspire local people's support for conservation by developing mechanisms to generate income from 'non-consumptive' use of *in situ* biodiversity (Borgerhoff Mulder & Coppolillo, 2005). It is now difficult to find conservation areas anywhere in the world that do include ecotourism development as part of their management plans.

This promise of ecotourism to support community-based sustainable development is based in large part on its particular character. First, ecotourism appears especially suited for development in low income areas of the Global South that have been largely neglected by previous development efforts as these are, by virtue of this very lack of industrialization, precisely the places that ecotourists characteristically seek. In addition, rural areas of low income societies are seen to hold a competitive advantage for ecotourism enterprises due to the fact that they are less expensive (as a result of lower wages and property costs, etc.) than rural areas of wealthier societies. Ecotourism is also considered ideally suited for small scale, locally-based development as ecotourists tend to seek less-developed areas and will go elsewhere if areas experience too much transformation. Hence the practice seems to embody an intrinsic incentive to limit the type of rampant overdevelopment associated with the mainstream tourism industry in many parts of the world (Mowforth & Munt, 2008).

## Refining the framework

Yet in actual practice it has proven much more difficult to develop ecotourism operations that approach this lofty potential than early proponents had hoped, leading to a large body of research highlighting the deficiencies of 'actually existing' ecotourism and the many obstacles in the face of its effective implementation (Duffy, 2002). This has caused some to become disillusioned with the concept and to argue for discarding it altogether (Mowforth & Munt, 2008), while others insist instead on the need to work harder to make sure that only operations that meet the standards of sound ecotourism are able to claim the label (Honey, 2008). Consequently, various organizations have developed lists of criteria that committed ecotourism operations should seek to fulfill. The following, from UNEP, is typical:

- Contribute to conservation of biodiversity
- Sustain well-being of local people
- Include interpretative/learning experience
- Promote responsible tourist action
- Delivered by small businesses to small groups
- Emphasize local participation and ownership

(Wood, 2002)

A key question for promoters of ecotourism, therefore, is how to encourage operators to implement these practices rather than merely greenwashing. One popular strategy is certification schemes, wherein operations are evaluated by a third party in terms of their fulfilment of sound ecotourism criteria (see Honey, 2002). Based on this, operators are certified and often promoted as genuine deliverers of quality services. In Costa Rica, for instance, a nationwide Certification for Sustainable Tourism (CST) programme confers on operators between one and five 'leaves' that they can showcase in promotional material (Bien, 2002). The common rationale for such programmes is that the extra market share operators can capture via certification will incentivize the adoption of sound social and environmental practices.

## Neoliberalizing environmental governance

In this sense, ecotourism can be understood as a particular form of environmental governance intended to influence the way users relate to natural resources (Fletcher, 2014). In particular, ecotourism has been described as a key component of a broader trend towards 'neoliberalization' within environmental governance in general since the 1980s, in line with the larger campaign to promote core neoliberal principles including privatization, decentralization, de-regulation, marketization, and commodification within the global political economy as a whole (Heynen et al., 2007; Büscher et al., 2014). Practices like ecotourism that seek not to extract resources and transform them into physical commodities like timber or gold, but rather to maintain them *in situ* as the basis of non-consumptive revenue generation, are considered a subset of this dynamic focused on 'neoliberal conservation' specifically (Büscher et al., 2014; Fletcher, 2014).

Key to this neoliberalization of environmental management is a conviction that governance in general functions most efficiently when it entails not direct regulation but rather a softer practice of creating incentive structures to influence how people choose among alternative courses of action (Fletcher, 2010). In terms of this perspective, individuals are commonly understood as 'rational actors' who calculate the costs and benefits of these different possible actions and choose that which maximizes their material utility. Effective governance, in this sense, entails providing incentives sufficient that individuals will choose the desired behavior, thus obviating the need for direct regulation. Within the promotion of ecotourism, Honey (2008: 14) terms this the 'stakeholder theory' asserting "that people will protect what they receive value from." This is, essentially, the rationale for ecotourism certification schemes and other voluntary strategies intended to motivate the adoption of sound ecotourism practices.

## Conclusion

The extent to which this strategy actually works remains a matter of contention. All in all, available research suggests that while effective ecotourism is difficult to implement it can be accomplished under certain circumstances (Krüger, 2005). Regardless, ecotourism remains a strong growth industry and key component of sustainable development policy worldwide and will likely continue to be so for the foreseeable future.

## Learning resources

Many resources and examples concerning ecotourism can be found through the Center for Responsible Tourism (CREST): www.responsibletravel.org.

For general information about ecotourism principles and initiatives, see The International Ecotourism Society (TIES): www.ecotourism.org.

For an example of an ecotourism certification program, visit Costa Rica Certification for Sustainable Tourism (CST): www.turismo-sostenible.co.cr.

For general information about sustainable tourism globally, visit the United Nations World Tourism Organization (UNWTO): www.unwto.org.

## Bibliography

Bien, A. (2002) 'Environmental certification for tourism in Central America: CST and other programs'. In Honey, M. (ed.) *Ecotourism and Certification: Setting Standards in Practice*, Washington, DC: Island Press.

Borgerhoff Mulder, M. & Coppolillo, P. (2005) *Conservation*. Princeton, NJ: Princeton University Press.

Büscher, B., Dressler, W. & Fletcher, R. (eds) (2014) *Nature™ Inc.: Environmental Conservation in the Neoliberal Age*. Tucson, AZ: University of Arizona Press.

Duffy, R. (2002) *Trip Too Far: Ecotourism, Politics, and Exploitation*. London: Earthscan.

Fletcher, R. (2010) 'Neoliberal environmentality: towards a poststructuralist political ecology of the conservation debate', *Conservation and Society*, 8(3): 171–181.

Fletcher, R. (2014) *Romancing the Wild: Cultural Dimensions of Ecotourism*. Durham, NC: Duke University Press.

Heynen, N., McCarthy, J., Prudham, S. & Robbins, P. (eds) (2007) *Neoliberal Environments: False Promises and Unnatural Consequences*. New York: Routledge.

Honey, M. (ed.) (2002) 'Environmental certification for tourism in Central America: CST and other programs'. In Honey, M. (ed.) *Ecotourism and Certification: Setting Standards in Practice*, Washington, DC: Island Press.

Honey, M. (2008) *Ecotourism and Sustainable Development: Who Owns Paradise?* 2nd edn. New York: Island Press.

Krüger, O. (2005) 'The role of ecotourism in conservation: Panacea or Pandora's box?', *Biodiversity and Conservation*, 14: 579–600.

Mowforth, M. & Munt, I. (2008) *Tourism and Sustainability: Development, Globalization, and New Tourism in the Third World*. 3rd edn. London: Routledge.

United Nations World Tourism Organization (UNWTO) (2016) *Tourism Highlights 2015*. Madrid: UNWTO.

Wood, M. (2002) *Ecotourism: Principles, Policies and Practices for Sustainability*. Nairobi, Kenya: UNEP.

# 6.4

# Ecological restoration

Matthias Gross

Since the early 1980s ecological restoration has been a rapidly growing field of ecological science and practice. As originally developed in the 1970s on the tall-grass prairies of the North American midwest and subsequently in other regions of the world, ecological restoration has proven value, not only as a way of reversing environmental damage, but also as a context for negotiating the relationship between human society and other parts of nature (cf. Egan et al., 2011; Jordan & Lubick, 2011). Most often ecological restoration refers to the practice of restoring, "renaturing," or re-designing a piece of land, whereas the term *restoration ecology* refers to a specialism within ecology as a science. *Ecological restoration*, in turn, is rendered the overall practice. Despite the etymology of the term restoration the practice does not strictly focus on returning degraded sites to conditions that prevailed prior to social influence. But whatever the exact goals of restoration and ecological design projects, these processes disrupt existing natural processes and ecosystems (cf. France, 2011; Higgs et al., 2014; Havlick et al., 2014).

The term restoration ecology originally was coined by William R. Jordan III (see Jordan et al., 1987; Court, 2012). However, neither the ecological foundations nor the practice of restoration are entirely new (cf. Gross, 2007). One could say that humans have practiced ecological restoration ever since farmers discovered shifting cultivation. However, in today's restoration projects, many of the practitioners employ strategies where major parts of the work and planning are undertaken by lay people and community organizations. In many areas, success in restoration projects is measured in the context of practitioner skills and community goals and less so by the manifestation of underlying scientific principles. At the very least, the restoration of new natural areas will be beyond that of any single contributing discipline or interest group.

Although ecological restoration has come to play an important role in discussions relating to the environment and environmental policy over the past decade, different authors and groups understand the terms quite differently. The most widely accepted definition is perhaps that ecological restoration is the active attempt to return an ecological system to a former condition following a period of alteration or disturbance through the reconstitution of processes, the reintroduction of species, and the removal or control of species that are inappropriate to the model system. The Society for Ecological Restoration (SER) defines ecological restoration as "the process of assisting the recovery of an ecosystem that has been degraded, damaged, or destroyed" (www.ser.org). Restoration is perhaps most readily understood as a form of environmental design and rehabilitation that is distinguished from other forms of rehabilitation by its commitment to the re-creation of different aspects of a model system, regardless of their value to humans. It furthermore emphasizes participatory processes aimed at a "re-wilding" of the landscape that is being restored. This also means that a piece of nature can develop in a direction that may not have been intended by the planners involved. This, indeed, can also include unexpected and even unwanted elements. Steven Packard, an early practitioner of ecological restoration from the late 1970s on, stated that "every restorationist knows the ecosystem will respond in unpredictable ways that rise out of itself. That's precisely what we want to liberate" (Packard, 1993: 14). For Jordan, this attitude goes so far that ecological restoration should represent "an act of self-abnegation and deference to the ecosystem that takes the restorationist beyond economics to establish the deeper relationship with the ecosystem" (Jordan, 2006: 26). Thus understood, the concept of ecological restoration promotes a special intimacy between nature and culture where the unexpected or self-organizational elements of the natural world are aimed to be included. Ecological restoration practice then is inseparable from cultural restoration. In many cases such as in Australia or Africa this includes the recovery of indigenous knowledge about the natural world as well as the cultural survival of indigenous peoples (cf. Martinez, 2003). Given that a lot of indigenous conceptions of nature, natural aesthetics, and ecological time (e.g. among North American or Australian natives) can differ among different parts of society, paradoxical situations based on "contradictory certainties" of "correct" scientific findings and different expert opinions.

Unlike in North America, in Europe, the term "renaturing" has traditionally been in use instead of restoration (cf. Westphal et al., 2010). It can be argued that the term "renature" moves beyond the time-honored debates about "restore to what time period?", and allows participants to more broadly envision what aspects of past, present, and future that they would like to be part of. It also seems to have the advantage of treating humans, and nonhumans, fire, wind, and so forth as participating in the shaping of this new space, thus avoiding another discussion on "what's natural and what's human?". However, although the terminology is different, in many respects both renaturing and restoration activities have both been focusing on the surprising aspects in the activities of "designing" or "shaping" a piece of land, addressing the negative influences of people on the "natural" landscape and the potential benefits, and the values attached to nature. In this way, many debates have started since the early 2000s about the novelty of ecosystems that have been "restored" into the future (Hobbs et al., 2013; Kueffer, 2015). Indeed, despite the "conservative" connotation of the term restoration, most ecological restoration projects have been projects of inventing new ecosystems and, given the dynamics of natural processes, all ecosystems, whether strictly designed along historical reference points or invented towards new forms of ecosystem compositions, eventually will all be novel since natural processes always move in unforeseen directions. Thus, even the early restoration projects today can be seen as projects that invented "new nature." This has been indicated by terms such as "invented landscapes" (Turner, 1994) or even as "synthetic ecology" (cf. Jordan et al., 1987). Thus, in a way, to restore and to invent something new in the actual practice of restoration always go hand in hand.

From the beginning in the later 1970s, the core strategy in restoration projects has been that hands-on practitioners who, although they may have little or no formal training in academic ecology, often achieve insights that contribute to and even challenge existing ideas about the ecology of the system being restored (Jordan, 2006). For a long time, the diffusion of restoration knowledge thus occurred primarily as the original practitioners moved to new problem contexts, rather than through reporting results that were reported in professional journals or at academic conferences. Communication links were maintained partly through formal and partly through informal channels (Gross, 2002). Through this kind of research, a good deal of often site-specific knowledge was lost after a few years. Until the mid-1980s, the state of science did not matter to the majority of ecological restoration practitioners. The practitioners learned only as much as they needed to restore or re-design a system. Robert Cabin (2011) even goes so far as to conclude that testing general scientific hypotheses and rigorous data in ecological restoration would be of limited use. In this vein, many ecological restoration projects from their beginnings were practice-oriented and were carried out by amateurs who learned as much ecology as they needed for restoring ecosystems. Restoration expertise thus is a form of expertise where the practitioners are not necessarily able to talk like academic ecologists, but they can actually do their ecology tests.

By treating humans as a mature part of nature whose existence in the age of the Anthropocene is even necessary for plants and animals to exist at all, ecological restoration as a nature-designing and novelty producing activity has little place for romantic notions of contemplation or humility. This questions the idea that nature is natural only when it is left untouched by humans.

## Learning resources

The website of the Society for Ecological Restoration (SER) includes an overview of forms of restoration in different cultural contexts. SER was founded in 1987 as a non-profit organization to connect individuals and organizations from around the world, utilizing many types of knowledge and cultural perspectives: www.ser.org.

The New Academy for Nature and Culture (USA) is promoting ecological restoration practice that is ecologically sound and culturally sensitive: http://environmentalprospect.org/new-academy.

The journal *Ecological Restoration* was founded in 1981 and serves as a forum for interdisciplinary aspects of restoring landscapes, including participatory collaborations between different stakeholders. This publication focusses in on education and reports restoration ecology progress all over the world: https://uwpress.wisc.edu/journals/journals/er.html.

To find out more about the key debates on ecological restoration, the following is an accessible overview: Egan, D., Hjerpe, E.E. & Abrams, J. (eds) (2011) *Human Dimensions of Ecological Restoration: Integrating Science, Nature, and Culture*. Washington, DC: Island Press.

An accessible history of the ecological restoration movement and its development into an academic discipline can be found in: Jordan, W.R. III & Lubick, G.M. (2011) *Making Nature Whole: A History of Ecological Restoration*. Washington, DC: Island Press.

## Bibliography

Cabin, R.J. (2011) *Intelligent Tinkering: Bridging the Gap between Science and Practice*. Washington, DC: Island Press.

Court, F.E. (2012) *Pioneers of Ecological Restoration: The People and Legacy of the University of Wisconsin Arboretum*. Madison, WI: University of Wisconsin Press.

Egan, D., Hjerpe, E.E. & Abrams, J. (eds) (2011) *Human Dimensions of Ecological Restoration: Integrating Science, Nature, and Culture*. Washington, DC: Island Press.

France, R.L. (ed.) (2011) *Restorative Redevelopment of Devastated Ecocultural Landscapes*. Boca Raton, FL: CRC Press.

Gross, M. (2002) 'New Natures and Old Science: Hands-on Practice and Academic Research in Ecological Restoration', *Science Studies*, 15(2): 17–35.

Gross, M. (2007) 'Restoration and the Origins of Ecology', *Restoration Ecology*, 15(3): 375–376.

Gross, M. (2016) 'Layered Industrial Sites: Experimental Landscapes and the Virtues of Ignorance'. In Hourdequin, M. & Havlick, D. (eds) *Restoring Layered Landscapes: History, Ecology, and Culture* (pp. 73–91). Oxford: Oxford University Press.

Havlick, D., Hourdequin, M. & John, M. (2014) 'Examining Restoration Goals at a Former Military Site: Rocky Mountain Arsenal, Colorado', *Nature & Culture*, 9(3): 288–315.

Higgs, E., Falk, D.A., Guerrini, A., Hall, M., Harris, J., Hobbs, R., Jackson, S.T., Rhemtulla, J.M. & Throop, W. (2014) 'The Changing Role of History in Restoration Ecology', *Frontiers in Ecology and the Environment*, 12(9): 499–506.

Hobbs, R.J., Higgs, E.S. & Hall, C. (eds) (2013) *Novel Ecosystems: Intervening in the New Ecological World Order*. New York: Wiley-Blackwell.

Jordan, W.R. III (2006) 'Ecological Restoration: Carving a Niche for Humans in the Classic Landscape', *Nature & Culture*, 1(1): 22–35.

Jordan, W.R. III & Lubick, G.M. (2011) *Making Nature Whole: A History of Ecological Restoration*. Washington, DC: Island Press.

Jordan, W.R. III, Gilpin, M.E. & Aber, J.D. (eds) (1987) *Restoration Ecology: A Synthetic Approach to Ecological Research*. Cambridge: Cambridge University Press.

Kueffer, C. (2015) 'Ecological Novelty: Towards an Interdisciplinary Understanding of Ecological Change in the Anthropocene'. In Greschke, H.M. & Tischler, J (eds) *Grounding Global Climate Change* (pp. 19–37). Heidelberg: Springer.

Martinez, D. (2003) 'Protected Areas, Indigenous Peoples, and the Western Idea of Nature', *Ecological Restoration*, 21(4): 247–250.

Packard, S. (1993) 'Restoring Oak Ecosystems', *Restoration & Management Notes*, 11(1): 5–16.

Turner, F. (1994) 'The Invented Landscape'. In Baldwin, A.D., de Luce, J. & Pletsch, C. (eds) *Beyond Preservation: Restoring and Inventing Landscapes* (pp. 35–66). Minneapolis: University of Minnesota Press.

Westphal, L.M., Gobster, P.H. & Gross, M. (2010) 'Models for Renaturing Brownfield Areas'. In Hall, M. (ed.) *Restoration and History: The Search for a Usable Environmental Past* (pp. 208–217). Abingdon: Routledge.

# Environmentalism

6.5

Marco Armiero

Environmentalism is a political movement whose aim, initially, was to protect valued aspects of the non-human world, expanding later to humans, especially to human health and its connections to the environment. John Muir, Chico Mendez, Rachel Carson, and Medha Patkar: all may be considered crucial figures in the development of environmentalism since the late nineteenth century. Nonetheless, it seems odd, if not impossible, to gather them under one single banner. A passionate advocate of wilderness protection, the union leader fighting to save the Amazon and the people living through it, the scientist who exposed the dangers of pesticides, and the activist opposing the construction of dams can represent the multiplicity of environmentalism; thereby, the difficulty of presenting it as a coherent and unified movement. In the book *Keywords for Environmental Studies* (Adamson et al., 2016) the economist and historian Joan Martinez-Alier has chosen to speak of environmentalisms, in the plural, to signal the irreducible diversity of the movement and its cultures. Although using the singular for the title of her book, environmental historian Carolyn Merchant (2005) traced the multiple trajectories of what she called 'radical ecology', individuating at least six main strands: deep ecology, spiritual ecology, social ecology, green politics, ecofeminism, and anti-globalization and sustainability. This chapter shares this vision of environmentalism as a plural movement. I will follow Martinez-Alier's and Guha's (1997) threefold organization of environmentalism, adding to it a few nuisances coming from Merchant's classificatory attempt.

## Three varieties of environmentalism

The difficulty in defining environmentalism goes together with the problematic definition of nature. Raymond Williams's observation (1983: 219) that nature is one of the most complex and controversial words in the English language remains the basis for any discussion on the topic. The complexity of nature is mirrored in the multifaceted character of environmentalism. The megafauna and a working class neighborhood, the human body and Gaia, the forest and the indigenous people living in it can all be part of what we call nature, thereby igniting different environmentalist cultures and practices. The Catalan economist Joan Martinez-Alier and the Indian historian Ramachandra Guha usefully identified three main branches of environmentalism: the cult of wilderness, the gospel of eco-efficiency, and the environmentalism of the poor. Although some branches seem to not fit in this classification—and I will elaborate on them later—Martinez-Alier's and Guha's proposal is still one of the most comprehensive.

## The cult of wilderness

The cult of wilderness includes the environmentalist cultures and practices aiming to preserve 'untouched' portions of nature, that is, what is considered wild. It presupposes a certain idea of wilderness and a cultural shift in public opinion, transforming it from wasteland to a space of beauty and appreciation. Historically, this kind of environmentalism has been associated with the creation of the first national parks and legislations aiming to protect special landscapes or iconic species. The champions of that strand of environmentalism were people such as John Muir (1838–1914) and John Ruskin (1819–1900), that is, intellectuals who contributed in proposing a positive vision of wilderness. For a long time mainstream environmentalist organizations stayed in that realm, concentrating their efforts in protecting wilderness and teaching people to appreciate it—something that did not go without contradictions, as in the controversial relationship between the protection of nature and promotion of mass outdoor cultures. In the last decades the cult of wilderness has been criticized for its blindness towards social issues, its colonial overtones in the Global South, and its obliteration of the rights of land-based indigenous societies in order to foster protection of some kind of untouched nature. The cult is radical in its own way. When nature is mostly seen as a mine for resources or a dump for waste, fighting to protect some parts of it from the worst ravages of the market economy might become an attempt to challenge the sacredness of economic growth.

## Eco-efficiency

By contrast, the gospel of eco-efficiency does not oppose economic growth and development. As the cult of wilderness is based upon separation, protection, and contemplation, eco-efficiency environmentalism has as keywords, wise-use, sustainability, and expertise. The origins of such environmentalist culture can be traced to eighteenth-century German forestry, which aimed at exploiting forests without destroying them. Undoubtedly, the rise of sustainability as a concept and policy-object has represented a turning point in the environmental movement. Sustainability embodies the most contemporary version of the eco-efficient environmentalism through its ambition to use nature without compromising its reproduction, thereby, its availability for future generations. A great trust has been placed in technology and science as possible solutions for environmental problems. This model of environmentalism has been criticized for its utilitarian view of nature, its faith in science, and its frequent disregard of social inequalities when considering the best use of the environment. Nonetheless, the gospel of efficiency has contributed to a better management of natural resources and waste and has enhanced the debate on the ecological limits of the planet.

## *Third class environmentalism*

It has been often argued that environmentalism is a luxury for rich people. Only those who have a full belly can care for less essential items and issues. That the environment is 'less essential' is, of course, questionable. While this might be the Global North vision of it, it does not work at all in the Global South where the 'environment' is manifestly a pillar of local livelihoods. Caring for the environment, Martinez-Alier and Guha have argued, is not an exclusive prerogative of the rich. Indigenous communities and subaltern groups around the world have fought against the destruction of their environments confronting extractivist industries, monocultures, enclosures, invasive infrastructures, and contamination. One can think of Chico Mendez (1944–1988), the well known Brazilian unions' leader and environmental activist, whose proposal of nature reserves protected from overseas mining firms represents a radically alternative way of thinking about nature conservation (Rodrigues, 2007; Sedrez, 2014). In the United States this subaltern environmentalism has become organized since the early 1990s under the banner of the environmental justice movement. Rooted in the civil rights movement, this form of environmentalism has mobilized minorities against what has been defined environmental racism, that is, the systematic targeting of those communities as the ultimate site for any kind of unwanted facilities. The fact that the environmental justice movement started to be organized in the early 1990s—the first National People of Color Environmental Leadership Summit was held in Washington DC in 1991—does not imply that it was born at that time. Robert Gottlieb (2005) has explored the longer history of what we might call subaltern environmentalism, unearthing the traces of it in the early twentieth-century reform movement and in the birth of occupational health fostered by the work of Alice Hamilton. In fact, some scholars, such as Stefania Barca (2012) and Chad Montrie (2008), have talked of a working-class environmentalism. According to those scholars, workers were often aware of the ecological risks produced by the organization of work, struggling to improve their conditions of work and life.

## Other environmentalisms

Although quite inclusive, the matrix proposed by Martinez-Alier and Guha might leave some aspects out of the picture. For instance, one might wonder what the advent of the Anthropocene has implied for environmentalism. While one might expect the idea of the Anthropocene to be connected to the most pessimistic strands of environmentalism, most strikingly it has given strength to the so-called ecomodernist trend. Though obviously connected to the eco-efficiency approach, the ecomodernists take the arguments of humans' management of nature to extremes, celebrating humans' total control over the entire planet (Shellenberger et al., 2015). While the ecomodernists propose a technocratic approach to the environmental crisis, the awareness of climate change has ignited a global environmental movement for climate justice (Dawson, 2010). In this case, more than for technical solutions, environmentalists have campaigned for social and political transformations, by and large informed by notions of social justice and fair distribution of harms and benefits.

Radically alternative to the ecomodernist wave, but also distant from the social ecologies of climate justice activism, is the deep ecology tradition. Coming from Norwegian philosopher Arne Naess (1912–2009), deep ecology has embodied the most radical critique of any kind of anthropocentrism, proposing a planetary egalitarianism among species and the overcoming of a utilitarian approach to nature. James Lovelock's Gaia theory—the idea of the Earth as a living organism—can been considered part of this broader movement.

Finally, eco-feminism is another branch of environmentalism which does not fit easily into Martinez-Alier's and Guha's systematization. Eco-feminism is a complex phenomenon which cannot be reduced to one single movement. Born in the 1970s with the writings of French writer Francoise d'Eaubonne and the foundation in Paris of the Ecology-Feminism Center, it consolidated in the US with conferences, publications, and the creation of the Center for Social Ecology in Vermont. Although eco-feminism shares a radical critique of patriarchy as the cause of the oppression of both women and nature, Carolyn Merchant has illustrated the significant differences among its various kinds (liberal, cultural, social, and socialist) in terms of aims and analysis.

## Conclusion

In this chapter Martinez-Alier's and Guha's interpretation has been used to present environmentalism as a threefold, plural historical phenomena. This plurality is connected to the complexity of the concept of nature which leaves unsolved the controversial question of what is included, or what is not. While for a long time, mainstream environmentalism has focused on wilderness, struggling for its preservation and mourning its disappearance, the raising of the eco-efficiency and sustainability discourse has shifted the attention from nature to 'the environment' and its wise use, rather than its conservation. Finally, I have shown that subaltern environmentalism aligns with the academic shift towards socio-natures and hybridity which overcomes the dichotomy nature vs. culture, or humans vs. environments. The chapter ended by including some currents of environmentalism which are not represented in this systematization: ecomodernism, climate justice and ecofeminism. One of the most challenging issues is to explore the rate and possibilities of alliance/conflict among those different branches of environmentalism. I believe this is one of the research lines which deserves more scrutiny.

## Learning resources

### Audio-visual materials

An excellent documentary on the history of environmentalism is *A Fierce Green Fire*: http://www.afiercegreenfire.com.

Robert Bullard is one of the most influential figures in the field of environmental justice. It is possible to watch one of his speeches at: https://www.youtube.com/watch?v=SYVvbs6XsNw.

Environmental historian Robert Gioielli has attempted a sketch of the history of the environmental movements since the mid-nineteenth century. His timeline is available at: https://www.academia.edu/23493428/Environmentalism_An_Incomplete_History.

Environmental racism explained in five minutes by the US geographer Laura Pulido for the KTH Environmental Humanities Laboratory VideoDictionary: https://www.kth.se/2.1231/historia/ehl/ehl-dictionary/environmental-racism-laura-pulido-1.516854.

## Useful texts

Bullard, R.D. (1990) *Dumping in Dixie: Race, Class, and Environmental Quality*. Boulder: Westview Press.

del Mar, D.P. (2014) *Environmentalism*. Abingdon: Routledge.

Martinez-Alier, J. (2005) *The Environmentalism of the Poor: A Study of Ecological Conflicts and Valuation*. New Delhi: Oxford University Press.

Nash, R. (1982) *Wilderness & the American Mind*. New Haven: Yale University Press.

Nixon, R. (2011) *Slow Violence and the Environmentalism of the Poor*. Cambridge: Harvard University Press.

Rootes, C. (2003) *Environmental Protest in Western Europe*. New York: Oxford University Press.

## Bibliography

Adamson, J., Gleason, W.A. & Pellow, D.N. (eds) (2016) *Keywords for Environmental Studies*. New York: New York University Press.

Barca, S. (2012) On working-class environmentalism: a historical and transnational overview, *Interface: A Journal for and About Social Movements*, 2: 61–80.

Dawson, A. (2010) Climate justice: the emerging movement against green capitalism, *South Atlantic Quarterly*, 109(2): 313–338.

Gottlieb R. (2005) *Forcing the Spring: The Transformation of the American Environmental Movement*. Washington, DC: Island Press.

Martinez-Alier, J. & Guha, R. (1997) *Varieties of Environmentalism: Essays North and South*. London: Earthscan.

Merchant, C. (2005) *Radical Ecology. The Search for a Liveable World*. New York: Routledge.

Montrie, C. (2008) *Making a Living: Work and Environment in the United States*. Chapel Hill: University of North Carolina Press.

Rodrigues, G. (2007) *Walking the Forest with Chico Mendes: Struggle for Justice in the Amazon*. Austin: University of Texas Press.

Sedrez, L. (2014) Rubber, trees and communities: Amazon in the twentieth century. In M. Armiero and L. Sedrez (eds) *A History of Environmentalism: Local Struggles, Global Histories*. London: Bloomsbury.

Shellenberger, M., Ellis, E. & Nordhaus, T. (2015) *An Ecomodernist Manifesto*. Online. Available at: http://www.ecomodernism.org/manifesto-english.

Williams, R. (1983) *Keywords*. New York: Oxford University Press.

# 6.6 Environmental art

## Harriet Hawkins and Anja Kanngieser

Art has, for millennia, offered artists both subject matter and materials through which to engage with and respond to the environment. Indeed art 'frames' environmental relations, whether this be prehistoric cave paintings with their hand-printed depictions of human–animal relations, or the institutionalized art forms of the Western Enlightenment era when landscape painting was grudgingly admitted to the artistic canon alongside portraiture and history painting. Within the scope of landscape painting alone environmental relations depicted range from romantic rural idylls (nature as spectacle reproducing establishment power relations) and productively ordered agricultural landscapes (environment as a resource organized by godly order), to vast and unsettling sublimes whose very force lies in the inhuman power of a nature that resists intelligibility. Throughout the nineteenth and twentieth centuries art forms can be conceived of as responses to environmental concerns stimulated through the effects of intensified technological development. We see this in work as varied as J.M.W. Turner's atmospheric oil paintings of nineteenth-century London's industrial smogs, or the landscape-scale sculptures of the mid-twentieth century American Land artists. The latter understood as a response to the 'fragile earth' pictured in the era's famous Earthrise and the Blue Marble images taken from space. Climate change, and the Anthropocene more generally, have become a defining force in early twenty-first-century art. The effectiveness, however, of a visual vocabulary dominated by icy destruction, flooded islands and dislocated people and animals has been questioned. Many have sought instead environmental art fit for an Anthropocene era in which the space-times of our environmental understandings are thrown into disarray (Clark, 2015).

Here we explore three forms of address that art works might make to such concerns around: i) who makes environmental knowledge, ii) the often abstract and insensible dimensions of environmental issues, and iii) the evolution of environmental encounters premised on entanglements of humans and other forms of life. Crucial throughout is a movement beyond conceiving of art as 'picturing' towards the possibilities various forms of artistic practice now offer in understanding, imagining and even bringing about alterative environmental futures.

## Remaking environmental expertise

Art not only produces certain forms of knowledge, it also offers the coordinates through which to critique privileged knowledge forms. We might think, for example, of the 'amateur' knowledge extrapolated from paintings, stories and photographs that have long offered source material – aesthetic proxies – for the reconstruction of past environments (Dixon et al., 2013). Or we might reflect on the role of arts practices as critical to the production and reiteration of indigenous knowledges about the environment and so of relations to it (Horton & Berlo, 2013). Arts practices, whether art-science ones or participatory arts practices, combine with forms of citizen science and competency groups not only to produce new environmental knowledge, but also to remake the terms of the production of science itself. Unseating the singular image of the climate expert, a broader sense emerges of who understands and can intervene within our changing environments (Ingram, 2013; Hawkins et al., 2015; Zurba & Berkes, 2014). The *Arctic Perspective Initiative* (2010–), which is a collaboration between artists Marko Peljhan and Matthew Biederman, and the Arts Catalyst and geographer Michael Bravo, offers an excellent example of this. It focuses on exploring and enabling the transmission of Inuit Arctic knowledges through free and accessible technologies (Bravo & Triscott, 2010; see learning resources for further information). Exploring such everyday icy inhabitations as well as the Inuit's specialist environmental knowledge challenges scientific imperialisms and remakes the coordinates through which this charismatic space of climate change is known. Another Arts Catalyst project entitled *Wrecked: On the Intertidal Zone* (2015, see Figure 6.6.1) created the Thames Estuary (UK) as experimental terrain across which myriad ways of knowing, from those of local bird watchers, fishermen and mud-walkers to scientific studies and information gleaned through political and corporate wrangling created a collage of environmental knowledges that refused any knowledge hierarchy and co-joined artistic, scientific and everyday experimentalisms (see learning resources for further information).

## Reimagining abstractions

The complexities of environmental change have a well-developed visual logic, rendering abstract ideas and complexities intelligible through the identification of tipping points, the diagramming of hockey stick warming curves, and the coloration of burning embers that co-join the symbolization of warming with an aesthetic schematics of danger (Liverman, 2009). Such scientific visualizations combine with the picturing of the charismatic species and spaces of climate change; from Arctic polar bears to the 'drowning' of Pacific small island states. For lay Western audiences, however, these work through a logic of distancing, situating environmental change and its effects in space-times other than those of the communities implicated within the

**Figure 6.6.1** Wrecked: On the Intertidal Zone – Graveyard of Lost Species.

www.yoha.co.uk, 2015.

creation of current environmental crisis, and those best positioned to begin mitigation and resilience measures. Given the worries environmental scientists and activists alike share over gaps between environmental knowledge and action caused (many posit) by such distancing, we might look to art for alternative environmental logics. One of the most common, and arguably successful of these, has been the everyday imaginaries of climate change cultivated by Western feminist artists. Juanita Schlaepfer-Miller, for example, worked alongside Swiss environmental scientists to create *Climate Hope Garden*, a project utilizing growing chambers to recreate the effects of climate change (under three different scenarios) on Swiss domestic gardens. The result demonstrated how many of the species of flowers, trees and vegetables the Swiss had grown up tending would no longer be viable under predicted climatic conditions. Another project that brought climate change 'home', revealing its effects across Western landscapes and on beloved species, was Deirdre Nelson's *Bird Yarns* (2013–). (See Figure 6.6.2; see also learning resources for more information).

Working with island communities in North Scotland, Nelson devised a knitting pattern for novice and experienced knitters alike to produce woolly Arctic terns. In crafting these avian bodies in knitting groups, and as part of virtual internet knitting networks across Europe and beyond, knitters were encouraged into collective reflection on the effects of climate change on tern food sources and mating patterns, which, in turn, affected their life span, bodily morphology and their very presence on once popular migration paths. Through these knitted bird bodies, climate change is at once a global phenomenon, but one with localized effects that are proximate to, rather than distant from, Western spaces and environmental practices.

**Figure 6.6.2** Deidre Nelson, *Bird Yarns* (2013–).

Photographer: Elizabeth Straughan.

## Environmental entanglements

For many environmental thinkers one of the challenges underpinning positive change in Western knowledge and environmental action is the separation between humans and other life forms. To appreciate the fundamental connections between humans and the stuff of life – whether that be animals that are 'big like us', teeming bacterial worlds, or geologic processes and forms – requires the building of alternative understandings of humanity that countermand logics of dominance that refuse the entanglement of human and non-human (Gibson-Graham, 2011). Such 'flat' ontologies, seeking commonalities and contiguities between human and non-human, often situate environmental connections as the building blocks for an alternative environmental ethics and politics. Over-easy celebrations of connection risk, however, proposing an under-defined and sentimental ethics of care. Complex environmental encounters offered through Arts practices can challenge this, one example being *You Are Variations* (2012–, see Figure 6.6.3) developed by Christina Della Giustina alongside scientists in the Swiss Federal Institute for Snow, Forest and Landscape Research. During a 9-month residency Della Giustina deployed a range of artistic and scientific visualization and sonification practices to explore the trees that were the subject of the Institute's research. An array of

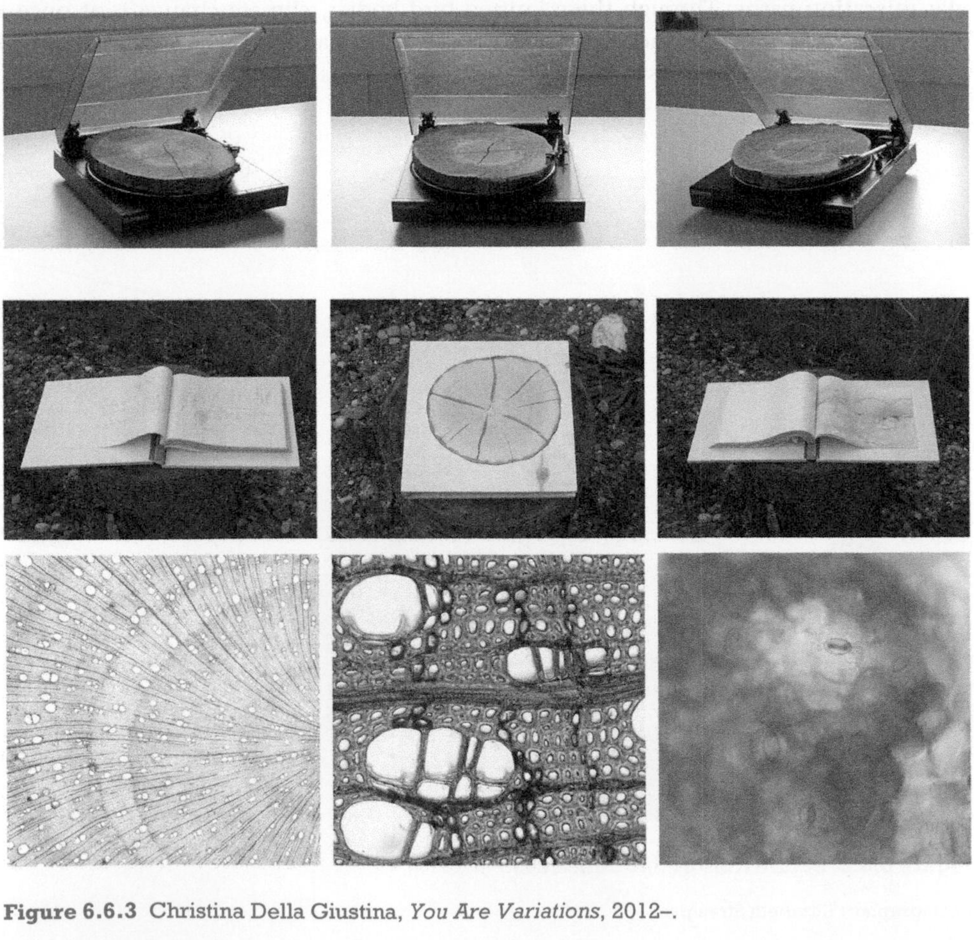

**Figure 6.6.3** Christina Della Giustina, *You Are Variations*, 2012–.

practices from microscopic films to taxonomic drawings, experiments with field instrumentation, as well as sketches and performative explorations enabled her to create multi-sensuous encounters for scientists, herself and her audiences.

A series of workshops saw scientists devising parameters to map biological and atmospheric characteristics onto functions of sound creating intricate soundings of trees from live streams of field data. For example, the atomic and molecular structures of the tree at a particular point (e.g. soil-root interface) became operations of rhythm, pulse and pitch, whilst temperature readings gave the time signature, and changing rates of evapo-transpiration rendered shifts in volume. As they worried over fidelity to data as much as what trees and their lively processes would sound like these scientists were encouraged into other relationships with their specimens than those premised on difference and distance.

## Conclusion: encountering the environment

Within many literatures on the environment, especially those concerned with the current environmental crisis, there is a clear need to foster new forms of environmental encounter. We would contend that this is precisely what arts practices enable; they make possible environmental understanding and action that countermands the knowledge hierarchies and the abstractions and distances that constitute much contemporary environmental knowledge. Instead, arts practices offer environmental encounters composed according to other logics; of proximity, of flat ontologies and epistemologies that entangle knowledge forms as much as they entangle forms of life. As we reflect on our earthly futures, we should heed the possibilities (whilst also remaining critical) of what arts practices can offer, especially when they come together with environmental scientists and local communities to remake our ways of being in, knowing and relating to the environments within which we live.

## Learning resources

There are lots of examples throughout the art work of individual artists, arts organizations and institutions as well as other wider groups or individuals who are engaging with environmental issues. The below offers some key ones:

The Arts Catalyst (interactive website of an arts-science organization who do a range of environmental work) www.artscatalyst.org.

Cape Farewell: The Cultural Response to Climate Change (interactive website of an arts organization dedicated to developing creative practices engaging with climate change) www.capefarewell.com.

The Pacific Storytellers Cooperative (a platform for place-based stories from the Pacific Islands) http://prel.org/programs/storytellers.

Glacier Hub (a platform collecting together multi-model knowledge on glaciers) http://glacierhub.org.

## Bibliography

Bravo, M.T. & Triscott, N. (eds) (2010) *Arctic Geopolitics and Autonomy*. London: Hatje Cantz.

Clark, T. (2015) *Ecocritism on the Edge: The Anthropocene as a Threshold Concept*. London: Bloomsbury.

Dixon, D., Hawkins, H. & Straughan, E. (2013) 'Sublimity, formalism and the place of art within geomorphology', *Progress in Physical Geography*, 37(2): 227–247.

Gibson-Graham, J.K. (2011) 'A feminist project of belonging for the Anthropocene', *Gender, Place and Culture*, 18(1): 1–21.

Hawkins, H., Marston, S., Ingram, M. & Straughan, E. (2015) 'The arts of socio-ecological transformation', *Annals of the Association of American Geographers*, 105(2): 331–341.

Horton, J.L. & Berlo, J.C. (2013) 'Beyond the mirror: indigenous ecologies and "new materialisms" in contemporary art', *Third Text*, 27(1): 17–28.

Ingram, M. (2013) 'Washing urban water: diplomacy in environmental art in the Bronx, New York City'. *Gender, Place, Culture*, 21(1): 105–122.

Liverman, D.M. (2009) 'Conventions of climate change: constructions of danger and the dispossession of the atmosphere', *Journal of Historical Geography*, 35(2): 279–296.

Zurba, M. & Berkes, F. (2014) 'Caring for country through participatory art: creating a boundary object for communicating indigenous knowledge and values', *Local Environment*, 19(8): 821–836.

# Environmental fiction

6.7

Joni Adamson

The failure of the 2009 United Nations Copenhagen talks (COP 15) on climate change illustrated some significant barriers to achieving sustainable futures, including bureaucratic inertia and the tendency of governments, university academics and other groups to work in isolation. To break through this inertia, there has been a notable movement among literature professors, engineers, architects, and urban planners to embrace a fast emerging genre of fiction referred to as "cli-fi," short for "climate fiction." In a 2014 "Room for Debate" feature in *The New York Times*, several writers, critics and scientists contributed opinion pieces addressing the efficacy of imagination and fiction for identifying and solving the problems associated with climate change. One of them, George Marshall, founder of the Climate Outreach Network, argued that fiction and imagination are vitally important for making sense of climate change. We need "the alchemy of stories" to turn cold data "into the emotional gold it needs to mobilize" (Marshall, 2014). Another, Sheree Renée Thomas, recalled the role of folktale and myth in building resistance to slavery in the Americas. "There is power in fantasy, especially in stories that urge us to face the impossible or find ways to survive" (2014).

In *The Guardian*, Rodge Glass observed that two decades ago it would have been difficult to identify even a handful of books that fell under the banner of "cli-fi." However, now an award-winning corpus of novels is aiming to warn readers of all the environmental nightmares to come (Glass, 2013). Books such as Barbara Kingsolver's *Flight Behavior*, George Turner's *The Sea and Summer*, Nathaniel Rich's *Odds Against Tomorrow*, Ian McEwan's *Solar*, and Paolo Bacigalupi's *The Windup Girl* are often mentioned as exemplars (Gunn, 2014).

In universities, there has been a parallel explosion in courses with titles such as "The Cultures of Climate Change," "The Political Ecology of the Imagination," and "Science Fiction to Science Fabrication" that focus on cli-fi and its parent genre, science fiction (Pérez-Peña, 2014; Gunn, 2014). Courses are designed not only by literature departments, but by departments in other disciplines also interested in exploring how fiction might be employed to break through the inertia of inaction on today's "wicked environmental problems." Interviewed about their goals, instructors express a hope that these novels will fire "inventors' imaginations" among their students who will then create, engineer, and build "plausible futures" that are socially, technologically, and ecologically agile enough to adapt to coming weather extremes (Pérez-Peña, 2014).

But not everyone is getting on the cli-fi bandwagon. Science fiction has a much longer literary history and, arguably, has already inspired space travel and a number of technologies that have changed modern cultures. In John Christopher's *The Death of Grass* (1956), a new virus infects grasses across the globe, causing mass famine. J.G. Ballard's *The Drowned World* (1962) is set in 2145, after solar radiation has melted the polar ice caps and London has become a tropical swamp. T.C. Boyle's *A Friend of the Earth* (2000) is set in 2025—in a hot, food-scarce U.S. plagued by mass extinction (Morin, 2014). With reference to her popular *MaddAddam* trilogy, internationally renowned novelist Margaret Atwood has also questioned the term cli-fi, noting that she prefers the term "speculative fiction" (Finn, 2015). "I think calling it climate change is rather limiting . . . I would rather call it *everything change*." Like other writers of speculative or science fiction, she writes to spark imaginations towards alternative possibilities that will allow both humans and nonhumans to survive, rather than face certain apocalypse (Finn, 2015).

Environmental fiction has been studied professionally at least since the late 1990s, when the field of environmental literary criticism, or ecocriticism, emerged as a vibrant new humanities endeavor. This field became one of the foundations of the environmental humanities. Ecocritics specifically, and environmental humanists more generally, have long been interested in how study of the "literature of nature" in diverse cultures (Murphy, 1998) can yield a better understanding of how narrative and storytelling relate to human attitudes and behaviors (Armbruster, 2016: 157).

The long relationship of the literatures of nature to what founding ecocritic Lawrence Buell has famously called the emergence of the modern "environmental imagination" (Buell, 1995) may make claims that cli-fi is "new," a bit premature. As cultural theorist Raymond Williams once noted, any full history of the uses of the keyword "nature," would "be a history of a large part of human thought" (1976[1983]: 219, 221). Nature has been the subject of imaginative narratives around the globe, among many peoples and for thousands of years, inspiring sustained attention as far and wide as the *Epic of Gilgamesh*, the Hebrew Bible, the *Tao Te ching*, Aristotle's *Physics*, Virgil's *Georgics*, and the *Bhagavad Gita* (Armbruster, 2016: 156). Many of the oldest recorded oral story cycles and texts written in Egypt, Rome, China, Europe, and the Americas illustrate that humans have long been telling stories that collect and archive sophisticated ethnobotanical and scientific information in narratives explaining human group relations to other human groups or to other species. For example, ancient mixed-genre texts referred to as "almanacs," intermingle practical advice on the agricultural arts, animals, mathematics, and astronomy with imaginative social commentary that often takes the form of cosmological creation or migration stories, songs, or prayers (Adamson, 2001: 133–134).

For example, in *Works and Days* (ca. 700 BCE), Greek poet Hesiod offers advice to farmers dealing with extreme weather and poor soils. Often considered one of the first written farmer's almanacs because of its advice on drought and poor soils, Hesiod weaves imaginative social commentary into his recitation of the stories of Prometheus, who created humans from clay, and Pandora, whose curiosity impels her to open a jar from which malicious spirits escape. These "evil" spirits, which Hesiod clearly links to Greek "seizure" of wealth from their "neighbors" (1983: 320–341), raise questions about human behaviors, especially in times of scarcity and change, and leads him to advocate ethical attention to the weather and the activities of animals (1983: 342–351).

Similarly, social commentary is woven into the first printed versions of the Mayan almanac, the *Popul Vuh* (c. 1500 CE), a corpus of mytho-historical-astronomical narratives about the gods of the sea and sky. Over the course of four attempts to create human beings capable of living ethically with other humans and animals, the gods create a being of mud that is washed away by the rain. Anthropologist Dennis Tedlock has suggested that Mayan elders were not copying or paraphrasing the Greek story of Prometheus or the Biblical story of Adam's creation from mud in this passage. They are negating the story of Adam and thus directly critiquing Spanish colonizers who did not believe the Maya to be human and who were plundering the resources of the Americas (Tedlock, 1983: 263–264, 270–271). Thus, both Hesiod and the Mayan elders made "visible" the connections between the weather, human activities, and cosmic and biogeochemical processes critical to the well-being of all life on the planet (Adamson, 2001: 145). The gods in these stories, much like the heroes in today's cli-fi, work as an imaginative force for encouraging humans to reevaluate human behaviors that unleash "malicious spirits" (or what environmental scholars today refer to as "unintended consequences").

The literary history of a "climate canon," to coin a phrase, might draw connections from these older mixed-genre narratives, farmers' almanacs, indigenous story cycles, to contemporary speculative and climate fictions. Indeed, texts such as the *Epic of Gilgamesh*, the *Popol Vuh*, the *Tao Te ching*, like cli-fi, muse on anthropocentric values and beliefs that result in unintended consequences. In *The Song of the Earth*, literary critic Jonathan Bate rereads some older fictions, and specifically that most iconic of eighteenth-century British literary works, Mary Shelley's *Frankenstein: Or, The Modern Prometheus* (1818), for what they can tell us about human understanding of the biogeophysical forces of the planet, and climate, as the "primary sign of the inextricability of culture and nature" (Bate, 2000: 102). He argues that, in the Anthropocene, there is much we can learn from literary works that attend "to the weather" about human hubris and its unintended consequences (Bate, 2000: 102). *Frankenstein* portrays a young scholar turning away from the understandings of "natural philosophy" found in ancient almanacs and embracing "modern science" as he creates an ungovernable monster from the "mud" of human flesh. Shelley began writing the novel, she explains in the novel's preface, during a "cold and rainy season" (2003[1818]: 14). As Bate establishes, the stormy weather of 1816, often referred to as the "Year with No Summer," was the result of the catastrophic 1815 explosion of Tambora, an Indonesian volcano, which affected the atmosphere of the entire northern hemisphere for the next three years, leading to worldwide crop failure, riots and starvation, and a global cholera epidemic (Bate, 2000: 97; Wood, 2014). Today it is known that a tropical eruption paradoxically cools the planet with a blanket of volcanic dust but drastically warms the Arctic owing to changes in wind circulation and north Atlantic ocean currents (Wood, 2014). As a result, the British Admiralty began receiving reports of a remarkable loss of sea ice around Greenland and planned a massive endeavor in polar exploration. This helps to explain why Shelley sets the opening scenes of *Frankenstein* in a frozen North, where a young ship's captain first encounters the monster and its creator. Famous for its laboratory scenes, *Frankenstein* has often been dismissed as a "romantic reaction" at odds with science, a categorization that veils the brilliance of Shelley's critique of science as a discourse that authorized colonial activities on a scale that today can be recognized as having altered planetary systems. Although Shelley may not herself have associated *Frankenstein* with a cataclysmic volcanic event, today the novel can be seen as associated with biogeophysical events that changed the weather. *Frankenstein,* then, offers a kind of proof that a "changing climate changes everything", from the weather to literary history (Wood, 2014).

As this illustrates, a literary "climate canon" would include ancient story cycles, classic texts and novels, and recent speculative fiction, or cli-fi. We might think of environmental fiction and narratives as powerful tools for reading "the signs of the times in the signs of the skies" (Bate, 2000: 102). Today, as they have for thousands of years, humans are employing the alchemy of stories as an imaginative force for thinking about the ongoing evolutionary transformations of the world and its inhabitants, especially as humans face the many complex challenges of a changing climate.

## Learning resources

Adamson, J., Gleason, W.A. & Pellow, D.N. (eds) (2016) *Keywords for Environmental Studies*. New York: New York University Press.

Bate, J. (2000) *The Song of the Earth*. London: Picador.

Buell, L. (1995) *The Environmental Imagination: Thoreau, Nature Writing, and the Formation of American Culture*. Cambridge, MA: Belknap Press of Harvard University Press.

## Bibliography

Adamson, J. (2001) *American Indian Literature, Environmental Justice, and Ecocriticism: The Middle Place*. Tucson, AZ: University of Arizona Press.

Adamson, J. (2016) 'Humanities' in Adamson, J., Gleason, W.A. & Pellow, D.N. (eds) *Keywords for Environmental Studies* (pp. 135–139). New York: New York University Press.

Armbruster, K. (2016) 'Nature writing' in Adamson, J., Gleason, W.A. & Pellow, D.N. (eds) *Keywords for Environmental Studies* (pp. 156–158). New York: New York University Press.

Bate, J. (2000) *The Song of the Earth*. London: Picador.

Bloom, D. (2014) 'Will fiction influence how we react to climate change? – Room for debate', *The New York Times*. Online. Available at: http://www.nytimes.com/roomfordebate/2014/07/29/will-fiction-influence-how-we-react-to-climate-change.

Buell, L. (1995) *The Environmental Imagination: Thoreau, Nature Writing, and the Formation of American Culture*. Cambridge, MA: Belknap Press of Harvard University Press.

Finn, E. (2015) 'An interview with Margaret Atwood', *Slate*. Online. Available at: http://www.slate.com/articles/technology/future_tense/2015/02/margaret_atwood_interview_the_author_speaks_on_hope_science_and_the_future.html.

Glass, R. (2013) 'Global warming: the rise of cli-fi', *The Guardian*. Online. Available at: https://www.theguardian.com/books/2013/may/31/global-warning-rise-cli-fi.

Gunn, E. (2014) 'How America's leading science fiction authors are shaping your future', *Smithsonian Magazine*, Online. Available at: http://www.smithsonianmag.com/arts-culture/how-americas-leading-science-fiction-authors-are-shaping-your-future-180951169/?page=2.

Hesiod (1983[ca. 700 BCE]) *Theogony, Works and Days and The Shield*. Baltimore: Johns Hopkins University Press.

Marshall, G. (2014) 'Climate fiction will reinforce existing views', *The New York Times*. Online. Available at: http://www.nytimes.com/roomfordebate/2014/07/29/will-fiction-influence-how-we-react-to-climate-change/climate-fiction-will-reinforce-existing-views.

Morin, J.L. (2014) 'Universities make "cli-fi" dreams come true', *HuffPost*. Online. Available at: http://www.huffingtonpost.com/j-l-morin/universities-make-clifi-d_b_5564491.html

Murphy, P. ed. (1998) *Literature of Nature: An International Sourcebook*. Chicago: Dearborn.

Pérez-Peña, R. (2014) 'College classes use arts to brace for climate change', *The New York Times* (March 31, 2014) A-12.

Shelley, M. (2003[1818]) *Frankenstein: Or, The Modern Prometheus*. London: Penguin Classics.

Tedlock, D. (1983) *The Spoken Word and the Work of Interpretation*. Philadelphia: University of Pennsylvania Press.

Thomas, S.R. (2014) 'Imagination will help find solutions to climate change' *The New York Times*. Online. Available at: http://www.nytimes.com/roomfordebate/2014/07/29/will-fiction-influence-how-we-react-to-climate-change/imagination-will-help-find-solutions-to-climate-change.

Williams, R. (1983 [1976]) *Keywords: A Vocabulary of Culture and Society*. New York: Oxford University Press.

Wood, G.D. (2014) 'The volcano that changed the course of history', *Slate*. (April 9, 2014). Online. Available at: http://www.slate.com/articles/health_and_science/science/2014/04/tam. . . ption_caused_the_year_without_a_summer_cholera_opium_famine_and.html.

# 6.8 Environmental celebrity

Mike Goodman

## Environmental star power

For contemporary media celebrities, the environment—and in particular, climate change—is a popular cause. A cursory gaze across the media-scape confirms this: Leonardo DiCaprio has produced and starred in *Before the Flood* (2016) which tells the tale of his journey as a UN Ambassador of Peace engaging powerful leaders about climate change. Olivia Munn, Harrison Ford, Arnold Schwarzenegger and Jessica Alba witness and 'emote' for audiences about the impacts of climate change on ordinary people and ecologies in the television programme *Years of Living Dangerously* (2014). Mark Ruffalo writes prominent columns about fracking and solar power in the Huffington Post, the millennial news-source of record. Even celebrity public intellectuals, such as Naomi Klein, are getting in on the act: she has starred—along with her six-year-old son—in the short film *Under the Surface* that shows the impacts of climate change on the Great Barrier Reef. For her, discussing her emotional responses to climate change and introducing her son in on the film, worked as a way for her to 'communicate in a visceral way, the intergenerational theft at the heart of this crisis' (Klein, 2016). Environmental politics, with this celebritization of climate change and ecological issues, have gone *spectacular*. Celebrities, as our witnesses and muses, now speak loudly for and about the environment in important ways that have impacts on what we know about nature, how we feel about it and what we should do to save it.

## The celebritized environment: a history of diverse personas and performances

Celebrity involvement and engagement with politics is not new, nor is it confined to the environment. Celebrity diplomacy can be traced back to the 1950s with the efforts of

Danny Kaye and later Audrey Hepburn who expanded the UN Goodwill Ambassador position (Cooper, 2008). Bono, Sting, Oprah, Madonna, George Clooney and Angelina Jolie have all taken up this mantle of celebritized, global humanitarian actor and often moonlight as environmental celebrities. Food has also been subject to its own celebritization: think of the wild growth of food celebrities telling us how to cook, what to eat and what others should eat. From Gwyneth Paltrow and her cleanses to Jamie Oliver and Michelle Obama with healthy eating to Chris Martin fronting for fair trade, 'good' food has been elevated to the level of celebrity icon. Here, too, there is celebrity cross-over with the likes of Miley Cyrus, a committed vegan, Paul McCartney rapping about Meatless Mondays and British chef Hugh Fernley-Whittingstall engaging European policy on sustainable fishing. Simply put, social and environmental activism are part of the job description and identities of contemporary celebrities (Littler, 2008) given their status as 'Big Citizens' able to confront the world's problems through what Chris Rojek (2014) calls *celanthropy*.

Nor are environmental celebrities all of one type. Indeed, there is an established and growing taxonomy of environmental celebrities. There are, for example, celebrity actors, musicians and entertainers, celebrity politicians (e.g. Al Gore), celebrity athletes, celebrity business people (e.g. Elon Musk) and celebrity public intellectuals. We could also add celebrity scientists like former NASA climatologist James Hansen, celebrity activists like environmentalist Bill McKibbin and celebrity religious leaders such as Pope Francis. Across this taxonomy, there are numerous types of roles that celebrity environmentalists take on and perform in the creation of the different media interventions they star in. For example, there are those who work to convince the public that climate change is real by translating the latest science, there are those who endorse 'brand' climate change to gain media coverage, there are those who work in the background orchestrating interventions and there those who express their emotions about climate change and show us its impacts to spur greater action. There are also those celebrities who remain in the role of the climate sceptic, the most famous of these being the US President Donald Trump who we might now dub as the celebrity 'denier-in-chief'. All of this points to the fact that there are a series of different characters, nodes and performances through which audiences might engage with celebrity environmentalists.

## Knowing the work of environmental celebrities

Environmental celebrities do their work through four processes. The first is that of *embodiment*. Environmental celebrities embody—make fleshy and lively—environmental politics. They are, at the very same time, commodities made flesh. As the celebrity studies scholar Graeme Turner (2004: 9) argues, 'the celebrity is . . . produced, traded and marketed by the media and publicity industries. . . . Celebrity's primary function is commercial and promotional' to which we might now add, also political. Environmental celebrities and their media interventions—much like solar panels and carbon credits—circulate and create value in the economy of green political markets. They also often embody responses to environmental issues in the ways they do, to make low-carbon lifestyles fashionable. That electric car and environmentally conscious dress must *simply* be seen at the Oscars! With this, the individualized responses of shifting to more sustainable lifestyles through high-tech means is greatly accentuated by environmental celebrities (Anderson, 2011).

Second, environmental celebrities *mediate* amongst the audience, different publics and nature. They are in between us and the environment, witnessing for us, educating us, telling us what to do and how to do it to save the planet. They are our go-betweens amid glaciers, tropical forests, scientists, world leaders and green capitalists. The interventions of celebrity environmentalists are designed to gain and maintain our attention through diverse performances and forms of 'enviro-tainment'. Yet, let's be clear here: mediation is not simply a passive exercise in environmental celebrities passing on the 'truth' about climate change or fracking. Rather, much like any other media form, environmental celebrities actively construct the representations, meanings and values of those everyday ecologies and communities affected by, for example, drought or polluted water.

Third, through this active mediation, environmental celebrities implicitly and/or explicitly *frame* how the public should not just think about the environment, but also how we should fundamentally relate to it. In this, environmental celebrities function through visual registers and performances—a bearded and sunglassed DiCaprio atop a melting glacier, a hard-hatted Schwarzenegger surrounded by the flames of a burning forest—as much or even more than they do the discursive. Given this emphasis on the visual, environmental celebrities are not simply just translating and transmitting facts, words and 'rational' knowledge about global environmental change, but are, now more than in past media interventions, working to foster emotional responses in the audience and wider public ecosystems. Environmental celebrities work to frame affect as much as they do cognition. Harrison Ford visibly rages at the President of Indonesia about deforestation, a jocular Elon Musk smiles to DiCaprio as he explains how green entrepreneurialism in the form of solar power and battery storage will solve the climate crisis, Don Cheadle visibly sympathises with a devote Christian Scientist doggedly trying to convince her church of the existence of climate change: Environmental celebrities work to give us visualized performances of emotion that attempt to frame our own affective responses to save the world.

Finally, with environmental celebrities, affect, imaginaries, mediation and embodiments are formulated through the processes of *relating*. Embodiments, mediation, ecological imaginaries, affects and—indeed—celebrity itself only come into being through relationships amongst actors, media forms and economies, technological platforms and audiences. Relating *to* environmental celebrity crosses the realms of the cognitive, the visceral, the material, the discursive and the visual. Relating *through* and *with* environmental celebrity, on the other hand, creates the potential for green politics that are conventional and/or radical, progressive and/or regressive. Social media, one of the core pillars of anthropologist Bram Büscher's (2013) concept of 'Nature 2.0' which describes the digitization of ecological politics, is clearly one of the key ways publics relate to environmental celebrities. Environmental celebrities' contribution to the increasing 'app-ification' of environmental politics—and how this might both hinder conservation through the deeper commodification of nature or further facilitate it through monetary contributions—is worthy of greater critical attention.

## Performing ecologies: pseudo-expertise, impact potential and inequality

This celebritization of the environment—and its seemingly unabated intensification—raises a number of further issues. First, celebrity environmentalists seem to be displacing the notions of 'expertise' and the figure of the 'expert'. Expertise, through environmental celebrities, is formulated by popularity, media profile and authenticity as much as by scientific training. Second, while environmental celebrities and their interventions have the potential to raise awareness, does this really matter? How and in what ways do these interventions impact the public in both material and behavioural terms? In political and policy terms? Indeed, as geographer Dan Brockington (2014) has argued, perhaps the key intervention that environmental celebrities might make is when they talk to other elites such as national and global policy makers. Or maybe, given that September, 2016 was the hottest on record, environmental celebrities have yet to gain real world traction. Or would things be even worse *without* environmental celebrities? A final concern is one about the privileged positionality of environmental celebrities, namely that celebrities—and the economic and cultural systems that support them—are fundamentally un-democratic (Kapoor, 2013). Here, we have yet another instance where the rich are speaking on behalf of poor and marginalized communities and ecologies in ways that maintain if not entrench inequalities of social and economic power and facilitate the impression that celebrities will be our saviours. If anything, this is a vitally critical time for more and greater cross-community resistance and public action, celebrity inspired or not.

## Learning resources

For more on environmental celebrity see the following:

Alexander, J. (2013) 'The case of the Green Vampire: eco-celebrity, Twitter and youth engagement', *Celebrity Studies*, 4(3): 353–368.

Boykoff, M. & Goodman, M. (2009) 'Conspicuous redemption? Reflections on the promises and perils of the "celebritization" of climate change', *Geoforum*, 40: 395–406.

Doyle, J., Farrell, N. & Goodman, M. (2018) Celebrities and Climate Change: History, Politics and the Promise of Emotional Witness, in Nisbet, M. (ed.) *The Oxford Encyclopedia of Climate Change Communication.* Oxford: Oxford University Press.

## Bibliography

Anderson, A. (2011) 'Sources, media, and modes of climate change communication: the role of celebrities', *WIREs Climate Change*, 2(4): 535–546.

Brockington, D. (2014) *Celebrity Advocacy and International Development*. Abingdon: Routledge.

Büscher, B. (2013) 'Nature 2.0', *Geoforum*, 44: 1–3.

Cooper, A. (2008) *Celebrity Diplomacy*. Boulder: Paradigm.

Kapoor, I. (2013) *Celebrity Humanitarianism: The Ideology of Global Charity*. Abingdon: Routledge.

Klein, N. (2016) 'Climate change is intergenerational theft. That's why my son is part of this story', *The Guardian*, 7 November.

Littler, J. (2008) 'I feel your pain: cosmopolitan charity and the public fashioning of the celebrity soul', *Social Semiotics*, 18(2): 237–251.

Rojek, C. (2014) 'Big citizen celanthropy and its discontents', *International Journal of Cultural Studies*, 17(2): 127–141.

Turner, G. (2004) *Understanding Celebrity*. London: Sage.

6.9

# Environmental certification and standards

Brooke Lahneman

Many organizations today adopt environmental standards and certifications in order to ensure their operations are more environmentally-friendly. Succinctly called 'environmental certified management standards' (ECMS), these programs are voluntary ones that specify sets of internal organizational management practices to be complied with across adopting firms (Terlaak, 2007). ECMS programs are designed to motivate firms to improve environmental performance through compliance with standardized practices that mitigate risk to the natural environment in key organizational systems. By design, ECMS do not specify the nature of a firm's output or its quality, nor how these firms are to implement these requirements. As ECMS permit firms the flexibility in how they implement the required practices, such programs cannot be directly linked to a firm's financial performance (Epstein & Roy, 2001). ECMS programs are typically formulated and monitored through collaborations among multiple stakeholders to determine requirements and enforcement mechanisms appropriate for a particular industry (Terlaak, 2007).

## Types of ECMS

There are diverse types of ECMS programs existing in the market. ECMS programs can develop industry-related standards, such as with the Forest Stewardship Council (FSC) that requires organizations in forestry-related industries to implement practices that mitigate environmental risks unique to forest management and wood-products manufacturing (Taylor, 2005). On the other hand, ECMS programs can standardize practices across industries, such as with ISO 14001 (Delmas, 2001; Delmas & Montiel, 2008). This general type of ECMS program often standardizes environmental management practices and systems that reach across manufacturing and services sectors, having firms comply with minimum environmental targets such as pollution emissions reduction and water conservation (Potoski & Prakash, 2005).

## Motivations to adopt ECMS

Prior studies have developed three primary reasons for why firms adopt an ECMS program:

### *Signaling commitment to stakeholders*

The 'stakeholder-signaling' motivation focuses on how firms can generate positive value through communicating their commitment to mitigating environmental risk to a wide range of stakeholders (King et al., 2005). A firm's adoption of an ECMS program creates transparency and accountability around a firm's compliance with a set of minimum standards in practices and processes regarding environmental risk mitigation (King et al., 2005). Accountability is key, as firms are typically required to be audited by a third party certifier as part of the ECMS program membership (Terlaak, 2007). In this way, adoption mitigates concerns on the part of regulators, suppliers, activists, and consumers, and employees, who might take action against a firm otherwise perceived to be engaging in activities harmful to the environment (Potoski & Prakash, 2005).

### *Competitive barriers to other firms*

Another primary reason why firms adopt an ECMS is to create competitive barriers for other firms, either through changing norms in how business is conducted or increasing operational costs (Bansal & Hunter, 2003). Some firms may have a greater involvement in developing environmental standards for an industry, and thus can design the standards and associated practices to closely fit their own current operations while creating the need for change for other firms (King et al., 2005). Compared to competitor firms, lower implementation costs give participating firms an operational advantage, and early-adopter status signals to stakeholders a strong commitment to environmental risk management.

### *Acquisition of 'green' capabilities*

Increasingly firms are adopting ECMS to 'learn' how to be greener in their operations. Research demonstrates that ECMS adoption can be associated with increased environmental performance, typically measured by lowered pollution emissions or decreased energy usage (King & Lenox, 2001). By their nature, ECMSs often require measurement and reporting, and so can prompt firms that desire to be more environmentally friendly to put that desire into action. The design of an ECMS influences how substantively a firm adopts required practices, with ECMS that impose more rigorous standards and oversight being associated with greater adoption of practices throughout key organizational systems (Lahneman, 2015). Indeed, firms can utilize the adoption of an ECMS primarily as a signaling tool, as mentioned above, to communicate commitment to environmental issues to a range of stakeholders, rather than as a direct means to enact large-scale practice changes (Delmas, 2001; Howard et al., 1999). Even so, research demonstrates the promise of ECMS in motivating change at deeper operational and cultural levels.

## Outcomes of adopting an ECMS

Firms hope to benefit in three primary ways from adopting an ECMS:

### *Financial profit*

Often called the 'business case for sustainability,' firms hope to create a profit margin from adopting an ECMS. Firms typically charge higher prices on products that are certified under the ECMS program. Called an 'eco-label premium' or 'eco-premium,' organizations hope that customers see products with eco-labels as being higher in value and thus are willing to pay more for those products (Bjørner et al., 2004). Firms also hope that by adopting greener practices, they will widen the profit margin by lowering the cost of operations. There is a cost to join ECMS programs as such programs often charge fees to join and require firms to change practices upon joining to comply with ECMS requirements. However, firms anticipate gaining 'eco-efficiency,' meaning that practices common across many ECMSs such as decreased waste and more efficient use of materials, as well as reduced water and energy use, will in fact decrease operating costs over the long term (Derwall et al., 2005).

### *Reputation boost*

The activities of firms are monitored by stakeholders such a regulators, governments, activists, and customers. Given the scale of operations and breadth of relationships necessary to maintain a low-cost, high-quality strategy, firms have the potential to impact the environment in a negative or positive way. Research has shown that it is often the negative impacts that receive the most attention from such influential stakeholders (Bansal & Clelland, 2004; Barnett & King, 2008). Thus, firms are motivated to adopt an ECMS to communicate a commitment to environmental protection, both avoiding sanctions from and boosting reputation with influential stakeholders (Bansal & Clelland, 2004; Barnett & King, 2008).

### *Mitigation of environmental risks in operational activities*

Firms that adopt an ECMS, even if the motivation was primarily symbolic at the start, often need to change practices to comply with a minimum set of standards

(Howard-Grenville et al., 2014). Overall, research demonstrates that firms that adopt an ECMS tend to report higher environmental performance (King & Lenox, 2001). Furthermore, firms that more readily implement ECMS practices if they adopt a program that provides a high level of detail in practice descriptions, sets demanding objectives to achieve, and tailors practices specific to an industry (Lahneman, 2015). Thus, although stakeholder-signaling might be a primary motivation of ECMS adoption, many firms find that the adoption of an ECMS does indeed result in a change in practices and mitigation of environmental risks.

## Moving forward with ECMS

Overall, there remains great promise in firms' adoption of ECMS in an effort to make impactful changes toward more environmentally-friendly practices. To accomplish this, ECMS must be intentionally designed to motivate substantive rather than symbolic adoption (Potoski & Prakash, 2005; Terlaak, 2007). This indicates a need to shift from a focus on monitoring environmental outputs, such as pollution and water or energy use, to a focus on inputs, such as practices (Howard et al., 1999). Research demonstrates that greater rigor in standards, certification oversight, and practices tailored to an industry are design factors associated with the implementation of a range of ECMS practices in key organizational systems (Lahneman, 2015). With such design, ECMS can be leveraged by organizations to implement real change toward becoming more environmentally-friendly in their operations.

## Learning resources

For additional information on environmental standards and certification programs in practice, please reference the following websites:

U.S. Environmental Protection Agency (EPA) on 'Environmental Management Systems (EMS)': https://www.epa.gov/ems.

U.S. Small Business Administration (SBA) on 'Green Certification and Ecolabeling': https://www.sba.gov/managing-business/running-business/green-business-guide/green-certification-and-ecolabeling.

Eco-Label Index website: http://www.ecolabelindex.com.

ISO 14001: http://www.iso.org/iso/iso14000.

Certified B Corporations ('B Corps'): https://www.bcorporation.net.

Bristol, F. (2015) 'Why using Fair Trade is good for your business': Online. Available at: http://gogreenbusiness.co.uk/blog/2015/02/using-fairtrade-good-business.

## Bibliography

Bansal, P. & Clelland, I. (2004) 'Talking trash: legitimacy, impression management, and unsystematic risk in the context of the natural environment', *Academy of Management Journal*, 47(1): 93–103.

Bansal, P. & Hunter, T. (2003) 'Strategic explanations for the early adoption of ISO 14001', *Journal of Business Ethics*, 46: 289–299.

Barnett, M.L. & King, A.A. (2008) 'Good fences make good neighbors: a longitudinal analysis of an industry self-regulatory institution', *Academy of Management Journal*, 51(6): 1150–1170.

Bjørner, T.B., Hansen, L.G. & Russell, C.S. (2004) 'Environmental labeling and consumers' choice—an empirical analysis of the effect of the Nordic Swan', *Journal of Environmental Economics and Management*, 47(3): 411–434.

Delmas, M. (2001) 'Stakeholders and competitive advantage: the case of ISO 14001', *Production and Operations Management*, 10(3): 343–358.

Delmas, M. & Montiel, I. (2008) 'The diffusion of voluntary international management standards: Responsible Care, ISO 9000, and ISO 14001 in the chemical industry', *The Policy Studies Journal*, 36(1): 65–93.

Derwall, J., Guenster, N., Bauer, R. & Koedijk, K. (2005) 'The eco-efficiency premium puzzle', *Financial Analysts Journal*, 61(2): 51–63.

Epstein, M.J. & Roy, M.J. (2001) 'Sustainability in action: identifying and measuring the key performance drivers', *Long Range Planning*, 34(5): 585–604.

Howard, J., Nash, J. & Ehrenfeld, J. (1999) 'Industry codes as agents of change: Responsible Care adoption by U.S. chemical companies', *Business Strategy and the Environment*, 8: 281–295.

Howard-Grenville, J., Bertels, S.P. & Lahneman, B. (2014) 'Sustainability: how it shapes organizational culture and climate', in Schneider, B. & Barbera, K.M. (eds) *Handbook of Organizational Climate and Culture: An Integrated Perspective on Research and Practice*. Oxford: Oxford University Press.

King, A.A. & Lenox, M.J. (2001) 'Lean and green? An empirical examination of the relationship between lean production and environmental performance', *Production and Operations Management*, 10(3): 244–256.

King, A.A., Lenox, M.J. & Terlaak, A. (2005) 'The strategic use of decentralized institutions: exploring certification with the ISO 14001 Management Standard', *The Academy of Management Journal*, 48(6): 1091–1106.

Lahneman, B. (2015) 'In vino veritas: understanding sustainability with environmental certified management standards', *Organization & Environment*, 28(2): 160–180.

Potoski, M. & Prakash, A. (2005) 'Covenants with weak swords: ISO 14001 and facilities' environmental performance', *Journal of Policy Analysis and Management*, 24(4): 745–769.

Taylor, P.L. (2005) 'In the market but not of it: fair trade coffee and forest stewardship council certification as market-based social change', *World Development*, 33(1): 129–147.

Terlaak, A. (2007) 'Order without law? The role of certified management standards in shaping socially desired firm behaviors', *Academy of Management Review*, 32(3): 968–985.

6.10

# Environmental insecurity

Peter Hough

Environmental insecurity has increasingly been invoked in political discourse around the world but subsequent political responses have proven to be highly diverse. In particular this reflects the fact that there is no consensus on what 'environmental security' actually means. Is the referent object to be secured the state, 'the human' or the environment? The question of whether environmental problems merit the politically significant label of 'security' is a complex one, and highly contested. In essence there are four positions that have evolved: 1. *Traditional International Relations Realists* reject the coupling together of the environment and security either or both because environmental degradation is not considered significant enough to merit such a label and the contention that the politics of 'security' is about the military defence of the state, not tackling things like pollution; 2. *Security Wideners* consider that environmental challenges can invoke the politics of security but only if they can be seen to cause wars or threaten the sovereignty of states; 3. *Traditional Ecologists* resist 'securitization' through concerns that this risks invoking inappropriate, militaristic 'national security' responses to complex environmental problems; 4. *Human/Critical Security Ecologists*, receptive to the ontological and epistemological challenges to the conventions of international relations that emerged following the end of the Cold War, contend that environmental problems can and should be 'securitized' by abandoning the traditional preoccupation with the state and military defence, and mobilizing global responses to different kinds of threats to life.

Designating an issue as a matter of security is not just a theoretical question but carries 'real world' significance since it is indicative of where political priorities lie. Security widening can be seen in the political practice of many governments since the early 1990s, when the post-Cold War landscape provided space for concerns other than nuclear war, such as anthropogenic environmental change. Especially influential in this was the 'Resource Wars thesis', associated with Canadian academic Thomas Homer-Dixon, which posited that environmental degradation would increasingly be a spark for armed conflict (Homer-Dixon, 1994). This thesis influenced the US Clinton administration and Homer-Dixon is known to have briefed Vice President Gore in the early 1990s (Floyd, 2010). This government then went on to create a Deputy Under Secretary for Environmental Security and cite environmental degradation as a security risk in their 1994 'National Security Strategy'. Other instances of governments making the environment the stuff of high politics subsequently emerged elsewhere in North America and Northern Europe, such as in defence and foreign policy statements from Finland, Canada, the Netherlands and the UK. In 2007 the UK government used their presidency of the UN Security Council to push through, with some resistance from some other members, the first discussion in that arena of on an overtly environmental topic: climate change. Perhaps, though, the clearest illustration of the embrace of environmental security comes from its adoption by members of NATO (the North Atlantic Treaty Organization):

> Based on a broad definition of security that recognizes the importance of political, economic, social and environmental factors, NATO is addressing security challenges emanating from the environment. This includes extreme weather conditions, depletion of natural resources, pollution and so on – factors that can ultimately lead to disasters, regional tensions and violence.
>
> (NATO, 2013)

As this statement indicates, securitizing environmental issues in practise has tended to be in the traditional national security manner of factoring such concerns into calculations of defence needs. An exception, though, comes from the low-lying states threatened with literal extinction under the waves of the rising oceans. Following the Security Council discussion of climate change two years earlier, the United Nations General Assembly in 2009 took up this theme with a resolution drafted by the government of low-lying Nauru called *Climate Change and its Possible Security Implications* (A64/350) calling on all UN agencies to prioritize global warming mitigation. Whilst the resolution was unanimously adopted, international political practice for the vast majority of governments has still tended not to couple together the environment and security without thinking in military terms. The discourse of environmental change in venues of intergovernmental 'high politics' invariably cites the resource wars thesis or the apparent threat posed by a rise in environmental migration. Environmental degradation is deemed important because it *might* be a cause of war and instability rather than because it *is* a threat to life in itself. The UK UN delegation pushing for the Security Council debate in 2007 cited the following security implications of climate change: border disputes due to the melting of ice sheets and rising sea levels; increased migration with "the potential for instability and conflict"; conflict over energy supplies; conflict due to scarcity; conflict due to poverty; and conflicts related to extreme weather events (UNSC, 2007). This tendency to treat environmental insecurity as a matter of military defence has alarmed many ecologists, wary that this serves to distract political attention from the more fundamental and multi-faceted social and economic challenges posed by issues like climate change. It was scientists not troops who saved the world from the threats posed by ozone depletion in the 1980s.

Human and Critical Security advocates share the ecologist's misgivings about widened security but, nevertheless, support ecological securitization because, for them, the concept is far more profound than the resource war thesis. The simple and unambiguous fact that casualties of pollution far outstrip those of war and terrorism is enough to merit the urgent prioritization of tackling environmental issues in international relations from this perspective. However, in practise, in cases where human security has come to inform government policy there has been a notable lack of consistency as to whether environmental change can be given the same sort of prioritization as non-military issues such as human rights or transnational crime. Governmentally, human security has been endorsed in very different ways. The 'freedom from fear' interpretation favoured by the Canadian government over recent years, for example, tends not to consider non-violent 'natural' threats as security matters. In contrast, the United Nations Development Programme and the Japanese government have championed a wider 'freedom from want' variant, which does not preclude threats with less direct human causation. However, this more expansive version of human security has generally served to emphasize poverty and development rather than environmental degradation. In a more general sense human security is still somewhat problematic from an ecological perspective since this is, by definition, an anthropocentric rather than ecocentric way of framing problems.

However, not all environmental securitization has been of the state-centric and militaristic form and ecocentric thinking has been evident in some other statements of government political priorities. The Netherlands' 2006 Foreign Policy Agenda refers to the role that environmental degradation may play in triggering conflicts but notably goes on to declare as one of eight goals a commitment 'to protect and improve the environment', without the addition of any clause indicating that this is another case of valuing the environment for instrumental rather than intrinsic reasons. (Netherlands, 2006) This ecocentric turn of making the environment the referent object of security has also been advanced in a different political form in recent years outside of the Western world as part of the 'pink tide' in Latin America. In 2008 Ecuador's new constitution declared that nature had the 'right to exist, persist, maintain and regenerate its vital cycles, structure, functions and its processes in evolution' and mandates the government to take 'precaution and restriction measures in all the activities that can lead to the extinction of species, the destruction of the ecosystems or the permanent alteration of the natural cycles' (Ecuador, 2008). Whilst many countries have cited environmental protection in their constitutions, none have done so in such unambiguously ecocentric terms. This 'rights of nature' approach has also been followed by the Morales government in Bolivia where the 'Law of Mother Earth' proclaims the right of nature 'to not be affected by mega-infrastructure and development projects that affect the balance of ecosystems' (Bolivia, 2011). For both of these Andean countries this idea of environmental rights comes from the impact of indigenous people identifying with nature against environmental pollution from oil and tin multinational corporations operating on their land. Ecocentric responses to environmental insecurities have increased in the discourse of international politics – as they have in the domestic politics of some states since the 1960s – but widened security thinking remains much more prevalent.

The rise of widened security – and particularly the resource wars thesis – means that environmental securitization for many still invokes a perception of militarization which jars with the pacifistic instincts of most ecologists. National securitization may be welcomed in terms of getting governments on board and giving environmental issues the spotlight they deserve, but old habits die hard and evidence suggests that this does tend to lead to the issues being framed in militaristic terms. Enquiry in International Relations (and particularly Security Studies) often, rightly, stands accused of being so pre-occupied with semantics, ontology and epistemology that matters of life and death are not addressed as fully as they deserve to be. However, determining how best to address insecurities related to environmental change does necessitate such reflection on what 'security' means and how it can be optimized. Leaving aside the ever-dwindling gaggle of (chiefly non-academic) 'environmental sceptics', a lack of consensus on the precise meaning of 'environmental insecurity' is hampering political efforts to tackle some of the most urgent threats facing the world today.

## Learning resources

The links between environmental change and security, in a range of ways, are explored in this UNEP wepage:

http://www.unep.org/roe/KeyActivities/EnvironmentalSecurity/tabid/54360/Default.aspx.

Articles on the human security implications of environmental change are collated on this United Nations University site:

http://ehs.unu.edu.

## Bibliography

Bolivia (2011) *Law of Mother Earth*, Law 071. Online. Available at: http://www.worldfuturefund.org/Projects/Indicators/motherearthbolivia.html (accessed 2.9.16).

Ecuador (2008) 'Rights for Nature', *Constitution*, adopted September 28th.

Floyd, R. (2010) *Security and the Environment. Securitisation Theory and US Environmental Security Policy*. Cambridge: Cambridge University Press.

Homer-Dixon, T. (1994) 'Environmental Scarcities and Violent Conflict: Evidence from Cases', *International Security*, 19(1): 5–40.

Hough, P. (2013) *Environmental Security. An Introduction*. Abingdon: Routledge.

NATO (2013) *Environmental Security*. Online. Available at: http://www.nato.int/cps/en/natolive/topics_49216.htm (accessed 6.5.12).

Netherlands (2006) *Policy Agenda 2006*. Ministry of Foreign Affairs.

United Nations Security Council (UNSC) (2007) Letter dated 5 April 2007 from the Permanent Representative of the United Kingdom of Great Britain and Northern Ireland to the United Nations addressed to the President of the Security Council April 5th S/2007/186.

# Environmental (in)justice

6.11

David Schlosberg

## A history of environmental justice

The idea of environmental justice – that some communities, or parts of communities, endure more environmental risks than others – is not a new concept. As the historian Martin Melosi (2004) has argued, things like sewage and municipal waste have been concentrated near the working poor and marginalized communities since ancient Greece, Rome, and Egypt – if not before.

Contemporary societies also have long histories of environmental injustices – and groups of people that have responded. In the US, the professions of public health and public planning were responses to progressive urban movements reacting to, and making demands about, the long-standing realities of how some are subject to the waste and impacts of others.

And yet it was not until the 1980s that the term 'environmental justice' came to prominence to address the reality of the inequity of environmental quality in everyday life. Initially, the term grew out of the confluence of two overlapping parts of the grassroots environmental movement in the US: the anti-toxics movement and the movement against environmental racism (Cole & Foster, 2001). It has since expanded to cover a wide range of environmental concerns in communities across the globe.

The anti-toxics movement began in the aftermath the events in Love Canal, New York, in the late 1970s and early 1980s, which led to a growth of awareness of the prevalence and dangers of toxics in poor or lower-middle class communities (Gibbs, 1982). Toxic emissions, waste dumpsites, and contaminated communities with threats to human health were the initial focus of the movement, but it grew to address a wide variety of issues relating to environmental threats to human health: from industrial waste sites, to municipal and hazardous waste dumps and incinerators, to nuclear waste, pesticides, and dioxin exposure.

But it was the ground-breaking struggle against a municipal hazardous waste landfill in Warren County, North Carolina in 1982 that led to the widespread use of the term 'environmental justice'. North Carolina needed to find a disposal site for up to 6000 truckloads of PCB-tainted soil, and proposed a facility that would continue to accept industrial waste on a commercial basis. The state ultimately chose the poorest county – where 65 percent of the population was African-American. Citizens saw it as an act of environmental racism, and resisted. Over 500 people were arrested when the trucks first rolled in, including members of Civil Rights organizations, along with members of environmental organizations.

Warren County forged a synthesis between civil rights discourse and environmental issues, and was the birthplace of what became the environmental justice movement (Bullard, 1993). Subsequent studies found evidence for the claim that a wide range of environmental wastes and impacts – toxic waste sites, air pollution, hazardous waste landfills, lead poisoning, occupational hazards, and more – were disproportionately impacting poor communities, communities of color, and indigenous communities. Studies also found unequal enforcement of environmental laws, illustrating that it is not just that people of color and the poor are more likely to be exposed to environmental bads of one sort or another, but also that the government does not protect those communities as well as richer, whiter communities.

Overall, the main point of environmental justice is that already vulnerable and overburdened communities face yet another symptom of social injustice – inequity and disparity in the distribution and experience of environmental damage. These inequities have been found based on both race and class, and spurred the growth of a broad social movement.

## The justice of environmental justice

So what does 'justice' mean when we talk about environmental justice – or environmental *in*justice? There are at least four ways that the movement and scholars have understood the concept.

### *Unjust distributions*

The earliest academic reflections on environmental justice originally focused on the existence of inequity in the distribution of environmental bads (Bullard, 1994). Clearly, one of the basic motivating factors for the environmental justice movement was a belief that environmental goods, environmental bads, and environmental protection where not distributed equally across the population. Clearly, some people got more waste, more lead, worse air, than others – and that is an injustice.

Distributional approaches are the dominant way to think about justice. Who gets what, and are the good (and bad) things in a society distributed equally? If not, are there good reasons for inequality, whether it is inequality of income or of toxic waste?

Justice theorist John Rawls (1971) famously argued that inequality could only be justified if, in some way, it helps the least well off and comes with some sort of larger social benefit. Any inequality that does *not* benefit the least well off – and especially an inequality that actually harms them – is unjust. Many environmental justice battles have been over this very point – does the inequity in the distribution of polluting factories or dumps help the least well off, by providing jobs or a way out of poverty (as many officials argue)? Or does the damage done, and the inequitable distribution of risk, unduly burden those who are already subject to a range of inequalities? Overall, inequity in the distribution of environmental goods and bads is one clear way that environmental injustice is experienced and understood.

## Recognition

But environmental justice is not simply about establishing the fact that more environmental bads and risks were being put on poor and minority communities – activists and scholars endeavored to explore the question of *why* those communities were devalued in the first place.

While equity has been a major issue for EJ, with racism and other forms of disrespect came a concern with a broader politics of *recognition*. Here, injustice comes with a lack of recognition of both individuals and their communities; such a lack of recognition can be understood both as a psychological experience and a status-based, institutional, and cultural structure.

On the one hand, recognition is a basic human need – something that gives us self-worth. We *feel* less without recognition from others, and disrespect is often specifically meant to harm. On the other hand, some theorists are dissatisfied with such an individualist and psychological approach; they focus on the lack of recognition as based in cultural, social, and political institutions. Nancy Fraser (1997), for example, identifies three status-based definitions and processes of misrecognition. First is a general practice of cultural domination; second is a pattern of non-recognition, which is the equivalent of being rendered invisible; and third is disrespect, or being routinely maligned or disparaged in stereotypical public and cultural representations. Racism and gender discrimination are clearly two forms of status-based misrecognition.

In practice, people experience both personal and institutional forms of discrimination and disrespect – and both are the focus of recognitional justice. In addition, and importantly, we see such disrespect at both the individual and community level. So, individually, activists criticize how they are represented as ignorant or hysterical, and dismissed as incapable of understanding science, the policy process, or economic development. At the community level, demands for collective and cultural recognition permeate the movement as well – as we see, for example, with indigenous responses to the desecration of sacred sites (Whyte, 2011).

## Participation

Crucially, both inequity and misrecognition create real structural obstacles to political participation. If you do not have resources, or recognition, it is difficult to be heard. In response, individuals and communities commonly call for participatory or procedural justice as an element of environmental justice.

Environmental justice has long articulated the demand to 'speak for ourselves', as well as a 'place at the table' – equal, informed, respectful participation. The movement exemplifies the idea of the 'all-affected principle' – where those who are impacted by a political decision have the right to participate in the decision-making process.

Environmental justice activists call for policy-making procedures that encourage active community participation, institutionalize public participation, recognize community knowledge, and utilize cross-cultural exchanges to enable the participation of the diversity that exists in a community. Movements focused on issues from local siting decisions to global climate change consistently articulate participatory demands as an element of environmental justice.

### *Capabilities*

Finally, and most broadly, what movements have meant by the 'justice' of environmental justice also broadly encompasses the basic needs and well-being of individuals and communities. Theoretically, this idea is most aligned with the 'capability' approach to justice developed by Amartya Sen (1999) and Martha Nussbaum (2011). The central argument is that justice should be understood in terms of how distributions affect our well-being and how we function.

Likewise, environmental justice advocates have long talked about community health, good jobs, and the basic capabilities necessary to live flourishing lives. Many environmental justice movements focus on various threats to the ability of a community to function. If you despoil the environment or undermine health, the argument goes, you also destroy the ability of a community to function and reproduce itself and its culture.

## The expanding reach of environmental justice

The idea of environmental justice may have begun with groups organizing around race, class, and environmental impacts in the United States, but the framework has expanded in application in the past 30 years. Environmental justice as an organizing frame is being applied to a wide range of issues, in an increasing number of places (Schlosberg, 2013).

Moving beyond the initial concerns with toxins and dumps, environmental justice has been applied to analyses of transportation, land use and smart growth policy, water quality and distribution, energy development and jobs, brownfields refurbishment, food justice, and questions of the role of scientific expertise.

The expansion of the discourse has occurred not only in issues, but in geographic location as well. Applications include cases of postcolonial environmental justice in India, waste management in the UK, agrarian change in Sumatra, nuclear waste in Taiwan, salmon farming and First Nations in Canada, gold mining in Ghana, oil politics in Ecuador, wind farm development in Wales, pesticide drift in California, Aboriginal water rights and co-management of lands in Australia, and many more. One of the key arguments is to bring environmental justice and sustainability into a single framework (Agyeman, 2013).

In addition to the expansion of issues, there has been a push to globalize environmental justice frames (Pellow, 2007). This is clearly illustrated by the use of environmental justice as an organizing theme by a number of global movements, such as climate justice, e-waste, food security, and indigenous rights. The idea has also been extended to the nonhuman realm, with claims for justice applied to animals, species, and ecosystems.

From its origins in the US, environmental justice has become a salient organizing idea for community movements worldwide, and an expanding analytical frame for scholars across the social sciences and humanities.

## Learning resources

There are a variety of websites that offer further and ongoing information on environmental justice issues; these include those of movement groups, academics, and government agencies.

http://www.weact.org. WeAct for Environmental Justice is based in New York City and addresses a range of environmental and climate concerns at the local level.
http://www.ienearth.org is the website of the Indigenous Environmental Network, which covers issues, groups, and conflicts across the globe.
http://www.ejolt.org. EJOLT is a global mapping project that catalogues and analyses environmental justice issues.

Government resources:
https://www.epa.gov/environmentaljustice is the home of the EPA Office of Environmental Justice, and has links to a range of environmental justice resources.

To learn more about environmental justice, here are two recent contributions that lay out both the history and scope of the movement and idea:

Agyeman, J., Schlosberg, D., Craven, L. & Matthews, C. (2016) 'Trends and Directions in Environmental Justice: From Inequity to Everyday Life, Community, and Just Sustainabilities'. In *The Annual Review of Environment and Resources*, 41: 6.1–6.20. doi: 10.1146/annurev-environ-110615-090052.
Holifield, R., Walker, G. and Chakraborty, J. (eds) (2017) *Routledge Handbook of Environmental Justice*. Abingdon: Routledge.

## Bibliography

Agyeman, J. (2013) *Just Sustainabilities: Policy, Planning, and Practice*. London: Zed Books.
Bullard, R. (ed.) (1993) *Confronting Environmental Racism: Voices from the Grassroots*. Boston: South End Press.
Bullard, R. (ed.) (1994) *Unequal Protection: Environmental Justice and Communities of Color*. San Francisco: Sierra Club Books.
Cole, L.W. & Foster, S.R. (2001) *From the Ground Up: Environmental Racism and the Rise of the Environmental Justice Movement*. New York: New York University Press.
Fraser, N. (1997) *Justice Interruptus: Critical Reflections on the 'Postsocialist' Condition*. New York: Routledge.
Gibbs, L. (1982) *Love Canal: My Story*. Albany, NY: SUNY Press.
Melosi, M.V. (2004) *Garbage in the Cities: Refuse, Reform, and the Environment*. Pittsburgh, PA: University of Pittsburgh Press.

Nussbaum, M.C. (2011) *Creating Capabilities: The Human Development Approach*. Cambridge, MA: Harvard University Press.

Pellow, D.N. (2007) *Resisting Global Toxics: Transnational Movements for Environmental Justice*. Boston, MA: MIT Press.

Rawls, J. (1971) *A Theory of Justice*. Oxford: Oxford University Press.

Schlosberg, D. (2013) 'Theorizing Environmental Justice: The Expanding Sphere of a Discourse', *Environmental Politics*, 22(1): 37–55.

Sen, A. (1999) *Development as Freedom*. New York: Anchor.

Whyte, K.P. (2011) 'The Recognition Dimensions of Environmental Justice in Indian Country', *Environmental Justice*, 4(4): 199–205.

# 6.12

# Environmental education

Sarah Burch

## Education in a changing environment

The environment is not a realm separate from human experience, but rather is central to the fabric of our communities, values, and aspirations. Once viewed as a wilderness worthy of fear or domination, then later as the subject of value-free scientific inquiry, the environment is increasingly understood to be a social and political space. The preservation and exploitation of natural systems is shaped by human values: what we feel entitled to consume, what we desire to preserve, and what we wish to create in the future all have a bearing on how we perceive 'the environment'. That environment never 'speaks for itself', as it were, but is always spoken for.

This changing landscape reveals a complex web of interactions between ecological and social systems. It highlights the central importance of excellent environmental education, but also calls into question the traditional view of education as being the unidirectional flow of 'truth' from the expert to the ignorant. Emerging approaches to environmental education acknowledge that possession of facts is not sufficient to change behaviour (Blake, 1999), there are many 'holders' of knowledge (not simply scientists) (Irwin, 2013), and myriad ethical implications are implicit in both what we choose to study and the solutions to environmental problems that we choose to pursue.

The sections that follow explore the powerful emerging trends that are reshaping the landscape of environmental education, with potentially transformative implications for environmental studies, policy, and practice.

## The rise of environmental education

In the late eighteenth and nineteenth centuries, Jean-Jacques Rousseau and Louis Agassiz laid the foundations of what would become known as 'nature study': an effort to nurture appreciation of the natural world while also imparting moral lessons and cultural values (McCrea, 2006). Nature study morphed over the decades that followed, adding layers of rigorous scientific training (as in the study of conservation) on top of the more traditional natural history, and ultimately being influenced by civil rights and environmental activism to encapsulate social movements as well. The environmental education that we see today often aims to expose students to nature, while tackling (from both biophysical and social perspectives) complex problems such as climate change, species extinction, and preservation of nature in urban spaces.

While there are many desired outcomes of effective environmental education, a common goal is responsible citizenship behaviour (Hungerford & Volk, 1990). In other words, by learning about the natural and social systems that give rise to environmental problems, we might create generations of students who feel both individually and collectively responsible for preserving natural spaces, consuming ethically-produced goods, and contributing to the creation of more sustainable communities.

Running parallel to the emergence of nature study and conservation education has been advances in pedagogy – the study of how we learn, and how best to teach. Since the early part of the twentieth century it has been argued that learning involves the whole person – not simply cognition but also thinking, feeling, perceiving and behaving (see for example, Dewey, 1916; Kolb, 1984). As such, experience plays a pivotal role in learning. Students can be exposed to concrete experiences, encouraged to reflect on these experiences, and taught to distil reflections into abstract concepts that can then shape future behaviour (Kolb & Kolb, 2005).

Experiential learning has long been central to environmental education. By engaging directly with nature, or with the social and political systems that shape our interaction with it (picture sitting in a City Council room as a climate change action plan is debated, and the costs and benefits considered), the complexity of social-ecological systems is made plain.

## New frontiers: collaboration and creativity

### *The many flavours of knowledge that comprise environmental education*

For several decades, environmental education has been treated as fundamentally interdisciplinary: soil chemistry, hydrology, and conservation biology collide in the study of climate change impacts on wetlands, and the implications of changing precipitation regimes for species at risk. Fisheries ecology intersects with ecotoxicology and environmental geochemistry as we study the rise and fall of commercially valuable fish species. But it is also clear that the social sciences and humanities are central to addressing environmental challenges: our emissions of the greenhouse gases that cause climate change, our demand for particular foods, and the waste produced by our cities, for instance, are explored in the domains of public policy, macro-economics, and urban planning.

This interdisciplinarity raises an important question in environmental education: do we build a deep skill-set within one particular discipline and then collaborate with others of similar depth, or do we gain reasonable skills in a host of domains and bring a more holistic, if shallow, view to these complex problems? Both approaches have merit, but still leave unexplored the prospect for knowledge held outside the realm of the 'expert' to hold sway. If one of the core purposes of environmental education is to inform and motivate transitions toward more environmentally, socially, and economically sustainable interactions with the natural world, then the way that nature is 'constructed' by each of us becomes central to the solutions that we design and implement.

For this reason, environmental scholars are increasingly engaged in processes of knowledge co-production. In the words of Sheila Jasanoff, "scientific knowledge, in particular, is not a transcendent mirror of reality. It both embeds and is embedded in social practices, identities, norms, conventions, discourses, instruments, and institutions – in short, all of the building blocks of what we consider 'social'" (Jasanoff, 2004).

Rather than being value-free and flawlessly objective, knowledge can be viewed as a social construct. Education is not simply an expert-driven process, designed to address a simple deficit in information and lead smoothly to a change in behaviour (Hargreaves et al., 2013). These insights lead us to consider environmental education through a new lens: knowledge about nature can be deepened, enriched, and expanded if co-created through an iterative and deeply interactive process of engagement.

## *Crowd-sourcing education and the advent of Massive Open Online Courses*

Traditional scientific inquiry tends to de-emphasize interests and values in favour of objective analysis and a scientific lens through which to view environmental problems. As with 'nature study' over a century ago, however, cultural values are a central strand in the story about the origins of (and possible solutions to) environmental problems. New tools are emerging that may facilitate a reframing, or at least the collaborative co-production, of environmental education.

Mail, videos, and later the internet have long been used to deliver traditional educational materials, but web-based social media have provided a new level of connectedness and capacity to collaborate (Li et al., 2016). This has now fed directly into the merging of the social and educational worlds in the form of Massive Open Online Courses (MOOCs): web-based, freely-available courses taught by university and college instructors, and offered to thousands of students at a time (Burch & Harris, 2014).

MOOCs, facilitated by social media, represent a pedagogical approach and technological platform that can facilitate learning in a variety of disciplines. Media attention, paired with both extreme enthusiasm and cutting criticism, has swelled since 2012, creating a rich and varied conversation about the future of education.

Ultimately, the goal of crowd-sourced environmental education may not be to push towards converging views or definitions of pressing issues, but rather to create a space within which varying perceptions of risk, cultural myths or narratives about the human relationship to nature, and diverse identity claims can be explicitly shared and negotiated (Burch & Harris, 2014).

### *Visioning, geo-visualization, and gamification*

While maps, graphs, and tables are the typical scientific tools for illustrating results, most humans are not moved to action by dots and lines. Stories, pictures, and games, however, trigger deep emotional responses (Shaw et al., 2009), allowing us to grasp vast complexity in a single glance and consider environmental change as a fundamentally personal phenomenon.

Three broad sets of tools have emerged that capitalize on this insight, and promise to transform the way in which environmental education (and engagement with the public) is undertaken. Visioning is a participatory process in which students imagine a variety of futures, exploring the implications of decisions (maybe build a wetland instead of a water purification plant so that biodiversity is enhanced while water is treated) and both the desirability and effectiveness of various solutions. Geo-visualization brings together models (hydrologic, climatic, socio-economic) with digital elevation, Light Detection and Ranging (LIDAR), and satellite data to create powerful images of the ways that landscapes shift over time. These visualizations can offer neighbourhood-scale experiences of the impacts associated with climate change, for instance, illustrating rising sea levels, shifting ecosystems, and a host of potential response options (Sheppard, 2012; Sheppard et al., 2011). Gamification, the newest tool on the block, offers students the opportunity to explore the intersection of ecological and social systems in real time (Cohen, 2011). Together, these tools hold significant potential to enhance the effectiveness of both environmental education and citizen engagement by triggering emotions, communicating uncertainty, and linking environmental problems to our everyday lives.

## Charting a path forward for environmental education

Education comes in many forms, from the passive receipt of information from an expert, to the collaborative solving of problems, to the critical exploration of whole systems including values, assumptions, and biophysical realities. Given the multitude of environmental challenges we face, the rapid pace of scientific inquiry, and the vast implications for the way our communities function, new tools are emerging that may reshape the landscape of environmental education. This chapter has sketched a brief history of environmental education, and presented several potentially powerful trends that may more effectively connect education with behaviour change.

## Learning resources

Coursera. This is a web-based platform for freely available Massive Open Online Courses, several of which deal with topics related to the environment such as climate change, biodiversity, and sustainable communities. www.coursera.org.

Environmental Education and Training Unit (United Nations Environment Programme). This unit provides publications, events, and tools for enhancing environmental education, including forging links between institutes of higher education and communities. www.unep.org/training.

'CCAFS scenarios engage regions to plan for uncertain futures': this blog post and associated publications, as part of the Research Program on Climate Change, Agriculture, and Food Security, illustrate how creative scenario and visioning techniques can be used to enhance learning. https://ccafs.cgiar.org/blog/ccafs-scenarios-engage-regions-plan-uncertain-futures#.WE77yHeZM_U.

## Bibliography

Blake, J. (1999) 'Overcoming the "value-action gap" in environmental policy: tensions between national policy and local experience', *Local Environment: The International Journal of Justice and Sustainability*, 4: 257–278.

Burch, S. & Harris, S. (2014) 'A Massive Open Online Course on climate change: the social construction of a global problem using new tools for connectedness', *Wiley Interdisciplinary Reviews: Climate Change*, 5: 577–585.

Cohen, A. (2011) 'The gamification of education', *The Futurist*, 45: 16–17.

Dewey, J. (1916) *Democracy and Education*. New York: Perigee Books.

Hargreaves, T., Nye, M. & Burgess, J. (2013) 'Keeping energy visible? Exploring how householders interact with feedback from smart energy monitors in the longer term', *Energy Policy*, 52: 126–134.

Hungerford, H. & Volk, T. (1990) 'Changing learner behaviour through environmental education', *The Journal of Environmental Education*, 21: 8–21.

Irwin, A. (2013) *Sociology and the Environment: A Critical Introduction to Society, Nature and Knowledge*, New Jersey: John Wiley & Sons.

Jasanoff, S. (2004) *States of Knowledge: The Co-production of Science and Social Order*. London: Routledge.

Kolb, A. (1984) *Experiential Learning: Experience as the Source of Learning and Development*. New Jersey: Prentice-Hall.

Kolb, A. & Kolb, D. (2005) 'Learning styles and learning spaces: enhancing experiential learning in higher eduction', *Academy of Management Learning and Education*, 4: 193–212.

Li, Y., Krasny, M. & Russ, A. (2016) 'Interactive learning in an urban environmental education online course', *Environmental Education Research*, 22: 111–128.

McCrea, E. (2006) *The Roots of Environmental Education: How the Past Supports the Future*. Stevens Point, WI: Environmental Education and Training Partnership.

Shaw, A., Sheppard, S., Burch, S., Flanders, D., Weik, A., Carmichael, J., Robinson, J. & Cohen, S. (2009) 'Making local futures tangible – synthesizing, downscaling, and visualizing climate change scenarios for participatory capacity building', *Global Environmental Change*, 19: 447–463.

Sheppard, S.R.J. (2012) *Visualizing Climate Change: A Guide to Visual Communication of Climate Change and Developing Local Solutions*. Abingdon: Earthscan/Taylor & Francis.

Sheppard, S.R.J., Shaw, A., Burch, S., Flanders, D., Wiek, A., Carmichael, J., Robinson, J. & Cohen, S. (2011) 'Future visioning of local climate change: a framework for community engagement and planning with scenarios and visualization', *Futures*, 43: 400–412.

# 6.13 Environmental markets

John O'Neill

Markets and market-mimicking procedures are increasingly used in environmental governance. Markets are institutional arrangements that involve the transfer of rights over goods and services between buyers and sellers using money as the medium of exchange. The direct use of markets is evident in the emissions trading regimes that have become the major approach to climate change mitigation policy and, more recently, in offsetting regimes for the protection of biodiversity. Market-mimicking procedures involve the monetary valuation of environmental goods and bads for the purposes of cost-benefit analysis (CBA).

## Why use markets to reduce environmentally harmful human activities?

Two lines of argument for the use of markets and market mimicking process have been particularly influential.

The first is a neo-classical economic argument that runs as follows: the absence of markets for environmental goods is the source of environmental problems, since preferences for those goods are not captured in market exchanges (Arrow, 1984: 155). Consequently, the solution to environmental problems lies in the extension of market prices to include preferences for environmental goods. The extension can be direct, through the definition of property rights for those goods so that they can be exchanged in actual markets. It can be indirect, through the identification of shadow prices of what people would pay were there a market. Shadow prices can be constructed through 'revealed preference methods' which infer individuals' willingness to pay at the margin for environmental goods from their market behaviour or through 'stated preference methods' in which individuals are asked their willingness to pay for a good in hypothetical markets.

Shadow prices enter into a CBA to include monetary valuations of all costs and benefits affected by some project.

The second line of argument does not require that all goods can be valued through markets, but claims markets provide the most efficient and effective procedure for realising environmental goods (Caney, 2010: 206). Schemes for trading emissions to pollute within a cap are sometimes defended on these grounds. The level of cap itself can be set by scientific and political debate without an attempt to place monetary values on all the goods and bads involved. However, once the cap has been set, trading in rights to emit pollutants offers the most efficient method of keeping total levels of pollution under a certain total. A policy of 'no net loss' of biodiversity has similarly appealed to market mechanisms. Credits are assigned to landowners to create, restore or enhance biodiversity. These can be bought by developers to offset losses to biodiversity caused by a development. The market transactions ensures biodiversity losses and gains are allocated efficiently while ensuring no net loss of biodiversity within a region or country.

## Criticisms of market-based environmental governance

The increasing monetisation and marketisation of environmental goods has come up against civil society resistance. This resistance draws upon very different traditions of argument that move in an opposite market-sceptical direction. Weak scepticism denies that markets and market mimicking mechanisms are a solution to environmental problems. Strong scepticism claims that their spread is itself a source of environmental problems.

Market scepticism is sometimes directed at the attempt to extend monetary measures to environmental goods. Three arguments are particularly influential.

### *Distributional*

The marginal value of a unit of income – a dollar or a euro, for instance – is higher for someone on a lower income. Hence, standard willingness to pay measures will put lower values on the preferences of the poor. Harms and benefits to those on lower incomes will count for less than those on higher incomes. One response is to add distributional weights to monetary valuations that reflect differences in income (Kolstad et al., 2014: 3.6.1.1). However, standard cost-benefit analysis in practice uses unweighted willingness to pay values. Hence policy will tend to benefit those on lower incomes less and adversely affect them more. A further distributional problem concerns future generations and non-humans who cannot express a monetary valuation. Their interests count only precariously to the extent current generations express concern for their interests in their willingness to pay. The problem of intergenerational justice is compounded by the practice of discounting the value of future benefits and costs.

### *Value incommensurability*

To say that values are commensurable is to claim that there is a common measure of value through which value of different options can be measured and ordered. Monetary valuation promises a cardinal scale of measurement for comparing the welfare value of different options, where welfare is understood in terms of preference satisfaction. Can monetary values deliver such a measure? One reason for holding that they cannot concerns constitutive incommensurability (Raz, 1986: 345ff.; O'Neill et al., 2008: Ch. 5). One response to requests to price environmental goods is to protest against the possibility of pricing. The response is rational. Monetary valuation has a social meaning. It is not an exercise in the neutral use of a measuring rod. Many relationships and ethical commitments are constituted by a refusal to price them. To show the strength of one's social and ethical commitments by saying how much it would take to give them up, would not be to show the extent of one's commitments, but rather to reveal their absence. Insofar as places and environmental goods embody relationships to past, present and future community or are the object of ethical concern, one way to rationally express that concern is to refuse to put monetary values on them.

## *Reason and deliberation*

Monetary valuations are reason-blind. They express the intensity of people's preferences for a marginal change in a bundle of goods. They do not reflect the soundness of the reasons for their judgements (O'Neill, 2007). Environmental decisions should be the outcome of public deliberation that reflect the judgements of individuals as citizens, not their private preferences as consumers (Sagoff, 2008). Since reasons must be able to survive being made public, participants are forced to offer reasons that appeal to general rather than particular interests. Publicness ensures that the interests of future generations and non-humans are better represented than through market expressions of private preferences where no publicness condition exists (Goodin, 1996: 846–847).

Market-scepticism is also directed against the claim that markets provide an effective and ethically defensible means to protect the environment.

## *Markets are not effective*

Existing emissions trading schemes have in practice been ineffective. Defenders of emissions trading claim this is a problem of market design. Typical design failures are the allocation of emission permits to those with the largest prior emissions (grandfathering) and the low levels of the cap on emissions. Proponents of emissions trading suggest improved design will resolve problems of effectiveness (Caney & Hepburn, 2011). Market-sceptics suggest that improved design alone will not resolve the problems (Aldred, 2012; Spash, 2010). The general assumption that emissions should be reduced where they are cheapest fails to address the current technological lock-in of carbon-dependent sources of energy. Carbon offsetting whereby emitters compensate emissions by financing projects, normally in developing countries, that would increase emissions elsewhere or capture carbon – for example through forestation, forest protection, or the replacement of polluting factories – raise additional problems. They assume counterfactual claims about what would have been emitted without offsetting which are difficult to verify. They turn polluting activities into assets and hence can create perverse incentives to maintain them for future trade (Lohmann, 2006).

## The markets are not ethically defensible

One central criticism of emissions trading is that it allows wealthy individuals and corporate actors to use their wealth to avoid making those sacrifices required to contribute to the solution of common problems (Goodin, 1994; Sandel, 2005). The wealthy are able to buy the benefits of emissions levels which would cause serious harms if all were permitted to act similarly (Goodin, 1994: 585). The burdens of action and responsibility are shifted to those who contribute least to the problem. A minor short-term responsibility for one group, e.g. to avoid a flight, can be displaced into a major long-term responsibility for another e.g. to grow and sustain trees over several decades.

## Biodiversity offset markets bring their own special problems

They are founded upon particular assumptions about the value of environmental goods. What is valued is not the specific good but rather the services it provides. One site can offset losses in another if it provides the same services. The approach fails to acknowledge the ways distinct places and sites matter as particulars that embody specific natural and cultural histories (O'Neill et al., 2008).

# Conclusion

Proponents of market modes of environmental governance claim that well-designed market instruments offer the solution to environmental problems. The minimal critical claim against environmental markets is that they are neither effective nor ethically defensible modes of environmental governance. A stronger sceptical claim is that the very spread of markets and market-mimicking modes of governance is a source of environmental problems, not their solution. There are two particularly influential forms of the stronger claim. One is Karl Polanyi's argument that markets in land and labour remove customary or ethical constraints on the unsustainable exploitation of environmental goods (Polanyi, 1957: 73). A second is that market economies structurally foster economic growth (Marx, 1970: Ch. 4) and this growth has become unsustainable. On this view the solution to environmental problems requires not more markets, but rather the development of non-market modes of economic and environmental governance.

# Learning resources

The major site advocating the valuation of environmental goods in market terms is *The Economics of Ecosystems and Biodiversity*. It has links to a number of reports and other resources: www.teebweb.org.

The following site contains links to a number of civil society organisations critical of the marketisation of environmental goods. It includes reports, films and other resources: naturenotforsale.org.

An excellent source of resources mapping environmental justice conflicts which includes reports on degrowth, non-market decision making and the limits of market modes of governance is this: www.ejolt.org.

Finally another fine source of critical reports is offered by The Corner House: www.thecornerhouse.org.uk.

## Bibliography

Aldred, J. (2012) "The Ethics of Emissions Trading", *New Political Economy*, 17(3): 339–360.

Arrow, K. (1984) "Limited Knowledge and Economic Analysis." In *The Economics of Information*. Cambridge, MA: Harvard University Press.

Caney, S. (2010) "Markets, Morality and Climate Change: What, If Anything, Is Wrong with Emissions Trading?", *New Political Economy*, 15: 197–224.

Caney, S. & Hepburn, C. (2011) "Carbon Trading: Unethical, Unjust and Ineffective?", *Royal Institute of Philosophy* Supplement, 69: 201–234.

Goodin, R. (1994) "Selling Environmental Indulgences", *Kyklos*, 47(4): 573–596.

Goodin, R. (1996) "Enfranchising the Earth, and its Alternatives", *Political Studies*, 44: 835–849.

Kolstad C., K. Urama, J. Broome, A. Bruvoll, M. Carino Olvera, D. Fullerton, C. Gollier, W. M. Hanemann, R. Hassan, F. Jotzo et al. (2014) "Social, Economic and Ethical Concepts and Methods". In Edenhofer, O., R. Pichs-Madruga, Y. Sokona, E. Farahani, S. Kadner, K. Seyboth, A. Adler, I. Baum, S. Brunner, P. Eickemeier et al. (eds) *Climate Change 2014: Mitigation of Climate Change. Contribution of Working Group III to the Fifth Assessment Report of the Intergovernmental Panel on Climate Change*. Cambridge: Cambridge University Press.

Lohmann, L. (ed.) (2006) *Carbon Trading* (Development Dialogue 48). Uppsala: Dag Hammarskjöld Foundation.

Marx, K. (1970) *Capital I*. London: Lawrence and Wishart.

O'Neill, J. (2007) *Markets, Deliberation and Environment*. London: Routledge.

O'Neill, J., Holland, A. & Light A. (2008) *Environmental Values*. London: Routledge.

Polanyi, K. (1957) *The Great Transformation*. Boston: Beacon Press.

Raz, J. (1986) *The Morality of Freedom*. Oxford: Clarendon.

Sagoff, M. (2008) *The Economy of the Earth*, 2nd edn. Cambridge: Cambridge University Press.

Sandel, M. (2005) "Should We Buy the Right to Pollute?". In *Public Philosophy: Essays on Morality in Politics*. Cambridge, MA: Harvard University Press.

Spash, C. (2010) "The Brave New World of Carbon Trading", *New Political Economy*, 15(2): 169–195.

# Environmental metaphor 6.14

Brendon M. H. Larson

Many environmental issues – including the suite of topics covered in this book – are couched in terms of metaphors, and hence, so are our responses to them. A metaphor is an everyday term that we apply to understanding something very different, such as when we conceptualize nature in terms of *systems* (e.g., ecosystems), our impact on ecosystems in terms of the *footprints* we leave after walking on a beach, and their potential to recover in terms of *resilience* (throughout this chapter, important environmental metaphors will be italicized the first time they occur).

On the one hand, these metaphors may seem trite and commonplace, yet on the other hand they are profound because many scholars have argued that attaining *sustainability* has everything to do with replacing existing metaphors ('metaphors we die by', see Goatly, 1996) with ones more conducive to sustainability (e.g., Larson, 2011; Stibbe, 2015). Further, the study of metaphors reveals that we never really understand things 'as they are', but only through our concepts and our language. We know this not least through historical studies that reveal changes in the metaphors that dominate people's conception of domains, such as nature, over time (Mills, 1982). Accordingly, they can be changed.

To appreciate the importance of metaphors one challenge is to notice them in the first place. Metaphors occur everywhere in discussions about the environment, though by their nature they often blend right into the background; unlike related similes, which state that something is *like* something else, metaphors are implicit, they are camouflaged (like the word 'blend', above). We simply talk about ecosystems and footprints and resilience or planetary *boundaries* and ecosystem *services* and *invasive* species, without noticing that they are metaphorical. If nothing else, this chapter should help to highlight the presence of such metaphors.

For you must first notice a metaphor before can you begin to consciously ask some questions about how it is influencing your thinking about an issue. Otherwise, you may not be aware of how it is swaying your perspective – a concern that applies not just to the influence of environmental metaphors, but to metaphors in media, politics, and your everyday life (e.g., the way that people speak of time in terms of money, such as "spending" and "saving" it).

In particular, consider asking the following four key questions of any metaphor you encounter. These questions encourage reflection that is an entrée to critical thinking about how environmental issues are constituted and whether we might respond to them differently.

1. Is this metaphor **apt**? This is a common place to begin when reflecting on a metaphor: does it appropriately fit what it is trying to describe? A metaphor provides a way for us to make sense of the world, as long as we seek to ensure it is not too misleading. If a metaphor does not fit well at all then you will probably never hear it – someone may have tried it out, only to see it die. This may lead us to assume that common metaphors are good descriptors, simply by the fact that they have become part of how we think and speak about something (a so-called 'dead' metaphor). Yet we must keep in mind that no metaphor 'fits' perfectly because it is always describing something in a way that it literally 'is not.'

Consider the planetary boundaries metaphor (Rockström et al., 2009). This phrase focuses attention on nine proposed biophysical limits to human activity, with the metaphor of boundaries implying that the consequences for humanity would be dire if we exceed them. Critics argue, however, that the metaphor is inaccurate because some of the boundaries are not really boundaries at all and because they are regional rather than global in scale (Nordhaus et al., 2012). Critics once raised similar issues about ecosystem *health*: is an ecosystem something that can have the trait of 'health,' like an individual human being?

2. What is **highlighted** and what is **hidden** by this metaphor? Every metaphor both highlights and hides perspectives on the world. When we think in terms of ecosystem services, for example, we highlight all the ways that nature provides valuable services to human beings (and in this sense it is apt). Some would say that this is a good thing, because for too long many of these benefits have gone unrecognized. But there are always, simultaneously, perspectives that a metaphor hides. The metaphor of services, for example, may cause us to neglect certain under-recognized or ineluctable values.

As another example, an insightful analysis by Moore and Moore (2013) considers the metaphor of ecological *restoration* in this regard. The metaphor highlights that we can to some extent fix our errors, by returning nature to an earlier state, yet it obscures how this belief may cause us to neglect questions about the scale of our needs and our wants. If we were less destructive of the natural world in the first place, perhaps restoration would be unnecessary – and the belief that we can restore may abet continued destruction.

3. What **values** does this metaphor communicate? This may seem like an odd question, especially since so many environmental metaphors derive from science. It may not seem that they have much to do with values at all. Consider an example, though. If you understand the natural world in terms of metaphors such as *stability* and *'the balance of nature'*, which is traditionally how people have understood it, then it would seem right to try to prevent *disturbance*. However, there is a tremendous philosophical schism between the former, what 'is,' and the latter, what 'ought to be'. By attending to metaphor, you may recognize that it is a value-laden interpretation of the former that leads to the latter. The problem here becomes even more pronounced once you recognize that ecologists now discount the metaphor of 'a balance of nature' and instead emphasize the importance of disturbance (e.g., in the form of fire) for the function of ecological systems.

In short, metaphors implicitly promote the view of the world that they highlight. This holds for the metaphors of environmental science, too, and in that sense they are value-laden and advocate for particular views. Accordingly, many environmental metaphors speak at two levels: with scientific authority *and* to people's values and preferences. Thus, a further reason to investigate the metaphors structuring an issue is to know whether you are being swayed by their value resonance rather than by pertinent evidence.

In this context, briefly reconsider the concept of aptness (or suitability). A metaphor may be apt, scientifically, yet we also need to ensure that it is appropriate in the domain of sustainability (Harré et al., 1999). Does it nurture positive relations among people and with the environment? If not, then in this era we should seriously consider alternatives.

4. What does this metaphor **do**? In the past, metaphors were often considered mere rhetoric, but cognitive science now recognizes that we think in terms of metaphor. Metaphors occur everywhere in human thought and language, not just when we use them consciously (the word 'use' there being an example). Consequently, metaphors are not simply words, they are (metaphorical) actors. Because humans understand the environment in terms of metaphors, our responses to environmental problems are shaped by metaphors, too.

Returning to the example of ecosystem services, notice how the metaphor tends to emphasize how to quantify services (as occurs in many contexts around the world at present), which raises tremendous issues related to whether this is appropriate – or whether the very act of trying to quantify something might actually devalue it (e.g., if it is essentially priceless or if there is a risk we may not understand its value). Economist Richard Norgaard (2010) questions whether the metaphor does more harm than good to the extent that it seduces us with a vision of continuing to do what we are already doing rather than enacting more radical change.

As another example, when you perceive some species as invasive, you act to prevent their spread (Larson, 2005, 2011). This may be appropriate, though the nearly imperceptible shift from value-laden metaphor to action here is noteworthy. For if these species are not so much the drivers of environmental change (as the implicit notion of 'invaders' implies), but merely passengers responding to change, then perhaps 'fighting' them misdirects resources that would be better utilized elsewhere.

Ultimately, then, such reflections about metaphors can contribute to humility in our actions. We have to act, yet we are rarely certain that our actions are correct. So it behooves us to continually re-evaluate our actions in the light of associated metaphors. Schön (1979) famously pointed out that we too quickly jump into a mode of 'problem-solving', before having reflected carefully enough on the 'problem-framing'; that is, whether the problem we are solving is really even a problem. When we carefully reflect on the metaphors shaping our actions, we have a new capacity to rethink our response to environmental issues.

## Learning resources

This engaging video, by a poet, helps to introduce metaphors and how they shape our understanding:

Hirshfield, J. (n.d.) The art of the metaphor. TEDEd Video: http://ed.ted.com/lessons/jane-hirshfield-the-art-of-the-metaphor.

The following book provides an introduction to the role of metaphors in attaining sustainability, with a focus on the role of scientific/ecological metaphors:

Larson, B.M.H. (2011) *Metaphors for Environmental Sustainability: Redefining Our Relationship with Nature*. New Haven: Yale University Press.

This book, by a leader in the field of ecolinguistics, provides an overview of the main 'stories' that influence our thinking about and interactions with the natural world:

Stibbe, A. (2015) *Ecolinguistics: Language, Ecology and the Stories We Live By*. New York: Routledge.

## Bibliography

Goatly, A. (1996) 'Green grammar and grammatical metaphor, or language and the myth of power, or metaphors we die by', *Journal of Pragmatics*, 25: 537–560.

Harré, R., Brockmeier, J. & Mühlhäusler, P. (1999) *Greenspeak: A Study of Environmental Discourse*. Thousand Oaks, CA: Sage.

Larson, B.M.H. (2005) 'The war of the roses: demilitarizing invasion biology', *Frontiers in Ecology and the Environment*, 3: 495–500.

Mills, W.J. (1982) 'Metaphorical vision: changes in Western attitudes to the environment', *Annals of the Association of American Geographers*, 72: 237–253.

Moore, K.D. & Moore, J.W. (2013) 'Ecological restoration and enabling behavior: a new metaphorical lens?', *Conservation Letters*, 6: 1–5.

Nordhaus, T., Shellenberger, M. & Blomqvist, L. (2012) *The Planetary Boundaries Hypothesis: A Review of the Evidence*. Oakland, CA: Breakthrough Institute.

Norgaard, R.B. (2010) 'Ecosystem services: from eye-opening metaphor to complexity blinder', *Ecological Economics*, 69: 1219–1227.

Rockström, J., Steffen, W., Noone, K., Persson, Å, Chapin III, F.S., Lambin, E.F., Lenton, T.M., Scheffer, M., Folke, C., Schellnhuber, H.J. et al. (2009) 'A safe operating space for humanity', *Nature*, 461: 472–475.

Schön, D. (1979) 'Generative metaphor: a perspective on problem-setting in social policy', in A. Ortony (ed.) *Metaphor and Thought*. Cambridge: Cambridge University Press.

# Environmental migrants and refugees

Romain Felli

The influence of environmental and climatic conditions on human life and actions (such as human mobility) is a perennial object of rumination for geographers, historians, anthropologists or political philosophers. The novelty of recent academic and political debates on environmental migrants and refugees (since the late 1980s), however, relates to the influence of human-made environmental transformations or degradations (such as anthropogenic climatic change, deforestation or desertification) on human migrations. Are the epoch-making modifications of human–nature interactions (in an era that some are calling the 'Anthropocene') significantly reshaping human agency, restricting, constraining, or enlarging the ability and the opportunities of different human groups to move across territories, borders and scales? And what, if anything, should be done about it?

## Environmental change, migration and conflict

A particularly poignant consequence of climate change is the displacement of populations because of a transformed environment, such as an increase in droughts or rising sea-levels. Very high numbers of these so-called 'climate refugees' (150–200 million) have been cited by environmentalists, journalists, and policy-makers since the early 1990s, sometimes largely exceeding the number of actual ('political') refugees (Gemenne, 2011). Stories of 'sinking islands' in the Pacific (such as Tuvalu) or submerged floodplains in South East Asia, for instance, led to the impression that dozens of millions of people would soon be forced to move away from their homes and livelihoods. But, is the relation between environmental transformations and human displacement so straightforward?

Some authors (sometimes called 'neo-Malthusians') have long used a rather crude environmental determinism to argue that environmental transformations, such as deforestation or climate change would lead to increased human migrations. According to them, a growing population puts pressure on limited natural resource (such as water or land) and thereby degrades them. This, in turn, leads to conflicts over resources (such as war) and/or possibly to out-migrations of population. These migrations may also put pressure on the receiving environments, such as those of refugee camps, and foster conflicts there (Reuveny, 2007). From this perspective, climate change, by further degrading natural resources (through increased drought or extreme weather events), 'multiplies' the possible outburst of violence and triggers migrations. The famed biologist Norman Myers was instrumental in producing reports forecasting hundreds of millions of people forced to move because of climate change – mainly in the Global South (see for instance Myers, 1993). These numbers were then circulated and broadcast by charitable organisations, non-governmental organisations and international organisations, such as the United Nations Environmental Programme (UNEP). More recently, the UNEP was caught in a controversy about its prediction of 50 million environmental refugees by 2010, which it failed to substantiate.

## Contesting the 'environmental refugees' thesis

Since the early 2000s, however, this sort of forecast has been strongly criticised by geographers, sociologists and specialists of migration. They have argued that no direct relation between environmental change and migration could be established, and that attempts at quantifying the 'number' of environmental or climate refugees were misleading. Different vulnerabilities (such as employment in a sector strongly affected by climate variations, for instance agriculture or tourism) and assets (social networks, money, education, etc.) mean that people in a given area are not similarly affected by a given natural hazard and that they do not respond similarly to it. On the contrary, migrations are complex social actions. They need to be explained by their economic, political, cultural and geographic contexts. They involve power relations and cannot simply be assumed to arise from the existence of a given natural hazard (Carr, 2005). The neo-Malthusian undertones of much of the early debate were criticised as playing into the hands of the Northern (especially US) security and military apparatus (Hartmann, 2010). Critics of the 'environmental refugee' or 'climate refugee' theory have pointed out that it neglects the ability of human beings to cope with changes, to adapt to environmental transformations by producing social, technological and economic responses. Furthermore, they objected to the use of the notion of 'refugee' in this context, arguing that it should remain a specific legal category for people needing protection because of persecutions. The 'climate refugee' thesis conjured unfounded fears about massive South-North migration in a context in which support for refugee protection in Northern countries was eroding (Bettini, 2013). Also, empirical research showed that international migration was not the option automatically favoured by supposedly 'sinking' islanders, or other people exposed to environmental change (Farbotko & Lazrus, 2012).

## Global management and local translation

The complexity of these debates was further compounded by the advocacy of scholars, policy-makers and activists demanding the creation of international schemes – possibly a new international convention – for protecting environmental, or climatic, 'refugees'. Various international organisations, such as the International Organization for Migration (IOM), and the UN Refugee Agency (UNHCR) are struggling over the definition, and jurisdiction, of the management of persons displaced by environmental disasters. For instance, a set of international norms, known as the Nansen Initiative, seeks to offer guidelines to manage people displaced "in the context of disasters and the effects of climate change". However, some concerns have arisen that contest the framing of environmental migration as an issue that needs to be governed or managed at a global level, and in relation to global climate policy. At national and regional levels, the question of environmental migration – and its recognition, or the absence thereof – raises new questions and conflicts in governance and the redefinition of administrative and political boundaries and scales.

## A more complex understanding of migrations . . .

Since the early 2000s, the sustained attention received by the issue of the environment and migration in international policy-making circles and various media, has led to a renewed academic interest in the topic (Piguet, 2013). Empirical research has flourished on the subject and major research projects – academic and policy driven – have been funded. Many scholars in the field also function as policy experts to governments, advocacy groups and international organisations.

These research projects, and their associated policy translations, have offered a more complex understanding of the relations between environmental transformations and human migrations. They have underlined that temporary or circular migrations have long been a strategy used to sustain rural livelihoods, that environmentally-related migrations tend to be regional rather than international, and that, often, only those with sufficient assets can actually migrate. One outcome of this research is the notion of 'migration as adaptation' that suggests viewing migration in a positive light and managing it (through international organisations, for instance), rather than trying to avoid it at all costs. According to a major research project undertaken by the UK's Government Office for Science (Foresight, 2011; Black et al., 2011), the focus of policymakers should be on those 'trapped' by environmental degradation, unable to move because they lack resources, networks, etc. However, the flourishing of the notions of 'migration as adaptation', of people helping themselves out of poverty, and of the idea that managed environmental migrations could build resilience have led some commentators to fear that environmental migration could be instrumentalised in entrenching existing inequalities of power and wealth (Felli & Castree, 2012)

## . . . in a postcolonial environment

Colonial enterprises from Europe have long shaped the will to manage populations in the South: their birth-rates, their places of residence and work, their labour, movements and (im)mobilities. In this context, planned resettlement – and transfers of population – have long been advocated, and sometimes implemented, by more powerful actors in the global system. These resettlements and transfers have often been justified in relation to environmental hazards (McAdam, 2015). They are still undertaken in some parts of the world, where climate change is used, for instance, as a justification to relocate poor peasants out of fertile floodplains, thereby freeing these spaces for capitalist agro-industries, a process known as 'primitive accumulation'.

Environmental migration has thus been pictured essentially as a process affecting poorer people from rural areas in the Global South. In comparison, less public attention and academic research has been devoted to environmental migrations in the North, in urban and industrial areas, and on 'amenity' migrations (where richer people move into more welcoming environments, such as expatriates and retirement communities). These sorts of migrations have not been pictured as resulting from global environmental crises and as processes that need to be managed.

## Learning resources

There are many institutions involved in the management of environmental migration, or that offer scientific analysis and advocacy for environmental migrants.

The International Organization for Migration (IOM) offers a very complete website with many written and documentary resources: http://environmentalmigration.iom.int.

The TransRe, a research project based at the University of Bonn in Germany, has a very lively numeric presence with various blogs and activities on social networks: http://transre.org/en/blog.

One important advocacy group on environment and migration is the UK based "Climate and Migration Coalition". Various podcasts on different aspects of the topic are on display on its website: http://climatemigration.org.uk.

A useful visual resource is the richly illustrated *Atlas of Environmental Migration* (edited by Dina Ionesco, Daria Mokhnacheva, François Gemenne. Abingdon: Earthscan/Routledge, 2017).

## Bibliography

Bettini, G. (2013) 'Climate Barbarians at the Gate? A Critique of Apocalyptic Narratives on "Climate Refugees"', *Geoforum*, 45: 63–72.

Black, R., Adger, N., Arnell, N.W., Dercon, S., Geddes, A. & Thomas, D.T (2011) 'The Effect of Environmental Change on Human Migration', *Global Environmental Change*, 21(1): 3–11.

Carr, E.R. (2005) 'Placing the Environment in Migration: Environment, Economy, and Power in Ghana's Central Region', *Environment and Planning A*, 37(5): 925–946.

Farbotko, C. & Lazrus, H. (2012) 'The First Climate Refugees? Contesting Global Narratives of Climate Change in Tuvalu', *Global Environmental Change*, 22(2): 382–390.

Felli, R. & Castree, N. (2012) 'Neoliberalising Adaptation to Environmental Change: Foresight or Foreclosure?', *Environment and Planning A*, 44(1): 1–4.

Foresight (2011) *Migration and Global Environmental Change*, Final Project Report. London: The Government Office for Science.

Gemenne, F. (2011) 'Why the Numbers Don't Add Up: A Review of Estimates and Predictions of People Displaced by Environmental Changes', *Global Environmental Change*, 21(1): 41–49.

Hartmann, B. (2010) 'Rethinking Climate Refugees and Climate Conflict: Rhetoric, Reality and the Politics of Policy Discourse', *Journal of International Development*, 22(2): 233–246.

McAdam, J. (2015) 'Relocation and Resettlement from Colonisation to Climate Change: The Perennial Solution to "Danger Zones"', *London Review of International Law*, 3(1): 93–130.

Myers, N. (1993) 'Environmental Refugees in a Globally Warmed World', *BioScience*, 43(11): 752–761.

Piguet, E. (2013) 'From "Primitive Migration" to "Climate Refugees": The Curious Fate of the Natural Environment in Migration Studies', *Annals of the Association of American Geographers*, 103(1): 148–162.

Reuveny, R. (2007) 'Climate Change-induced Migration and Violent Conflict', *Political Geography*, 26(6): 656–673.

# 6.16 Environment and the news media

Anders Hansen

Research on the media and the environment since the 1970s has overwhelmingly focused on the news and the reporting of environmental controversies and problems. Most of this research has focused on news coverage of specific environmental issues, problems or disasters over a limited period of time, providing valuable evidence on the processes involved in the short-term public construction and representation of particular issues or problems. A smaller, but growing body of research, has examined the longer-term ups and downs of news media attention to the environment and contributed evidence on the 'drivers' and processes impacting on the extent and nature of news reporting on the environment and its roles in relation to public opinion and political action.

## News coverage of the environment

The 1960s saw the emergence of a more holistic environmental or ecological paradigm and the 'environment' as a clearly distinguishable category of news coverage first arose in the late 1960s. Environmental disasters – whether 'natural' or specifically caused by human activity – are inherently news-worthy and thus inevitably attract considerable news coverage. Beyond disasters, analyses of news coverage have encompassed a wide range of individual environmental issues and concerns, including pesticides and chemical pollution, soil/water/air pollution and quality, population-growth, unsustainable resource exploitation and depletion, nuclear power, carbon-based energy production, genetically modified crops and food production, etc. Since the late 1980s, global warming/climate change has become the single most prominent focus of environmental news coverage, and this is also reflected in where the main emphases of environmental communication research have moved to.

Analyses of news coverage of environmental events, disasters, and issues have shown how such coverage is influenced by a wide range of factors, including key news values (e.g. geographical proximity), the practices and routines of journalists, the ownership and political outlook of news organizations, the type of news organization, and the publicity practices and communications activities of sources. National and cultural differences have also been shown to impact on how the environment and environmental issues are reported on.

Comparative studies of news coverage of environmental issues have shown two important characteristics: one, that news attention to environmental issues goes up and down in issue-attention cycles that bear little relationship to the severity of these issues, as measured by, for example, scientific, economic or other indicators; and two, an overall firmly upward trend in the amount of news attention to the environment and confirmation of 'the environment' as a distinctive and indeed institutionalized category of news coverage (Hansen, 2015).

While a multitude of factors impact on the extent and framing of news coverage, studies have confirmed that key influences are political attention, the activities of campaigning groups and non-governmental organizations, and major international summit events. By contrast, scientific publications, weather events, or indeed disasters are not by themselves major contributors to the overall fluctuations in media attention to the environment (Schäfer et al., 2014).

The consolidation of environmental news as a distinct category of news coverage is confirmed by research showing that – rather than different environmental issues

**Figure 6.16.1** Climate summits: participants in the Paris COP21 Climate Change Conference 2015.

*Source*: https://commons.wikimedia.org/wiki/File%3ACOP21_participants_-_30_Nov_2015_(23430273715).jpg. By Presidencia de la República Mexicana [CC BY 2.0 (http://creativecommons.org/licenses/by/2.0)], via Wikimedia Commons.

**Figure 6.16.2** Environmental protest: Global Climate March Berlin.

*Source*: https://commons.wikimedia.org/wiki/File%3AGlobal_Climate_March_Berlin_-56_(22799527953).jpg. By mw238 (Global Climate March Berlin – 56) [CC BY-SA 2.0 (http://creativecommons.org/licenses/by-sa/2.0)], via Wikimedia Commons.

competing for space on the news agenda – enhanced news attention to one environmental issue often brings with it increased coverage of other environmental issues (Djerf-Pierre, 2013). News studies have also shown that it often takes a 'threshold event' or 'tipping point' for news attention to increase significantly. Pioneering research in the early 1980s (e.g. Mazur, 1984), for example, showed how it took a major nuclear accident like Three Mile Island to significantly sensitize and alert the news media to nuclear issues, but once alerted, news reporting of all things nuclear increased dramatically and continued to remain high for years following the accident. Set against other established news categories such as terrorism, war reporting, economic news, international conflict news, etc., environmental news often, however, loses out in competition for the limited space of traditional news formats of print and broadcast news.

## Environmental journalism

The sociology of media organizations and news journalism has provided much evidence relevant to understanding how the news media report on the environment. Early studies (Schoenfeld et al., 1979) noted how the establishment in the late 1960s and early 1970s within many news organizations of designated environmental news 'beats' staffed by specialist environment reporters was key to the rise of the environment on the public agenda. The rise of 'environmental journalism' as a specialty type of news reporting brought with it the benefits of a professional journalistic culture with implications not only for the extent to which the news media drew attention to environmental problems, but importantly for the complexity, accuracy, investigative and critical nature of news reporting on major environmental issues. However, even during its comparatively short history, environmental journalism has been the subject of enormous changes and fluctuations, experiencing significant cycles of expansion and contraction in news organizational arrangements such as the designation of specialist beats and specialist reporters.

Surveying the history of environmental journalism, Sharon Friedman (2015) notes how media convergence, downsizing and the rise of the internet and digital media technologies have had both positive and negative effects on environmental news coverage. Negative effects include the reduction or elimination across print and broadcast news media of specialist environment beats and designated environmental reporters, with 'environmental news' often being dispersed across more mainstream news categories (business, politics, technology, health, etc.) and covered by reporters, who would not necessarily have accumulated the breadth of insight or the range of reliable sources that tends to be characteristic of specialty environmental or science reporters (Dunwoody, 2015). Positive influences of the rise of the internet and digital media include the vast increase in the availability and accessibility of environmental information, speed of communication, and enlargement of the range of 'voices' participating in public debate about the environment.

## Source influence in environmental news

The relationship between journalists and their sources is a key factor shaping environmental news coverage. Environmental journalism has been repeatedly shown to be highly 'authority-oriented', relying on government and official institutions, experts and representatives for statements about environmental issues, whilst affording much less prominence or credibility to environmental pressure groups or non-governmental organizations critical of environmental policies or practices. At the same time, sources of all types have become increasingly adept at news management and public relations approaches to influencing public communication about controversial environmental issues. Studies have begun to uncover the range and effectiveness of news management strategies deployed. These range from those simply seeking to take advantage of knowledge of news values, media organizational arrangements and journalistic values and practices, to the more sinister approaches aimed at confusing public debate, 'sowing doubt' about the scientific consensus on, for example, climate change, and hiding particular vested economic and political agendas behind a veneer of seemingly independent 'expert' institutions and think-tanks (Miller & Dinan, 2015) ostensibly concerned with offering impartial evidence-based contributions to public debate.

The relationship between sources and journalists has been characterized as a dance, where one or the other may lead at various points in time, but the evidence from communication research shows that the balance of power has shifted increasingly in favour of sources. This is due to a combination of economic pressures on news organizations resulting in re-structuring and downsizing (often with particular detrimental impact on environmental and other specialty journalisms), enhanced public relations and news management practices of sources, and the new opportunities and challenges of digital and multi-platform media. The convergence of these pressures and developments mean that environmental journalists are under increasing pressure to produce more textual and visual reporting, across multiple news platforms and with tighter timelines, due not least to the round-the-clock nature of digital news reporting. These pressures have led to journalism that is increasingly desk-bound, which in turn has increased the scope for pro-active news sources and news-providers to 'subsidise' the work of news organizations and their journalists with ready-packaged and advantageously framed information, while at the same time significantly reducing the scope for journalists to exercise traditional journalistic networking skills and fact/accuracy-checking strategies (Williams, 2015).

## News reporting and public/political agendas

Research into the influence of news reporting on public opinion and political decision making regarding the environment has drawn on a wide range of models from communications theory and many other disciplines. Agenda-setting and framing research (Trumbo & Kim, 2015; Nisbet & Newman, 2015) have provided valuable insights into the news media's role in influencing not just which environmental topics get discussed in the public sphere, but also how they are framed in terms of defining causes, apportioning blame and identifying solutions. Above all, research on news reporting of the environment has confirmed the complex influences on how the environment becomes 'news', how environmental issues are defined and framed in news media, and the complex and dynamic nature of how news media, political and public agendas interact. The key challenge for research on the environment and news media is to map and understand the changing nature of these complex interactions.

## Learning resources

For fascinating continuous monitoring of world news media coverage of climate change, see the website of the International Collective on Environment, Culture and Politics: http://sciencepolicy.colorado.edu/icecaps/research/media_coverage/index.html.

*The Routledge Handbook of Environment and Communication* (https://www.routledge.com/The-Routledge-Handbook-of-Environment-and-Communication/Hansen-Cox/p/book/9780415704359) offers state-of-the-art chapters by leading environmental communication scholars on the production, representation and social/political implications of environmental news.

## Bibliography

Djerf-Pierre, M. (2013) Green metacycles of attention: reassessing the attention cycles of environmental news reporting 1961–2010, *Public Understanding of Science*, 22(4): 495–512.

Dunwoody, S. (2015) Environmental scientists and public communication. In A. Hansen & R. Cox (eds) *The Routledge Handbook of Environment and Communication*. Abingdon: Routledge.

Friedman, S. (2015) The changing face of environmental journalism in the United States. In A. Hansen & R. Cox (eds) *The Routledge Handbook of Environment and Communication* (pp. 144–157). Abingdon: Routledge.

Hansen, A. (2015) News coverage of the environment: a longitudinal perspective. In A. Hansen & R. Cox (eds) *The Routledge Handbook of Environment and Communication* (pp. 209–220). Abingdon: Routledge.

Mazur, A. (1984) The journalists and technology: reporting about Love Canal and Three Mile Island, *Minerva, 22*(Spring): 45–66.

Miller, D. & Dinan, W. (2015) Resisting meaningful action on climate change: think tanks, 'merchants of doubt' and the 'corporate capture' of sustainable development. In A. Hansen & R. Cox (eds) *The Routledge Handbook of Environment and Communication* (pp. 86–99). Abingdon: Routledge.

Nisbet, M.C. & Newman, T.P. (2015) Framing, the media, and environmental communication. In A. Hansen & R. Cox (eds) *The Routledge Handbook of Environment and Communication* (pp. 325–338). Abingdon: Routledge.

Schäfer, M.S., Ivanova, A. & Schmidt, A. (2014) What drives media attention for climate change? Explaining issue attention in Australian, German and Indian print media from 1996 to 2010, *International Communication Gazette*, 76(2): 152–176.

Schoenfeld, A.C., Meier, R.F. & Griffin, R.J. (1979) Constructing a social problem – the press and the environment, *Social Problems*, 27(1): 38–61.

Trumbo, C. & Kim, S.-J.S. (2015) Agenda-setting with environmental issues. In A. Hansen & R. Cox (eds) *The Routledge Handbook of Environment and Communication* (pp. 312–324). Abingdon: Routledge.

Williams, A. (2015) Environmental news journalism, public relations and news sources. In A. Hansen & R. Cox (eds) *The Routledge Handbook of Environment and Communication* (pp. 197–205). Abingdon: Routledge.

# 6.17 Environment and popular culture

Alison Anderson

## What is popular culture and why is it important?

Popular culture encompasses a very wide range of everyday cultural forms that have a large following among ordinary citizens including films, video games, music, and social media sites. It can also be seen as extending to lived experience via popular lifestyle activities involving shopping, food and sports. Defining popular culture, however, is far from straightforward since it is historically variable and always defined in relation to other conceptual categories such as 'high culture' or 'folk culture' (Fiske, 2010; Storey, 1993). Contemporary 'green popular culture' has to be understood in the context of globalisation and the ideological power of global capitalism. Through what Goodman et al. (2016) refer to as 'spectacular environmentalims', it increasingly turns nature into a spectacle with striking visual icons, often involving mega-fauna such as the drowning polar bear, that tend to favour indvidualised rather than collective responses to environmental harm. Through chasing the 'money shot', entertainment media focusing on nature frequently seek to maximise profit and hold onto the capitalist dream yet embrace ecological sustainability. This is bound up with the inherent contradictions of green popular culture reflecting our opposing desires for consumerism and for ecological sustainability.

## Analysing nature and the environment in popular culture

Popular culture can exert a powerful force in the communication of environmental issues, yet it has often been overlooked by scholars; it is so pervasive it is easy to ignore or dismiss as trivial. The constant stream of images and ideas from popular culture in Western society plays an important role in how we construct and maintain our understanding of environmental issues, especially when we have little direct contact with nature. As Julia Corbett observes, "recognizing multiple levels of everyday communication – individual belief and ideology, popular culture, and discourse by social institutions – is vital for understanding the subtle complexities of environmental communication" (2006: 8).

In recent years, however, an increasing number of studies have begun to analyse how nature and the environment is variously represented. These range from studies on advertising campaigns, red top newspapers, blockbuster films and documentaries, to cartoons and greetings cards (see Meister & Japp, 2002). Much of this work focuses on US popular culture – for example, research on portrayals of nature in *The Simpsons*, Hollywood films such as *The Lorax, Wall-E,* and *Ice Age 2* – and a significant amount of attention has focussed on the American climate science fiction disaster *The Day After Tomorrow* (Brereton, 2005, 2015; Moore, 2015; Svoboda, 2016).

However, in line with a more general tendency in media and cultural studies, the bulk of research has tended to focus on text rather than images (Anderson, 2009). Despite the recent advances that have been made in this area, it still remains comparatively under-researched (Hansen & Doyle, 2011). Also, relatively little work has examined how audiences make sense of green popular culture and the extent to which it impacts on perceptions and behaviour. As Parham notes, "the question of reception – of how, exactly, green media or popular cultural texts affect audiences, cultivate environmental awareness, or engender activism – remains largely untested" (2016: 3). To some extent, the focus on textual representations reflects the practical difficulties in undertaking extensive ethnographic audience research and the complexities of examining visual materials.

Likewise, until recently the production of green popular culture has tended to be neglected. An example of a recent study is Maxwell and Miller (2012) who draw attention to the numerous ways that media technology consumes and wastes natural resources – including the environmental impact of the Hollywood film industry.

## Dominant themes in green popular culture

Environmental problems are frequently portrayed as the result of natural forces, rather than connected to issues of social injustice, which tends to result in piecemeal solutions that do not get to the root of the problems. Representations of nature can naturalise and justify existing relationships of power and privilege in society. Noel Sturgeon (2009) argues that environmentalism in popular culture fails to recognise that environmental problems are bound-up with social injustices linked to gender, race and sexuality. A mainstream strand of environmentalism since the late 1980s has been a common narrative framework to legitimate particular aspects of US ways of life including: family values, American history, global military power and consumerism.

Dominant themes in green popular culture include:

- Human mastery over nature
- Mother nature

- Nature as commodity
- Nature and national identity

Alternative discourses are emerging, however, and digital media offers new possibilities to challenge dominant themes. New media activism is based on a sophisticated grasp of popular culture and its modes of expression (Anderson, 2014).

## New media activism and the subversion of popular culture

In the battle to win hearts and minds some transnational environmental NGOs have become masters of subversion and 'brandjacking'. For example, Greenpeace's innovative and humorous 'Barbie It's Over' 2011 campaign against Mattel, the largest toy company in the world, plays on all the ingredients that characterise contemporary popular culture – celebrity, scandal, gossip and conflict.

The environmental NGO sought to expose the destruction of rainforests for pulp paper used in the toy's cardboard packaging through a high profile social media campaign involving two of Mattel's most well-known toys – Barbie and Ken. When examining the packaging from Asia Pulp and Paper (APP), Greenpeace discovered fibres from trees found on the Indonesian island of Sumatra. The forests here are disappearing more rapidly than anywhere else in Indonesia and it is home to an indigenous tiger.

In June 2011 Greenpeace released a spoof video on the internet in which Ken storms out of an interview and dramatically dumps Barbie when he is shown evidence of her cutting down forests. The video, which was translated into 18 different languages, was posted on YouTube and Vimeo and gained over 2 million views by the end of the campaign. Greenpeace's viral video was a play on Mattel's own advertising campaign where Ken wins Barbie back after a 7 year split (Stine, 2011).

Greenpeace created a fake Twitter account for Ken, Barbie's long-standing boyfriend, with an @ken_talks Twitter handle, tweeting messages such as, "Did you know there are only about 400 Sumatran tigers left in the wild? Feel a bit sick :{ ". In response an @Barbie account was created by a fan and began tweeting back with messages such as, "Off to bed. Hope @ken_talks will reply to my texts. Maybe I'll take up recycling and win him back!" (Sullivan, 2012).

Greenpeace also created a Facebook page for Ken and a stunt involving Barbie in a pink bulldozer, and released inappropriate photos of Barbie with a chainsaw. The social media campaign was accompanied by offline action too, including a banner of Ken protesting against Mattel's actions being hung on the wall of the company's headquarters in California (see Figure 6.17.1).

In excess of half a million emails were sent by members of the public to Mattel calling for them to end their association with APP. The company bowed to pressure and announced in October 2011 that they had instructed their suppliers to avoid wood fibre from controversial sources, including companies "that are known to be involved in deforestation".

Popular culture then, despite its inherent contradictions, is a potentially important vehicle for transmitting and contesting environmental meanings and understandings. It offers a rich source of material for analysing how environmental issues are constructed and perceived, and for reflecting on our own relationship with nature.

**Figure 6.17.1** Environmental protest advertising: an example.



## Learning resources

Environmental Justice/Cultural Studies: This website provides a number of useful resources for studying representations of environment/nature in popular culture: http://culturalpolitics.net/environmental_justice/popular_culture.

To find out more about the literature on popular culture and climate change see: http://www.kcl.ac.uk/sspp/departments/geography/research/Research-Domains/Contested-Development/BoykoffetalWP28.pdf.

## Bibliography

Anderson, A. (2009) 'Media, politics and climate change: towards a new research agenda', *Sociology Compass*, 3(2): 166–182.

Anderson, A. (2014) *Media, Environment and the Network Society*. Basingstoke: Palgrave Macmillan.

Brereton, P. (2005) *Hollywood Utopia: Ecology in Contemporary American Cinema*. Bristol: Intellect Press.

Brereton, P. (2015) *Environmental Ethics and Film*. Abingdon: Routledge.

Corbett, J. (2006) *Communicating Nature: How we Create and Understand Environmental Messages*. Washington, DC: Island Press.

Fiske, J. (2010) *Understanding Popular Culture*. London: Routledge.

Goodman, M.K., Littler, J., Brockington, D. & Boykoff, M.T. (2016) 'Spectacular environmentalisms: media, knowledge and the framing of ecological politics', *Environmental Communication*, 10(6): 677–688.

Hansen, A. & Doyle, J. (2011) 'Communicating the environment: guest editors' introduction', *International Communication Gazette*, 73(1–2): 3–6.

Maxwell, R. & Miller, T. (2012) *Greening the Media*. New York: Oxford University Press.

Meister, M. & Japp, P.M. (eds) (2002) *Enviropop: Studies in Environmental Rhetoric and Popular Culture*. Westport: Praeger.

Moore, E.E. (2015) 'Green screen or smokescreen? Hollywood's messages about nature and the environment', *Environmental Communication*, 10: 539–555.

Parham, J. (2016) *Green Media and Popular Culture: An Introduction*. Basingstoke: Palgrave Macmillan.

Stine, R. (2011, August 5th) 'Social media and environmental campaigning: brand lessons from Barbie', *Ethical Corporation*. Online. Available at: http://www.ethicalcorp.com/supply-chains/social-media-and-environmental-campaigning-brand-lessons-barbie.

Storey, J. (1993) *An Introductory Guide to Cultural Theory and Popular Culture*. New Jersey: Prentice-Hall.

Sturgeon, N. (2009) *Environmentalism in Popular Culture: Gender, Race, Sexuality and Politics of the Natural*. Tucson: University of Arizona Press.

Sullivan, E. (2012, 1st January) 'Greenpeace facilitates Ken-Barbie breakup for the rainforest's sake', *PR Week*. Online. Available at: http://www.prweek.com/article/1280545/greenpeace-facilitates-ken-barbie-breakup-rainforests-sake#TlzRo6Ve2dTrlzy4.99.

Svoboda, M. (2016) 'Cli-fi on the screen(s): patterns in the representations of climate change in fictional films', *Wiley Interdisciplinary Reviews: Climate Change*, 7(1): 43–64.

# Environmental policy

6.18

James Palmer

## What is environmental policy?

Environmental policy refers to any deliberate intervention in the world that is designed to improve the conditions under which natural systems exist and operate. Environmental policies normally seek to reduce or reverse the harm inflicted by human activity upon a natural system or process, or alternatively on non-human species. They can be formulated and implemented at multiple scales, assume a diversity of forms, and increasingly involve the action of non-state actors and organisations alongside, or instead of, national governments. The formal emergence of environmental policy, at least in the US and UK, can be traced to 1970 (although some earlier policies in both countries can be viewed clearly today as environmental). Since this time, several transformations have occurred in the forms taken by environmental policy, but also more fundamentally in the understandings of the environment – and, in a closely related sense, of 'nature' – on which those policies are based.

Especially notable shifts can be associated with the invention of the concept of sustainable development in the late 1980s, the rise to prominence of global climate change as an 'emblematic' environmental issue in the 1990s and 2000s (Hajer, 1995), and the advent of the concept of the Anthropocene in the twenty-first century. Yet a recurring theme throughout this period is one of limits to human development and prosperity, evidenced by the Club of Rome's famous *Limits to Growth* report in 1972, and by more recent efforts to identify a series of 'planetary boundaries' within which society may operate safely (Steffen et al., 2015). This chapter summarises the basis for shifting approaches to environmental policy making across the period from 1970 to the present, focusing on questions of discourse and definition – concerning how the goals of policy are delineated – as well as of practice and implementation.

## A shifting environment for policy: from pollution control to sustainable development and the end of (a single) nature?

Whilst 1970 marks the beginning of environmental policy in Anglo-American contexts – this year witnessed the passing of the US *National Environmental Policy Act* and the formation of the first UK *Department of the Environment* – such formal interventions find their roots in a societal environmentalism which had mushroomed in the 1960s, catalyzed by Rachel Carson's publication, in 1962, of seminal research into the effects of pesticide spraying on birds and other species. Carson's *Silent Spring* had a profound impact on public perceptions of the potential for human activities to cause irreversible damage to the environment, and combined with other high-profile incidents (such as the 1967 *Torrey Canyon* oil spill, and the proliferation of nuclear weapons testing) to galvanise a combative and well-organised environmental movement. Numerous environmental protection societies were founded on both sides of the Atlantic in this period, including the *Environmental Defense Fund, Friends of the Earth,* the *World Wildlife Foundation* and the *Conservation Society*.

A notable feature of the environmental movement, at least in this period, was its adherence to a strict view of nature as separate from society, and of the environment, therefore, as a sphere to be protected and defended from contamination, pollution or other damage caused by human activities. In policy terms, this translated into a 'pollution control' paradigm, prioritising measures designed to limit the issuance of various contaminants – whether to the air, water or land – principally through retrospective, 'end-of-pipe' modifications to industrial processes. By the 1980s, however, this approach was giving way to *ecological modernisation*, a central-European school of thought which interprets environmental problems as structurally embedded in existing modes of human economic development, as well as their associated forms of social and political organisation. From this perspective, the purpose of environmental policy is not simply to reduce gross pollution, but to restructure prevailing socio-economic systems in ways mutually beneficial to human prosperity and to ecological entities and processes. Whilst it places more emphasis on questions of equity than ecological modernisation, the closely related concept of *sustainable development*, emerging from the 1987 *Bruntland Report* of the UN World Commission on Environment and Development, also advocates a more structural view of environmental problems, calling for forms of technology and social organisation that will not compromise "the environment's ability to meet present and future needs" (WCED, 1987). As new guiding principles for environmental policy making, these concepts enabled regulatory attention to move beyond the preservation of a pristine, untarnished natural world, permitting in addition the construction of novel imaginaries about the kinds of environments humans should seek to establish for future generations.

With the emergence of climate change as an emblematic environmental issue in the 1990s, and more recently still the coining of the concept of the Anthropocene, understandings of what is at stake in the formulation of environmental policy have shifted still further, blurring the boundaries between society and nature. As the average level of carbon dioxide in the atmosphere passes 400 parts per million, the goal of eradicating human influence from this sphere of the environment becomes increasingly unattainable, and discussions have instead shifted to the questions of how to limit change, and – crucially – how to *adapt* to an altered global climate system. More broadly, the environment is increasingly conceptualised as a hybrid construction in which human and non-human influences are inextricably intertwined – this realisation has arguably permitted the flourishing of a rewilding agenda in the sphere of biodiversity conservation, for instance. As increasing numbers of commentators declare the onset of the Anthropocene epoch, the object of environmental policy can no longer so easily be defined in relation to a single idea of nature as pristine wilderness, throwing open the prospect of a multi-, or perhaps even post-natural, environmental agenda for the twenty-first century.

## Shifting policies for the environment: from regulation to markets, networks and collaboration

Following the pollution control paradigm, early environmental policies routinely consisted of direct regulation imposed by national governments upon particular groups within society, often industrial organisations involved in large-scale manufacturing or energy generation. Figure 6.18.1 demonstrates that environmental policy was initially restricted to laws in just a handful of countries, designed either to restrict air and water pollution, to enable access to environmental information, or to promote nature protection as a whole.

By the turn of the century, this landscape had undergone profound transformation, with a raft of new environmental policy instruments (NEPIs) having been adopted to tackle problems across the sector. Alongside conventional regulatory tools based upon control and enforcement of norms and standards, policy instruments based upon the provision of clearer environmental information to citizens (e.g. eco-labelling and energy labels), upon cooperation between state and non-state actors (e.g. voluntary agreements and technology transfer commitments), and – perhaps most significantly – upon market principles (e.g. emissions trading schemes and feed-in tariffs), have all proliferated in number and scope.

Paralleling a broader shift towards neoliberal modes of political-economic governance in advanced democracies, the rise of market-based environmental policy instruments has not been without controversy. One source of unease is that many of these tools imply the 'neoliberalisation of nature' (Castree, 2008), and bring to bear upon environmental governance a set of economic rationalities that, in other spheres of human activity, have arguably been responsible for much environmental degradation in the first place. Certainly, the interpretation of environmental problems through an economic lens has sometimes raised their profile, as in the examples of the publication of the Stern Review on the Economics of Climate Change in the UK, and the development of conservation policies based upon the principle of securing payments for ecosystem services. Yet there remain doubts over the ability of market-based measures to offer reliable incentives, or indeed deterrents, to actors and organisations engaged in environmentally-consequential activities; not to mention over the capacity of these measures to account adequately for the diverse – and not merely monetary – ways in which humans value nature.

Market-based policy innovations have also motivated additional actors and organisations to participate in efforts to remediate environmental problems. Indeed, with the development of such instruments, responsibility for implementing, and in some cases for designing, environmental policy has moved beyond the nation state, as civil society

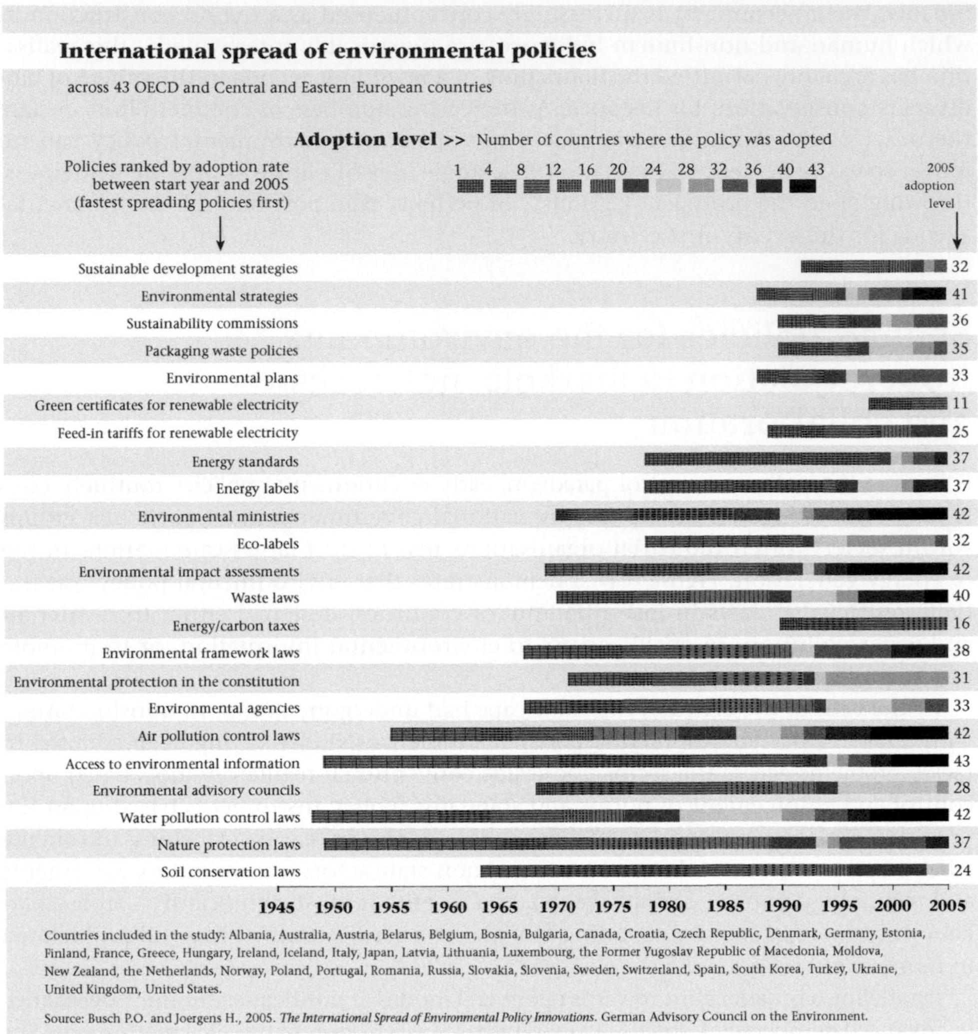


**Figure 6.18.1** Graph showing the spread of environmental policy innovations across 43 OECD and Central and Eastern European countries, 1945–2005.

*Source*: Busch & Joergens, 2005.

actors, private sector organisations, and supranational and intergovernmental bodies all contribute to the production of knowledge and ideas relevant to the policy-making process, and in some cases, gain legislative powers. Complex, multi-stakeholder, multi-level governance is now the new 'normal' in many spheres of environmental policy – for example, responses to climate change continue to evolve internationally (through the United Nations Framework Convention on Climate Change), nationally (through state targets for greenhouse gas emissions reductions and other legislation), at the urban level (through experimental forms of climate governance such as congestion charging, as well as collaboration amongst cities), and locally (through community renewable energy schemes and local climate change adaptation strategies). Whilst they have widened participation however, these new forms of environmental governance have also been diagnosed by some as agents of 'post-politicisation', whereby an over-riding concern with producing consensus around technical, managerial and administrative arrangements necessarily extrudes more fundamental dissensus and contestation over both the prevailing socio-political order, and the framings of nature, both in its present and future form, to which that order grants primacy (Swyngedouw, 2013).

## Learning resources

Institute for European Environmental Policy (2012) *Manual of European Environmental Policy.* Online. Available at: http://www.ieep.eu/understanding-the-eu/manual-of-european-environ mental-policy.

Jordan, A. Adelle, C. (eds) (2013) *Environmental Policy in the EU: Actors, Institutions, Processes.* Abingdon: Routledge.

Owens, S. (2015) 'The Environment in Politics', in *Knowledge, Policy, and Expertise: The UK Royal Commission on Environmental Pollution 1970–2011* (pp 23–44). Oxford, Oxford University Press.

## Bibliography

Busch, P.O. & Joergens, H. (2005) *The International Spread of Environmental Policy Innovations.* Berlin: German Advisory Council on the Environment.

Castree, N. (2008) 'Neoliberalising Nature: Processes, Effects, Evaluations', *Environment and Planning A*, 40: 153–173.

Hajer, M. (1995) *The Politics of Environmental Discourse: Ecological Modernization and the Policy Process.* Oxford: Oxford University Press.

Steffen, W., Richardson, K., Rockström, J., Cornell, S.E., Fetzer, I., Bennett, E.M., Biggs, R., Carpenter, S.R., de Vries, W., de Wit, C.A. et al. (2015) 'Planetary Boundaries: Guiding Human Development on a Changing Planet', *Science*, 347(6223): 1259855.

Swyngedouw, E. (2013) 'The Non-political Politics of Climate Change', *ACME: An International E-Journal for Critical Geographies*, 12(1): 1–8.

World Commission on Environment and Development (WCED) (1987) *Bruntland Report: Our Common Future.* Report of the World Commission on Environment and Development. New York: United Nations.

# 6.19 Environmental values

## Alan Holland

It is sometimes said that you only find out whether a thing really matters to you when you are in danger of losing it. Certainly, the project of articulating and defending environmental values has only come to the fore since the realisation dawned that we might be witnessing what has been called the 'end of nature' (McKibben, 1990). For some, no doubt, this might be of no consequence. And even those who have regard for nature might respond in very different ways. We cannot here canvas all possible attitudes to the environment or even all the reasons people have for valuing it. Our aim is simply to identify what it is about the environment – that is, the non-human world – that makes it valuable *in its own right*. Some believe nature has what is called *intrinsic value*.

The initial contrast here is with *instrumental value*. If we should value the environment purely for the health, wealth and happiness that we derive from it, then we would be valuing it for instrumental reasons rather than valuing it in its own right. At the same time it is possible, of course, to value the environment for both instrumental and non-instrumental reasons. A second contrast is with *extrinsic value*. If we should value a particular landscape purely for the reason that it was where we spent our childhood holidays, then again we would not be valuing that landscape as such, but valuing it for a reason that was accidental or 'extrinsic'. Again, it is of course possible to value the environment for both intrinsic and extrinsic reasons.

More controversially, it has been claimed that the environment has an intrinsic value that is 'objective' in the sense that it would exist irrespective of whether anyone recognised it. We can see what is true about the claim if we compare what is meant by saying that gold is valuable with what is meant by saying that the environment is valuable. Gold is valuable just in so far as people value it. If they ceased to do so we should no longer say that gold is valuable. But when we say that the environment is valuable we have made a judgement to the effect that there is good reason to value it; and this might still be true even if no one happened to value it. Our next question therefore must be: what reasons might there be for judging that the environment *as such* has value?

## Values and ethics

Before addressing that question we need to clarify briefly the relation between values and ethics. Values and ethics are distinct, but interconnected. For several decades now environmental philosophers have devoted themselves to identifying and clarifying the various ways in which the environment and its components make ethical claims upon us (Attfield, 1983; Taylor, 1986; Rolston, 1988). And the fact that something makes an ethical claim upon us can itself be a reason for valuing it. Conversely, many would claim that the presence of value in a thing makes it something that we ought (ethically) to protect and maintain. Thus, environmental value can both generate ethical claims and be generated by them. But it can also rest on a much broader range of considerations.

Most of the reasons that people have found for attributing value to the environment as a whole, or to various features of the environment, can be subsumed under three categories: stand-alone reasons, relational reasons and constitutive or contributory reasons. These categories are not exclusive, and many features may be valued for reasons that fall under more than one category.

## Stand-alone reasons

By 'stand-alone' reasons we mean reasons that are independent of relational or contributory considerations. Among the most frequently cited factors that meet this description are the health, well-being or flourishing of non-humans. For in general – and setting aside, for example, the well-being of the single-celled parasite that causes malaria – we regard well-being and flourishing as welcome states, and have reason therefore to cherish and foster their presence in living things. Again, and over and above the well-being of individuals, we mourn the 'loss' of species (the term 'loss' already indicative of a prior evaluative commitment), when this loss is untimely rather than part of the natural course of events. The reasons are varied and complex and will include reasons from all three of our categories; they are at any rate strong enough to lead E.O. Wilson to declare that "we should not knowingly allow any species or race to go extinct" (1992: 335). We also mourn the loss of the habitats and ecosystems to which these species were integral. But stand-alone reasons are not only to be located in the living biosphere. We have reason to value deserted places for their tranquillity and the night sky for its beauty, serenity and scale – valuations revealed by our talk of 'noise pollution' or again, of 'light pollution' when, for example, this taints our view of the stars. More generally, both living and non-living components of the non-human world attract a range of aesthetic valuations based, it may be, on their grandeur or delicacy, their dramatic or haunting qualities, their intricacy or ingenuity.

## Relational reasons

A great many features of our environment are judged valuable on account of the relations that obtain between them and human social, economic and cultural activities.

It has become common of late to refer to these features as 'services' and to reconfigure valuable environmental features as 'ecosystem services', subdividing these 'services' into provisioning, regulating, supporting and cultural (MA, 2003: Summary p. 5). Different concepts of course suit different purposes but suffice it to say here that the language of 'ecosystem service', with its distinctly instrumental flavour, does little justice to the variety and richness of these relationships. The notion of 'mother earth' may seem fanciful, but it seems a lot nearer the mark in capturing these relationships than the notion of 'ecosystem services'. The earth is at once our birthplace, our dwelling place and the (co)-provider of our livelihoods ('co-provider' since there is very little that literally drops into our laps). To speak of these as services seems about as appropriate as referring to a mother as a 'service provider' for her baby. As our birthplace, the earth yields the opportunity to interpret and understand our geological and biological inheritance, the kinship that we feel – and that exists – with our fellow voyagers in 'the odyssey of evolution' (Leopold, 1949: 109), and the seasonal beat and rhythm that is the backdrop of our lives. As our dwelling place the earth is source of our sense of belonging and our sense of identity, the context within which our personal and social histories are embodied and the setting for our familiar everyday existence. As our co-provider the earth supplies opportunities for developing the crafts and skills that are exercised in husbandry and horticulture, for the associated exercise of nurture and care and for the design and creativity deployed in the making and shaping of things. If we consider in this connection the reason why there is such deep concern over climate change we might think that this has at least as much to do with the rupture of our environmental relationships as with the disruption of environmental services (Gardiner, 2006).

## Constitutive or contributory reasons

When out for a morning walk, we might be lucky enough to experience the heady scent of gorse, the richly varied song of a nightingale and the spider's web jewelled with dewdrops. These enhance the value of our walk not in an instrumental sense but by being constituents of it. More generally, we can readily approach the notion of constitutive or contributory reasons by considering a confidence that Aldo Leopold shares with his reader in the preface to his *Sand County Almanac*: 'There are some who can live without wild things', he remarks, 'and some who cannot. These essays are the delights and dilemmas of one who cannot'. For Leopold, a life that lacks some significant engagement with 'wild things' is without worth. John O'Neill is another who has developed a version of this perspective, arguing that the flourishing of non-humans, besides being valuable in its own right, is valuable because it is partly constitutive of a flourishing human life (1992: 133). It is a thought that resonates equally with the 'deep ecology' of Arne Naess and John Donne's meditation – 'No man is an Island'. While it is true that Donne's empathy extends only as far as our fellow humans, what all three have in common is opposition to the view that beings of any kind exist independently of each other. As Naess expresses it: 'we seek what is best for ourselves, but through the extension of the self, our own best is also that of others . . . when we harm others we also harm ourselves' (1989: 174–175). In a word, in an impoverished environment, we could live only impoverished lives.

## Last word

Values and ideals, where these come into conflict, have been and continue to be responsible for some of the most appalling episodes of human history. Yet it bears reflection that the environment is a site of common and shared concern where we have every reason to make common cause. An increasing focus on environmental values, therefore, has the potential, at least, to be a unifying rather than divisive force in our fractured world.

## Learning resources

Go outside instead of reading indoors. Recommended sites include: a river bank; a coastal path; a woodland clearing; open moorland; a wetland reedbed. Sit quietly for at least one hour and focus on your surroundings with all five senses, aiming to identify and distinguish the stand-alone, relational and contributory values to be found there.

## Bibliography

Attfield, R. (1983) *The Ethics of Environmental Concern*. Oxford: Blackwell.
Gardiner, S. (2006) 'A perfect moral storm: climate change, intergenerational ethics and the problem of moral corruption', *Environmental Values*, 15: 397–413.
Leopold, A. (1949) *A Sand County Almanac*. New York: Oxford University Press.
McKibben, B. (1990) *The End of Nature*. New York: Viking.
Millenium Ecosystem Assessment (MA) (2003) *Ecosystems and Human Well-Being: A Framework for Assessment*. Washington, DC: Island Press.
Naess, A. (1989) *Ecology, Community and Lifestyle* (trans. and ed. D. Rothenberg). Cambridge: Cambridge University Press.
O'Neill, J. (1992) 'The varieties of intrinsic value', *The Monist*, 75: 119–137.
Rolston III, H. (1988) *Environmental Ethics: Duties to and Values in the Natural World*. Philadelphia: Temple University Press.
Taylor, P. (1986) *Respect for Nature*. Princeton, NJ: Princeton University Press.
Wilson, E.O. (1992) *The Diversity of Life*. Harmondsworth: Penguin.

# 6.20 Environmental science and public policy

Maria Carmen Lemos and Katherine Browne

## Why do we need environmental policy?

The proliferation of environmental problems and threats to the conservation of natural resources and human well-being has continuously challenged governments and communities. Because these challenges include the management of resources that are fundamental for the survival of all life on Earth, such as air and water, action to avoid and address environmental threats and risk is paramount. Yet despite the unprecedented mobilization of communities, governments and knowledge systems in implementing programs such as the Sustainable Development Goals (SDGs), water and air continue to be polluted, forests and wildlife disappear before our eyes and extreme poverty persists.

Many characteristics make environmental problems particularly complex when compared to other public policy issues. First, environmental problems can be irreversible, as with species loss or desertification that make large swaths of land unsuitable for agriculture. Second, they may affect systems that cannot directly advocate for themselves policy-wise such as fauna and flora and future generations that will suffer from depletion of natural resources today. Third, there may be a disconnection between the causes and solutions of environmental problems; people and communities who have not caused a problem may still be negatively affected by it. For example, in the case of global warming, poor communities in the global south that contributed little to historical emissions are nevertheless disproportionally affected by a changing climate. Fourth, environmental problems have scale-implications in that they often cross political jurisdictions and national borders and solutions in one scale can cause problems in another. For example, storing water to solve a drought problem can affect fisheries, or damming water in one country upriver can affect supply to another downriver.

To solve environmental problems we need a broad array of governance mechanisms that include different institutions, organizations and actors. We also need different kinds of knowledge that inform both the construction of environmental depletion and natural resource use as public problems, as well as the design of solutions to these problems. Among different kinds of knowledge that play that role, scientific knowledge is critical, especially in the process of public policy making (Kingdon, 1984). In principle, scientific knowledge that is relevant, credible and legitimate makes decisions better and informs the design, implementation and evaluation of solutions of environmental problems (Cash et al., 2006). Yet, in practice, the use of scientific knowledge in environmental decision making is more complex and less straightforward than scientists and policy-makers expect and hope.

## The science–policy connection and divide

While environmental problems are often ubiquitous and transboundary, policy systems tend to be very specific, spatially defined and contextual. At the state level, environmental action is generally shaped by formal institutions, such as laws and regulations. At the international level, however, environmental policy often takes the form of agreements and conventions. And because the general process of making environmental policy involves a myriad of actors (governmental and non-governmental), mechanisms, and knowledges, its contextual character has challenged attempts from environmental scholars to conceptualize a general model able to describe all its facets. Yet a number of models have emerged which seek to understand what mechanisms and institutional arrangements are the most successful in helping governments to prepare and respond to environmental challenges worldwide (Sabatier, 1991). In many of these models, science is expected to play an important role. For example, in Kingdon's (1984) classic three streams model – broadly comprised of a crisis, a political and a policy stream – scientific knowledge informs all of them, but especially the policy stream, where scientific expertise supports the design of solutions and many times, policy entrepreneurs rely on science to push for action (Kingdon, 1984). Scientists have also been instrumental in framing environmental issues as public problems, such as the ozone layer depletion and climate change, and helping to place them in governmental policy agendas.

Yet for all the expectation that science has a great role to play in informing and supporting policy, the connection between the two is complex and at times uncomfortable. On the one hand, there is a widespread perception among scholars studying the science–policy connection that not enough knowledge makes its way into policy making, and that special attention needs to be paid to narrow the gap between science production and its use in support of policy and decision making (Kirchhoff et al., 2013; Weiss, 1978). Traditionally, the expectation has been that once scientific knowledge is made available and disseminated through publication in peer-review journals and participation in scientific conferences – an approach scholars have equated to a 'loading dock' (Cash et al., 2006) – knowledge would be taken up by policy-makers and other stakeholders. However, in reality, research shows that the process of scientific knowledge creation and use is often not straightforward or easy. For example, in climate-related policy making, scholars have found that common barriers for the use of climate knowledge are the fact that policy makers believe the available knowledge does not 'fit' their needs (e.g. because it is too uncertain or not available at the spatial and temporal scale they need), that it interferes with their well-established practices (e.g. when new knowledge forces them to change their current decision-making processes), or that it is not relevant, credible and legitimate (Cash et al., 2006; Lemos et al., 2012; Rayner et al., 2002).

On the other hand, historically, the way in which science production has been organized advocates that science production should be unfettered and neutral to avoid being politicized (Sarewitz, 1996). This model, referred as Mode 1 of science, relies on disciplinary settings in universities and research organizations as the main producers of knowledge (Gibbons et al., 1994) and assumes that the science produced is expected and presumed to be useful to solve problems (Dilling, 2007). In addition, influential scholars such as Bruno Latour and Sheila Jasanoff have convincingly argued and shown that the separation between science, policy and society is artificial and that scientific knowledge is not neutral or objective but co-produced in the day-to-day interaction between scientists and their social environment. (Jasanoff, 1990; Latour & Wolgar, 1979).

## New models of scientific knowledge production and use

Proponents of alternative models of science–policy intersection argue that the complexity of contemporary problems requires more than one disciplinary view to solve them (Gibbons et al., 1994), and that the process of creating scientific knowledge to solve problems needs to be more interactive and include other actors, especially intended users of that knowledge such as policy makers (Klenk et al., 2015). This is especially the case, for example, when problems are very uncertain and risks very high. Here scholars have proposed that we need to go beyond 'normal science' to consider a post-normal approach, in which scientific knowledge alone is not enough and decision-makers need their own forms of knowing to better evaluate their risk situation (Funtowicz & Ravetz, 1993).

Mechanisms to improve the interaction between scientists and policy makers (and with other stakeholders) include the intentional design of specific processes of co-production of science and decision making. These processes include consulting policy makers trying to solve a specific problem about their knowledge needs; involving stakeholders from the very beginning of a scientific project design; interacting with them regularly and continuously throughout the process of creating new knowledge, and evaluating the results in terms of how usable knowledge has been in supporting decision making (Lemos et al., 2012, Meadow et al., 2015).

In this model, interaction with policy makers (as well as other stakeholders) through co-production require that scientists make significant changes in the way they produce science including engaging in better communication and visualization of their scientific findings, interacting with stakeholders and making significant effort to understand their needs and decision-making processes. To facilitate these processes, specific organizations – known as boundary organizations – have emerged that engage and bridge the work of scientists and potential users of their science. These organizations broker knowledge between producers and users, add value to scientific products and disseminate knowledge to broader audiences than those scientists could reach through direct interaction. Knowledge produced at the interface of science and society is expected to be more relevant for solving problems and supporting management; more likely to be 'bought in' by stakeholders and be more legitimate in their eyes; and more likely to build trust (Kirchhoff et al., 2013).

Scientific knowledge produced in this fashion has been shown to be more used, is expected to be more democratic, inclusive and transparent by avoiding inequity often introduced by scientific expertise and more likely to consider and integrate across different kinds of knowledge (e.g. scientific, lay and indigenous knowledge) (Wesselink & Hoppe, 2011; Yearley, 2000). In contrast, the creation of participatory knowledge production and governance processes in itself does not guarantee knowledge democracy, especially when the use of scientific knowledge becomes a source of authority of some groups over others and an instrument of inequity in the distribution of power across participant groups (Blok, 2007).

## Learning resources

Climate.gov is a webpage from the US National Oceanographic and Atmospheric Administration that focuses on making climate science accessible and usable. It includes teaching and youth-driven materials and user-friendly climate information such as the US Third Climate Assessment.

https://www.climate.gov/news-features.

The United Nations Sustainable Development Goals Knowledge Platform focuses on communicating the goals of the SDGs, including opportunities for stakeholder engagement.

https://sustainabledevelopment.un.org/mgos.

The International Council for Science (ICSU) offers different resources at the science-decision interface, including a Science for Policy Program.

http://www.icsu.org.

## Bibliography


Blok, A. (2007) Experts on public trial: on democratizing expertise through a Danish consensus conference, *Public Understanding of Science*, 16(2): 163–182.

Cash, D.W., Borck, J.C. & Patt, A.G. (2006) Countering the loading-dock approach to linking science and decision making. Comparative analysis of El Niño/Southern Oscillation (ENSO) Forecasting systems, *Science, Technology and Human Values*, 31: 465–494.

Dilling, L. (2007) Towards science in support of decision making: characterizing the supply of carbon cycle science, *Environmental Science & Policy*, 10(1), 48–61.

Funtowicz, S.O. & Ravetz, J.R. (1993) The emergence of post-normal science. In Von Schomberg, R. (ed.) *Science, Politics and Morality: Scientific Uncertainty and Decision Making* (pp. 85–123). New York: Springer.

Gibbons, M., Limoges, C., Nowotny, H., Schartzman, S., Scott, P. & Trow, M. (1994) *The New Production of Knowledge*. London: Sage Publications.

Jasanoff, S. (1990) *The Fifth Branch: Science Advisers as Policymakers*. Cambridge, MA: Harvard University Press.

Kingdon, J.W. (1984) *Agendas, Alternatives, and Public Policies*. Boston: Little, Brown.

Kirchhoff, C., Lemos, M.C. & Desai, S. (2013) Actionable knowledge for environmental decision making: broadening the usability of climate science. *Annual Review of Environment and Resources*, 38: 393–414.

Klenk, N., Meehan, K., Pinel, S.L., Mendez, F., Torres Lima, P. & Kammen, D. (2015) Stakeholders in climate science: beyond lip service? *Science*, 350(6262): 743–744.

Latour, B. & Wolgar, S. (1979) *Laboratory Life*. Beverly Hills, CA: Sage.

Lemos, M.C., Kirchhoff, C. & Ramparasad, V. (2012) Narrowing the climate information usability gap, *Nature Climate Change*, 2(11): 789–794.

Meadow, A.M., Ferguson, D.B., Guido, Z., Horangic, A., Owen, G. & Wall, T. (2015) Moving toward the deliberate coproduction of climate science knowledge, *Weather, Climate, and Society*, 7(2): 179–191.

Rayner, S., Lach, D., Ingram, H. & Houck, M. (2002) *Weather forecasts are for wimps: Why water resource managers don't use climate forecasts*. Final report to NOAA Office of Global Programs. Oregon: Center for Water and Environmental Sustainability, Oregon State University.

Sabatier, P.A. (1991) Towards better theories of the policy process. *PS: Political Science and Politics*, 24(2): 147–156.

Sarewitz, D. (1996) *Frontiers of Illusion: Science, Technology, and the Politics of Progress*. Philadelphia: Temple University Press.

Weiss, C.H. (1978) Improving the linkage between social research and public policy. In L.E. Lynn, Jr. (ed.) *Knowledge and Policy: The Uncertain Connections* (pp. 23–81). Washington, DC: National Academy of Sciences.

Wesselink, A. & Hoppe, R. (2011) If post-normal science is the solution, what is the problem?: The politics of activist environmental science, *Science Technology & Human Values*, 36(3): 389–412.

Yearley, S. (2000) Making systematic sense of public discontents with expert knowledge: two analytical approaches and a case study, *Public Understanding of Science*, 9: 105–122.

# Geoengineering



Jack Stilgoe

Responses to global climate change are conventionally seen as either mitigation through reducing greenhouse gas emissions or adapting to the changing environment. As concerted global action on climate has stalled, national science advisers and others have augmented this to press their case. US Presidential science adviser John Holdren told a 2010 climate change conference that 'We only have three options . . . It's really that simple: mitigation, adaptation, and suffering'.

However, another more radical option has sometimes been discussed: geoengineering (or 'climate engineering'). The idea of intentional technological interference in the climate system in order to cool the planet has a long history, but has only recently emerged to become a topic of mainstream scientific discussion.

## History: rethinking the unthinkable

Geoengineering is often discussed as an 'emerging technology'. But it is not a technology at all. It is not even a basket of technologies, or potential technologies, although some proposals are better developed than others. Geoengineering is an idea, perhaps 'a bad idea whose time has come' (Kintisch, 2010, p. 13).

As an idea, geoengineering has a history as long as that of modern science. Francis Bacon saw the control of weather as an important part his natural philosophy (Horton, 2014). Three centuries later, J.D. Bernal (1939) claimed that diverting ocean currents could melt the Arctic and so improve the climate of the Northern hemisphere.

Such speculations would continue throughout the twentieth century. At the same time, as described by James Fleming (2010), enthusiasts and engineers of varying credibility promised control of the weather to desperate farmers and others who had fallen victim to climatic whims. The story of these rainmakers moved slowly from mythology to respectable science during the twentieth century. As the growth of computing power and global climate models promised greater predictive power over the weather and potent, world-changing technologies emerged during the two World Wars and the Cold War, some began to construct more detailed schemes. John Von Neumann wanted to take 'the first steps toward influencing the weather by rational, human intervention' (quoted in Harper, 2008). Such visions of control lay behind the rapid growth of meteorology as a science, although in public most scientists would emphasise that their aims were merely predictive.

Some of the earliest thinking on reflecting sunlight to cool the planet came from Mikhail Budyko, a Russian who was one of the leading figures in the quantification of meteorology, previously ridiculed as a 'guessing science' (Harper, 2006). In the 1970s, Budyko (1977) sketched a plan for increasing the reflectivity of the planet's upper atmosphere using sulphate particles, dispensed from aeroplanes. It was this idea that, having lain all-but dormant for 40 years, inspired the intervention of Paul Crutzen in 2006. Crutzen, a highly respected Nobel Laureate for his work on atmospheric ozone, argued that the problem of climate change was an intractable Gordian knot. Geoengineering with a stratospheric sunshade provided a sword, or, as he put it, 'a contribution to resolve a policy dilemma' (Crutzen, 2006).

Crutzen's paper brought a veneer of respectability to what had previously been considered by many to be either taboo or a Cold War joke. He argued that the technologies with which to geoengineer were cheap and readily available. Even scientists who hated the idea could not ignore the possibility that it might be put into action at some point.

## Assessing geoengineering options

Oliver Morton (2015) begins his analysis of geoengineering by asking two questions:

1. Do you believe the risks of climate change merit serious action aimed at lessening them?
2. Do you think that reducing an industrial economy's carbon dioxide emissions to near zero is very hard?

If we answer yes to both, Morton argues, we should take geoengineering seriously. As twenty-first-century climate scientists reflected inadequate policy responses to the problem they had elucidated, they reluctantly began to agree. For many of these scientists, geoengineering has aroused particular concerns. International negotiations on climate change mitigation have been fragile and geoengineering seems to present a 'moral hazard': if insurance against the risks of climate change were on offer, people in power would surely become less interested in reducing greenhouse gas emissions. Scientists were not only concerned that geoengineering would be seen as a 'get out of jail free' card, they also worried that its deployment would have unintended consequences on global weather. Alan Robock (2008), one of the first wave of natural scientists to seriously explore geoengineering, offered what he called a 'fairly comprehensive list of reasons why geoengineering might be a bad idea'. The list was wide-ranging, encompassing politics, ethics and effects on local weather. Robock concludes that, in addition, 'there is reason to worry about what we don't know'.

As the idea moved towards the scientific mainstream, right-wing pundits in the USA who were eager for hassle-free solutions to the question of climate change grew increasingly interested. The Royal Society (2009), the UK's national academy of sciences, began an assessment that aimed to bring a cool scientific rationality to what had become a heated discussion.[1]

The Royal Society divided geoengineering options into two proposed mechanisms of intervention. The first, carbon dioxide removal, involves the reduction of greenhouse gas concentrations in the atmosphere with machines or by enhancing natural systems. The second, solar radiation management, bounces a proportion of sunlight back into space by making the Earth's surface, clouds or upper atmosphere more reflective (see Figure 6.21.1)

The Royal Society assessed the various options on multiple criteria, including effectiveness, speed, cost, safety (see Figure 6.21.2), and concluded, as Paul Crutzen had done, that stratospheric particle injection was the most potent, the cheapest, but also the riskiest option available. Scientific interest in stratospheric particle injection continued to grow (Oldham et al., 2015), in part because of an assumption that it would be 'cheap and technically easy' (Keith, 2013). Economist William Nordhaus (1992) was among the first to argue that, compared with decarbonisation of industrial society, geoengineering offered the potential for 'costless mitigation of climate change'. The history of similar sociotechnical systems suggests that the complexities and uncertainties associated with such cost estimates are, in fact, vast.

For some (including an early reviewer of The Royal Society's report) the profundity of ethical and safety concerns raised by stratospheric particle injection warranted ruling it out altogether. Geographer Mike Hulme (2014) has argued that it represents 'an illusory solution to the wrong problem' and should therefore be taken off the table altogether. Whether in spite of, or because of its Promethean connotations, stratospheric particle injection continues to dominate geoengineering discussions.

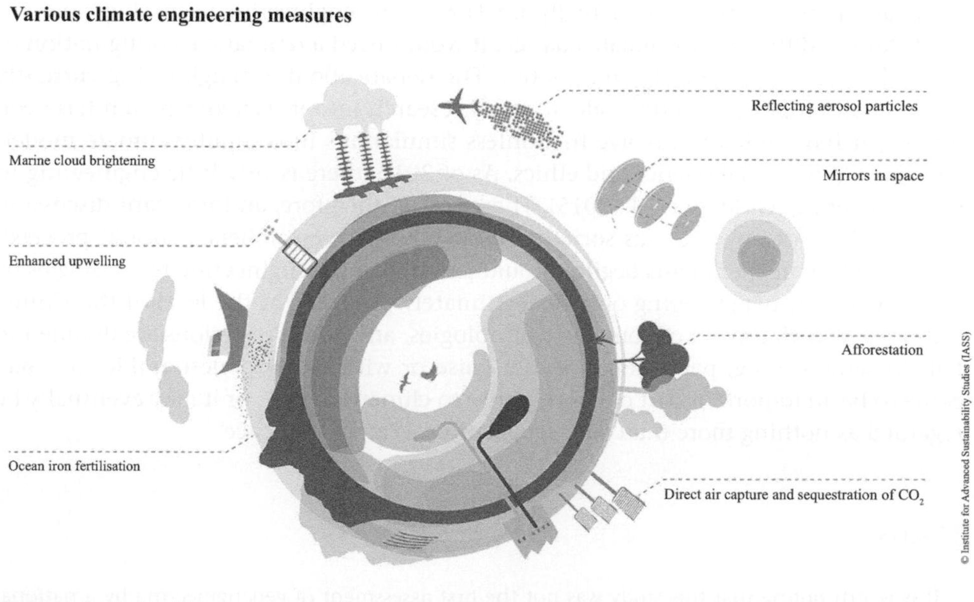


**Figure 6.21.1** Climate engineering proposals, reproduced from the IASS, Potsdam (permission to reproduce granted).

http://www.tyndall.ac.uk/sites/default/files/climate_engineering_infographic_iass_berlin.jpg.

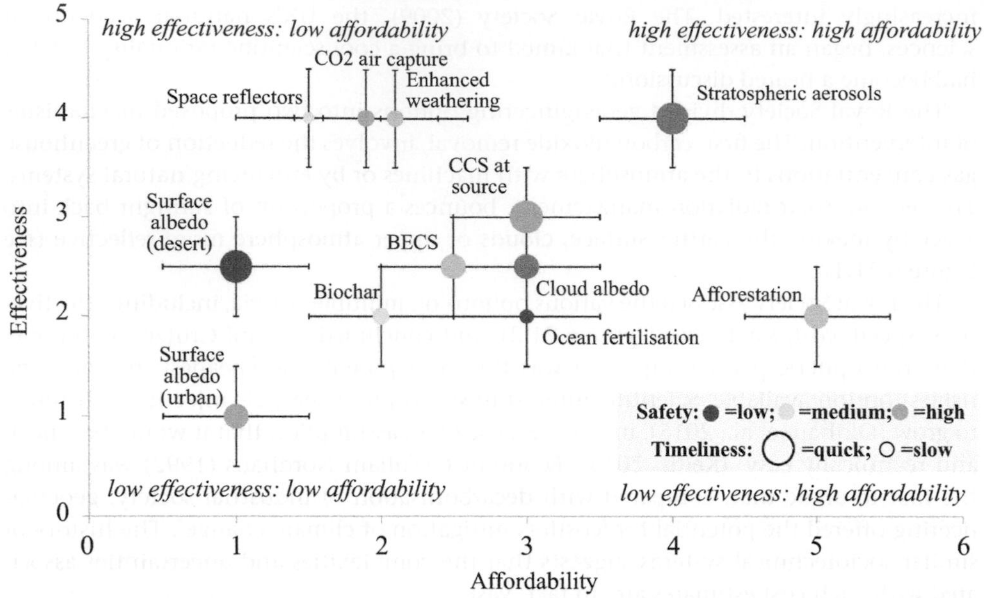


**Figure 6.21.2** Preliminary overall evaluation of the geoengineering techniques (reprinted from The Royal Society, 2009) (reproduced with permission).

## Conclusion

The debate about geoengineering has tended to make technologies and ideas appear closer and more real than they actually are. For any geoengineering technology to make a substantial difference to climate change, it would need a dramatic reconfiguration of research, technology, society and politics. The debate about geoengineering currently is out of all proportion to the scale of actual research into it. Where research has been funded, it has tended to involve frictionless simulations in computer climate models and speculative social science and ethics. As of 2016, there is very little engineering in geoengineering (Oldham et al., 2015). There is still, therefore, an important discussion to be had about how we – as society and as scientific researchers – should proceed: Should outdoor experiments begin? Should patents on geoengineering technologies be allowed? Can geoengineering only be legitimately governed at the level of the United Nations? As with any set of complex technologies, any hard predictions are doomed to fail. Geoengineering, perhaps in another guise or with more modest ambitions, may come to be an important part of the response to climate change, or it may eventually be regarded as nothing more than wild speculation. Watch this space.

## Note

1 It is worth noting that this study was not the first assessment of geoengineering by a national academy. The US National Academies addressed the issue, albeit in politically unsophisticated terms, in 1992 as part of an assessment of options for tackling climate change (NAS (1992) *Policy Implications of Greenhouse Warming*. National Academy Press, Washington). In 2015, the US

National Academies revisited geoengineering, renaming it 'climate intervention' because of the committee's opinion that the previous label 'implies a greater level of precision and control than might be possible'.

## Learning resources

Hulme, M. (2014) *Can Science Fix Climate Change? A Case Against Climate Engineering.* Chichester: John Wiley & Sons.

Morton, O. (2015) *The Planet Remade: How Geoengineering Could Change the World.* Princeton, NJ: Princeton University Press.

Stilgoe, J. (2015). *Experiment Earth: Responsible Innovation in Geoengineering.* Abingdon: Routledge.

## Bibliography

Bernal, J.D. (1939/2010) *The Social Function of Science.* London: Faber & Faber.

Buchanan, R.A. (2006) *Brunel: The Life and Times of Isambard Kingdom Brunel.* New York: Continuum.

Budyko, M.I. (1977) *Climatic Changes* (transl. Izmeniia Klimata Leningrad: Gidrometeoizdat). Washington, DC: American Geophysical Union.

Crutzen, P.J. (2006) Albedo enhancement by stratospheric sulfur injections: a contribution to resolve a policy dilemma?, *Climatic Change,* 77: 211–219.

Fleming, J. (2010) *Fixing the Sky: The Checkered History of Weather and Climate Control.* New York: Columbia University Press.

Harper, K. (2006) Meteorology's struggle for professional recognition in the USA (1900–1950), *Annals of Science,* 63(2): 179–199.

Harper, K. (2008) *Weather by the Numbers: The Genesis of Modern Meteorology.* Cambridge, MA: MIT Press.

Holdren, J. (2010) Text of remarks by Obama science adviser John Holdren to the National Climate Adaptation Summit, May 27, 2010.

Horton, Z. (2014) 'Collapsing scale: nanotechnology and geoengineering as speculative media'. In K. Konrad, C. Coenen, A.B. Dijkstra, C. Milburn & H. van Lente (eds) *Shaping Emerging Technologies. Governance, Innovation, Discourse* (pp. 203–218): Bristol: IOP Press.

Hulme, M. (2014) *Can Science Fix Climate Change? A Case Against Climate Engineering.* Chichester: John Wiley & Sons.

Keith, D. (2013) *A Case for Climate Engineering.* Cambridge, MA: MIT Press.

Kintisch, E. (2010) *Hack the Planet: Science's Best Hope – or Worst Nightmare – for Averting Climate Catastrophe.* Chichester: John Wiley & Sons.

Morton, O. (2015) *The Planet Remade,* Princeton, NJ: Princeton University Press.

Nordhaus, W.D. (1992) An optimal transition path for controlling greenhouse gases, *Science,* 258(5086): 1315–1319.

Oldham, P., Szerszynski, B., Stilgoe, J., Brown, C., Eacott, B. & Yuille, A. (2015) Mapping the landscape of climate engineering, *Philosophical Transactions of The Royal Society of London A,* 372(2031): 1–20.

Robock, A. (2008) 20 reasons why geoengineering may be a bad idea, *Bulletin of the Atomic Scientists,* 64(2): 14–18.

The Royal Society (2009) *Geoengineering the Climate: Science, Governance and Uncertainty.* London: The Royal Society.

# 6.22 Green consumption

Ivan R. Scales

## Mass consumption and its environmental impacts

The last 100 years have seen a dramatic increase in humanity's capacity to produce and consume goods, particularly in affluent industrialised countries. While global human population doubled between 1960 and 2000, household consumption expenditure quadrupled (Gardner et al., 2004). Mass consumption societies have emerged where 'not a few individuals, nor a thin upper class, but the majority of families enjoy the benefits of increased productivity and constantly expand their range of consumer goods' (Matsuyama, 2002: 1035–1036).

It is important to note that these global trends hide considerable variation, both between and within countries (Gardner et al., 2004; Ivanova et al., 2015). While 12 per cent of the world's population lives in North America and Europe, it accounts for over 60 percent of global consumer spending (Gardner et al., 2004). In contrast, while South Asia and sub-Saharan Africa together account for over a third of the world's population, they make up less than 4 per cent of global consumer spending (ibid.). Consumption in China and India has grown rapidly over the last decade, albeit unevenly, with affluent urban consumers accounting for a significant proportion of growth (Gardner et al., 2004; Hubacek et al., 2007).

The simplest proximate drivers of the growth of mass consumption have been rises in income and reductions in the costs of goods and services (Matsuyama, 2002). However, these have been underpinned by a diverse and interlinked set of technological, economic, political and cultural factors, with considerable geographical and historical variation (Matsuyama, 2002; Cohen, 2007). The transition to fossil fuels underpinned the mechanisation of manufacturing and mass production, helping to produce standardised goods in large volumes, increase efficiency, and reduce costs. Fossil fuels, combined with the development of intermodal transportation (where containers can be used across different modes of transport) also transformed the movement of goods. In particular, container shipping has helped to create global commodity networks and to vastly increase the quantities and types of goods available to consumers (Knowles, 2006).

In addition to income and the price of goods, household consumption is strongly influenced by daily routines, consumer behaviour, and lifestyle, which are in turn embedded in social norms and expectations (Ivanova et al., 2015). Mass consumption has been enabled and encouraged through the proliferation of consumer credit, as well as the expansion of advertising and branding (Cohen, 2007; Gardner et al., 2004).

The growth of mass consumption societies has had significant environmental implications. Raw materials and energy are required to produce, transport, store and sell consumer goods. Household consumption also generates considerable waste. A global assessment estimated that household consumption contributes more than 60 per cent of global greenhouse gas emissions and between 50 and 80 per cent of total land, material, and water use (Ivanova et al., 2015). Again, it is important to stress geographical differences and unevenness. Ecological 'footprints' (the amount of productive land required to produce resources and assimilate waste) range from averages of 9.7 hectares per person in the United States to 0.47 hectares in Mozambique (Gardner et al., 2004).

## The basic principle of green consumption

Green consumption, or green consumerism, seeks to reduce the impact of household consumption on the environment by changing the way consumers buy goods and services, use energy, and produce and deal with waste. It is part of a broader phenomenon of 'ethical consumption', where consumer decisions are based not only on the quality and price of goods or services on offer but also on their moral attributes (Carrier, 2010). Green consumption emphasises the power of 'thinking globally, acting locally', where influencing individual choices is seen as the best way of dealing with environmental problems.

The possibility of 'greening' consumption gained popularity in the 1980s and early 1990s. During this period, ideas of 'limits to growth' (Meadows et al., 1972), which argued for radical curbs on global production and consumption, began to give way to ideas of 'sustainable development'. Environmental thinking also began to move away from a focus on top-down 'command and control' regulation by states and international bodies to bottom-up participatory and market-based approaches.

Green consumption is based on some key assumptions (Scales, 2014):

1. Consumers can be provided with clear information about the environmental conditions of production and the impacts of their consumptive behaviour.
2. This information enables consumers to make choices about the way they consume and to change their consumptive behaviour.
3. Producers of goods and services will react to changing consumer demands in the marketplace by changing the ways in which they make goods and provide services.

In theory, information campaigns, labelling schemes and branding enable consumers to make informed choices about what they consume. Examples of 'eco' labelling include 'dolphin friendly' tuna and the Marine Stewardship Council's (MSC) 'sustainably sourced' seafood, as well as the International Resin Identification Coding System, which sets global standards to facilitate plastic recycling.

Green consumption's emphasis on consumer power in the marketplace is grounded in the principles of neoclassical economics, where free markets (in other words interactions between buyers and sellers where states and other external actors do not interfere) enable resources to be distributed according to the forces of supply and demand. Green consumerism also fits well with neoliberal economic values, which emphasise individual liberty, minimal involvement of the state, and see free markets as the most efficient way to coordinate the diverse needs of people (Cohen, 2007; Scales, 2017).

In addition to changes in the *types* of goods purchased, green consumption also needs to deliver changes in the *quantity* of goods purchased, owned, and consumed, as well as changes in the ways consumers deal with the waste generated by their consumption. Efforts to encourage consumers to reduce consumption include various 'switch off' campaigns to draw attention to energy use; government policies to enable and incentivise consumers to improve house insulation and install solar panels; and more coercive measures such as fees charged for the collection and disposal of domestic waste. Some proponents of green consumerism envisage a move away from economies based on the ownership of goods to 'service and flow' economies where goods are primarily rented (Hawken et al., 1999).

## Criticisms of green consumption

While the last decade has seen a proliferation of 'eco' labels and brands, as well as attempts to inform and educate consumers, there is a growing body of research that has revealed the limitations of green consumption and argues that it is incapable of addressing the environmental degradation that results from mass consumption.

One of the biggest problems facing green consumption is the scale and complexity of environmental impacts. Many of the commodity chains that connect producers with consumers are vast, involving numerous processes and steps (Ivanova et al., 2015; Knowles, 2006). These include extracting raw materials, manufacturing, transportation and retail. Each step can involve a wide variety of environmental impacts, from greenhouse gas emissions to various pollutants and waste products (Ivanova et al., 2015). Once these impacts are calculated, the next challenge is to communicate the information clearly to consumers. However, there is a tendency for labelling schemes to confuse consumers (Carrier, 2010; West, 2010). For example, West (2010), carrying out research on coffee consumption in New York found that despite efforts to label and certify coffee, consumers had little idea of where coffee was produced, under what social and environmental conditions or even what the different labels and certifications corresponded to.

The need to simplify complex environmental problems into easily identified 'eco' labels, metrics and messages can also lead to unintended and undesirable environmental outcomes. For example, a focus on reducing carbon emissions to tackle climate change has already led to an increase in the production, marketing and consumption of 'green' biofuels. However, the production of some biofuel feedstock (especially palm oil) has involved clearing tropical rainforest, threatening biodiversity in the process (Groom et al., 2008).

Critics of green consumption have also challenged the idea of consumer power. They point out that in industrialised societies power tends to lie with large producers and retailers, as well as states (who have the power to regulate, subsidise or tax particular industries and activities). Large retailers are able to dominate markets and dictate the conditions of production, often pressuring labelling schemes to relax standards to allow them to capture the price premiums of 'eco' products (Gould et al., 2008; Hartwick, 2000; Ponte & Gibbon, 2005). Manufacturers and retailers also play a key role in promoting mass consumption through branding, advertising, planned obsolescence and rapid product turnover to encourage consumers to replace goods that are still functional (Gould et al., 2008; Hartwick, 2000). Some critics argue that any attempt to reduce human environmental impacts will need more radical political, economic and cultural changes that are incompatible with mass consumption capitalism (see for example Foster, 2000; Fridell, 2007; Gould et al., 2008, Hartwick, 2000; Jackson, 2009).

## Learning resources

There are now a large number of websites and apps to help you calculate and reduce your environmental footprint. Here are some suggestions:

The Centre for Alternative Technology (http://learning.cat.org.uk/en/resources) has a wide range of resources, from apps to games, to help individuals and groups better understand the impacts of their lifestyles. It covers food, energy, building and transport.

The World Wildlife Fund for Nature has produced an ecological footprint calculator: http://footprint.wwf.org.uk.

The 10:10 website (http://1010uk.org) provides advice on how to reduce your carbon footprint.

For more information on how policy can encourage behaviour change:

Dobson, A. (2010) *Environmental Citizenship and Pro-environmental Behaviour: Rapid Research and Evidence Review*. London: Sustainable Development Research Network. Online. Available at http://www.sd-research.org.uk/sites/default/files/publications/SDRN%20Environmental%20Citizenship%20and%20Pro-Environmental%20Full%20Report_0.pdf.

Jackson, T. (2005) *Motivating Sustainable Consumption*. London: Sustainable Development Research Network. Online. Available at http://www.sustainablelifestyles.ac.uk/sites/default/files/motivating_sc_final.pdf.

## Bibliography

Carrier, J.G. (2010) 'Protecting the environment the natural way: ethical consumption and commodity fetishism', *Antipode*, 42: 672–689.

Cohen, M.J. (2007) 'Consumer credit, household financial management, and sustainable consumption', *International Journal of Consumer Studies*, 31: 57–65.

Foster, J.B. (2000) *Marx's Ecology: Materialism and Nature*. New York: Monthly Review Press.

Fridell, G. (2007) 'Fair-trade coffee and commodity fetishism: the limits of market driven social justice', *Historical Materialism*, 15: 79–104.

Gardner, G., Assadourian, E. & Sarin, R. (2004) 'State of the world: the state of consumption today'. In Starke, L. (ed.) *State of the World*. New York: W.W. Norton and Company.

Gould, K., Pellow, D.N. & Schnaiberg, A. (2008) *The Treadmill of Production*. Boulder: Paradigm Publishers.

Groom, M.J., Gray, E.M. & Townsend, P.A. (2008) 'Biofuels and biodiversity: principles for creating better policies for biofuel production', *Conservation Biology*, 22: 602–609.

Hartwick, E.R. (2000) 'Towards a geographical politics of consumption', *Environment and Planning A*, 32: 1177–1192.

Hawken, P., Amory, L. & Hunter, L. (1999) *Natural Capitalism: Creating the Next Industrial Revolution*: Boston: Little, Brown and Company.

Hubacek, K., Guan, D. & Barua, A. (2007) 'Changing lifestyles and consumption patterns in developing countries: a scenario analysis for China and India', *Futures*, 39: 1084–1096.

Ivanova, D., Stadler, K., Steen-Olsen, K., Wood, R., Vita, G., Tukker, A. & Hertwich, E.G. (2015) 'Environmental impact assessment of household consumption', *Journal of Industrial Ecology*, 20(3): 526–536.

Jackson, T. (2009) *Prosperity Without Growth: Economics for a Finite Planet*. London: Routledge.

Knowles, R.D. (2006) 'Transport shaping space: differential collapse in time–space', *Journal of Transport Geography*, 14: 407–425.

Matsuyama, K. (2002) 'The rise of mass consumption societies', *Journal of Political Economy*, 110: 1035–1070.

Meadows, D.H., Meadows, D.L., Randers, J. & Behrens III, W.W. (1972) *The Limits to Growth: A Report for the Club of Rome's Project on the Predicament of Mankind*. London: Pan Books.

Ponte, S. (2008) 'Greener than thou: the political economy of fish ecolabeling and its local manifestations in South Africa', *World Development*, 36: 159–175.

Ponte, S. & Gibbon, P. (2005) 'Quality standards, conventions and the governance of global value chains', *Economy and Society*, 34: 1–31.

Scales, I.R. (2014) 'Green consumption, ecolabelling and capitalism's environmental limits', *Geography Compass*, 8(7): 477–489.

Scales, I.R. (2017) 'Green capitalism'. In Richardson, D., Castree, N., Goodchild, M.F., Kobayashi, A., Liu, W. & Marston, M.A. (eds) *The International Encyclopedia of Geography: People, the Earth, Environment, and Technology*. New York: Wiley-AAG.

West, P.C. (2010) 'Making the market: speciality coffee, generational pitches, and Papua New Guinea', *Antipode*, 42: 690–718.

# Greenwashing

6.23

Frances Bowen

## Greenwashing and the challenge of environmental communication

Communication challenges are at the centre of understanding human responses to the environment. Decades ago, the general public had relatively low environmental awareness, so companies, governments, NGOs, the media and other social actors usually did not need to find ways to communicate about environmental issues. But as concerns about human impacts on the environment have increased, societies have had to develop new language and symbols to communicate about environmental issues. We are now surrounded by ecolabels, green marketing, corporate sustainability reports, environmental standards, industry pledges and new environmental terminology that are all designed to communicate environmental credibility. However, it is very difficult for individuals as consumers or voters to differentiate misleading corporate spin from authentic environmental improvements.

Greenwashing is "any communication that misleads people into adopting overly positive beliefs about an organization's environmental performance, practices or products" (Lyon & Montgomery, 2015). The term was first used by NGOs in the environmental justice movement in the 1990s (e.g. Greer & Bruno, 1996). Among these early activists, the term 'greenwashing' is derived from a combination of 'green' and 'brainwash', which emphasizes the role of elite power and propaganda in successfully influencing how people think of companies' environmental performance.

By the time the term 'greenwashing' entered the *Oxford English Dictionary* in 1999, it had acquired its now more common meaning derived from 'green' and 'whitewashing': deliberately attempting to conceal unpleasant or incriminating facts about a person or organization in order to protect their reputation. Since then, growing academic literatures in marketing, economics and organizational theory on greenwashing have developed ever more precise definitions of greenwashing.

## Varieties of greenwash

Bowen (2014) reviewed and compared many definitions of greenwash, and argued that the idea of greenwashing had become too narrow in the recent academic literature. She argued that while common uses of the term included a wide variety of phenomena – including corporate tactics such as 'confusion', 'fronting' and 'posturing' (Laufer, 2003) – academic research had begun to focus too much on a narrower idea of greenwashing as an information disclosure strategy that is (1) deliberate, (2) initiated by companies, and is (3) beneficial to firms and costly to society. She explores a much broader idea of 'symbolic corporate environmentalism', which recognizes that greenwashing may not always be deliberate, and that greenwashing can be undertaken by NGOs, governments or others as well as companies. Also, although traditionally greenwashing is understood to benefit companies (e.g. through improving their reputation or brand image) and be costly to society (e.g. through stimulating less efficient decision making because stakeholders are given inaccurate information), she pointed out that this is not always the case. One of the central controversies in environmental communication is to understand if the wasteful corporate spin of greenwashing can ever be outweighed by the benefits of raising awareness of environmental issues through using new language or through exposing and publicizing examples of greenwashing behaviour.

Lyon and Montgomery (2015) build on Bowen's (2014) critiques of greenwashing, and identify several broader 'varieties of greenwash'. Broadly, these can be categorized into two families: (1) greenwashing through active communication, and (2) greenwashing through association.

The former includes *selective disclosure*, that is, choosing which areas of activity to communicate about and highlighting strong performance areas. This variety of greenwash is most prominent in the academic literature, as it is relatively easy to measure whether large samples of companies communicate on only a narrow selective range of their activities, or whether they disclose across a wide variety of issues. For example, they may voluntarily report about water saving initiatives, but without disclosing how much water they have used over time or compared with other companies. In contrast, researching *empty green claims and policies* – that is, making claims or stating policies without living up to the promises – is much more difficult, as it can be hard to directly measure the difference between what firms say and what they do. Identifying empty claims and policies is thus more the domain of investigative journalists and case studies. The third variety of greenwash in category (1) above is *misleading narrative, discourse and visual imagery*. This is even harder to detect as it involves using environmental images to reinforce brands, logos or corporate messages in reports. Famous examples include when the oil company BP rebranded itself as 'Beyond Petroleum' in 2000, and introduced its green and yellow flower or sunburst logo. There is now increasing academic interest in the more subtle dimensions of this variety of greenwash, including for example, analysing the types of pictures used in corporate reports and websites.

In contrast, greenwashing through association involves communicating an environmental message by connecting environmental activities with other third parties such as certification schemes, governments or NGOs. The most prominent variety of greenwash by association is *dubious certifications and labels* that are not worth the paper they are written on. The ecolabelindex.com website now lists over 450 different ecolabeling schemes, but it is hard for consumers to know which ones are credible. A growing stream of academic research now compares the characteristics of different certification and label schemes in industries such as coffee, forestry or toy manufacturing. Researchers emphasize the stringency (or strictness) of the design of the scheme, such as whether there are ways to cheat the system, or whether the label is actually based on a high level of environmental performance. So far, this literature is leading to the disappointing conclusion that the environmental performance of firms with certifications and labels is often no better and is sometimes worse than those without (see for example Blackman & Rivera (2011), based on 46 studies of sustainability certification).

Greenwashing by association can also arise through *ineffective public voluntary programmes*. Delmas and Montes-Sancho's (2010) study of the US EPA's Climate Leaders Programme and Kim and Lyon's (2011) study of the DoE's Voluntary Greenhouse Gas Registry scheme both showed that associating with government-backed schemes can provide a false reassurance that companies have good environmental performance through participating in the schemes, despite lower actual emissions reductions. The final variety of greenwash involves association with NGOs in *co-opted NGO endorsements and partnerships*. Involving NGOs in an environmental programme or initiative can be an important source of credibility for companies, but has also given rise to concerns that NGOs are being co-opted into providing environmental legitimacy for companies when this is not warranted.

## Drivers of greenwash

Delmas and Burbano (2011) provide a comprehensive overview of why companies greenwash. The usual explanation is that companies are deliberately seeking to gain some external advantage from misleading consumers, investors, regulators or others. Delmas and Burbano divide these into non-market drivers (such as a lax regulatory environment, or very active attention from NGOs of the media), and market drivers (such as consumers' willingness to pay for green products, competitive pressures or investor demand). However, greenwashing can also be caused by internal organizational drivers and even by the individual psychology of managers. For example, organizations can be resistant to change due to their cultures, processes and structures, so that even well-intentioned corporate greening statements can be difficult to implement in practice inside the organization. This can be particularly challenging for large companies, or where there is poor communication inside the firm. At the individual level, greenwashing can arise from psychological biases such as optimism bias; that is, the tendency for individual managers to over-estimate the likelihood of positive results from environmental communication, and under-estimate the likelihood of inaccurate communication being exposed.

Thus, greenwashing can arise because of deliberate attempts to manipulate environmental information, or it can arise because large, complex organizations are 'muddling through'. Crilly, Zollo and Hansen (2012) found that multinational companies that faced the same external pressures greenwashed in different ways and for different reasons. Greenwashing could be deliberate or emerge from a pattern of activities across large organizations. Identifying when greenwashing is deliberate, and when it is an unintended consequence of managing a variety of environmental issues within complex organizations is a challenge for both researchers and activists.

## Limiting greenwashing

Closing the gap between green talk and green action requires monitoring both companies' environmental communications and actions. In some countries, regulators issue guidelines on fair environmental communications – for example the US Federal Trade Commission's *Guides for the Use of Environmental Marketing Claims*, or the UK's *Green Claims Guide* published by DEFRA. Websites such as the greenwashingindex.com allow users to upload examples of environmental advertising and for others to evaluate and rate them. Investigative journalists run exposés on companies that overstate their environmental claims. Indeed, there is some optimism that the rise of social media, with its distributed surveillance of companies' activities and claims will limit corporate greenwashing (Lyon & Montgomery, 2012).

However, this traditional way of limiting greenwashing through exposing deliberately misleading information will only work for the more obvious and active greenwash. Lots of greenwash is subtle, and not even necessarily intended by those committing it. Limiting this deeper form of greenwash requires a more sophisticated understanding of the organizational and individual-level reasons why greenwashing occurs, and why consumers and other stakeholders are often blind to it. Society needs new symbols and language to talk about environmental issues; but scholars agree that we also need to look at who is shaping this new green talk.

## Learning resources

www.greenwashingindex.com. This website allows users to post examples of environmental advertisements and green claims for others to rate and rank on a greenwashing scale from 'authentic' to 'bogus'.

www.ecolabelindex.com. This website is the largest global directory of ecolabels. Established in 2007, it is maintained by a non-profit organization based in the USA. As of November 2017, the index includes entries on 465 ecolabels in 199 countries in 25 industry sectors.

www.sinsofgreenwashing.com. This interactive website includes games, tools and social media to educate about the main types of greenwashing, including examples of greenwashing from a wide variety of consumer products industries.

## Bibliography

Blackman, A. & Rivera, J. (2011) 'Producer-level benefits of sustainability certification', *Conservation Biology*, 25(6): 1176–1185.

Bowen, F.E. (2014) *After Greenwashing: Symbolic Corporate Environmentalism and Society*. Cambridge: Cambridge University Press.

Crilly, D., Zollo, M. & Hansen, M. (2012) 'Faking it or muddling through? Understanding decoupling in response to stakeholder pressures', *Academy of Management Journal*, 55(6): 1429–1448.

Delmas, M. & Burbano, V. (2011) 'The drivers of greenwashing', *California Management Review*, 54(1): 64–87.

Delmas, M. & Montes-Sancho, M. (2010) 'Voluntary agreements to improve environmental quality: symbolic and substantive cooperation', *Strategic Management Journal*, 31(6): 575–601.

Greer, J. & Bruno, K. (1996) *Greenwash: The Reality Behind Corporate Environmentalism*. Lanham, MD: Apex Press.

Kim, E-H. & Lyon, T. (2011) 'Strategic environmental disclosure: evidence from the DOE's Voluntary Greenhouse Gas Registry', *Journal of Environmental Economics and Management*, 61: 311–326.

Laufer, W.S. (2003) 'Social accountability and corporate greenwashing', *Journal of Business Ethics*, 43(3): 253–261.

Lyon, T.P. & Montgomery, W. (2012) 'Tweetjacked: the impact of social media on corporate greenwash', *Journal of Business Ethics*, 118(4): 747–757.

Lyon, T.P. & Montgomery, W. (2015) 'The means and ends of greenwash', *Organization and Environment*, 28(2): 223–249.

# 6.24 Green technology

James Meadowcroft

Technology – including tools and other artifacts, as well as techniques, skills and processes – represents the application of knowledge to attain practical outcomes. 'Green technology' is generally understood as technology that is kind to the environment; that avoids, reduces or remediates harm humans are doing to their natural surroundings. But defining what counts as green technology, and the role such technologies can play in addressing global environmental problems, remain hotly contested. This is because what is 'harmful' cannot be defined in a timeless, universal way. Ultimately, the notion of green technology raises profound questions about the relationship between humans and the non-human natural world.

## What counts as green technology?

Technologies to manage air and water pollution, generate energy from renewable sources (like solar or wind), improve energy efficiency, conserve natural resources and protect biodiversity, or clean up contaminated sites can all be presented as 'green'. Like related expressions – including 'clean technology', 'environmental technology' and 'sustainable technology' – the green technology label is applied liberally, subject to considerable hype and open to continual challenge. Smart meters, modular nuclear power, organic farming, solar water heaters, chemical dispersants to manage oil-spills and straw-bale home construction, can all be presented as green technology. Such claims involve judgements about the environmentally destructive character of dominant technological practices and the relative merit of alternatives. But they also reflect implicit assumptions about the environmental impacts that are given weight, the boundaries within which assessments are carried out, and the underlying causes of environmental problems. What appears 'green' is place and time sensitive: a technology that looks good for the environment at one location or historical juncture may seem a bad choice in other circumstances.

Consider electric passenger vehicles (EVs) like those promoted by the company Tesla and (with varying degrees of enthusiasm) by mainstream automotive manufacturers. On the one hand, EVs are a perfect illustration of green technology. They are more energy efficient than automobiles with internal combustion engines (which waste about 60 per cent of the energy in gasoline as heat). They have no tailpipe emissions to create local air pollution or drive climate change. On the other hand, their green credentials can be challenged. In some jurisdictions where coal dominates the generation mix, consumers could be kinder to the climate by driving efficient gasoline powered vehicles. There are issues related to life cycle emissions, including GHGs released during vehicle fabrication, and impacts from battery production and disposal. Above all, many environmentalists argue emphasis should be placed on transforming our cities, and how we live and work, to break car-dependence and encourage reliance on public transit, biking and walking for personal mobility. From this perspective EVs are more about perpetuating consumerist lifestyles than a breakthrough in green technology.

Carbon capture and storage (CCS), which can trap GHG emissions from fossil power plants and other industrial process, provides another example. Its green technology credentials are challenged by those who reject reliance on 'end of pipe' solutions that encourage continued reliance on coal-fired power (which causes other environmental problems in addition to climate change), rather than accelerating the transition to renewables. On the other hand, climate models suggest the availability of CCS in the mitigation toolbox would dramatically reduce the costs of low carbon pathways; the technology might be needed to control non-combustion industrial emissions; or it could one day be used to secure 'negative' carbon emissions, removing carbon dioxide from the atmosphere when combined with air capture or bio-fuel based electricity generation. So, is CCS a green technology or not?

Similar controversies swirl around all sorts of technologies, from big hydroelectric generation projects to genetically modified crops or consumer recycling. To a large extent the tension boils down to whether it is enough that green technologies reduce some of the impacts of existing lifestyles and systems of production, or whether they should contribute more actively to the transformation of current ways of living (promoting environmental and perhaps other progressive values) and to healing the planet.

## What can green technology accomplish?

Over recent decades technological innovation has allowed developed societies to address some of the more visible adverse environmental impacts of industrialisation. In particular it has helped reduce pollutant emissions to air and water, improve waste management, and increase resource efficiencies. 'Relative decoupling' of economic growth from specific environmental impacts (a declining burden per unit of GDP) is relatively common across The Organisation for Economic Co-operation and Development (OECD), while 'absolute decoupling' (a real fall in a specific environmental harm even as GDP grows) has been observed in some countries, on particular issues, for limited periods of time. Sulphur dioxide emissions are a success story here, as absolute emissions continue to fall in developed states and may even have peaked internationally (Smith et al.,

2011). Novel technologies – such as flue gas desulphurisation and cleaner generation alternatives (including wind and solar) – have played an important role here, although other factors, such as a switch to low sulphur coal for electricity production and structural economic change, have also contributed.

Government policies, public concern about environmental issues, and business efforts to secure material and energy efficiencies and reputational advantage, have all encouraged a shift to cleaner technologies. Regulatory measures, that force performance improvements – for example, industrial discharge limits or consumer appliance standards – provide the foundation of modern environmental policy. But government incentives to support research and development (for example, by funding university research or providing support to firms) and deployment of clean technologies (for example, through public procurement policies or feed-in tariffs) are also important. Increasingly, state support for the clean technology sector is seen as an economic development opportunity, with the potential for business expansion, job creation and developing export markets. Promoting green technology is a central plank of the 'green growth' or 'green economy' initiatives supported by international organisations such as the OECD and UNEP (OECD, 2011; UNEP, 2011).

Barriers to the emergence and uptake of green technologies are substantial. Path dependence makes it risky and costly to depart from an established industrial trajectory. Technologies are bound into interdependent complexes, and linked with dominant social practices and institutions. Economic interests associated with incumbent technologies resist change – think of the response of fossil energy producers to the climate issue. There is now a rapidly growing scholarly literature on promoting innovation (Nelson, 1993; Hekkert et al., 2007), understanding socio-technical transitions (Geels, 2005), and encouraging the development of more sustainable technology pathways (Foxon, 2013).

Yet the question remains: to what extent can technology provide the answer to environmental problems? Some analysts focus almost exclusively on the potential of technological transformation: so, environmental policy becomes largely a matter of inducing the required technological shifts. Others point to underlying socio-economic and political drivers (population increase, economic growth, rising material consumption, affluent lifestyles and consumerist culture, capitalist economic relationships, and so on), arguing *mere* technological change cannot prevent an ever-expanding sphere of environmental harm. Clearly any serious response to the issue of climate change will *necessarily* be technological, since we cannot continue to rely on existing (GHG emitting) energy technologies to power human development. And yet, such a response may not be forthcoming or prove adequate unless broader socio-economic and political issues are confronted, and institutional reform secured. Moreover, since many technologies, and socio-political arrangements, could contribute to a low carbon future, there is no avoiding political arguments over which technological and socio-economic trajectories (with their different distributions of costs and benefits) should be pursued. And of course, climate change is not the only environmental problem contemporary societies confront.

## Humans, technology and the biosphere

Discussions about the nature and potential of green technology raise broader issues about the character of social progress, the human capacity to orient societal and technological advance, and our stewardship of the biosphere. Questions about green technology are ultimately questions about how human beings are to make their living in the world. About what sorts of eco-social transformations and systems are possible or desirable. For millennia human societies have been altering their natural surroundings. The deliberate, but also the unintended, consequences of the deployment of human technologies – from the mastery of fire, agriculture and the domestication of animals, through the harnessing of fossil energy, to genetic engineering – have literally remade the world around us. And today questions about the consequences of climate change and biodiversity loss can no longer be avoided.

Technological options continue to expand and in the not-too-distant future humans may be able to manipulate their genome, dramatically extend life spans, develop artificial intelligence, transform ecosystems through the deployment of biotechnology, fight battles with combat robots, conduct large-scale geoengineering, or establish settlements on other planets. Yet to say we could do these things does not tell us whether it would be wise or ethical to do them. So, issues of green technology, about what kind of a world we want to make, will not go away.

## Learning resources

The following offer interesting insights into why green technology both reflects and stands to alter the wider social, economic and political context:

MacKay, D.J.C. (2009) *Sustainable Energy – Without the Hot Air*. Cambridge: UIT.

Mulvaney, D.R. (ed.) (2011) *Green Technology: An A-to-Z Guide*. Los Angeles: Sage Publications.

Winner, L. (1978) *Autonomous Technology: Technics-out-of-Control as a Theme in Political Thought*. Massachusetts: The MIT Press.

## Bibliography

Foxon, T. (2013) 'Transition pathways for a UK low carbon electricity future', *Energy Policy*, 52(1): 10–24.

Geels, F. (2005) *Technological Transitions and System Innovations: A Co-evolutionary and Socio-technical Analysis*. Cheltenham: Edward Elgar.

Hekkert, M., Suurs, R., Negro, S., Kuhlmann, S. & Smits, R. (2007) 'Functions of innovation systems: a new approach for analysing technological change', *Technological Forecasting & Social Change*, 74(4): 413–432.

Nelson, R. (1993) *National Innovation Systems: A Comparative Analysis*. Oxford: Oxford University Press.

The Organisation for Economic Co-operation and Development (OECD) (2002) *Indicators to Measure the Decoupling of Environmental Pressure from Economic Growth*. Paris: OECD.

The Organisation for Economic Co-operation and Development (OECD) (2011) *Towards Green Growth*. Paris: OECD.

Smith, S.J., Van Aardenne, J., Klimont, Z., Andres, R.J., Volke, A. & Delgado Arias, S. (2011) 'Anthropogenic sulfur dioxide emissions: 1850–2005', *Atmospheric Chemistry and Physics*, 11: 1101–1116.

United Nations Environment Programme (UNEP) (2011) *Towards a Green Economy: Pathways to Sustainable Development and Poverty Eradication*. Nairobi: United Nations Environment Programme.

# 6.25 Indigenous territorial rights

Philippe Hanna and Frank Vanclay

## Defining Indigenous peoples

In around 90 countries of the world, there are people who have a cultural identity separate to the dominant culture in the country in which they live, and who typically have a strong attachment to the land. The United Nations estimates there are over 370 million such individuals in the world, speaking over 4,000 languages (UNDESA, 2009). Because of their strong connection to the land, Indigenous peoples are typically very vulnerable to any activity or project that impacts on their territory, or on the natural resources on which they depend and/or to which they are culturally attached. Indigenous peoples are over-represented amongst the world's poor and are worse off on most health and well-being indicators (UNDESA, 2009).

The generic term, 'Indigenous peoples', is widely used to describe these peoples, although it is difficult to define, sometimes resisted, and much contested. A range of largely equivalent terms are sometimes used, including: tribal groups, first peoples, first nations, Aboriginal peoples, ethnic minorities, adivasi, and traditional peoples. Occupational and geographical terms can also be used, such as hunter-gatherers, fishing communities, reindeer herders, nomads, peasants, and hill people. Occasionally, official designations, such as scheduled tribes, might be used (UNDESA, 2008).

Because of their diversity, and the variety of contexts in which they live, a precise definition that applies universally has been difficult to develop. Nevertheless, a general understanding is now established within the United Nations. Indigenous people are regarded as having at least some (but not necessarily all) of the following characteristics (UNDESA, 2008, p. 8):

- They identify themselves as indigenous peoples and are, at the individual level, accepted as members by their community.
- They have historical continuity or association with a given region or part of a given region prior to colonization or annexation.
- They have strong links to territories and surrounding natural resources.
- They maintain, at least in part, distinct social, economic and political systems.
- They maintain, at least in part, distinct languages, cultures, beliefs and knowledge systems.
- They are resolved to maintain and further develop their identity and distinct social, economic, cultural and political institutions as distinct peoples and communities.
- They form non-dominant sectors of society.

Arguably, two criteria are paramount: self-identification and connection to land. Human rights concerns mean that nobody can be forced to be Indigenous against their will, therefore self-identification is essential. Connection to land is also important, because it is the basis of Indigenous or collective rights (United Nations, 2007; UNOHCHR, 2013).

Many ethnic or cultural groups may have a special relationship to land (and in some cases to water). However, for various reasons related to identity politics, some not necessarily identify themselves as being Indigenous, or may not be accepted as Indigenous by national authorities. Therefore, alternative wordings, such as 'traditional communities' or 'tribal groups', are used in some countries because they allow more flexibility. For example, an international NGO, Minority Rights Group International, advocates on behalf of 'ethnic, religious and linguistic minorities'. In Brazil, in addition to the over 252 recognized Indigenous groups, about 27 categories of 'traditional communities' are recognized, including traditional fishermen, maroons (descendants of escaped slaves) and collectors of various natural products, for example rubber tappers and nut collectors (Hanna & Vanclay, 2013).

Continued access to their territories is particularly important so that Indigenous peoples can transmit their traditional knowledge to younger generations and engage in some of their cultural practices. To teach about their environment, they need to be in the place where they belong. Culture, that is the shared knowledge, meanings, values and practices of a particular group, is strongly related to the environment in which people live. This Indigenous, local, or traditional knowledge is intrinsically related to the territory in which it has been developed. Moreover, this knowledge influences and protects the environment because, in many cases, it regulates the environmental practices of the group by imposing restrictions on the collection of certain plants or the hunting of certain animals. Most, if not all Indigenous groups, have a strong ethic of stewardship to protect the environment. This Indigenous knowledge, which has been built up over generations, provides a deep understanding of their environment and actively shapes that environment. Activities that impinge on Indigenous people or their territories need to respect this Indigenous knowledge and their culture (Convention on Biological Diversity, 2004).

## The struggle for territorial rights and recognition

It took decades of struggle for Indigenous peoples to be recognized as culturally-differentiated groups and for their rights to be established (Chartres & Stavenhagen, 2009). Historically, they were discriminated against and denied their human rights. In the worst cases, they were denied full citizenship of the country in which they lived. There is now an established international legal framework of Indigenous rights reflected in two key documents, the United Nations *Declaration on the Rights of Indigenous Peoples* (United Nations, 2007) and the Indigenous and Tribal Peoples Convention of the International Labour Organization (1989). These documents emphasize that Indigenous peoples are entitled to the same human rights as all other peoples of the world; however, because of their special connection to land and their vulnerability, they deserve special consideration to ensure their rights are respected.

Two key concepts form the basis of this legal framework: *collective rights* and the *right to self-determination*. While most human rights accrue to individuals by virtue of being human, collective rights are intended to protect those characteristics of Indigenous peoples that only exist by virtue of being a member of a specific cultural group. Collective rights ensure that Indigenous peoples have the right to live in freedom, peace and security as a distinct group, to maintain their culture, and not be subjected to any action that seeks to, or could, obliterate their identity as a distinct cultural group (i.e. ethnocide). The right to self-determination establishes that Indigenous peoples can freely pursue the economic, social and cultural development of their choosing (Tauli-Corpuz et al., 2010).

The principle of Free, Prior and Informed Consent (FPIC) arose as a mechanism to ensure that Indigenous peoples can implement their right to self-determination. Any project, policy or activity which has the potential to impact on the lives and/or territories of Indigenous peoples must not proceed without their approval. Engagement activities to gain their consent must be undertaken with enough time for them to consider the issue in culturally-appropriate ways and to seek external advice as they wish. All relevant information about the project must be fully and fairly disclosed in accessible language. The objective of FPIC is that local communities must be able to meaningfully influence the decision-making processes of any planned intervention that may affect their lives (Hanna & Vanclay, 2013; FAO, 2016).

Gaining formal title to the land they inhabit is an important part of establishing the security of tenure of Indigenous peoples and protecting their resources. Communal land subject to only informal traditional title tends not to be secure. Around the world, Indigenous peoples have often had this land taken away from them. Gaining ownership of their land and resources is important for their survival as well as for their autonomy.

## Conflicts over the use of natural resources on Indigenous territories

The existence of natural resources on Indigenous territories frequently leads to conflict with companies and governments that want to exploit the resources, and within the communities themselves. Typically, the territorial rights of affected communities are not fully respected, especially in the planning and implementation stages, violating the principle of FPIC. Furthermore, the negative environmental and social impacts are not properly addressed creating grave consequences for Indigenous communities, including (1) landlessness; (2) joblessness; (3) homelessness; (4) marginalization: (5) food insecurity; (6) loss of access to common property resources; (7) increased morbidity and mortality (i.e. declining health); and (8) community disarticulation (Cernea, 1997). This is frequently the case with resource extraction (e.g. oil and gas mining) and energy generation, including renewable energy sources such as hydroelectric powerplants, windfarms and biofuel. However, it can also apply to biodiversity or conservation projects, such as the REDD scheme (Reducing Emissions from Deforestation and Degradation), which provides payments to communities for the preservation of forests for carbon sequestration. REDD projects have been criticised for being forced on communities, misrepresenting the deal, and not providing fair terms.

Some best-practice examples can be identified where the affected Indigenous group has had enough influence in the negotiation process to contribute to project design and participate in the monitoring of impacts. Indigenous communities might even own the project, in full alignment with their ethnodevelopment aspirations.

## Indigenous activism

Indigenous peoples' rights do not always ensure that self-determination is respected when their territories are targeted by resource extraction or other projects. There are many forms of political action Indigenous peoples pursue when their rights are not respected, or when their territories, culture or way of life are at risk (Hanna et al., 2016a, 2016b). Indigenous activism, which is usually aligned to environmental causes, has become increasingly vocal since the 1980s, and particularly prominent in recent years, for example with the Belo Monte Dam in Brazil and the North Dakota Access Pipeline in the United States. Both cases gained a worldwide audience through the use of social media to share information about the projects, their impacts, and resistance strategies.

## Conclusion

Indigenous territorial rights, supported by protest actions, are crucial to ensure the self-determination of Indigenous peoples. Where national states have inadequate legal frameworks, or lack the resources to ensure protection, Indigenous territorial rights can be violated. This happens by the illegal exploitation of natural resources, coercion or physical violence against Indigenous groups. Such violations often occur with the complicity of governments, for example by approving resource extraction from Indigenous territories without the Free, Prior and Informed Consent of the affected groups. Comprehending that Indigenous cultures and knowledge are strongly related to the territories where they have been developed over generations, and better understanding this symbiotic relationship will ensure the respect and protection of Indigenous territories and cultures into the future.

## Learning resources

*United Nations Permanent Forum on Indigenous Issues* is the UN coordinating body on Indigenous rights and Indigenous peoples' concerns: https://www.un.org/development/desa/indigenouspeoples.

*Cultural Survival* is one of the first international NGOs to focus on Indigenous peoples' rights, cultures, self-determination and political resilience. Resources include a quarterly publication and an online radio: https://www.culturalsurvival.org.

*International Working Group for Indigenous Affairs* is an international NGO with extensive educational materials on Indigenous people's rights and their situation: http://www.iwgia.org.

## Bibliography

Cernea, M. (1997) The Risks and Reconstruction Model for resettling displaced populations, *World Development*, 25(10): 1569–1587.

Chartres, C. & Stavenhagen, R. (eds) (2009) *Making the Declaration Work: The United Nations Declaration on the Rights of Indigenous Peoples*. Copenhagen: International Work Group for Indigenous Affairs.

Convention on Biological Diversity (2004) *Akwé: Kon – Voluntary guidelines for the conduct of cultural, environmental and social impact assessment regarding developments proposed to take place on, or which are likely to impact on, sacred sites and on lands and waters traditionally occupied or used by Indigenous and local communities*. Montreal: Secretariat of the Convention on Biological Diversity. Online. Available from: https://www.cbd.int/doc/publications/akwe-brochure-en.pdf.

Food and Agriculture Organization (FAO) (2016) *Free, Prior and Informed Consent: An Indigenous Peoples' Right and a Good Practice for Local Communities – Manual for Project Practitioners*. Rome: Food and Agriculture Organization. Online. Available from: http://www.fao.org/3/a-i6190e.pdf.

Hanna, P. & Vanclay, F. (2013) Human rights, Indigenous peoples and the concept of free, prior and informed consent, *Impact Assessment and Project Appraisal*, 31(2): 146–157.

Hanna, P., Langdon, E.J. & Vanclay, F. (2016a) Indigenous rights, performativity and protest, *Land Use Policy*, 50: 490–506.

Hanna, P., Vanclay, F., Langdon, E.J. & Arts, J. (2016b) Conceptualizing social protest and the significance of protest actions to large projects, *The Extractive Industries and Society*, 3(1): 217–239.

International Labour Organization (1989) *Convention Concerning Indigenous and Tribal Peoples in Independent Countries* (C169). Online. Available from: http://www.ilo.org/dyn/normlex/en/f?p=NORMLEXPUB:12100:0::NO::p12100_instrument_id:312314.

Tauli-Corpuz, V., Enkiwe-Abayao, L. & de Chavez, R. (eds) (2010) *Towards an Alternative Development Paradigm: Indigenous People's Self-Determined Development*. Baguio City (Philippines): TEBTEBBA. Online. Available from: http://www.tebtebba.org/index.php/all-resources/category/8-books?download=28:indigenous.

United Nations (2007) *Declaration on the Rights of Indigenous Peoples*. Online. Available from: http://www.un.org/esa/socdev/unpfii/documents/DRIPS_en.pdf.

United Nations Department of Economic and Social Affairs (UNDESA) (2008) *Resource Kit on Indigenous Peoples Issues*. New York: United Nations Department of Economic and Social Affairs. Online. Available from: http://www.un.org/esa/socdev/unpfii/documents/resource_kit_indigenous_2008.pdf.

United Nations Department of Economic and Social Affairs (UNDESA) (2009) *State of the World's Indigenous Peoples*. United Nations Department of Economic and Social Affairs. UN Document ST/ESA/328. Online. Available from: http://www.un.org/esa/socdev/unpfii/documents/SOWIP_web.pdf.

United Nations Office of the High Commissioner for Human Rights (UNOHCHR) (2013) *Indigenous Peoples and the United Nations Human Rights System*. Fact Sheet No.9 (Rev 2). Geneva: United Nations Office of the High Commissioner for Human Rights. Online. Available from: http://www.ohchr.org/Documents/Publications/fs9Rev.2.pdf.

# Indigenous knowledge systems

Deborah MacGregor

6.26

We take wisdom where we can find it. No system has all the answers.

Borrows, 2005: 5.

For thousands of years, Indigenous peoples all over the world developed complex and sophisticated knowledge systems that facilitated sustainable relationships with the environment, or 'Mother Earth'. While these cultures are highly diverse in nature, they share among them certain common philosophical foundations. Integral to Indigenous knowledge systems (IKS) is a responsibility to maintain and enhance relationships with the Earth as a living entity. Such notions are evident in various international Indigenous declarations pertaining to the environment, including the *Universal Declaration on the Rights of Mother Earth* developed at the World Peoples' Conference on Climate Change in Cochabamba, Bolivia in 2010. Article 1 of the *Declaration* states that "Mother Earth is a living being." Article 3.1 states that "Every human being is responsible for respecting and living in harmony with Mother Earth" (Global Alliance, 2010).

Having been developed and practiced for countless generations, Indigenous knowledge systems are certainly not new. What *is* new is the interest in and recognition of these systems by governments, academics, ENGOs and others (Johnson, 1992; Menzies, 2006). This interest has emerged in part due to recognition of the fact that Western science and technology alone may be unable to effectively address the global environmental challenges faced by humanity. IKSs offer alternative approaches, although they are often presented in opposition to Western knowledge (Kermoal & Altamirano-Jiménez, 2016). Still, the international community has for over three decades recognized the value of IKSs in helping to address global environmental issues and has been applying them in various contexts (Mason et al., 2012; Shackeroff & Campbell, 2007; Weiss et al., 2013). In addition to this, Indigenous peoples themselves have been calling for the recognition of IKSs in the decision-making processes that impact their lives, lands, and waters.

## What are Indigenous knowledge systems?

Different terms for Indigenous knowledge systems are used interchangeably in the scholarly and international literature. Such terms include 'local knowledge', 'folk knowledge', 'people's knowledge', 'traditional wisdom', 'ethnoscience', 'native science', 'traditional science', 'traditional knowledge', and 'traditional ecological knowledge (TEK)'. Many commentators utilize the term TEK, as it allows for an emphasis on a Western scientific understanding of 'ecology', even though this may differ significantly from Indigenous views of the topic (Berkes, 2012; Houde, 2007). 'Traditional knowledge' (TK) is also frequently used in the literature, in part due to the usage of the term in international agreements and conventions, such as the Convention on Biological Diversity (CBD, 1992). Definitions of TK in the international context vary, yet may be captured as "a living body of knowledge that is developed, sustained and passed on from generation to generation within a community, often forming part of its cultural or spiritual identity" (WIPO, 2016: 10). TEK is similarly defined as "a cumulative body of knowledge, practice, and belief, evolving by adaptive processes and handed down through generations by cultural transmission about the relationships between living beings (including humans) with one another and their environment" (Berkes, 2012: 8).

Indigenous activist Winona LaDuke describes TEK as "the culturally and spiritually based way in which Indigenous peoples relate to their ecosystems" (LaDuke, 1994: 127). The major difference between Indigenous and non-Indigenous conceptions of TEK is that Indigenous people see TEK as much more than a "body of knowledge"; rather, it is a "way of living", it is about how one relates to Mother Earth (Kimmerer, 2012). An important specific example of this existing divide between how Indigenous and Western knowledge systems approach sustainability is seen in the area of water governance, which may be thought more of as 'water justice' from an Indigenous viewpoint. In an IKS approach, water is conceptualized and related to as a living entity, an actual *relative* of humanity, with humanity's responsibilities to the waters forming a key emphasis for governance (McGregor, 2014). Failure to live up to these responsibilities, and working to correct such failures, is thus a water *justice* issue. In Western systems, water is viewed as a critical resource or commodity required for human consumption and basic needs. Conflicts around supply and demand for water are thus seen as commercial and/or human rights issues (McGregor, 2015).

Whatever terms are used in referring to Indigenous knowledge systems, Indigenous perceptions of them do not 'fit' within Eurocentric concepts. Furthermore, IKS/TEK/TK do not represent a "uniform concept across all Indigenous peoples; it is diverse knowledge that is spread throughout different peoples in many layers" (Battiste & Henderson, 2000: 35). The concepts convey different meanings across cultures. It has also been argued that because of the tendency to privilege some definitions over others, it is perhaps more functional to consider a term such as TEK as a collaborative concept "that bridges cross-cultural and cross-situational divides" (Whyte, 2013: 8).

Indigenous knowledge systems are also embedded within broader societal systems. We may be familiar with how Western science or knowledge is embedded within institutional settings (usually educational), with norms around who has knowledge, who is regarded as an expert, what knowledge is taught or not taught to students, how knowledge is validated, what counts as knowledge, etc. Indigenous knowledge is similarly situated within Indigenous society and its institutions (Kimmerer, 2012). However, Western knowledge continues to dominate the environmental governance landscape (Latulippe, 2015; Nadasdy, 2006) and the struggle for respect of IKSs continues accordingly.

## International recognition

Recognition of Indigenous peoples' unique perspectives, knowledge systems and concerns with respect to environmental issues goes back to the early 1980s when the International Union for the Conservation of Nature (IUCN) established a Working Group on Traditional Ecological Knowledge, chaired by Graham Baines (Williams & Baines, 1993). These early international initiatives were supported by a series of workshops and symposia examining the value of TEK for natural resource management. The 1987 Report of the World Commission on Environment and Development (the Brundtland Report), which emphasized the important role of Indigenous peoples in sustainable development, served as a catalyst for increased recognition of TEK worldwide (WCED, 1987).

Since the first Rio 'Earth Summit' in 1992, the United Nations has promoted global recognition of traditional knowledge systems in achieving various environmental goals. This support has taken the form of intergovernmental guidance for the use of traditional knowledge, including its protection, access and benefits sharing, its potential as a complement to science, and the need for on-the-ground support to ensure its continued propagation and vitality (Nakashima, 2010).

In 1997, at the United Nations Conference on Environment and Development, the *Convention on Biodiversity* (CBD) was signed, one of two legally-binding agreements to arise out of that conference. The CBD reiterated the vital role of Indigenous people and their knowledge in achieving sustainability. The CBD has had significant influence in terms of putting TK on the international environmental and sustainable development landscape. For countries that are signatory to the CBD, the consideration of TK in environmental management has been one of the main drivers compelling governments to include TK in environmental and resource management decision-making regimes. The recognition of the unique role that Indigenous peoples have in achieving sustainable development goals in such agreements also obligates governments and other actors (e.g., ENGOs) to strive to work more co-operatively with Indigenous peoples. In many respects these trends over the past three decades have represented opportunities for the involvement of Indigenous peoples in global environmental governance.

The worldwide climate change agenda is the most recent initiative to recognize the importance of Indigenous knowledge in addressing global challenges.

> Research on indigenous environmental knowledge has been undertaken in many countries, often in the context of understanding local oral histories and cultural

> attachment to place. . . . With the increased interest in climate change and global environmental change, recent studies have emerged that explore how indigenous knowledge can become part of a shared learning effort to address climate-change impacts and adaptation, and its links with sustainability.
>
> (IPCC, 2007)

Canada, among other countries, has been engaged in TEK and climate change research for some time (Nakashima et al., 2012). South of the border, a recent initiative within the US Forest Service has been the creation of *The Guidelines for Considering Traditional Knowledges in Climate Change Initiatives*. These guidelines were developed by the Climate and Traditional Knowledges Workgroup (CTKW) for understanding TEK in the context of climate change.

Internationally, one of the most important undertakings in recent years has been the adoption of The *United Nations Declaration on the Rights of Indigenous Peoples* (UNGA, 2007) by the United Nations General Assembly on September 13th, 2007, after decades of advocacy by Indigenous peoples. UNDRIP explicitly recognizes the importance of Indigenous knowledge as having a key role in realizing a sustainable future. As Article 31 of UNDRIP states:

> Indigenous peoples have the right to maintain, control, protect and develop their own cultural heritage, traditional knowledge and traditional cultural expressions, as well the manifestation of their sciences, technologies and cultures, including human and genetic resources, seeds, medicines, knowledge of the properties of fauna and flora, oral traditions, literatures, designs, sport and traditional games and visual and performing arts. They also have the right to maintain, control, protect, and develop their intellectual property over such cultural heritage, traditional knowledge and traditional cultural expressions.
>
> (UNGA, 2007)

The recognition of traditional knowledge (IKS, TEK, TK) in UNDRIP underscores the importance of its application in realizing a sustainable future and self-determination for Indigenous peoples. We are moving beyond the binary understanding of IKS and Western science as being opposing entities, to understanding how different knowledge systems can be utilized collaboratively and cooperatively to address environmental challenges faced by all.

## Conclusion

> Since Rio 1992, we as Indigenous Peoples see that colonization has become the very basis of the globalization of trade and the dominant capitalist global economy. The exploitation and plunder of the world's ecosystems and biodiversity, as well as the violations of the inherent rights of Indigenous Peoples that depend on them, have intensified.
>
> (Kari-Oca 2 Declaration, 2012)

Indigenous rights, IKSs and environmental sustainability are interwoven concepts, as expressed by Indigenous peoples in international fora in various environmental declarations over the past three decades. Despite improvements made internationally, nationally and locally in recognizing IKSs as valuable in addressing challenges, IKSs continue to be marginalized in relation to Western-based knowledge. Recognition of IKSs has not necessarily resulted in their robust application on the ground, and Indigenous people continue to see their territories exploited, their rights ignored and planetary destruction ongoing. Indigenous peoples, among others, have observed that "Imperialist globalization exploits all that sustains life and damages the Earth. We need to fundamentally reorient production and consumption based on human needs rather than for the boundless accumulation of profit for a few" (Kari-Oca 2 Declaration, 2012). IKSs in their various forms challenge the dominant political and economic world order and call for fundamental change in order to achieve sustainability for all beings on Mother Earth (McGregor, 2016).

## Learning resources

The following resources include further information for understanding IKSs and how they are being applied in various contexts:

### *Websites*

These sites contain information on how to engage Western science with IKSs in addressing key environmental issues such as climate change:

Alaska Native Knowledge Network: http://www.ankn.uaf.edu.
Guidelines for Considering Traditional Knowledges (TKs) in Climate Change Initiatives: https://climatetkw.wordpress.com/guidelines.
Center for World Indigenous Studies: http://cwis.org.
Local and Indigenous Knowledge Systems (LINKS), UNESCO: http://www.unesco.org/new/en/jakarta/natural-sciences/small-islands-and-indigenous-knowledge/local-and-indigenous-knowledge-systems-links.

### *Book*

This important text by Ellen Simmons looks at how IKSs are being applied in various contexts (forestry, biodiversity, food sovereignty) across the globe:

Simmons, E. (2013) *Indigenous Earth: Praxis and Transformation.* Penticton, BC: Theytus Books.

## Bibliography

Battiste, M. & Henderson, J.Y. (2000) *Protecting Indigenous Knowledge and Heritage: A Global Challenge.* Saskatoon, Saskatchewan: Purich.
Berkes, F. (2012) *Sacred Ecology* (3rd edn.). New York: Routledge.
Borrows, J. (2005) *Crown and Aboriginal Occupations of Land: A History & Comparison* (Rep.). Toronto, ON: Ministry of the Attorney General. Online. Available at: http://www.attorneygeneral.jus.gov.on.ca/inquiries/ipperwash/policy_part/research/pdf/History_of_Occupations_Borrows.pdf.

Convention on Biological Diversity (1992) Convention on Biodiversity Office. United Nation Environment. Retrieved March 22, 2018, from https://www.cbd.int/convention/text/default.shtml.

Global Alliance for the Rights of Nature (2010) *Universal Declaration of Rights of Mother Earth*. Online. Available at: http://therightsofnature.org/universal-declaration.

Houde, N. (2007) *Ecology and Society: The Six Faces of Traditional Ecological Knowledge: Challenges and Opportunities for Canadian Co-Management Arrangements*. Online. Available at: http://www.ecologyandsociety.org/vol12/iss2/art34.

Intergovernmental Panel on Climate Change (IPCC) (2007) *Indigenous Knowledge for Adaptation to Climate Change*. Online. Available at: https://www.ipcc.ch/publications_and_data/ar4/wg2/en/xccsc4.html.

Johnson, M. (1992) *Lore: Capturing Traditional Environmental Knowledge*. Hay River, NWT: Dene Cultural Institute.

Kari-Oca 2 Declaration (2012) *Indigenous Peoples Global Conference on Rio 20 and Mother Earth*. Online. Available at: http://www.forestpeoples.org/tags/rio20-united-nations-conference-sustainable-development.

Kermoal, N.J. & Altamirano-Jiménez, I. (2016) *Living on the Land: Indigenous Women's Understanding of Place*. Edmonton, AB: Athabaska University Press.

Kimmerer, R.W. (2012) Searching for synergy: integrating traditional and scientific ecological knowledge in environmental science education, *Journal of Environmental Studies and Sciences*, 2(4): 317–323.

LaDuke, W. (1994) Traditional ecological knowledge and environmental futures. In Colorado Journal of International Environmental Law & Policy (ed.) *Endangered Peoples: Indigenous Rights and the Environment* (pp. 126–148). Niwot, CO: University Press of Colorado.

Latulippe, N. (2015) Bringing governance into the conversation: introducing a typology of traditional knowledge literature, *AlterNative*, 11(2): 118–131.

Mason, L., White, G., Morishima, G., Alvarado, E., Andrew, L., Clark, F., Durglo, M., Durglo, J., Eneas, J., Erickson, J. et al. (2012) Listening and learning from traditional knowledge and Western science: a dialogue on contemporary challenges of forest health and wildfire, *Journal of Forestry*, 111(4): 193–197.

McGregor, D. (2014) Traditional knowledge and water governance: the ethic of responsibility. *AlterNative*, 10(5): 493–507.

McGregor, D. (2015) Indigenous women, water justice and zaagidowin (love). Women and water. *Canadian Woman Studies/Les Cahiers de la Femme*, 30(2/3): 71–78.

McGregor, D. (2016) Living well with the Earth: Indigenous rights and environment. In C. Lennox & D. Short (eds) *Handbook of Indigenous Peoples Rights* (pp. 167–180). Abingdon: Routledge.

Menzies, C.R. (ed.) (2006) *Traditional Ecological Knowledge and Natural Resource Management*. Lincoln: University of Nebraska Press.

Nadasdy, P. (2006) The case of the missing sheep. In Menzies, C.R. (ed.) *Traditional Ecological Knowledge and Natural Resource Management* (pp. 127–152). Lincoln: University of Nebraska Press.

Nakashima, D. (2010) *Indigenous Knowledge in Global Politics and Practice for Education, Science and Culture*. Natural Sciences Sector. Paris: UNESCO. Online. Available at: http://www.unesco.org/new/fileadmin/MULTIMEDIA/HQ/SC/temp/LINKS/sc_LINKS-UNU-TKinGPP.pdf.

Nakashima, D.J., Galloway M.K., Thulstrup, H.D., Ramos C.A. & Rubis, J.T. (2012) *Weathering Uncertainty: Traditional Knowledge for Climate Change Assessment and Adaptation*. Paris, UNESCO, and Darwin, UNU. Online. Available at: http://unesdoc.unesco.org/images/0021/002166/216613e.pdf.

Shackeroff, J. & Campbell, L. (2007) Traditional ecological knowledge in conservation research: problems and prospects for constructive engagement, *Conservation and Society*, 5(3): 343–360.

United Nations General Assembly (UNGA) (2007) *United Nations Declaration on the Rights of Indigenous Peoples*. Online. Available at: http://www.dd-rd.ca/site/_PDF/un/A_61_L67eng.pdf.

Weiss, K., Hamann, M. & Marsh, H. (2013) Bridging knowledges: understanding and applying indigenous and Western scientific knowledge for marine wildlife management, *Society and Natural Resources*, 26(3): 285–302.

Whyte, K.P. (2013) On the role of traditional ecological knowledge as a collaborative concept: a philosophical study, *Ecological Processes*, 2(7).

Williams, N. & Baines, G. (1993) *Traditional Ecological Knowledge: Wisdom for Sustainable Development.* Canberra: Centre for Resources and Environmental Studies.

World Commission on Environment and Development (WCED) (1987) *Our Common Future*. New York: Oxford University Press.

World Intellectual Property Organization (WIPO) (2016) *Traditional Knowledge and Intellectual Property*. Online. Available at: http://www.wipo.int/edocs/pubdocs/en/wipo_pub_tk_1.pdf.

# 6.27 Institutions and natural resource management

Tim Forsyth

Institutions are rules or behaviors shared by a group of individuals. They are important for natural resource management because they show how societies and communities work together rather than as competing individuals. They also give a framework for how to reduce the degradation of resources that are being overused. Institutions can be formal and visible structures, such as quotas that govern commercial fishing in the North Sea. They can also be informal and tacit, such as the traditional rules governing agricultural land or forests in developing countries.

Institutions are usually portrayed as the solution to the problem known as the 'Tragedy of the Commons'. The Tragedy refers to the hypothetical situation where people such as shepherds use a common land where there is no mechanism to prevent overuse. Each shepherd uses the pasture excessively because it is in his or her individual self-interest to do so. As a result, the pasture becomes degraded. The Tragedy was considered a useful metaphor for how human societies are based on individualistic behavior that can degrade natural resources.

Theorists of institutions have criticized the assertions of the commons tragedy. Most influentially, Elinor Ostrom (1990) argued that the Tragedy mistakenly labeled pastureland as an open access resource, or one that is open to all users without restriction. Instead, pastureland should be called a common pool resource because it is governed by rules (known as property rights) that allow local users to exclude other people, otherwise known as a common property regime. A common property regime is an institution because it provides a framework of shared rules and expectations that refers to how people use a resource as a group rather than as unregulated individuals. Ostrom and others produced evidence from around the world that showed how people had already built institutions to protect natural resources without state intervention.

Ostrom's work inspired a new field of identifying design principles crucial for building successful institutions, such as the clear communication of information; graduated sanctions for people who broke rules; and mechanisms for monitoring and dispute resolution (Poteete et al., 2010). These rules also offered ways to 'upscale' institutions, or connect institutions with similar objectives at different spatial (and temporal) scales—such as by applying the lessons of one successful community to new locations, or by connecting forest protection at the local level with national and international policies. International rules on carbon trading also allow the global system of accounting to connect to smaller scales, and vice versa.

But there is also a debate about how to build institutions, and how they come into being. Analysts from disciplines such as anthropology have argued that Ostrom's approach to institutions relied too heavily upon economics and rational-choice forms of politics, which assume that individuals always act rationally in order to maximize their self-interest (Bardhan & Ray, 2008; Forsyth & Johnson, 2014). As an alternative, these other analysts have argued that institutions can be based on a wider range of motivations. For example, studies of community-based natural resource management (CBNRM) have shown a wide variety of cultural codes and practices that govern how shifting cultivators clear land for new agriculture or leave it fallow; how rice farmers allocate land and labor between families; or engage in communal practices to protect soil from erosion in the Sahel (Netting, 1993). For example, in Java in Indonesia, farmers are engaged in long-term sharing of labor to cultivate irrigated rice fields, and then to share the rice harvest. In Burkina Faso in the Sahel, men and women collaborate to build lines of stones to protect agricultural land, as well as take communal decisions about increasing livelihood opportunities for local people. In addition, some analysts also showed that governments could change individualistic behavior, and encourage local people to adopt institutions, by long-term education and involvement in decision making (Agrawal, 2005).

In contrast to Ostrom's work, these explanations argue that rational behavior is not always based on profit maximization. For example, one study of irrigation tanks in southern India showed that people did not only administer tanks to protect the supply of water; they also use tanks for social activities such as meeting people, or negotiating caste roles (Mosse, 1997).

Second, these studies also acknowledged a broader range of environmental problems. For example, forest institutions do not only govern the extraction of timber, but also diverse other activities such as agriculture, hunting, or the collection of non-timber forest products such as herbs and leaves.

Third, institutions can also demonstrate the concept of legal pluralism, or the simultaneous existence of different jurisdictions over property rights (Sikor & Lund, 2009). For example, one village might comprise traditional (or customary) rules for using forests and pasture. Simultaneously, religious leaders might also determine the location of sacred groves used for burials. Local governments (such as *panchayats* in South Asia) might delimit and tax land used for public purposes. And national governments might impose other rules such as declaring certain land reserve forest or national parks, or even part of international agreements such as the Convention on Biological Diversity.

Consequently, institutions frequently involve a combination of formal and informal institutions, and comprise deliberative processes for negotiating rules and access between diverse stakeholders. The process of using multiple and complex common property regimes simultaneously is sometimes called institutional bricolage (Cleaver, 2012).

Institutions can also be transformed or created in order to change existing resource use. Some analysts and development organizations, for example, seek to understand current institutions in order to see which social groups are included and excluded in resource use. For example, in Nepal, some informal institutions forbid women from selling land, or from entering commercial trade of logs (Nightingale, 2011). Reforming institutions can sometimes mean building the political space to criticize or increase participation in existing practices. Some analysts have also used the concept of environmental entitlements to refer to how institutions allow or deny poorer people access to resources that are valuable to them (Leach et al., 1999). For example, national parks usually require villagers to restrict commercial activities. An environmental entitlements approach would assess how these institutions benefit some groups more than others, and then highlight how to make institutions more inclusive and empowering.

These approaches to institutions also show ways of making multi-scale environmental governance more inclusive. For example, the framework known as Reduced Emissions from Deforestation and forest Degradation (or REDD+) bears many similarities with Ostrom's ideas for institutions because it offers incentives and penalties for individuals to increase forest cover. REDD+, however, has been criticized for overlooking more diverse objectives such as enhancing biodiversity and local livelihoods. New approaches to governance such as the Landscapes Approach attempt to diversify REDD+ by representing multiple stakeholders and diverse objectives in land use policy, rather than aiming to increase forest cover alone (Sayer et al., 2013).

## Learning resources

The International Association for the Study of the Commons is an organization that offers information, meetings, and open-access to the *International Journal of the Commons*. https://www.iasc-commons.org.

Ostrom, E. (2005) *Understanding Institutional Diversity*. Princeton: Princeton University Press. This is a useful summary of institutions from a rational-choice perspective and summarises the seminal work of Elinor Ostrom.

Vatin, A. (2006) *Institutions and the Environment*. Cheltenham: Edward Elgar. This is a useful introductory text that summarises debates about institutions and the environment.

## Bibliography

Agrawal, A. (2005) *Environmentality: Technologies of Government and the Making of Subjects*. Durham, NC: Duke University Press.

Bardhan, P. & Ray, I. (2008) *The Contested Commons: Conversations Between Economists and Anthropologists*. Malden, MA: Blackwell Publishing.

Cleaver, F. (2012) *Development Through Bricolage: Rethinking Institutions for Natural Resource Management*. Abingdon: Routledge.

Forsyth, T. & Johnson, C. (2014) Elinor Ostrom's legacy: governing the commons and the rational choice controversy, *Development and Change*, 45(5): 1093–1110.

Leach, M., Mearns, R. & Scoones, I. (1999) Environmental entitlements: dynamics and institutions in community-based natural resource management, *World Development*, 27(2): 225–247.

Mosse, D. (1997) The symbolic making of a common property resource: history, ecology and locality in a tank irrigated landscape in South India, *Development and Change*, 28(3): 467–504.

Netting, R.M. (1993) *Smallholders, Householders: Farm Families and the Ecology of Intensive, Sustainable Agriculture*. Stanford: Stanford University Press.

Nightingale, A.J. (2011) Bounding difference: intersectionality and the material production of gender, caste, class and environment in Nepal, *Geoforum*, 42(2): 153–162.

Ostrom, E. (1990) *Governing the Commons: The Evolution of Institutions for Collective Action*. New York: Cambridge University Press.

Poteete, A.R., Janssen, M. & Ostrom, E. (2010) *Working Together: Collective Action, the Commons, and Multiple Methods in Practice*. Princeton: Princeton University Press.

Sayer, J., Sunderland, T., Ghazoul, J., Pfund, J.-L., Sheil, D., Meijaard, E., Venter, M., Boedhihartono, A.K., Day, M., Garcia, C. et al. (2013) Ten principles for a landscape approach to reconciling agriculture, conservation, and other competing land uses, *Proceedings of the National Academy of Sciences (PNAS)*, 110(21): 8349–8356.

Sikor, T. & Lund, C. (2009) Access and property: a question of power and authority, *Development and Change*, 40, 1–22.

# 6.28 Privatizing environmental assets

## Wim Carton

Privatization has, in recent decades, become a common approach to the management of natural resources and the resolution of environmental problems. To clarify the logic and implications of this trend, it is useful to start with a brief elaboration of the concept. Broadly speaking, privatization can be defined as the transfer of ownership of an asset (a company, resource, service, etc.) from the public domain to the private sector. It implies that the legal right over a previously open access or public asset is assigned to a private individual, group or company, and therefore also the control and management of that asset and any benefits that accrue from it. Because this process often excludes certain groups of people from the resources they formerly had access to (and might be dependent on), privatization is commonly described as a form of 'enclosure', 'resource grabbing' or 'dispossession' (Fairhead et al., 2012; Peluso, 2007). Privatization serves important economic functions because it often marks a first step towards commodification, or the process of bringing new resources or services into capitalist markets by making them into tradeable goods that can sold for a profit (Castree, 2003). As part of a wider turn to neoliberal policies since the 1970s, it has been aggressively promoted by many economists, governments and international organizations as a way to open up new markets and stimulate free trade.

The ongoing move towards the privatization of environmental goods and services (or 'nature') should be understood in this context. At one level, the logic behind this trend is straightforward capital accumulation, based on the neoclassical idea that overall maximization of utility and profit provides the greatest benefit for all. More than just serving economic interests, however, privatization increasingly also came to be seen as a means to instil environmental responsibility in people and organizations, and promote the 'rational' use of environmental resources and ecosystem services. Already in the 1950s and 1960s, scholars like Scott Gordon (1954) and Garrett Hardin (1968) had argued that overexploitation of resources arises when people strive to maximize their economic benefits – a central but problematic assumption in neoclassical economics about the natural behaviour of humans – under 'common property' or 'open access' resource regimes. In the absence of private property rights, they claimed, there are no incentives to manage resources sustainably because the costs of environmental degradation are socialized and therefore of marginal interest to each individual or company exploiting those resources. As Gordon (1954) puts it: "Wealth that is free for all is valued by none because he [sic] who is foolhardy enough to wait for its proper time of use will only find that it has been taken by another" (p. 135). It is important to note that this claim, later popularized in terms of Hardin's (1968) 'tragedy of the commons', has been extensively questioned and problematized. Institutional economist Elinor Ostrom's work, for example, shows that there are plenty of empirical cases where Hardin's thesis does not hold, and where complex social structures are perfectly able to manage common property resources sustainably (Ostrom, 1990). Neoclassical understandings of private property relations thus emerge from this critique as above all a distinctly Western construction (Turner, 2016).

Neither Gordon nor Hardin actually made a case for the privatization of environmental resources as a way to overcome the problem that they identified. Hardin (1968) ultimately argued for the abolishment of the commons through some form of coercive measures, while Gordon (1954) envisioned that the issue could also be resolved through government control. Despite this, and despite substantial criticism against the idea, the 'tragedy of the commons' has, over the past decades, been used as a central argument for the privatization of nature, either as a way to counteract environmental problems or simply to optimize resource allocation (Harvey, 2010). In this reinterpretation of the argument, private property relations are portrayed as the best and most efficient way to manage environmental resources and prevent their degradation. The reasons for this wedlock between neoliberal policies and environmentalism are complex but probably involve a mix of contingency, opportunism and concerted action by neoliberal organizations (Castree, 2010). As geographers James McCarthy and Scott Prudham (2004) point out, neoliberalism very much emerged in tandem with modern environmentalism, and in many ways should be seen as a distinctly environmental project from the start.

The literature on this is vast and includes discussions on a wide range of resources in various contexts (for an overview see, for example Castree, 2010; Fairhead et al., 2012; Heynen et al., 2007). To give just one example, debates about overfishing have commonly drawn on the 'tragedy' narrative to explain the overexploitation of fisheries and consequently prescribe privatized fishing rights as the logical solution (Carothers & Chambers, 2012; Mansfield, 2004). In an illustration of the influence of this narrative, many governments have, since the 1970s, turned to 'individual tradable quotas' (ITQs) – also known as individual fishing quotas or 'catch shares' – as a way to manage their fisheries. Essentially, this system allocates the rights to part of a total fishing quota for a specific species to eligible actors, who can then often lease or sell these rights to others. Apart from enabling the generation of rents through fisheries, this system is claimed to "foster a conservation ethic among participants who, as owners, want the fishery resources to remain healthy over the long term for their own benefit" (Carothers & Chambers, 2012, p. 42). James Fairhead et al. (2012) have proposed the term 'green grabbing' to describe how the privatization and appropriation of environmental resources is increasingly framed within such conservation narratives. Their own work reveals how such narratives are used to justify land acquisitions by private actors in the name of biodiversity conservation, carbon sequestration, biofuel development and investments in ecotourism. This development is part of what has been called a global 'land grab', where states and private actors are scrambling to acquire land, often in developing countries, for reasons of 'green' investments, food production and financial speculation (Borras et al., 2011).

Despite what the above discussion might suggest, however, privatization processes are not straightforward or monothetic, making generalizations about social and environmental implications difficult. ITQs can take many forms, not all of which bestow the same kind of rights on their owners. They can, moreover, be used for other purposes than the maximization of private profits, as in the case of the Western Alaska Community Development Quota system, which uses ITQs as a way to redistribute part of the proceeds from lucrative fisheries to coastal communities (Mansfield, 2007). Similarly, the appropriation of land and natural resources for 'green ends' does not necessarily mean that existing users are entirely closed off, or that land or resources change ownership (e.g. are 'sold') in line with commonplace Western understandings of property relations (Fairhead et al., 2012). Both proponents and critics commonly use the term 'privatization' to describe processes that do not necessarily result in 'pure' private property, but in hybrid forms that still involve certain public rights and responsibilities (Turner, 2016). To deny privatization this hybrid characteristic would be to bestow too much power onto the abstracting concepts of neoclassical economics, and to belie the many ways in which the restructuring of resource access and control is ultimately articulated through relations with the state and more informal resource management regimes.

That being said, the impossibility of simple generalizations does not diminish the often disruptive impacts that privatization has had in concrete cases. While specific outcomes ultimately depend on socioeconomic contexts, as well as on the biophysical characteristics of the environmental goods or services that are targeted (Bakker, 2005; Fairhead et al., 2012), some clear tendencies can be discerned. The privatization of fishing rights has often meant a trend towards the consolidation of power in the hands of a few large actors, and the loss of jobs for small-scale fishermen, who essentially are pushed out of traditional income activities. Meanwhile, there's little evidence that privatized fishing rights help conserve fish stocks more than ordinary total catch quotas would (Carothers & Chambers, 2012). In the case of land acquisitions, privatization has often meant a restructuring of historically established access and user rights of indigenous groups and local communities. Private sector investments in tree plantations for carbon sequestration purposes, for example, commonly enclose existing forests or public lands, preventing locals from using forest resources and from activities such as hunting, cattle grazing, or the gathering of firewood (Leach & Scoones, 2015). In some cases, land acquisitions have involved violent evictions and open conflicts (Cavanagh & Benjaminsen, 2014). Moreover, in both of the above examples the discursive bias towards the 'tragedy' narrative and the *de facto* criminalization of traditional subsistence activities in order to 'protect' natural resources implicitly or explicitly reframes locals as 'irrational' resource users, and the ultimate drivers behind, for example, deforestation or soil degradation (Fairhead et al., 2012; Leach & Scoones, 2015).

If nothing else, such examples show that the proposed virtuous marriage between privatization and environmental conservation is much more problematic and contested than proponents commonly acknowledge. Privatization often involves a highly unequal distribution of benefits and disadvantages, to the detriment of economically marginalized groups. Common criticism and widespread opposition to the privatization of nature bear witness to this.

## Learning resources

For an overview of debates on the privatization (and neoliberalization) of nature, see:

Castree, N. (2010) 'Neoliberalism and the biophysical environment: a synthesis and evaluation of the research', *Environment and Society: Advances in Research*, 1(1): 5–45.

Heynen, N., McCarthy, J., Prudham, S. & Robbins, P. (2007) *Neoliberal Environments: False Promises and Unnatural Consequences*. New York: Routledge.

Mansfield, B. (ed.) (2008) *Privatization: Property and the Remaking of Nature-Society Relations*. Malden & Oxford: Wiley-Blackwell.

For an overview of recent debates on the privatization of nature for purported environmental purposes, see:

Fairhead, J., Leach, M. & Scoones, I. (2012) 'Green Grabbing: a new appropriation of nature?', *The Journal of Peasant Studies*, 39(2): 237–261.

## Bibliography


Bakker, K. (2005) 'Neoliberlizing nature? Market enviromentalism in water supply in England and Wales', *Annals of the American Geographers*, 95(3): 542–565.

Borras, S.M., Hall, R., Scoones, I., White, B. & Wolford, W. (2011) 'Towards a better understanding of global land grabbing: an editorial introduction', *The Journal of Peasant Studies*, 38(2): 209–216.

Carothers, C. & Chambers, C. (2012) 'Fisheries privatization and the remaking of fishery systems', *Environment and Society: Advances in Research*, 3(1): 39–59.

Castree, N. (2003) 'Commodifying what nature?', *Progress in Human Geography*, 27(3): 273–297.

Castree, N. (2010) 'Neoliberalism and the biophysical environment: a synthesis and evaluation of the research', *Environment and Society: Advances in Research*, 1(1): 5–45.

Cavanagh, C. & Benjaminsen, T.A. (2014) 'Virtual nature, violent accumulation: the "spectacular failure" of carbon offsetting at a Ugandan National Park', *Geoforum*, 56: 55–65.

Fairhead, J., Leach, M. & Scoones, I. (2012) 'Green grabbing: a new appropriation of nature?', *The Journal of Peasant Studies*, 39(2): 237–261.

Gordon, H.S. (1954) 'The economic theory of a common-property resource: the fishery', *The Journal of Political Economy*, 62(2): 124–142.

Hardin, G. (1968) 'The tragedy of the commons', *Science*, 162(3859): 1243–1248.

Harvey, D. (2010) 'The future of the commons', *Radical History Review*, 2011(109): 101–107.

Heynen, N., McCarthy, J., Prudham, S. & Robbins, P. (2007) *Neoliberal Environments: False Promises and Unnatural Consequences*. New York: Routledge.

Leach, M. & Scoones, I. (2015) *Carbon Conflicts and Forest Landscapes in Africa*. New York: Routledge.

Mansfield, B. (2004) 'Rules of privatization: contradictions in neoliberal regulation of North Pacific fisheries', *Annals of the Association of American Geographers*, 94(3): 565–584.

Mansfield, B. (2007) 'Property, markets, and dispossession: the western Alaska community development quota as neoliberalism, social justice, both, and neither', *Antipode*, 39(3): 479–499.

McCarthy, J. & Prudham, S. (2004) 'Neoliberal nature and the nature of neoliberalism', *Geoforum*, 35(3): 275–283.

Ostrom, E. (1990) *Governing the Commons: The Evolution of Institutions for Collective Action*. Cambridge: Cambridge University Press.

Peluso, N.L. (2007) 'Enclosure and privatization of neoliberal environments'. In N. Heynen, J. McCarthy, S. Prudham & P. Robbins (eds) *Neoliberal Environments: False Promises and Unnatural Consequences* (pp. 89–93). New York: Routledge.

Turner, M.D. (2016) 'Political ecology III: the commons and commoning', *Progress in Human Geography*, 41(6) 795–802.

# Sustainability transitions

6.29

Frank W. Geels

## The need for sustainability transitions and increasing interest

Many environmental problems, such as climate change, loss of biodiversity, and resource depletion, are formidable societal challenges, which relate to unsustainable consumption and production patterns in electricity, heat, buildings, mobility and agro-food systems. Addressing these problems requires reductions in environmental burdens by a factor of four or five (e.g. 80 per cent reduction in greenhouse gas emissions by 2050). Such reductions require shifts to new kinds of systems, shifts which are called 'sustainability transitions' (Markard et al., 2012).

These transitions are not only about new technologies, but also about changes in user practices, cultural discourses, policies and institutions, infrastructure and business models. As shorthand, these configurations are labelled 'socio-technical systems' (Elzen et al., 2004). Sustainability transitions and system innovation are multi-actor processes, involving interactions between firms, consumers, policymakers, wider publics, NGOs, and scientists.

Academic research on sustainability transitions started in the early 2000s, initially pioneered by scholars in the field of innovation studies, who combined insights from history, evolutionary economics and sociology of technology. The field gradually attracted more attention, diversified, and institutionalized with the creation of the Sustainability Transitions Research Network (STRN) in 2009. The field now has more than 1000 members from a range of disciplines, e.g. innovation studies, political science, geography, cultural studies, consumption studies, business studies and economics. Political attention has also increased as countries like the UK launched a Low-Carbon Transition Plan (2009) and Germany embarked on the Energy Transition (2011).

The core challenge for research and policy on sustainability transitions is the tension between stability and change. On the one hand, there is a flurry of sustainability initiatives (e.g. eco-cities, community energy, transition towns) and the launch of 'green' technologies (e.g. wind turbines or electric vehicles). On the other hand, existing systems are deeply entrenched and appear to change very slowly in sustainable directions, if at all. Transitions research aims to understand this tension and the associated struggles on techno-economic and socio-political dimensions.

## Multi-level perspective on transitions

A prominent framework is the multi-level perspective (Rip & Kemp, 1998; Geels, 2002), which conceptualizes transitions as non-linear processes that result from the interplay of developments within and between three analytical levels: niches (the locus for radical innovations), socio-technical regimes (the locus of established practices and associated rules) and an exogenous sociotechnical landscape. These 'levels' refer to heterogeneous configurations of increasing stability, which can be seen as a nested hierarchy (Figure 6.29.1).

*Niches* are 'protected spaces' such as research and development laboratories, subsidized demonstration projects or small market niches where users have special demands and are willing to support emerging innovations. Niche actors work on radical innovations that deviate from existing regimes. The literature on niche-innovation (Schot & Geels, 2008) distinguishes three social processes within niches: 1) The articulation of *expectations* or *visions*, which provide directionality to innovation activities and attract attention and funding from external actors. 2) The building of social *networks* and the enrollment of more actors, which bring in new knowledge and resources. 3) *Learning processes* on various dimensions, e.g. technical performance, consumer demand, infrastructure requirements, policy instruments, symbolic meanings and business models.

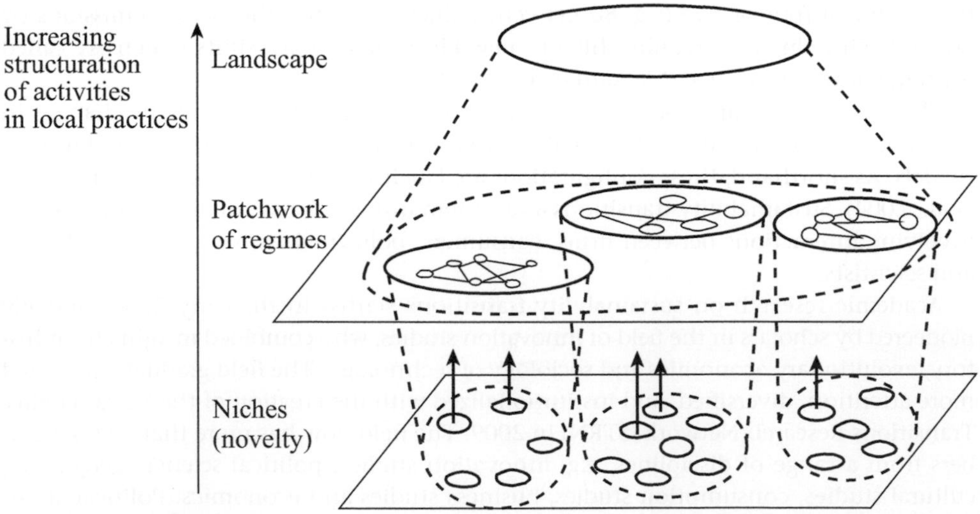


**Figure 6.29.1** Multiple levels as a nested hierarchy.

*Source*: Geels (2002: 1261).

The alignment of these three processes stimulates niche development and may enhance internal momentum.

Niche-innovations usually struggle against existing *socio-technical systems* and *regimes*, which are stabilized by many lock-in mechanisms (Unruh, 2000) such as sunk investments (in machines, people and infrastructure); regulations and laws that create market entry barriers; resistance from vested interests; low costs because of economies of scale and consumer lifestyles. Innovation within existing regimes tends to be incremental, giving rise to predictable trajectories.

The *socio-technical landscape* is the wider exogenous context that influences niche and regime dynamics. It includes spatial structures, macro-economic trends, political ideologies, demographic trends, societal values and beliefs, and environmental change. Landscape changes usually unfold gradually, beyond the direct influence of niche and regime actors.

The multi-level perspective (MLP) suggests that transitions come about through the interplay between processes at these different levels: (a) niche-innovations build up internal momentum, (b) changes at the landscape level create pressure on the regime, and (c) destabilization of the regime creates windows of opportunity for niche-innovations (Figure 6.29.2). Instead of single 'causes' or 'drivers', the MLP emphasizes 'circular causality' in which multi-dimensional processes at different levels link up and reinforce each other.

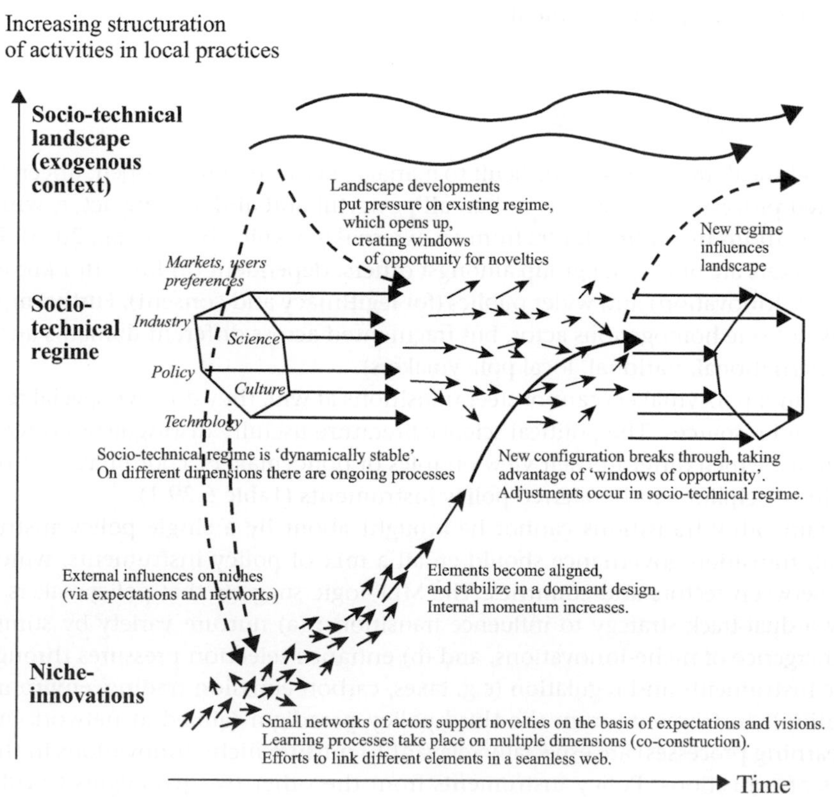


**Figure 6.29.2** Multi-level perspective on transitions.

*Source*: Adapted from Geels (2002: 1263).

Although each transition is unique, the complexity can be stylized by distinguishing different phases in transitions. In the first phase, radical innovations emerge in niches, often outside or on the fringe of existing regimes. The social network of innovators is unstable and fragile. There is much uncertainty, various design options co-exist, actors improvise and engage in experiments. The niche-innovations do not (yet) form a threat to the existing regime.

In the second phase the innovation enters small market niches, which provide resources for further development and specialization. The radical innovation develops a trajectory of its own and rules begin to stabilize. Users build up experience with the new technology and further articulate their preferences. Radical innovations may remain stuck in market niches for a long time, especially when they face a mis-match with the existing regime. As long as the regime remains stable, niche-innovations often have little chance to diffuse more widely.

The third phase is characterized by wider breakthrough of the niche-innovation and competition with the established regime. On the one hand, this process depends on niche-internal drivers such as price/performance improvements, scale economies, development of complementary technologies, and support from powerful actors. On the other hand, external landscape developments exert pressures on the regime, leading to tensions and windows of opportunity for the diffusion of niche-innovations.

The fourth phase is characterized by substitution and broader adjustments in infrastructures, regulations and views on normality. The new system may (eventually) influence wider landscape developments.

## Policy implications

Socio-technical transitions are difficult to manage, because they are open, uncertain and contested processes. The state is not an all-powerful and all-knowing actor, which can steer transitions by pulling levers from an external 'cockpit' (Hajer et al., 2015). Rather, policymakers are one social group amongst others, dependent on firms (for knowledge, resources, innovation) and wider publics (for legitimacy and consent). Furthermore, the state is not one homogenous actor, but fragmented across different domains and levels (e.g. international, national, local policymakers).

Although policymakers cannot steer transitions at will, they do have special responsibilities and resources. The political science literature usefully distinguishes three policy paradigms, which differ in their view on roles of policymakers, coordination processes, scientific disciplines and preferred policy instruments (Table 6.29.1).

Sustainability transitions cannot be brought about by a single policy instrument. Instead, transition governance should entail a mix of policy instruments, which may differ between sectors and countries. The MLP-logic suggests that policymakers should follow a dual-track strategy to influence transitions: (a) nurture variety by stimulating the emergence of niche-innovations, and (b) enhance selection pressures through economic instruments and regulation (e.g. taxes, carbon emission trading, environmental legislation). Instruments from the third policy paradigm (aimed at network building and learning processes) are especially relevant to nurture niche-innovations in the early phases of transitions. Policy instruments from the other two paradigms (regulations, standards, taxes, financial incentives) are more suited to create pressure on existing regimes and stimulate the wider diffusion of niche-innovations, which is more important in later phases of transitions.

**Table 6.29.1** Different policy paradigms

| | *Market model (bottom-up)* | *Classic steering (top-down)* | *Interactive network governance* |
|---|---|---|---|
| **Characterization of relationships** | Autonomous: government sets 'rules of the game', but leaves autonomous actors free to decide on actions. | Hierarchical command-and-control: government sets goals or tells actors what to do (e.g. via standards). | Mutually dependent interactions. |
| **Characterization of coordination processes** | Incentives and price signals coordinate self-organizing actors through markets. | Government coordinates through regulations, goals, and targets. | Coordination happens through social interactions and exchange of information and resources. |
| **Foundational scientific disciplines** | Neo-classical economics. | Classic political science. | Sociology, innovation studies, neo-institutional theory. |
| **Policy instruments** | Financial incentives (subsidies, taxes). | Formal rules, regulations and laws. | Demonstration projects and experiments, knowledge transfer policies, network management, vision building through scenario workshops, strategic conferences, and public debates. |

*Source*: Geels et al. (2015: 8).

## Learning resources

The research manifesto of STRN (available at http://www.transitionsnetwork.org) provides useful references and information about specific transition related research themes, including 'governance, power and politics', 'civil society, culture and social movements', 'the role of firms and industries in transitions', and 'sustainable consumption, and the geography of transitions'.

The STRN website also provides access to all newsletters since 2011, which contain lists of recent publications on empirical topics, countries, and conceptual frameworks.

The following special issues also offer useful introductions into current debates:

'Innovation and sustainability transitions: the allure of the multi-level perspective and its challenges', *Research Policy*, 2010, Vol. 39, No. 4.

'Sustainability transitions: an emerging field of research and its prospects', *Research Policy*, 2012, Vol. 41, No. 6.

'Sustainability transitions in the making: a closer look at actors, strategies and resources', *Technological Forecasting and Social Change*, 2012, Vol. 79, No. 6.

'Past and prospective energy transitions: insights from history', *Energy Policy*, 2012, Vol. 50.

'Transition pathways to a low carbon economy', *Energy Policy*, 2013, Vol. 52.

## Bibliography

Elzen, B., Geels, F.W. & Green, K. (eds) (2004) *System Innovation and the Transition to Sustainability: Theory, Evidence and Policy*. Cheltenham: Edward Elgar.

Geels, F.W. (2002) 'Technological transitions as evolutionary reconfiguration processes: a multi-level perspective and a case-study', *Research Policy*, 31(8–9): 1257–1274.

Geels, F.W., McMeekin, A., Mylan, J. & Southerton, D. (2015) 'A critical appraisal of sustainable consumption and production research: the reformist, revolutionary and reconfiguration positions', *Global Environmental Change*, 34: 1–12.

Hajer, M., Nilsson, M., Raworth, K., Bakker, P., Berkhout, F., de Boer, Y., Rockström, J., Ludwig, K. & Kok, M. (2015) 'Beyond cockpit-ism: four insights to enhance the transformative potential of the sustainable development goals', *Sustainability*, 7: 1651–1660.

Markard, J., Raven, R. & Truffer, B. (2012) 'Sustainability transitions: an emerging field of research and its prospects', *Research Policy*, 41(6): 955–967.

Rip, A. & Kemp, R. (1998) 'Technological change', in Rayner, S. & Malone, E.L. (eds) *Human Choice and Climate Change* (pp. 327–399). Columbus, Ohio: Battelle Press.

Schot, J.W. & Geels, F.W. (2008) 'Strategic niche management and sustainable innovation journeys: theory, findings, research agenda and policy', *Technology Analysis & Strategic Management*, 20(5): 537–554.

Unruh, G.C. (2000) 'Understanding carbon lock-in', *Energy Policy*, 28(12): 817–830.

6.30

# The Sustainable Development Goals

David Griggs

On 1 January 2016 'Transforming our World: The 2030 Agenda for Sustainable Development (the 2030 Agenda)' came into effect. This United Nations document sets out 17 Sustainable Development Goals (UN, 2016). Its objective is to end poverty, protect the planet, and ensure prosperity for all. The 17 goals and 169 targets aim to be achieved by 2030. The 2030 Agenda was unanimously adopted by the 193 Member States of the United Nations and is acting as a framework for nations, the UN and all sectors of society to end poverty by 2030 and pursue a sustainable future.

The 17 SDGs are as follows:

- Goal 1. End poverty in all its forms everywhere
- Goal 2. End hunger, achieve food security and improved nutrition, and promote sustainable agriculture
- Goal 3. Ensure healthy lives and promote well-being for all at all ages
- Goal 4. Ensure inclusive and equitable quality education and promote lifelong learning opportunities for all
- Goal 5. Ensure access to affordable, reliable, sustainable and modern energy for all
- Goal 6. Ensure availability and sustainable management of water and sanitation for all
- Goal 7 Achieve gender equality and empower all women and girls
- Goal 8. Promote sustained, inclusive and sustainable economic growth, full and productive employment and decent work for all

- Goal 9. Build resilient infrastructure, promote inclusive and sustainable industrialization and foster innovation
- Goal 10. Reduce inequality within and among countries
- Goal 11. Make cities and human settlements inclusive, safe, resilient and sustainable
- Goal 12. Ensure sustainable consumption and production patterns
- Goal 13. Take urgent action to combat climate change and its impacts
- Goal 14. Conserve and sustainably use the oceans, seas and marine resources for sustainable development
- Goal 15. Protect, restore and promote sustainable use of terrestrial ecosystems, sustainably manage forests, combat desertification, and halt and reverse land degradation and halt biodiversity loss
- Goal 16. Promote peaceful and inclusive societies for sustainable development, provide access to justice for all and build effective, accountable and inclusive institutions at all levels
- Goal 17. Strengthen the means of implementation and revitalize the Global Partnership for Sustainable Development

They are often represented using the icons in Figure 6.30.1.

They follow on from, and expand upon, the Millennium Development Goals (MDGs), which started a global effort in 2000 to tackle key development issues. The MDGs established measurable, universally-agreed objectives for tackling extreme poverty and hunger, preventing deadly diseases, and expanding primary education to all children, among other development priorities.

**Figure 6.30.1**

*Source*: United Nations website, https://sustainabledevelopment.un.org/?menu=1300.

The MDGs were as follows:

- Eradicate extreme poverty and hunger
- Achieve universal primary education
- Promote gender equality and empower women
- Reduce child mortality
- Improve maternal health
- Combat HIV/AIDS, malaria, and other diseases
- Ensure environmental sustainability
- Develop a global partnership for development

## What did the Millennium Development Goals achieve?

The MDGs have been called 'the most successful anti-poverty movement in history', but what achievements were made and can those achievements be attributed to the MDGs?

**MDG 1:** The number of people living on less than $1.25 a day was reduced from about 1.9 billion in 1990 to 836 million in 2015, but the target of halving the proportion of people suffering from hunger was narrowly missed.

**MDG 2:** Primary school enrolment figures have greatly increased, but the goal of achieving universal primary education was narrowly missed, with the net enrolment rate increasing from about 83 per cent in 2000 to 91 per cent in 2015.

**MDG 3:** By 2015, women made up 41 per cent of paid workers outside the agricultural sector, an increase from 35 per cent in 1990, about two-thirds of developing countries have achieved gender parity in primary education and 90 per cent of countries now have more women in parliament.

**MDG 4:** The child mortality rate reduced by more than half over the MDG period, from about 90 to 43 deaths per 1,000 live births, but it missed the MDG target of a drop of two-thirds.

**MDG 5:** The global maternal mortality ratio fell by nearly half, missing the target of a two-thirds reduction.

**MDG 6:** The number of new HIV infections fell by around 40 per cent between 2000 and 2013, but the target of halting and beginning to reverse the spread of HIV/AIDS by 2015 was not met. The global malaria incidence rate fell by an estimated 37 per cent and the mortality rate by 58 per cent.

**MDG 7:** The target of integrating the principles of sustainable development into country policies and programmes and reversing the loss of environmental resources was not met and in many areas the situation has deteriorated, hence the need for the SDGs. However, about 2.6 billion people have gained access to improved drinking water since 1990, so the target of halving the proportion of people without access to improved sources of water was achieved in 2010, although over 650 million people across the world still do not have access to improved drinking water.

**MDG 8:** Between 2000 and 2014, overseas development increased by 66 per cent in real terms, and in 2013 reached the record figure of US $134.8bn.

So it is clear that, while many of the goals were not met, significant progress was achieved against many targets. While some of the gains can simply be attributed to economic development gains in countries such as China, it is widely accepted that the MDGs played some role in many of the achievements.

## Key differences between the MDGs and the SDGs

There are many differences between the MDGs and the SDGs but here are a few of the key ones:

### *A much broader scope*

The SDGs are broader in scope because they address the interconnected elements of sustainable development: economic growth, social inclusion and environmental protection, whereas the MDGs focused primarily on the social aspects.

### *Apply to all countries*

The MDGs applied to developing countries with funding coming from rich countries. However, the SDGs apply to all countries equally, meaning that it is equally important and relevant for a developed country to implement the SDGs as a developing country.

### *Integrated*

In bringing together the so-called three pillars of sustainable development, social inclusion, economic prosperity and environmental protection in a much more holistic way, the SDGs recognize the interlinkages between goals, and the fact that implementation action must be carried out in an integrated way.

### *Bottom-up*

The MDGs were created through a top-down process largely determined by OECD countries and international donor agencies. The SDGs, by contrast, were created in one of the most inclusive participatory processes the world has ever seen with face-to-face consultations in more than 100 countries, and millions of citizen inputs through social media.

### *Half way to all the way*

The MDG targets for 2015 were set to get us "half way" to the goal of ending hunger and poverty, with similar proportional goals in other fields. The SDGs are designed to eliminate extreme poverty and hunger.

### *No one will be left behind*

Human development, human rights and equity are deeply rooted in SDGs, enshrined in an overarching principle that "no one will be left behind".

## *The role of the private sector and civil society*

All sectors of society, including business and civil society are far more engaged in the SDGs than they were in the MDGs, which were seen as more of a government-to-government development aid agenda.

## How will the SDGs be implemented?

Implementation of the SDGs and reporting on progress is 'State led', i.e., responsibility lies at the national level. The United Nations' role is to be the forum where nations come together to talk about implementation, share data, and report on progress. The UN High Level Political Forum (HLPF) is the central platform for follow-up and review of the 2030 agenda and the SDGs. The HLPF meets every 4 years at the level of Heads of State and Government under the auspices of the General Assembly, and in other years under the auspices of the Economic and Social Council.

The HLPF will be informed by the Global Sustainable Development Report (GSDR) which aims to strengthen the science–policy interface at the HLPF. The GSDR will adopt an assessment-of-assessments approach, documenting and describing information on specific issues that are policy-relevant in the field of sustainable development. The report is designed not to compete with new assessments and other reports and assessments being prepared by the UN system and outside.

A robust follow-up and review mechanism for the implementation of the 2030 Agenda requires a solid framework of indicators and statistical data to monitor progress, inform policy and ensure accountability of all stakeholders. A global indicator framework was developed by the Inter-Agency and Expert Group on SDG Indicators (IAEG-SDGs) and agreed to, as a practical starting point, at the 47th session of the UN Statistical Commission in March 2016. The list includes 230 indicators on which general agreement has been reached. Each country must report on progress at least twice over the 15-year period.

Although formal responsibility for implementation of the SDGs and reporting on progress lies with national governments, the SDGs are a universal agenda, meaning that everyone has a role to play in implementation, including business, civil society, sub-national and local government, and even individuals. International organizations enable implementation of the SDGs; examples include the Sustainable Development Solutions Network (SDSN), which aims to mobilize global scientific and technological expertise to promote practical problem solving for sustainable development, and Future Earth which is a major international research platform providing the knowledge and support to accelerate transformations to a sustainable world. In the business sphere the UN Global Compact and the Business Council on Sustainable Development drive sustainable development by creating sustainable business solutions.

## Dealing with interactions between SDGs

In UN (2016) it describes the SDGs as being "integrated and indivisible". Implicit in this is that the goals depend upon each other and these dependencies need to be taken into account when implementing the SDGs. Failure to do so could result in perverse outcomes where action to address one goal leads to negative outcomes for another goal. For example, using coal to improve energy access (goal 7) would accelerate climate change (goal 13) and acidify the oceans (goal 14). Alternatively, synergies could be missed where, if action to address one goal was undertaken in a certain way, it could also lead to positive outcomes for another goal.

Another complicating factor is that the interactions between goals and targets take many different forms. One framework that is used to classify these interactions is that of Nilsson et al., 2016, in which interactions are rated on a seven-point scale.

**Table 6.30.1** The interactions among the SDGs: an analytical framework

**GOALS SCORING**

The influence of one Sustainable Development Goal or target on another can be summarized with this simple scale.

| *Interaction* | *Name* | *Explanation* | *Example* |
|---|---|---|---|
| +3 | Indivisible | Inextricably linked to the achievement of another goal. | Ending all forms of discrimination against women and girls is indivisible from ensuring women's full and effective participation and equal opportunities for leadership. |
| +2 | Reinforcing | Aids the achievement of another goal. | Providing access to electricity reinforces water-pumping and irrigation systems. Strengthening the capacity to adapt to climate-related hazards reduces losses caused by disasters. |
| +1 | Enabling | Creates conditions that further another goal. | Providing electricity access in rural homes enables education, because it makes it possible to do homework at night with electric lighting. |
| 0 | Consistent | No significant positive or negative interactions. | Ensuring education for all does not interact significantly with infrastructure development or conservation of ocean ecosystems. |
| –1 | Constraining | Limits options on another goal. | Improved water efficiency can constrain agricultural irrigation. Reducing climate change can constrain the options for energy access. |
| –2 | Counteracting | Clashes with another goal. | Boosting consumption for growth can counteract waste reduction and climate mitigation. |
| –3 | Cancelling | Makes it impossible to reach another goal. | Fully ensuring public transparency and democratic accountability cannot be combined with national-security goals. Full protection of natural reserves excludes public access for recreation. |

*Source*: Nilsson et al. (2016) with permission.

In addition, there are other considerations that need to be taken into account, such as, if the interaction is reversible, e.g. species loss is irreversible, does the interaction go in both directions, how important are the consequences of the interaction, and how certain or uncertain is the interaction? Context is also important, such as differences in geography, technology, timescale and governance.

## Will the SDGs be successful?

Having only come into force on 1 January 2016 it is too early to tell if the SDGs will be successful. There are some encouraging signs as some governments and businesses align behind the goals. But they are incredibly ambitious, so the reality is almost certain that they will not be fully implemented. However, if, as with the MDGs, they help to drive significant progress then that in itself will be enough to deem them a success.

## Learning resources

The following are useful web resources for understanding the SDGs and their realization in the years ahead:

### *UN*

- Official portal: http://www.un.org/sustainabledevelopment.
- Knowledge platform: https://sustainabledevelopment.un.org/index.php?menu=1300.
- Goal images: http://www.un.org/sustainabledevelopment/news/communications-material.

### *IISD*

- SDGs Knowledge Hub: http://sdg.iisd.org/

### *SDSN*

- Home page: http://unsdsn.org.
- SDGs Index & Dashboards: http://www.sdgindex.org.

### *Business*

- SDGs Compass: http://sdgcompass.org.
- IN Global Compact: https://www.unglobalcompact.org.

## Bibliography

Nilsson, M., Griggs, D. & Visbeck, M. (2016) Policy: map the interactions between Sustainable Development Goals, *Nature*, 534: 320–322.

UN (2016) A/RES/70/1 – Transforming our world: the 2030 Agenda for Sustainable Development. Online. Available at: https://sustainabledevelopment.un.org/post2015/transformingourworld.

# Part 7

## Key debates

# Part 7 Introduction

One of the signature features of environmental studies is dispute and debate. Perceptions of, and feelings about, the non-human world vary considerably. Even in science, where perceptions are guided by 'objective' methods and where feelings are supposedly irrelevant, there are disagreements about the 'facts' of environmental change and about the implications of these facts for policy action on phenomena like climate change. In the wider social sciences and the humanities, lively arguments persist about how far humans can 'control' the non-human world, about the proper goals of nature conservation, and about whether changing consumer behaviour is really the best response to the 'environmental crisis' of our time – among many other topics. The point of these debates, for some involved, is to quest for the 'right', or at least the 'best', answer. For others, by contrast, the debates are a symptom of a clash of perspectives that cannot be resolved by recourse to evidence or reason. Take the use of government regulations or markets as ways of managing environmental harms like the illegal killing of endangered species. Is one intrinsically better than the other? Advocates of 'market-based environmental policies' believe so and have argued their case for 40 years. Yet others point out that the 'markets versus regulation' debate is false since context matters to the effectiveness of either approach. In this part, readers are introduced to several ongoing debates in different areas of environmental studies. These debates reveal something of the vibrancy, but also continued contention and dispute, that characterizes environmental studies.

# Anthropocentrism

7.1

Eileen Crist

Before the word 'anthropocentrism' existed, poet William Blake wrote: 'Man has closed himself up, till he sees all things thro' narrow chinks of his cavern. If the doors of perception were cleansed everything would appear to man as it is, infinite' (1977: 258). That narrowness of view and related constriction of thought are what the concept of anthropocentrism primarily intends to convey. It is an inherently normative concept in aiming to expose the 'human-centered' approach to the natural world as profoundly limited—inherently wanting in a breadth of understanding that expresses gratitude, appreciation, kinship, embeddedness, and compassion within the more-than-human world.

Anthropocentrism has been associated with orthodox Christianity, and more generally with monotheistic religions, which conventionally elevated man above worldly nature, representing him in the image of a God who granted him dominion over Earthly life. This perspective is most indebted to historian Lynn White's classic essay 'The Historical Roots of our Ecologic Crisis', in which he submitted that Christianity is one of the most anthropocentric religions of the world and therefore bearing a huge responsibility for nature's destruction (1967). White's paper generated sustained controversy and has been countered by ecological thinkers of a Christian persuasion (Berry, 1981). His argument has been invaluable in inspiring new and more nuanced readings of the Bible and driving changes within faith communities. To mention an example of paramount significance, Pope Francis's recent renunciation of an anthropocentric reading of scripture is a direct response to White's thesis. The Pope's 2015 *Encyclical* inaugurated an official shift within Catholicism toward a more ecological standpoint and a non-anthropocentric Biblical hermeneutics.

While White's paper was a landmark on the topic of anthropocentrism, historically it did not reach far enough. Political scientist John Rodman traced the discursive roots of anthropocentrism to classical antiquity. Classical philosophical and political thought, he argued, riveted by an interest in the essence of 'the human', framed its inquiry through the question: 'How are humans *different* from all other life-forms?'. Rodman aptly termed this ubiquitous question of Western thought 'the Differential Imperative' (1987). The Differential Imperative has haunted Western discourses ever since antiquity, and there has been no shortage of proposed supposedly unique human traits (e.g., reason, logos, free will, morality, technology, personhood, and others). The detailed scholarship of philosopher Gary Steiner shows just how prevalent, enduring, and entrenched the obsession with the human difference from the nonhuman realm has been in the Western canon (Steiner, 2005).

The question of the human difference gave rise to the longest-lived cosmology of Western civilization: the Great Chain of Being which posited a hierarchy within which Man was positioned on the highest Earthly rung. This hierarchy was both an ontological scheme and a moral order, sanctioning human use of whatever was ostensibly beneath him—which, on the Earth plane, was everything (Thomas, 1983). Christianity did not invent the Great Chain of Being but was undoubtedly responsible for disseminating its narrative to the masses via uncountable pulpits over the course of many centuries.

The cosmology of the Great Chain of Being underwrote a disinclination to merge human and nonhuman narratives, to treat animals and plants as kin, to respect the self-integrity of wild places, and to develop traditions of reciprocity with the natural world—all quintessential features of indigenous cultures. The natural world became, to oversimplify with a justifiable generalization, a place to use. This blazed the way toward the establishment of Western instrumentalism and the eventual conceptualization of the world as 'natural resources'. The combination of cognizing the world as resources and acting upon it instrumentally defines the legacy of anthropocentrism, ongoing in our time.

Defining anthropocentrism in precise terms is anything but simple. Etymologically, anthropocentrism refers to the deluded placement of humans at the center with everything else existing to serve them, much like the geocentric perspective positioned Earth at the center as the most exalted celestial body of the universe (Crist, 2014). On the ground, anthropocentrism is often understood to mean that human interests always come first. But this is where defining the term gets tricky, because it is far from clear that anthropocentrism actually serves human interests. Thus, eco-psychologist David Kidner has cogently argued that 'anthropocentrism is not anthropocentric,' meaning that human beings are conditioned into, and victimized by, the very perspective that fuels nature's destruction as though it were nothing but an instrumental field populated by useful stuff (Kidner, 2014).

While the spatial metaphor of humans-at-the-center conveys the standard import of anthropocentrism, Blake was arguably on a better track when he compared its perspective to seeing through blinkers. After all, it is at least theoretically possible for humans to position themselves at a center while still taking in the panoramic beauty, majesty, and mystery of the cosmos—of the Earth included, since it is the cosmos humanity inhabits. Blake's 'seeing thro' narrow chinks' nicely nails the fact that human self-elevation undermines an inclination (or ability) to see grandeur elsewhere and in other beings.

The Romantics created the first Western intellectual tradition to decry the narrowness of anthropocentrism and its deadening of the natural world. For example, in his celebrated poem *Lines Written in Early Spring,* William Wordsworth sung out the numinous character of the more-than-human world, wherein 'Nature's holy plan' is redolent with joyful awareness. Juxtaposing that 'plan' to man's blindness to perceive and celebrate it, Wordsworth ended his sonnet with a pointed question: 'Have I not reason to lament what man has made of man?'

In North America, John Muir and Henry Thoreau (as well as Ralph Emerson to a certain extent) carried the Romantic sensibility forward, unpacking it in more systematic analyses and extending it to activism on nature's behalf. Invoking Christian imagery and language, Muir sought to couple entities long cleaved asunder in Western culture: unifying the natural world with the Creation and wilderness with Eden. His activism for preserving portions of wild nature virtually untouched by people gave rise to the now global institution of the National Park and other types of protected areas (1901). Thoreau was a more complex thinker. His magnum opus, *Walden*, defies facile description: while caustically anti-anthropocentric, it is simultaneously a paean to human achievement and potential (1991). Thoreau's tender, sensual, and detailed attention to the nonhuman world weaves the human standpoint in such a way as to reveal that immersion within, and appreciation for, the larger birthplace, context, and support of human life is what gives depth to, and actually elevates, the human being. Most of the real humans (of Concord, Massachusetts) who populate *Walden* emerge through Thoreau's eyes as cognitively and emotionally stunted: they are unable to place themselves, *to ground themselves*, within an existential plane that is infinitely greater than they are, and they remain unresponsive to a world that objectively commands awe.

A century after Thoreau's time, Arne Naess was the first philosopher to systematically dissect anthropocentrism. His approach echoed Blake and Thoreau in that he exposed the limitations of human identity and being-in-the-world that the anthropocentric perspective instills. Naess argued that we underestimate who we are when we identify with the separate sense of self—otherwise known as the ego (1995). In Naess's ecosophy, anthropocentrism is essentially the human ego writ large. 'Human nature is such,' he maintained, that 'with sufficient comprehensive (all sided) maturity we cannot help but "identify" our self with all living beings' (Naess, 1995: 225). The ecological self is able to see himself and herself reflected in other people, animals, and plants, as well as in other kinds of living beings such as rivers, forests, and mountains. The identification of the self with the broader world fosters what Naess called (after Immanuel Kant) 'beautiful acts' in the world: meaning, we do the right thing not because we 'ought to', or because we are driven by 'enlightened self-interest', but because we see our own selves in all beings and places. The evolution from ego to ecological self was a process of enrichment for Naess. Like Thoreau, but more tactfully, Naess pointed out how anthropocentrism diminishes human identity and ultimately undermines human life.

After the birth of the environmental movement, and in the wake of Naess's deep ecological writings (as well as White's seminal essay), anthropocentrism became a central topic of investigation in environmental philosophy in the 1980s and 1990s (Fox, 1989; Katz et al., 2000). Following Naess's lead, deep ecologists argued that anthropocentrism constitutes the deepest causal layer of the ecological crisis (Devall & Sessions, 1985; Sessions, 1995). This claim generated controversy, and intellectual sparring soon ensued between proponents of deep ecology, social ecology, and ecofeminism. Social ecologists argued for the primacy of human–human inequality and noted the paramount role of capitalism in environmental destruction (Bookchin, 1996). Ecofeminists emphasized how patriarchical oppression subjugated women, disempowered others, and 'feminized' nature (Warren, 1990).

Differences are bound to persist with respect to how diverse thinkers diagnose ecological and social maladies. It remains all the more important, therefore, to focus on the common threads between different forms of oppression. For example, 'resourcism' (a view of nature affiliated with anthropocentrism) tends to promote social inequities, for when the world is reduced to a resource-base in the service of amassing power and wealth, exploitation of fellow humans easily follows (Foreman, 2007; Crist, 2014). What's more, as Matthew Calarco has elaborated, anthropocentrism has never been a hierarchical schema for categorizing humans and nonhumans alone. (Anthropocentrism is thus different from speciesism.) Rather, anthropocentrism has elevated certain humans as representing the highest expression of 'the human' (predictably white educated males), over purportedly inferior others (predictably nonwhites, women, children, laborers, and so-called savages). Inequality is comprehensively built into the anthropocentric worldview (Calarco, 2014).

While the idea of anthropocentrism has served a valuable purpose in environmental thought and activism, it may be that the term has come to outlive its usefulness. Shortcomings of the term include the dubious metaphor of 'centeredness' (rather than narrowness), unclear meaning (who, if anyone, is actually benefitted by it?), and an overly circumscribed affiliation with academia. In the quest for new conceptual venues to express the core normative insights of anthropocentrism, certain authors are using the idea of 'human supremacy', which has the advantage of being immediately obvious in meaning and critical valence (Crist, 2012; Jensen, 2016). Additionally, an important recent development carrying critiques of anthropocentrism in promising directions has been the emergence of the environmental humanities. Pioneers within this field are exploring the vast terrain of alliances that exist and can be formed between human and nonhuman domains and endeavoring to articulate the hopeful possibility of creating a world of multispecies flourishing (Rose, 2011; van Dooren, 2014; Haraway, 2016).

## Learning resources

Three journals are especially focused on publishing anti-anthropocentric (or ecocentric) writings and promoting multispecies flourishing:

*The Ecological Citizen:* http://www.ecologicalcitizen.net.
*The Trumpeter: Journal of Ecosophy:* http://trumpeter.athabascau.ca/index.php/trumpet.
*Environmental Humanities:* http://environmentalhumanities.org.

## Bibliography

Berry, W. (1981) *The Gift of the Good Land*. New York: North Point Press.

Blake, W. (1977) *The Portable Blake*. London: Penguin Classics.

Bookchin, M. (1996) *The Philosophy of Social Ecology: Essays on Dialectical Naturalism*. Montreal: Black Rose Books.

Calarco, M. (2014) "Being toward meat: anthropocentrism, indistinction, and veganism", *Dialectical Anthropology*, 38(4): 415–429.

Crist, E. (2012) "Abundant Earth and the population question". In P. Cafaro & E. Crist (eds) *Life on the Brink: Environmentalists Confront Overpopulation* (pp. 141–153). Athens: University of Georgia Press.

Crist, E. (2014) "Ptolemaic environmentalism". In G. Wuerthner, E. Crist & T. Butler (eds) *Keeping the Wild: Against the Domestication of Earth* (pp. 16–30). Washington DC: Island Press.

Devall, B. & Sessions, G. (1985) *Deep Ecology: Living as if Nature Mattered*. Layton, Utah: Gibbs Smith.

Foreman, D. (2007) "The arrogance of resourcism", *Around the Campfire* Issue 5. Albuquerque: Rewilding Institute.

Fox, W. (1989) "The deep ecology-ecofeminism debate and its parallels", *Environmental Ethics*, 11(1): 5–25.

Haraway, D. (2016) *Staying with the Trouble: Making Kin in the Cthulucene*. Durham: Duke University Press.

Jensen, D. (2016) *The Myth of Human Supremacy*. New York: Seven Stories.

Katz, E., Light, A. & Rothenberg, D. (eds) (2000) *Beneath the Surface: Critical Essays in the Philosophy of Deep Ecology*. Cambridge, MA: MIT Press.

Kidner, D. (2014) "Why 'Anthropocentrism' is not Anthropocentric", *Dialectical Anthropology*, 38(4): 465–480.

Muir, J. (1901) *Our National Parks*. San Francisco: Sierra Club Books.

Naess, A. (1995) "Self-realization: an ecological approach to being in the world". In G. Sessions (ed.) *Deep Ecology for the 21st Century*. Boston: Shambhala.

Pope Francis (2015) *Encyclical on Climate Change and Inequality: On Care of Our Common Home*. New York: Melville House.

Rodman, J. (1987) "Paradigm change in political science: an ecological perspective", *American Behavioral Scientist*, 24(1): 49–78.

Rose, D. (2011) *Wild Dog Dreaming: Love and Extinction*. Charlottesville: University of Virginia Press.

Sessions, G. (ed.) (1995) *Deep Ecology for the 21st Century*. Boston: Shambhala.

Steiner, G. (2005) *Anthropocentrism and its Discontents: The Moral Status of Animals in the History of Western Philosophy*. Pittsburgh: University of Pittsburgh.

Thomas, K. (1983) *Man and the Natural World: Changing Attitudes in England 1500–1800*. Oxford: Oxford University Press.

Thoreau, H. (1991) *Walden, or, Life in the Woods*. New York: Vintage Books.

van Dooren, T. (2014) *Flight Ways: Life and Loss at the Edge of Extinction*. New York: Columbia University Press.

Warren, K.J. (1990) "The power and the promise of ecological feminism", *Environmental Ethics*, 12(2): 125–146.

White, L. (1967) "The historical roots of our ecologic crisis", *Science*, 155(3767): 1203–1207.

# 7.2 Biology and culture

Maurizio Meloni

The distinction between biology and culture became a twentieth-century consensus view as a result of an immensely influential article by anthropologist Alfred Kroeber in 1917. Titled 'The Superorganic', the contribution appeared in the journal *American Anthropologist* and in an elegant way established what still today appears, if not the only, certainly the most obvious way to conceive the difference between biology and culture: one transmitted by blood, operating at the organic level, the other belonging to a transcendental realm of symbols and information, culture or 'civilization' as Kroeber called it.

Take, for instance, Kroeber's argument about flying: birds and humans can both 'fly', however they do so in radically contrasting ways. The point of contrast is not only that for birds flying is the result of a somatic feature (having wings) while for humans it depends on an extra-somatic arrangement (airplane technology as a civilizational resource). More importantly and crucially, the contrast is between the radically different ways in which this know-how is transmitted among birds and humans. For birds it is written, so to speak, in blood. It is the result of hereditary forces (i.e. biological and instinctive), whereas for humans it is a learned acquisition independent of blood relationships: 'People who have not the slightest blood kinship to the first designers of airplanes can fly and are flying today', Kroeber points out. Not only that, social learning can go in the reverse genealogical direction, so that any invention of a child (for instance an airplane) can be 'used, enjoyed and profited [from]' by a parent (1917: 166).

Take a second scenario in Kroeber's text: 'a French baby, born in France of French parents, themselves descended for numerous generations from French-speaking ancestors' via a 'mute nurse', is sent just after birth to the interior of China and there adopted by a Chinese couple. If we meet this baby again in 3, 10 or 30 years, Kroeber says, it is obvious to expect that this person will speak 'not a word of French; and absolutely pure

Chinese, without a trace of accent and with Chinese fluency'. As in the case of flying, a radical contrast is established between organic and superorganic traits, this time with regard to their degree of changeability. After 3 or 30 years, regardless of the 'amount of association with Chinese' people, we certainly can't expect to find this person's nose or eyes or hair to have adopted a Chinese outlook. For language, however, it is exactly the amount of association (that is, learning) that makes this individual a perfectly fluent Chinese speaker. Building on these and other intuitively convincing examples, Kroeber established his notion of the superorganic as a boundary-generating machine. His key goal is to neatly separate what was still confused in the mind of many fellow anthropologists, sociologists and even scientists: namely, organic from civilizational causes, biology from culture, the somatic from the extra-somatic.

Kroeber's transcendent view of culture was neither unchallenged among fellow anthropologists, nor radically unprecedented, given how similar his approach was to the Durkheimian view (preceding Kroeber by nearly 20 years) of a radically discontinuous relationship between biological and social facts. These two great traditions, the Kroeberian and the Durkheimian, undoubtedly set the stage for the debate between biology and society/culture that took place over the 'long twentieth-century of the gene' (Keller, 2000; Meloni, 2016a). The reference to *genetics* is not casual here, nor is the statement that the century of the gene was *a long one*. These are both important points to consider. Firstly, the radically dichotomizing move performed by both Kroeber and Durkheim was closely related to the triumph of genetics. More accurately, it was strictly associated with the 'hard' view of biological heredity that denied (against Lamarckian views) the porosity of the hereditary substance to sociocultural influences. Heredity was conceived as impermeable to environmental influences: this view could not be entirely attributed to Darwin (who was a bit shaky on the matter) but to Darwin's half-cousin Francis Galton (1875), and more crucially to German zoologist August Weismann (1892). It was upon this late nineteenth-century assumption that genetics was made possible a decade later. It is often unnoticed or neglected among mainstream histories of sociology and anthropology that both Durkheim and Kroeber, the proclaimed masters of the independence of the sociocultural, in fact 'purified' their disciplines by relying heavily on this hard-hereditarian view of biology that was dominant at the time. Durkheim had made use of Weismann already in the *Division of Labor in Society* (1893), just one year after the publication of Weismann's major compendium *The Germplasm* (1892), a fact made all the more significant considering the bitter French-German scientific rivalry of the time and the largely Lamarckian context in which Durkheim operated. Durkheim found in Weismann exactly what he needed: his radical decoupling of individual behaviour from social facts mirrored perfectly Weismann's severing of individual lifetime modifications from race heredity. Both social facts and the germplasm now transcended individual acquisitions (Meloni, 2016b).

Kroeber was much more explicit about borrowing from Galton, Weismann and genetics in order to construct culture as a transcendent force operating above the level of individuals and outside of their bodies. 'The Superorganic' is actually the culmination of a series of minor texts in which Kroeber came to terms with the Mendelian revolution as the basis for a new, independent social science (1916a, 1916b). What Weismann and Galton have given us, Kroeber claims, is an utter separation between biological heredity and social inheritance: the first occurring within the body, the latter out of it, and with no way to influence the first (against Lamarckian confusions). Kroeber claimed that the Weismannian phase of biology is its mature phase, the one in which 'organic phenomena must be interpreted solely by organic processes' (1916a:

369); he later added in his 1917 text that superorganic phenomena can be explained only superorganically.

If Kroeber and Durkheim both decided to break with this confused way of thinking which mixed the organic and the socio-cultural one might ask what the legacy of this gesture is today. The two dichotomizing claims made by Kroeber in his Superorganic piece are not wrong by themselves, but they would be certainly seen in a much more problematic light today. For instance, there is a growing awareness of the existence of cultural traditions in non-human animals that are profoundly problematic for Kroeber's sharp separation of blood and culture (Avital & Jablonka, 2000). Also regarding his second claim, things look more confusing. While it is obviously true that the baby sent to China would still not look Chinese, microbiologists and epigeneticists would easily show how something crucial within her epigenome or gut microbiome has changed: for instance, new enzymes and bacteria and even genes (via horizontal gene transfer) may have been acquired from a different diet (Hehemann et al., 2010); or novel epigenomic markings may have been established because of the exposure to different toxins, air-quality, etc. Some of these acquired changes may even be transferrable to the next generation. Actually, one of the ironies of Kroeber's text is that just while he was arguing that only extra-organic features are locally-dependent and changeable, Franz Boas had just recognized that environmental influences could alter even the supposedly unchangeable cephalic index in the children of new immigrants to the US (1910). This Boasian line is certainly more in tune with contemporary findings that are made up of a novel constellation of new studies and theorizations that problematize any neat biological/cultural distinction (Lock, 2015). These findings use frameworks like the 'biosocial' or the 'biocultural' to challenge an implausible partitioning between what is biologically given and what culturally acquired in a human body (Ingold & Palsson, 2013; Frost, 2016). Furthermore, they aim at a double destabilization: both toward the biological, now increasingly considered to be socially shaped from the very beginning (gene functioning, see Landecker & Panofsky, 2013); but also toward the cultural, by 'refusing the idea that culture is something that humans do or create by virtue of transcending, overcoming, or exceeding the stodgy, staid limitations of their fleshy incarnation' (Frost, 2016). Notions like inscription, embodiment, embedment, nature–culture, and entanglement, are part of this emerging conceptual repertoire that destabilizes twentieth-century dichotomies of biological versus social causation. It is not by chance that all this is occurring at a time of intense crisis for the Weismannian idea of a lack of communication between the hereditary substance and its socio-environmental determinants: this crisis is punctuated with overt claims of a return to 'soft heredity', though now dressed in a much more sophisticated molecular framework (Gissis & Jablonka, 2011).

## Learning resources

Two great introductions to the mutated understanding of heredity (compared to what was available in Kroeber's time) can be found in:

1. BBC, 'The Ghost in Your Genes', 2006 (English): http://www.dailymotion.com/video/x1f27b5_bbc-the-ghost-in-your-genes-2006xvid_tech.
2. Les nouveaux secrets de notre hérédité ARTE #Documentaire, 2016 (French): https://www.youtube.com/watch?v=5XPcUdNH5Ho.

# Bibliography

Avital, E. & Jablonka, E. (2000) *Animal Traditions: Behavioural Inheritance in Evolution*. Cambridge: Cambridge University Press.

Boas, F. (1910) *Changes in Bodily Form of Descendants of Immigrants*. Washington, DC: Government Printing Office.

Durkheim, E. (1893/1997) *The Division of Labor in Society*, W.D. Halls (trans.). New York: Free Press.

Frost, S. (2016) *Biocultural Creatures: Toward a New Theory of the Human*. Durham, NC: Duke University Press.

Gissis, S. & Jablonka, E. (2011) *Transformations of Lamarckism: From Subtle Fluids to Molecular Biology*. Cambridge, MA: MIT Press.

Hehemann, J.H., Correc, G., Barbeyron, T., Helbert, W., Czjzek, M. & Michel, G. (2010) 'Transfer of carbohydrate-active enzymes from marine bacteria to Japanese gut microbiota', *Nature*, 464: 908–912.

Ingold, T. & Palsson, G. (eds) (2013) *Biosocial Becomings: Integrating Social and Biological Anthropology*. Cambridge: Cambridge University Press.

Keller, E. (2000). *The Century of the Gene*. Cambridge, MA: Harvard University Press.

Kroeber, A. (1916a) 'The cause of the belief in use inheritance', *The American Naturalist*, 50: 367–370.

Kroeber, A. (1916b) 'Inheritance by magic', *American Anthropologist*, 18: 19–40.

Kroeber, A. (1917) 'The superorganic', *American Anthropologist*, 19: 163–213.

Landecker, H. & Panofsky, A. (2013) 'From social structure to gene regulation, and back: a critical introduction to environmental epigenetics for sociology', *Annual Review of Sociology*, 39: 18–25.

Lock, M. (2015) 'Comprehending the body in the era of the epigenome', *Current Anthropology*, 56(2): 151–177.

Meloni, M. (2016a) *Political Biology: Science and Social Values in Human Heredity from Eugenics to Epigenetics*. London/New York: Palgrave Macmillan.

Meloni, M. (2016b) 'The transcendence of the social: Durkheim, Weismann and the purification of sociology', *Frontiers in Sociology*, 1(11): 1–13.

# 7.3 Environmental science and politics

Tim Forsyth

Environmental science is authoritative knowledge about biophysical processes, events and entities that underpins environmental concern and policy. Science is usually defined as information and explanations that have been made in rigorous and unbiased ways. It is usually portrayed as separate from social and political influences: science is supposedly insulated from the hurly burly of public debate and political decision making. Debates within social science, however, argue that scientific information nonetheless reflects social values and is always political. Consequently, there is a need to make science more transparent and openly governed.

The most common definition of science is information generated using the scientific method, a method founded on the philosophy known as positivism. Positivist science is a process of testing hypotheses through experimentation that ensures that samples are selected without bias, and that scientists do not influence results. Positivists assume that the world is knowable objectively so long as great care is taken to measure and interpret it. It is worth noting, however, that scientific methods have changed over time. The earliest positivists looked for generalizations inside datasets. This practice was replaced in the early twentieth century by seeking to verify generalizations in different datasets. Verification was then replaced by a new process of identifying hypotheses and conducting experiments to falsify them. This last style of science was the most ambitious because it treated hypotheses as accurate theories about reality until they could be falsified and replaced by newer theories.

Social analysts of science, however, have criticized positivist approaches to science. In particular, the field known as Sociology of Scientific Knowledge (SSK) has argued that theories and datasets always reflect the values or the problems of the societies or scientific networks that create them. This is a problem for environmental policy because it means that scientific knowledge is treated as though it is universally relevant and detached from society, but in fact is based only on selected, and often historic, values. Accordingly, it is important to ask: whose values, or which political concerns, started the search for scientific explanations? And what social conditions need to be reproduced in order for a scientific generalization to appear true in different times and places?

Analysts within the so-called Strong Program of SSK argued that scientific truth claims exist with social structures or networks that uphold those beliefs because they 'dovetail' with those structures or networks. Others, such as Bruno Latour, argued that the work of the biologist Louis Pasteur was considered true partly because France created laboratories and official tests in order to reproduce appropriate conditions to generalize Pasteur's discoveries. Latour used the term 'inscription' to refer to the specific process by which information about a phenomenon was collected, analyzed, and then presented as a universal scientific truth (Latour, 1993). Making science more transparent, therefore, requires looking at when, how, and by whom scientific truths were made. Over time, these ideas led to the development of Actor-Network Theory (ANT), which seeks to show how scientific explanations co-exist with social networks involving scientists, people and biophysical objects. The objective of ANT is to show how scientific explanations and the ordering of social and natural things co-constitute each other. Accordingly, ANT argues that distinctions between nature and society are blurred because they are based on distinct, and social, networks. Moreover, scientific progress through experimentation or positivism is unlikely to succeed because scientific explanations both reflect, and are upheld by, broader social orderings of society and nature (Braun & Whatmore, 2010).

A further question asks why do different societies frame, or perceive, environmental problems in different ways? The field known as Cultural Theory (with a capital C and T) followed the work of Mary Douglas (1921–2007) and argued that beliefs about the fragility or resilience of nature cannot be separated from deep-set cultural beliefs about how individuals and societies work together. Consequently, scientific facts and explanations both reflected and helped to maintain these worldviews. For example, Cultural Theorists showed that projections of deforestation in the Himalayas varied by a factor of 67 because of the different perspectives and objectives of expert organizations and scientists generating knowledge (Thompson et al., 1990). The implication is that it is pointless asking, 'what are the facts' without also asking what different people or organizations would like the facts to be.

Other analysts, instead, have used the ideas of Michel Foucault (1926–1984) to demonstrate how scientific knowledge has grown over time within specific contexts. In particular, these analysts use the terms environmental narratives or storylines to show how commonly-heard explanations of environmental problems contain causes, effects, and blame that are simplistic and politically convenient (Hajer, 1995). For example, researchers in Africa have argued that many debates about deforestation and desertification are narratives because they allocate blame onto smallholders in simplistic ways that also justify certain solutions favored by different states (Forsyth, 2003). Narratives are problematic because policymakers and activists often use them to support political objectives, even if scientific research questions their accuracy. Indeed, some analysts have suggested the word desertification is itself a problem because it frames environmental science and policy around biophysical changes rather than enhancing social development in the face of complex change (Davis, 2016).

The field known as Science and Technology Studies (STS) advances on these studies by showing how scientific practices and expert organizations rationalize authoritative knowledge, and by offering insights for governing science. A key concern is to identify which values or assumptions are left unchallenged, and which social groups might be excluded from framing environmental science and policy. STS scholars use the term boundary work to refer to the mechanisms by which scientists or policymakers reduce public debate about some truth claims, and instead present these as unchallengeable. Another important concept is co-production, which refers to the mutual constitution of knowledge with visions of social order. Together, these terms indicate how political and epistemic authorities arise together, such as in expert organizations, and to indicate which alternative forms of knowledge and social participation might be possible (Jasanoff, 2004).

The case of climate change highlights these concerns. Many climate scientists wish to fight back against 'climate change deniers' by demonstrating how science is authoritative and free from social influence. Social analysts of science do not deny the existence of anthropogenic climate change, but seek ways to make information about it more diverse, democratic, and accountable to different stakeholders. There are different visions of how to live with climate change (Hulme, 2009). Yet information about climate change is governed by one large, expert organization, the Intergovernmental Panel on Climate Change (IPCC), which has developed its own style of epistemic and political authority. For example, the IPCC defines the risks associated with climate change as linked to each additional unit of atmospheric greenhouse gas concentrations, rather than the social and economic conditions that create vulnerability to those gases (Ribot, 2010). In addition, the IPCC has adopted a style of communication—of speaking with one voice, in a way that fails to acknowledge diverse experiences or values. Critics suggest that these characteristics mean that the IPCC might be mistrusted as a source of information because it lacks meaningful engagement with how climate change raises problems for people, and places too much emphasis on measurements of global temperatures as an epistemic justification for addressing climate change (Beck et al., 2014). Alternative forms of scientific governance might include greater reliance on citizen science—or involving local people in the generation and framing of knowledge about risks, rather than relying on atmospheric models alone.

The study of environmental science and politics does not imply that environmental problems do not exist, or that scientific information is unhelpful. Instead, it aims to make scientific explanations and proposed solutions more diverse and effective by analyzing how, and for what purpose, scientific statements were made.

## Learning resources

The Centre for Science and Environment (http://www.cseindia.org) in Delhi, India, is an example of a think tank that produces critical information about environmental science and what this means for policy in developing countries.

Chilvers, J. & Kearnes, M. (eds) (2015) *Remaking Participation: Science, Environment and Emergent Publics*. Abingdon: Routledge. This is a useful review of how debates about social science and knowledge are relevant to environmental science.

Felt, U., Fouché, R., Miller, C. & Smith-Doerr, L. (eds) (2017) *The Handbook of Science and Technology Studies*, 4th Edition. Cambridge MA: MIT Press. This is an overview of Science and Technology

Studies (STS) that refers to various disciplines and policy debates. In particular, this chapter is a useful introduction of how STS can be applied to environmental politics: Beck, S., Forsyth, T., Kohler, P., Lahsen, M. & Mahony, M. (2017) 'The making of global environmental science and politics'.

Keller, A.C. (2009) *Science in Environmental Policy: The Politics of Objective Advice*. Cambridge, MA: MIT Press. This is a useful introductory textbook that considers scientific information from a political science perspective.

## Bibliography

Beck, S., Borie, M., Chilvers, J., Esguerra, A., Heubach, K., Hulme, M. & Görg, C. (2014) 'Towards a reflexive turn in the governance of global environmental expertise: the cases of the IPCC and the IPBES', *Gaia*, 23(2): 80–87.

Braun, B. & Whatmore, S. (2010) *Political Matter: Technoscience, Democracy, and Public Life*. Minneapolis: University of Minnesota Press.

Davis, D.K. (2016) *The Arid Lands: History, Power, Knowledge*. Cambridge, MA: MIT Press.

Forsyth, T. (2003) *Critical Political Ecology: The Politics of Environmental Science*. London: Routledge.

Hajer, M.A. (1995) *The Politics of Environmental Discourse: Ecological Modernization and the Policy Process*. Oxford: Clarendon Press.

Hulme, M. (2009) *Why we Disagree about Climate Change: Understanding Controversy, Inaction and Opportunity*. Cambridge: Cambridge University Press.

Jasanoff, S. (2004) *States of Knowledge: The Co-production of Science and Social Order*. London: Routledge.

Latour, B. (1993) *We Have Never Been Modern*. New York: Harvester Wheatsheaf.

Ribot, J. (2010) 'Vulnerability does not just come from the sky: framing grounded pro-poor cross-scale climate policy'. In R. Mearns & A. Norton (eds) *Social Dimensions of Climate Change: Equity and Vulnerability in a Warming World* (pp. 47–74). Washington, DC: World Bank.

Thompson, M., Ellis, R. & Wildavsky, A.B. (1990) *Cultural Theory*. Boulder: Westview Press.

# 7.4 Environmental behaviour change

Stewart Barr

## The behaviour change debate

It's likely that nearly everyone reading this chapter will have experienced a publicity campaign designed to persuade them to change their habits (see Figure 7.4.1 for an example). Whether it's eating more healthily, taking more exercise, avoiding driving whilst under the influence of alcohol or recycling more household waste, we all regularly encounter campaigns to encourage us to alter our everyday behaviours. Environmental behaviour change has become a particularly popular tool for policy makers throughout the developed and, more recently, the developing world, not least because it supports a shift towards encouraging individual responsibility and a reduced reliance on regulations and laws to enforce change. Indeed, there has been considerable academic and policy-funded research into the factors that influence individual behavioural choices, whether these are related to water use, energy consumption, travel habits or sustainable consumption behaviours (Jackson, 2005). However, there is considerable debate in the social science community about two key aspects of environmental behaviour change (Dobson, 2010). First, researchers disagree about whether promoting individual behaviour change is an appropriate tool for managing the environment and dealing with grand challenges such as anthropogenic climate change. Second, there is much debate about both the theoretical and methodological underpinnings of behaviour change research, in particular whether scholars should focus their attention on individual behaviours or, alternatively, what are termed wider 'social practices'. This chapter will elaborate on each of these issues in turn and will conclude by outlining some of the most recent developments in the 'behaviour-practice' debate.

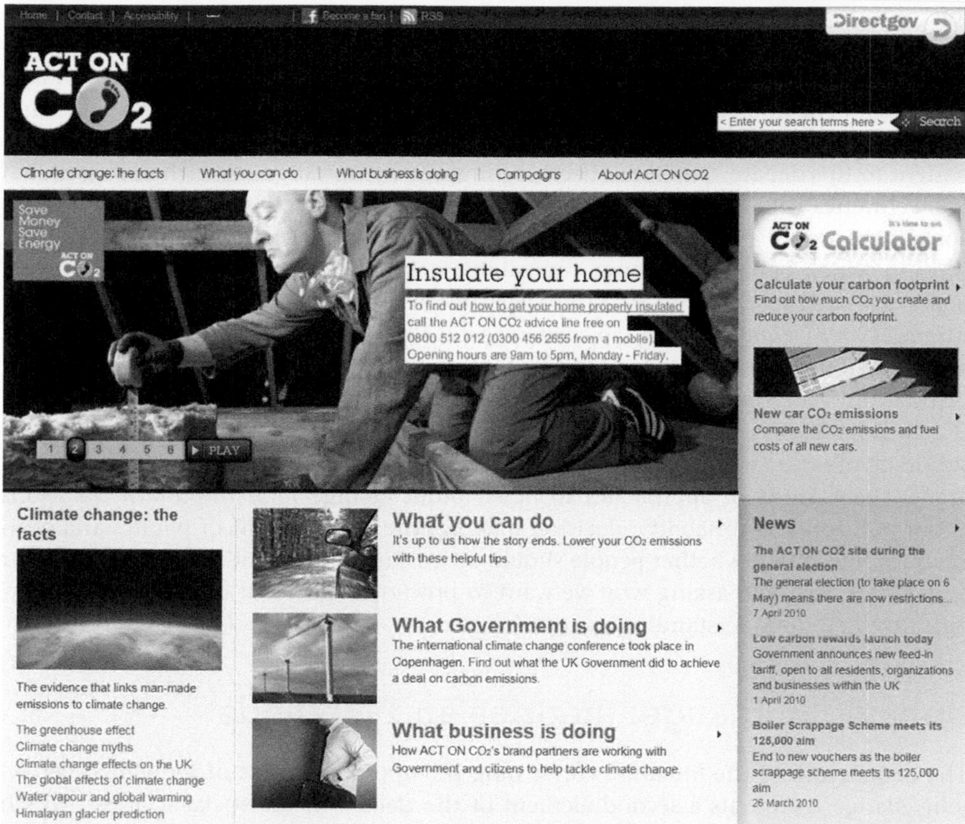


**Figure 7.4.1** The UK government's Act on $CO_2$ campaign from 2010.

*Source*: Act on $CO_2$ Website: http://webarchive.nationalarchives.gov.uk/20101007164856/http://actonco2.direct.gov.uk/home.html. Image covered by Crown copyright.

## Why behaviour change?

Perhaps the most important question about environmental behaviour change is why it is apparently such an assumed orthodoxy amongst the majority of policy makers in both developed and developing world nations (Jones et al., 2013). The corpus of literature that has been built up since the 1970s on environmental behaviour change (Stern, 2000) has largely avoided questions of whether encouraging individuals to change their behaviours is either a worthwhile or indeed morally and politically appropriate strategy for tackling resource scarcity, environmental degradation and anthropogenic climate change. Yet scholars from political science, sociology and human geography have progressively challenged the assumption that focusing on behaviour change is an appropriate strategy (Jones et al., 2013). This challenge has two main components. First, political scientists (see Clarke et al., 2007) argue that promoting individual behaviour change is part of a wider political project to drive down responsibility for social and environmental issues to individuals, as nation states continue to uphold the role of individual choice and aspire to 'less government' as the state 'rolls back' (Jessop, 2002). Commonly referred to as neo-liberalisation, this process is about creating a market-driven economy, where consumer choice is key. Accordingly, these scholars argue that behaviour change is a manifestation of neo-liberal approaches to governing nations and promoting choice amongst free consumers. Yet there is also a second component to this debate, which has been highlighted by investigators such as Barr (2014) and Whitehead et al. (2011). They argue that although the nation state has been rolled back, there are still ways in which the state exercises influence, and this is through the ways in which behaviour change campaigns seek to govern the choices that people make. For example, in the UK, the Department for Environment, Food and Rural Affairs' (2008) *Framework for Pro-environmental Behaviours* very specifically identifies key behaviours as a priority for change. More recently the UK Government has taken inspiration from the highly influential book *Nudge* (Thaler & Sunstein, 2008), in which the authors discuss how consumers can be influenced to change their behaviours by subtle changes in what they refer to as 'choice architectures'. As Jones et al. (2013) argue, these kinds of approaches to behavioural change raise important questions about political accountability, decision making, the involvement of publics and ethical questions about whether people should be un-knowingly 'nudged' to change their behaviours. As such, asking why we want to promote behaviour change is an important first step in understanding its context.

## Behaviour change: theories and methods

The disjuncture in the literature concerning the appropriateness of individual behaviour change represents a second element of the debate in the social sciences, which relates to the epistemological and methodological basis for behaviour change research. Sociologist Elisabeth Shove (2010) has sought to crystalise this debate through a provocative paper on what she views as two perspectives on the ways social scientists theorise and practice behaviour change research. In essence, these two major traditions relate to, first, a psychological understanding of behaviour change and, second, a sociologically-informed perspective that rejects the focus on individual behaviour, but rather advocates a focus on wider social practices.

To begin with what has undoubtedly been the dominant field in environmental behaviour change research, those adopting a psychological approach have been concerned with utilising a range of social-psychological frameworks to understand the influences of various factors on specific individual behaviours. Commonly applied frameworks have been Fishbein and Ajzen's (1975) Theory of Reasoned Action (TRA) and Ajzen's (1991) Theory of Planned Behaviour (TPB), both of which explore how behavioural intensions, social norms, attitudes and, in the case of the TPB, perceived behavioural control influence reported behaviours (De Groot & Steg, 2007). There has also been a considerable amount of research that has sought to correlate and statistically model psychological constructs (such as attitudes, perceptions and intensions) with reported individual behaviours (see Stern, 2000). Such research is therefore theory-driven and utilises existing psychological understandings of behaviour to explore environmental behaviours. As Owens (2000) has noted, this research is also characterised by its focus on individual cognition (how people process information), and makes certain assumptions about the linear basis for decision making;

most importantly that if a particular factor influencing behaviour can be manipulated, changes should occur in response to such interventions. As such, methodologically this kind of behavioural research is driven by a traditional scientific method, with a major focus on quantitative methods (such as large social surveys), analysed statistically to produce models of behaviour that should be scientifically repeatable and replicable (Spaargaren & Mol, 2008).

In contrast to the psychological approach to understanding behaviour change, many of those from a sociological and human geography background have progressively questioned both the theoretical and methodological assumptions made by those using a psychological approach (Shove, 2010). Instead, these researchers argue that the primary focus of researchers should not be on behaviour, but what they term 'social practice' (Reckwitz, 2002). Social practices are routines manifested in everyday life and which can be traced across individuals and connected to underlying social and economic structures (Shove et al., 2012). A good example would be to consider car driving to and from work. From a psychological perspective, this might be considered an issue of an individual's decision to drive a car and not to use other modes of transport; from a social practice perspective, this would be referred to as the practice of daily commuting by car that has become an expectation or even desired outcome for millions of people who live in auto-based suburbs and who live far away from work. As such, social practices are linked much more to infrastructures of living around us. The sociological approach therefore leads us to examine how people behave in a particular social context, and therefore the unit of analysis is not the individual, but social groupings. Indeed, methodologically there is the assertion that intensive qualitative methods (such as ethnographic methods, observation and interviews) are more likely to reveal the complexity of how people engage in environmentally significant practices, given the important link to underlying social contexts (Spaargaren & Mol, 2008).

## Conclusion: bridging the gap in behaviour change research

The debate amongst the social science community on environmental behaviour change has recently been re-energised by interventions from various scholars (Shove, 2010; 2011; Whitmarsh et al., 2011; Wilson & Chatterton, 2011) who have debated whether or not there is the potential for utilising the knowledge from both the psychological and sociological perspectives with a view to facilitating policy making and practical action. Wilson and Chatterton (2011) have argued that there are pragmatic benefits to be gained from appreciating the role of social practices with a structured framework that can connect to certain policy goals for behaviour change, which recognise the role of underlying structural influences. However, these are first attempts at resolving what is a major area of contention in the social sciences. Indeed, it is worth reflecting on why this is so; the psychological and sociological perspectives on behaviour change are drawn from radically different epistemological and methodological starting points. There is, as yet, no agreement on how we can define 'environmental behaviour' (at the individual or practice scale), nor on the role of broader economic and political factors. In this way, the behaviour change debate is emblematic of wider debates in environmental studies concerning the role of theory, method and practice.

## Learning resources

There are some very useful examples of typical behaviour change campaigns, all of which attempt to engage individuals to encourage them to reduce specific kinds of consumption:

10:10: this website is designed to help people lead a lower carbon lifestyle: http://1010uk.org.

My 2050: This is an interactive way to explore the relationship between consumption behaviours and carbon emissions: http://my2050.decc.gov.uk.

Travelwise (Utah): This is a good example of a website dedicated to promoting low carbon travel and congestion reduction: http://travelwise.utah.gov.

Waterwise: This exemplifies ways that individuals can save resources in the home, focussing on water: http://www.waterwise.org.uk.

To find out more about the key debates on behaviour change and environment, the following are very accessible and succinct reviews:

Dobson, A. (2010) *Environmental Citizenship and Pro-environmental Behaviour: Rapid Research and Evidence Review*. London: Sustainable Development Research Network.

Available at: http://www.sd-research.org.uk/sites/default/files/publications/SDRN%20Environmental%20Citizenship%20and%20Pro-Environmental%20Full%20Report_0.pdf.

Jackson, T. (2005) *Motivating Sustainable Consumption*. London: Sustainable Development Research Network.

Available at: http://www.sustainablelifestyles.ac.uk/sites/default/files/motivating_sc_final.pdf.

## Bibliography

Ajzen, I. (1991) "The theory of planned behavior", *Organizational Behavior and Human Decision Processes*, 50: 179–211.

Barr, S. (2014) "Practicing the cultural green economy: where now for environmental social science?", *Geografiska Annaler: Series B, Human Geography*, 96: 231–243.

Clarke, J., Newman, J., Smith, N., Vidler, E. & Westmarland, L. (2007) *Creating Citizen-Consumers: Changing Publics and Changing Public Services*. London: Sage.

De Groot, J. & Steg, L. (2007) "General beliefs and the theory of planned behavior: the role of environmental concerns in the TPB", *Journal of Applied Social Psychology*, 37: 1817–1836.

Department for Environment, Food and Rural Affairs (DEFRA) (2008) *Framework for Pro-environmental Behaviours*. London: DEFRA.

Dobson, A. (2010) *Environmental Citizenship and Pro-environmental Behaviour: Rapid Research and Evidence Review*. London: Sustainable Development Research Network.

Fishbein, M. & Ajzen, I. (1975) *Belief, Attitude, Intention and Behavior: An Introduction to Theory and Research*. Reading, MA: Addison-Wesley.

Jackson, T. (2005) *Motivating Sustainable Consumption*. London: Sustainable Development Research Network.

Jessop, B. (2002) "Liberalism, neoliberalism, and urban governance: a state–theoretical perspective", *Antipode*, 34: 452–472.

Jones, R., Pykett, J. & Whitehead, M. (2013) *Changing Behaviours: On the Rise of the Psychological State*. Cheltenham: Edward Elgar.

Owens, S. (2000) "Engaging the public: information and deliberation in environmental policy", *Environment and Planning A*, 32: 1141–1148.

Reckwitz, A. (2002) "Toward a theory of social practices: a development in culturalist theorizing", *European Journal of Social Theory*, 5: 243–263.

Shove, E. (2010) "Beyond the ABC: climate change policy and theories of social change", *Environment and Planning A*, 42: 1273–1285.

Shove, E. (2011) "On the difference between chalk and cheese—a response to Whitmarsh et al's comments on 'Beyond the ABC: Climate change policy and theories of social change'", *Environment and Planning A*, 43: 262–264.

Shove, E., Pantzar, M. & Watson, M. (2012) *The Dynamics of Social Practice: Everyday Life and How it Changes*. London: Sage.

Spaargaren, G. & Mol, A.P.J. (2008) "Greening global consumption: redefining politics and authority", *Global Environmental Change*, 18: 350–359.

Stern, P. (2000) "New environmental theories: toward a coherent theory of environmentally significant behaviour", *Journal of Social Issues*, 56: 407–424.

Thaler, R.H. & Sunstein, C.R. (2008) *Nudge: Improving Decisions about Health, Wealth and Happiness*. New Haven, CT: Yale University Press.

Whitehead, M., Jones, R. & Pykett, J. (2011) "Governing irrationality, or a more than rational government? Reflections on the rescientisation of decision making in British public policy", *Environment and Planning A*, 43: 2819–2837.

Whitmarsh, L., O'Neill, S. & Lorenzoni, I. (2011) "Climate change or social change? Debate within, amongst, and beyond disciplines", *Environment and Planning A*, 43: 258–261.

Wilson, C. & Chatterton, T. (2011) "Multiple models to inform climate change policy: a pragmatic response to the 'beyond the "ABC" debate", *Environment and Planning A*, 43: 2781–2787.

# 7.5 Environmental citizenship

Bronwyn Hayward

In the last two decades, philosophers, citizens and activists alike have begun to talk about 'environmental citizenship' and the related idea of 'ecological citizenship'. This interest has been driven by questions about how humans should respond to increasingly complex and multi-scale (local, regional, national and global) environmental risks. For example, what does it mean to be a responsible citizen when the actions of people in one place or time have increasingly far-reaching environmental implications for distant others, including future generations and non-human nature (Dobson, 2003)?

Citizenship is traditionally defined as a legal *status* or everyday *practice* of members of a defined polity. By contrast, environmental citizenship argues that citizens are not only members of territorially defined communities like nation states, but are also members of interconnected ecosystems. Environmental citizenship claims that citizens have rights and responsibilities within this wider ecosystem. For example, environmental citizens have rights to equitable access to environmental goods, like clean drinking water and public parks, and have responsibilities to act in thoughtful, virtuous ways to address growing environmental problems (for example, by conserving water, eating less meat, lowering use of fossil fuels and consuming fewer resources).

As public awareness about complex environmental challenges like climate change and the loss of biodiversity has grown, there has been greater debate about how citizens should respond to these problems. Many of these debates raise complex moral questions about how we should live and what it means to be a 'good' citizen. Since the seventeenth century, liberalism has inspired a vision of citizenship as a legal status which protects the rights and freedoms of individuals who live within politically defined borders. However, growing concerns about environmental degradation has expanded liberal ideas of rights and injustices. For example, the 1972 UN Conference on Human Environments and the 1992 UN Earth Summit's publication of Agenda 21, both recognize access to good quality environments as an important citizen right (Christoff, 1996).

Civic republicanism has also inspired discussion about environmental citizenship, particularly the duties or responsibilities citizens have to participate in the life of their community and to contribute to collective self-government (see, for example, works by Aristotle, Machiavelli, or Rousseau). In the face of global environmental threats such as climate change, the ideas of civic republicanism have also inspired citizens to take responsibility to act collectively across time and space to reduce environmental degradation and risk in a variety of novel ways, from grass roots communitarian movements (like transition towns or local climate justice movements), to civic participation through new methods of decision making including mini-public assemblies, citizen juries, and digital deliberation (Barry, 1999; Barber & Bartlett, 2015).

But how far can environmental rights and responsibilities be accommodated within existing theories of liberal citizenship and civic republicanism?

## The challenges of environmental citizenship

Complex and multi-scale, environmental problems have highlighted five challenging questions for theories of citizenship:

1. *Political agency*. How should citizens think and act to enhance environmental security and promote ecological well-being?
2. *Justice*. If citizens are embedded in eco-systems, what ideas of justice should inform decisions and actions over the long term regarding the distribution of environmental harms and benefits? In addition, how should the needs, rights and claims of non-human nature and future generations be properly considered?
3. *Knowledge*. Given the complexity of environmental problems, what information do citizens need to be able to live well and contribute to the environmental sustainability of places and communities they inhabit?
4. *Decision making*. What are the most appropriate methods for decision making when communities and non-human nature are densely interconnected?
5. *Transformation*. What are the best ways to make significant changes that enhances the well-being of people and ecosystems?

## An ecological turn

By the mid-1990s, growing concerns about the failure of successive governments to adequately protect the environment as a shared or common good led some theorists and activists to argue democracy was 'failing ecology' (Plumwood, 1995). As a consequence, citizenship debates took an increasingly 'ecological turn'.

Andrew Dobson (2003) has offered one of the most comprehensive accounts of ecological citizenship. He argued that it is not only informed by the rights-based rhetoric of liberal democracy, but also by the civic republic tradition of concern for the protection and promotion of common goods and public duties and ideas of non-contractual justice (Dobson, 2003: 96–97). As Dobson notes, advocates of an *ecological* citizenship go further than liberal citizenship and civic republicanism. In recognizing that citizens live in ecosystems in conditions of mutual interdependence, ecological citizenship places a particularly high value on principles of environmental justice. Virtuous ecological citizens should live their lives in ways that reduce environmental harm both to the planet and others, including those who live in distant places, not because of market incentives or contractual obligations, but simply because it is right to do so (Dobson, 2003). For example, advocates of ecological citizenship draw attention to the 'asymmetrical' obligations that arise between citizens when lives are interconnected and the impact of high carbon consumer lifestyles, for example, have serious long-term consequences for distant others who will suffer disproportionate impacts of climate change.

Questions of how to live fairly and justly are very significant concerns for ecological citizens. Ecological citizenship also embraces conceptions of citizenship that recognize the rights of children, non-human nature and future generations (Clark, 2011). Children, for example, are recognized as citizens in the wider ecosystems they inhabit, in the sense that they identify with, participate in and make claims upon these communities (Lister, 2007). Ecological citizenship also places significant emphasis on justice as responsibility for non-human nature (MacGregor, 2006).

Concerns about environmental injustice as the unfair distribution of environmental goods and bads, and a desire to promote and participate in inclusive and equitable environmental decision procedures are features of both environmental and ecological citizenship. However, ecological citizenship places significant emphasis on environmental justice as everyday actions to enhance a fairer distribution of material resources, improve opportunities for inclusive decision making and encourage compassion in interactions between people and the non-human world (Hayward, 2012; Wolf et al., 2009).

Given that physical environments and access to resources are unevenly shared, ecological citizenship places a high expectation on individuals to live sustainably and to occupy ecological space in a way that reduces the ability of individuals to compromise the opportunities of others to pursue options that are important to them. In this context, the ecological turn in the environmental citizenship debate has also highlighted unresolved but important tensions about the locus and drivers of positive environmental change. Liberalism's focus on individuals as the appropriate agents of social transformation, has encouraged much research into the way behavioral change can advance sustainability. However, drawing as it does on traditions of civic republicanism, ecological citizenship also highlights wider questions about how the social-economic structures and power relationships which lock in environmental degradation can be transformed (Kenis, 2016). In the face of a global environmental crisis, the actions of individuals are inadequate. Sustained, coordinated collaborative citizen action is also vital. Achieving a more sustainable future will require individual environmental citizens who have the capacity and moral concern to think critically and act responsibly on environmental issues, but it will also require significant, concerted collective action, of ecologically aware and politically motivated citizens who have the capacity and opportunity to act together to address the underlying social, economic and political drivers of environmental degradation.

## Learning resources

The website of The Centre for Environmental Philosophy (now closed) introduces the ideas of environmental citizenship in the North American context: http://www.cep.unt.edu/citizen.htm.

Another website (2013) hosted by the Royal Geographical Society details examples of conference discussions that are beginning to explore ecological citizenship in the global south: http://conference.rgs.org/AC2013/3.

## Bibliography

Barber, W. & Bartlett, R. (2015) *Consensus and Global Environmental Governance: Deliberative Democracy in Nature's Regime*. Cambridge, MA: MIT Press.

Barry, J. (1999) *Rethinking Green Politics*. London: Sage.

Christoff, P. (1996) 'Ecological citizens and ecologically guided democracy', in Doherty, B. & De Geus, M. (eds) *Democracy and Green Political Thought: Sustainability, Rights and Citizenship*. London: Routledge.

Clark, N. (2011) *Inhuman Nature: Sociable Life on a Dynamic Planet*. London: Sage.

Dobson, A. (2003) *Citizenship and the Environment*. Oxford: Oxford University Press.

Hayward, B. (2012) *Children, Citizenship and Environment: Nurturing a Democratic Imagination in a Changing World*. Abingdon: Routledge.

Kenis, A. (2016) 'Ecological citizenship and democracy: communitarian versus agonistic perspectives', *Environmental Politics*, 25(6): 949–970.

Lister, R. (2007) 'Why citizenship? Where, when and how children?', *Theoretical Enquiries in Law*, 8(2): 693–717.

MacGregor, S. (2006) *Beyond Mothering Earth. Ecological Citizenship and the Politics of Care*. Vancouver: UBC Press.

Plumwood, V. (1995) 'Has democracy failed ecology? An ecofeminist perspective', *Environmental Politics*, 4(4): 134–168.

Wolf, J., Brown, K. & Conway, D. (2009) 'Ecological citizenship and climate change: perceptions and practice', *Environmental Politics*, 18(4): 503–521.

7.6

# Environmental conservation and restoration

Jamie Lorimer

There is a long history of human efforts to both conserve and restore the environment in the face of undesired changes. Religious groups have protected sacred groves. Farmers and foresters have tended the soil and nurtured beneficial plants and animals. Hunters have defended forests for game. Artists have valued aesthetic animals and landscapes. And scientists have designed interventions to manage in the interests of particular species and ecological systems. All of these groups might talk of conserving and restoring the environment. But they often have very different ideas about what this ought to involve. It is rare that they agree on what the environment is, how best it might be protected and who ought to benefit from their interventions. The conservation and restoration of the environment is thus a contentious field that has been subject to much debate in environmental studies. This chapter introduces some of the key debates by reflecting on the diverging answers environmental studies offers to four central questions: *What, where and when to conserve? Who knows best? What works? And who and what benefits?*

## What, where and when to conserve?

Debates about environmental conservation and restoration are often underpinned by different understandings of *what* the environment is. Perhaps the most significant tension here relates to whether or not people are seen to be part of the environment, or whether we are separate from it. There is a powerful strand of thought in Western environmentalism that holds people apart from the environment, and argues that the quality of an environment can be measured by the absence of human impacts (Cronon, 1995). Here conservation involves the protection of untouched wilderness. In contrast, many farmers and hunters (and some conservationists) see humans as stewards of the environment. They argue that some low-intensity or traditional human land uses can be reconciled with (or are even necessary for) environmental protection.

A second set of debates concern the geographies of *where* conservation should focus its energies. The tendency to equate nature with the absence of modern humans has led to the consistent relegation of urban environments as the target for conservation (Hinchliffe, 1999). Conservation has tended to focus on the countryside, on nature reserves and National Parks. But a growing interest in ecological connectivity – or the movement of plants and animals within the environment – has challenged an understanding of the environment as composed of bounded sites or patches. Instead, we are encouraged to think in terms of ecological networks and corridors that cut across political borders and intersect with human infrastructure and cities.

A further conservation debate concerns *when* the environment is understood to be most natural. Wilderness enthusiasts tend to value prehistoric landscapes. Those seeking the reconciliation of humans with the environment look to premodern landscapes. While others concerned with post-industrial settings have more freedom to offer more futuristic ecological baselines. These debates over what, where and when to conserve have been amplified by recent discussions amongst conservationists of the Anthropocene and its implications for environmental management (Lorimer, 2015). This proclamation of a new anthropogenic epoch highlights how the entire environment has been modified by people. Some suggest that it undermines approaches to conservation that try to preserve bounded places without human impacts as emblematic of past ecologies. Instead some ecologists suggest that humans will need to assist our 'novel ecosystems' (Hobbs et al., 2013) to adapt to fast changing future scenarios through proactive management.

## Who knows best?

Debates in conservation and restoration about what the environment is are often related to how the environment is known. Artists, farmers, hunters and scientists all have very different knowledge practices. There are many different forms of expertise that come to shape the practices and the politics of conservation. Perhaps the most common cause of tension stems from the differences between those who adopt a scientific approach to the environment and those whose expertise emerges from their everyday experience. Anthropologist Tim Ingold (2000) explains how, for natural science, the environment is best conceived as an object viewed from the outside. Objective science tests hypotheses and seeks generalizable findings that might apply to environments everywhere. It elevates reason above emotion and seeks to purify environmental knowledge of personal, political and economic interests. In contrast, Ingold argues, for farmers, hunters and artists the environment is a 'lifeworld'. Here knowledge emerges out of frequent embodied encounters in a particular place. This knowledge is local, heart-felt and is strongly shaped by the personal, cultural and economic context in which it is produced. The supposed differences between lay and scientific knowledge are a central concern for science and technology studies. Academics and practitioners in this field have developed methods for understanding and bringing together different forms of expertise. In the context of conservation and restoration this often involves forms of public engagement and participatory, or action, research.

## What works?

Environmental conservation and restoration are established professions whose practitioners have gained significant first-hand experience. Debates in the field are more focused on what works best. For example, there is a great deal of discussion about the effectiveness of National Parks and nature reserves. This conservation strategy involves marking out a territory and designating it for legal protection. In a park, borders may be fenced to exclude people and contain wildlife. Access to the site can be controlled and commodified in the form of tourism or hunting permits. National Parks have been a popular strategy. They have helped conserve large areas of land and stem the loss of biodiversity. However, critics dub this approach 'fortress conservation' (Brockington, 2002). They suggest that parks can curtail the movement of wildlife, risk ignoring the environmental history of indigenous land use, and help to privatize land and resources that were once common property. Instead, they propose forms of community conservation, or argue for common ownership and the maintenance of low intensity agriculture.

A second debate amongst practitioners and academics concerns the best means of valuing the environments that are to be conserved or restored. Early European and North American conservation was often a public activity supported by the State. The land was held as common property for the Nation, or for the colonial settler elite. In the last 30 years, there has been a growing move to make conservation into a more capitalist or neoliberal enterprise. Public lands have been privatized and markets have been developed for ecotourism, carbon offsets and a variety of ecosystem services. Critics of this shift suggest that capitalism is ecologically irrational and perpetuates social injustice. The practical, ecological and social implications of these different modes of political economy are hotly contested in environmental studies.

## Who and what benefits?

These debates about what works are linked to a broader set of discussions about who and what benefits from conservation and restoration. Hunting, tourism and scientific conservation are closely entangled with histories of Western imperialism and colonialism (Adams & Mulligan, 2003). During the eighteenth and nineteenth centuries, large numbers of people were cleared from their land (especially in Africa and North America) to create spaces for wildlife. During this period the social benefits of conservation accrued to a small elite. Critics argue that postcolonial conservation has failed to systematically address this history and find means of better distributing the economic and social benefits of conservation. Others argue that the creation of nature reserves in affluent countries must be understood in the context of the globalization of agriculture and forestry, through which land can be spared in the North as a consequence of deforestation and intensification in the tropics. It is argued that conservation, in both

North and South, is still frequently practiced and supported by wealthier outsiders at the expense of more local and marginal groups.

Attention has also focused on which nonhumans benefit. Conservation has been driven by public affections for particular species and places, and people have made great sacrifices to protect valued plants, animals and landscapes. But concerns have been expressed that conservation and restoration are too concerned with a narrow set of charismatic species and aesthetic landscapes. Popular 'flagship' species (like the panda) often gain much more attention than ecologically significant 'keystone' species (Caro & Girling, 2010). National Parks more commonly feature sublime upland landscapes than swamps or oceans. Critics link these partialities to the dependence of conservation on amateur enthusiasms and public donations. They suggest that these trends are likely to be exacerbated by the shift to market conservation.

## Conclusions

Environmental conservation and restoration are established, heterogeneous and growing fields within environmental studies. Conservation and restoration are animated by the contemporary environmental crisis. They are thus the subjects of especially heated debates. This is a healthy situation, indicating a rich politics and mature intellectual capacity. This review offers readers a brief route map for navigating these debates and contributing to their future vitality and development.

## Learning resources

Policy reports and summaries of the current state of conservation are available from the websites of the Convention on Biological Diversity (CBD), the International Union for the Conservation of Nature (IUCN) and the Intergovernmental Science-Policy Platform on Biodiversity and Ecosystem Services (IPBES). More information on the current practices of conservation is available on the websites of the major conservation NGOs. See for example the Nature Conservancy, Conservation International, the World Wildlife Fund, the National Trust, the Royal Society for the Protection of Birds, and the Wildlife Trusts.

The BBC has extensive television and radio resources on the science, history and politics of conservation. These include documentaries produced by BBC Nature and by BBC Radio 4. See, for example, a series entitled *Unnatural Histories* on the environmental and social histories of famous nature reserves. Accessible magazine articles on the history, politics and current debates in conservation can be found in *National Geographic*, *The Conversation*, and *The New York Times*.

## Bibliography

Adams, W.M. (2004) *Against Extinction: The Story of Conservation*. London: Earthscan.

Adams, W.M. & Mulligan, M. (2003) *Decolonizing Nature: Strategies for Conservation in a Post-Colonial Era*. London: Earthscan.

Brockington, D. (2002) *Fortress Conservation: The Preservation of the Mkomazi Game Reserve, Tanzania*. Bloomington: Indiana University Press.

Brockington, D. & Duffy, R. (2011) *Capitalism and Conservation*. Malden, MA: Wiley-Blackwell.

Caro, T. & Girling, S. (2010) *Conservation by Proxy: Indicator, Umbrella, Keystone, Flagship, and Other Surrogate Species*. Washington, DC: Island Press.

Cronon, W. (1995) The Trouble with Wilderness; Or, Getting Back to the Wrong Nature. In Cronon W. (ed.) *Uncommon Ground: Rethinking the Human Place in Nature* (pp. 69–90). New York: W.W. Norton & Co.

Hinchliffe, S. (1999) Cities and Natures: Intimate Strangers. In Allen J.E. (ed.) *Unsettling Cities* (pp. 137–180). London: Open University Press.

Hobbs, R.J., Higgs, E.S. & Hall, C. (eds) (2013) *Novel Ecosystems: Intervening in the New Ecological World Order*. New York: Wiley-Blackwell.

Ingold, T. (2000) *The Perception of the Environment: Essays on Livelihood, Dwelling and Skill*. London: Routledge.

Lorimer, J. (2015) *Wildlife in the Anthropocene: Conservation after Nature*. Minneapolis: University of Minnesota Press.

Marris, E. (2011) *Rambunctious Garden: Saving Nature in a Post-Wild World*. New York: Bloomsbury USA.

Primack, R.B. (2014) *Essentials of Conservation Biology*. Sunderland, MA: Sinauer Associates.

Wilson, E.O. (1992) *The Diversity of Life*. Cambridge, MA: Belknap Press of Harvard University Press.

# 7.7

# Environment and economy


## Richard B. Norgaard


People hold widely different perspectives on how environment and economy interrelate. The differences frequently boil down to distinct perceptions of the nature of the environment and the role new technology can play in how people interact with the environment. These differences underlie policy debates over environmental pollution, economic growth, environmental sustainability, and anthropogenic climate change.

## Historical context

For the vast majority of human history, people knew that their survival was intimately linked to the fecundity of nature. The interplay between human life and the environmental system – the sun, rain, and soils that sustained other species on which humans depended – was intimately understood. As agriculture arose some 15,000 years ago, societies became organized hierarchically. Nevertheless, the kings and priests at the top were well aware of humanity's dependence on the environment. Eco, the root of both the words economy and ecology, stems from the Greek word *ecos* referring to the household as an organized productive unit. In Western thought, nature's economy and the human economy were seen as conceptually parallel and functionally intertwined. Carl Linnaeus, the eighteenth-century Swedish botanist who established modern taxonomy, also wrote with authority on agriculture, or 'practical economy'. Early economists – the Physiocrats of the mid-eitheenth century as well as Malthus and Ricardo in the early nineteenth century – grounded their understandings of economics on the productivity of soil (Schabas, 2005). In the second half of the nineteenth century, British economist William Stanley Jevons sounded the alarm that the coal that drove Britain's industry was limited.

The idea that agriculture, forestry, fishing, and mining constitute the primary economic sector, the underlying source of economic well-being and wealth, has remained central to the mindsets of rural people. Environmental scientists have also understood the human economy as dependent on the environmental system. By mid-twentieth century, however, economists declared this primary sector idea to be dead. In a modern economy, they argued, agriculture and other so-called primary industries are fully dependent on the technologies and inputs of the industrial sectors. Furthermore, farmers need doctors and lawyers, even grocery stores, the same as city dwellers. The economy should be understood as an interdependent whole with no sector more important than another.

Economic thinking after 1950 swung even further, severing its historical roots to sun, water, and soil. In an analysis of the factors supporting economic growth in the United States, the role of land was deemed insignificant, *a priori*, and not included in the analysis (Denison, 1962). Harold Barnett and Chandler Morse furthermore argued in *Scarcity and Growth* (1963) that technological progress was making resources more and more abundant. Thus, there was no reason to worry that natural resource scarcity would limit endless economic growth. Human ingenuity overcomes all earthly barriers now and forever.

By the late 1960s, however, people became increasingly aware, for example, that pesticides were killing more than agricultural pests; that using more and more petroleum was tied to oil spills spoiling nature and air pollution in cities. Among economists, the dominant response to such environmental problems was that their costs were external to the economy and merely needed to be made internal to the economy. Environmental markets or taxes would ensure that the costs of pollution were included in economic decisions. Environmental problems were a matter of fine-tuning the economic system to include everything that is important about the environmental system. This is the essence of environmental economics, which today informs a good deal of environmental policy worldwide (such as markets in greenhouse gas emissions).

A few economists, however, pleaded for much greater caution. Kenneth Boulding suggested the metaphor of Spaceship Earth (1966) to stress that our planet was, except for the energy from the Sun, a system with limited supplies. Nicholas Georgescu-Roegen invoked basic physics, arguing that the economy is not immune to the second law of thermodynamics (1971). Herman Daly argued for a Steady-State Economy that stayed within environmental limits (1973). These economists, with the support of ecologists and philosophers, as well as system theorists, were among the founders of ecological economics, a field formalized in the 1980s.

## Environment

It is widely understood that physical and biotic components of environmental systems are interconnected. For practical purposes, however, western and westernized people have treated the environment as consisting of separate parts. The basic social concept of property, for example, presumes that land is divisible, its separate parts can be owned, and the owners of land can do what they want with their property. In practice, property law has long recognized that there are limits to what activities an individual can engage in on his or her property, if those activities affect the health of other people or their ability to use their property. Thus, in practice, environmental interconnections are acknowledged. Strong property rights advocates, however, contest whether any such limits should be imposed on the use of property or that earlier limits are sufficient. People with a more ecological understanding of the environment accept such limits and advocate for new restrictions as new understandings of the connections between people's activities, such as climate change, become known.

Environmental economists argue that the appropriate level of restrictions on rights can be economically determined. Ecological economists tend to see questions of rights as requiring ethics beyond economic reasoning, particularly ethics that include future generations, people who are not represented in today's markets.

## Technology

New technologies come from advances in knowledge, and advances in knowledge are surely good. New technologies replace old technologies because they produce goods more cheaply. The material advance of humankind has been driven by the increased productivity of new technologies. Countering this 'technology is good' view, skeptics – since the time of E.F. Schumacher's *Small is Beautiful* (1973) – argue that the new knowledge used to create new technologies derives from disciplines with only very partial knowledge of the environment's complex interrelationships. If we had known that agrichemical technologies would pollute the environment, and had had a system of property rights that protected people from the pollution, the technologies would not be used as much. If we had known that combustion technologies would lead to climate change as soon as they did, we would have taken preventive measures sooner and restricted the use of combustion technologies. The central point here is that new technologies have never been only good and the details are in how the unintended consequences of new technologies should be handled. The current arguments for large-scale 'geoengineering' of the atmosphere and oceans to mitigate global warming alarm many people because they may have large-scale unintended consequences.

## Economic growth

Increases in economic activity, or growth, make people richer, and as a society becomes richer, it can more easily invest in education, research, and the invention and implementation of new technologies including technologies that work with the environment or correct environmental problems. It is also argued that as people become wealthier, they care more about environmental quality and future generations. These are the positions taken by many environmental economists. The evidence is mixed on these points. Ecological economists argue that growth can further disrupt the environment and that there needs to be clear guidelines with respect to the ways in which the economy should grow. Furthermore, some aspects of the economy should shrink. When income is highly unequal, the rich should degrow in order to make opportunities for the poor (D'Alisa et al., 2015).

## The current situation

History since the mid-twentieth century documents the strong links between environment and economy through economic growth driven by the use of modern technologies. Combustion technologies pollute the air, runoff carrying agricultural fertilizers pollute the water, and agricultural pesticides contaminate food and farm workers, for example. Greenhouse gases from combustion technologies and diverse other human activities drive climate change. Environmental scientists argue that humanity is exceeding several planetary boundaries already and is on course to exceed others (Rockström et al., 2009). Technological optimism still reigns among most economists who argue that with more growth, greater investment in better technologies will be possible. Slowing growth, or even degrowth as some ecological economists now advocate, will hurt the poor according to the dominant economic perspective. From this perspective a more palatable alternative is arguments for a 'Green New Deal' which bases new growth and new jobs on taking the economy in a more sustainable direction under government guidance but with considerable business input and benefit. These differences in perceptions on environment and economy will likely stir political controversy for decades to come.

## Bibliography

Barnett, H.J. & Morse, C. (1963) *Scarcity and Growth: The Economics of Resource Availability*. Baltimore: RFF Press, Johns Hopkins University Press.

Boulding, K. (1966) *The Economics of the Coming Spaceship Earth. Environmental Quality in a Growing Economy*. Baltimore: RFF Press, Johns Hopkins University Press.

D'Alisa, G., Demaria, F. & Kallis, G. (eds) (2015) *Degrowth: A Vocabulary for a New Era*. New York: Routledge.

Daly, H.E. (ed.) (1973) *Toward a Steady-State Economy*. San Francisco: W.H. Freeman.

Denison, E.F. (1962) *The Sources of Economic Growth in the United States and the Alternatives Before Us*. New York: Committee for Economic Development.

Georgescu-Roegen, N. (1971) *The Entropy Law and the Economic Process*. Cambridge, MA: Harvard University Press.

Rockström, J., Steffen, W., Noone, K.J., Persson, Å., Chapin III, F.S., Lambin, E.F., Lenton, T.M., Scheffer, M., Folke, C., Schellnhuber, H.J. et al. (2009) A safe operating space for humanity, *Nature*, 461: 472–475.

Schabas, M. (2005) *The Natural Origins of Economics*. Chicago: University of Chicago Press.

Schumacher, E.F. (1973) *Small is Beautiful: A Study of Economics as if People Mattered*. New York: Blond & Briggs.

# Expert and lay environmental knowledges

Carol Morris

Environmental knowledge is important because it is only through processes of knowing that we can identify what we understand as an environmental 'problem', such as climate change, pollution or biodiversity loss, and then begin to develop solutions to these problems. The same can be said about environmental 'resources' and the opportunities they afford us. Although this might seem all too obvious, the point of this chapter is to highlight that environmental knowledge is not singular, hence the title. Recognising that there are multiple ways of knowing the environment might better equip us to tackle the phenomena we think of as environmental problems (and opportunities). In short, environmental knowledge is central to how the environment is governed and managed. The chapter firstly outlines some of the reasons why a range of ways of knowing, including but not limited to science, have come to be studied by environmental social scientists. It then discusses lay environmental knowledge and some of the ways that this has been examined before turning its attention to studies of formally accredited environmental 'experts'.

## Science and environmental knowledge production

Science is widely understood to be an important, if not the preeminent, domain of environmental knowledge production. For example, it is scientists (from many different academic disciplines) that have brought climate change to popular and political attention and, for many people, science is key to helping address this environmental problem.

However, the relationship between science and society has been changing in recent years, opening up opportunities for other ways of knowing the environment. First, following a series of high profile scientific controversies, publics have become more sceptical and less trusting of scientific knowledge and have begun to assert that other ways of knowing the environment are valid and important. One example of this is the clash between scientists working for the United Kingdom's Ministry of Agriculture, Fisheries and Food and sheep farmers in the management of the Cumbrian hills following the radioactive fallout from the Chernobyl nuclear plant accident in 1986 (Wynne, 1996). The farmers drew on their practical and experience based knowledge of mountain ecology and sheep behaviour to challenge the inappropriateness of the management prescriptions developed by the scientists. Second, there have been changes in how we govern ourselves, notably the introduction of more participatory forms of democracy. This is illustrated in the use of consultative processes within government policy making and citizen's juries (Petts, 2001). This has provided new routes into decision making for these other ways of environmental knowing. Third, in the context of reduced levels of government funding of science, it is beginning to be recognised, including by scientists themselves, that 'volunteered' environmental data gathered by amateurs and enthusiasts can usefully and legitimately help to produce environmental knowledge. This is being realised through crowd sourcing initiatives such as the Royal Society for the Protection of Birds' (RSPB) 'Big Garden Bird Watch' and formal organisations that monitor the environment such as the Climatological Observers Link, a UK based amateur weather society that produces weather data from its network of member's weather stations (Morris & Endfield, 2012). People other than scientists, therefore, also produce environmental knowledge which is variously labelled as 'unaccredited', 'non-certified', 'experiential', 'practical', 'lay', 'local', and 'indigenous'.

## Lay environmental knowledges

The study of environmental knowledge that is produced outside formal, official or professional processes and institutions of knowledge production such as universities or research institutes has become a popular theme within environmental research. The places and peoples of interest here include 'indigenous' communities such as 'First Nations' peoples in Canada or Aboriginal groups in Australia, specific residential, occupational and recreational communities such as farmers, anglers and amateur naturalists and meteorologists (e.g. Eden, 2012; Endfield & Morris, 2012). In this type of research it is typically assumed that livelihood dependencies, recreational activities embedded in specific localities, or long-term residence, result in a very intimate relation between humans and (aspects of) the nonhuman world. Such is the closeness and sustained nature of this relationship that people develop a very detailed knowledge of local environmental conditions, but this is knowledge that is produced outside of any scientific methodology and typically in association with a set of ongoing *practices*. Such knowledge accumulates and can be handed down through generations by cultural transmission but is also understood to be dynamic, as evolving over time through adaptive processes.

Researchers have drawn attention to how these lay environmental knowledges are often ignored or marginalised in modernisation and development processes, both in the global north and south, but also within the design and implementation of environmental policy. They have also explored the differences between lay and expert knowledges and how these can lead to tension and struggles over the identification and definition of environmental problems and, by extension, the formulation of solutions to those problems. This illustrates research interest in the politics of environmental knowledge which is also evident in work that examines the treatment by professionals of amateur knowledge producers (Ellis & Waterton, 2005; Lawrence & Turnhout, 2010). Here, the exploitation of volunteered environmental data and the volunteers who provide them is seen as a threat to their continued contribution to environmental knowledge production. It is asserted that lay environmental knowledges need to be taken more seriously by environmental 'professionals' including policy makers and scientists as these knowledges can and should make a valuable contribution to the governance and management of natural environments, including the resolving of specific environmental problems.

At the same time, however, there has been a questioning of the implied essentialism (i.e. fixed or 'given') of the prefixes 'expert' or 'lay'. Focusing on types or forms of knowledge, so it is argued, runs the risk of reifying and fixing that knowledge. Instead, there is a need to conceptualise knowledge as dynamic and fluid, being continuously re-produced through ongoing social interactions, e.g. between farmers or anglers, and their associated practices. Moreover, the content and epistemology of knowledge is believed to be less significant than the links between power and knowledge. As such, it is recommended that the emphasis of research efforts should be the examination of the social, historical and institutional relations in which a particular way of environmental knowing develops and is recognised as valid or legitimate, while other claims to knowledge are de-legitimised (Leach & Fairhead, 2002; Robbins, 2000).

## Expert environmental knowledges

A distinct form of research on environmental knowledge focuses on formally accredited environmental knowledge and the activities of associated experts, such as environmental advisors. In this context, environmental experts are identified as such through their scientific training and/or occupation. Some of this work involves historical investigations of expert environmental knowledge domains (e.g. Hulme, 2010; Bell & Sheail, 2005).

Another theme in this body of work is the 'doing' or 'practising' of formal environmental expertise in 'the field' (as opposed to spaces such as the laboratory, or zoological/botanical collections) by particular types of 'knowledge worker'. Here, the focus of concern is the 'how' of environmental knowledge production. The particular 'fields' of interest have included forests being managed under certification schemes that promote sustainability (e.g. Eden, 2008), and farms participating in agri-environmental schemes (e.g. Proctor et al., 2012). The knowledge actors investigated are those who mediate between institutional science and actors such as farmers and foresters, e.g. field level advisors, and who work with specialised environmental knowledge to facilitate the implementation of certification standards and policy prescriptions.

This research has used concepts from science and technology studies to reveal the heterogeneity of field-based environmental expertise, i.e. how it is produced in association with many other human actors, but also nonhuman, material objects and ecologies. Also revealed by this work, is that 'intermediary' knowledge workers are vital actors in environmental governance and management, and yet their role can be constrained significantly. For example, in the public policy context of agri-environmental schemes it has been shown that field advisors are limited in the extent to which they can encourage a wider environmental learning amongst farmers as opposed to simply ensuring that scheme objectives are met.

## Conclusion

Studies of environmental knowledge have been conducted in an ever-increasing variety of environmental contexts. Researchers have explored how, where and why environmental knowledge comes into being and drawn attention to the context-specific but also dynamic nature of both 'expert' and 'lay' environmental knowledge. Although the labelling of particular forms of knowledge has been questioned, researchers continue to pay particular attention to lay environmental knowledges, and to argue that these need to be engaged more fully in environmental governance and management through a variety of participatory mechanisms. One recent development in this respect is the idea of 'joint' or 'co-production' of environmental knowledge by multiple groups of stakeholders, both 'expert' and 'lay'. This perspective avoids the possible limitations of naming and distinguishing at the start of a study particular forms of environmental knowledge and how these might be brought together. The ongoing commitment to participatory forms of decision making amongst policy makers suggests that the challenges involved in co-producing environmental knowledge will continue to be of interest to environmental social science.

## Learning resources

The following two websites are for initiatives that enable amateurs and citizens to contribute to the production of environmental knowledge:

The Weather Observations website: http://wow.metoffice.gov.uk.

Established in 2010 by the UK's Meteorological Office, in collaboration with the Royal Meteorological Society, the Weather Observations website enables anyone to submit weather observations. This development signals a formal recognition of amateur meteorology by two of the organisations representing meteorology's professional interests.

The Big Garden Bird Watch website:

https://ww2.rspb.org.uk/discoverandenjoynature/discoverandlearn/birdwatch.

Big Garden Birdwatch is an initiative that is described by the RSPB as enabling this environmental NGO to "build a picture of garden wildlife across the UK. Armed with this information, we can identify what is in danger – and how we can help".

The Rural Economy and Land Use (RELU) programme was a major interdisciplinary research programme in the UK. One of the projects funded under RELU, entitled 'Understanding environmental knowledge controversies', explored how environmental knowledge (specifically about flooding and pollution) could be produced differently, through scientists and social scientists working together with non-certified environmental experts. The project website can be found at:

http://knowledge-controversies.ouce.ox.ac.uk.

The following article provides a review of how global environmental knowledge is produced:

Turnhout, E., Dewulf, A. & Hulme, M. (2016) What does policy-relevant global environmental knowledge do? The cases of climate and biodiversity. *Current Opinion in Environmental Sustainability*, 18: 65–72.

Available as a pdf at:

http://www.mikehulme.org/wp-content/uploads/2014/03/Turnhout-Dewulf-Hulme-FINAL.pdf

## Bibliography

Bell, M. & Sheail, J. (2005) 'Experts, publics and the environment in the UK: twentieth-century translations', *Journal of Historical Geography*, 31: 496–512.

Eden, S. (2008) 'Being fieldworthy: environmental knowledge practices and the space of the field in forest certification', *Environment and Planning A*, 26: 1018–1035.

Eden, S. (2012) 'Counting fish: performative data, anglers' knowledge-practices and environmental measurement', *Geoforum*, 43: 1014–1023.

Ellis, R. & Waterton, C. (2005) 'Caught between the cartographic and the ethnographic imagination: the whereabouts of amateurs, professionals, and nature in knowing biodiversity', *Environment and Planning D: Society and Space*, 23: 673–693.

Endfield, G. & Morris, C. (2012) 'Exploring the role of the amateur in the production and circulation of meteorological knowledge', *Climatic Change*, 113: 69–89.

Hulme, M. (2010) 'Problems with making and governing global kinds of knowledge', *Global Environmental Change*, 20: 558–564.

Lawrence, A. & Turnhout, E. (2010) 'Personal meaning in the public sphere: the standardisation and rationalisation of biodiversity data in the Netherlands', *Journal of Rural Studies*, 26: 353–360.

Leach, M. & Fairhead, J. (2002) 'Manners of contestation: "citizen science" and "indigenous knowledge" in West Africa and the Caribbean', *International Social Science Journal*, 53: 299–311.

Morris, C. & Endfield, G. (2012) 'Exploring contemporary amateur meteorology through a historical lens', *Weather*, 67: 4–8.

Petts, J. (2001) 'Evaluating the effectiveness of deliberative processes: waste management case-studies', *Journal of Environmental Planning and Management*, 44(2): 207–226.

Proctor, A., Donaldson, A., Phillipson, J. & Lowe, P. (2012) 'Field expertise in rural land management', *Environment and Planning A*, 44: 1696–1711.

Robbins, P. (2000) 'The practical politics of knowing: state environmental knowledge and local political economy', *Economic Geography*, 76: 126–144.

Wynne, B. (1996) 'May the sheep safely graze? A reflexive view of the expert–lay knowledge divide', in Lash S., Szerszynski, B & Wynne, B (eds) *Risk, Environment and Modernity: Towards a New Ecology* (pp. 44–83). London: SAGE.

# 7.9 Gender and environment

Sherilyn MacGregor

Gender broadly refers to the collection of characteristics used to categorize people according to dominant understandings of what it means to be masculine and feminine. Just as 'the environment' did not enter popular discourse until the late 1960s, the importance of 'gender' in human societies was popularized in the global north only as recently as the 1970s. Each connected to political movements for social change – feminism and environmentalism respectively. They are both contested concepts that attract no shortage of controversy. But what does gender have to do with the environment? This question puzzles scholars of environmental studies. Surely all humans share the blame and suffer the effects of environmental problems, so why does it matter what gender they are? Answers to this question, and the debates it provokes, can be found in the field of feminist environmental studies, which is dedicated to understanding the gender-environment nexus.

## The gender of nature and the nature of gender

One response to the 'how are gender and environment connected?' question is that, historically and in many different societies, nature has been gendered feminine and women have been associated with nature in ways that justifies their subordination. To support this claim, scholars have examined the history of European cultures whose systems and values dominate the globe.

For example, the historian Carolyn Merchant has documented how nature and the feminine have been closely connected in science and philosophy from the Ancient Greeks (Aristotle, Plato) to the Renaissance poets (Donne, Spenser) to the Enlightenment men of science (Bacon, Boyle). A recurring theme depicts the Earth as alive in the form of "a beneficent and kindly mother" who should be protected from ecologically-damaging activities such as mining (Merchant, 1990: 29–41). With the end of feudalism and the rise of capitalism in the sixteenth century, this view – and its related constraints on the manipulation of 'mother nature' – was undermined to make way for techno-scientific progress. In came "a new ethic sanctioning the exploitation of nature" (ibid.: 164) to drive the scientific revolution. Merchant explored the role of Francis Bacon, founder of both the scientific method and The Royal Society (and Lord Chancellor during King James I's witch trials in England), in shaping a world-view in which nature was cast as at once inert matter and a feminized slave to be dominated. Bacon described the scientific method as a means of interrogating nature (described as a 'common harlot') that had much in common with courtroom techniques for extracting the truth out of women accused of witchcraft, including torture, penetration and restraint.

Feminist environmental scholars have argued that the centuries-old binary structures of Western thought are responsible for the devaluation of all things feminine (Plumwood, 1993). The association of men with mind, rationality and culture and women with their opposites – body, emotion and nature – has formed the lens through which differences between the sexes have been viewed. Aristotle believed women to be naturally defective and inferior to men, akin to tame animals, and best confined to the home along with slaves. Defining 'what women are like' in relation to 'what men are like' has been a focus of scientific and philosophical debate for centuries. A common argumentative strategy has been to make claims that what is 'natural' and 'normal' is determined by biology. Although scientific reasons for women's alleged natural inferiority have been proven wrong, they were for a long time powerful enough to bar women from university education and to deny them the right to vote.

This discussion highlights an important position in feminist environmental studies, namely that ideas about gender have shaped scientific knowledge about nature and vice versa. It demonstrates that, as Donna Haraway (1988) argues, scientific knowledge is always partial, never innocent or neutral. Contrary to beliefs in scientific objectivity, feminist scholars assert that human observers inevitably see the world through a prism of their own values, assumptions and interests. Although all sorts of sexualities and gendered behaviours are observable 'in nature', they have been either interpreted to fit into a heteronormative and binary gender framework or else labelled 'abnormal'. For example, in *Evolution's Rainbow* Joan Roughgarden (2013) presents a litany of evidence from humans and other species to cast a long shadow of doubt on the veracity of traditional, Darwinian understandings of evolution and biologically-based claims about gender. The sheer diversity of observable sexual behaviour in living beings – from reptiles and birds to primates and humans – makes it nearly impossible to see ideas about 'normal' as unmediated truths.

## Earth mothers and cool dudes

A second way of thinking through the gender-environment connection is to consider how people, as gendered beings, think about the natural world and how they respond to ecological challenges. Some have asserted that women care more and worry more about the state of the planet than men because they give birth. It is also common to hear that men are more likely to take risks and trust technological progress to solve environmental problems than women because they can 'do science' and are expected to be breadwinners. Are these claims valid?

In what is often referred to as the 'essentialism debate', those who make such claims have been criticized for generalizing a specific set of traits to all women and men as if there were a set of 'hard wired' qualities (or essences) that lead them to act and think as they do. Often these claims have more to do with ideology than evidence, and so are difficult to defend. The dominant approach in feminist environmentalism today is to reject the virtuous 'Earth mother' narrative in favour of intersectional analyses that take a wide range of power relations into account. Most scholars look to sociological explanations and empirical research for insight into gender differences in environmental impacts, values and practices.

There is evidence to suggest that environmental problems are experienced differently by men and women and that environmental attitudes are shaped by gender roles and identities. Epidemiological research shows that women are more vulnerable to environmental toxins than men, due (among other reasons) to high levels of exposure to chemicals in feminized labour (Scott, 2015). Reports about the impacts of climate change on human populations have documented that women are more severely hurt by extreme weather events than men, due to their relative poverty, domestic roles, and lower social status (Nagel, 2016). Environmental sociologists have found that women in affluent societies tend to be more 'pro-environmental' in their beliefs and practices than men. Cross-national studies have found that women express higher levels of concern for climate change and other environmental problems than men (Buckingham, 2010). They also are more likely than men to engage in pro-environmental practices in the household, such as recycling and making eco-friendly purchasing decisions (Johnsson-Latham, 2007). Although debateable, possible reasons for these feminized tendencies are the facts that women still do a disproportionate amount of housework and are socialized from the time they are girls to be altruistic.

On the other hand, researchers have found men generally to be lagging behind women in their take-up of the green agenda. Men as a group eat more meat, drive more and consume more leisure goods than women, perhaps in order to conform to hegemonic forms of masculinity (Hultman, 2013). Opinion poll data show that men tend to be more sceptical of climate change than any other demographic group. Referring to them as 'cool dudes', Aaron McCright and Riley Dunlap (2011) have suggested that conservative white men are driving the climate denial movement that has contributed to the US's laggard status in the global climate policy arena. Other scholars have pointed to dominance of elite men in decision-making positions in government, industry and science to suggest that hegemonic masculinity could be a significant, but invisible, factor in climate change (Nelson, 2012).

## Conclusion

An examination of these questions and debates yields many interesting insights into academic research on gender and environment. A long-standing criticism of environmental studies has been that it is blind to gender (Banerjee & Bell, 2007). There is substantially more research on women and environment, conducted mostly by feminist scholars, than there is on men and environment. Unfortunately, this imbalance has resulted in the equation 'gender = women' and the lack of a more inclusive approach to gender that could lead to greater attention in environmental studies to genderqueer and LGBT+ people. What is perhaps more problematic is that it has also enabled men to exist as an unmarked category in environmental studies, when they have always held the greatest power to shape the natural environment. There is much more work to be done to develop a comprehensive body of knowledge about the connections between gender and environment.

## Learning resources

This chapter has focused on research on gender-environment connections in the global north. An interdisciplinary collection of essays exploring a wide range of topics can be found in:

MacGregor, S. (ed.) (2017) *The Routledge Handbook of Gender and Environment*. Abingdon: Routledge.

A useful website maintained by a network of German feminist environmental researchers and activists is Genanet: gender-environment-sustainability: http://www.genanet.de/en/home.html.

There is also a wealth of academic, policy and activist material linking gender-environment-development in the global south. A comprehensive report can be found at:

United Nations Environment Programme (UNEP) (2016) *Global Gender and Environment Outlook: The Critical Issues*. Online. Available at: http://web.unep.org/ggeo.

## Bibliography

Banerjee, D. & Bell, M.M. (2007) 'Ecogender: locating gender in environmental social science', *Society and Natural Resources*, 20(1): 3–19.

Buckingham, S. (2010) 'Call in the women', *Nature*, 468(25): 502.

Haraway, D. (1988) 'Situated knowledges: the science question in feminism and the privilege of partial perspective', *Feminist Studies*, 14(3): 575–599.

Hultman, M. (2013) 'The making of an environmental hero: a history of ecomodern masculinity, fuel cells and Arnold Schwarzenegger', *Environmental Humanities*, 2: 83–103.

Johnsson-Latham, G. (2007) 'A study on gender equality as a prerequisite for sustainable development'. Report to the Environment Advisory Council, Sweden.

McCright, A. & Dunlap, R. (2011) 'Cool dudes: the denial of climate change among conservative white males in the United States', *Global Environmental Change*, 21(4): 1163–1172.

Merchant, C. (1990) *The Death of Nature: Women, Ecology and the Scientific Revolution*. New York: Harper and Row.

Nagel, J. (2016) *Gender and Climate Change: Impacts, Science, Policy*. Abingdon: Routledge.

Nelson, J. (2012) 'Is dismissing the precautionary principle the manly thing to do? Gender and the economics of climate change'. Research Note #013. New York: Institute for New Economic Thinking. Online. Available at: http://ineteconomics.org/sites/inet.civicactions.net/files/Note-13-Nelson.pdf.

Plumwood, V. (1993) *Feminism and the Mastery of Nature*. London: Routledge.

Roughgarden, J. (2013) *Evolution's Rainbow: Diversity, Gender, and Sexuality in Nature and People* (10th anniversary edition). Berkeley: University of California Press.

Scott, D. (ed.) (2015) *Our Chemical Selves: Gender, Toxics and the Environment*. Vancouver: University of British Columbia Press.

# 7.10 Interdisciplinary environmental inquiry

Lauren Rickards

Interdisciplinarity is core to the existence and performance of the field of 'environmental studies'. But while interdisciplinarity is well known, it is poorly understood. This chapter describes the political character of the context and content of interdisciplinarity. It outlines how interdisciplinarity is driven by calls for more innovative and socially accountable research that reflects the complex boundary-crossing character of the contemporary world, including the way the environment is multiply entangled with society. It distinguishes interdisciplinarity from similar terms and critiques the idealized 'integration' model, presenting alternatives. To end, it points to the challenges and joys of actually doing interdisciplinary research.

## Why interdisciplinarity?

Interdisciplinarity is an increasingly idealized mode of research (Jacobs & Frickel, 2009). Across the world, most research funding agencies now require research proposals to have an interdisciplinary element. Underlying this drive are arguably three main motivations (Barry et al., 2008). The first is creativity, or its more commercialized synonym, 'innovation' (Barry et al., 2008). Being interdisciplinary is envisaged to push researchers out of their insular comfort zones into a necessarily more challenging, creative space. A second, greater motivation is social accountability. Under the umbrella of the 'impact agenda' and calls for academics to more directly address the world's 'grand challenges', researchers are being pushed to work at the interface of academia and the 'real world'. In this light, interdisciplinarity is valued for producing knowledge shaped more to the contours of practical concerns than disciplinary borders.

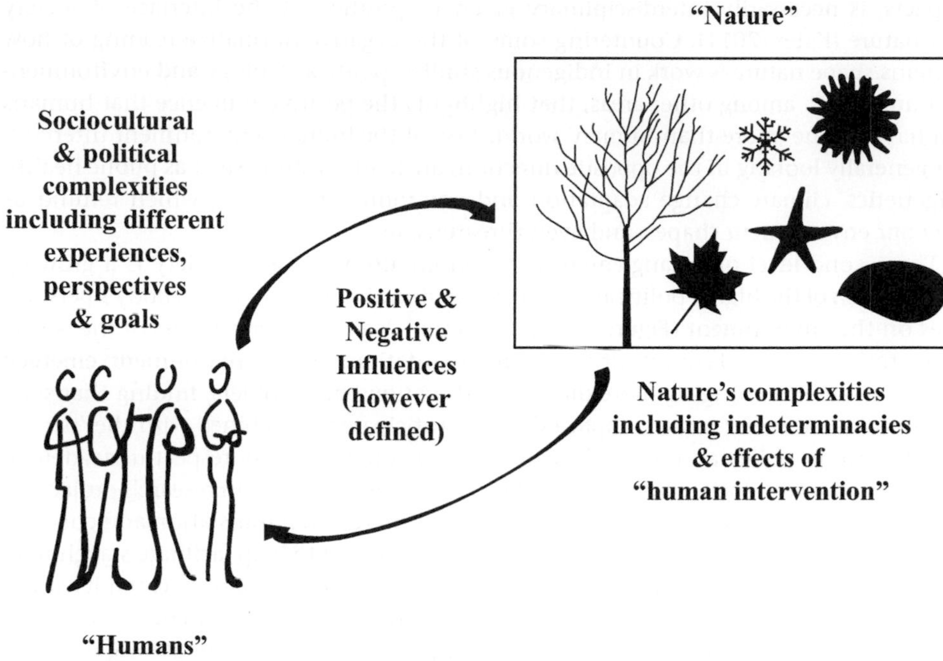


**Figure 7.10.1** Overview of main rationales for why environmental inquiry needs to be interdisciplinary.

Perhaps the greatest motivation for interdisciplinarity on environmental questions is, third, an ontological one: the object of inquiry – 'the environment' – does not make sense, arguably does not even exist, from any single disciplinary perspective. To begin with, the increasingly apparent complexity of 'pure nature' itself seems to demand that disciplines join forces (Figure 1). As Weszkalnys and Barry (2013) put it:

> it is the environment itself – on account of its complexity, heterogeneity, or the range of its impacts – and not just society or the growing social concern with the environment that [. . . requires] what is referred to as the integration of different methods and concepts, seemingly forcing researchers to move beyond their normative disciplinary commitments.
>
> (Weszkalnys & Barry, 2013: 179)

Beyond the environment's increasingly appreciated – and arguably deepening – intrinsic complexity, mutability and indeterminacy is the fact that it is (now) fundamentally entangled with society at multiple levels, so that expertise on 'the social' must be brought to bear on all 'environmental' questions.

Three levels of entanglement between environment and society have stimulated the turn to interdisciplinary environmental research. The first is the multi-directional physical interactions between humanity and environment. On the one hand, the very existence of environmental studies speaks to strong awareness of human influences upon the environment, as conventionally studied by environmental *impact* studies. The related field of 'sustainability science', which tries to address and ameliorate these impacts, is necessarily interdisciplinary given its position at the interface of society and nature (Kates, 2011). Countering some of the negative normative reading of how humans shape nature is work in indigenous studies, political ecology and environmental humanities, among other areas, that highlights the positive influence that humans can have on the 'more-than-human' world. Also at the human–environment interface, but generally looking in the opposite direction, are fields of study such as public health, epigenetics, climate change adaptation and environmental justice, which remind us how our environment shapes, and even threatens, us.

The second level of entanglement between environment and society is a growing appreciation of the highly political, partial, and cultural character of everybody's perspectives on the environment (Figure 7.10.1). To some degree, this awareness of the 'social character' and political complexity of our representations of 'the environment' emerged out of researchers engaging more directly with various stakeholders, finding that such conversations were far more complicated than anticipated. Combined with the 'participatory turn' in policy and research, which has encouraged more respectful academia-society encounters, social scientists have been frequently enrolled in research projects to not only design more effective uni-linear processes of 'research translation and communication' once projects are complete (Fox et al., 2006), but to help facilitate stakeholder engagement more broadly (Phillipson et al., 2009). Interdisciplinary research itself has underlined the importance of interpretive differences between researchers, extending awareness that such differences exist not only among stakeholders but among researchers, thanks in part to their habitual disciplinary lenses.

Third, interdisciplinary research on environmental issues has been driven, unsettled, and succeeded by growing questions about whether environment (or Nature) and society (or Humans) are even meaningful as isolated entities (Figure 7.10.1). Underlying this ontological critique is a belief that the distinction is now outdated (thanks to the profound physical entanglement of the two mentioned above), or a conviction that the basic modernist ontology of Human and Nature categories is a partial and dangerous Western fiction (Head & Gibson, 2012). The upshot of all this is that, combined with the drive for more innovative solutions to real world environmental problems discussed above, environmental research is now seen as 'necessarily interdisciplinary' (Weszkalnys & Barry, 2013: 179).

## What is interdisciplinarity?

Besides bringing together more than one discipline, what is *interdisciplinary* research? Three key elements of the concept are presented here. The first is that interdisciplinary inquiry includes academics only as they try to work through the challenges of marrying different disciplinary perspectives. In this sense, it can reinforce the imagined division between those inside and outside the Academy and is often positioned as a kind of academic house-keeping, a necessary precursor to later academic engagements with social stakeholders. In contrast, 'transdisciplinarity' generally means the production of knowledge across the academia-society boundary, moving beyond or transcending any ('internal') concerns with disciplinarity (Pohl, 2008) (Table 7.10.1).

The second aspect of the concept of interdisciplinarity is that it presumes the existence of disciplines. Many disciplines are demarcated according to the part, scale or aspect of the world they study (e.g. atoms, numbers, language, social relations, animals, plants, planets). The desire to bring them together is often motivated by a perceived

**Table 7.10.1** Main relationships between types of collaborative, cross-disciplinary knowledge production

| | Academics only | Academics + Non-academics |
|---|---|---|
| Juxtaposed knowledges | **Multidisciplinarity** *Jigsaw model* | |
| ↕ | ↕ *Dialogical model* ↕ | **Transdisciplinarity** |
| Integrated knowledges | **Interdisciplinarity** *Blender model* | |

need to generate a more comprehensive, realistic picture of the world. But besides carving up the world, disciplines are also characterized by different paradigms, concepts, methods and assumptions, such as a preference for reductionist or holistic forms of knowledge, inductive versus deductive or abductive knowledge production, or applied or 'pure' research. More than just areas of knowledge (what Becher (1989) calls 'territories') or even 'epistemic cultures' (Knorr-Cetina, 1999), disciplines are also 'tribes' of people: groups with their own networks, cultures, norms, expectations and dialects (e.g. Bracken & Oughton, 2006). Thus, as discussed further below, interdisciplinary research can necessitate complex inter-personal interactions and challenging institutional negotiations (including, for example, via peer-review processes), even if one conducts interdisciplinary research on one's own.

What counts as a discipline is itself a matter of debate. Historically, recognition of an area of knowledge as a discipline conferred institutional kudos. But with the contemporary enthusiasm for interdisciplinarity has come enthusiasm for more hybrid and open-ended areas of knowledge, as the popularity of environmental studies illustrates. Ironically, such fields can find it difficult to respond to funders' calls for 'radical' interdisciplinarity: the bringing together of clearly 'very different', non-overlapping research areas, as opposed to facilitating conversations between 'cognate' areas of knowledge (whether within one inter-discipline or between closely related sub-disciplines) (Evans & Marvin, 2006). Often the interdisciplinary research sought on environment-society matters is of a 'radical' type, notably across the STEM-HASS divide. For example, arts-science initiatives are proliferating; some to enroll art to help communicate science, but others in order to unsettle, interpose and rework each area of knowledge in relation to the other (Barry et al., 2008).

The third aspect of interdisciplinarity is the extent to which and the way in which different knowledges are imagined to combine. 'Integration' is the dominant metaphor (e.g. Stock & Burton, 2011), giving rise to an imagined spectrum. At the low integration end, cross-disciplinary knowledge forms are often labeled 'multidisciplinary' (Klein, 2010) (Table 7.10.1). For example, the 'triple bottom line' approach to environmental sustainability – in which environmental science, economics, and other social sciences are all taken into account – is generally multi- not interdisciplinary because the three forms of knowledge are viewed side-by-side, as complementary perspectives that need to be traded-off more than integrated into something new. In this 'jigsaw' model of research, the more pieces amass together, the more comprehensive and impressive the result is considered to be. At the high integration end of the spectrum is interdisciplinarity, where separate knowledges are 'blended' into a new perspective, irreducible to any of the 'ingredient' disciplines (Huutoniemi et al., 2010). In keeping with its broadly normative feel, such interdisciplinarity tends to be promoted as more desirable than multidisciplinarity on account of its more complete elimination of disciplinary boundaries.

But, as discussed further below, this 'blender model' of interdisciplinarity, as we could call it, tends to be based on the assumption that the source knowledges are commensurate. This works when all disciplines involved share basic assumptions about the world, such as the existence of a single reality. A shared positivist paradigm or at least quantitative approach to information, for instance, can enable different inputs to be integrated 'on the one plane' in a common model, table or calculation. But when the knowledges involved in cross-disciplinary endeavours involve more interpretivist forms of knowledge – those about social relations, values, discourse and power, for example, which reject the notion that 'data' or 'information' can be separated from place or knowers – they do not have an easy, shared basis on which to integrate. Rather than just differing in *what* they are looking at, such disciplines differ fundamentally in *how* they look at the world. Some even assert that we do not just see the (one) world in different ways but that we often enact it differently, creating different realities (Mol, 2002). In such knowledge situations, the task of 'integration' becomes the far more exploratory one of sharing and juxtaposing different stories and perspectives, not in order to sew them tightly into a more comprehensive whole, but to learn about different experiences and concerns, find the strengths and weaknesses of different perspectives, identify appropriate synergies, seek inspiration, and generate new insights. Indicative of this 'dialogical model' of interdisciplinarity is Castree's (2016) call for global change research that is 'not [an] ever more integrated, accurate analysis of dynamic, coupled human-environment interactions – as if we live in just one world requiring ever more "joined-up" and granular description, explanation and prediction' (p. 328), but one more akin to 'engaged pluralism' (*cf* Barnes & Sheppard, 2010; see also Miller et al., 2008). Here, epistemological and ontological differences and tensions are taken not as a problem but rather as a resource for creative and ethical work.

## Challenges and joys of interdisciplinary research

Debate over the basic model and meaning of interdisciplinary research points to some of the profound challenges it involves. As in any research encounter, power relations are central. The tension between a jigsaw and dialogical model of interdisciplinarity, for example, reflects struggles between STEM and HASS researchers, or positivist and interpretivist paradigms, as each favours a model commensurate with their own assumptions about knowledge.

At a more prosaic (but for individuals, often equally important) level, challenges of interdisciplinarity include various 'border troubles' (Petts et al., 2008) such as different assumptions about the process, purpose, politics, quality and outputs of research, above and beyond the need to simply generate a productive degree of shared intellectual understanding. Given rising reflexivity about knowledge production, research on *doing* interdisciplinary research on the environment, and the related topic of teaching about the environment in an interdisciplinary way, are now research topics in their own right (e.g. Harris et al., 2009; Gosselin et al., 2016; Lélé & Norgaard, 2005; Walshe, 2016). This literature underlines that while interdisciplinary research can pose risks and costs for academics (especially young academics) (e.g. Rhoten & Parker, 2004), notably in contexts that do not genuinely value or understand it (Lowe & Phillipson, 2009), it can be enormously enlivening and rewarding in an intellectual, interpersonal and institutional sense (e.g. Rickards, 2014). Not only can it help you find that your work as a geologist, for example, has far more in common with poets than you ever imagined (see Frodeman, 2011), but you can help open up new ways of thinking and acting on the undeniably serious environmental issues that we know now (partly thanks to interdisciplinary research) that we are inescapably part of.

## Learning resources

Brown, R.R., Deletic, A. & Wong, T.H. (2015) 'Interdisciplinarity: how to catalyse collaboration', *Nature*, 525(7569): 315–317.

This recent paper on how to foster interdisciplinary research outlines five principles for achieving effective collaboration. It is part of a special issue of *Nature* on interdisciplinarity.

Frodeman, R. (ed.) (2010) *The Oxford Handbook of Interdisciplinarity*. Oxford: Oxford University Press.

This comprehensive edited collection is edited by one of the leading thinkers on interdisciplinarity. It contains valuable chapters on tricky questions such as how to evaluate the quality of interdisciplinary research and the challenge it poses to normal peer review.

Klein, J.T. (1990) *Interdisciplinarity: History, Theory, and Practice*. Detroit: Wayne State University Press.

This seminal book is by one of the leading scholars on interdisciplinarity and still provides a useful point of departure.

http://learningforsustainability.net/integration-interdisciplinary.

This portal provides a wealth of web-based resources for those working on sustainability, including a set of readings on interdisciplinarity.

## Bibliography

Barnes, T.J. & Sheppard, E. (2010) 'Nothing includes everything: towards engaged pluralism in Anglophone economic geography', *Progress in Human Geography*, 34: 193–214.

Barry, A., Born, G. & Weszkalnys, G. (2008) 'Logics of interdisciplinarity', *Economy and Society*, 37: 20–49.

Becher, T. (1989) *Academic Tribes and Territories: Intellectual Enquiry and the Cultures of Disciplines*. Buckingham: Open University Press and the Society for Research into Higher Education.

Bracken, L.J. & Oughton, E.A. (2006) 'What do you mean? The importance of language in developing interdisciplinary research', *Transactions of the Institute of British Geographers*, 31: 371–382.

Castree, N. (2016) 'Geography and the new social contract for global change research', *Transactions of the Institute of British Geographers*, 41: 328–347.

Evans, R. & Marvin, S. (2006) 'Researching the sustainable city: three modes of interdisciplinarity', *Environment and Planning A*, 38: 1009–1028.

Fox, H.E., Christian, C., Nordby, J.C., Pergams, O.R., Peterson, G.D. & Pyke, C.R. (2006) 'Perceived barriers to integrating social science and conservation', *Conservation Biology*, 20: 1817–1820.

Frodeman, R. (2011) 'Interdisciplinary research and academic sustainability: managing knowledge in an age of accountability', *Environmental Conservation*, 38: 105–112.

Gosselin, D., Vincent, S., Boone, C., Danielson, A., Parnell, R. & Pennington, D. (2016) 'Introduction to the special issue: negotiating boundaries: effective leadership of interdisciplinary environmental and sustainability programs', *Journal of Environmental Studies and Sciences*, 6: 268–274.

Harris, F., Lyon, F. & Clarke, S. (2009) 'Doing interdisciplinarity: motivation and collaboration in research for sustainable agriculture in the UK', *Area*, 41: 374–384.

Head, L. & Gibson, C. (2012) 'Becoming differently modern: geographic contributions to a generative climate politics', *Progress in Human Geography*, 36(6): 699–714.

Huutoniemi, K., Klein, J.T., Bruun, H. & Hukkinen, J. (2010) 'Analyzing interdisciplinarity: typology and indicators', *Research Policy*, 39: 79–88.

Jacobs, J.A. & Frickel, S. (2009) 'Interdisciplinarity: a critical assessment', *Annual Review of Sociology*, 35: 43–65.

Kates, R.W. (2011) 'What kind of a science is sustainability science?', *Proceedings of the National Academy of Sciences*, 108: 19449–19450.

Klein, J.T. (2010) 'A taxonomy of interdisciplinarity'. In Frodeman, R. (ed.) *The Oxford Handbook of Interdisciplinarity* (pp. 15–30). Oxford: Oxford University Press.

Knorr-Cetina, K. (1999) *Epistemic Cultures: How the Sciences Make Knowledge*. Cambridge, MA: Harvard University Press.

Lélé, S. & Norgaard, R. (2005) 'Practicing interdisciplinarity', *BioScience*, 55: 967–975.

Lowe, P. & Phillipson, J. (2009) 'Barriers to research collaboration across disciplines: scientific paradigms and institutional practices', *Environment and Planning A*, 41: 1171–1184.

Miller, T.R., Baird, T.D., Littlefield, C.M., Kofinas, G.P., Chapin III, F.S. & Redman, C.L. (2008) 'Epistemological pluralism: reorganizing interdisciplinary research', *Ecology & Society*, 13: 46.

Mol, A. (2002) *The Body Multiple: Ontology in Medical Practice*. Durham, NC: Duke University Press.

Petts, J., Owens, S. & Bulkeley, H. (2008) 'Crossing boundaries: interdisciplinarity in the context of urban environments', *Geoforum*, 39: 593–601.

Phillipson, J., Lowe, P. & Bullock, J.M. (2009) 'Navigating the social sciences: interdisciplinarity and ecology', *Journal of Applied Ecology*, 46: 261–264.

Pohl, C. (2008) 'From science to policy through transdisciplinary research', *Environmental Science & Policy*, 11: 46–53.

Rhoten, D. & Parker, A. (2004) 'Risks and rewards of an interdisciplinary research path', *Science*, 306: 2046–2046.

Rickards, L. (2014) *Interdisciplinary Collaboration in Context: Academics and Agendas*. Melbourne: University of Melbourne.

Stock, P. & Burton, R.J. (2011) 'Defining terms for integrated (multi-inter-trans-disciplinary) sustainability research', *Sustainability*, 3: 1090–1113.

Walshe, N. (2016) 'An interdisciplinary approach to environmental and sustainability education: developing geography students' understandings of sustainable development using poetry', *Environmental Education Research*, 23(8): 1130–1149.

Weszkalnys, G. & Barry, A. (2013) 'Multiple environments: accountability, integration and ontology'. In Barry, A. & Born, G. (eds) *Interdisciplinarity: Reconfigurations of the Social and Natural Sciences* (pp. 178–208). Abingdon: Routledge.

# Multi-level environmental governance



Andrew Jordan and David Benson

In recent years, multi-level environmental governance has become a standard term in environmental studies. Its growing popularity reflects a profound shift that has occurred in the ways that environmental issues are considered and acted upon – that is to say, 'governed'. As a result, many centralised, sovereign state-led systems of governing have been replaced and/or complemented with new forms of governance in which power is shared with non-state actors such as businesses and NGOs, often functioning at (many) other institutional levels.

The term multi-level environmental governance draws our attention to two critical aspects. First, the notion of *governance,* or increased reliance on a wider array of governing modalities such as networks, markets and communities. Second, the fact that much contemporary governing occurs through and across *multiple levels,* i.e. international, supra-national, national, regional and local.

This chapter describes the emergence of multi-level environmental governance and examines the various means through which it is enacted. As governors have realised that steps are required to steer policy through greater multi-level complexity, they have adopted a fresh set of institutional devices. These include new packages of policy instruments (Wurzel et al., 2013), cross sectoral policy integration strategies (Jordan & Lenschow, 2010) and policy innovations (Jordan & Huitema, 2014a, 2014b). The final sections describe some of these governance institutions, the purpose of which is to 'govern governance'.

## From environmental government to governance

Governing environmental issues is a relatively recent phenomenon. It is strongly associated with the big or 'strong state' of the mid-twentieth century. One of the first countries to respond was the USA where intervention by the federal government occurred via a series of landmark legislative measures including the Clean Air (1970) and Clean Water (1972) Acts. The early 1970s witnessed a rapid diffusion of environmental protection norms across the globe, with many industrialised countries introducing environmental legislation and the machinery of government to implement it (Jänicke & Weidner, 1997). As in the US, the approach adopted was 'command and control', i.e. governments and other state institutions regulating within their geographical borders.

Over time, this relatively rapid centralisation of environmental government functions began to morph into more diffuse systems of multi-level environmental governance as non-state actors began to assume or be granted greater control. These adaptations resulted from several interconnected processes: internationalisation, regionalisation, marketisation, decentralisation, and neo-liberalism. To give some examples, even during the early 1970s, governments formulated *international* agreements to address transboundary environmental problems, thereby shifting powers upwards to the global level. For example, marine pollution problems led to the signing of MARPOL which controls pollution of the oceans from oil and other hazardous substances. Further national powers were ceded to supranational bodies through the development of *regional markets* – such as that in Europe after the establishment of the European Economic Community (Jordan & Adelle, 2013). Mirroring wider devolutionary thrusts in governance, environmental powers were also *devolved* downwards to lower governmental levels. Finally, in the 1980s and 1990s the emergence of *neo-liberal* ideas weakened state control, particularly in the USA (Andrews, 2006) and parts of Europe (Gravey & Jordan, 2016). Although the wholesale dismantling of environmental regulations that had emerged in the 1970s did not occur, business actors became more influential in shaping and delivering environmental governance. The delivery of sustainable development through multi-stakeholder partnerships perfectly illustrates this trend.

## Modes and levels of governing

The net result of these interconnected shifts was the emergence of a more complex landscape of *multi-level environmental governance* – one that embraces many different modes and levels of governing (Hooghe & Marks, 2001). Thus today, environmental governing through hierarchical regulation is undertaken at the international level under legally binding regimes on climate and biodiversity. In Europe, the EU has adopted a broad-ranging suite of policies which shape actions in its Member States (see Jordan & Adelle, 2013). And at the sub-national scale, regional and local governments are important sources of environmental law and policy (Benson et al., 2013).

Crucially, Pierre and Peters (2000: 14–22) identify three additional modes of governance that have become more common in the era of multi-level governing: markets, networks, and communities. A defining feature of multi-level environmental governance is the importance of governance through *markets*. Governments seek to employ market forces to implement their environmental policy objectives on the basis that, given the appropriate financial incentives, they exert a powerful steer on unsustainable practices. Another emblematic feature of governance is the growth of policy *networks*, both within and beyond the state (Rhodes, 1997). Multi-level environmental governance is therefore characterised by networks of actors, often spanning different institutional levels and policy sectors.

Finally, *communities*, and more specifically individual citizens, play greater roles in environmental decision making via collaborative partnerships between local actors and state agencies (Benson et al., 2013). Environmental governance also increasingly involves citizens directly in decision making. Public participation has been promoted through initiatives such as the UN's Agenda 21 and the Aarhus Convention 1998, with a plethora of institutional mechanisms to facilitate this interaction.

## Institutional innovations in governing

Another way of understanding multi-level environmental governance is in terms of process. Thus, governors employ a variety of *policy instruments* and institutions to 'steer' society (Pierre & Peters, 2000) towards more environmentally sustainable outcomes. Although the term institution can be interpreted differently (Peters, 2012), in environmental studies it generally denotes the rules that coordinate societal action (for example, Ostrom, 2015) or, in its more literal sense, the *organisations* that are responsible for steering.

*Regulation* is still the most widely used policy instrument globally, invariably involving setting environmental standards and enforcement, often supported by punitive action against transgressors. An example *par excellence* is the US federal Clean Water Act, which sets uniform standards for water quality that must be implemented by states under central agency enforcement. The EU also relies heavily on regulation (Jordan & Adelle, 2013).

But governments also employ many other environmental policy instruments, including *market-based approaches*, *voluntary agreements* and various *informational* devices such as eco-labels (Wurzel et al., 2013). Market-based instruments have also proliferated as coordinating mechanisms under multi-level environmental governance. A good example is the EU's emissions trading scheme – developed to reduce carbon pollution through forcing industry to purchase emissions allowances. Governors have also entered into voluntary agreements with private actors to secure environmentally responsible behaviour, for example the so-called 'covenants' agreed by the Dutch government as part of their national sustainability plan. Finally, information is provided to social actors to influence their actions in more environmentally friendly ways. Ecolabel schemes, that provide information on the environmental efficiency or performance of specific products, are used in several countries (Wurzel et al., 2013).

Multi-level environmental governance has also witnessed the emergence of different *organisational* forms. From the 1970s onwards, many countries created dedicated environmental ministries and implementing agencies (Jänicke & Weidner, 1997). The most famous example is the US Environmental Protection Agency, which formed the blueprint for similar institutions worldwide. In the UK, the Department of the Environment was established in 1970. At the European level, the EU established a dedicated Directorate-General within the European Commission in 1973 to develop environmental policy (Jordan & Adelle, 2013). At the international level, international secretariats, often located in United Nations bodies, work alongside NGOs and so-called 'epistemic communities' of experts to govern (Haas, 1989). The latter are comprised of scientific or knowledge-based actors, such as the Intergovernmental Panel on Climate Change. Ranged against them, sometimes in direct opposition, are industry groups and federations, who have learned to mobilise lobbying support for their interests at various levels. At lower levels, federal states in countries such as the USA, Germany and Australia have their own environmental agencies while local governments directly perform some environmental functions such as waste collection and recycling.

As these instruments, institutions and organisations of environmental governance have proliferated, so too has the perceived need to ensure their coordination to support sustainability. To give one example of how governance interventions themselves have to be governed, consider the idea of environmental policy integration. This policy principle seeks to ensure that all policy sectors (rather than just the environmental sector) adopt environmentally sensitive polices (Jordan & Lenschow, 2010). Meanwhile, strategic environmental assessment is a policy-making tool which aims at ensuring that higher level plans and policy programmes account for their environmental impacts. Finally, as different types of policy instrument have proliferated, policy makers have realised that they have to 'mix' or package them to ensure that they complement one another (Howlett, 2011).

## Conclusion

The shift from the 'strong state' to systems of multi-level governance has fundamentally altered the ways in which the environment and its interaction with human society is understood and ultimately governed. Governing now engages many more levels, actor types (private, public, etc.) and operates across complex vertical spaces. Steering through this complexity requires greater reliance on designing sophisticated packages of modes and instruments. The perceived need to govern systems of governance has, in turn, necessitated new institutional and organisational mechanisms. Indeed, as governance becomes ever more diffuse and interconnected, some analysts have pointedly asked who or what is ultimately in charge of, and thus accountable for, multi-level governing (Meadowcroft, 2007).

## Learning resources

For succinct summaries of the wider debate about governance and sustainability, see:

Biermann, F. (2007) '"Earth system governance" as a crosscutting theme of global change research', *Global Environmental Change*, 17(3–4): 326–337.

Jordan, A.J. (2008) 'The governance of sustainable development: taking stock and looking forwards', *Environment and Planning C*, 26(1): 17–33.

There is a rich seam of work on the wider theme of multi-level governance. A good place to start is:

Bache, I. & Flinders, M. (eds) (2004) *Multi-level Governance*. Oxford: Oxford University Press.

For a much fuller compendium of the most impactful work on multi-level governance, see:

Bache, I. & Flinders, M. (eds) (2015) *Multi-level Governance: Essential Readings (Volumes I and II)*. Cheltenham: Edward Elgar.

Finally, the academic discussion of multi-environmental governance is taking place within a broader debate about what is known as earth system governance. The Earth System Governance project is a good place to find out more about opportunities (conferences, visiting fellowships, summer schools) in this area: http://www.earthsystemgovernance.org.

## Bibliography

Andrews, R.N.L. (2006) *Managing the Environment, Managing Ourselves*. New Haven: Yale University Press.

Benson, D., Jordan, A. & Smith, L. (2013) 'Is environmental management really more collaborative? A comparative analysis of putative "paradigm shifts" in Europe, Australia and the USA', *Environment and Planning A*, 45(7): 1695–1712.

Gravey, V. & Jordan, A.J. (2016) 'Does the European Union have a reverse gear? Policy dismantling in a hyperconsensual polity', *Journal of European Public Policy*, 23(8): 1180–1198.

Haas, P.M. (1989) 'Do regimes matter? Epistemic communities and Mediterranean pollution control', *International Organization*, 43(03): 377–403.

Hooghe, L. & Marks, G. (2001) *Multi-level Governance and European Integration*. Lanham: Rowman & Littlefield.

Howlett, M. (2011) *Designing Public Policies: Principles and Instruments*. Abingdon: Routledge.

Jänicke, M. & Weidner, H. (eds) (1997) *National Environmental Policies: A Comparative Study of Capacity-building*. Berlin: Springer.

Jordan, A. & Adelle, C. (eds) (2013) *Environmental Policy in the EU: Actors, Institutions and Processes*. Abingdon: Routledge.

Jordan, A.J. & Huitema, D. (2014a) 'Innovations in climate policy: the politics of invention, diffusion and evaluation', *Environmental Politics*, 23(5): 715–734.

Jordan, A.J. & Huitema, D. (2014b) 'Policy innovation in a changing climate: sources, patterns and effects', *Global Environmental Change*, 29: 387–394.

Jordan, A. & Lenschow, A. (2010) 'Environmental policy integration: a state of the art review', *Environmental Policy and Governance*, 20(3):147–158.

Meadowcroft, J. (2007) 'Who is in charge here? Governance for sustainable development in a complex world', *Journal of Environmental Policy & Planning*, 9(3–4): 299–314.

Ostrom, E. (2015) *Governing the Commons: The Evolution of Institutions for Collective Action*. Cambridge: Cambridge University Press.

Peters, B.G. (2012) *Institutional Theory in Political Science*. New York: Continuum.

Pierre, J. & Peters, B.G. (2000) *Governance, Politics and the State*. Basingstoke: Macmillan.

Rhodes, R.A.W. (1997) *Understanding Governance*. Buckingham: Open University Press.

Wurzel, R.K.W., Zito, A.R. & Jordan, A.J. (2013) *Environmental Governance in Europe: A Comparative Analysis of New Environmental Policy Instruments*. Cheltenham: Edward Elgar.

# 7.12 International environmental institutions

Frank Biermann

## What are institutions for global environmental governance?

Since global ecological processes do not respect political borders, governments have to work together to achieve global sustainability. Major issues of global environmental change – such as global climate change, ocean pollution or stratospheric ozone depletion – require the collaboration of many countries to be effectively resolved. Key instruments of global environmental governance are thus intergovernmental agreements that set binding standards for participating countries. At present, over 1100 environmental treaties are in force (see Figure 7.12.1 for examples). Some of these agreements involve almost all nations and address issues of vital importance for humankind. The United Nations Framework Convention on Climate Change and its more detailed sub-agreements of Kyoto (1997) and Paris (2015) is one of the most prominent examples. The international agency to support this treaty today employs several hundred international civil servants, and the annual conferences of the parties to the convention have evolved into huge gatherings of world leaders, public officials, environmentalists, business leaders and so forth. Other intergovernmental agreements, however, are rather limited in scope and membership. For example, the Agreement on the Conservation of Polar Bears of 1973 has only five countries as parties, and its institutional apparatus is minimal. To keep an overview of all environmental treaties that are in force, several database projects have been set

- Barcelona Convention for the Protection of the Mediterranean Sea Against Pollution, 1976
- Basel Convention on the Control of Transboundary Movements of Hazardous Wastes and their Disposal, 1989
- Convention on Biological Diversity, 1992
- Convention on International Trade in Endangered Species of Wild Fauna and Flora, 1973
- International Convention for the Prevention of Pollution of the Sea by Oil, 1954
- International Convention for the Regulation of Whaling, 1946
- Montreal Protocol on Substances that Deplete the Ozone Layer, 1987
- United Nations Framework Convention on Climate Change, 1992

**Figure 7.12.1** Examples of international environmental treaties.

See https://iea.uoregon.edu for all treaties.

up that allow researchers and students to compare and study the myriad international treaties. One such database is the International Environmental Agreements Database at the University of Oregon (see https://iea.uoregon.edu).

Importantly, international environmental institutions cannot function without continuous policy making at national and subnational levels. Global standards need to be implemented at the local level, and global norm setting requires local decision making to set the frames for global decisions. As a result, policy making at the subnational, national, regional and global levels coexists, with the potential of both conflicts and synergies between different levels of regulatory activity.

## Are international institutions effective?

Are these intergovernmental agreements effective in resolving global environmental problems, and if not, what could be improved? This is a vital question that has been studied extensively since the 1980s (overviews in Jordan, 2008; Young et al., 2008). Key questions have been the conditions that allow for quick and effective negotiations of new agreements, rules and regulations, and the key factors that explain high levels of implementation of such agreements – or their failure. Still today, there is no clear-cut answer regarding the overall achievements of international institutions in saving the global environment. Some agreements are undoubtedly major success stories that have helped to save humanity from self-inflicted destruction. For instance, the Montreal Protocol on Substances that Deplete the Ozone Layer, agreed in 1987, has provided for a binding phase-out schedule for chlorofluorocarbons and other gases that had begun to destroy the life-protecting ozone layer in the stratosphere. Most governments have now implemented this phase-out plan, and the ozone layer is slowly recovering. Similar success stories are the international treaties on whaling, trade in endangered species and some types of marine pollution.

Concerning other, often more complex issues, however, international treaties have been slow to be accepted, or still lack clear binding regulations. The slow process of international negotiations to find common ground on combatting climate change is a key example. Also, deforestation, as another example, is progressing rapidly without any strong international agreements in place to halt this loss. Political scientists have developed numerous propositions to explain such variation in the effectiveness of international institutions, with key factors being, among others, the complexity of the problem, the existence of normative consensus, the costs of regulation, and the distribution of burdens and benefits among countries, to name just a few.

## What are the alternatives?

Especially in areas where governments could not agree on strong international rules, or where their implementation was slow and weak, other actors have stepped up and tried to take over from governments, notably by creating global institutions that rely on civil society organizations, global corporations and other non-governmental actors. Examples are the Forest Stewardship Council or the Marine Stewardship Council, two standard-setting bodies that have been set up by major corporations and environmental advocacy groups without direct involvement of governments. It sometimes seems as if traditional intergovernmental policy making through diplomatic conferences is now replaced by such networks of non-state actors, which some also see as being more efficient and transparent. On the other hand, the distribution of such networks is often linked to the particular interests of private actors who have to respond to their particular constituencies, and serious questions about the legitimacy of private standard setting remain. For example, policy making by private organizations, such as corporations and environmentalists, cannot relate back to democratic elections or other forms of direct representation, and is often perceived as being biased in favour of the interests and concerns of the rich countries in the North.

## Are international institutions fair?

International institutions have to operate in a nation-state system that is marked by substantial differences in wealth and power. The richest 20 per cent of humanity account for 76.6 per cent of total private consumption, while the poorest 20 per cent consume merely 1.5 per cent. One billion people lack sufficient access to water, and 2.6 billion have no basic sanitation. International institutions have to address this context of vast inequalities among people and countries.

One consequence has been that in many international environmental institutions, developing countries insisted on various exemptions. For instance, a general rule of international environmental governance widely accepted today, is the principle of 'common but differentiated responsibilities and capabilities' that has been agreed upon by governments in 1992. This principle is operationalized, for example, in a provision of the Montreal Protocol on Substances that Deplete the Ozone Layer that allows developing countries to delay implementation of the phase-out of such substances by 10 years. Industrialized countries also agreed to compensate the full agreed incremental costs that developing countries would incur in complying with this treaty. In the UN Framework Convention on Climate Change, industrialized countries agreed 'to take the lead' in addressing global warming, and quantified binding obligations to reduce emissions of greenhouse gases apply at present only to industrialized countries. In the last 20 years, a variety of international funding mechanisms have been set up to support poorer countries in their efforts to support international environmental governance. One example is the Global Environment Facility that is jointly operated by the World Bank, the UN Development Programme and the UN Environment Programme and that has allocated over the last two decades 9.2 billion US dollars to developing countries, along with over 40 billion US dollars in co-financing. However, the overall funding levels committed to by industrialized countries are not sufficient to carry all additional costs that developing countries will incur. Such conflicts continue to result in substantial deadlocks in many international negotiations, with no easy compromises in sight.

## What needs to be done to make international institutions more effective?

In sum, international environmental governance is today one of the most important domains of international relations. Its overall effectiveness, however, remains insufficient. This calls for renewed efforts by decision-makers in both public and private agencies to strengthen the overall system of international environmental governance in order to maintain stable conditions on our planet that allow for a sustainable co-evolution of human and natural systems.

A multitude of concrete reform proposals have been published in recent years, drawing on broader assessments of the social-science research in this field. Yet the implementation of such reforms is lagging behind. More political will and efforts in this policy domain are thus urgently needed. In this sense, effective international environmental institutions need to be transformative. Business as usual will hardly prevent critical transitions in the earth system. Technological revolutions and efficiency gains might provide a partial solution within the current systems of distribution and consumption. Yet this will likely not be sufficient. Instead, international environmental institutions, to be effective, will need to address directly key social concerns. Notably, in a highly divided world, international environmental governance poses fundamental questions of equity and allocation within and among nations. The entire system of global governance and world politics is likely to be affected by the changes that are required in international environmental governance or, as it is increasingly referred to, 'Earth system' governance (Biermann, 2014). In times of progressing climate change, accelerating depletion of biological diversity or growing levels of human-made pollutants all over the world, the current system of independent decision making by 190 nation states seems increasingly outdated. The success of international environmental governance is thus inherently interlinked with the evolution of the overall architecture of world politics. More political engagement and progress towards a 'constitutional moment' in international politics (Biermann et al., 2012) seems thus the need of the hour.

## Learning resources

*Earth Negotiations Bulletin.* This Bulletin is a balanced, timely and independent reporting service on United Nations environment and development negotiations. It provides daily coverage at selected UN environment and development negotiations. See http://www.iisd.ca/enb.

*Earth System Governance Project*. This global research programme brings together hundreds of scholars from industrialized and developing countries to study sustainability governance, from local to global levels. The Project's website has useful information on international conferences, experts, regional and country-based activities, open events, and so forth. See www.earthsystemgovernance.org.

*The International Environmental Agreements Database Project*. This project seeks to foster analysis of international environmental agreements by providing a repository for most information related to such agreements and the evaluation of their influence. Initiated in 2002, the Database seeks to provide negotiators, treaty secretariats, scholars, students and interested citizens with a reliable list of all historic and current international environmental agreements. See https://iea.uoregon.edu.

*United Nations Environment Programme*: This UN programme, serving as a catalyst and coordinator of environmental policy in the UN system, also maintains a special programme to support international environmental institutions. See http://www.unep.org/environmentalgovernance.

## Bibliography

Biermann, F. (2014) *Earth System Governance: World Politics in the Anthropocene*. Cambridge, MA: MIT Press.

Biermann, F., Abbott, K., Andresen, S., Bäckstrand, K., Bernstein, S., Betsill, M.M. Bulkeley, H., Cashore, B., Clapp, J., Folke, C. et al. (2012) 'Navigating the Anthropocene: improving earth system governance', *Science*, 335(6074): 1306–1307.

Jordan, A.J. (2008) 'The governance of sustainable development: taking stock and looking forwards', *Environment and Planning C*, 26(1): 17–33.

Young, O.R., King, L.A. & Schroeder, H. (2008) *Institutions and Environmental Change. Principal Findings, Applications and Research Frontiers*. Cambridge, MA: MIT Press.

7.13

# Markets and governments in environmental policy

Edward B. Barbier

The most common rationale cited for government intervention in markets for the purposes of environmental policy is the presence of *environmental externalities* (Barbier & Markandya, 2012; Baumol & Oates, 1988; OECD, 1975; Pearce et al., 1989). These are uncompensated environmental effects of production and consumption that affect consumer utility and enterprise cost outside the market mechanism. Common examples are pollution damages, soil erosion that leads to sedimentation of water bodies, oil and toxic waste spills, loss of biodiversity from deforestation, and human-induced global warming.

If the environmental damages caused by pollution or over-exploitation of natural resources are not fully reflected in the market prices for the goods and services, then there will be a divergence between these prices and the true social costs of producing and consuming those goods and services. That is, because the individuals and firms affected by environmental damages are not compensated, these damages are ignored in market transactions. The result is that too much of the goods and services causing environmental damages will be produced and consumed, and the environmental damages arising from pollution or over-exploitation of natural resources will continue. This outcome is a classic example of *market failure* – markets not reflecting the full costs of producing and consuming goods and services.

There are essentially two ways in which governments can intervene to correct the market failures associated with environmental damages. First, public authorities can impose regulations that restrict or require changes in the way that goods and services are produced and consumed so that any environmental damages are reduced or eliminated. This approach is referred to as the *regulatory approach* to environmental policy, or simply 'command and control'. Alternatively, environmental damages can be rectified through the creation of markets for environmental resources or through the introduction of taxes or other charges on producers that use the resources or pollute the environment. The latter approach is referred to as the use of *market-based incentives* to control environmental impacts.

All these approaches are said to 'internalize' any uncompensated environmental costs into the final prices paid. They are also consistent with the Polluter Pays Principle, which stipulates that the costs of achieving the desired environmental standards should be borne by those responsible for creating the environmental burdens, as long as no subsidies are given to the producers in implementing the command and control regulations or in the way that market rights are allocated (OECD, 1975). As the regulatory approach to environmental policy has been well established for decades, the remainder of this chapter will discuss the key issues and debates concerning the removal of environmentally perverse subsidies and the use of market-based incentives across many economies. Though 'market-based' implies the absence of government, the truth is that most markets in environmental goods and services have been designed by governments. The question then arises: are governments using markets as an effective tool to manage human uses of the environment?

## Phasing out environmentally harmful subsidies

Eliminating or reducing environmentally harmful subsidies has proved to be a difficult task. In recent years, fossil fuel subsidies have ranged around $500 billion globally, or approximately 0.7 per cent of global GDP and 2 per cent of all government revenues (Clements et al., 2013; IEA, 2014). Support for fossil fuel exploration in the 20 largest and richest economies amounts to an additional $90 billion per year (Bast et al., 2014). According to one estimate, phasing out all fossil fuel consumption and production subsidies by 2020 could result in a 5.8 per cent reduction in global primary energy demand and a 6.9 per cent fall in greenhouse gas emissions (IEA/OPEC/OECD/World Bank, 2010). Environmentally harmful subsidies are also pervasive in other sectors, notably agriculture and fishing (Barbier, 2015).

## Implementing market-based incentives

Various market-based incentives are currently used to correct any under-pricing of environmental damages. These include: taxes and charges for environmental services, tradable permit schemes, payments for environmental service (PES), and voluntary mechanisms.

A variety of environmental taxes and charges are currently employed in many high-income economies, including emissions charges, product charges (including deposit/refund schemes for recycling), and user fees for environmental services, natural resource use or waste disposal (Barbier & Markandya, 2012). However, most of these incentives are currently set at low rates or subject to many exemptions, so they do not reduce environmental damages significantly or provide sufficient incentives for economy-wide green investments. In addition, many taxes and charges are designed principally for raising revenues rather than for generating incentive effects. There are fewer cases of environmental taxes and charges in developing countries. Although taxes on petroleum products are the most common form of instrument, there is growing use of charges for pollution, solid waste and water supply use (Barbier & Markandya, 2012).

The benefits to many economies of more effectively implementing taxes, charges and other environmental market-based incentives could be substantial. For example, the environmental damages imposed by under-pricing fossil fuels are substantial, and include the costs of climate change, local pollution, traffic congestion, accidents and road damage (Barbier, 2015; Clements et al., 2013; Parry et al., 2014). The economic costs are especially high in rapidly developing, large emerging market economies, such as China, India, Indonesia and Russia. Moreover, the burden of the environmental damages caused by under-pricing of fossil fuels is also borne by the poorer and less advantaged populations within countries (Barbier, 2015). This is especially the case for local pollution effects, such as deteriorating air quality.

Similar inefficient pricing exists in markets for key goods and services throughout many economies, including agriculture, water supply and use, natural resources and transport. To take one example – inefficient water pricing – this problem may best be tackled by a variety of market-based incentives, such as establishing water markets, tradable permit schemes, taxes and charges for water use and pollution, and payments for watershed services. Although the use of water markets and market-based reforms for a wide range of water sector applications is growing, there remain fundamental barriers to instigating such market-based incentives in many countries. Nevertheless, water pricing and institutional reforms, especially in agriculture, will be essential to controlling the growing problem of freshwater availability in many developing economies (Dinar & Saleth, 2005; Dosi & Easter, 2003; Easter & Archibald, 2002; Schoengold & Zilberman, 2007). In addition, greater efforts need to be made in all economies on developing and implementing water pollution charges, including permit fees, discharge levies and fines, as a means to discouraging excessive effluent discharges from point and non-point sources. The use of water quality trading schemes is occurring in some river basins in Canada and the United States, but the geographic coverage remains small (Horan & Shortle, 2011).

Payments for environmental services (PES) are also emerging worldwide, but are still underutilized. PES are agreements whereby a user or beneficiary of an environmental service provides payments to individuals or communities whose management decisions influence the provision of that service. The main purpose of introducing payments for environmental services is to influence land-use decisions by enabling landholders to capture more of the value of these services than they would have done in the absence of the mechanism. However, existing PES schemes have largely focused on four services: carbon sequestration, watershed protection, biodiversity benefits and landscape beauty. Hydrological services from watershed protection tend to predominate, although land use and geological carbon sequestration schemes have expanded in recent years (OECD, 2010).

Major developments have also taken place in the use of voluntary mechanisms to promote higher environmental standards. The growing use of such mechanisms is directly related to the greater public availability of information on environmental quality. As a result, numerous schemes across a variety of products and countries, including developing economies, have flourished in recent decades (Barbier & Markandya, 2012). One notable area of advance is in product labelling. This has been a powerful force in promoting products that are environmentally more benign, and there is evidence that it has led to changes in consumer behaviour with respect to purchasing less environmentally damaging goods and services, and even impacted international trade (Zarilli et al., 1997). Environmental labelling has also been credited with having a positive environmental impact through changing consumer demand in the case of organic agriculture (DEFRA, 2003).

## Conclusion

These examples illustrate the growing popularity of market-based incentives to control environmental damages. The increased use of taxes and charges has brought flexibility and reduced the cost of meeting environmental standards. However, more could be done to use these policy instruments to reduce environmental damages than just raise revenues. Other market-based incentives that have grown in popularity include tradable permits, PES schemes and voluntary mechanisms. Unfortunately, environmentally damaging subsidies still contribute significantly to environmental impacts worldwide and, as the example of fossil fuel subsidies indicates, there is little evidence that there is the political will to remove such disincentives to environmental improvement.

## Learning resources

The Organization for Economic Cooperation and Development tracks a variety of environmental market-based instruments for a number of high-income and emerging market economies at http://www2.oecd.org/ecoinst/queries/#, which is frequently updated.

There are several websites that give an overview of the economic approach to environmental management. See, for example, http://www.env-econ.net and http://greeneconomics.blogspot.in for a good introduction and key debates.

For broader debates about markets and governments in environmental policy, see http://triplecrisis.com/author/edward-barbier.

## Bibliography


Barbier, E.B. (2015) *Nature and Wealth: Overcoming Environmental Scarcity and Inequality*. London: Palgrave Macmillan.

Barbier, E.B. & Markandya, A. (2012) *A New Blueprint for a Green Economy*. Abingdon: Routledge.

Bast, E., Makhijani, S., Pickard, S. & Whitley, S. (2014) *The Fossil Fuel Bailout: G20 Subsidies for Oil, Gas and Coal Exploration*. London: Overseas Development Institute, and Washington, DC: Oil Change International.

Baumol, W. & Oates, W. (1988) *The Theory of Environmental Policy*. Cambridge: Cambridge University Press.

Clements, B., Coady D., Fabrizio, S., Gupta, S., Alleyne, T. & Sdalevich, C. (eds) (2013) *Energy Subsidy Reform: Lessons and Implications*. Washington, DC: International Monetary Fund.

Department for Environment, Food and Rural Affairs (DEFRA) (2003) 'An assessment of the environmental impacts of organic farming. A review for Defra-funded project OF0405. DEFRA, ADAS', Elm Farm Research Centre and IGER. Online. Available at: http://www.defra.gov.uk/farm/organic/policy/research/pdf/env-impacts2.pdf.

Dinar, A. & Saleth, R.M. (2005) 'Water institutional reforms: theory and practice', *Water Policy*, 7: 1–19.

Dosi, C. & Easter, K.W. (2003) 'Water scarcity: market failure and the implications for water markets and privatization', *International Journal of Public Administration*, 26: 265–290.

Easter, K.W. & Archibald, S. (2002) 'Water markets: the global perspective', *Water Resources Impact*, 4: 23–25.

Horan, R.D. & Shortle, J.S. (2011) 'Economic and ecological rules for water quality trading', *Journal of the American Water Resources Association*, 47: 59–69.

IEA/OPEC/OECD/World Bank (2010) *Analysis of the Scope of Energy Subsidies and the Suggestions for the G-20 Initiative*. Joint Report Prepared for Submission to the G-20 summit Meeting Toronto (Canada), 26–27 June 2010.

International Energy Agency (IEA) (2014) *World Energy Outlook 2014*. Paris: IEA.

Organization for Economic Cooperation and Development (OECD) (1975) *The Polluter Pays Principle: Definition, Analysis, Implementation*. Paris: OECD.

Organization for Economic Cooperation and Development (OECD) (2010) *Paying for Biodiversity: Enhancing the Cost-Effectiveness of Payments for Ecosystem Services*. Paris: OECD.

Parry, I., Heine, D., Lis, E. & Li, S. (2014) *Getting Prices Right: From Principle to Practice*. Washington, DC: International Monetary Fund.

Pearce, D.W., Markandya, A. & Barbier, E.B. (1989) *A Blueprint for a Green Economy*. London: Earthscan.

Schoengold, K. & Zilberman, D. (2007) 'The economics of water, irrigation, and development.' In R. Evenson & P. Pingali (eds) *Handbook of Agricultural Economics*, vol. III (pp. 2933–2977). Amsterdam: Elsevier,.

Zarrilli, S., Jha, V. & Vossenaar, R. (1997) *Ecolabelling and International Trade*. London: Macmillan Press.

# 7.14 Nature and nurture

## Peter J. Taylor

Proponents of sustainability seek to change themselves and others, so environmental studies might well concern itself with the development and changeability of behavior. On this issue, the question *how much is nature, how much is nurture* is common in popular debate and announcements about developments in science—even though scientists and other commentators repeatedly assert that nature versus nurture is an ill-framed formulation. This entry provides, in four steps, a basis for critical interpretation of the nature–nurture opposition and for moving beyond it.

## Interpreting ideas of nature

First step: Ideas of nature underlie a great deal of social thought, present and past. The changing meanings of "nature" and tensions among co-existing meanings show a history of the social order being defended or promoted (Williams, 1980). The romantic ideal, for example, of unspoiled places and sentiments (nature *separate* from "man") arose in response to industrialization under early capitalism, which was rapidly escalating exploitation of people and natural resources (producing unprecedented *interdependencies* among peoples and nature), exploitation underwritten by the removal of traditional checks in the name, ironically, of *natural* principles of individual autonomy and unconstrained pursuit of utility in economic transactions. Moreover, in this history a multiplicity of processes and specificity of qualities tends to be eclipsed by the abstract and singular (Table 7.14.1). Following Williams, discussions of nature invite interpretation that attends to "the history of human labour" (p. 78) suppressed in promoting abstract ideas about nature.

**Table 7.14.1** General historical trends in ideas of nature*

| *From* | *Towards* |
|---|---|
| 1. Multiplicity of natural processes, attributes, spirits | Organized around a single principle (Unitary, Singular Nature) |
| 2. Specific (nature = essential quality of something) | Abstract (nature = the whole material world) Personified (nature = inherent force) |
| 3. Fixed place | Changing place; increased intervention |

* Williams acknowledges that a multiplicity of meanings persist at all times.

**Table 7.14.2** Examples of nature–culture analogues in environmental discourse

| *Arena* | *Nature, natural, universal, environment* | *Culture, artificial, particular, society* |
|---|---|---|
| Environmentalism | (Non-human) nature, which holds lessons for us | People, out of touch with (non-human) nature |
| Anti-toxics campaigns | Human bodies | Chemicals ("toxic wastes") |
| Environmental justice | Political-economic processes common to all societies and fractions of society | Communities, having specific identities, are specifically targeted |
| Globalization | Global integration | Local resistance |
| Economics | Markets | Non-market residue in social relations ("market failure") |
| Human bodies and non-human nature | Dominated by or resisting technology | Dependent on or integrating technology |
| Ethics | Self-interested rationality | Particular, norm-governed behavior |
| Language | Oral | Text |
| Anthropology | Female | Male |
| Anthropology | Other cultures, primitives | Euro-, white, advanced |
| Discourse | Unmarked (by race, gender, . . .) | Marked |
| Systems | Adaptive, self-regulating | Disordered, crisis-prone |
| Determinants of behavior | Nature | Nurture |

## Interpreting nature–culture oppositions

Second step: Culture is placed in opposition to nature in a variety of ways (Table 7.14.2) that warrant critical interpretation. As well as the same historical trends to the abstract and singular, there are also tensions or contradictions among the oppositions. Sometimes the nature side is positioned as a counter to changes or interventions made by human societies. At other times the culture side provides the checks that allow humans to set themselves apart from what is deemed natural.

These oppositions can be criticized from several angles: 1. Taking note of resonances and contradictions among the oppositions, question whether one wants, say, the naturalness of human bodies to have associations with the naturalness of markets; 2. Dispute either the naturalness or universality of the left-hand side or the effectiveness of the checks from the right-hand side; 3. Deny the divide, asserting that both sides contribute to, or interact in, the situation at hand; and 4. Move beyond the divide, with its emphasis on the abstract and singular, noting, for example, that real markets have always been embedded in political-economic relations. All four angles of criticism can be applied to nature–nurture, a key instance of the nature–culture opposition. In what follows, the last two angles are considered.

## Five nature–nurtures

Third Step: Five different things to be understood should be distinguished (Table 7.14.3). The common response to the nature–nurture opposition—"of course, it is some of both" or "it is both in interaction" (i.e., third angle of criticism above)—implies a single nature–nurture opposition. That suggestion discounts the control of materials and

**Table 7.14.3** Five distinct approaches in the study of nature–nurture

| *To be understood* | *Scientific approach* |
|---|---|
| A. Variation in observable trait among varieties and locations | Analysis of variance: Compare how much variation in the trait is associated with differences among means for varieties, locations, variety-location combinations, and residual contributions* |
| B. Variation in trait in relation to measurable factors | Compare how much variation is associated with differences in measured genetic factors, environmental factors, combinations (gene-environment interaction), and residuals |
| C. Differences between group averages | Associate the difference between the averages for groups (e.g., Euro-Americans and African-Americans) with measurable factors |
| D. Changeability of individual development | Piece together a picture of the processes of development of a trait; on that basis, speak to fixity versus flexibility of traits |
| E. Adaptiveness of trait | Look for basis of current human traits in natural selection for the trait in the past |

*Source*: Taylor (2015).

* These differences are conventionally but ambiguously labeled genetic (genotypic), environmental, genotype-environment interaction, and error variance. Heritability is the technical name for the first component as a fraction of the variation in the trait.

conditions needed to establish reliable knowledge under the different approaches and to translate, if that is possible, between them. It is as if finding (or disputing) support for the relative weighting or interaction of nature and nurture through one approach providing (or removing) support for a similar conclusion in another. These issues are not widely recognized and require careful attention.

To illustrate the apparent mutual reinforcement: The idea that IQ test scores have 60–80 per cent heritability (i.e., type A analysis) makes it seem plausible that researchers will find the genes involved in differences among people in general (B). (The ambiguity of the term "genetic variance" [A] allows many people to think, incorrectly, that heritability measures the strength of the influence of those yet-to-be-identified genes.) Those genes would surely be good candidates for investigating whether genes are involved in the persistent difference in average test scores for African- and Euro-Americans (C). In turn, a genetic basis for between-racial-group averages, in combination with the basic XX-XY chromosomal difference for most males and females, renders plausible that genes are involved in between-sex differences for some given trait, e.g., modes of sexual arousal or sexual aggression. Then, to the extent that genes are involved, development would show more fixity than flexibility (D). Finally, a trait (or differences among groups for the trait) that can be linked to genes invites attention to the evolutionary past in which this trait must have been advantageous over alternatives (at least in the situations where each group lived) (E).

To appreciate how the apparent reinforcements overlook the differences between the different nature–nurtures, note that, even if heritability estimated from the similarity between twins or close relatives (A) were associated with the similarity of yet-to-be-identified genetic factors (B), the factors may be heterogeneous, i.e., not the same from one set of relatives to the next, or from one environment to the next. Moreover, the measured factors (B) need not be modifiable (e.g., chromosomal sex is a commonly measured but non-modifiable genetic factor). And, if the factors were modifiable, it does not follow that modifying them would generate the differences observed in the original data set. Lower income level, to use another example, is a significant factor associated with smoking rates, but there is no reason to expect that disbursing $10,000 to poor smokers would lead them to quit. After all, the dynamics through which a person develops a low income and those through which a person becomes a smoker are separately and jointly far more complex than any static statistical model can capture. In short, conflation of different nature–nurture sciences discounts, borrowing Williams's words, the labour involved in knowing nurture as well as nature as "varied and variable, as the changing conditions of a human world." To understand the development and changeability of human behavior requires going beyond the third, interactionist angle of criticism.

## Intersecting processes

Final step: Build on the well-developed alternatives to the nature–culture opposition found in political-economic studies of environment that analyze situations, such as soil erosion or decline in a fishery, as the outcome of *intersecting processes*—from climate and geo-morphology, to disputes over roles and responsibilities within local institutions of production, through to changes in national political economies and international debt relations (Taylor & García-Barrios, 1995). By exposing multiple points of engagement that together would help modify the situation, such studies provide models for understanding and engaging with processes through which organisms or agents construct themselves from diverse resources (whose qualities are obscured when lumped under the labels nature and culture or nurture). Even in phenylketonuria, genetic medicine's poster-child, individuals' life courses are shaped by not only diagnosis and the special diet—genes *then* environment—but also the presence (or absence) of: insurance coverage for that diet; family, peer and cultural support for compliance; counseling and support groups; access and responsiveness to the drug BH4; and more.

## Conclusion

Use of the abstract nature–nurture opposition and conflation of quite different things that are being studied invites, on the critical side, interpretation of the social order being defended or promoted and, on the constructive side, examination of specific intersecting processes informed through interdisciplinary research that erases the differences between social science and natural science.

## Learning resources

On repeated assertions that nature versus nurture is an ill-framed formulation, read then look for updates of Paul, D.B. (1998) *The Politics of Heredity: Essays on Eugenics, Biomedicine, and the Nature-nurture Debate*. Albany, SUNY Press.

## Bibliography

Taylor, P.J. (2015) 'What difference does it make? An essay review of James Tabery "Beyond versus: The struggle to understand the interaction of nature and nurture"', *Working Papers on Science in a Changing World*. Online. Available at: http://scholarworks.umb.edu/cct_sicw/1.

Taylor, P.J. & García-Barrios, R. (1995) 'The social analysis of ecological change: from systems to intersecting processes', *Social Science Information*, 34(1): 5–30.

Williams, R. (1980) 'Ideas of nature', in *Problems in Materialism and Culture* (pp. 67–85). London: Verso.

# Population numbers and global demography

Stephen G. Warren

## The importance of population in environmental problems

Each of the various environmental problems facing us, including global warming, air pollution, water pollution, deforestation, and species extinction, may be addressed by targeting relevant factors that are different for each problem. But population is a multiplicative factor common to all of them, so it deserves special attention. And it is a multiplier not only for the causes of environmental problems but also for their impacts. For example, the harm caused by sea-level rise depends on the number of people living in the coastal areas subject to flooding.

## What exponential growth can do

Exponential growth occurs when the change of a quantity is proportional to the quantity itself. Populations (of plants, animals, and bacteria) are capable of exponential growth if the resources necessary for life are available, because the number of offspring in a population is proportional to the existing population. For humans, the "total fertility rate" (*TFR*) of a defined group (e.g. a nation) is the average number of children produced by a woman in her lifetime. The *TFR*, the population doubling time, and the annual growth rate, are all related by simple equations (Warren, 2015).

Two key examples (assuming for simplicity that all children live to reproductive age) are (a) *TFR* = 2 children per woman (cpw), resulting in a stable population and called "replacement-level fertility" because the number of children is just enough to replace their two parents; and (b) *TFR* = 4 cpw, for which the population doubles in one generation. At present, the highest *TFR*s are in Africa, 6–7 cpw (e.g. Niger, Angola, Congo), and the lowest are in Eastern Europe and East Asia, 1.2 cpw (e.g. Poland, Greece, Taiwan, Singapore).

If exponential growth continues for several generations, the population can grow by an astonishing amount. In 1900 the population of the Philippines was 7 million, but then it doubled every 28 years, reaching 100 million in 2014.

## Recent population history

The world population doubled in 40 years, from 3 billion in 1960 to 6 billion in 2000 (now in 2016 it is 7.4 billion). This growth rate, a doubling in just 40 years, was unprecedented. Now the world is experiencing a slower growth rate in percent per year, but a constant number added per year, because the base population is growing. For example, in 1977 the population of 4.3 billion, growing at 1.9 per cent per year, meant an addition of 82 million people per year. In 2008 the growth rate was only 1.2 per cent per year, but building on a population of 6.7 billion it caused the population to grow, again, by 82 million per year.

## Is our understanding of the demographic transition backwards?

At various times in the past two centuries the fertility rate in country after country declined dramatically, from ~6 to ≤2. In many developing countries, the fertility rate is inversely correlated with affluence and with education of girls, suggesting that the way to reduce *TFR* in poor countries is to promote economic and educational development; then women will choose to have fewer children. Expressing this view is the slogan "Development is the best contraceptive."

But in this correlation the causality could be in the other direction. In countries that recently transitioned from poor to rich and from high to low fertility, the fertility transition *preceded* the rapid economic development (O'Sullivan, 2015). In East Asia in 1970, Thailand and the Philippines both had high fertility (6 cpw) and similar per-capita GDPs. Because of differences in their family-planning policies, fertility declined more rapidly in Thailand than in the Philippines (now 1.6 and 2.9 cpw, respectively), and now Thailand's per-capita GDP is $6000, versus Philippines' $2400. With fewer children to support, resources are not spread so thin, and the children can be better educated. The outcome of this debate has implications for the focus of development aid; more attention might be paid to providing and promoting contraceptives, so as to reduce the fertility rate now rather than waiting for economic development to (hopefully) lead to smaller families. Currently, only 1 per cent of foreign development aid is allocated to family planning; a doubling to 2 per cent has been advocated to slow population growth (Bongaarts, 2016).

## Is immigration necessary to maintain a prosperous economy in low-fertility countries?

Many developed countries, particularly in Europe, now have below-replacement-level fertility, yet their populations are still growing because of immigration from nearby high-fertility countries. Immigration of young workers is widely seen as necessary to maintain public finances and standards of living in countries whose native populations are aging, such as Germany. This view has been challenged by Lee et al. (2014); their survey of 40 diverse countries found that, on the contrary, "Fertility below replacement, and modest population decline, favor higher material standards of living." Japan is taking this alternative path. Its fertility has been low for long enough that its population is already declining. But instead of bringing in immigrants, which now constitute only 2 per cent of Japan's population, alternative solutions are sought: recruiting women and retirees into the work force, and developing robots to lessen the demand for human workers. For example, robots in Japan are taking on some tasks in the nursing-home care of patients with dementia (Weisman, 2013).

## Debates about population projections

The U.N. Population Division, in collaboration with the Statistics Department at the University of Washington (Gerland et al., 2014), projects world population reaching 11±1.3 billion at 2100 and still growing. Their work was challenged by Lutz et al. (2014), who expect world population to peak in 2075 at 9 billion and then decline. Both projections have *TFR* declining in the high-fertility countries; their debate is about how fast it will decline. For example, Gerland et al. assume Nigeria's *TFR* to drop from today's 5.5 cpw, down to 2.2 by 2100; this trajectory would result in growth of Nigeria's population from today's 182 million to 900 million in 2100. Lutz et al. expect a more-rapid drop in fertility, resulting in a smaller population in 2100.

Most of the projections have incorporated an assumption that all nations will eventually converge to a uniform value of $TFR \approx 2.1$ cpw, which is the replacement-level fertility for developed countries (it is slightly higher than 2.0 to account for childhood deaths). The U.N. has experimented with alternative limiting *TFR*s, but then returned to 2.1 for its most recent projections. But no stabilizing feedbacks have been found that can maintain *TFR* within a narrow range (Demeny, 1997), so why has this assumption, precise to two significant figures, been so persistent? We can get a clue by noting that any constant value of *TFR* greater than 2.1 leads to a population growing indefinitely toward infinity, and any value less than 2.1 results in the population declining eventually to zero (Figure 7.15.1). The assumption of 2.1 may therefore be based on the U.N.'s desire to avoid disaster.

What the projections do not incorporate is any increase in mortality (decrease of life expectancy) due to famine, disease, or war, that might result from overpopulation, and the projections have been criticized for this neglect.

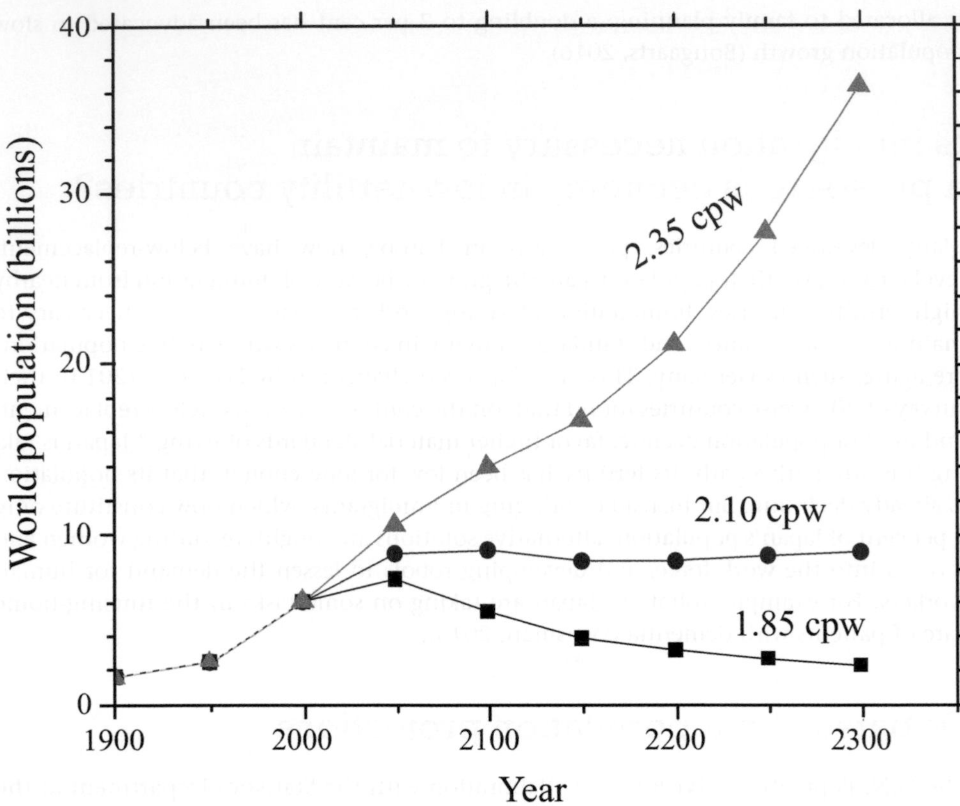


**Figure 7.15.1** Projections of global population, assuming long-term fertility at the replacement level of 2.1 children per woman (cpw) as well as two variants of constant fertility, higher or lower than replacement level by 0.25 children.

*Source*: Redrawn from World Population to 2300, by Population Division, ©2004 United Nations. Reprinted with the permission of the United Nations.

## The coming population decline *versus* the problem of outliers

The worldwide trend to smaller families has led many policy makers to conclude that "population will take care of itself", allowing them to focus efforts for environmental protection on other factors. This complacency has also been fostered by the projections of the U.N. and others that project fertility to stabilize at replacement level in all countries.

In opposition to this view is the expectation that the world will continue to exhibit cultural diversity, so that nations cannot all be expected to converge to identical social behavior. Even if most countries do limit their fertility to replacement level or below, the few that do not can grow to dominate numerically, because of the nature of exponential growth (Warren, 2015).

## Women effectively control their own fertility, *versus* an "unmet need" for contraception

There is a widespread belief that women find ways to limit their fertility so that they in fact produce the number of children they say they want. This belief, along with the declining fertility projections of the U.N., has hindered efforts to make availability of birth control a priority in development aid. But if a woman in the Sahel has seven children and says that this is the number she desires, is she telling the truth? Women in patriarchal societies are often not free to choose; they may be married off in their early teens and then pressured by their parents and husbands to have more children than they desire. When contraceptives become available, these women eagerly start using them (Campbell et al., 2013).

The U.N.'s International Conference on Population and Development in 1994 resulted in diminished support for family planning, which led to a reversal of the ongoing fertility decline in East Africa, causing the projected 2050 population of Kenya alone to be 34 million more than if family planning had continued at its pre-1994 level (Figure 2 of Ezeh et al., 2009).

## Is abortion necessary for fertility reduction?

The debate about legalizing abortion has mostly been between opponents who argue that abortion is murder, and proponents who argue for a woman's right to control her fertility and who are dismayed by the thousands of maternal deaths caused each year by unsafe illegal abortions. A different point of view is taken by those concerned about the environmental effects of overpopulation, who see abortion as an essential backup method of birth control for when contraception fails. It is rare to find a country whose fertility has dropped to replacement level without legal abortion.

## Will birth control stop population growth, or will population be limited by the food supply?

The debate here is whether projections of human populations are useful if they do not include feedbacks from the Earth system. The projections by the U.N. and others assume that life expectancy will slowly increase everywhere but that fertility will decline so much that world population will eventually peak and then begin to decrease. Those projections do not consider possible resource shortages that would increase mortality, particularly by famine. Earth System models, by contrast, model the effects of population on environmental measures such as greenhouse gas emissions, and the resulting effects of climate change on agriculture, disease, and sea level, and then the feedback effects of this environmental damage on population and on national economies. Modeling of population as part of the Earth System began 40 years ago with the "Limits to Growth" study, but has only recently begun to attract renewed research effort.

## Learning resources

The following are all insightful resources that allow exploration of key themes covered in this chapter:

Bongaarts, J. (2016) 'Development: slow down population growth', *Nature*, 530(7591): 409–412.

Cohen, J.E. (1995), *How Many People Can the Earth Support?* New York: Norton.

Population Reference Bureau (2016) *World Population Data Sheet*. Online. Available at: http://www.prb.org.

Rosling, H. (2016) *The Truth about Population*. Online. Available at: https://www.youtube.com/watch?v=eA5BM7CE5–8&feature=youtu.be.

Weeks, J.R. (1999) *Population: An Introduction to Concepts and Issues*. Belmont, CA: Wadsworth Publishing Co.

Weisman, A. (2013) *Countdown: Our Last, Best Hope for a Future on Earth?* London: Little, Brown and Company.

## Bibliography

Bongaarts, J. (2016) 'Development: slow down population growth', *Nature*, 530(7591): 409–412.

Campbell, M.M., Prata, N. & Potts, M. (2013) 'The impact of freedom on fertility decline', *Journal of Family Planning and Reproductive Health Care*, 39: 44–50.

Demeny, P. (1997) 'Replacement-level fertility: the implausible endpoint of the demographic transition', in Jones, G.W., Douglas, R.M., Caldwell, J.C. & D'Souza, R.M. (eds) *The Continuing Demographic Transition* (pp. 94–110). Oxford: Clarendon Press.

Ezeh, A.C., Mberu, B.U. & Emina, J.O. (2009) 'Stall in fertility decline in Eastern African countries: regional analysis of patterns, determinants and implications', *Philosophical Transactions of the Royal Society B*, 364: 2991–3007.

Gerland, P., Raftery, A.E., Ševčiková, H., Li, N., Gu, D., Spoorenberg, T., Alkema, L., Fosdick, B.K., Chunn, J., Lalic, N. et al. (2014) 'World population stabilization unlikely this century', *Science*, 346: 234–237.

Lee, R., Mason, A., Amporfu, E., An, C-B., Rosero-Bixby, L., Bravo, J., Bucheli, M., Chen, Q., Comelatto, P., Coy, D. et al. (2014) 'Is low fertility really a problem? Population aging, dependency, and consumption', *Science*, 346: 229–234.

Lutz, W., Butz, W., Samir, K.C., Sanderson, W. & Scherbov, S. (2014) 'Population growth: peak probability', *Science*, 346: 561.

O'Sullivan, J. (2015) 'The infrastructure dividend: conceptualising and quantifying the cost of providing capacity for additional people'. Seventh African Population Conference, Johannesburg, South Africa, December 2015. Online. Available at: http://uaps2015.princeton.edu/abstracts/150981.

United Nations Population Division (2004) *World Population to 2300*. New York: UN Department of Economic and Social Affairs.

Warren, S.G. (2015) 'Can human populations be stabilized?', *Earth's Future*, 3: 82–94.

Weisman, A. (2013) *Countdown: Our Last, Best Hope for a Future on Earth?* London: Little, Brown and Company.

# Public engagement with environmental science

## Helen Pallett

Approaches to public engagement with environmental science have been varied and contested. While often presented as a twentieth-century invention, public engagement dates back to the inception of the environmental sciences as formal disciplines, and has evolved alongside them. Public engagement projects and programmes have not only drawn from advances in the environmental sciences, but have also contributed significantly to them. Formal public engagement with science activities have often been based on misconceptions about the public, presenting them as often 'irrational' and lacking in understanding. However, publics in different countries have resisted and defied this narrow categorisation through the variety of ways in which they have engaged with the environmental sciences.

## Public engagement at the establishment of environmental sciences

The early foundations of the study of the environmental sciences were laid by the gentleman explorers of eighteenth- and nineteenth-century Europe. Access to this social world was as much to do with one's background and resources as it was about having formal scientific training. These explorers, led by figures such as Alexander von Humbolt, began to systematically chart and record the flora and fauna they encountered in far-flung corners of the globe, creating the outlines of classificatory systems still used today by biologists and ecologists. They also began to map and richly describe these regions as a precursor to the development of geography as a discipline.

Early conservationists, too, were not professional scientists, but rather were concerned and often wealthy citizens responding to the environmental destruction they were witnessing as western states modernised and industrialised. These conservationists set about systematically monitoring important populations and habitats to provide evidence of their decline and to test their efforts at conservation. The evidence they amassed was also bolstered by the work of often less wealthy amateurs such as bird watchers, anglers and plant enthusiasts who had been doing monitoring of their own. They also created associations and trusts, such as the Sierra Club in America and the National Trust in the UK, to provide long-term funding for these efforts, creating resources and laying the foundations for field sciences like ecology. While the development of colonial field sciences beyond the West was often a story of domination, sometimes these disciplines also advanced by learning from indigenous peoples.

The development of the meteorological sciences was also enabled in no small part by the labours of lay observers, in creating large data sets to work from and beginning to infer connections between different aspects of the weather. Large networks of lay observers remained central to the work of predicting and understanding weather systems well into the twentieth century, and indeed are still vital in some areas of meteorology.

The professionalisation of the environmental sciences and their formalisation as disciplines through the eighteenth and nineteenth centuries was driven in part by concerns about the rationality and level of understanding of the lay public, which have been echoed in debates about public engagement ever since. At this time it was public outrage about emerging theories of evolution, as well as new ways of thinking about environmental change from geology, which conflicted with accepted religious teachings, which animated these discussions. Therefore, the founding of the environmental sciences, though dependent on much work by amateurs, was also predicated on the exclusion of much of the public from the practice of science in order to create a new elite of professional scientists who practiced their craft in a rigorous and systematic way. However, while the lay public were removed from the day-to-day practice of the environmental sciences, a particular kind of public – wealthy, respectable, male – still had an important role in witnessing and therefore validating scientific claims and findings through the emergent experimental method. It was at this very point when the role of the public in the environmental sciences had been judiciously curtailed that discussions about the 'popularisation of science' – namely the need to be more engaged with this lay public – began (Shapin, 1990).

## Communicating the environmental sciences

The Victorian passion for the popularisation of science through travelling road shows, educational books, museums, public gardens and more has left a lasting legacy. After 1945, in response to widespread public unease with certain developments in science such as new energy technologies or the role of scientific advances in warfare, this project was reframed in terms of the public understanding of science, rather than its popularisation. Both governments and scientific institutions saw public unease and distrust of science as a threat, and believed it was a consequence of a lack of understanding of scientific facts and principles. They believed that better communication and education of science would improve public acceptance of controversial scientific advances and policies.

Science communication was the burgeoning academic field and industry set up to respond to this crisis in public confidence, using many of the same tools as the popularisation of science movement. This was also supported by the creation of new institutions, such as the Government's Committee on the Public Understanding of Science and the British Science Association in the UK. It remains a thriving field encompassing activities as diverse as documentaries, corporate social responsibility agendas, and the increasing onus placed on academics to communicate their findings beyond the academy.

However, cracks have also emerged in this project, revealing the limits of this one-way public engagement. Around high profile environmental science controversies towards the end of the twentieth century, such as the British BSE (Mad Cow disease) crisis, continuing debates about nuclear power and waste, and the introduction of genetically modified organisms (GMOs) into the food system, science communication has failed to diffuse controversy and opposition as had been hoped. One of the reasons for this is that public distrust of and unease with science was not only down to a lack of scientific understanding, but rather resulted from people holding fundamentally different values and visions of the future to governing institutions (Jasanoff, 2005). Therefore, repeatedly telling people that they were wrong to be concerned about issues like BSE or GMOs – and in some cases exaggerating the level of certainty about the scientific facts – was not enough to convince people who also had concerns about social justice, regulation and other ethical dimensions. Furthermore, public groups continually demonstrated that in many cases they were not ignorant of the scientific facts, but rather were basing their positions on forms of counter expertise, sometimes showing superior knowledge about environmental impacts on their local areas, or choosing to monitor and measure different elements.

Two-way public engagement has been proposed as a response to this apparent impasse, allowing for much more active public participation around environmental issues and the environmental sciences rather than merely passive acceptance. This has emerged from a recognition of the substantive role lay publics could play in debating the ethical dimensions of environmental issues and contributing relevant situated knowledges to the discussion. More instrumentally, governing institutions have realised that two-way engagement is also more likely to obtain the public 'buy-in' necessary to enable the radical societal transformations which will be required to address environmental problems such as climate change, or restructuring food systems. As a result public participation has become an institutionalised and in many cases routine part of environmental science and governance, encompassing local government planning decisions, government agencies, scientific projects and national policy making.

## Citizen science and environmental activism

The term 'citizen science', often attributed to science analyst Alan Irwin (1995), describes the involvement of lay publics in the actual practice of science, which as this chapter has shown, has been a long-running endeavour. Contemporary citizen science projects are perhaps the most high-profile and widespread instances of public engagement with environmental sciences. These projects are often orchestrated by professional scientists, but then involve large numbers of the lay public in carrying out scientific tasks such as species counts, measuring environmental quality, and even analysing parts of large data sets. The exponential increase in these kinds of projects has revolutionised certain areas of the environmental sciences by providing a large and willing labour force without which some kinds of data collection and analysis would not be possible. Furthermore, social media platforms have made it increasingly easy to recruit and communicate with this labour force. However, these projects have sometimes been criticised for being extractive and failing to properly acknowledge the valuable contributions made by their citizen scientists.

A more foundational criticism made of these projects is that they tend to remain firmly rooted in the conventional scientific paradigm, and therefore fail to recognise the value of other 'situated knowledges' (Haraway, 1991) such as local understandings about a particular area, or entirely different epistemic systems like indigenous knowledges. This mismatch limits the contributions of citizen scientists, as well as potentially restricting the pool of people who can legitimately participate in these projects. Furthermore, as citizen scientists are not usually involved in the formulation of research problems and questions – merely at the data collection and initial analysis stage – citizen science projects are not responsive to these diverse public values and knowledge systems.

Another significant area of lay public engagement around the environmental sciences is around environmental activism and environmentalism. Environmentalists themselves have long played a significant role in raising money to fund scientific projects and environmental protection efforts, and in raising broader public consciousness about environmental issues from habitat degradation, to climate change and recycling. Often environmental activists have used environmental science as a direct support to their causes. For example, the UK-based Climate Camp protesters famously proclaimed 'we are armed only with peer-reviewed science'. However, other environmental activists have had more ambiguous relationships with established environmental science, sometimes marshalling their own counter expertise to challenge the dominant scientific view. For example, GMO activists challenged biologists in this way, and threatened the safe conduct of biological field trials. Around the issue of fracking, activists have recently challenged dominant narratives of geologists around the safety of fracking by gathering ecological and other forms of evidence about the impacts of the practice.

## Conclusion

Public engagement with the environmental sciences has been as diverse and longstanding as the disciplines themselves. The ongoing challenge for scientific and governing institutions is to be able to recognise and capture this diversity, in order to respond to the useful insights and challenges which public engagement generates. The instigators of public engagement also need to reflect on and challenge their own assumptions about the public and the issues at hand to create more meaningful public engagement initiatives.

## Learning resources

The article and report below give an overview of the still growing and diversifying field of citizen science:

Buchan, K. (2016) Citizen science: how the net is changing the role of amateur researchers. *The Guardian*, 3 July. Online. Available at: https://www.theguardian.com/science/2016/jul/03/citizen-science-how-internet-changing-amateur-research (last accessed 8 September, 2016).

UK Environmental Observation Network (2012) *Understanding Citizen Science and Environmental Modelling: Final Report*. Online. Available at: https://www.ceh.ac.uk/sites/default/files/citizensciencereview.pdf (last accessed 8 September 2016).

This film follows three environmental activists trying to address environmental problems in very different contexts, showing the contested and ambiguous role of environmental science in their struggles:

*Elemental* (film) (2012) Directed by Emmanuel Vaughan-Lee and Gayatri Roshan. More information here: http://www.elementalthefilm.com.

## Bibliography

Haraway, D.J. (1991) *Simians, Cyborgs, and Women: The Reinvention of Nature*. London: Routledge.

Irwin, A. (1995) *Citizen Science: A Study of People, Expertise and Sustainable Development*. London: Routledge.

Jasanoff, S. (2005) *Designs on Nature: Science and Democracy in Europe and the United States*. Princeton: Princeton University Press.

Shapin, S. (1990) Science and the public. In Olby, R.C., Cantor, G.N., Christie, J.R.R. & Hodge, M.J.S. (eds) *Companion to the History of Modern Science* (pp. 990–1007). London: Routledge.

# 7.17 Race, nature and society

Peter Wade

The concept of race is associated most frequently with European colonialism. The race concept has divided humans into a hierarchy of kinds of people, based on a combination of elements in three domains: aspects of physical appearance (e.g. skin colour, hair texture), heritable essences (e.g. "blood"), and behaviour (including mental and moral qualities). These three have been perceived as linked such that elements in one domain give clues to elements in another (e.g. skin colour indicates behaviour). The balance between the three components has varied over time, as described below (see Table 7.17.1 for a schematic summary). The race concept has been used to justify social hierarchies – especially those arising from European colonialism, such as "civilised" versus "barbaric" people – by rooting them in things deemed to be "natural".

The natural environment has long been a central component of racial thinking insofar as it has been seen to shape the human body (its surface and its internal essences) and human behaviour: human physical and social diversity has long been considered in part a product of environmental difference. Historically, however, the role attributed to

**Table 7.17.1** Schematic indication of the changing balance of the component dimensions of racial thinking.

| | *Key aspects of surface appearance* | *Underlying hereditary essence* | *Behaviour or "culture"* |
|---|---|---|---|
| Pre-18th-century environmentalism | Secondary | Secondary | Primary |
| Scientific biological racism | Secondary | Primary | Secondary |
| Post-WWII social constructionism | Secondary | Tertiary | Primary |

environmental influences has varied in terms of their supposed weight and effects: how permanently or irreversibly does the environment shape a humankind? What exactly does the environment shape: just superficial aspects of human bodies (e.g. skin colour) or more significant characteristics (intelligence, moral qualities, etc.)? These questions continue to vex researchers across the disciplines, and answers to them animate wider debates about peoples' appearance and behaviour.

## Early racial thinking

In the ancient world (roughly eighth century BC to seventh century AD), thinkers and physicians in Europe and Asia related human diversity and moral worth to climate and geography, although historians disagree on how important the natural environment was considered to be. Some think that, in differentiating among humans, the ancient Greeks attributed more importance to forms of political order. Accordingly, there is disagreement about whether "proto-racism" existed in antiquity (Dikötter, 1992; Eliav-Feldon et al., 2009).

In medieval Europe (fifth to fifteenth centuries), ideas about religion contributed to the development of racial thinking, but with an emphasis on "blood" and breeding, rather than environment. Jews and Moors (Muslims) were seen by Christians as infidels, but religious conflict, especially in Iberia, created a context in which they were also increasingly seen as tainted by *raza*, understood as lineage or blood, an internal essence that predisposed them to continue practising Jewish and Muslim rites even after apparent conversion to Christianity.

These ideas influenced the colonisation of the Americas, which was the crucible for the consolidation of racial thinking and racism, although some scholars see these as emerging already from the early sixteenth century, while others think that racial theory did not develop fully until the eighteenth or even nineteenth century (Hannaford, 1996; Mignolo, 2011). European colonialism encouraged a social hierarchy in which Europeans dominated native Americans, Africans and Asians, and racial thinking emerged as a way to understand and justify this hierarchy and its attendant practices of conquest, enslavement and exploitation. Climate and geography were important components in ideas about why these kinds of humans differed naturally in terms of blood or "humours" (bodily substances influenced by the environment and thought to condition human being) and behaviour. Aspects of physical appearance, supposedly linked to blood and behaviour, became increasingly significant over time as an indicator for racial classification into global categories such as white, black, red, yellow, etc.

## Scientific racism

During the nineteenth century, racial thinking depended increasingly on the emerging science of biology in formulating ideas about human diversity and hierarchy. The era of "scientific racism" saw theories that divided humans into four or five permanent "races", different in anatomy, biology, civilisational qualities, intelligence and moral worth. The environment was thought to have played a formative role in creating these races in the distant past and, although the underlying racial type was fairly permanent, the environment continued to exert a superficial influence on how these types manifested (Banton, 1987).

In the late nineteenth century, Darwinian evolutionary theory set the scene for modern understandings of human physical diversity, in the process undermining the idea of permanent biological races (see below). However, it took some time for the full implications of the theory to overturn scientific racism, which continued into the early decades of the twentieth century, for example in the guise of eugenics. Eugenics promoted the idea that "races" (understood as biological-cultural entities) could be improved by controlling human reproduction and development; in most countries, this involved policies to improve the environment in which children grew up and attempts to encourage disease-free marriages, but in some countries it resulted in the forcible sterilisation of people deemed "unfit". The latter approach reached an awful apogee under the Nazis in the 1930s and 40s, when Jews and some others were deemed racially inferior and massacred.

## Race as a social construction

Meanwhile, some scientists began to suggest the alternative idea that biology and behaviour – or "culture" – were not linked. Human physical diversity had nothing to do with intelligence and much less with moral worth (Barkan, 1992; Smedley, 1993). After the horrors of Nazism and World War II, there was a global shift to this kind of anti-racist ideology that divorced culture from biology. This was helped by research showing that people are all biologically very similar – recent figures say humans are about 99 percent genetically identical – a fact that entirely refutes the notion of clearly distinct biological races.

Social science thus affirms that "race is a social construction" – racial thinking interweaves ideas of nature and culture to construct narratives of human difference, seen as deeply engrained; it typically deploys variants of colonial and post-colonial racial categories, such as black, white, African, Asian, Jew, etc., although it is not limited to these categories. To complicate matters, global anti-racist orthodoxy has meant the language of race has become politically awkward or even toxic in many contexts, so that racial thinking often hides behind references to "ethnicity" or "culture" (Wade, 2015).

## Environment and race today

The environment still plays an important role in current thinking about human physical diversity and the idea of race. Evolutionary theory, developing Darwinian ideas, affirms that *Homo sapiens* evolved in east Africa and, about 100,000 years ago, started to migrate across the world, finally peopling the last continent – the Americas – from about 20,000 years ago. During this process, humans became physically differentiated – genetically and in body form – through evolutionary mechanisms such as natural and sexual selection, genetic drift and founder effects. This differentiation seems to have broadly continental dimensions, but geneticists disagree on how biologically significant those dimensions are (Chakravarti, 2014). Some believe that continental genetic diversity is important in understanding human health – e.g. predispositions to certain diseases – and that familiar racial categories can be a useful short-hand for comprehending that diversity. Others say that medically significant diversity does not follow continental, much less racial, lines, and that to use racial categories represents a dangerous regression to scientific racism. Such fears are exacerbated by forensic genetic technologies that claim to predict a suspect's "racial" appearance on the basis of a DNA sample (Duster, 2015).

The environment is important in a different way too. Race may be a social construction but it has powerful effects, insofar as people use it to guide actions such as social exclusion. These effects can include impacts on people's biology – environmental racism can create toxic neighbourhoods that affect a racialised minority's health; racially segregated environments may shape the acquisition of bodily dispositions and skills (Wade, 2002). This embodiment of racial experience can create negative cycles: a pregnant woman who experiences stress (such as racism) may give birth to an underweight girl who is therefore predisposed to high blood pressure later in life, which makes it more likely she herself will give birth to an underweight baby (Hartigan, 2013). The way social race becomes biology can even lead to genetic effects, through epigenetic mechanisms whereby environmental impacts shape gene expression, including in ways that can be passed on in genetic inheritance (Richardson & Stevens, 2015).

## Continuing debates

Claims that we are now in a "post-racial" era are based on the rise of multiculturalist and anti-racist policies, and on iconic events such as Barack Obama's election as US president. Such claims are premature, given continuing racial disparities in health and social inclusion, the resurgence of geneticised versions of human diversity that resonate with familiar racial categories, and the recent rise in segments of US and European societies of a nationalist xenophobia that thinly veils an underlying racism. The powerful and mercurially adaptable mix of nature and culture represented by racial thinking looks set to persist in mutating forms into the future.

## Learning resources

*Race: Are We So Different?* A project of the American Anthropological Association. http://www.understandingrace.org/home.html

Online Supplement for *How Real is Race? A Sourcebook on Race, Culture and Society* (by Carol C. Mukhopadhyay, Rosemary Henze, and Yolanda T. Moses, Rowman and Littlefield Publishers, 2nd edition, 2013). https://sites.google.com/a/sjsu.edu/how-real-is-race-sourcebook/

*Black in Latin America*. A website supporting the TV documentary series and book of the same name by Henry Louis Gates. http://www.pbs.org/wnet/black-in-latin-america/

City-Data.com (www.city-data.com/) gives open access to detailed information about the United States (racial demographics, unemployment, sexual diversity, etc.) at levels ranging from the entire country to individual census tracts. Users can explore the degree of racial segregation of spatial units at varying degrees of resolution.

## Bibliography

Banton, M. (1987) *Racial Theories*. Cambridge: Cambridge University Press.

Barkan, E. (1992) *The Retreat of Scientific Racism: Changing Concepts of Race in Britain and the United States between the World Wars*. Cambridge: Cambridge University Press.

Chakravarti, A. (ed.) (2014) *Human Variation: A Genetic Perspective on Diversity, Race, and Medicine*. Cold Spring Harbor, NY: Cold Spring Harbor Laboratory Press.

Dikötter, F. (1992) *The Discourse of Race in Modern China*. Stanford: Stanford University Press.
Duster, T. (2015) 'A post-genomic surprise. The molecular reinscription of race in science, law and medicine', *British Journal of Sociology*, 66(1): 1–27.
Eliav-Feldon, M., Isaac, B. & Ziegler, J. (eds) (2009) *The Origins of Racism in the West*. Cambridge: Cambridge University Press.
Hannaford, I. (1996) *Race: The History of an Idea in the West*. Washington, DC: Woodrow Wilson Center Press.
Hartigan, J. (ed.) (2013) *Anthropology of Race: Genes, Biology, and Culture*. Santa Fe, NM: School for Advanced Research Press.
Mignolo, W. (2011) *The Darker Side of Western Modernity: Global Futures, Decolonial Options*. Durham, NC: Duke University Press.
Richardson, S.S. & Stevens, H. (eds) (2015) *Postgenomics: Perspectives on Biology after the Genome*. Durham, NC: Duke University Press.
Smedley, A. (1993) *Race in North America: Origin and Evolution of a Worldview*. Boulder and Oxford: Westview Press.
Wade, P. (2002) *Race, Nature and Culture: An Anthropological Perspective*. London: Pluto Press.
Wade, P. (2015) *Race: An Introduction*. Cambridge: Cambridge University Press.

# Representation and reality

7.18

## Zoë Sofoulis

Representation involves presenting a set of differences that has some analogies or homologies to a set of differences in a different medium, site, or state. This permits one thing, event or gesture to stand in for, indicate, or comment about another thing, event or gesture. Whoever claims the power to define reality, and the rules of evidence for it, also has the power to marginalise alternative world-views as unrealistic, unthinkable or unrepresentative.

The concept of 'semiotic modality' helps distinguish different kinds of relationship between representations and reality or truth. A statement beginning 'I suspect that . . .' is of lower modality (certainty) than the assertion 'I'm sure that . . .', but of higher modality than the speculation 'I wonder if . . .'. Techniques of mechanical reproduction allow photographs to present close homologies to the pre-camera reality, so they are often considered more 'realistic' and of 'higher modality' than drawing or a cartoon. But sometimes, as in botanical illustration, hand drawing or painting is preferred for clarity.

How relations between representation and reality are understood has implications for theories of knowledge (epistemology).

## Representation *versus* reality

Representation is opposed to an ideal reality in the philosopher Plato's influential metaphor of the cave, where what initiates took for reality was revealed as a mere shadow play of forms. To early modern scientists, Nature was a book that scientists read to reveal God's laws of creation. The current ideal is of universal integrated knowledge into which each bit of scientific knowledge slots like a jigsaw piece.

## Representation *of* reality

Many discussions of representation hinge on the binaries of constructed or real: artificial or natural, artistic or realistic, invented or discovered, interpretive or factual, subjective or objective, scripted or spontaneous. Scientific realism hopes for representations offering transparent windows onto a reality that would ideally be experienced without mediation. Proponents of the arts counter that artifice and poetics enhance our understanding of reality by moving us to respond to unmeasurable dynamics and affects that connect us and other elements of our lifeworlds.

### *Scientific representations of reality*

Much of science's political and cultural authority rests upon claims to represent reality in the highest possible modality (certainty, accuracy, facticity). Counting and measuring observable phenomena (such as the amount of $CO_2$ in the atmosphere) is a key technique for anchoring scientific representations in reality. Scientific reports often feature visual signs like photographs, pictures, diagrams, graphs and charts, that help convey a realist effect. While the unfunded, voluntary labour of peer review helps assure the validity of scientific reports, the biggest investments in scientific truth-making go to scientific facilities and instruments that allow real-world phenomena to be detected, measured and translated into visual, acoustic or mathematical signs with (purportedly) close homologies to happenings in the world itself. These include innumerable specialised cameras and micro- and macroscopes, sonic and electromagnetic wave technologies, and other devices for detecting rare phenomena like gravitational waves. Operators of these instruments can sometimes forget their outputs are not reality itself, but representations of phenomena fitting the device's parameters.

The old modern disciplinary divide can be mapped as a binary of the high-modality representations of technoscience ('the facts') versus the lower modality expressions in the humanities and social sciences that rely on meanings and interpretations rather than machinic algorithms to bridge the gap between reality and representation. As Bruno Latour argues, such binaries are produced in acts of 'purification' that attempt to erase traces of the messy in-betweens. The high modality of scientific facts is exaggerated by downplaying the roles of interpretive factors like standpoint, context, judgement, narrative, expectation and emotion in producing scientific knowledge. The authorial 'I' is banished from the official science report, along with all traces of scientist's labours to get the equipment, the critter, the chemical, the colleague or the Dean to all behave properly for the sake of the project.

Misrepresentation of how scientific knowledge is produced leads to problems. The crisis of scientific replicability could well be an outcome of idealised representations of scientific knowledge production, an over-enthusiastic 'purification' of accounts of procedures that leave out background details, like how the technician tweaked the set-up to yield results, or the outlier results omitted so as to not upset the calculation of averages. Climate denialism could, paradoxically, be fuelled by insistence that science representations are factual truths with no underlying agenda; such claims mean little to cynical media- and communications-savvy audiences who assume every image can be photo-shopped, every fact 'spun', and every research dollar turned to outcomes favourable to investors.

In Haraway's (1988) updated model of scientific objectivity, the disengaged god-like stance is replaced by a more humble and realistic 'situated knowledge', where the knower acknowledges instead of denying the limits and biases of their own standpoint. Whereas positivism recognises just one valid methodology (science), situated knowledges support epistemological pluralism and recognise that every body of knowledge illuminates some aspects of the world while making others harder to see. 'Truth' about a phenomenon here is not captured in a unified explanatory field but understood from multiple perspectives and modalities. The positivist fantasy of integrated knowledge typically requires cultural and social research findings to be presented within a positivist framework as 'social data', while in a situated knowledge ecology a diversity of knowledge types is valued.

## Representation *as* reality

Countering positivist ideas of the transparent medium, the high-modality representation, and the disembodied scientist are theories of knowledge informed by twentieth- and twenty-first-century developments in linguistics, anthropology, communications studies, sociology, and the history and sociology of science. These regard representations as active components of shared social and material realities. Not only are signs part of an everyday lifeworld (via spoken language and on phones, screens and advertising signage, including clothing), they alter how we see and interpret that world.

Marxist and feminist critiques of Enlightenment philosophy rejected the possibility of an unlocated and unbiased standpoint in a real world where social inequality ensures knowledge has politics, and ideology influences what kinds of knowledge are prioritised or ignored. Linguistic relativity – articulated in the Sapir-Whorf hypothesis that 'the limits of my language are the limits of my world', and variants of the theme that language particularities in different cultures influence perceptions of reality – undermines claims that any representation system is unbiased and superior to all others. Science is not free of cultural or gender bias, as shown in anthropologist Emily Martin's (1991) analysis of heterosexist clichés of courtship and marriage in textbook accounts of the egg and the sperm in human reproduction.

New kinds of theories about representational activity creating reality have come from science studies. Donna Haraway's material-semiotic actors, Latour's socio-technical networked actants and Karen Barad's intra-actions that inaugurate the 'mutual constitution of entangled agencies' are all products of an ontological levelling that treats humans and non-humans, events and entities, representations and material realities as being of similar status, and thus potentially able to constitute and act upon each other.

## Reality as representation

Modern mass production processes reduced the distinction between original and copy, while the ubiquity of media means people inhabit worlds disconnected from most lifeforms but full of symbols, signs, and simulacra. When the difference between the genuine and the fake no longer matters, argued influential French philosopher Jean Baudrillard, the simulacrum can take precedence over the real. Tourists experience real places like the Grand Canyon as already over-familiar icons of themselves. The map now precedes the territory, which may be crumbling away, like the Great Barrier Reef, now best viewed through a David Attenborough virtual reality experience. More generally in the so-called 'Anthropocene' era, fossil fuel-dependent extractive and construction industries and global trade networks have rearranged the substances of the world and climatic phenomena to such a profound degree as to break down the old distinctions between what was natural or real and what was artificial or a product of human activity, further troubling the disciplinary and epistemological divisions based on those differences.

## Non-representational theory and methods

Non-representational approaches re-enliven accounts of research by recycling the discards of both realism and constructionism, starting with phenomena considered too unrepresentative to integrate into universal scientific knowledge. Approaches like actor-network theory, practice theory, performance studies, and certain kinds of human geography seek out novelty and irregularity instead of averages, and learn from unique events as they unfold. Attention goes to the tangled interactive network rather the isolated specimen; to practices and performances not rules and scripts; to affects and intercorporeality instead of reading bodies as texts; and to the atmospheres, backgrounds, routines and spaces that facilitate production of the objects that normally capture all attention. Instead of the bland emotionless language of 'objective' science writing, researchers try to embody more of the liveliness of the experienced world into their own performances, responses and reports, leaving explanations open and multiple rather than folded into a single neat package. One site where non-representational aspects of scientific work might be glimpsed is popular science shows, where telegenic and mellifluous scientists passionately enthuse about topics directly to camera from laboratories, observatories, or more exotic locales, occasionally making melancholy reflections on extinction, habitat loss or climate change.

## Learning resources

Chandler, D. (2017) *Semiotics for Beginners*. Online. Available at: http://visual-memory.co.uk/daniel/Documents/S4B/sem02a.html. Online resource with introduction to theories of semiotics, including examples of visual and verbal modality.

Hall, S. (1997) *Representations: Cultural Representations and Signifying Practices*. London: Sage/Open University. The Introduction and Chapter 1 provide a useful summary of cultural theory of representation. Copies can be found online as course readings, for example at: https://faculty.washington.edu/pembina/all_articles/Hall1997.pdf.

Kerr, S. 'Three Minute Theory: What is Intra-Action?'. Online. Available at: https://www.youtube.com/watch?v=v0SnstJoEec. This short animation gives a helpful condensed explanation of Barad's concept, as does the website Concepts in STS: https://conceptsinsts.wikispaces.com/Agential+Realism+(Weiss).

Latour, B. & Woolgar, S. (1979) *Laboratory Life: The Social Construction of Scientific Facts*. Princeton: Princeton University Press. Although now dated and still overly long, this remains a standard demonstration of science in the making, and what aspects of reality do and don't make it into the 'facts' constructed though interactions among scientists, apparatus, and objects. Key points from *Laboratory Life* can be gleaned from reviews such as:

Haraway, D. (1980) '*Laboratory Life: The Social Construction of Scientific Facts*. Bruno Latour, Steve Woolgar', *Isis*, 71(3): 488–489.
Shapin, S. (1981) 'Laboratory Life. The Social Construction of Scientific Facts', *Medical History*, 3: 341–342. Online. Available at: https://afinetheorem.wordpress.com/2014/07/25/laboratory-life-b-latour-s-woolgar-1979.

Sofoulis, Z. (2009). 'Social construction for the twenty-first century: a co-evolutionary makeover', *Australian Humanities Review*, 46: 81–98. Online. Available at: http://www.australianhumanitiesreview.org/archive/Issue-May-2009/sofoulis.htm. Charts some points of connection and divergence between social constructionism as found in cultural studies and feminism dating from the 1970s, and contemporary actor-network perspectives.

## Bibliography

Barad, K. (2003) 'Posthumanist performativity: toward an understanding of how matter comes to matter', *Signs*, 28(3): 801–831.
Baudrillard, J. (1983) *Simulations*, trans. Paul Foss et al. Los Angeles: Semiotext(e). https://archive.org/details/Simulations1983.
de Sousa Santos, B. (2009) 'A non-occidentalist West? Learned ignorance and ecology of knowledge', *Theory Culture Society*, 26: 103–125.
Haraway, D. (1988) 'Situated knowledges: the science question in feminism and the privilege of partial perspective.' *Feminist Studies*, 14(3): 575–599.
Hodge, R. & Kress, G. (1988) *Social Semiotics*. Cambridge: Polity Press.
Latour, B. (1993) *We Have Never Been Modern*, trans. Catherine Porter. Cambridge, MA: Harvard University Press.
Martin, E. (1991) 'The egg and the sperm: how science has constructed a romance based on stereotypical male-female roles.' *Signs*, 16(3): 485–501.
Sofoulis, Z. (2015) 'A knowledge ecology of urban Australian household water consumption', *ACME: An International E-Journal for Critical Geographies*, 14(3): 765–785. Online. Available at: http://acme-journal.org/index.php/acme/article/view/1232.
Vannini, P. (2015) 'Non-Representational Research methodologies: an introduction' in Vannini, P. (ed.) *Non-Representational Methodologies: Re-Envisioning Research*. New York: Routledge.

# 7.19 Rewilding

## Steve Carver

Attempts to conserve and protect wildlife and natural habitats have been around for hundreds of years. This usually involves placing a fence (either real or metaphorical) around an area to protect it. This is supported by conservation measures targeted at preserving species and habitats. Rewilding has recently emerged as a more proactive approach by creating space and freedom for wild nature rather than simply protecting that which is left and restricting wildlife to parks and reserves.

The idea of rewilding has been around for more than 20 years, first appearing in print in *Newsweek* magazine in 1990. The term was later clarified by Dave Foreman of the US Wildlands Network and by Michael Soulé and Reed Noss in 1998 to refer to 'the scientific argument for restoring big wilderness based on the regulatory roles of large predators'. The rewilding movement has since gathered momentum and caught the public imagination with especially strong interest in Britain and Europe. This may be seen as a reaction partly to the parlous state of wild nature in Europe and the stagnation of traditional conservation approaches in the face of neoliberal views of natural capital.

## Rewilding: definitions and philosophical origins

There is as yet no universally accepted definition of rewilding. However, rewilding may be defined as restoring and protecting natural processes, often in core wild areas, providing connectivity between such areas, and protecting or reintroducing keystone species. Rewilding may require active intervention through ecological restoration, particularly to restore ecological connectivity between fragmented protected areas, and the reintroduction of key species where these are no longer present. In essence, rewilding is about providing space for nature to determine its own trajectory and outcomes with minimal human intervention.

## *Philosophical roots*

Although rewilding is relatively new, its philosophical underpinnings go back many years, not least because of our long relationship with nature and landscape. In Cicero's *De Natura Deorum* he talks about 'first nature' as wilderness and 'second nature' as its bountification.

> We sow corn, we plant trees, we fertilize the soil by irrigation, we dam the rivers and direct them where we want. In short, by means of our hands we try to create as it were a second nature within the natural world.

'Third nature' refers to the development of the human nature aesthetic, usually expressed through the arts, gardens, landscape architecture and the appreciation of wild, untamed nature that evolved during the Romantic movement of the mid-nineteenth century. Rewilding creates a shift in the nature–culture paradigm toward a 'fourth nature' based around the (re)creation of wildness through the process of ecological restoration. This brings us back full circle to Cicero's first nature by reducing human influence such that the primary dynamic is that of natural processes leading to more natural form and function (Figure 7.19.1).

**Figure 7.19.1** The cycle of nature–culture.

*Source*: after Carver (2013).

## A typology of rewilding

There are many different types of rewilding. The form and shape of any rewilding project will depend on its landscape context, the people and organisations involved and the intended outcomes. Rewilding is just one set of approaches that sits somewhere on a land management continuum between the wholly artificial (i.e. indoor spaces) and the wholly wild (i.e. pristine wilderness). This is known as the wilderness continuum or environmental modification spectrum (Figure 7.19.2).

Rewilding can be categorised according to levels of human intervention, the role of keystone species, spatial scale and intended outcomes.

## *Trophic and abiotic rewilding*

Rewilding often involves the reintroduction of 'keystone species' (predators and large herbivores) that are able to determine the course of ecological processes through their behaviour and actions. This is referred to as trophic rewilding. A popular example is how the reintroduction of wolves to Yellowstone National Park had far-reaching and unforeseen effects on the ecosystem through their modification of elk herd behaviour which had previously prevented tree regeneration in riparian areas. The wolves' return reinstated a 'landscape of fear' among the elk which then no longer congregated in large numbers on floodplains where they were vulnerable, thus reducing grazing pressure and allowing

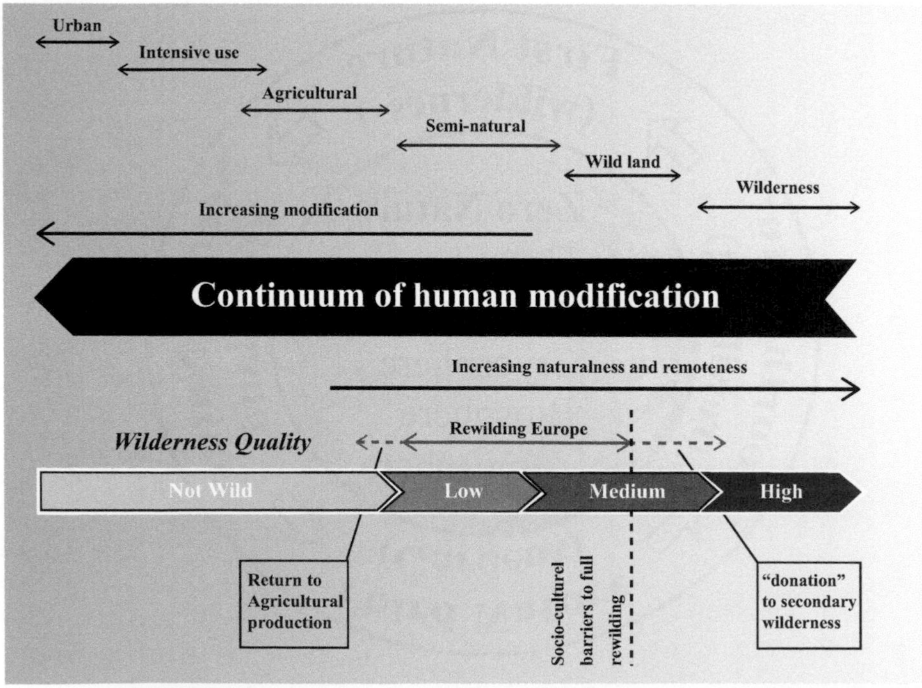


**Figure 7.19.2** The wilderness continuum.

*Source*: reproduced from Lesslie (1993) with permission.

the trees to re-grow. This in turn encouraged more beaver, which altered the river geomorphology which then encouraged more fish, birds and amphibians. This is known as a trophic cascade wherein the reintroduction of one missing element to an ecosystem can have cascading effects throughout, leading to beneficial, if unforeseen, outcomes.

Other projects have to start from a more basic position of needing to reinstate more natural vegetation patterns before species can be reintroduced. Carrifran Wildwood is aimed at restoring native tree cover to the over-grazed sheep-trod within a valley in the southern uplands of the Scottish Borders. It is recognised that the natural tree cover needs to be re-established to provide suitable habitat before species can be reintroduced. In heavily modified landscapes, this often means actively re-establishing the native vegetation that would be present under current climatic and edaphic conditions (soil, geology, etc.) in the absence of human land use. Only after natural vegetation cover is established can large herbivores and their predators be reintroduced to establish trophic cascades and modify the land cover.

Rewilding can additionally focus on restoring physical dynamics in the landscape, such as seasonal flooding and wildfire, where these are part of the natural ecosystem. This is termed abiotic rewilding. Examples include river restoration and natural flood management (NFM). The distinction between passive and abiotic rewilding can be blurred, for example, using beaver as a natural vector in NFM. Abiotic rewilding can also include the removal of human structures and artefacts such as tracks, fences and bridges from a landscape to make it feel wilder and increase remoteness.

## Active and passive rewilding

Rewilding can be either active or passive. Active rewilding implies the maintenance of some level of human intervention in land and species management. Passive rewilding is where human management is withdrawn and the processes of natural succession and interruption are allowed to take their course. Rewilding initiatives often have an active phase of trophic and abiotic rewilding, followed by a passive phase. Initially some level of human intervention is often required to establish a natural vegetation pattern (e.g. via tree planting) and a representative population of keystone species (e.g. via species reintroductions) before the processes of natural succession, herbivory, predation, etc. are allowed to take over and determine the future direction of the ecosystem.

## Spatial scale and connectivity

All these approaches to rewilding are landscape and context specific, requiring careful consideration of the geography of each project. Spatial scale and setting will determine which approach is appropriate under which set of circumstances. For example, reintroductions of large predators requires large areas of land and therefore is not appropriate for small-scale projects. Spatial and ecological connectivity is another important consideration since rewilding in isolation can leave species and habitats vulnerable to the effects of climate change, genetic simplification, etc. For this reason rewilding should be planned with the wider geographical setting in mind and as part of a wider network of spatially and ecologically connected core areas. This is illustrated in Figure 7.19.3 using the widely accepted Cores, Corridors and Carnivores (CCC) model wherein keystone

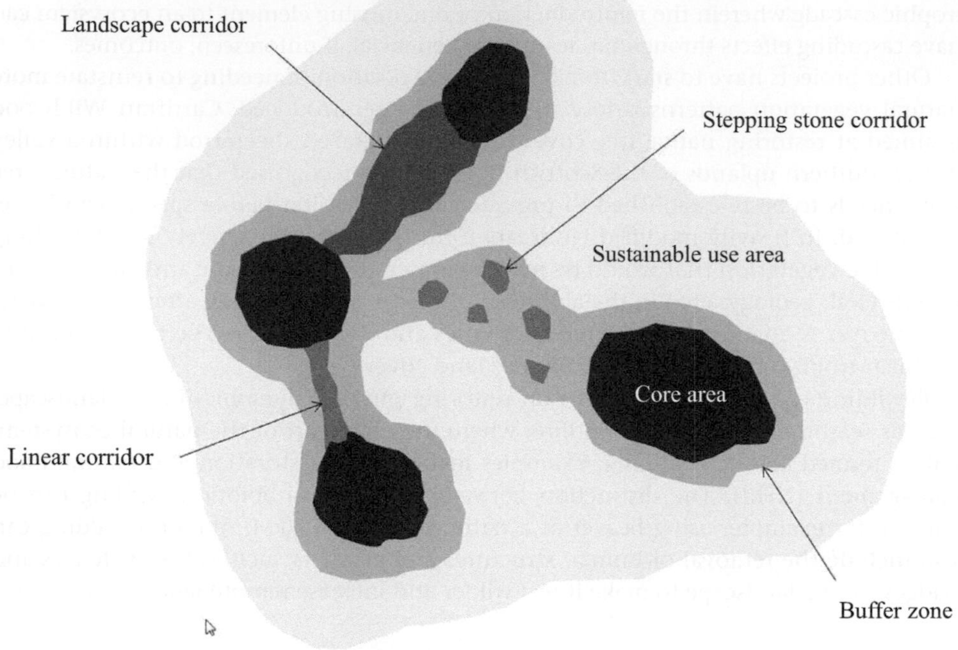


**Figure 7.19.3** The 'Cores, Corridors and Carnivores' model.

*Source*: reproduced with permission from Worboys et al. (2010).

species driving trophic cascades (carnivores) can move freely between existing wilderness and rewilded areas (cores) through rewilded landscapes (corridors).

## Issues: costs and benefits

There are many potential benefits associated with rewilding. These include benefits for wildlife and natural processes resulting in improved biodiversity and resilience. Landscape aesthetics can be enhanced through removal of human artefacts. Downstream benefits can be realised through improvements to water quality and flood protection, while nearby urban populations can benefit in terms of health and well-being from having wild, green space in the vicinity for recreation and enjoyment. This in turn can lead to local economic benefits from wildlife tourism and environmental stewardship.

Despite obvious benefits for ecosystems and wildlife, rewilding can be controversial. Conflicts can arise between rewilding and traditional human land uses including agriculture, forestry and game management. Conflicting demands on land resources and spill-over effects from rewilded areas mean that rewilding is unpopular with some people. Issues such as predation on domestic livestock and game species, spread of disease and crop damage are among the concerns raised. Conflicts are also present within the conservation sector with changes brought about by rewilding often blamed for loss of rare species dependent on traditional land management practices.

This mix of cost and benefits, conflict and controversy make rewilding an interesting topic. The future of many species and ecosystems will depend very much what we do today and this in turn will dictate the quality of our world and its resilience to climate change. Rewilding is likely to be part of our response to these challenges.

## Learning resources

The following websites contain much of relevance and interest:

Rewilding Britain. http://www.rewildingbritain.org.uk.
Rewilding Institute. http://rewilding.org/rewildit.
Self-Willed-Land. http://www.self-willed-land.org.uk.
Trees for Life. http://treesforlife.org.uk/forest/missing-species-rewilding/rewilding.
Wildland Research Institute. http://www.wildlandresearch.org.
Wildlands Network. http://www.wildlandsnetwork.org.

## Bibliography

Carver, S. (2013) 'Rewilding and habitat restoration', in Howard, P., Thompson, I. & Waterton, E. (eds) *The Routledge Companion to Landscape Studies*. Abingdon: Routledge.
Carver, S. (2014) 'Making real space for nature: a continuum approach to UK conservation', *ECOS*, 35(3–4): 4–14.
Castree, N. (2005) 'De-naturalisation: bringing nature back in', in Castree, N. *Nature* (pp. 108–176). London: Routledge.
Cicero, M.T. (1896) *De Natura Deorum* (*On the Nature of the Gods*), trans. Francis Brooks. London: Methuen.
Foote, J. (1990) 'Trying to take back the planet', *Newsweek*, 115(6): 24.
Foreman, D. (1993) 'Around the campfire', *Wild Earth*, 2(3).
Lesslie, R. (1993) 'The National Wilderness Inventory: wilderness identification, assessment and monitoring in Australia'. In Martin, V. & Hendee, J.C. (eds) *International Wilderness Allocation, Management and Research*. Proceedings of the 5th World Wilderness Congress (p. 3136). Fort Collins, CO: International Wilderness Leadership (WILD) Foundation.
Navarro, L.M. & Pereira, H.M. (2012) 'Rewilding abandoned landscapes in Europe', *Ecosystems*, 15(6): 900–912.
Soulé, M. & Noss, R. (1998) 'Rewilding and biodiversity: complementary goals for continental conservation', *Wild Earth*, 8: 18–28.
Spens, M. (2004) 'The garden at Portrack designed and created by Charles Jencks (1986–2004): entrapment and release', in Conan, M. (ed.) *Contemporary Garden Aesthetics, Creations and Interpretations*. Cambridge, MA: Harvard University Press.
Whatmore, S. (1999) 'Culture-nature', in Cloke, P., Crang, P. & Goodwin, M. (eds) *Introducing Human Geographies*. London: Hodder Arnold.
Worboys, G., Francis, W.L. & Lockwood, M. (2010) *Connectivity Conservation Management: A Global Guide*. Abingdon: Earthscan.

# Index

*Notes*
*Italics* are used to indicate figures and **bold** for tables, however when interspersed with related text these are included in regular page spans.
Mc is filed as Mac.
UK spelling is used throughout.